Cellular and Molecular Mechanisms in Gynecological Disorders

Cellular and Molecular Mechanisms in Gynecological Disorders

Guest Editor

Qiwei Yang

MDPI

Basel • Beijing • Wuhan • Barcelona • Belgrade • Novi Sad • Cluj • Manchester

Guest Editor
Qiwei Yang
Department of Obstetrics and
Gynecology
University of Chicago
Chicago
USA

Editorial Office
MDPI AG
Grosspeteranlage 5
4052 Basel, Switzerland

This is a reprint of the Special Issue, published open access by the journal *Cells* (ISSN 2073-4409), freely accessible at: https://www.mdpi.com/journal/cells/special_issues/HH76Q9R85A.

For citation purposes, cite each article independently as indicated on the article page online and as indicated below:

Lastname, A.A.; Lastname, B.B. Article Title. *Journal Name* **Year**, *Volume Number*, Page Range.

ISBN 978-3-7258-5413-4 (Hbk)
ISBN 978-3-7258-5414-1 (PDF)
https://doi.org/10.3390/books978-3-7258-5414-1

Contents

About the Editor

Qiwei Yang

Qiwei Yang works in the Department of Obstetrics and Gynecology at the University of Chicago, where he leads research programs at the intersection of cell biology, molecular biology, epigenetics, and tumor biology. His work is driven by the overarching goal of understanding how genomic, epigenomic, and epitranscriptomic dysfunction contributes to the onset and progression of women's reproductive diseases. In recent years, his research has focused heavily on epigenetic regulators, RNA modifications, and signaling pathways underlying uterine disorders. His studies have centered on uterine fibroids, the most common benign tumors of the female reproductive tract, which impose a significant health and economic burden worldwide. At the same time, he has advanced work on leiomyosarcoma, a rare but highly aggressive uterine malignancy with limited treatment options and poor patient outcomes. His investigations aim to bridge the gap between mechanistic understanding and translational applications, with the ultimate objective of improving women's health. Dr. Yang has served as a Principal Investigator and Co-Investigator on multiple NIH-funded grants examining the molecular basis of uterine disease pathogenesis. His scholarly productivity is reflected in more than 110 peer-reviewed publications and over 125 conference abstracts. Beyond his research, he has demonstrated leadership in the scientific community, serving as an Academic Editor for two journal Special Issues, Associate Editor for three international journals, and Chair of international scientific symposia. He is also an active reviewer for numerous international journals, contributing his expertise to advancing the field. In addition, Dr. Yang is deeply committed to mentorship, having supervised and trained MD, MD/PhD, and PhD students, as well as postdoctoral researchers and clinical fellows, providing guidance in both basic and translational research.

Preface

Gynecological disorders represent a diverse group of conditions affecting the female reproductive system, including the uterus, ovaries, fallopian tubes, cervix, vagina, and external genitalia. These disorders encompass both benign diseases, such as endometriosis, uterine fibroids, and polycystic ovary syndrome, and malignant tumors, including ovarian and uterine cancers. Together, they impose profound physical, reproductive, psychological, and societal burdens. Despite significant advances, many of these conditions remain poorly understood at the cellular and molecular levels, and effective diagnostic, preventive, and therapeutic strategies are still limited.

This Reprint aims to provide a comprehensive and up-to-date overview of emerging discoveries in gynecological disease research, with a particular focus on the molecular, cellular, and physiological mechanisms underlying both benign and malignant pathologies. The collection brings together review and original research articles covering key topics such as autophagy in reproductive aging, the roles of inflammation and oxidative stress, protease regulation in ovarian function, molecular drivers of uterine fibrosis and senescence, immune–tumor interactions in ovarian cancer, and novel therapeutic targets such as BET proteins in uterine leiomyosarcoma and AURKA in ovarian cancer. By addressing processes that cut across multiple disease entities, these contributions highlight the interconnected nature of gynecological pathophysiology.

The motivation for assembling this collection stems from the urgent need to bridge basic scientific advances with clinical applications. Recent progress in genomics, epigenomics, single-cell transcriptomics, and organoid modeling has transformed our understanding of reproductive tissues and disease mechanisms. These technological advances now offer unprecedented opportunities to identify biomarkers, redefine disease taxonomies, and develop precision-based diagnostic and therapeutic strategies. However, major gaps remain in translating these insights into improved outcomes for women worldwide, particularly for aggressive or understudied malignancies such as uterine leiomyosarcoma.

The purpose of this work is to stimulate further research and interdisciplinary dialogue by providing a platform that integrates diverse perspectives—from molecular biology and immunology to translational and clinical studies. It is intended for a wide audience, including basic scientists, translational researchers, clinicians, oncologists, reproductive biologists, and healthcare professionals dedicated to advancing women's health. Graduate students and early-career investigators will also find this collection to be a valuable resource for understanding current challenges and future directions in the field.

We hope that the articles presented here will not only expand scientific knowledge but also inspire new strategies for early detection, fertility-preserving interventions, and mechanism-driven treatments that ultimately improve the reproductive health and quality of life of women worldwide.

Qiwei Yang
Guest Editor

cells

MDPI

Editorial

Uncovering the Cellular and Molecular Landscape of Gynecological Disorders

Qiwei Yang

Department of Obstetrics and Gynecology, University of Chicago, Chicago, IL 60637, USA; yangq@bsd.uchicago.edu

1. Introduction

Gynecological disorders encompass a diverse array of conditions that affect the female reproductive system, including the uterus, ovaries, fallopian tubes, cervix, vagina, and external genitalia. These disorders range from common benign conditions, such as uterine fibroids, endometriosis, and polycystic ovary syndrome (PCOS), to malignancies, including ovarian, cervical, uterine leiomyosarcoma, and endometrial cancers. Collectively, they impose a significant burden on women's physical, reproductive, and psychological health, often leading to chronic pain, infertility, and impaired quality of life [1–4]. Importantly, gynecological cancers remain a leading cause of cancer related morbidity and mortality among women worldwide, with late-stage diagnoses and limited treatment options contributing to poor long-term outcomes in many cases [5,6]. Beyond the individual impact, both benign and malignant gynecological diseases contribute to substantial societal and economic costs through healthcare expenditures, loss of productivity, and long-term disability [7,8]. In recent years, breakthroughs in molecular biology, genomics, epigenomics, and advanced imaging technologies have profoundly expanded our understanding of the cellular and molecular underpinnings of these conditions. These advances have uncovered both distinct and overlapping mechanisms driving the pathogenesis of benign and malignant gynecological diseases, suggesting shared regulatory networks that span the spectrum of reproductive pathology [9,10]. For example, intricate crosstalk between the hormonal and immune systems has emerged as a central feature influencing tissue homeostasis, inflammation, and disease progression. Likewise, the role of cellular senescence in shaping the microenvironment [11], as well as dynamic remodeling of the extracellular matrix (ECM) [12,13], is now recognized as a key determinant of disease onset and chronicity.

Moreover, emerging evidence implicates autophagy and inflammasome activation as important modulators of cellular stress responses and immune activation in gynecologic tissues. These insights are reshaping our conceptual frameworks for how gynecological disorders originate and evolve. At the same time, innovative technologies such as single-cell RNA sequencing (scRNA-seq), epigenetic-based omics analysis, spatial transcriptomics, and patient-derived organoid models are revolutionizing the study of reproductive diseases. These platforms enable high-resolution dissection of cellular heterogeneity, allow for more refined disease classification, and accelerate the discovery of novel biomarkers and therapeutic targets. Together, these developments are transforming gynecological disease research, paving the way toward more precise, mechanism-based diagnostic and therapeutic strategies.

This collection of reviews and original research articles offers a comprehensive and timely overview of emerging insights into the cellular and molecular mechanisms underlying ovarian and uterine pathologies. The included studies span a diverse yet intercon-

 https://doi.org/10.3390/cells14171399

nected spectrum of gynecologic conditions, highlighting the growing recognition of shared pathogenic processes across both benign and malignant diseases. Topics covered include mesothelial clearance by endometriosis spheroids, impaired decidualization, autophagy in reproductive aging, protease regulation in ovarian function, immune–tumor interactions in ovarian cancer, and the molecular drivers of fibrosis, senescence, transformation, and tumorigenesis in uterine tissues. Collectively, these contributions underscore the urgent need for integrated research approaches to advance novel diagnostics, fertility-preserving interventions, and precision therapies for gynecologic diseases.

2. An Overview of Published Articles

2.1. Female Fertility and Cellular Mechanisms

Two articles in this Special Issue focus on gynecological disorders that affect multiple organs, including both the ovaries and the uterus. Autophagy, a highly conserved cellular process vital for maintaining intracellular homeostasis, plays a pivotal role in female reproductive function [14]. Recent studies underscore the essential role of autophagy in embryo implantation, particularly within the endometrium, where it supports optimal fetal development. However, despite these advances, critical gaps remain in our understanding of how autophagy affects oocyte quality, follicular development, and overall reproductive capacity at the molecular level. Autophagy contributes to oocyte maturation and developmental competence and is implicated in reproductive aging by influencing age-related declines in fertility. The review article by Harrath et al. (contribution 1) explores the physiological roles of autophagy across the female reproductive system and its involvement in reproductive toxicity. The authors further summarize the functional links between autophagy, endometrial receptivity, and embryo development. Moreover, this review article highlights the intricate relationship between autophagy and various aspects of female reproductive health, including pregnancy, ovarian function, infertility, endometriosis, and gynecologic malignancies. Emerging therapeutic strategies aimed at modulating autophagy through both natural compounds and synthetic agents are also discussed, with the goal of enhancing female fertility and improving reproductive outcomes. By synthesizing current findings and identifying existing knowledge gaps, this review aims to stimulate further research into the multifaceted roles of autophagy in reproductive biology. A comprehensive understanding of autophagy's impact on female fertility holds promise for advancing diagnostics, therapeutics, and the broader field of reproductive medicine.

Oxidative stress plays a key role in accelerating the age-related decline of reproductive function and fertility [15]. Recent research has identified human amniotic membrane-derived mesenchymal stem cells (AMSCs) as a promising tool for reducing oxidative damage and promoting tissue repair. In an original study by Ra (contribution 2), intravenous administration of AMSC-conditioned medium (AMSC-CM) was shown to reduce oxidative stress and improve reproductive function in aged female mice. The treatment modulated reproductive hormone levels to resemble those of younger mice and induced beneficial gene expression changes in the ovaries and uterus, supporting anti-aging, energy metabolism, and reproductive functions. These results suggest that AMSC-CM holds therapeutic potential for combating age-related female infertility through its antioxidative and regenerative properties.

The ovary plays a central role in gynecological disorders, including polycystic ovary syndrome (PCOS), ovarian cysts, endometriosis, and ovarian cancer [16,17], all of which can disrupt hormone balance, fertility, and reproductive health. Proteases are key regulators of ovarian folliculogenesis, guiding processes from primordial follicle activation to ovulation and corpus luteum formation. These enzymes mediate essential cellular and tissue-level functions such as extracellular matrix (ECM) remodeling, cell proliferation, apoptosis, and intercellular signaling—each vital for proper follicular development and ovarian function [18]. The review article by Kushawaha and Pelosi (contribution 3) provides a comprehensive synthesis of current knowledge on the classification, regulation, and function of proteases in ovarian physiology and pathology. The authors highlight several major protease families, including matrix metalloproteinases (MMPs), tissue inhibitors of metalloproteinases (TIMPs), the plasminogen activator (PA) system, and cathepsins, discussing their specific contributions to follicle development. These proteases tightly regulate ECM turnover and cellular communication, facilitating the dynamic structural changes required for follicular growth, oocyte maturation, and luteal phase transition. Beyond normal ovarian physiology, dysregulated protease activity has been implicated in several ovarian pathologies. The authors explore the involvement of aberrant proteolytic signaling in the development and progression of ovarian cancer, polycystic ovary syndrome (PCOS), and primary ovarian insufficiency (POI), highlighting how altered expression or activity of specific proteases contributes to disease mechanisms. Finally, by integrating findings from clinical genomic studies and mechanistic investigations using animal models, the authors underscore the translational potential of targeting proteases for therapeutic intervention. This review provides a foundation for future research into protease-based diagnostics and treatments in reproductive medicine and ovarian disease.

Chronic inflammation is increasingly recognized as a key contributor to female reproductive dysfunction [16], affecting both natural conception and the success of assisted reproductive technologies such as in vitro fertilization (IVF). As a central mediator of innate immune responses, the NLR family pyrin domain-containing 3 (NLRP3) inflammasome plays a pivotal role in inflammatory signaling and is closely associated with oxidative stress. Activation of the NLRP3 inflammasome in response to oxidative stress has been linked to reduced oocyte quality, compromised fertilization potential, and impaired embryo development. Within the ovarian microenvironment, the interplay between oxidative stress and NLRP3 activation exerts a substantial impact on oocyte competence and reproductive outcomes. The review article by Moustakli et al. (contribution 4) elucidates the molecular mechanisms underlying NLRP3 inflammasome activation and its interaction with oxidative stress in the context of female fertility. Additionally, the authors explore emerging therapeutic strategies that target these pathways to enhance oocyte quality, support embryo development, and ultimately improve reproductive success.

2.2. Ovarian Cancer: Molecular Pathways and Biomarkers

Hypoxia-Inducible Factor-1α (HIF-1α) plays a pivotal role in the progression of multiple cancers by promoting cellular adaptation, survival, angiogenesis, metabolic reprogramming, immune evasion, and metastasis [19]. The review article by Rahman et al. (contribution 5) elucidates the molecular functions of HIF-1α in ovarian cancer and infertility, highlighting innovative therapeutic strategies aimed at both disease control and fertility preservation. Under hypoxic conditions, HIF-1α translocates to the nucleus, where it dimerizes with HIF-1β and binds to hypoxia-response elements (HREs) in the promoter regions of target genes. This activation drives the expression of genes involved in angiogenesis, metabolic reprogramming, epithelial-to-mesenchymal transition, therapeutic resistance, and immune evasion. HIF-1α–mediated angiogenesis and glycolytic reprogramming support tumor proliferation, survival, and metastasis. Concurrently, dysregulation of HIF-1α disrupts ovarian homeostasis by impairing follicular development, hormone production, and vascular integrity, processes essential for female fertility. Additionally, HIF-1α promotes chronic inflammation and oxidative stress, further compromising reproductive health. Given its dual role in tumor progression and infertility, HIF-1α is an attractive therapeutic target. Emerging strategies, including small-molecule inhibitors and nanoparticle-based drug delivery systems, hold promise for reducing HIF-1α activity, thereby limiting cancer progression while safeguarding reproductive function.

Epithelial ovarian cancer (EOC) is the most common and aggressive form of ovarian cancer and represents the leading cause of gynecological cancer-related deaths in women. It is often diagnosed at an advanced stage, contributing to its poor prognosis. Unlike many other cancers, EOC spreads through a unique metastatic mechanism involving the formation of multicellular spheroids within the peritoneal cavity [20]. The study by Tomas et al. (contribution 6) offers valuable insights into the signaling plasticity driving this process, using an EOC spheroid model. The authors propose that EOC cells undergo biological shifts between tumor and spheroid states during cancer dormancy. By directly comparing cultured EOC spheroids and organoids, they found increased AMPK activity and decreased Akt signaling in spheroids, indicating distinct phenotypes. RNA sequencing further revealed key differences, particularly in cell cycle regulation in organoids. The inhibition of the G2/M checkpoint confirmed its essential role in organoid viability. Notably, both spheroids and organoids showed equal sensitivity to Alisertib, highlighting AURKA as a potential therapeutic target. These findings underscore the need to understand EOC cell adaptations, as therapeutic vulnerabilities may vary across disease stages.

In addition to investigating signaling plasticity in ovarian cancer using two 3D models by Tomas (contribution 6), Kang et al. (contribution 7) identified prognostic biomarkers for high-grade serous carcinoma (HGSC), a highly lethal malignancy characterized by initial chemotherapy sensitivity and frequent recurrence. Unlike prior studies limited to bulk RNA sequencing or a few genetic markers, this study utilized spatial transcriptomics and multispectral immune cell immunofluorescence (IF) to uncover markers associated with disease progression post-first-line treatment. Notable findings included elevated NKG7 expression in $CD45^+$ immune regions and increased TFPI2 and PIGR in tumor areas, all linked to improved progression-free survival. Multispectral IF further revealed higher Treg/$CD8^+$ T cell ratios and closer proximity of Tregs to cancer cells and macrophages in recurrent tumors—correlating with poor outcomes. Integrated analyses demonstrated that immune cell density and immune pathway scores in recurrent tumors were positively correlated with cancer pathway activity, excluding NF-κB. These results offer new insights into immune–tumor interactions and potential biomarkers for predicting HGSC recurrence.

2.3. *Uterine Aging, Fibrosis, and Genetic Mutations*

The uterus is the central organ linked to the highest prevalence and diversity of gynecological conditions. Fibrosis in the uterus often arises from chronic inflammation, repeated injury, hormonal dysregulation—particularly involving estrogen and progesterone—or aberrant wound-healing responses. Characterized by excessive accumulation of extracellular matrix (ECM) proteins, uterine fibrosis disrupts normal reproductive function and contributes to infertility in humans. The review article by Matsuno and Imakawa (contribution 8) explores the role of uterine fibrosis and biological aging as underrecognized yet potentially critical contributors to declining fecundity in cattle, which serve as valuable research models. The authors detail the mechanisms of fibrosis in the female reproductive tract of cattle, highlighting key pathways such as TGF-β, β-catenin, and Hippo signaling, as well as hormonal interactions that form a complex cross-regulatory network amplifying fibrotic responses. Additionally, the review addresses the impact of biological aging—including molecular drivers, epigenetic aging, and cellular senescence—on uterine function and infertility. Targeting fibrotic signaling pathways may offer promising avenues for the development of non-surgical therapies for uterine fibrotic disorders.

Uterine fibroids, or leiomyomas (UFs), are often associated with excessive extracellular matrix (ECM) deposition and fibrotic remodeling, contributing to their firm, fibrous nature and disease progression [13]. UFs are common benign tumors affecting up to 70–80% of women by age 50. Arising from the myometrium, they cause symptoms like abnormal bleeding, pelvic pain, infertility, and pregnancy complications, and are a major reason for hysterectomy. Genomic studies have identified *MED12* gene mutations—especially in exon 2—as the most frequent alteration, found in 60–70% of UFs. These mutations, which disrupt mediator-associated CDK activity, transcription, and chromatin regulation [21,22], likely drive the clonal expansion of myometrial stem cells but are rarely seen in malignant uterine leiomyosarcomas (ULMS), suggesting a benign trajectory. The cause of *MED12* mutations remains unclear. The article by Li et al. (contribution 9) demonstrated that *MED12* mutations can be detected in myometrial cells from hysterectomy specimens using duplex deep sequencing (DDS), which offers higher accuracy and lower error rates compared to standard next-generation sequencing (NGS) methods. The authors found that oxidative stress promotes *MED12* mutations in myometrial cells and is associated with increased DNA damage, suggesting that it may be a significant risk factor for the acquisition of these mutations. This study provides important insights into the molecular basis underlying the high incidence of uterine fibroids in women of reproductive age.

The endometrium is essential for reproductive success, facilitating implantation and pregnancy through tightly regulated processes involving hormonal responsiveness, immune modulation, and tissue regeneration. However, aging disrupts these functions, with cellular senescence—a state of irreversible cell cycle arrest induced by DNA damage and oxidative stress—emerging as a key contributor. The review article by Kobayashi et al. (contribution 10) summarizes the molecular mechanisms underlying cellular senescence in age-related endometrial dysfunction. Although senescence plays beneficial roles in tumor suppression and tissue repair, its dysregulation can impair endometrial integrity and compromise fertility. Central to the regulation of senescence are signaling pathways involving p53, AMPK, and mTOR, which coordinate cellular stress responses, metabolic regulation, and proliferation control. Specifically, p53 activates AMPK and inhibits mTOR, thereby promoting energy conservation and limiting excessive senescence. AMPK further suppresses mTOR activity, mitigating age-associated cellular dysfunction. This p53–AMPK–mTOR axis, together with autophagy, plays a pivotal role in determining cell fate in response to stress and nutrient availability. While moderate levels of senescence may support endometrial function, excessive accumulation of senescent cells can disrupt tissue homeostasis and reduce fertility. A deeper understanding of these molecular interactions may guide the development of targeted strategies to counteract reproductive aging and improve fertility outcomes.

2.4. Endometriosis: Immune and Molecular Pathophysiology

Endometriosis is a gynecologic disorder characterized by the presence of endometrium-like glandular and stromal tissue outside the uterine cavity. The role of hormonal dysregulation, particularly estrogen, is well established in the initiation, progression, and persistence of the disease [23]. Emerging evidence also underscores the involvement of immune system dysfunction in the pathophysiology of endometriosis. The human endometrium is a highly dynamic tissue that undergoes cyclical remodeling driven by hormonal cues throughout the menstrual cycle. Similarly, ectopic endometrial lesions exhibit hormone-responsive behavior, though their pathogenic mechanisms remain incompletely understood. This complexity poses significant challenges to fully elucidating the etiology and pathology of the disease. The review article by Greygoose et al. (contribution 11) explores the intricate interplay between estrogen and both the innate and adaptive immune systems during the menstrual cycle, focusing on their roles in the development and progression of endometriosis. Estrogen has been shown to modulate numerous inflammatory and immunoregulatory processes, influencing the behavior of both tissue-resident and circulating immune cells. The authors highlight key estrogen-mediated interactions with specific myeloid and lymphoid cell populations, emphasizing their contributions to the immunopathogenesis of endometriosis. To investigate the pathology and lesion spread in endometriosis, Kloeckner and Walker (contribution 12) developed an in vitro mesothelial clearance assay using spheroids of endometriotic and normal epithelial cells cultured on a mesothelial monolayer. They found that both cell types could mediate mesothelial clearance, with endometriotic spheroids showing enhanced clearance ability. This assay offers a useful platform for testing potential therapies, though further research is needed to understand the mechanisms of mesothelial clearance and whether endometriotic lesions can invade the underlying extracellular matrix, similar to ovarian cancer.

Defective decidualization has been mechanistically linked to uterine conditions that impair successful conception, such as endometriosis [24]. However, the presence and underlying causes of decidualization defects in the eutopic endometrium of women with endometriosis remain largely unclear. In this review, Retis-Resendiz et al. (contribution 13) integrate and critically evaluate molecular evidence from both in vivo and in vitro studies that have investigated alterations in decidualization within the eutopic endometrium of endometriosis patients. Multiple studies have reported impaired decidualization, implicating dysregulation in key genes such as *IGFBP1, PRL, connexin 43, HOXA10,* and *FOXO1.* Disruptions have also been observed in several signaling pathways, including sex steroid hormone signaling via genomic and non-genomic mechanisms, TGF-β signaling through SMAD2/3/4, bone morphogenetic protein signaling through SMAD1/5/4, and Notch signaling. Additionally, pro-inflammatory cytokines and epigenetic mechanisms, such as DNA methylation and histone modifications, have been shown to contribute to the dysregulation of the decidualization process. Despite these insights, further functional studies are needed to determine whether these molecular defects directly contribute to endometriosis-associated infertility. A deeper understanding of the decidualization process and its disruption in endometriosis may not only inform fertility-targeted therapies but also support the development of more effective treatments for this complex and chronic condition.

Increasing evidence implicates immune system dysregulation, particularly innate immunity and Toll-like receptors (TLRs), in the development and progression of diseases [25]. The study by Sobstyl et al. (contribution 14) investigated the expression of specific TLRs (TLR2, TLR3, TLR4, TLR7, TLR8, and TLR9) on peripheral blood lymphocyte subpopulations ($CD4^+$, $CD8^+$, and $CD19^+$ cells) in women with endometriosis. The authors also measured the levels of their soluble forms in serum and urine. The results revealed significantly increased TLR expression in lymphocyte subpopulations of endometriosis patients compared to healthy controls, indicating active involvement of TLR pathways in both systemic and localized immune responses. Receiver operating characteristic analysis further demonstrated the potential of TLR profiles to distinguish endometriosis patients from controls and possibly differentiate among disease subtypes. These findings reinforce the role of TLRs in the immunopathogenesis of endometriosis and highlight their promise as diagnostic biomarkers and therapeutic targets. Further studies with larger cohorts and mechanistic investigations of TLR signaling are warranted to confirm and expand upon these observations.

2.5. Decoding Benign and Malignant Uterine Pathologies

Recent advances in transcriptomics, especially single-cell RNA sequencing (scRNA-seq), have transformed our understanding of uterine biology by revealing the cellular diversity and molecular mechanisms driving these conditions [26]. The review article by Boldu-Fernandez et al. (contribution 15) summarizes current research on mapping the cellular architecture of the several gynecological diseases through scRNA-seq technology. In the endometrium and myometrium, scRNA-seq and spatial transcriptomics uncovered key cell types, signaling pathways, and dynamic changes across the menstrual cycle. These technologies have also highlighted functional abnormalities in stromal and immune cells, such as fibroblast-to-myofibroblast transitions and impaired macrophage function, that contribute to fibrosis, chronic inflammation, and lesion persistence in endometriosis. In endometrial cancer, scRNA-seq has revealed the complexity of the microenvironment, identifying distinct cancer-associated fibroblast subtypes and immune cell profiles that influence disease progression and therapy resistance. Similarly, studies of adenomyosis have identified disrupted signaling pathways, including Wnt and VEGF, and novel progenitor cell populations associated with tissue invasion and neuroinflammation. In UFs, single-cell analyses have characterized smooth muscle and fibroblast subpopulations involved in extracellular matrix remodeling and key pathways such as ERK and mTOR. While challenges like scalability and reproducibility remain, the authors concluded that multi-omics high-resolution single-cell analyses hold great promise for biomarker discovery, therapeutic target identification, and the development of personalized treatments for gynecological diseases.

ULMS is a rare but highly aggressive malignancy arising from the smooth muscle cells of the myometrium. Representing approximately 1–2% of all uterine cancers, ULMS is clinically and biologically distinct from the far more common benign UFs. ULMS is characterized by rapid growth, high recurrence rates, and poor prognosis, with a five-year survival rate of less than 40% in advanced stages [9]. Emerging evidence highlights the role of epigenetic dysregulation in ULMS [27–31]. However, the contribution of BET (bromodomain and extra-terminal domain) proteins to ULMS pathogenesis has not been fully elucidated. In this study by Yang et al. (contribution 16), members of the BET protein family, including BRD2, BRD3, and BRD4, were found to be aberrantly overexpressed in ULMS tissues compared to normal myometrium. Pharmacological inhibition of BET proteins using small-molecule inhibitors JQ1 and I-BET762 significantly suppressed ULMS cell proliferation in a dose-dependent manner by inducing cell cycle arrest. Notably, high-throughput RNA-sequencing analysis revealed that targeted inhibition of BET proteins led to alterations in several critical signaling pathways, including the Hedgehog pathway, epithelial–mesenchymal transition (EMT), and transcription factor-driven regulatory networks. In addition, BET inhibition impacted multiple epigenetic regulators, including DNA methyltransferases, histone-modifying enzymes, and N6-methyladenosine RNA methylation machinery, suggesting broader epigenetic remodeling because of BET targeting. These findings not only highlight BET proteins as potential biomarkers of disease progression but also suggest that BET inhibition could serve as a promising therapeutic strategy for targeting dysregulated transcriptional and epigenetic programs in ULMS.

3. Future Perspectives

As our understanding of gynecological disorders continues to deepen, future research must leverage emerging cellular and molecular insights to address the persistent clinical challenges these conditions present. The integration of advanced omics technologies, high-resolution imaging, and patient-derived models offers unprecedented opportunities to dissect disease complexity at single-nucleotide, single-cell, and spatial resolution. These cutting-edge approaches will be pivotal in redefining disease taxonomies, shifting from traditional histopathological classifications toward molecularly informed subtypes that more accurately capture the underlying biology and therapeutic vulnerabilities.

A critical next step in gynecological disease research is the integration of multi-omic datasets, including genomic, transcriptomic, epigenomic, proteomic, and metabolomic information, to build comprehensive molecular atlases of gynecologic tissues across developmental stages, physiological states, and disease progression. These integrative frameworks will be essential for uncovering early pathogenic drivers, identifying dynamic biomarkers of disease onset and progression, and revealing novel therapeutic targets. In parallel, the expanded application of patient-derived organoids, advanced ex vivo systems, and in vivo imaging technologies will enhance the translational pipeline, enabling more predictive and physiologically relevant preclinical testing, particularly for understudied or aggressive malignancies such as uterine leiomyosarcoma.

Clinical translation research should focus on developing diagnostic approaches grounded in underlying disease mechanisms, such as non-invasive biomarkers and risk stratification assays, augmented by Artificial Intelligence (AI) technologies to enable earlier detection and personalized management of gynecological diseases [32,33]. Efforts to develop low-cost, high-throughput platforms and to ensure the inclusion of diverse populations in biomarker discovery studies are also crucial. Leveraging AI-driven analytics alongside these strategies will be essential to translating scientific advances into accessible, equitable solutions that improve early detection, guide treatment decisions, and ultimately enhance health outcomes for all women globally.

4. Concluding Remarks

This Special Issue presents a diverse, multidisciplinary collection of original research and review articles that advance our understanding of the molecular, cellular, and physiological mechanisms underlying a broad spectrum of gynecological disorders. The contributions address ovarian and uterine diseases, infertility, endometriosis, age-related reproductive decline, and both benign and malignant tumors, offering critical insights into their pathogenesis and identifying potential avenues for therapeutic intervention.

Key molecular players and pathways include the impact of autophagy, senescence, and proteolytic signaling on reproductive aging and fertility; the identification of critical regulators such as HIF-1α and NLRP3 in cancer and infertility; and the discovery of promising therapeutic targets such as BET proteins in uterine leiomyosarcoma and AURKA in ovarian cancer. Notably, recent advances in single-cell transcriptomics and spatial technologies have begun to redefine our cellular and molecular understanding of reproductive tissues, opening new opportunities for precision diagnostics and targeted therapies.

Collectively, the articles underscore the complexity and interconnectivity of biological systems that govern female reproductive health. They also reveal substantial knowledge gaps, particularly in the translation of basic discoveries to clinical practice, and underscore the importance of continued interdisciplinary research to develop effective, fertility-preserving treatments for women affected by these conditions.

By integrating current findings and exploring emerging directions, this Special Issue not only enhances our understanding of gynecological diseases but also serves as a catalyst for future studies aimed at improving the reproductive health and overall well-being of women worldwide.

Funding: This study was partially supported by the National Institutes of Health (NIH) grant RO1 HD106285.

Institutional Review Board Statement: Not applicable.

Informed Consent Statement: Not Applicable.

Data Availability Statement: No new data were created or analyzed in this study. Data sharing is not applicable to this article.

Acknowledgments: I would like to extend my sincere gratitude to all the authors for their outstanding contributions to this Special Issue. Their high-quality work has greatly enriched the scientific discourse on gynecological disorders. I also thank the dedicated team at the *Cells* Editorial Office for their invaluable support in coordinating a rigorous peer review process and ensuring seamless communication with contributors. Finally, I am deeply appreciative of the expert reviewers, whose timely and constructive evaluations were instrumental in maintaining the scientific integrity and success of this Special Issue.

Conflicts of Interest: The author declare no conflict of interest.

List of Contributions:

1. Harrath, A.H.; Rahman, M.A.; Bhajan, S.K.; Bishwas, A.K.; Rahman, M.H.; Alwasel, S.; Jalouli, M.; Kang, S.; Park, M.N.; Kim, B. Autophagy and Female Fertility: Mechanisms, Clinical Implications, and Emerging Therapies. *Cells* **2024**, *13*, 1354. https://doi.org/10.3390/cells13161354.
2. Ra, K. Therapeutic Potential of Human Amniotic Membrane-Derived Mesenchymal Stem Cell Conditioned Medium in Combating Oxidative Stress and Age-Related Female Infertility. *Cells* **2025**, *14*, 801. https://doi.org/10.3390/cells14110801.
3. Kushawaha, B.; Pelosi, E. Spotlight on Proteases: Roles in Ovarian Health and Disease. *Cells* **2025**, *14*, 921. https://doi.org/10.3390/cells14120921.
4. Moustakli, E.; Stavros, S.; Katopodis, P.; Skentou, C.; Potiris, A.; Panagopoulos, P.; Domali, E.; Arkoulis, I.; Karampitsakos, T.; Sarafi, E.; et al. Oxidative Stress and the NLRP3 Inflammasome: Focus on Female Fertility and Reproductive Health. *Cells* **2025**, *14*, 36. https://doi.org/10.3390/cells14010036.
5. Rahman, M.A.; Jalouli, M.; Bhajan, S.K.; Al-Zharani, M.; Harrath, A.H. The Role of Hypoxia-Inducible Factor-1α (HIF-1α) in the Progression of Ovarian Cancer: Perspectives on Female Infertility. *Cells* **2025**, *14*, 437. https://doi.org/10.3390/cells14060437.
6. Tomas, E.J.; Valdes, Y.R.; Davis, J.; Kolendowski, B.; Buensuceso, A.; DiMattia, G.E.; Shepherd, T.G. Exploiting Cancer Dormancy Signaling Mechanisms in Epithelial Ovarian Cancer Through Spheroid and Organoid Analysis. *Cells* **2025**, *14*, 133. https://doi.org/10.3390/cells14020133.
7. Kang, H.; Joung, J.-G.; Park, H.; Choi, M.C.; Koh, D.; Jeong, J.-Y.; Lee, J.; Kim, S.-Y.; Jung, D.; Hwang, S.; et al. Identification of Prognostic Biomarkers of Ovarian High-Grade Serous Carcinoma: A Preliminary Study Using Spatial Transcriptome Analysis and Multispectral Imaging. *Cells* **2025**, *14*, 681. https://doi.org/10.3390/cells14100681.
8. Matsuno, Y.; Imakawa, K. Biological Aging and Uterine Fibrosis in Cattle: Reproductive Trade-Offs from Enhanced Productivity. *Cells* **2025**, *14*, 955. https://doi.org/10.3390/cells14130955.
9. Li, Y.; Asif, H.; Feng, Y.; Kim, J.J.; Wei, J.-J. Somatic MED12 Mutations in Myometrial Cells. *Cells* **2024**, *13*, 1432. https://doi.org/10.3390/cells13171432.
10. Kobayashi, H.; Umetani, M.; Nishio, M.; Shigetomi, H.; Imanaka, S.; Hashimoto, H. Molecular Mechanisms of Cellular Senescence in Age-Related Endometrial Dysfunction. *Cells* **2025**, *14*, 858. https://doi.org/10.3390/cells14120858.
11. Greygoose, E.; Metharom, P.; Kula, H.; Seckin, T.K.; Seckin, T.A.; Ayhan, A.; Yu, Y. The Estrogen–Immune Interface in Endometriosis. *Cells* **2025**, *14*, 58. https://doi.org/10.3390/cells14010058.
12. Kloeckner, A.A.; Walker, S.R. Endometriosis Cell Spheroids Undergo Mesothelial Clearance in a Similar Manner to Ovarian Cancer Cell Spheroids. *Cells* **2025**, *14*, 742. https://doi.org/10.3390/cells14100742.
13. Retis-Resendiz, A.M.; Gómez-Suárez, S.K.; García-Gómez, E.; Vázquez-Martínez, E.R. Molecular Basis of Impaired Decidualization in the Eutopic Endometrium of Endometriosis Patients. *Cells* **2025**, *14*, 326. https://doi.org/10.3390/cells14050326.
14. Sobstyl, A.; Mertowska, P.; Mertowski, S.; Tarkowski, R.; Dudzi 'nski, D.; Kotowski, M.; Bojarski, K.; Stelmach, B.; Chermuła, B.; Br.azert, M.; et al. Expression of Toll-like Receptors

on Lymphocyte Subpopulations and Their Soluble Forms in Serum and Urine of Women with Endometriosis. *Cells* **2025**, *14*, 1273. https://doi.org/10.3390/cells14161273.

15. Boldu-Fernández, S.; Lliberos, C.; Simon, C.; Mas, A. Mapping Human Uterine Disorders Through Single-Cell Transcriptomics. *Cells* **2025**, *14*, 156. https://doi.org/10.3390/cells14030156.
16. Yang, Q.; Falahati, A.; Khosh, A.; Lastra, R.R.; Boyer, T.G.; Al-Hendy, A. Unraveling the Role of Bromodomain and Extra-Terminal Proteins in Human Uterine Leiomyosarcoma. *Cells* **2024**, *13*, 1443. https://doi.org/10.3390/cells13171443.

References

1. Zondervan, K.T.; Becker, C.M.; Koga, K.; Missmer, S.A.; Taylor, R.N.; Vigano, P. Endometriosis. *Nat. Rev. Dis. Primers* **2018**, *4*, 9. [CrossRef]
2. Lagana, A.S.; Garzon, S.; Gotte, M.; Vigano, P.; Franchi, M.; Ghezzi, F.; Martin, D.C. The Pathogenesis of Endometriosis: Molecular and Cell Biology Insights. *Int. J. Mol. Sci.* **2019**, *20*, 5615. [CrossRef] [PubMed]
3. Culley, L.; Law, C.; Hudson, N.; Denny, E.; Mitchell, H.; Baumgarten, M.; Raine-Fenning, N. The social and psychological impact of endometriosis on women's lives: A critical narrative review. *Hum. Reprod. Update* **2013**, *19*, 625–639. [CrossRef] [PubMed]
4. Hazimeh, D.; Coco, A.; Casubhoy, I.; Segars, J.; Singh, B. The Annual Economic Burden of Uterine Fibroids in the United States (2010 Versus 2022): A Comparative Cost-Analysis. *Reprod. Sci.* **2024**, *31*, 3743–3756. [CrossRef]
5. Torre, L.A.; Trabert, B.; DeSantis, C.E.; Miller, K.D.; Samimi, G.; Runowicz, C.D.; Gaudet, M.M.; Jemal, A.; Siegel, R.L. Ovarian cancer statistics, 2018. *CA Cancer J. Clin.* **2018**, *68*, 284–296. [CrossRef]
6. Ginsburg, O.; Bray, F.; Coleman, M.P.; Vanderpuye, V.; Eniu, A.; Kotha, S.R.; Sarker, M.; Huong, T.T.; Allemani, C.; Dvaladze, A.; et al. The global burden of women's cancers: A grand challenge in global health. *Lancet* **2017**, *389*, 847–860. [CrossRef]
7. Yang, Q.; Ciebiera, M.; Bariani, M.V.; Ali, M.; Elkafas, H.; Boyer, T.G.; Al-Hendy, A. Comprehensive Review of Uterine Fibroids: Developmental Origin, Pathogenesis, and Treatment. *Endocr. Rev.* **2022**, *43*, 678–719. [CrossRef]
8. Guo, T.; Liu, H.; Xu, B.; Qi, Y.; Xu, K.; Wu, X.; He, X.; Qin, Y.; Chen, Z.J. Epidemiology, genetic etiology and intervention of premature ovarian insufficiency. *Endocr. Rev.* **2025**, bnaf011. [CrossRef]
9. Yang, Q.; Madueke-Laveaux, O.S.; Cun, H.; Wlodarczyk, M.; Garcia, N.; Carvalho, K.C.; Al-Hendy, A. Comprehensive Review of Uterine Leiomyosarcoma: Pathogenesis, Diagnosis, Prognosis, and Targeted Therapy. *Cells* **2024**, *13*, 1106. [CrossRef] [PubMed]
10. Yang, Q.; Carvalho, K.C.; Shirazi, R.; Falahati, A. Editorial: Emerging researchers in frontiers in pharmacology: Obstetric and pediatric pharmacology 2022. *Front. Pharmacol.* **2023**, *14*, 1215954. [CrossRef] [PubMed]
11. Coppe, J.P.; Desprez, P.Y.; Krtolica, A.; Campisi, J. The senescence-associated secretory phenotype: The dark side of tumor suppression. *Annu. Rev. Pathol.* **2010**, *5*, 99–118. [CrossRef]
12. Alshehri, S.; Pavlovic, T.; Farsinejad, S.; Behboodi, P.; Quan, L.; Centeno, D.; Kung, D.; Rezler, M.; Lee, W.; Jasinski, P.; et al. Extracellular Matrix Modulates Outgrowth Dynamics in Ovarian Cancer. *Adv. Biol.* **2022**, *6*, e2200197. [CrossRef]
13. Yang, Q.; Al-Hendy, A. Update on the Role and Regulatory Mechanism of Extracellular Matrix in the Pathogenesis of Uterine Fibroids. *Int. J. Mol. Sci.* **2023**, *24*, 5778. [CrossRef]
14. Komarla, A.; Dufresne, S.; Towers, C.G. Recent Advances in the Role of Autophagy in Endocrine-Dependent Tumors. *Endocr. Rev.* **2023**, *44*, 629–646. [CrossRef]
15. Broekmans, F.J.; Soules, M.R.; Fauser, B.C. Ovarian aging: Mechanisms and clinical consequences. *Endocr. Rev.* **2009**, *30*, 465–493. [CrossRef] [PubMed]
16. Sadeghi, H.M.; Adeli, I.; Calina, D.; Docea, A.O.; Mousavi, T.; Daniali, M.; Nikfar, S.; Tsatsakis, A.; Abdollahi, M. Polycystic Ovary Syndrome: A Comprehensive Review of Pathogenesis, Management, and Drug Repurposing. *Int. J. Mol. Sci.* **2022**, *23*, 583. [CrossRef] [PubMed]
17. Akter, S.; Rahman, M.A.; Hasan, M.N.; Akhter, H.; Noor, P.; Islam, R.; Shin, Y.; Rahman, M.D.H.; Gazi, M.S.; Huda, M.N.; et al. Recent Advances in Ovarian Cancer: Therapeutic Strategies, Potential Biomarkers, and Technological Improvements. *Cells* **2022**, *11*, 650. [CrossRef] [PubMed]
18. Curry, T.E., Jr.; Osteen, K.G. The matrix metalloproteinase system: Changes, regulation, and impact throughout the ovarian and uterine reproductive cycle. *Endocr. Rev.* **2003**, *24*, 428–465. [CrossRef]
19. Shi, R.; Liao, C.; Zhang, Q. Hypoxia-Driven Effects in Cancer: Characterization, Mechanisms, and Therapeutic Implications. *Cells* **2021**, *10*, 678. [CrossRef]
20. Capellero, S.; Erriquez, J.; Battistini, C.; Porporato, R.; Scotto, G.; Borella, F.; Di Renzo, M.F.; Valabrega, G.; Olivero, M. Ovarian Cancer Cells in Ascites Form Aggregates That Display a Hybrid Epithelial-Mesenchymal Phenotype and Allows Survival and Proliferation of Metastasizing Cells. *Int. J. Mol. Sci.* **2022**, *23*, 833. [CrossRef]
21. Turunen, M.; Spaeth, J.M.; Keskitalo, S.; Park, M.J.; Kivioja, T.; Clark, A.D.; Makinen, N.; Gao, F.; Palin, K.; Nurkkala, H.; et al. Uterine leiomyoma-linked MED12 mutations disrupt mediator-associated CDK activity. *Cell Rep.* **2014**, *7*, 654–660. [CrossRef]

22. Dannappel, M.V.; Zhu, D.; Sun, X.; Chua, H.K.; Poppelaars, M.; Suehiro, M.; Khadka, S.; Lim Kam Sian, T.C.; Sooraj, D.; Loi, M.; et al. CDK8 and CDK19 regulate intestinal differentiation and homeostasis via the chromatin remodeling complex SWI/SNF. *J. Clin. Invest.* **2022**, *132*, e158593. [CrossRef] [PubMed]
23. Bulun, S.E.; Yilmaz, B.D.; Sison, C.; Miyazaki, K.; Bernardi, L.; Liu, S.; Kohlmeier, A.; Yin, P.; Milad, M.; Wei, J. Endometriosis. *Endocr. Rev.* **2019**, *40*, 1048–1079. [CrossRef] [PubMed]
24. Chuffa, L.G.A.; Lupi, L.A.; Cucielo, M.S.; Silveira, H.S.; Reiter, R.J.; Seiva, F.R.F. Melatonin Promotes Uterine and Placental Health: Potential Molecular Mechanisms. *Int. J. Mol. Sci.* **2019**, *21*, 300. [CrossRef] [PubMed]
25. Elia, P.P.; Tolentino, Y.F.; Bernardazzi, C.; de Souza, H.S. The role of innate immunity receptors in the pathogenesis of inflammatory bowel disease. *Mediat. Inflamm.* **2015**, *2015*, 936193. [CrossRef]
26. Gonzalez-Fernandez, J.; Zaragozano, S.; Monteagudo-Sanchez, A.; Simon, C.; Vilella, F. Single-cell technology: The key to an improved understanding of the human endometrium in health and disease. *Am. J. Obstet. Gynecol.* **2025**, *232*, S43–S53. [CrossRef]
27. de Almeida, B.C.; Dos Anjos, L.G.; Dobroff, A.S.; Baracat, E.C.; Yang, Q.; Al-Hendy, A.; Carvalho, K.C. Epigenetic Features in Uterine Leiomyosarcoma and Endometrial Stromal Sarcomas: An Overview of the Literature. *Biomedicines* **2022**, *10*, 2567. [CrossRef]
28. Yang, Q.; Bariani, M.V.; Falahati, A.; Khosh, A.; Lastra, R.R.; Siblini, H.; Boyer, T.G.; Al-Hendy, A. The Functional Role and Regulatory Mechanism of Bromodomain-Containing Protein 9 in Human Uterine Leiomyosarcoma. *Cells* **2022**, *11*, 2160. [CrossRef]
29. Yang, Q.; Al-Hendy, A. The Functional Role and Regulatory Mechanism of FTO m(6)A RNA Demethylase in Human Uterine Leiomyosarcoma. *Int. J. Mol. Sci.* **2023**, *24*, 7957. [CrossRef]
30. Omran, M.M.; Vafaei, S.; Alkhrait, S.; Yang, Q.; Al-Hendy, A. Dual Targeting EZH2 and Histone Deacetylases in Human Uterine Sarcoma Cells Under Both 2D and 3D Culture Conditions. *J. Cell Mol. Med.* **2025**, *29*, e70626. [CrossRef]
31. Yang, Q.; Falahati, A.; Khosh, A.; Mohammed, H.; Kang, W.; Corachan, A.; Bariani, M.V.; Boyer, T.G.; Al-Hendy, A. Targeting Class I Histone Deacetylases in Human Uterine Leiomyosarcoma. *Cells* **2022**, *11*, 3801. [CrossRef] [PubMed]
32. Yu, H.; Luo, S.; Ji, J.; Wang, Z.; Zhi, W.; Mo, N.; Zhong, P.; He, C.; Wan, T.; Jin, Y. A Deep-Learning-Based Artificial Intelligence System for the Pathology Diagnosis of Uterine Smooth Muscle Tumor. *Life* **2022**, *13*, 3. [CrossRef] [PubMed]
33. Nakagawa, M.; Nakaura, T.; Namimoto, T.; Iyama, Y.; Kidoh, M.; Hirata, K.; Nagayama, Y.; Yuki, H.; Oda, S.; Utsunomiya, D.; et al. Machine Learning to Differentiate T2-Weighted Hyperintense Uterine Leiomyomas from Uterine Sarcomas by Utilizing Multiparametric Magnetic Resonance Quantitative Imaging Features. *Acad. Radiol.* **2019**, *26*, 1390–1399. [CrossRef] [PubMed]

cells

MDPI

Review

Autophagy and Female Fertility: Mechanisms, Clinical Implications, and Emerging Therapies

Abdel Halim Harrath [1], Md Ataur Rahman [2], Sujay Kumar Bhajan [3], Anup Kumar Bishwas [3], MD. Hasanur Rahman [3], Saleh Alwasel [1], Maroua Jalouli [4], Sojin Kang [5], Moon Nyeo Park [5] and Bonglee Kim [5,6,*]

1 Zoology Department, College of Science, King Saud University, Riyadh 11451, Saudi Arabia; hharrath@ksu.edu.sa (A.H.H.); salwasel@ksu.edu.sa (S.A.)
2 Department of Neurology, University of Michigan, Ann Arbor, MI 48109, USA; ataur1981rahman@hotmail.com
3 Department of Biotechnology and Genetic Engineering, Faculty of Life Sciences, Bangabandhu Sheikh Mujibur Rahman Science and Technology University, Gopalganj 8100, Bangladesh; sujaybge@gmail.com (S.K.B.); anupkumarbishwas21@gmail.com (A.K.B.); hasanurrahman.bge@gmail.com (M.H.R.)
4 Department of Biology, College of Science, Imam Mohammad Ibn Saud Islamic University (IMSIU), Riyadh 11623, Saudi Arabia; mejalouli@imamu.edu.sa
5 Department of Pathology, College of Korean Medicine, Kyung Hee University, 1-5 Hoegidong Dongdaemun-gu, Seoul 02447, Republic of Korea; bmb1994@khu.ac.kr (S.K.); mnpark@khu.ac.kr (M.N.P.)
6 Korean Medicine-Based Drug Repositioning Cancer Research Center, College of Korean Medicine, Kyung Hee University, Seoul 02447, Republic of Korea
* Correspondence: bongleekim@khu.ac.kr

Abstract: Autophagy, an evolutionarily conserved cellular mechanism essential for maintaining internal stability, plays a crucial function in female reproductive ability. In this review, we discuss the complex interplay between autophagy and several facets of female reproductive health, encompassing pregnancy, ovarian functions, gynecologic malignancies, endometriosis, and infertility. Existing research emphasizes the crucial significance of autophagy in embryo implantation, specifically in the endometrium, highlighting its necessity in ensuring proper fetal development. Although some knowledge has been gained, there is still a lack of research on the specific molecular impacts of autophagy on the quality of oocytes, the growth of follicles, and general reproductive health. Autophagy plays a role in the maturation, quality, and development of oocytes. It is also involved in reproductive aging, contributing to reductions in reproductive function that occur with age. This review explores the physiological functions of autophagy in the female reproductive system, its participation in reproductive toxicity, and its important connections with the endometrium and embryo. In addition, this study investigates the possibility of emerging treatment approaches that aim to modify autophagy, using both natural substances and synthetic molecules, to improve female fertility and reproductive outcomes. Additionally, this review intends to inspire future exploration into the intricate role of autophagy in female reproductive health by reviewing recent studies and pinpointing areas where current knowledge is lacking. Subsequent investigations should prioritize the conversion of these discoveries into practical uses in the medical field, which could potentially result in groundbreaking therapies for infertility and other difficulties related to reproduction. Therefore, gaining a comprehensive understanding of the many effects of autophagy on female fertility would not only further the field of reproductive biology but also open new possibilities for diagnostic and treatment methods.

Keywords: autophagy; female fertility; oocyte quality; follicular development; reproductive aging; therapeutic interventions

1. Introduction

Autophagy is crucial to cellular homeostasis in the complex network of cellular biology and controls self-degradation and recycling of cellular components that are essential to cell health [1]. Although its significance in cellular homeostasis is well established, recent investigations have demonstrated its complex participation in physiological and pathological processes, including female reproductive health [2]. Autophagy-related genes (ATGs) lead to complicated autophagic machinery. Beclin-1, ATG5, and ATG7 are key elements of this ensemble, whereby they guide autophagosome formation and cellular cargo breakdown [3]. Recently, it has become well known that autophagy is crucial for understanding its complex effects on female reproductive health [4]. With a delicately coordinated network of organs and hormonal signals, the female reproductive system depends on cellular equilibrium [5]. Recent investigations have linked autophagy to oocyte maturation, embryonic development, and placental function. Infertility and pregnancy difficulties have been linked to autophagy dysregulation [6]. Autophagy has far-reaching effects beyond cellular maintenance, and the complex role of autophagy in female reproductive health has not yet been well studied.

The aim of this study was to explore the functional aspects of autophagy in the female reproductive system beyond molecular details. This finding may explain the complex interaction of ATG proteins and signaling networks that regulate autophagy. An understanding of these molecular details is crucial for elucidating the reproductive health effects of autophagy [7]. Autophagy protects reproductive systems and processes, ranging from oocyte quality control to early embryonic development [8]. Autophagy also affects neurological illnesses, carcinogenesis, diabetes, development, life expectancy, and fertility [9]. During the process of follicular atresia, the remaining follicles undergo a process of degeneration and finally disintegration. The quantity of remaining oocytes, as well as their quality, continues to decrease with increasing age, which is the cause of the age-related decrease in female fertility [10]. There is a correlation between the quality of the oocyte and the development of the embryo and the fetus, as evidenced by the fact that older women have a greater chance of miscarriage [11]. To investigate the effects of autophagic dysregulation on polycystic ovary syndrome (PCOS), endometriosis, and recurrent pregnancy loss, in this review, we explored autophagy and its effects on female reproductive health in depth.

2. Crucial Relationship between Autophagy and Follicular Development

Follicular development is a complex process that is crucial for the fertility and reproduction of females. The follicles within the ovaries progress through a sequence of stages, thus culminating in ovulation [12]. Autophagy has a notable impact on multiple physiological processes, such as follicular growth [13]. An understanding of the complex interplay between autophagy and follicular development is crucial for comprehending ovarian function and fertility. Autophagy has multiple functions at various stages of folliculogenesis in ovarian follicles, ranging from the primordial to the preovulatory stages [14–17]. Primordial follicles, which are inactive follicles, depend on autophagy to maintain their dormancy and ensure their survival [18]. Autophagy inhibits the early activation of primordial follicles, hence controlling both the quantity and quality of the follicular pool [19]. During the shift from the dormant stage to the developing stage, autophagy becomes crucial for the functioning of granulosa cells and the quality of oocytes [19,20]. The granulosa cells that surround the oocyte play a vital role in supporting its development [21]. Autophagy in granulosa cells is essential for the appropriate turnover of organelles, the generation of energy, and the secretion of hormones, which are all crucial for the growth and maturation of follicles [22]. Moreover, autophagy protects against damage caused by oxidative stress throughout the process of follicular growth [23]. Oxidative stress caused by reactive oxygen species (ROS) produced in the follicle can negatively affect the quality of oocytes and reduce fertility [24]. Autophagy facilitates the reduction of oxidative stress by eliminating impaired cellular elements, thereby protecting the health of follicles and enhancing the competence of oocytes [25]. However, the regulation of autophagy in the ovaries is rigorously governed by many signaling pathways and molecular mechanisms. Autophagy is controlled by important regulators such as mTOR and AMPK [26,27]. These regulators use environmental cues and hormone signals to adjust autophagic activity based on the availability of nutrients, energy levels, and stressful conditions. The control of autophagy in the ovary is crucial, especially during the formation of follicles [28]. Gonadotropins, which include follicle-stimulating hormone (FSH) and luteinizing hormone (LH), control the process of autophagy in granulosa cells and theca cells, which correspondingly affects the formation of follicles and the production of steroids [19]. Moreover, ovarian hormones such as estrogen and progesterone have distinct impacts on autophagy in a specific manner according to the stage, thus emphasizing the complex interaction between autophagy and ovarian physiology [29].

Autophagy has diverse functions during folliculogenesis, including the regulation of primordial follicle quiescence and the support of granulosa cell function and oocyte quality [24]. An understanding of the control and abnormal function of autophagy in the ovary will provide valuable insights into the underlying causes of ovarian diseases and possible approaches for treating reproductive diseases [30]. Moreover, the recognition of the essential connection between autophagy and follicular growth has important practical implications for the fields of reproductive medicine, as well as fertility preservation [31]. Impaired autophagy has been associated with ovarian disorders, including premature ovarian insufficiency (POI) and polycystic ovary syndrome (PCOS), thus highlighting the significance of regulating autophagy as a potential treatment approach [32] (Figure 1). Future studies should prioritize investigating the molecular processes that govern the control of autophagy in the ovary and its influence on follicular development. Furthermore, the exploration of the therapeutic possibilities of directing autophagic pathways toward the treatment of ovarian diseases and infertility has potential for enhancing reproductive outcomes and the well-being of women.

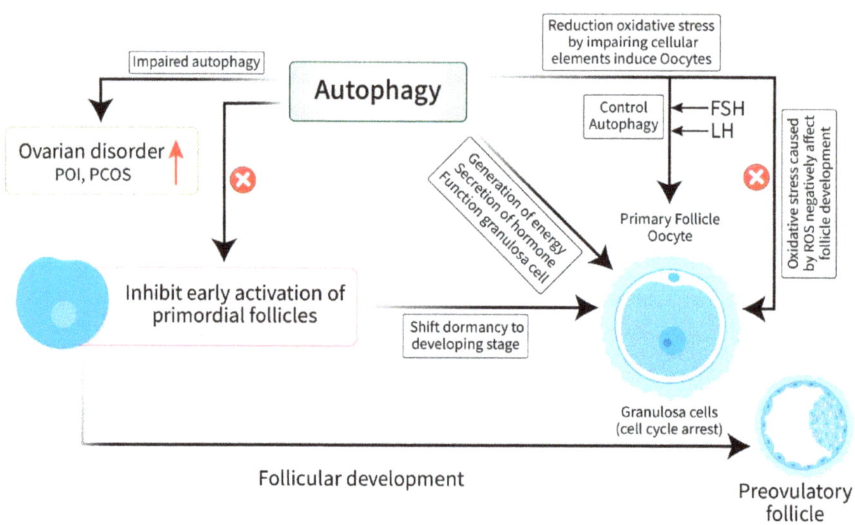

Figure 1. Relationship between autophagy and follicular development. Autophagy has a significant impact on follicular development ranging from primordial to preovulatory stages. Autophagy inhibits early activation of primordial follicles thus controlling the quality and quantity of it. It acts on granulosa cells resulting in the generation of energy and secretion of hormones controlled by gonadotropins. Moreover, autophagy inhibits ROS, which is produced by follicles and negatively affects the quality of oocytes. On the other hand, impaired autophagy may result in POI and PCOS.

3. Emerging Role of Autophagy in Oocyte Quality Control

Egg cells, which are also known as oocytes, are essential for reproduction and have a significant influence on both embryonic development and fertility [33]. It is essential to understand the complex processes regulating oocyte quality (including the function of autophagy) to advance reproductive medicine and fertility treatments [34]. Autophagy in oocytes is involved in many phases of early embryonic growth, folliculogenesis, and oogenesis [19]. Oocyte quality has been shown to be significantly impacted by autophagy, which is a cellular process that is known for its ability to recycle and preserve cellular homeostasis. Oogenesis is a highly regulated process in which oocyte maturation occurs through complex molecular and cellular processes [35]. Autophagy plays a crucial role in various important processes of oocyte development, such as removing excessive or damaged organelles, such as mitochondria, which are necessary for maintaining oocyte quality [36]. Malfunctioning mitochondria can result in oxidative stress, as well as impaired oocyte quality, thus ultimately impacting the process of fertilization and the development of embryos [37]. Autophagy is a cellular process that selectively eliminates damaged mitochondria to maintain their integrity [38]. This process is known as mitophagy.

Moreover, autophagy also impacts the survival and development of oocytes by controlling energy metabolism and nutrient availability [39]. Oocytes rely heavily on energy, thus necessitating adequate levels of ATP and metabolic substrates to undergo appropriate maturation and fertilization [40]. Autophagy is a process that breaks down and recycles cellular components to provide nutrients, which helps in the metabolism and development of oocytes [41]. Autophagy plays a significant role not only in the maturation of oocytes but also in the early stages of embryonic development [42]. After fertilization, the zygote undergoes a sequence of cellular divisions along with morphogenetic processes, thus resulting in the creation of a blastocyst [43]. Autophagy plays a role in embryonic development by facilitating the removal of paternal mitochondria, preserving the integrity of the genome, and promoting cellular differentiation [44]. Autophagy impairment during the initial stages of embryonic development can result in irregularities in development and growth, as well as difficulties during pregnancy [45].

The regulation of autophagy in oocytes is controlled by a complex system of signaling routes and molecular processes. Crucial controllers of autophagy, such as the mammalian target of rapamycin (mTOR) pathway and the AMP-activated protein kinase (AMPK) pathway, have essential functions in organizing autophagic activity in response to different environmental signals and cellular stresses [19,46]. Autophagy in oocytes is influenced by hormonal cues, growth factors, and nutrient availability, thus demonstrating the close relationship between metabolic conditions and reproductive function [47]. Ultimately, autophagy is crucial for controlling the quality of oocytes by participating in mitochondrial dynamics, energy metabolism, and cellular homeostasis [48]. An understanding of the complex relationship between autophagy and oocyte development has significant potential for developing reproductive medicine and fertility treatments, which can ultimately help people with infertility and reproductive diseases [49]. Although there have been notable advancements in comprehending the function of autophagy in oocyte quality (Figure 2), there are still several unresolved questions. Moreover, investigating prospective treatment strategies that focus on autophagy has potential for enhancing reproductive outcomes and addressing disorders associated with infertility.

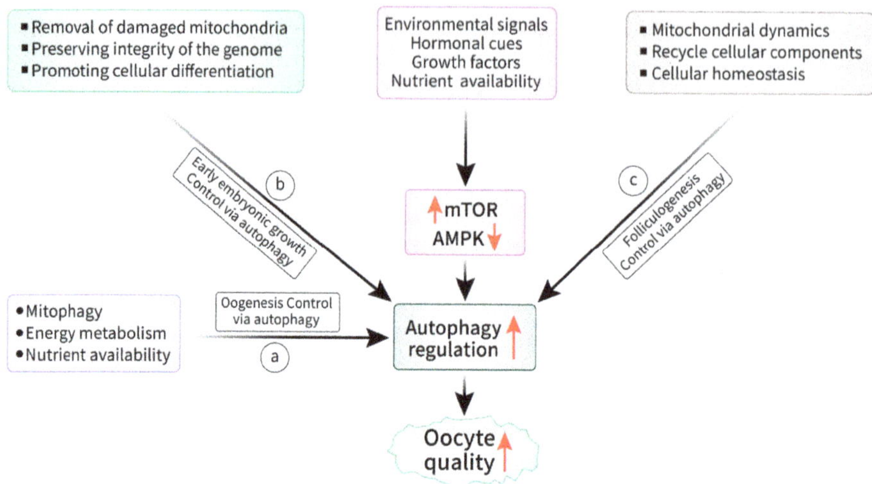

Figure 2. Autophagy control in oocyte quality via oogenesis, early embryonic growth and folliculogenesis. Specific environmental, external, and internal factors activate mTOR and AMPK pathways that result in the regulation and control of autophagy. Autophagy in oocytes is broadly involved in three interconnected phases: (a) In oogenesis: Malfunctioning mitochondria is eliminated, a process known as mitophagy, to avoid oxidative stress and impaired oocyte quality. In addition, autophagy provides adequate levels of ATP via metabolic substrates. Furthermore, it breaks down and recycles cellular components for nutritional requirements. (b) In early embryonic growth: Autophagy removes parental mitochondria since the mixture of parental mitochondria can lead to mitochondrial dysfunction. On the other hand, it preserves the integrity of the genome through removing damaged DNA, clearing oxidative stress, and regulating the repair of DNA. It also promotes cellular differentiation. (c) In folliculogenesis: Autophagy maintains cellular homeostasis by breaking down and recycling cellular components or eliminating damaged organelles.

Recent study has emphasized the crucial significance of autophagy in the process of oocyte maturation and follicle growth [50]. Investigations involving the deactivation of autophagic genes in oocytes have yielded valuable knowledge on the processes that determine the quality and developmental potential of oocytes [51]. For example, when important autophagic genes like Atg5 and Beclin1 are deactivated, it has been demonstrated that oocyte maturation is hindered, resulting in decreased fertility and worse embryonic development. Autophagy is a crucial process for eliminating impaired organelles and ensuring the stability of cellular conditions in oocytes. Nevertheless, despite these progressions, there is still a lack of literature that particularly focuses on the elimination of autophagic genes in oocytes. To close this gap, additional study is required to clarify the specific functions of several autophagic pathways and their interconnections in the process of oocyte development. These investigations will not only improve our understanding of reproductive biology but also potentially result in new therapeutic approaches for enhancing oocyte quality and fertility outcomes.

4. Role of Autophagy in Reproductive Aging

Reproductive aging is an inherent process marked by a gradual decrease in reproductive capacity as individuals grow older, and it impacts both males and females [52]. Although significant focus has been given to the hormonal and physiological transformations that occur throughout this period, recent studies have provided insight into the involvement of autophagy in reproductive aging [53]. A comprehensive understanding of the role of autophagy in the process of reproductive aging could provide valuable knowledge about the fundamental mechanisms that are involved and potential approaches for treatment [54]. The quality of oocytes is a crucial factor in determining fertility and reproductive success in females [55]. Oocytes are more susceptible to age-related deterioration because of the gradual increase in cellular damage. Research has demonstrated that the activity of autophagy declines in oocytes as they age, thus resulting in diminished quality and decreased ability to develop [56]. Impaired autophagy can lead to the buildup of impaired cell structures, such as mitochondria, and harmful clumps of proteins, which can contribute to the aging of egg cells and infertility [57]. The use of pharmacological or genetic methods to increase autophagic activity has been proposed as being a viable technique to enhance oocyte quality and fertility in elderly females [58]. The decrease in reproductive capacity due to aging is affected not only by internal factors but also by external factors such as lifestyle, nutrition, and environmental exposures [59]. Recent research indicates that dietary therapies and lifestyle changes that stimulate autophagy activation may have positive impacts on reproductive aging [60]. Studies have demonstrated that restricting caloric intake, practicing intermittent fasting, and supplementing the diet with chemicals that promote autophagy can increase autophagic activity and enhance reproductive outcomes in animal models [61,62]. Moreover, current research has demonstrated specific molecular pathways and regulatory systems that control autophagy during the process of reproductive aging (Figure 3). Autophagy is controlled by important regulators such as mTOR and AMPK. These regulators have been found to play a role in controlling the function of the ovaries and testes [63]. The exploration of the relationship between these communication routes and autophagy may result in the identification of new treatment targets for age-related loss of reproductive function [64].

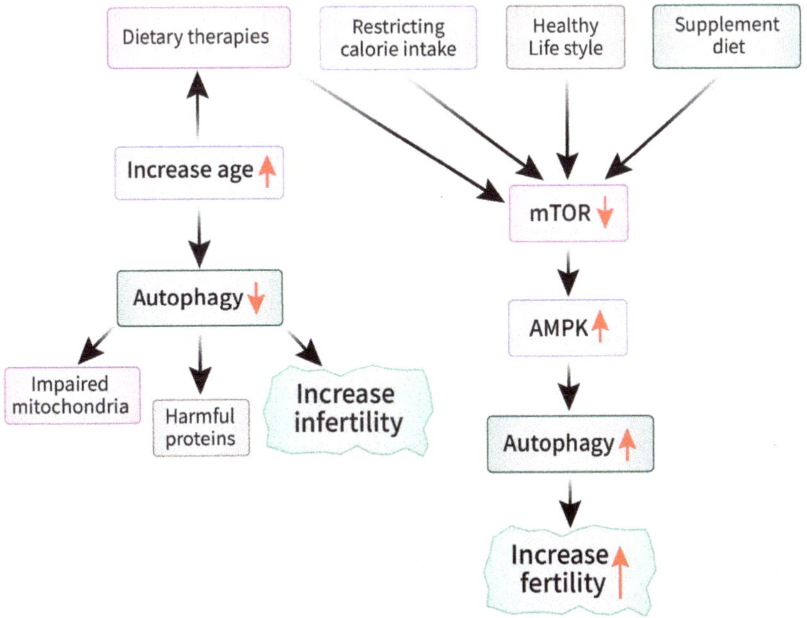

Figure 3. Role of autophagy in reproductive aging. Several internal and external factors upregulate autophagy by decreasing mTOR and increasing AMPK. Increased autophagy maintains cellular homeostasis that consequently inhibits reproductive aging. In contrast, impaired autophagy has been linked to lower oocyte and sperm quality, infertility, impaired cell structure, and production of harmful proteins that accelerate reproductive aging.

mTOR serves as a pivotal controller of cell growth, proliferation, and metabolism. It combines signals from nutrients, growth factors, and cellular energy levels [65]. More precisely, mTOR signaling plays a vital role in initiating the growth of primordial follicles, an essential process in the formation of ovarian follicles [66]. Primordial follicle activation refers to the transformation of inactive primordial follicles into developing primary follicles [67]. This process is carefully controlled by an intricate network of signaling channels. mTOR facilitates this transition by augmenting protein synthesis, food absorption, and energy generation, thereby supplying the essential resources for follicle growth and development [68]. Furthermore, the mTOR signaling system interacts with the PI3K-Akt pathway to guarantee accurate follicle activation and growth. Imbalance in mTOR activity can result in conditions like premature ovarian failure or PCOS, underscoring its crucial involvement in preserving ovarian function and fertility [69].

Autophagy is crucial in reproductive aging because it helps to maintain cellular balance and protects reproductive function [64]. Impaired autophagy has been linked to the deterioration of oocyte and sperm quality, thus resulting in age-related infertility in both males and females [19]. Strategies focused on increasing autophagic activity show potential for maintaining reproductive health and fertility in elderly people [53]. Additional investigations are required to clarify the precise molecular pathways that underlie the involvement of autophagy in reproductive aging. Therefore, further research is necessary to identify specific therapies that can alleviate the reduction in fertility associated with aging.

5. Physiological Role of Autophagy in the Female Reproductive System

Autophagy has a complex impact on female reproductive physiology, whereby it affects several elements of fertility, pregnancy, and reproductive health [6]. An understanding of the complex relationship between autophagy and reproductive processes has the potential to identify the mechanisms that cause infertility, pregnancy problems, and reproductive disorders [7] (Figure 4). Furthermore, directing attention toward autophagic pathways may demonstrate innovative therapeutic approaches for the treatment of reproductive diseases and the enhancement of reproductive outcomes in women. Additional investigations are necessary to clarify the exact processes and therapeutic possibilities involved in manipulating autophagy in female reproductive disorders. Recently, scientists examined the role of autophagy in the complex processes of female reproductive physiology. This research has provided insights into how autophagy affects fertility, pregnancy, and reproductive diseases [7,70]. Autophagy plays a significant role in oogenesis, which is the process of egg production and a crucial aspect of female reproductive physiology [33]. Autophagy is essential for the growth and maintenance of oocytes, thus ensuring that they mature correctly and are capable of fertilization [71]. Research has shown that when the process of autophagy in oocytes is not properly regulated, it may result in a reduced ability for oocytes to develop normally and increased vulnerability to abnormalities [72]. This can potentially cause infertility or difficulties during pregnancy.

Moreover, autophagy plays a crucial role in maintaining ovarian function and folliculogenesis, which refers to the recurring process of follicle growth and the release of eggs [36]. The delicate equilibrium between autophagy and apoptosis in granulosa cells, which encircle and provide support to the maturing oocyte within the follicle, is essential for the growth and selection of follicles [28,73]. Dysregulation of autophagy in granulosa cells has been linked to ovarian dysfunction, polycystic ovary syndrome (PCOS), and premature ovarian failure (POF), thus underscoring its importance in reproductive illnesses [74]. Autophagy plays diverse roles in sustaining the health of both the mother and the fetus during pregnancy, thus helping them adapt to constant physiological changes [75]. It promotes the growth and development of the placenta, invasion of trophoblasts, and delivery of nutrients, which are essential for healthy fetal growth and development [76]. In addition, autophagy assists in reducing oxidative stress and inflammation, both of which are elevated during pregnancy [6,77]. This correspondingly provides protection against pregnancy problems such as pre-eclampsia and gestational diabetes.

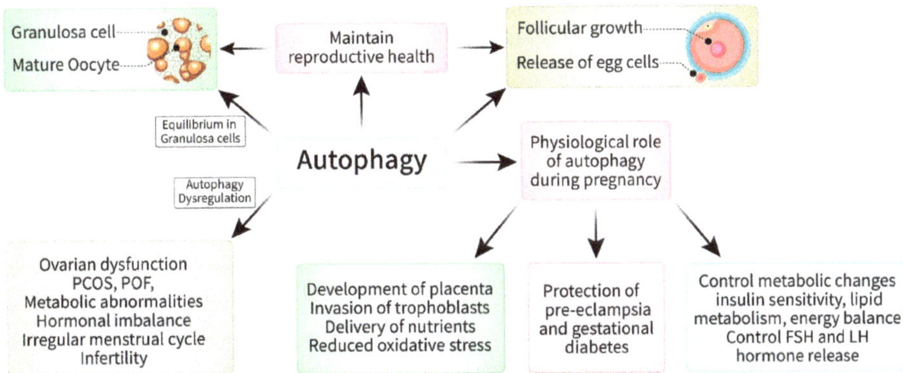

Figure 4. Physiological role of autophagy in female reproductive system. Autophagy has been linked to playing a crucial role in four areas: fertility, reproductive health, pregnancy, and metabolic changes. In fertility: autophagy helps in growth and maintenance of oocytes where impaired autophagy may result in abnormal development of oocytes. In reproductive health: autophagy keeps equilibrium in granulosa cell and impaired one causes ovarian distinction, PCOS, POF, cellular damage and oxidative stress. In pregnancy: autophagy develops placenta, helps in invasion of trophoblasts, etc., where impaired autophagy is associated with preeclampsia and gestational diabetes. In metabolic changes: normal autophagy regulates insulin sensitivity, lipid metabolism, and energy balance. Infraction of autophagy leads to irregular menstrual cycle.

Furthermore, autophagy plays a role in controlling maternal metabolic changes during pregnancy, including insulin sensitivity, lipid metabolism, and energy balance [78]. Abnormal autophagy has been linked to metabolic abnormalities during pregnancy, thus indicating that it could be a useful target for treating difficulties related to metabolic dysfunction in pregnant women [79]. In addition to its role in reproduction, autophagy also has an impact on hormone release, including effects on gonadotropins such as follicle-stimulating hormone (FSH) and luteinizing hormone (LH), which play crucial roles in regulating ovarian function and the menstrual cycle [80,81]. Imbalanced autophagy in the hypothalamic–pituitary–gonadal (HPG) axis can disturb the equilibrium of hormones, thus resulting in irregular menstrual cycles, infertility, or reproductive problems [81,82]. Moreover, there is growing evidence indicating a connection between autophagy and reproductive aging, which refers to the gradual decrease in fertility and ovarian function as one ages [83]. Decreases in autophagic activity associated with aging can lead to cellular damage, oxidative stress, and genetic instability in the ovary [19,84]. This can ultimately affect fertility and reproductive outcomes in older women.

6. Role of Autophagy in Female Reproductive System Toxicity

Granulosa cells, being the largest and most essential functional cells in the follicle, produce various signaling factors, growth factors, and cytokines [85]. They proliferate, differentiate, and regulate the microenvironment surrounding the follicle, which is crucial for its development and maturation. During follicle development, the number of granulosa cells increases, starting from early sinusoid follicles with only a few thousand cells to antral follicles with a minimum of 100,000 cells [86,87]. Numerous toxicological experiments have been conducted on ovarian granulosa cells to investigate the impact of toxic chemical pollutants such as bisphenol A, acrylamide, pesticides, mycotoxins, zearalenone, BTEX, and microplastics on granulosa cell proliferation and follicle development [15,17,88–92]. Reports showed that chemicals' toxic effects on the female reproductive system can lead to disorders in the ovarian cycle, increased spontaneous abortion rates, abnormal development of offspring, and a decline in fertility [92,93]. The developmental toxicological mechanisms of these chemicals involve overproduction of reactive oxygen species (ROS), cellular oxidative damage, apoptosis, and autophagy ultimately leading to follicular atresia [17,92,94]. Studies have demonstrated the significance of autophagy in granulosa cells for follicular atresia, as excessive ROS-induced oxidative stress is a major inducer of autophagy in various ovarian cells [17,95]. In fact, toxicants can induce reproductive toxicity through autophagy by increasing the levels of autophagy-related proteins such as Beclin-1, LC3B-II, and the LC3B-II/LC3BeI ratio.

Autophagy and apoptosis play significant roles in hormonally regulated follicular atresia [96]. The former serves as a dual-function mechanism that regulates the survival and death of ovarian cells, resulting in different effects, such as cytoprotection or cytotoxicity [97,98]. Under normal circumstances, autophagy levels in cells are low, promoting cell survival. It acts as an intracellular degradation system that encapsulates unnecessary or malfunctioning cellular components within double-membraned structures called autophagosomes, which are then degraded and recycled [99]. Transmission electron microscopy is the gold standard technique commonly used to observe double-membraned autophagic vacuoles, providing convincing evidence of cell autophagy [17,98]. In TEM images, ovarian cells exposed to toxic chemicals exhibit numerous autophagic vacuoles, that in a later stage fuse with lysosomes for degradation and recycling (Figure 5). Increased autophagy is accompanied by elevated levels of malondialdehyde (MDA) and reduced levels of SOD, GSH, and CAT, indicating that oxidative stress induces autophagy in chemical-exposed ovaries [17,92]. In addition, ROS reportedly activate autophagy in follicular granulosa cells via the mTOR pathway, highlighting the close relationship between oxidative stress and autophagy in ovarian cells [15–17,95]. This autophagy's activation serves as an initial step in maintaining the balance between ROS and antioxidant scavengers, protecting cell viability and playing a cytoprotective role against chemical-induced apoptosis in granulosa cells [100,101]. However, excessive ROS production can result in cell death due to autophagy's failure to repair ovarian cells. Thus, autophagy can act as an antagonist against apoptosis but can also contribute to cell apoptosis under different conditions [102,103]. This autophagy-induced reproductive toxicity is mediated through the inhibition of the Akt-mTOR signaling pathway, which plays a significant role in autophagic cell death [17,104]. It has also been reported that when autophagy is disrupted in ovarian cells under the effect of toxicants, the transcription factor WT1 accumulated in granulosa cells leads to the inhibition of their differentiation [96].

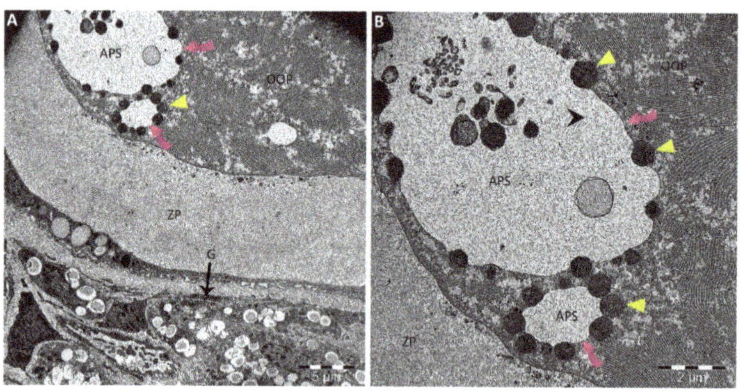

Figure 5. Transmission electron microscopy (TEM) images providing direct visual evidence of the autophagic process occurring within the oocyte cytoplasm. (**A**) The ultrastructure of an ovary from a rat shows the presence of autophagosomes (pink arrows) within the cytoplasm of an oocyte, which appears to be a protective mechanism after in utero treatment with acrylamide. (**B**) A higher-magnification microphotograph of the image in Figure (**A**) reveals large autophagosomes (APS) engulfing extracellular material. These autophagosomes are surrounded by many lysosomes (yellow arrowheads) at different stages of fusion, forming autophagolysosomes. APS: autophagosomes; G: granulosa cells; OOP: ooplasm; ZP: zona pellucida.

7. Crosstalk between Role of Autophagy on Endometrium and Embryo/Endometrium

Autophagy is a vital cellular process that is responsible for breaking down and reusing cellular components. It has a notable impact on the reproductive system, specifically in the endometrium and its interaction with the embryo [105]. The endometrium, which is the inner lining of the uterus, experiences periodic alterations that are controlled by hormonal signals to prepare for the attachment of an embryo [106]. Autophagy regulates cellular turnover and preserves endometrial homeostasis, creating a favorable environment for the embryo [107]. The interaction between autophagy in the endometrium and the embryo is essential for the effective attachment of the embryo and the occurrence of pregnancy [108]. Autophagy in the endometrium plays a crucial role in mitigating cellular stress, eliminating dysfunctional organelles, and controlling inflammation [109]. These processes are essential for establishing a favorable environment for the embryo. Dysfunction in autophagic mechanisms in the endometrium can result in compromised receptivity, which can contribute to disorders including infertility and recurrent pregnancy loss [110]. Autophagy plays a crucial role in the survival of the embryo during the pre-implantation phase by breaking down cellular components to provide nutrients and energy [111]. The embryonic signals and endometrial autophagy combine to achieve synchrony, facilitating effective implantation [112]. Factors derived from the embryo can influence the process of autophagy in the endometrium, which in turn improves the ability of the endometrium to receive the embryo and facilitates its invasion and attachment [113]. On the other hand, when endometrial autophagy is suitable, it can impact the release of signaling molecules that aid in the growth of an embryo [114]. Gaining insight into the interaction between autophagy in the endometrium and embryo is crucial for the development of therapeutic approaches to tackle reproductive diseases [115]. By specifically focusing on autophagic pathways, it may be feasible to enhance the ability of the endometrium to accept and support embryos, hence improving reproductive outcomes.

8. Recent Therapeutic Targets for Autophagy in Female Fertility and Treatment

The process of female fertility is complex and controlled by multiple molecular pathways, one of which is autophagy [58,72]. This process is essential for maintaining the balance and functioning of cells. Recently, there has been a growing emphasis on the understanding of the significance of autophagy in female fertility and its potential as a target for therapeutic intervention in diseases associated with fertility. Recent progress has been made in discovering therapeutic targets of autophagy in female fertility and their implications for treatment options [116]. Due to the significant role of autophagy in female reproductive physiology, a focus on this pathway has become a promising strategy for addressing infertility and associated diseases. Multiple recent studies have demonstrated specific targets within the autophagy pathway that could be utilized for reproductive treatment [53,58]. An example of a target is mTOR, which is a crucial regulator of autophagy and is responsible for controlling cellular development and metabolism. Preclinical studies have demonstrated that suppressing mTOR enhances autophagy and leads to improved egg quality, ovarian function, and reproductive outcomes [117]. AMPK is a cellular energy sensor that controls autophagy based on nutrient availability. AMPK activation has been shown to induce autophagy and enhance oocyte maturation and embryo development [118,119] (Figure 6). Furthermore, it has been suggested that manipulating the expression or function of autophagy-related genes, such as Beclin-1 and Atg7, could be a method to improve fertility in women who are undergoing assisted reproductive technology (ART) [120]. Moreover, recent findings indicate that the disruption of autophagy may play a role in the development of certain fertility problems, such as polycystic ovary syndrome (PCOS), endometriosis, and premature ovarian insufficiency (POI) [121,122]. The targeting of the autophagy mechanisms associated with these diseases could provide new treatment strategies for controlling infertility in affected individuals [123]. Moreover, more investigations are required to clarify the exact mechanisms through which autophagy controls female fertility and to discover other therapeutic targets within this system. The effective translation of these results into clinical practice and the development of autophagy-based medicines for infertility will require collaborative efforts among fundamental scientists, doctors, and pharmaceutical businesses.

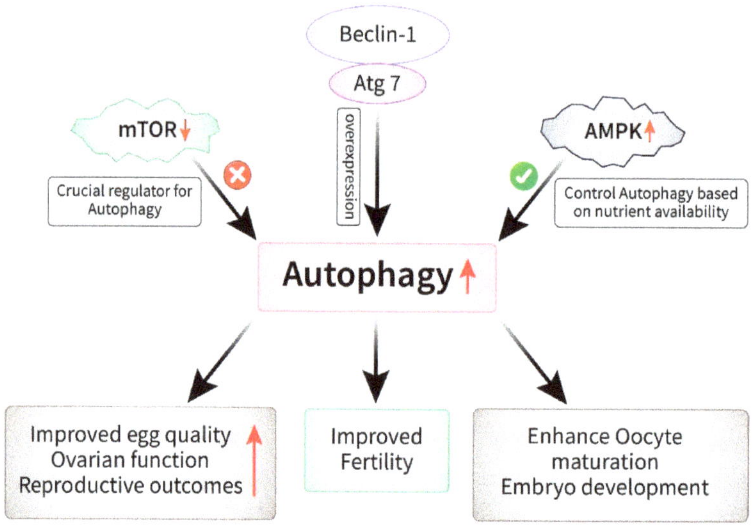

Figure 6. Recent therapeutic targets for autophagy in female fertility and treatment. Genes and proteins: Beclin-1 protein and Atg7 gene have an impact on autophagy that leads to improved fertility in those undergoing ART. Pathways: downregulated mTOR increases autophagy that results in improved egg quality, ovarian function, and reproductive outcomes. On the other hand, the upregulation of AMPK is associated with oocyte maturation and embryo development.

8.1. Therapeutic Targets for Autophagy Modulation in Female Fertility and Embryo Implantation via Natural Products

The modulation of autophagic activity using drugs or natural substances is a promising strategy for enhancing fertility and optimizing reproductive outcomes. Various substances have been found to be regulators of autophagy, with the ability to either inhibit or stimulate autophagy. Plants, herbs, and traditional medical systems produce natural products that contain a wide range of bioactive chemicals with various pharmacological actions. Several of these substances have demonstrated the ability to regulate autophagy and have potential therapeutic advantages for female fertility (Figure 7). Resveratrol, which is a polyphenolic molecule present in red grapes and berries, has been shown to boost autophagic activity and promote ovarian function in animal models of ovarian aging [124]. Curcumin, which is a beneficial component found in turmeric, has been shown to trigger autophagy and perhaps safeguard against oxidative stress-induced ovarian damage [125]. Furthermore, scientific studies have examined the effects of herbal extracts, including ginseng, green tea, and chasteberry, on the regulation of autophagy and the enhancement of female reproductive health [126,127]. These naturally occurring substances contain phytochemicals that possess antioxidant, anti-inflammatory, and hormone-modulating properties. These features may play a role in their positive impacts on fertility. By specifically focusing on autophagic pathways, these natural substances may provide a safe and efficient option for women who want to improve their fertility and overcome difficulties related to reproduction [128]. Additional investigations are necessary to clarify the mechanisms by which natural compounds impact autophagy and their potential use in reproductive medicine. By utilizing the medicinal properties of organic substances, we can create new methods to enhance women's reproductive well-being and pregnancy outcomes.

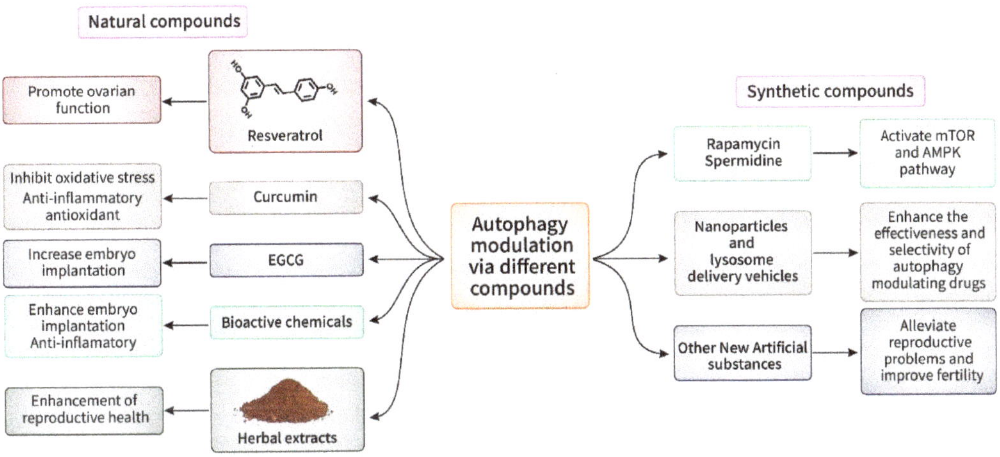

Figure 7. Autophagy modulation via natural and synthetic compounds. Both natural and synthetic compounds play a significant role in autophagy modulation. Natural compounds like resveratrol, curcumin, EGCG, bioactive chemicals and herbal extracts have numerous positive impacts on autophagy modulation including promotion of ovarian function, inhibition of oxidative stress, and enhancement of female reproductive health. They also possess anti-inflammatory, antioxidant, and hormone-modulating properties. One the other hand, synthetic compounds like rapamycin, spermidine, and other new artificial substances with their carriers (nanoparticles and lysosome) activate mTOR and AMPK pathways, lessen reproductive problems, and improve fertility by enhancing the effectiveness and selectivity of autophagy-modulating drugs.

Plants, marine organisms, and microorganisms have a wide range of bioactive chemicals with potential for use in therapy [129]. Various natural substances have been shown to act as regulators of autophagy, thus presenting potential opportunities for improving the process of embryo implantation [130]. Resveratrol, which is a type of polyphenol present in red grapes and berries, has gained significant recognition for its ability to enhance autophagy [131]. Research has shown that it can enhance the growth of embryos and their attachment to the uterus by stimulating autophagic pathways in both embryos and endometrial cells [132]. Curcumin, which is a biologically active molecule extracted from turmeric, has strong anti-inflammatory and antioxidant effects and can regulate autophagy [133]. Curcumin has been demonstrated to enhance autophagic flux, thus leading to improved endometrial receptivity and facilitating embryo implantation in animal models [134]. Moreover, studies have shown that epigallocatechin gallate (EGCG), which is a type of catechin found in green tea, may enhance autophagy and facilitate the process of embryo implantation by exerting its antioxidative and anti-inflammatory effects [135]. Recent research breakthroughs have demonstrated numerous natural substances that can effectively regulate autophagy to improve the process of embryo implantation [132]. Ginsenosides, which are active components extracted from ginseng, have been demonstrated to enhance autophagy and enhance the responsiveness of the endometrium, thereby promoting the successful attachment of embryos [136]. Quercetin, which is a flavonoid found in fruits and vegetables, has a positive impact on implantation by influencing autophagy-related pathways [137]. Natural compounds can enhance embryo implantation by modifying autophagy, and their therapeutic benefits extend beyond this specific property. These chemicals have several effects, thereby affecting different signaling pathways that are involved in the process of implantation and the formation of early pregnancy. For example, polyunsaturated fatty acids, such as omega-3 fatty acids found in fish oil, not only control autophagy but also have anti-inflammatory effects, which help with endometrial receptivity and embryo implantation [138]. To fully utilize the therapeutic benefits of natural products in improving embryo implantation, it is crucial to have a thorough understanding of their molecular mechanisms and pharmacological qualities. By incorporating sophisticated methods such as omics approaches and bioinformatics, we can discover the complex relationships among natural products, autophagy, and implantation. Moreover, it is crucial to conduct thorough preclinical investigations and clinical trials to confirm the effectiveness and safety of therapies based on natural products in enhancing reproductive outcomes. The utilization of the combined benefits of natural products on autophagy and other signaling pathways has great potential for enhancing fertility therapies and reducing difficulties related to implantation failure.

8.2. Targeting Autophagy Modulation in Female Fertility and Embryo Implantation via Synthetic Compounds

Synthetic substances that specifically affect the process of autophagy show great potential for use in medical treatments to improve female fertility and enhance the success of embryo implantation (Figure 7). Preclinical studies have demonstrated that small compounds, such as rapamycin and spermidine, can regulate autophagic activity and enhance reproductive outcomes [139]. These drugs exert their effects by activating several signaling pathways, including the mTOR and AMP pathways, which play crucial roles in regulating autophagy. Furthermore, the development of new artificial substances that particularly target autophagy in the female reproductive system has significant potential for clinical application [140]. By specifically controlling the flow of autophagy in oocytes, granulosa cells, and endometrial cells, these substances have the potential to alleviate reproductive problems and improve fertility. In addition, the use of delivery vehicles, such as nanoparticles and liposomes, could enhance the effectiveness and selectivity of autophagy-modulating drugs [141]. This scenario would reduce unintended effects and maximize therapeutic advantages. Nevertheless, there are still obstacles to overcome regarding the application of autophagy regulation in clinical settings for the purpose of addressing female infertility and facilitating embryo implantation. Additional investigations are required to clarify the exact mechanisms that cause autophagy dysregulation in reproductive diseases and to discover new targets for treatments. Furthermore, comprehensive preclinical and clinical investigations are important for evaluating the safety, effectiveness, and enduring impacts of synthetic drugs that specifically target autophagy in the female reproductive system [142]. By optimizing autophagic activity in the oocyte and the endometrium, these chemicals provide a new and effective treatment method for reproductive diseases and may improve assisted reproductive procedures [143]. Continued research endeavors in this domain are crucial for fully harnessing the potential of autophagy modulation in female reproductive health.

8.3. Potential Diagnostic and Therapeutic Applications of Autophagy in Female Fertility

Monitoring autophagy indicators can offer vital information about the condition of the ovaries and their ability to reproduce [4]. Autophagy-related genes (ATGs) and proteins, including LC3, Beclin-1, and p62, play a vital role in the quality of oocytes and the development of follicles [72,144]. Anomalous expression of these markers can suggest hindered egg maturation or diminished ovarian reserve, assisting in the identification of diseases such as PCOS and POF [145]. Moreover, assessing autophagic activity in granulosa cells, which provide support for oocyte development, can function as a prognostic indicator for the achievement of fertilization and the implantation of embryos in ART [146].

Targeting the mechanisms involved in autophagy has potential for improving female fertility. Regulating autophagy can enhance the quality of oocytes and their ability to mature, which is essential for achieving optimal pregnancy results [147]. Autophagy inducers, like resveratrol and spermidine, have demonstrated promise in stimulating oocyte maturation and safeguarding against oxidative stress [148]. On the other hand, inhibiting autophagy could be advantageous in situations such as endometriosis, where excessive autophagy promotes the survival of cells in abnormal endometrial tissues [149]. Moreover, the manipulation of autophagy can aid in maintaining ovarian function while undergoing cancer treatments [50]. Chemotherapy and radiotherapy frequently result in ovarian impairment and infertility [150]. Autophagy activators can alleviate these consequences by safeguarding the ovarian reserve and bolstering tissue repair processes [151]. The dual function of autophagy in both facilitating and impeding reproductive processes highlights its potential as a diagnostic and therapeutic target in female fertility [152]. Further investigation is crucial to completely understand the mechanisms involved in autophagy in reproductive biology, which would facilitate the development of new therapies that could enhance fertility outcomes and promote ovarian health in women.

9. Limitations and Future Perspectives on Autophagy in Female Fertility

The connection between autophagy and female fertility has received considerable attention in recent years. Autophagy plays a role in multiple reproductive processes, such as egg development, folliculogenesis, and embryo implantation [153]. Nevertheless, despite the progress made in comprehending this subject, there are still several constraints in elucidating the complex interaction between autophagy and female fertility. The constraints and deliberations on potential prospects and limitations for advancing research on the role of autophagy in female fertility are subsequently discussed.

Insufficient comprehension of autophagic mechanisms in reproductive tissues is a substantial obstacle due to the intricate control of autophagy. Although the importance of autophagy in preserving oocyte quality and controlling follicular growth has been acknowledged, the exact mechanisms underlying these processes are not fully understood. Further research is needed to understand the specific molecular processes that control autophagy in relation to female fertility. The investigation of autophagy in reproductive tissues presents significant technological obstacles. Existing techniques for evaluating autophagic activity, such as immunoblotting and immunofluorescence, suffer from a lack of specificity and can produce inconsistent findings. It is crucial to develop more precise and reliable methods that are specifically designed for evaluating autophagy in reproductive tissues. Although preclinical studies have indicated a connection between the disruption of autophagy and female infertility, there is a lack of clinical data to substantiate this correlation. There is a need for extensive clinical investigations to examine biomarkers related to autophagy in individuals who are unable to conceive to confirm the significance of autophagy in human fertility and discover possible targets for diagnosis and treatment. The impact of the reduction in autophagic activity associated with aging has been linked to reproductive aging and age-related infertility [154]. Nevertheless, the specific consequences of a decrease in autophagy on the quality of oocytes and the results of reproduction are not yet fully understood. To understand the age-related alterations in autophagy and their consequences for female fertility, it is necessary to conduct longitudinal research involving animal models and human cohorts.

To overcome this limitation, advanced methods in molecular biology are used to evaluate autophagy. Novel molecular methodologies, such as single-cell RNA sequencing and proteomics, have the potential to decipher the intricate regulatory networks of autophagy in reproductive organs. By combining high-throughput methods with functional assays, we can gain a thorough understanding of the dynamics of autophagy during oogenesis, folliculogenesis, and embryogenesis [154]. Advances in targeted autophagy modulators, which specifically target certain elements of the autophagic machinery, exhibit considerable potential as therapeutic approaches for addressing female infertility. The development of small compounds or biologics that selectively stimulate or suppress autophagy in certain tissues can boost the quality of oocytes, improve ovarian function, and optimize reproductive results [155]. The utilization of CRISPR/Cas9-mediated genome editing has exceptional potential for exploring the functional importance of autophagy-related genes in female fertility [156]. Genetically modified animal models with precise modifications in autophagy-related pathways will allow researchers to analyze the specific functions of autophagy in reproductive processes [78,157]. Furthermore, the performance of longitudinal studies that monitor autophagic activity over the entire lifespan of females is crucial for comprehending the loss in fertility that occurs with age. By integrating autophagy assessments with clinical outcomes in various groups, we can clarify the causal connection between autophagy malfunction and reproductive aging.

10. Conclusions

To understand female fertility, this review explored autophagy and its enormous effects on reproductive health. Our study provides important insights into the intricacy and potential of autophagy and female fertility. Dynamic autophagy modulation during reproductive stages is a key discovery. Autophagy maintains cellular homeostasis throughout follicular growth, fertilization, and implantation, thus allowing for adaptation to the reproductive microenvironment [158]. This multifaceted overview of autophagy challenges its static image and highlights its responsiveness to female fertility. The elucidation of this link suggests new paths for therapies to improve oocyte quality, which is a key component in age-related decreases in fertility and recurrent pregnancy loss [10]. An understanding of the molecular mechanisms of autophagy and regulatory pathways involved in female fertility provide opportunities for new treatments. Those with reproductive issues may find hope in the targeting of autophagic systems to reduce age-related fertility. This extensive research provides the groundwork for reproductive health-tailored treatments. The tailoring of reproductive therapies to the individual's autophagic profile can enhance success rates and reduce dangers and negative effects. Its potential applications extend beyond assisted reproductive technology, which can affect fertility-related diseases. However, the complex relationship between autophagy and female fertility provides a promising scientific and clinical picture. The dynamic nature of autophagy, its impact on oocyte quality, and its translational implications necessitate further investigations. At the intersection of basic science and practical innovation, the elucidation of autophagy in female fertility has encouraged researchers, clinicians, and policymakers to work together to develop customized, effective reproductive treatments. We must acknowledge the gaps in our understanding of autophagy in female reproductive health as we begin this exploration. Future research will encounter obstacles and possibilities in understanding the role of autophagy in reproduction. By identifying these gaps, we hope to motivate future research into the function of autophagy in reproductive biology and provide novel treatment possibilities.

Author Contributions: A.H.H.: Conceptualization, Writing—original draft, Writing—review and editing, Visualization. M.A.R.: Writing—original draft, Data curation, Visualization. S.K.B.: Writing—review and editing, Figure drawing. A.K.B.: Writing—review and editing, Figure drawing. M.H.R.: Writing—review and editing, Figure drawing. M.J.: Writing—review and editing, Visualization. S.A.: Writing—review and editing, Visualization. S.K.: Writing—review and editing, Visualization. M.N.P.: Writing—review and editing, Visualization. B.K.: Writing—review and editing, Visualization, Funding. All authors have read and agreed to the published version of the manuscript.

Funding: This research was supported by Basic Science Research Program through the National Research Foundation of Korea (NRF) funded by the Ministry of Education (NRF-2020R1I1A2066868), the National Research Foundation of Korea (NRF) grant funded by the Korea government (MSIT) (No. 2020R1A5A2019413), and Basic Science Research Program through the National Research Foundation of Korea (NRF) grant funded by the Korea government (MSIT) (RS-2024-00350362).

Data Availability Statement: The datasets supporting the conclusions of this study are included within the article.

Conflicts of Interest: There are no conflicts of interest to declare.

References

1. Voronina, M.V.; Frolova, A.S.; Kolesova, E.P.; Kuldyushev, N.A.; Parodi, A.; Zamyatnin, A.A. The Intricate Balance between Life and Death: ROS, Cathepsins, and Their Interplay in Cell Death and Autophagy. *Int. J. Mol. Sci.* **2024**, *25*, 4087. [CrossRef] [PubMed]
2. Salmeri, N.; Viganò, P.; Cavoretto, P.; Marci, R.; Candiani, M. The kisspeptin system in and beyond reproduction: Exploring intricate pathways and potential links between endometriosis and polycystic ovary syndrome. *Rev. Endocr. Metab. Disord.* **2024**, *25*, 239–257. [CrossRef]

3. Nematisouldaragh, D. Autophagy Gene Regulation in Cardiac Myocytes and Cardiac Fibroblasts. 2024. Available online: http://hdl.handle.net/1993/38063 (accessed on 15 July 2024).
4. Zhao, Y.; Zhao, X.; Jiang, T.; Xi, H.; Jiang, Y.; Feng, X. A Retrospective Review on Dysregulated Autophagy in Polycystic Ovary Syndrome: From Pathogenesis to Therapeutic Strategies. *Horm. Metab. Res.* **2024**, *56*, 547–558. [CrossRef] [PubMed]
5. Britto, T.I. *Endocrine Equilibrium: Navigating the Hormonal Seas: Navigating the Whispers of Hormones: A Delicate Dance of Balance*; BookRix, 2023; ISBN 3755446073, 9783755446071.
6. Komijani, E.; Parhizkar, F.; Abdolmohammadi-Vahid, S.; Ahmadi, H.; Nouri, N. Autophagy-mediated immune system regulation in reproductive system and pregnancy-associated complications. *J. Reprod. Immunol.* **2023**, 103973. [CrossRef] [PubMed]
7. Goudarzi, S.T.; Vousooghi, N.; Verdi, J.; Mehdizadeh, A.; Aslanian-Kalkhoran, L.; Yousefi, M. Autophagy genes and signaling pathways in endometrial decidualization and pregnancy complications. *J. Reprod. Immunol.* **2024**, 104223. [CrossRef]
8. Jiang, G.; Zhang, L.; Zhao, J.; Li, L.; Huang, Z.; Wang, Z. Dynamic autophagy map in mouse female germ cells throughout the fetal to postnatal life. *Reprod. Sci.* **2023**, *30*, 169–180. [CrossRef]
9. Pavlova, J.A.; Guseva, E.A.; Dontsova, O.A.; Sergiev, P.V. Natural activators of autophagy. *Biochemistry* **2024**, *89*, 1–26. [PubMed]
10. Huang, W.; Li, X.; Yang, H.; Huang, H. The Impact of Maternal Age on Aneuploidy in Oocytes: Reproductive Consequences, Molecular Mechanisms, and Future Directions. *Ageing Res. Rev.* **2024**, *97*, 102292. [CrossRef]
11. Farquhar, C.M.; Bhattacharya, S.; Repping, S.; Mastenbroek, S.; Kamath, M.S.; Marjoribanks, J.; Boivin, J. Female subfertility. *Nat. Rev. Dis. Primers* **2019**, *5*, 7. [CrossRef]
12. Hernandez-Medrano, J.; Belmpa, M. Folliculogenesis and Oogenesis. In *Mastering Clinical Embryology*; CRC Press: Boca Raton, FL, USA, 2024; pp. 75–81.
13. Han, X.-H.; Wang, M.; Pan, Y.-Y.; Fan, J.-F.; Wang, J.-L.; Zhao, L.; Wang, Y.-Y.; Zhang, T.-X.; Zhao, T.; Ding, T.-Y.; et al. Effect of follicle-stimulating hormone and luteinizing hormone on apoptosis, autophagy, and the release and reception of some steroid hormones in yak granulosa cells through miR-23a/ASK1 axis. *Cell. Signal.* **2024**, *115*, 111010.
14. Lecot-Connan, T.; Boumerdassi, Y.; Magnin, F.; Binart, N.; Kamenický, P.; Sonigo, C.; Beau, I. Anti-Müllerian hormone induces autophagy to preserve the primordial follicle pool in mice. *FASEB J.* **2024**, *38*, e23506. [CrossRef] [PubMed]
15. Harrath, A.H.; Alrezaki, A.; Jalouli, M.; Aldawood, N.; Aldahmash, W.; Mansour, L.; Alwasel, S. Ethylbenzene exposure disrupts ovarian function in Wistar rats via altering folliculogenesis and steroidogenesis-related markers and activating autophagy and apoptosis. *Ecotoxicol. Environ. Saf.* **2022**, *229*, 113081. [CrossRef]
16. Harrath, A.H.; Alrezaki, A.; Jalouli, M.; Al-Dawood, N.; Dahmash, W.; Mansour, L.; Sirotkin, A.; Alwasel, S. Benzene exposure causes structural and functional damage in rat ovaries: Occurrence of apoptosis and autophagy. *Environ. Sci. Pollut. Res.* **2022**, *29*, 76275–76285. [CrossRef] [PubMed]
17. Jalouli, M.; Mofti, A.; Elnakady, Y.A.; Nahdi, S.; Feriani, A.; Alrezaki, A.; Sebei, K.; Bizzarri, M.; Alwasel, S.; Harrath, A.H. Allethrin promotes apoptosis and autophagy associated with the oxidative stress-related PI3K/AKT/mTOR signaling pathway in developing rat ovaries. *Int. J. Mol. Sci.* **2022**, *23*, 6397. [CrossRef]
18. Zhou, X.-Y.; Lai, Y.-H.; Zhang, J.; Li, Y.; Wu, X.-M.; Yang, Y.-Z.; Zhang, X.-F.; Ma, L.-Z.; Zheng, K.-M.; Liu, Y.-D. Advanced oxidation protein products attenuate the autophagy-lysosome pathway in ovarian granulosa cells by modulating the ROS-dependent mTOR-TFEB pathway. *Cell Death Dis.* **2024**, *15*, 161. [CrossRef]
19. Samare-Najaf, M.; Neisy, A.; Samareh, A.; Moghadam, D.; Jamali, N.; Zarei, R.; Zal, F. The constructive and destructive impact of autophagy on both genders' reproducibility, a comprehensive review. *Autophagy* **2023**, *19*, 3033–3061. [CrossRef]
20. Jiang, Y.; He, Y.; Pan, X.; Wang, P.; Yuan, X.; Ma, B. Advances in oocyte maturation in vivo and in vitro in mammals. *Int. J. Mol. Sci.* **2023**, *24*, 9059. [CrossRef]
21. Sugiura, K.; Maruyama, N.; Akimoto, Y.; Matsushita, K.; Endo, T. Paracrine regulation of granulosa cell development in the antral follicles in mammals. *Reprod. Med. Biol.* **2023**, *22*, e12538. [CrossRef]
22. Stringer, J.M.; Alesi, L.R.; Winship, A.L.; Hutt, K.J. Beyond apoptosis: Evidence of other regulated cell death pathways in the ovary throughout development and life. *Hum. Reprod. Update* **2023**, *29*, 434–456. [CrossRef]
23. Xi, H.; Hu, Z.; Han, S.; Liu, X.; Wang, L.; Hu, J. FSH-inhibited autophagy protects against oxidative stress in goat Sertoli cells through p62-Nrf2 pathway. *Theriogenology* **2023**, *195*, 103–114. [CrossRef] [PubMed]
24. Gao, Z.; He, W.; Liu, Y.; Gao, Y.; Fan, W.; Luo, Y.; Shi, X.; Song, S. Perinatal bisphenol S exposure exacerbates the oxidative burden and apoptosis in neonatal ovaries by suppressing the mTOR/autophagy axis. *Environ. Pollut.* **2024**, *349*, 123939. [CrossRef] [PubMed]
25. Bai, J.; Wang, X.; Chen, Y.; Yuan, Q.; Yang, Z.; Mi, Y.; Zhang, C. Nobiletin Ameliorates Aging of Chicken Ovarian Prehierarchical Follicles by Suppressing Oxidative Stress and Promoting Autophagy. *Cells* **2024**, *13*, 415. [CrossRef] [PubMed]
26. Rahman, M.A.; Cho, Y.; Nam, G.; Rhim, H. Antioxidant compound, oxyresveratrol, inhibits APP production through the AMPK/ULK1/mTOR-mediated autophagy pathway in mouse cortical astrocytes. *Antioxidants* **2021**, *10*, 408. [CrossRef] [PubMed]
27. Rahman, M.A.; Hannan, M.A.; Dash, R.; Rahman, M.H.; Islam, R.; Uddin, M.J.; Sohag, A.A.M.; Rahman, M.H.; Rhim, H. Phytochemicals as a complement to cancer chemotherapy: Pharmacological modulation of the autophagy-apoptosis pathway. *Front. Pharmacol.* **2021**, *12*, 639628. [CrossRef] [PubMed]

28. Cacciottola, L.; Camboni, A.; Cernogoraz, A.; Donnez, J.; Dolmans, M.-M. Role of apoptosis and autophagy in ovarian follicle pool decline in children and women diagnosed with benign or malignant extra-ovarian conditions. *Hum. Reprod.* **2023**, *38*, 75–88. [CrossRef] [PubMed]
29. Kozieł, M.J.; Piastowska-Ciesielska, A.W. Estrogens, estrogen receptors and tumor microenvironment in ovarian cancer. *Int. J. Mol. Sci.* **2023**, *24*, 14673. [CrossRef] [PubMed]
30. Shi, Y.-Q.; Zhu, X.-T.; Zhang, S.-N.; Ma, Y.-F.; Han, Y.-H.; Jiang, Y.; Zhang, Y.-H. Premature ovarian insufficiency: A review on the role of oxidative stress and the application of antioxidants. *Front. Endocrinol.* **2023**, *14*, 1172481. [CrossRef] [PubMed]
31. Singh, A.K.; Kumar, S.L.; Beniwal, R.; Mohanty, A.; Kushwaha, B.; Prasada Rao, H. Influence of the Ovarian Reserve and Oocyte Quality on Livestock Fertility. In *Sustainable Agriculture Reviews 59: Animal Biotechnology for Livestock Production 3*; Springer: Cham, Switzerland, 2023; pp. 201–240. [CrossRef]
32. Li, P.; Dou, Q.; Zhang, D.; Xiang, Y.; Tan, L. Melatonin regulates autophagy in granulosa cells from patients with premature ovarian insufficiency via activating Foxo3a. *Aging* **2024**, *16*, 844. [CrossRef]
33. Yildirim, R.M.; Seli, E. The role of mitochondrial dynamics in oocyte and early embryo development. In *Seminars in Cell & Developmental Biology*; Academic Press: Cambridge, MA, USA, 2024; pp. 52–61.
34. Fogliano, C.; Carotenuto, R.; Cirino, P.; Panzuto, R.; Ciaravolo, M.; Simoniello, P.; Sgariglia, I.; Motta, C.M.; Avallone, B. Benzodiazepine Interference with Fertility and Embryo Development: A Preliminary Survey in the Sea Urchin *Paracentrotus lividus*. *Int. J. Mol. Sci.* **2024**, *25*, 1969. [CrossRef]
35. Zhai, F.; Ma, X.; Yan, L.; Qiao, J. The molecular genetics of oogenesis. In *Human Reproductive and Prenatal Genetics*; Elsevier: Amsterdam, The Netherlands, 2023; pp. 145–163.
36. Zhang, T.; He, M.; Zhang, J.; Tong, Y.; Chen, T.; Wang, C.; Pan, W.; Xiao, Z. Mechanisms of primordial follicle activation and new pregnancy opportunity for premature ovarian failure patients. *Front. Physiol.* **2023**, *14*, 1113684. [CrossRef] [PubMed]
37. Costa, J.; Braga, P.C.; Rebelo, I.; Oliveira, P.F.; Alves, M.G. Mitochondria quality control and male fertility. *Biology* **2023**, *12*, 827. [CrossRef] [PubMed]
38. Ahmed, K.R.; Rahman, M.M.; Islam, M.N.; Fahim, M.M.H.; Rahman, M.A.; Kim, B. Antioxidants activities of phytochemicals perspective modulation of autophagy and apoptosis to treating cancer. *Biomed. Pharmacother.* **2024**, *174*, 116497. [CrossRef] [PubMed]
39. El-Sheikh, M.; Mesalam, A.A.; Kang, S.-M.; Joo, M.-D.; Soliman, S.S.; Khalil, A.A.K.; Ahn, M.-J.; Kong, I.-K. Modulation of apoptosis and autophagy by melatonin in juglone-exposed bovine oocytes. *Animals* **2023**, *13*, 1475. [CrossRef] [PubMed]
40. Fan, J.; Liu, C.; Zhao, Y.; Xu, Q.; Yin, Z.; Liu, Z.; Mu, Y. Single-Cell RNA Sequencing Reveals Differences in Chromatin Remodeling and Energy Metabolism among In Vivo-Developed, In Vitro-Fertilized, and Parthenogenetically Activated Embryos from the Oocyte to 8-Cell Stages in Pigs. *Animals* **2024**, *14*, 465. [CrossRef] [PubMed]
41. He, H.; Wang, J.; Mou, X.; Liu, X.; Li, Q.; Zhong, M.; Luo, B.; Yu, Z.; Zhang, J.; Xu, T. Selective autophagic degradation of ACLY (ATP citrate lyase) maintains citrate homeostasis and promotes oocyte maturation. *Autophagy* **2023**, *19*, 163–179. [CrossRef] [PubMed]
42. Adel, N.; Abdulghaffar, S.; Elmahdy, M.; Nabil, M.; Ghareeb, D.; Maghraby, H. Autophagy-related gene and protein expressions during blastocyst development. *J. Assist. Reprod. Genet.* **2023**, *40*, 323–331. [CrossRef]
43. Kljajic, M.; Saymé, N.; Krebs, T.; Wagenpfeil, G.; Baus, S.; Solomayer, E.-F.; Kasoha, M. Zygote Diameter and Total Cytoplasmic Volume as Useful Predictive Tools of Blastocyst Quality. *Geburtshilfe Und Frauenheilkd.* **2023**, *83*, 97–105. [CrossRef]
44. Palozzi, J.M.E. The Role of Programmed Germline Mitophagy in mtDNA Quality Control. Ph.D. Thesis, University of Toronto, Toronto, ON, Canada, 2023.
45. Zhou, P.; Wang, J.; Wang, J.; Liu, X. When autophagy meets placenta development and pregnancy complications. *Front. Cell Dev. Biol.* **2024**, *12*, 1327167. [CrossRef]
46. Wang, Z.; Quan, W.; Zeng, M.; Wang, Z.; Chen, Q.; Chen, J.; Christian, M.; He, Z. Regulation of autophagy by plant-based polyphenols: A critical review of current advances in glucolipid metabolic diseases and food industry applications. *Food Front.* **2023**, *4*, 1039–1067. [CrossRef]
47. Telfer, E.E.; Grosbois, J.; Odey, Y.L.; Rosario, R.; Anderson, R.A. Making a good egg: Human oocyte health, aging, and in vitro development. *Physiol. Rev.* **2023**, *103*, 2623–2677. [CrossRef]
48. Leyria, J. Endocrine factors modulating vitellogenesis and oogenesis in insects: An update. *Mol. Cell. Endocrinol.* **2024**, *587*, 112211. [CrossRef]
49. Carrageta, D.F.; Pereira, S.C.; Ferreira, R.; Monteiro, M.P.; Oliveira, P.F.; Alves, M.G. Signatures of metabolic diseases on spermatogenesis and testicular metabolism. *Nat. Rev. Urol.* **2024**, *21*, 477–494. [CrossRef]
50. Afzal, A.; Zhang, Y.; Afzal, H.; Saddozai, U.A.K.; Zhang, L.; Ji, X.-Y.; Khawar, M.B. Functional role of autophagy in testicular and ovarian steroidogenesis. *Front. Cell Dev. Biol.* **2024**, *12*, 1384047. [CrossRef]
51. Zhang, Q.; Zhang, J.; Chang, G.; Zhao, K.; Yao, Y.; Liu, L.; Du, Z.; Wang, Y.; Guo, X.; Zhao, Z. Decoding molecular features of bovine oocyte fate during antral follicle growth via single-cell multi-omics analysis. *Biol. Reprod.* **2024**, ioae114. [CrossRef]
52. Lee, R.; Chu, C.C. Theoretical perspectives on reproductive aging. *Front. Ecol. Evol.* **2023**, *10*, 934732. [CrossRef]
53. Raee, P.; Tan, S.C.; Najafi, S.; Zandsalimi, F.; Low, T.Y.; Aghamiri, S.; Fazeli, E.; Aghapour, M.; Mofarahe, Z.S.; Heidari, M.H. Autophagy, a critical element in the aging male reproductive disorders and prostate cancer: A therapeutic point of view. *Reprod. Biol. Endocrinol.* **2023**, *21*, 88. [CrossRef]

54. Wang, X.; Wang, L.; Xiang, W. Mechanisms of ovarian aging in women: A review. *J. Ovarian Res.* **2023**, *16*, 67. [CrossRef]
55. Fair, T.; Lonergan, P. The oocyte: The key player in the success of assisted reproduction technologies. *Reprod. Fertil. Dev.* **2023**, *36*, 133–148. [CrossRef]
56. Zhang, Y.; Bai, J.; Cui, Z.; Li, Y.; Gao, Q.; Miao, Y.; Xiong, B. Polyamine metabolite spermidine rejuvenates oocyte quality by enhancing mitophagy during female reproductive aging. *Nat. Aging* **2023**, *3*, 1372–1386. [CrossRef]
57. Vertika, S.; Singh, K.K.; Rajender, S. Mitochondria, spermatogenesis, and male infertility–An update. *Mitochondrion* **2020**, *54*, 26–40. [CrossRef]
58. Peters, A.E.; Mihalas, B.P.; Bromfield, E.G.; Roman, S.D.; Nixon, B.; Sutherland, J.M. Autophagy in female fertility: A role in oxidative stress and aging. *Antioxid. Redox Signal.* **2020**, *32*, 550–568. [CrossRef]
59. D'Argenio, V.; Dittfeld, L.; Lazzeri, P.; Tomaiuolo, R.; Tasciotti, E. Unraveling the balance between genes, microbes, lifestyle and the environment to improve healthy reproduction. *Genes* **2021**, *12*, 605. [CrossRef] [PubMed]
60. Sciorio, R.; Tramontano, L.; Adel, M.; Fleming, S. Decrease in Sperm Parameters in the 21st Century: Obesity, Lifestyle, or Environmental Factors? An Updated Narrative Review. *J. Pers. Med.* **2024**, *14*, 198. [CrossRef]
61. Strilbytska, O.; Klishch, S.; Storey, K.B.; Koliada, A.; Lushchak, O. Intermittent fasting and longevity: From animal models to implication for humans. *Ageing Res. Rev.* **2024**, *96*, 102274. [CrossRef] [PubMed]
62. Wang, R.; Lv, X.; Xu, W.; Li, X.; Tang, X.; Huang, H.; Yang, M.; Ma, S.; Wang, N.; Niu, Y. Effects of the periodic fasting-mimicking diet on health, lifespan, and multiple diseases: A narrative review and clinical implications. *Nutr. Rev.* **2024**, nuae003. [CrossRef]
63. Zhou, Y.; Yan, J.; Qiao, L.; Zeng, J.; Cao, F.; Sheng, X.; Qi, X.; Long, C.; Liu, B.; Wang, X. Bone Marrow Mesenchymal Stem Cell-Derived Exosomes Ameliorate Aging-Induced BTB Impairment in Porcine Testes by Activating Autophagy and Inhibiting ROS/NLRP3 Inflammasomes via the AMPK/mTOR Signaling Pathway. *Antioxidants* **2024**, *13*, 183. [CrossRef]
64. Chang, Y.; Deng, H.; He, Y.; Zhou, B.; Yuan, D.; Wu, J.; Zhang, C.; Zhao, H. Wuzi Yanzong administration alleviates Sertoli cell injury by recovering AKT/mTOR-mediated autophagy and the mTORC1–mTROC2 balance in aging-induced testicular dysfunction. *J. Ethnopharmacol.* **2024**, *318*, 116865. [CrossRef]
65. Zhao, T.; Fan, J.; Abu-Zaid, A.; Burley, S.K.; Zheng, X.S. Nuclear mTOR Signaling Orchestrates Transcriptional Programs Underlying Cellular Growth and Metabolism. *Cells* **2024**, *13*, 781. [CrossRef]
66. Xie, Q.; Liao, Q.; Wang, L.; Zhang, Y.; Chen, J.; Bai, H.; Li, K.; Ai, J. The dominant mechanism of cyclophosphamide induced damage to ovarian reserve: Premature activation or apoptosis of primordial follicles? *Reprod. Sci.* **2024**, *31*, 30–44. [CrossRef]
67. Castrillon, D.H. Development and Maldevelopment of the Ovaries. In *Gynecologic and Obstetric Pathology*; Springer: Berlin/Heidelberg, Germany, 2024; pp. 1–23.
68. Athar, F.; Karmani, M.; Templeman, N.M. Metabolic hormones are integral regulators of female reproductive health and function. *Biosci. Rep.* **2024**, *44*, BSR20231916. [CrossRef]
69. Moruzzi, M.; Martinelli, I.; Micioni Di Bonaventura, M.V.; Giusepponi, M.E.; Gabrielli, G.; Fruganti, A.; Marchegiani, A.; Dini, F.; Marini, C.; Cuccioloni, M. Metabolic effects of Tart Cherries supplementation in an animal model of obesity. *Ital. J. Anat. Embryol.* **2017**, *122*, 150.
70. Wang, B.; Gao, M.; Yao, Y.; Shen, H.; Li, H.; Sun, J.; Wang, L.; Zhang, X. Enhancing endometrial receptivity: The roles of human chorionic gonadotropin in autophagy and apoptosis regulation in endometrial stromal cells. *Reprod. Biol. Endocrinol.* **2024**, *22*, 37. [CrossRef] [PubMed]
71. Li, C.; Zhang, H.; Wu, H.; Li, J.; Liu, Q.; Li, Y.; Pan, M.; Zhao, X.; Wei, Q.; Peng, S. Intermittent fasting improves the oocyte quality of obese mice through the regulation of maternal mRNA storage and translation by LSM14B. *Free Radic. Biol. Med.* **2024**, *217*, 157–172. [PubMed]
72. Zhu, J.-R.; Zhu, W.-J.; Li, T.; Ou, J.-P. Autophagy activity is increased in the cumulus cells of women with poor ovarian response. *Taiwan. J. Obstet. Gynecol.* **2024**, *63*, 205–213. [PubMed]
73. Ojo, O.A.; Nwafor-Ezeh, P.I.; Rotimi, D.E.; Iyobhebhe, M.; Ogunlakin, A.D.; Ojo, A.B. Apoptosis, inflammation, and oxidative stress in infertility: A mini review. *Toxicol. Rep.* **2023**, *10*, 448–462. [PubMed]
74. Usman, M.; Li, A.; Wu, D.; Qinyan, Y.; Yi, L.X.; He, G.; Lu, H. The functional role of lncRNAs as ceRNAs in both ovarian processes and associated diseases. *Non-Coding RNA Res.* **2023**, *9*, 165–177.
75. Ren, H.; Dai, R.; Nabil, W.N.N.; Xi, Z.; Wang, F.; Xu, H. Unveiling the dual role of autophagy in vascular remodelling and its related diseases. *Biomed. Pharmacother.* **2023**, *168*, 115643.
76. Burton, G.J.; Jauniaux, E. The human placenta: New perspectives on its formation and function during early pregnancy. *Proc. R. Soc. B* **2023**, *290*, 20230191. [CrossRef]
77. Chaudhary, M.R.; Chaudhary, S.; Sharma, Y.; Singh, T.A.; Mishra, A.K.; Sharma, S.; Mehdi, M.M. Aging, oxidative stress and degenerative diseases: Mechanisms, complications and emerging therapeutic strategies. *Biogerontology* **2023**, *24*, 609–662.
78. Wang, L.; O'Kane, A.M.; Zhang, Y.; Ren, J. Maternal obesity and offspring health: Adapting metabolic changes through autophagy and mitophagy. *Obes. Rev.* **2023**, *24*, e13567. [CrossRef]
79. Fang, H.; Li, Q.; Wang, H.; Ren, Y.; Zhang, L.; Yang, L. Maternal nutrient metabolism in the liver during pregnancy. *Front. Endocrinol.* **2024**, *15*, 1295677. [CrossRef]
80. Ma, L.; Shen, W.; Zhang, J.; Ma, L.; Shen, W.; Shen, W.; Ma, L.; Zhang, J.; Zhang, J. The Life Cycle of the Ovary. In *Ovarian Aging*; Springer: Berlin/Heidelberg, Germany, 2023; pp. 7–33.

81. Ramya, S.; Poornima, P.; Jananisri, A.; Geofferina, I.P.; Bavyataa, V.; Divya, M.; Priyanga, P.; Vadivukarasi, J.; Sujitha, S.; Elamathi, S. Role of hormones and the potential impact of multiple stresses on infertility. *Stresses* **2023**, *3*, 454–474. [CrossRef]
82. Fan, Y.; Jiang, X.; Xiao, Y.; Li, H.; Chen, J.; Bai, W. Natural antioxidants mitigate heavy metal induced reproductive toxicity: Prospective mechanisms and biomarkers. *Crit. Rev. Food Sci. Nutr.* **2023**, 1–13. [CrossRef]
83. Reiter, R.J.; Sharma, R.; Romero, A.; Manucha, W.; Tan, D.-X.; Zuccari, D.A.P.d.C.; Chuffa, L.G.d.A. Aging-Related Ovarian Failure and Infertility: Melatonin to the Rescue. *Antioxidants* **2023**, *12*, 695. [CrossRef] [PubMed]
84. Xiang, W.; Qin, Y.; Feng, H.L.; Li, J. *Ovarian Aging and Reproduction*; Frontiers Media SA: Lausanne, Switzerland, 2023.
85. Cui, J.; Li, Y.; Zhang, W.; Qian, H.; Zhang, Z.; Xu, K. Alginic acid induces oxidative stress-mediated hormone secretion disorder, apoptosis and autophagy in mouse granulosa cells and ovaries. *Toxicology* **2022**, *467*, 153099. [CrossRef]
86. Munakata, Y.; Kawahara-Miki, R.; Shiratsuki, S.; Tasaki, H.; Itami, N.; Shirasuna, K.; Kuwayama, T.; Iwata, H. Gene expression patterns in granulosa cells and oocytes at various stages of follicle development as well as in in vitro grown oocyte-and-granulosa cell complexes. *J. Reprod. Dev.* **2016**, *62*, 359–366. [CrossRef] [PubMed]
87. Havelock, J.C.; Rainey, W.E.; Carr, B.R. Ovarian granulosa cell lines. *Mol. Cell. Endocrinol.* **2004**, *228*, 67–78. [CrossRef] [PubMed]
88. An, R.; Wang, X.; Yang, L.; Zhang, J.; Wang, N.; Xu, F.; Hou, Y.; Zhang, H.; Zhang, L. Polystyrene microplastics cause granulosa cells apoptosis and fibrosis in ovary through oxidative stress in rats. *Toxicology* **2021**, *449*, 152665. [CrossRef]
89. Shi, J.; Liu, C.; Chen, M.; Yan, J.; Wang, C.; Zuo, Z.; He, C. The interference effects of bisphenol A on the synthesis of steroid hormones in human ovarian granulosa cells. *Environ. Toxicol.* **2021**, *36*, 665–674. [CrossRef]
90. Liu, X.-L.; Wu, R.-Y.; Sun, X.-F.; Cheng, S.-F.; Zhang, R.-Q.; Zhang, T.-Y.; Zhang, X.-F.; Zhao, Y.; Shen, W.; Li, L. Mycotoxin zearalenone exposure impairs genomic stability of swine follicular granulosa cells in vitro. *Int. J. Biol. Sci.* **2018**, *14*, 294. [CrossRef] [PubMed]
91. Aldawood, N.; Alrezaki, A.; Alanazi, S.; Amor, N.; Alwasel, S.; Sirotkin, A.; Harrath, A.H. Acrylamide impairs ovarian function by promoting apoptosis and affecting reproductive hormone release, steroidogenesis and autophagy-related genes: An in vivo study. *Ecotoxicol. Environ. Saf.* **2020**, *197*, 110595. [CrossRef] [PubMed]
92. Alhelaisi, A.; Alrezaki, A.; Nahdi, S.; Aldahmash, W.; Alwasel, S.; Harrath, A.H. Early-Life Exposure to the Mycotoxin Fumonisin B1 and Developmental Programming of the Ovary of the Offspring: The Possible Role of Autophagy in Fertility Recovery. *Toxics* **2023**, *11*, 980. [CrossRef] [PubMed]
93. He, H.; Wan, N.; Li, Z.; Zhang, Z.; Gao, Z.; Liu, Q.; Ma, X.; Zhang, Y.; Li, R.; Fu, X. Short-term effects of exposure to ambient PM2. 5 and its components on hospital admissions for threatened and spontaneous abortions: A multicity case-crossover study in China. *Chemosphere* **2024**, *350*, 141057. [CrossRef]
94. Al-Gubory, K.H. Environmental pollutants and lifestyle factors induce oxidative stress and poor prenatal development. *Reprod. Biomed. Online* **2014**, *29*, 17–31. [CrossRef] [PubMed]
95. Choi, J.Y.; Jo, M.W.; Lee, E.Y.; Yoon, B.-K.; Choi, D.S. The role of autophagy in follicular development and atresia in rat granulosa cells. *Fertil. Steril.* **2010**, *93*, 2532–2537. [CrossRef] [PubMed]
96. Shao, T.; Ke, H.; Liu, R.; Xu, L.; Han, S.; Zhang, X.; Dang, Y.; Jiao, X.; Li, W.; Chen, Z.-J. Autophagy regulates differentiation of ovarian granulosa cells through degradation of WT1. *Autophagy* **2022**, *18*, 1864–1878. [CrossRef] [PubMed]
97. Kumariya, S.; Ubba, V.; Jha, R.K.; Gayen, J.R. Autophagy in ovary and polycystic ovary syndrome: Role, dispute and future perspective. *Autophagy* **2021**, *17*, 2706–2733. [CrossRef]
98. Gozuacik, D.; Kimchi, A. Autophagy as a cell death and tumor suppressor mechanism. *Oncogene* **2004**, *23*, 2891–2906. [CrossRef]
99. Klionsky, D.J. Autophagy revisited: A conversation with Christian de Duve. *Autophagy* **2008**, *4*, 740–743. [CrossRef]
100. Liu, T.; Di, Q.-N.; Sun, J.-H.; Zhao, M.; Xu, Q.; Shen, Y. Effects of nonylphenol induced oxidative stress on apoptosis and autophagy in rat ovarian granulosa cells. *Chemosphere* **2020**, *261*, 127693. [CrossRef] [PubMed]
101. Lou, Y.; Yu, W.; Han, L.; Yang, S.; Wang, Y.; Ren, T.; Yu, J.; Zhao, A. ROS activates autophagy in follicular granulosa cells via mTOR pathway to regulate broodiness in goose. *Anim. Reprod. Sci.* **2017**, *185*, 97–103. [CrossRef] [PubMed]
102. Sun, Y.; Shen, J.; Zeng, L.; Yang, D.; Shao, S.; Wang, J.; Wei, J.; Xiong, J.; Chen, J. Role of autophagy in di-2-ethylhexyl phthalate (DEHP)-induced apoptosis in mouse Leydig cells. *Environ. Pollut.* **2018**, *243*, 563–572. [CrossRef] [PubMed]
103. Meng, L.; Jan, S.Z.; Hamer, G.; van Pelt, A.M.; van der Stelt, I.; Keijer, J.; Teerds, K.J. Preantral follicular atresia occurs mainly through autophagy, while antral follicles degenerate mostly through apoptosis. *Biol. Reprod.* **2018**, *99*, 853–863. [CrossRef] [PubMed]
104. Hasanain, M.; Bhattacharjee, A.; Pandey, P.; Ashraf, R.; Singh, N.; Sharma, S.; Vishwakarma, A.; Datta, D.; Mitra, K.; Sarkar, J. α-Solanine induces ROS-mediated autophagy through activation of endoplasmic reticulum stress and inhibition of Akt/mTOR pathway. *Cell Death Dis.* **2015**, *6*, e1860. [CrossRef]
105. Xiao, L.; Zuo, Z.; Zhao, F. Microbiome in Female Reproductive Health: Implications for Fertility and Assisted Reproductive Technologies. *Genom. Proteom. Bioinform.* **2024**, *22*, qzad005. [CrossRef] [PubMed]
106. Ahmadi, F.; Eshrati, B.; Hassani, F.; Hosseini, R.; Pooransari, P.; Ramazanali, F.; Tavalaee, M.; Barra, F. International journal of fertility and sterility. *Int. J. Fertil. Steril.* **2024**, *18*.
107. Liu, C. *Analysis of the microRNAs MiR-34a/b/c as Mediators of the Effects of Curcumin and Aspirin on Colorectal Cancer*; LMU: Munich, Germany, 2024. [CrossRef]
108. Jiang, L.; Fernando, S.R.; Kodithuwakku, S.P.; Cao, D.; Yeung, W.S.; Lee, K.-F. Stromal cell decidualization and embryo implantation: A vulnerable step leading to successful pregnancy. *Reprod. Dev. Med.* **2024**, *8*, 101–110. [CrossRef]

109. Arora, R.; Baldi, A. Revolutionizing Neurological Disorder Treatment: Integrating Innovations in Pharmaceutical Interventions and Advanced Therapeutic Technologies. *Curr. Pharm. Des.* **2024**, *30*, 1459–1471. [CrossRef]
110. Muhammad, T.; Wan, Y.; Lv, Y.; Li, H.; Naushad, W.; Chan, W.Y.; Lu, G.; Chen, Z.J.; Liu, H. Maternal obesity: A potential disruptor of female fertility and current interventions to reduce associated risks. *Obes. Rev.* **2023**, *24*, e13603. [CrossRef]
111. Lee, S.H.; Rinaudo, P.F. Metabolic Regulation of preimplantation Embryo Development in vivo and in vitro: Molecular Mechanisms and insights. *Biochem. Biophys. Res. Commun.* **2024**, *726*, 150256. [CrossRef] [PubMed]
112. Chettiar, V.; Patel, A.; Chettiar, S.S.; Jhala, D.D. Meta-analysis of endometrial transcriptome data reveals novel molecular targets for recurrent implantation failure. *J. Assist. Reprod. Genet.* **2024**, *41*, 1417–1431. [CrossRef] [PubMed]
113. Zipponi, M.; Cacciottola, L.; Dolmans, M.-M. Overview of crosstalk between stromal and epithelial cells in the pathogenesis of adenomyosis and shared features with deep endometriotic nodules. *Hum. Reprod.* **2024**, *39*, deae116. [CrossRef] [PubMed]
114. Liu, Y.; Sun, Y.; Cheng, S. Advances in the use of organoids in endometrial diseases. *Int. J. Gynecol. Obstet.* **2024**, *166*, 502–511. [CrossRef] [PubMed]
115. Brinca, A.T.; Peiró, A.M.; Evangelio, P.M.; Eleno, I.; Oliani, A.H.; Silva, V.; Vicente, L.F.; Ramalhinho, A.C.; Gallardo, E. Follicular Fluid and Blood Monitorization of Infertility Biomarkers in Women with Endometriosis. *Int. J. Mol. Sci.* **2024**, *25*, 7177. [CrossRef]
116. Wang, X.; Li, J.; Lu, W.; Gao, F.; Zhang, S.; Li, J. Therapeutic roles of platelet-rich plasma to restore female reproductive and endocrine dysfunction. *Front. Endocrinol.* **2024**, *15*, 1374382. [CrossRef] [PubMed]
117. Liang, J.; Huang, F.; Song, Z.; Tang, R.; Zhang, P.; Chen, R. Impact of NAD+ metabolism on ovarian aging. *Immun. Ageing* **2023**, *20*, 70. [CrossRef] [PubMed]
118. Zhang, T.; Wang, L.; Pan, Y.; He, H.; Wang, J.; Zhao, T.; Ding, T.; Wang, Y.; Zhao, L.; Han, X. Effect of rapamycin treatment on oocyte in vitro maturation and embryonic development after parthenogenesis in yaks. *Theriogenology* **2022**, *193*, 128–135. [CrossRef]
119. Lin, M.; Hua, R.; Ma, J.; Zhou, Y.; Li, P.; Xu, X.; Yu, Z.; Quan, S. Bisphenol A promotes autophagy in ovarian granulosa cells by inducing AMPK/mTOR/ULK1 signalling pathway. *Environ. Int.* **2021**, *147*, 106298. [CrossRef]
120. Wu, C.; Du, X.; Liu, H.; Chen, X.; Ge, K.; Meng, R.; Zhang, Z.; Zhang, H. Advances in polychlorinated biphenyls-induced female reproductive toxicity. *Sci. Total Environ.* **2024**, *918*, 170543. [CrossRef]
121. Gao, M.; Shen, H.; Li, Q.; Gu, X.; Jia, T.; Wang, Y. Perfluorooctane sulfonate (PFOS) induces apoptosis and autophagy by inhibition of PI3K/AKT/mTOR pathway in human granulosa cell line KGN. *Environ. Pollut.* **2024**, *344*, 123333. [CrossRef] [PubMed]
122. Xiao, S.; Du, J.; Yuan, G.; Luo, X.; Song, L. Granulosa Cells-Related MicroRNAs in Ovarian Diseases: Mechanism, Facts and Perspectives. *Reprod. Sci.* **2024**, 1–16. [CrossRef] [PubMed]
123. Li, L.; Zhang, Y.; Zhou, J.; Wang, J.; Wang, L. A systematic review of the mechanistic actions of microRNAs within integrated traditional Chinese medicine and western medical treatment for endometriosis. *Drug Discov. Ther.* **2024**, *18*, 1–9. [CrossRef] [PubMed]
124. Podgrajsek, R.; Ban Frangez, H.; Stimpfel, M. Molecular Mechanism of Resveratrol and Its Therapeutic Potential on Female Infertility. *Int. J. Mol. Sci.* **2024**, *25*, 3613. [CrossRef] [PubMed]
125. Hao, M.; Chu, Y.; Lei, J.; Yao, Z.; Wang, P.; Chen, Z.; Wang, K.; Sang, X.; Han, X.; Wang, L. Pharmacological mechanisms and clinical applications of curcumin: Update. *Aging Dis.* **2023**, *14*, 716. [CrossRef]
126. Song, L.; Zhang, S. Anti-Aging Activity and Modes of Action of Compounds from Natural Food Sources. *Biomolecules* **2023**, *13*, 1600. [CrossRef] [PubMed]
127. Rahman, M.A.; Hwang, H.; Nah, S.-Y.; Rhim, H. Gintonin stimulates autophagic flux in primary cortical astrocytes. *J. Ginseng Res.* **2020**, *44*, 67–78. [CrossRef] [PubMed]
128. Shen, Y.-L.; Cheng, L.-Y.; Meng, X.-R.; Li, Q.; Du, L.-Y.; Wang, E.-P.; Chen, C.-B. Effects of ginseng continuous soil crop on growth development and antioxidant system of ginseng at different fertility stages. *Chin. J. Appl. Chem.* **2023**, *40*, 109–115.
129. Rahman, M.A.; Dash, R.; Sohag, A.A.M.; Alam, M.; Rhim, H.; Ha, H.; Moon, I.S.; Uddin, M.J.; Hannan, M.A. Prospects of marine sterols against pathobiology of Alzheimer's disease: Pharmacological insights and technological advances. *Mar. Drugs* **2021**, *19*, 167. [CrossRef]
130. Wang, C.-R.; Ji, H.-W.; He, S.-Y.; Liu, R.-P.; Wang, X.-Q.; Wang, J.; Huang, C.-M.; Xu, Y.-N.; Li, Y.-H.; Kim, N.-H. Chrysoeriol improves in vitro porcine embryo development by reducing oxidative stress and autophagy. *Vet. Sci.* **2023**, *10*, 143. [CrossRef]
131. Buljeta, I.; Pichler, A.; Šimunović, J.; Kopjar, M. Beneficial effects of red wine polyphenols on human health: Comprehensive review. *Curr. Issues Mol. Biol.* **2023**, *45*, 782–798. [CrossRef] [PubMed]
132. Park, H.; Cho, M.; Do, Y.; Park, J.-K.; Bae, S.-J.; Joo, J.; Ha, K.-T. Autophagy as a therapeutic target of natural products enhancing embryo implantation. *Pharmaceuticals* **2021**, *15*, 53. [CrossRef] [PubMed]
133. Peng, Y.; Ao, M.; Dong, B.; Jiang, Y.; Yu, L.; Chen, Z.; Hu, C.; Xu, R. Anti-inflammatory effects of curcumin in the inflammatory diseases: Status, limitations and countermeasures. *Drug Des. Dev. Ther.* **2021**, *15*, 4503–4525. [CrossRef] [PubMed]
134. Temre, M.K.; Kumar, A.; Singh, S.M. An appraisal of the current status of inhibition of glucose transporters as an emerging antineoplastic approach: Promising potential of new panGLUT inhibitors. *Front. Pharmacol.* **2022**, *13*, 1035510. [CrossRef] [PubMed]
135. Hung, S.W.; Li, Y.; Chen, X.; Chu, K.O.; Zhao, Y.; Liu, Y.; Guo, X.; Man, G.C.-W.; Wang, C.C. Green tea epigallocatechin-3-gallate regulates autophagy in male and female reproductive cancer. *Front. Pharmacol.* **2022**, *13*, 906746. [CrossRef] [PubMed]

136. Tao, F.; Zhai, Q.; Cao, Y.; Gao, H.; Cai, Y.; Jia, W.; Ma, H.; Xue, P. Inhibition of p38 MAPK/NF-κB p65 signaling pathway activity by rare ginsenosides ameliorates cyclophosphamide-induced premature ovarian failure and KGN cell injury. *J. Ethnopharmacol.* **2024**, *326*, 117944. [CrossRef] [PubMed]
137. Xiong, Y.; Huang, C.-W.; Shi, C.; Peng, L.; Cheng, Y.-T.; Hong, W.; Liao, J. Quercetin suppresses ovariectomy-induced osteoporosis in rat mandibles by regulating autophagy and the NLRP3 pathway. *Exp. Biol. Med.* **2023**, *248*, 2363–2380. [CrossRef] [PubMed]
138. Sengupta, P.; Dutta, S.; Hassan, M.F. Polycystic ovary syndrome (PCOS) and oxidative stress. *J. Integr. Sci. Technol.* **2024**, *12*, 752. [CrossRef]
139. Jiang, D.; Ji, C.; Zhou, X.; Wang, Z.; Sun, Q.; Wang, X.; An, X.; Ling, W.; Kang, B. Pathway analysis of spermidine anti-oxidative stress and inducing autophagy in granulosa cells of Sichuan white geese. *Theriogenology* **2024**, *215*, 290–301. [CrossRef]
140. Tenchov, R.; Sasso, J.M.; Wang, X.; Zhou, Q.A. Antiaging Strategies and Remedies: A Landscape of Research Progress and Promise. *ACS Chem. Neurosci.* **2024**, *15*, 408–446. [CrossRef]
141. Liu, Z.; Lu, T.; Qian, R.; Wang, Z.; Qi, R.; Zhang, Z. Exploiting Nanotechnology for Drug Delivery: Advancing the Anti-Cancer Effects of Autophagy-Modulating Compounds in Traditional Chinese Medicine. *Int. J. Nanomed.* **2024**, *19*, 2507–2528. [CrossRef] [PubMed]
142. Islam, M.R.; Rahman, M.M.; Dhar, P.S.; Nowrin, F.T.; Sultana, N.; Akter, M.; Rauf, A.; Khalil, A.A.; Gianoncelli, A.; Ribaudo, G. The role of natural and semi-synthetic compounds in ovarian cancer: Updates on mechanisms of action, current trends and perspectives. *Molecules* **2023**, *28*, 2070. [CrossRef] [PubMed]
143. Dash, S.; Kanungo, S.K.; Mishra, B.; Sahoo, A.C.; Senapati, A.K.; Pattnaik, S. Evaluation of classical ayurvedic medicine "Sukumaram Kasayam" in a rat model of letrozole induced polycystic ovaries. *Adv. Tradit. Med.* **2021**, *21*, 675–683. [CrossRef]
144. Zheng, Y.; Qiu, Y.; Wang, Q.; Gao, M.; Cao, Z.; Luan, X. ADPN Regulates Oxidative Stress-Induced Follicular Atresia in Geese by Modulating Granulosa Cell Apoptosis and Autophagy. *Int. J. Mol. Sci.* **2024**, *25*, 5400. [CrossRef] [PubMed]
145. Kusumaningtyas, I.; Dasuki, D.; Harjana, S.M.; Sadewa, A.H.; Sweety, M.C.; Septiani, L. Unraveling the microRNAs, key players in folliculogenesis and ovarian diseases. *Middle East Fertil. Soc. J.* **2024**, *29*, 13. [CrossRef]
146. Meirelles, L.v.M.; Klamt, F.; Smitz, J. Redox biology of human cumulus cells: Basic concepts, impact on oocyte quality, and potential clinical use. *Antioxid. Redox Signal. Larchmt.* **2020**, *32*, 522–535.
147. Ren, C.; Zhang, S.; Chen, Y.; Deng, K.; Kuang, M.; Gong, Z.; Zhang, K.; Wang, P.; Huang, P.; Zhou, Z. Exploring nicotinamide adenine dinucleotide precursors across biosynthesis pathways: Unraveling their role in the ovary. *FASEB J.* **2024**, *38*, e23804. [CrossRef] [PubMed]
148. Khan, M.Z.; Khan, A.; Huang, B.; Wei, R.; Kou, X.; Wang, X.; Chen, W.; Li, L.; Zahoor, M.; Wang, C. Bioactive Compounds Protect Mammalian Reproductive Cells from Xenobiotics and Heat Stress-Induced Oxidative Distress via Nrf2 Signaling Activation: A Narrative Review. *Antioxidants* **2024**, *13*, 597. [CrossRef] [PubMed]
149. Ji, J.; Cheng, X.; Du, R.; Xie, Y.; Zhang, Y. Advances in research on autophagy mechanisms in resistance to endometrial cancer treatment. *Front. Oncol.* **2024**, *14*, 1364070. [CrossRef]
150. Guo, Y.; Xue, L.; Tang, W.; Xiong, J.; Chen, D.; Dai, Y.; Wu, C.; Wei, S.; Dai, J.; Wu, M. Ovarian microenvironment: Challenges and opportunities in protecting against chemotherapy-associated ovarian damage. *Hum. Reprod. Update* **2024**, dmae020. [CrossRef]
151. Wang, X.; Liu, Y.; Wang, J.; Lu, X.; Guo, Z.; Lv, S.; Sun, Z.; Gao, T.; Gao, F.; Yuan, J. Mitochondrial Quality Control in Ovarian Function: From Mechanisms to Therapeutic Strategies. *Reprod. Sci.* **2024**. [CrossRef] [PubMed]
152. Li, M.-Y.; Shen, H.-H.; Cao, X.-Y.; Gao, X.-X.; Xu, F.-Y.; Ha, S.-Y.; Sun, J.-S.; Liu, S.-P.; Xie, F.; Li, M.-Q. Targeting a mTOR/autophagy axis: A double-edged sword of rapamycin in spontaneous miscarriage. *Biomed. Pharmacother.* **2024**, *177*, 116976. [CrossRef] [PubMed]
153. Kobayashi, H.; Yoshimoto, C.; Matsubara, S.; Shigetomi, H.; Imanaka, S. Altered energy metabolism, mitochondrial dysfunction, and redox imbalance influencing reproductive performance in granulosa cells and oocyte during aging. *Reprod. Sci.* **2024**, *31*, 906–916. [CrossRef] [PubMed]
154. Feuz, M.B.; Nelson, D.C.; Miller, L.B.; Zwerdling, A.E.; Meyer, R.; Meyer-Ficca, M.L. Current Insights and a Potential Role of NAD in the Reproductive Health of Aging Fathers and Their Children. *Reproduction* **2024**, *167*, e230486. [CrossRef]
155. Zhu, Y.; Liu, H.; Zheng, L.; Luo, Y.; Zhou, G.; Li, J.; Hou, Y.; Fu, X. Vitrification of Mammalian Oocytes: Recent Studies on Mitochondrial Dysfunction. *Biopreservation Biobanking* **2024**. [CrossRef]
156. Tek, M.; Çalış, Ö. Gene-drive based strategy for engineering disease resistance with creating heritable mutation by using CRISPR/Cas9. In Proceedings of the 2nd International Molecular Plant Protection Congress, Bursa, Turkey, 15–18 May 2023; Volume 1, p. 111.
157. Mao, M.; Song, S.; Li, X.; Lu, J.; Li, J.; Zhao, W.; Liu, H.; Liu, J.; Zeng, B. Advances in epigenetic modifications of autophagic process in pulmonary hypertension. *Front. Immunol.* **2023**, *14*, 1206406. [CrossRef]
158. Díez-Juan, A.; Vladimirov, I.K. Perspective Chapter: Ovarian Reproductive Aging and Rejuvenation Strategies. In *IVF Technologies and Infertility-Current Practices and New Perspectives*; IntechOpen: London, UK, 2023.

cells

MDPI

Article

Therapeutic Potential of Human Amniotic Membrane-Derived Mesenchymal Stem Cell Conditioned Medium in Combating Oxidative Stress and Age-Related Female Infertility

Kihae Ra

Department of Theriogenology and Biotechnology, College of Veterinary Medicine and Research Institute for Veterinary Science, Seoul National University, Seoul 08826, Republic of Korea; ragh1102@snu.ac.kr

Abstract: Oxidative stress is a crucial factor accelerating the age-related functional deterioration of reproductive organs and fertility. Recently, human amniotic-membrane-derived mesenchymal stem cells (AMSCs) have emerged as a promising source with notable potential to reduce oxidative damage and support tissue regeneration. This study investigates the impact of the intravenous administration of human amniotic membrane-derived mesenchymal stem cell conditioned medium (AMSC-CM) on oxidative stress and reproductive competence in aged female mice. Antioxidative effects of human AMSC-CM were found in the reproductive organs of aged mice at both the RNA and protein levels. Human AMSC-CM positively regulated age-dependent changes in reproductive hormones, comparable to those observed in younger mice. RNA sequencing analysis revealed alterations in ovarian and uterine gene expression in aged mice showing that AMSC-CM treatment promotes the expression of genes essential for anti-aging, energy metabolism, and female reproductive processes. These findings highlight the potential of human AMSC-CM as a therapeutic strategy with anti-aging and antioxidative effects against age-related female infertility.

Keywords: amniotic membrane-derived mesenchymal stem cell; conditioned medium; female reproductive system; oxidative stress; reproductive aging; RNA sequencing

1. Introduction

Reproductive aging is assumed to be a major concern in terms of both the individual and socioeconomic aspects associated with infertility and adverse pregnancy-related complications. Age-related loss of ovarian and uterine function hampers the normal reproductive competence of aged females [1]. Female reproductive aging was once thought to be primarily caused by the deterioration of oocyte quality rather than uterine aging. However, increasing evidence indicates that both oocyte and uterine factors are involved in the decline in fertility in aged females [2]. Ovarian aging involves the gradual loss of ovarian follicles, a decrease in reproductive hormone production that is required for normal physiological functions, and the degradation of oocyte quality, all of which result in the decline in fertility [3]. The uterus, along with the ovaries, forms the essential reproductive organ where biological events for pregnancy occur, including embryonic development, elongation, implantation, placentation, and fetal development. The endometrial environment in the uterus determines the reproductive competence [4]. With aging, the number of implantation sites decrease during early pregnancy, and an abnormal uterine response disturbs the development of placenta decidualization, leading to a considerable reduction in fertility [5]. Age-dependent changes in the ovaries and uterus are known, together, to reduce the probability of a successful reproductive outcome [6]. Oxidative stress driven

https://doi.org/10.3390/cells14110801

by excessive reactive oxygen species (ROS) increases with aging, bringing detrimental effects to essential ovarian and uterine functions necessary for fertility [7]. For instance, oxidative stress in the ovaries is linked to reproductive disorders, such as abnormal ovulation, polycystic ovary syndrome, and premature ovarian failure [8]. Oxidative damage disrupts progesterone production in the corpus luteum and its modulatory function in the endometrium, which are crucial for embryo development, implantation, and the maintenance of pregnancy [9]. Furthermore, an impaired immune system within the uterus caused by oxidative stress during gestation can potentially lead to early pregnancy loss [10]. Consequently, managing oxidative stress, particularly for aged females, is essential for maintaining successful reproductive outcomes.

Mesenchymal stem cell (MSCs) isolated from various source tissues have been widely used as a therapy to resolve age-associated reproductive disorders. Perinatal tissue including the placenta, amniotic membrane, and fluid, was considered to be unnecessary and was generally discarded post-partum; however, it is now acknowledged as a competent source of stem cells that have an intermediate stemness between pluripotency and multipotency [11–13]. The injection of human amniotic fluid-derived MSCs into the ovary has been shown to improve the function of aged ovaries, increase follicle numbers, and improve hormone levels in aged mice [14]. The intravenous injection (IV) of human umbilical cord-derived MSCs stabilizes the reproductive hormone levels and improves ovarian structure by upregulating genes involved in ovarian function in perimenopausal rats [15]. In mice showing natural ovarian aging, the injection of human amniotic membrane-derived MSCs (AMSC) into the ovary improves the number of ovarian follicles and the proliferation of ovarian granular cells, and inhibits apoptosis by increasing the secretion of growth factors [16].

The use of CM, the processed culture medium containing the secreted factors collected by culturing stem cells, offers significant advantages over the original cells in terms of safety and convenience [7]. A previous study demonstrated the anti-apoptotic and anti-oxidative effects of human ASC-CMs in the ovaries and uteruses of pregnant mice of advanced maternal age following IV administration under the optimal conditions [17]. However, studies were required to elucidate the undiscovered effects of other cell sources of human MSC-CM for advanced efficiency. As the findings of the previous study revealed the improved effects of AMSC-CM compared with that of ASC-CM on embryo development [18], this study further aimed to identify the physiological implications and potential mechanisms of AMSC-CM on the reproductive organs, with particular emphasis on its antioxidative and antiaging properties, by molecular-level analyses. Based on this, the present study aimed to investigate the changes in ovarian and uterine gene expression following the IV injection of human AMSC-CM using RNA sequencing analysis in association with its beneficial effects on oxidative stress and reproduction in aged female mice.

2. Materials and Methods

2.1. Ethics Approval

Human AMSCs were provided from the R Bio Stem Cell Research Center under GMP conditions. All of the cell donors provided informed consent for inclusion before participation in the study. The study was conducted in accordance with the Declaration of Helsinki, and the protocol was approved by the Ethics Committee of Biostar Stem Cell Technology (IRB NO. 2019-03).

2.2. Chemicals and Reagents

All chemicals and reagents were purchased from Sigma-Aldrich (St. Louis, MO, USA), unless stated otherwise.

2.3. Culture and Characterization of AMSCs

Isolation and establishment of AMSCs were performed as previously described [18]. Briefly, cryopreserved AMSCs (1×10^6) from the amnion tissue of three donors were thawed and seeded in T-175 flasks (175 cm^2) containing RPME-P (R BIO, Seoul, Republic of Korea) supplemented with an antibiotic–antimycotic solution (1%) in a humidified incubator at 37 °C and 5% CO_2. AMSCs were cultured in AMSC medium (R BIO) until the cells were confluent. The immunophenotypic characterization of cultured AMSCs was performed using flow cytometry. The method and result of characterization were previously stated [18].

2.4. Preparation of AMSC-CM

AMSCs from a donor were cultured in RPME-P until 90% confluency, and then the medium was changed to Dulbecco's modified Eagle's medium (DMEM) after washing twice with PBS. The culture medium was collected every 24 h, and fresh DMEM was added to the original flask. Supernatants were collected for 5 d and then pooled. The pooled medium was centrifuged at 1700 rpm for 5 min and filtered using a 0.22 μm filter to be prepared as CM. CM from three donors was equally mixed in a total volume of 20 mL and concentrated to a final volume of 2 mL by centrifugation at $3000\times g$ for 90 min at 4 °C using a 3 kDa cut-off filter tube (Vivaspin 20; GE healthcare, Chicago, IL, USA).

2.5. Experimental Animals

The 7- and 28-week-old female C57BL/6J mice (Central Lab Animal Inc., Seoul, Republic of Korea) were housed under conventional conditions in an animal facility at the Seoul National University. Experimental animals were kept in a room under controlled temperature (23 °C) and humidity conditions (60%) with a 12 h light/dark cycle. Mice were humanely sacrificed by cervical dislocation as required. All experimental animal procedures were approved by the Institutional Animal Care and Use Committee of Seoul National University (SNU-200928-2-2).

2.6. Intravenous Administration of AMSC-CM

The 7-week-old mice were categorized into the 'young control' group (YC) and 28-week-old mice were randomly divided into the 'old control' (OC) and 'old treatment' (OT) groups. The OT group was intravenously injected with human AMSC-CM via the tail vein six times with 4 d intervals, which was previously confirmed as the optimal condition for IV CM treatments [17]. The YC and OC groups were intravenously injected with normal saline (0.9%) (JW Pharmaceutical, Seoul, Republic of Korea). The single dose amount was determined based on the weight of each mouse (1 μL/g). Prior to the treatment, pregnant mare serum gonadotropin (10 IU), followed by human chorionic gonadotropin (10 IU) after 48 h, were intraperitoneally injected into every group for estrous cycle synchronization.

2.7. Oxidative Stress and Antioxidant Marker Assays

The ovaries and uterine tissue lysates from the YC, OC, and OT groups were used to measure total antioxidant capacity (TAC), hydrogen peroxide (H_2O_2), and catalase (CAT) activities using OxiSelect™ assay kits (Cell Biolabs Inc., San Diego, CA, USA) according to the manufacturer's instructions. The collected ovaries and uterine tissue samples were homogenized and centrifuged, and the supernatant was collected according to the manufacturer's protocol, as per the recommendations in the respective assay kit. The results of each colorimetric assay were calculated using absorbance values measured at 490 nm for TAC activity, 520 nm for CAT activity, and 570 nm for H_2O_2 activity.

2.8. Serum Hormone Measurement

Serum was collected from mice in the YC, OC, and OT groups, and the levels of follicle-stimulating hormone (FSH) and anti-mullerian hormone (AMH) were measured using an enzyme immunoassay kits (Mybiosource, San Diego, CA, USA) according to the manufacturer's instructions. The optical density of the samples was measured using a spectrophotometer at 450 nm.

2.9. RNA Sequencing Analysis

RNA sequencing was performed on the ovaries and uterine tissues collected from the mice in the OC (n = 3) and OT (n = 4) groups. The RNA library was prepared using an Illumina TruSeq Stranded Total RNA Sample Prep Kit (Illumina, San Diego, CA, USA) according to the manufacturer's instructions. Briefly, after removing rRNA from the total RNA (700 ng) using an rRNA removal kit, RNA was purified using beads. After rRNA depletion, a fragment mix (EPH) was used for the remaining RNA at 94 °C for 6 min. Fragmented RNA primed with random hexamers was transcribed to first-strand cDNA using reverse transcriptase and random primers at 25 °C for 10 min, 42 °C for 15 min, and 70 °C for 15 min. Subsequently, to produce double-stranded cDNA, the replacement strand was synthesized by incorporating dUTP in place of dTTP using polymerase at 16 °C for 1 h. After cDNA was cleaned using sample purification beads, a single "A" nucleotide was attached to the 3′ ends of the blunt fragments using the A-Tailing mix reagent by incubation at 37 °C for 30 min, and then at 70 °C for 5 min. Indexing adapters were ligated to the ends of the DNA fragments using ligation mix 2 reagents at 30 °C for 10 min. PCR was conducted to enrich the DNA fragments with adapter molecules on both ends after washing twice with sample purification beads under the thermocycler conditions, as follows: 95 °C for 3 min, followed by eight cycles of 98 °C for 20 s, 60 °C for 15 s, and 72 °C for 30 min, with a final extension at 72 °C for 5 min. The quality and band size of the libraries were evaluated using an Agilent 2100 Bioanalyzer (Agilent, Santa Clara, CA, USA). Libraries were quantified by qPCR using the CFX96 Real-Time System (Bio-Rad, Hercules, CA, USA) and normalized. Sequencing of the prepared library was conducted using the NextSeq 500 system (Illumina). Potential sequencing adapters in the raw reads were trimmed using Skewer (version 0.2.2). The trimmed reads were mapped to the reference genome using STAR (version 2.5) software [19]. Since the sequencing libraries were strand specific, prepared using Illumina's strand-specific library preparation kit, the mapping process was performed using the strand-specific library option (--library-type = fr-firststrand). Cufflinks software (version 2.2.1) [20] was used with the strand-specific library option (--library-type = fr-firststrand) and other default options for the quantification of the mapped reads on the reference genome to the gene expression values. The gene annotation of the reference genome (mm10) from UCSC Genome Browser (https://genome.ucsc.edu) in the GTF format was used as a gene model, and the expression values were calculated in the fragments per kilobase of transcript per million fragments mapped (FPKM) unit. Differentially expressed genes (DEGs) between the OC and OT groups were analyzed using Cuffdiff software in the Cufflinks package [21] using the strand-specific library option (--library-type = fr-firststrand) and other default options. The normalized expression values of the selected DEGs were clustered unsupervised and scatter plots for the gene expression values between the two groups were drawn using in-house R scripts to compare the expression profiles.

2.10. Functional Category Analysis

To investigate the physiological function of the DEGs in the ovarian and uterine tissues between the OC and OT groups, the gene set overlapping test was conducted between the analyzed DEGs and functional categorized genes regarding biological processes of Gene Ontology by g: Profiler version 0.6.7 [22]. Gene set enrichment analysis was performed using the GSEA software v4.1.0 (https://www.gsea-msigdb.org/gsea/index.jsp, accessed on 11 August 2020).

2.11. Statistical Analyses

Statistical analyses were performed using GraphPad Prism version 5 (GraphPad; San Diego, CA, USA) using a one-way ANOVA followed by Tukey's post-hoc test. Data are presented as the mean ± standard error of the mean. A *p*-value of less than 0.05 was considered statistically significant among the groups. All experiments were conducted in at least three replicates.

3. Results

3.1. AMSC-CM Treatment Restrains Oxidative Stress and Improves Antioxidant Levels

To evaluate the oxidative effect of AMSC-CM administration in the reproductive organs of aged female mice, oxidative stress and antioxidant biomarker levels were compared in the ovarian and uterine lysates from the YC, OC, and OT groups.

In the ovarian lysates, the TAC of the OT group was significantly higher than that of the OC group ($p < 0.01$), but was lower than that of the YC group ($p < 0.001$, Figure 1A). The H_2O_2 content of the OT group was significantly decreased compared to OC group, but not as compared to the YC group ($p < 0.001$, Figure 1B). CAT activity showed no differences among all of the groups (Figure 1C).

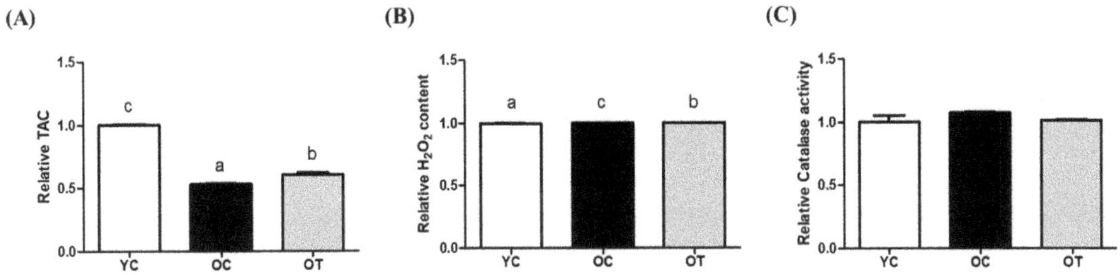

Figure 1. Comparison of oxidative stress and antioxidant biomarker levels in ovaries of young control (YC), old control (OC), and old treatment groups injected with AMSC-CM. The relative total antioxidant capacity (TAC) (**A**), hydrogen peroxide (H_2O_2) content (**B**), and catalase activity (**C**) were expressed. Data are normalized to average value of ASC-CM and presented as the mean ± SEM. Different superscript letters in each column indicate significant difference ($p < 0.05$).

In the uterine lysate, the TAC of the OC group was the lowest ($p < 0.001$), and that of the OT group was significantly improved compared to the YC ($p < 0.001$) and the OC groups ($p < 0.001$, Figure 2A). The H_2O_2 content of the OT group was significantly lower than that that of the OC group, but not higher than that of the YC group ($p < 0.0001$, Figure 2B). However, in contrast to the results with the ovarian lysate, the CAT activity was significantly increased in the OT group compared with both the YC ($p < 0.05$) and OC groups ($p < 0.01$, Figure 2C).

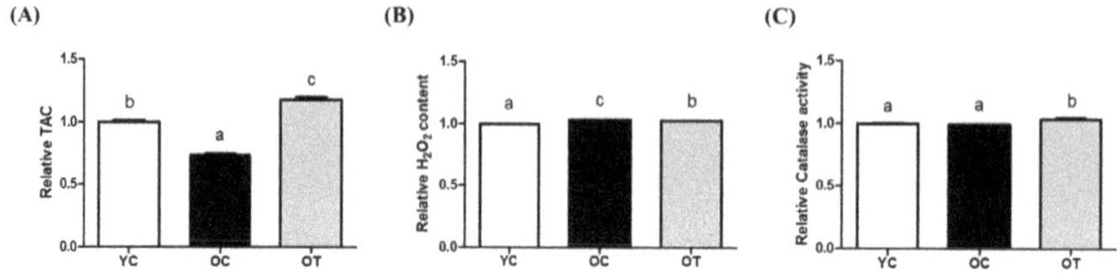

Figure 2. Comparison of oxidative stress and antioxidant biomarkers level in uterus of young control (YC), old control (OC), and old treatment groups injected with AMSC-CM. The relative total antioxidant capacity (TAC) (**A**), hydrogen peroxide (H_2O_2) content (**B**), and catalase activity (**C**) were expressed. Data are normalized to average value of ASC-CM and presented as the mean ± SEM. Different superscript letters in each column indicate significant difference ($p < 0.05$).

3.2. AMSC-CM Treatment Restores Serum Reproductive Hormone Levels

In addition to antioxidation, the physiological effects of AMSC-CM administration in aged female mice reproduction were validated by the serum hormone levels of FSH and AMH. The serum levels of FSH were significantly higher in the OC group than in the YC and OT groups ($p < 0.01$). FSH levels in the YC and OT groups were similar (Figure 3A). AMH levels in the OC and OT groups were lower than in the YC group but significantly increased in the OT group compared to the OC group ($p < 0.001$, Figure 3B).

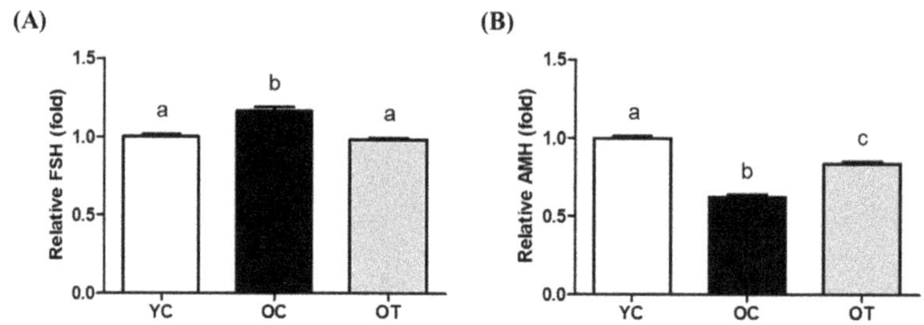

Figure 3. Relative serum hormone levels in young control (YC), old control (OC), and old treatment (OT) groups injected with AMSC-CM. (**A**) Follicle stimulating hormone (FSH); (**B**) anti-mullerian hormone (AMH). Data are presented as the mean ± SEM. Different superscript letters in each column indicate significant difference ($p < 0.05$).

3.3. AMSC-CM Treatment Alters the Ovary and Uterus Transcriptomes

For molecular understanding of the antioxidative and physiological effects above, transcriptome analysis was conducted in the ovaries and uterus of aged female mice according to AMSC-CM administration. The quality of reads from each sample was assessed by the rate of reads whose sequencing quality was over 30 in the total number of reads. The percentage of samples showing over Q30 bases was higher than 80%, thus validating the read quality. The normalized distribution of transcript expression in each sample of the OC and OT groups was identical. DEGs between the OC and OT groups were identified using a fold-change cutoff of 2 and a p-value cutoff of 0.05. Among a total of 75 DEGs, the unsupervised hierarchical clustering analysis found that 67 genes were upregulated and eight genes were downregulated in the OT-group ovaries compared to that in the OC-group ovaries (Figure 4A). As compared to the OC-group uteruses, 431 genes were upregulated and 163 genes were downregulated among a total of 594 DEGs (Figure 4B) in the OT-group uteruses (Figure 4B). Additionally, 54 of the 75 ovarian DEGs overlapped with the uterine DEGs between the OC and OT groups. Except for three genes, all genes were upregulated in both the ovary and uterus (Table 1).

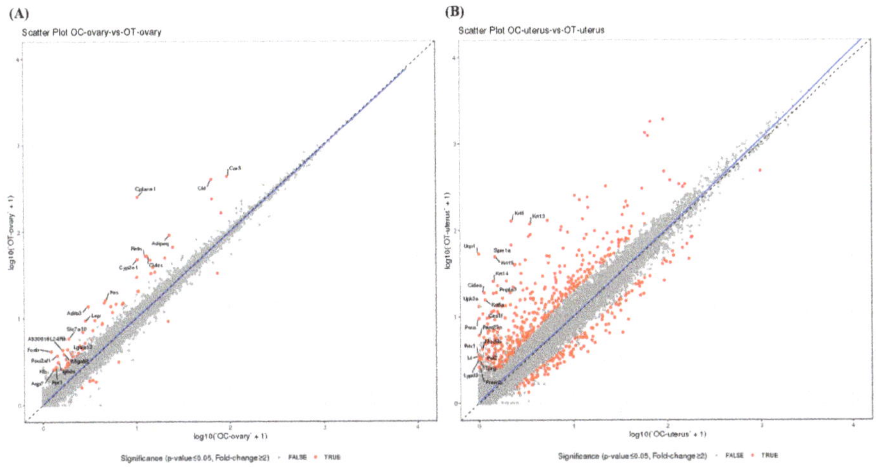

Figure 4. Scatter plot based on gene expression profiles to show the similarities among samples. Differentially expressed genes between groups were identified using a fold-change cutoff of 2 and a *p*-value cutoff of 0.05. (**A**) OC-group ovaries vs. OT-group ovaries; (**B**) OC-group uteruses vs. OT-group uteruses.

Table 1. Differentially expressed genes that are commonly upregulated in the ovary and uterus between the OC and OT groups.

OC vs. OT					
Official Gene Symbol	**Log_2 (Fold Change)**		**Official Gene Symbol**	**Log_2 (Fold Change)**	
	Ovary	**Uterus**		**Ovary**	**Uterus**
Slc7a10	2.406	5.256	*Ffar2*	1.260	3.868
A530016L24Rik	2.35	5.092	*Lpl*	1.362	3.758
Cfd	2.66	4.759	*Rbp4*	1.689	3.423
Adrb3	2.61	4.793	*Fosb*	3.741	1.284
Klb	2.35	5.009	*Retnla*	1.219	3.601
Pnpla3	1.819	5.391	*Abcd2*	1.184	3.334
Pou2af1	2.623	4.517	*Trarg1*	1.421	3.089
Aqp7	2.163	4.924	*Cdo1*	1.829	2.349
Cyp2e1	2.327	4.736	*Cd36*	1.48	2.575
Car3	2.211	4.396	*Irs3*	1.481	2.518
Cidec	2.102	4.377	*Jchain*	1.653	2.341
Plin1	1.438	4.895	*Fcor*	1.21	2.779
Retn	2.147	4.182	*Ces1d*	1.022	2.706
Fabp4	1.889	4.293	*Mzb1*	1.359	2.265
B3galt2	2.034	4.098	*Cd79b*	1.544	2.017
Lgals12	2.000	4.037	*Acp5*	1.14	2.398
Adipoq	2.071	3.909	*Ifi27l2a*	1.022	2.368
Adig	1.407	4.516	*Hcar1*	1.023	2.289
Lep	2.163	3.73	*Fos*	2.016	1.125
Acvr1c	1.483	4.181	*Cd209g*	1.029	1.871
Pck1	1.545	4.068	*Pxmp2*	1.279	1.493
Apoc1	1.154	4.404	*Tmem45b*	1.023	1.676
Igfals	2.183	3.335	*Ccl2*	1.284	1.384
Orm1	1.36	4.096	*Dgat2*	1.166	1.411
Serpina3c	1.284	4.144	*Thrsp*	1.265	1.218
Myl1	1.039	4.181	*Pla2g2d*	1.066	1.218
Sncg	1.979	3.201			

3.4. Functional Analysis of the Ovaries in Singular Enrichment Analysis Terms

Enriched Gene Ontology (GO) terms in the biological process category were analyzed using the g:Profiler (version 0.6.7) R package. In the biological process, 114 terms were significantly enriched in the OT-group ovaries compared to those in the OC group, including response to stimulus, immune system process, metabolic process, homeostatic process, and female pregnancy. As is shown in Figure 5A, the top ten enriched terms were lipid metabolic process, response to organonitrogen compound, lipid catabolic process, response to peptide, response to peptide hormone, response to nitrogen compound, cellular lipid metabolic process, lipid localization, response to insulin, and acylglycerol metabolic process.

(A)

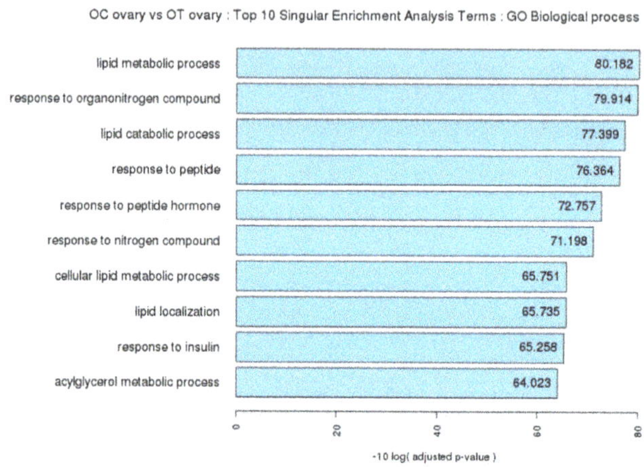

(B)

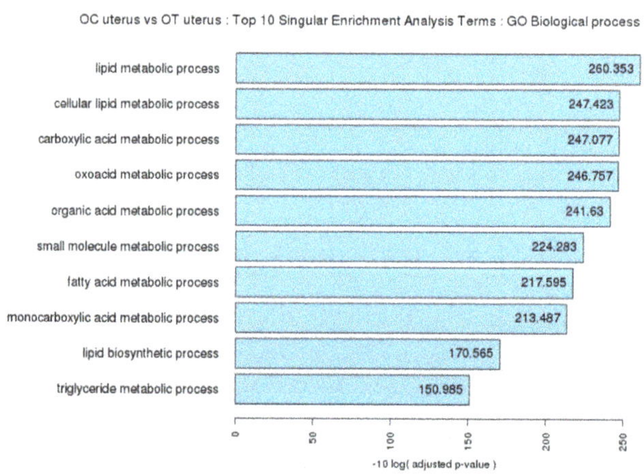

Figure 5. Top 10 singular enrichment analysis terms in GO biological process. (**A**) OC-group ovaries vs. OT-group ovaries, the enriched biological terms from the 75 DEGs selected by a fold-change cutoff of 2 and a p-value cutoff of 0.05. (**B**) OC-group uteruses vs. OT-group uteruses, the enriched biological terms from the 594 DEGs selected by a fold-change cutoff of 2 and a p-value cutoff of 0.05.

3.5. Functional Analysis of the Uterus in Singular Enrichment Analysis Terms

Among the enriched GO terms in biological processes, 348 terms were significantly enriched in the OT-group uteruses compared to those in the OC group, including cell differentiation, structure development, regulation of peptide secretion, response to stimulus, metabolic process, and homeostatic process. As is shown in Figure 5B, the top ten enriched terms were lipid metabolic process, cellular lipid metabolic process, carboxylic acid metabolic process, oxoacid metabolic process, organic acid metabolic process, small molecule metabolic process, fatty acid metabolic process, monocarboxylic acid metabolic process, lipid biosynthetic process, and triglyceride metabolic process.

3.6. Gene Set Enrichment Analysis of the Ovaries

GSEA was performed to compare the gene enrichment of selected terms in the ovaries of the OC and OT groups. The list of terms analyzed (Figure 6A,B) include the following: cell death in response to oxidative stress (normalized enrichment score: 0.814224, nominal p-value: 0.912), oxidation reduction process (−1.1907881, 0.026286965), reactive oxygen species metabolic process (−1.6204555, 0), developmental process involved in reproduction (1.0326562, 0.29069766), ovarian follicle development (0.99881315, 0.4673913), ovulation (0.96755165, 0.5), and ovulation from ovarian follicles (−0.8856735, 0.6181172).

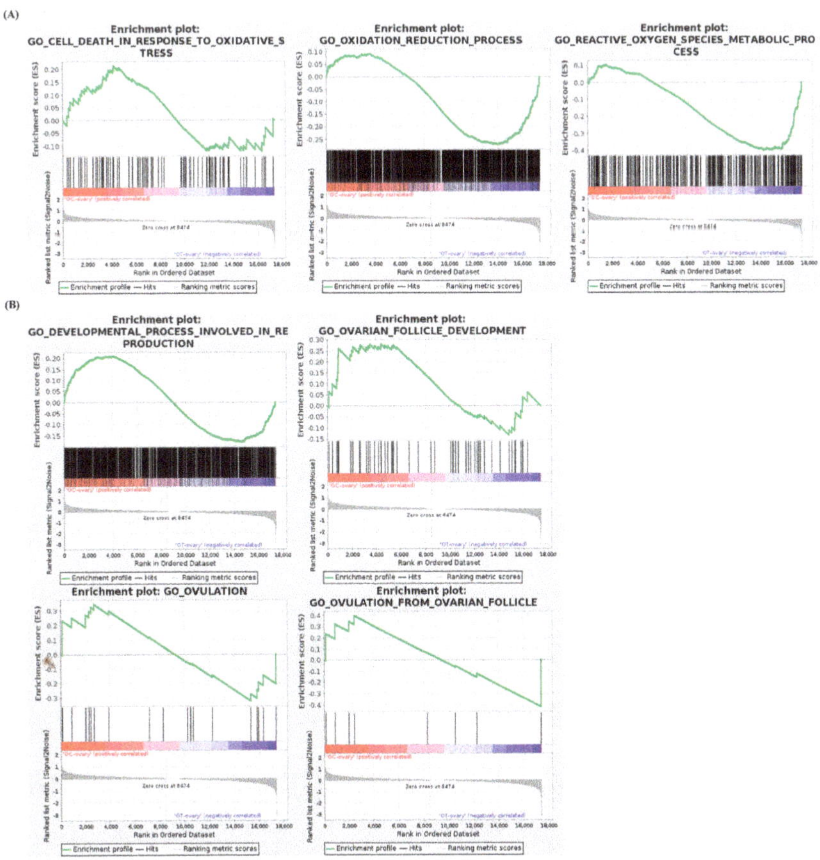

Figure 6. Gene set enrichment analysis between OC-group ovaries and OT-group ovaries. GO terms related to oxidative stress (**A**) and reproductive processes (**B**) in ovaries were presented.

3.7. Gene Set Enrichment Analysis of the Uterus

As is shown in Figure 7A,B, the list of terms analyzed by GSEA between the uteruses of the OC and OT groups include the following: cell death in response to oxidative stress (normalized enrichment score: −1.0534321, nominal p-value: 0.35905045), oxidation reduction process (−2.3681388, 0), reactive oxygen species metabolic process (−1.6318362, 0.0012919897), developmental process involved in reproduction (1.176085, 0), uterine development (1.5872613, 0.017721519), and embryo implantation (0.9732308, 0.4863222).

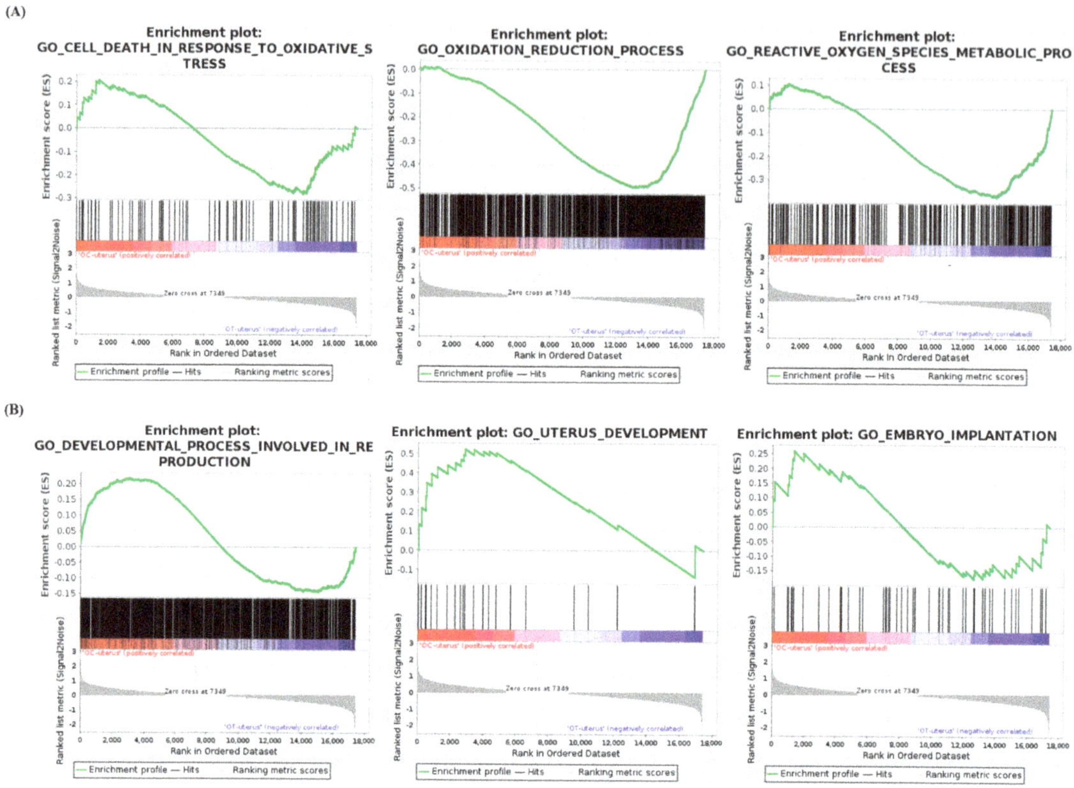

Figure 7. Gene set enrichment analysis between OC-group uteruses and OT-group uteruses. GO terms related to oxidative stress (**A**) and reproductive process (**B**) in uteruses were presented.

4. Discussion

In this study, the ovarian and uterine gene expression changes in aged mice were identified following the IV administration of AMSC-CM, which could demonstrate antioxidative and physiological effects in reproduction. These findings build on the previously reported antioxidant effect of AMSC-CM [18], thereby confirming the beneficial effects of AMSC-CM on female reproduction. First of all, oxidative stress and antioxidant biomarkers in the ovarian and uterine lysates were measured to confirm the antioxidative effect of AMSC-CM at the protein level. As there is a cumulative ability of antioxidants to counteract oxidative stress, TAC provides an overall measure of the antioxidant defense system maintaining cellular functions and preventing damage caused by ROS in various physiological and pathological conditions [23]. The TAC of reproductive organs in aged mice was lower than that in the young mice, as predicted; but, remarkably, AMSC-CM treatment improved the TAC in both the ovary and uterus (Figures 1A and 2A). In particular, the TAC of old mice uteruses treated with AMSC-CM was observed to be higher than that of YC mice (Figure 2A). The contents of H_2O_2 in both the ovary and uterus were reduced after the IV administration of AMSC-CM to aged mice (Figures 1B and 2B), showing its antioxidative effects on a key ROS that contributes to oxidative stress by causing damage to cellular components, including DNA, proteins, and lipids, associated with pathological conditions such as inflammation and aging [24]. Of note, the activity of CAT, which was one of the DEGs upregulated specifically in the OT group (in the uterus), was enhanced after AMSC-CM treatment in aged mice (Figure 2C). Moreover, this is consistent with the previous finding that adipose-derived MSC (ASC)-CM treatment as conventional antioxidant interventions in aged female mice increased CAT expression in the uterus by 7-fold in 4-month-old mice and 1.5-fold in 6-month-old mice [25].

Next, the reproductive hormones, namely FSH and AMH levels, were evaluated in the serum of the YC, OC, and OT groups. FSH and AMH are markers of aging as an increase in the FSH level and a decrease in the AMH level indicate ovarian aging [26], depletion of the follicular reserve, and uterine atrophy [27]. While aged mice showed increased FSH levels (Figure 3A), AMSC-CM treatment reduced FSH levels in aged mice, similar to the levels in the YC mice. Furthermore, as a biomarker to assess reproductive age [28], the AMH level was upregulated in aged mice by AMSC-CM treatment (Figure 3B). As the YC, OC, and OT groups were equally injected with superovulation-induction hormones for estrous cycle synchronization, the possibility that the difference in the hormone levels among groups is due to individual estrous cycles can be excluded. Thus, the results demonstrate the anti-aging effect of AMSC-CM on age-dependent changes in serum hormone levels.

In the transcriptome analysis, among the 54 genes differentially expressed in common in the AMSC-CM-treated ovaries and uteruses of aged mice compared to the control group (Table 1), the most remarkable genes were klotho beta (*klb*) and *adipoq*. Klotho has been identified as an aging-suppressor gene, and the its overexpression in mice leads to an extension of lifespan by 20–30% through regulation of the aging process [29]. Klotho helps to preserve cellular health during aging by supporting redox balance and counteracting oxidative stress, thereby contributing to mitochondrial function and overall metabolic stability [30–32]. *klb*, as a crucial element of endocrine fibroblast growth factor (FGF) receptor complexes, controls multiple metabolic processes including energy metabolic processes, glucose uptake, and fatty acid metabolism [29,33]. Interacting with FGF signaling, recent studies have demonstrated that *klb* functions as a critical molecular integrator of metabolism and reproductive aging. Its expression is associated with both hypothalamic gonadotropin-releasing hormone signaling and peripheral redox balance, linking systemic metabolic cues to reproductive axis regulation and ovarian homeostasis [34,35]. In the present study, treatment with AMSC-CM led to a marked upregulation of *klb* expression in the aged ovary and uterus, suggesting AMSC-CM may regulate the metabolic–reproductive crosstalk that is disrupted by aging, and its enhancement of *klb*-associated signaling could be a potential strategy for restoring age-related reproductive decline through both metabolic and redox-mediated mechanisms.

Adipoq encodes the protein adiponectin, one of the adipokine hormones that act as critical regulators in female reproductive organs regarding ovarian steroidogenesis, oocyte maturation, development, and the embryo implantation environment [36]. The effect of adiponectin on the ovary includes the activation of ovarian follicle remodeling genes, oocyte maturation, corpus luteum development, and the secretion of ovarian hormones [37]. Furthermore, adiponectin affects the uterine expression of genes involved in uterine tissue development, proliferation, apoptosis, and the regulation of implantation [37]. The interaction of *adipoq* and its receptors is associated with preimplantation embryo development and uterine receptivity through autocrine and paracrine signaling [38]. Mechanistically, *adipoq* plays a critical role in oxidative regulation, a key contributor to reproductive aging, by activating the AMP-activated protein kinase (*AMPK*) signaling pathway [36–38]. *AMPK* serves as an upstream modulator of the Forkhead box O (*FoxO*) transcription factors and their downstream genes involved in oxidative stress resistance, ultimately modulating cellular antioxidant defense [39–41]. Interestingly, in the previous study, *FoxO*-mediated expression of antioxidant genes were increased in blastocysts that developed in vitro in AMSC-CM–supplemented culture medium [18]. While the previous study focused on preimplantation embryos and the present study investigates ovarian and uterine tissues in the context of reproductive aging, the shared upregulation of genes involved in *AMPK* pathway indicates a broad anti-oxidative effect of AMSC-CM on reproductive competence. Taken together, these findings suggest that AMSC-CM may modulate oxidative stress not only in somatic reproductive tissues but also throughout the oocyte-to-blastocyst developmental trajectory via the *Adipoq–AMPK–FoxO* axis.

In both ovarian and uterine genes, the top ten singular enrichment analysis terms in biological processes, between the AMSC-CM-untreated and -treated groups, were primarily categorized under metabolic processes (Figure 5A,B). Among the various metabolic processes, energy metabolism, a common function of *klb* and *adipoq*, declines progressively with age [42]. The resulting metabolic problems may induce reproductive disorders, including ovarian disease or infertility [43]. The lipid-metabolism-related terms were predominant in the top ten singular enrichment analysis terms (Figure 2B). Lipid metabolism is highly involved in the regulation of aging, as demonstrated by studies showing that interventions modulating lipid metabolism increase life expectancy [44]. In the aspect of reproduction, abnormal lipid metabolism is a leading risk factor for infertility in aged women [45]. In the perimenopausal state, the administration of antioxidants can be applied to modulate disrupted lipid metabolism and improve ovarian function [46]. Furthermore, a disorder of lipid metabolism is considered to induce an adverse pregnancy outcome, considering the essential roles of lipids in various cellular processes, including cell apoptosis, differentiation, and growth [47]. In particular, uterine lipid metabolism is critical for uterine receptivity [48]. It was hypothesized that the IV administration of AMSC-CM in aged mice would alleviate aging-induced damage to reproduction and oxidative stress. Indeed, this study found that the genes related to female pregnancy were differentially expressed in the OT-group ovaries compared to those in the OC group. Moreover, based on the analysis of the GO biological process terms, the genes related to detoxification, the defense response, the oxidation–reduction process, and the reactive oxygen species metabolic process were differentially expressed in the OT-group uteruses compared to those in the OC group. Further, GSEA for specific terms was performed to evaluate the effect of AMSC-CM treatment on the ovaries and uteruses of aged mice, focusing on reproduction and oxidative stress. GSEA suggested that genes involved in the regulation of oxidation were enriched in the uterus following AMSC-CM treatment compared to those in the control group (Figure 7). In addition to oxidation reduction and ROS metabolism, genes engaged in ovulation were also enriched in the ovaries of the treatment group as compared to those of the control group (Figure 6).

These findings bridge the molecular effects of AMSC-CM with functional improvements in antioxidant modulation and endocrine regulation, providing a robust foundation for future applications to age-related infertility. Although the present study remains at the preclinical stage, existing findings suggest that AMSC-CM could be a promising therapeutic agent and an effective solution for female reproductive aging, advancing from bench to bedside [7,18]. As recent studies have pointed out, AMSC-CM has emerged as a potential candidate with advantages over cell-based approaches in terms of safety and practical aspects, such as being free from immunogenicity risks, having a standardized production protocol, and offering storage stability and manageability. From a legal standpoint, regulatory guidelines for manufacturing and clinical trial standards will be required to establish AMSC-CM as a translational medicine after thorough preclinical research investigating its pragmatic applications. In this regard, expanding on the outcomes of the present study, further research could evaluate additional metrics of reproductive performance, such as implantation rates and live birth outcomes, to provide a more comprehensive understanding of functional improvements. Additionally, identifying specific factors within AMSC-CM that contribute to these antioxidative and anti-aging effects would clarify the underlying mechanisms and support its therapeutic applications.

5. Conclusions

The IV of human AMSC-CM resulted in upregulated genes with their main functions being modulating aging and endocrine effects in the female reproductive system and biological metabolic processes in the ovaries and uteruses of aged female mice. Moreover, the antioxidative effect of human AMSC-CM was confirmed at both the RNA and protein levels in the ovary and uterus as the level of indicative markers of antioxidant capacity was recovered and age-dependent alteration of reproductive hormones was restored, all of which resulted in the improvement of reproductive competence in aged mice.

Funding: This research was supported by Nature Cell (#550-20200076), R Bio (#550-20230116), and the Basic Science Research Program, through the National Research Foundation of the Republic of Korea (NRF), funded by the Ministry of Education (RS-2023-00271197), BK21 plus program, and Research Institute for Veterinary Science.

Institutional Review Board Statement: The study was conducted according to the guidelines of the Declaration of Helsinki, and approved by the Ethics Committee of Biostar Stem Cell Technology (IRB NO. 2019-03).

Informed Consent Statement: All experiments using experimental animals in this study were approved by the Institutional Animal Care and Use Committee of Seoul National University (SNU-200928-2-2).

Data Availability Statement: The RNA sequencing data presented in this study are available in the Gene Ex-pression Omnibus (GEO) under the accession number GSE297514.

Acknowledgments: The authors would like to thank Eun Young Kim, Sung Keun Kang, and Jeong Chan Ra from Nature Cell and R Bio (Seoul, Republic of Korea) for the management and provision of Human AMSCs, and Chae Yeong Choi and Sung Yeoun Lee from LAS Inc. (Gimpo, Republic of Korea) for their kind assistance with sequencing data interpretation.

Conflicts of Interest: The author declares no conflicts of interest.

Abbreviations

The following abbreviations are used in this manuscript:

MSCs	Mesenchymal stem cells
AMSCs	Amniotic membrane-derived mesenchymal stem cells
CM	Conditioned medium
ROS	Reactive oxygen species
IV	Intravenous injection
EGs	Differentially expressed genes
FSH	Follicle-stimulating hormone
AMH	Anti-mullerian hormone
TAC	Total antioxidant capacity
SOD	Superoxide dismutase
CAT	Catalase
GSEA	Gene set enrichment analysis

References

1. Beshay, E.; Rehman, K.U.; Carrier, S. Impact of Aging on Reproduction and Sexual Function. In *Autonomic Nervous System in Old Age*; Kuchel, G.A., Hof, P.R., Eds.; Interdiscipl Top Gerontol; Karger: Basel, Switzerland, 2004; Volume 33, pp. 94–106.
2. Meldrum, D.R. Female reproductive aging—Ovarian and uterine factors. *Fertil. Steril.* **1993**, *59*, 1–5. [CrossRef] [PubMed]
3. Velarde, M.C.; Menon, R. Positive and negative effects of cellular senescence during female reproductive aging and pregnancy. *J. Endocrinol.* **2016**, *230*, R59–R76. [CrossRef]
4. Shirasuna, K.; Iwata, H. Effect of aging on the female reproductive function. *Contracept. Reprod. Med.* **2017**, *2*, 23. [CrossRef] [PubMed]
5. Woods, L.; Perez-Garcia, V.; Kieckbusch, J.; Wang, X.; DeMayo, F.; Colucci, F.; Hemberger, M. Decidualisation and placentation defects are a major cause of age-related reproductive decline. *Nat. Commun.* **2017**, *8*, 352. [CrossRef]
6. Nelson, S.M.; Telfer, E.E.; Anderson, R.A. The ageing ovary and uterus: New biological insights. *Hum. Reprod. Update* **2013**, *19*, 67–83. [CrossRef] [PubMed]
7. Ra, K.; Park, S.C.; Lee, B.C. Female Reproductive Aging and Oxidative Stress: Mesenchymal Stem Cell Conditioned Medium as a Promising Antioxidant. *Int. J. Mol. Sci.* **2023**, *24*, 5053. [CrossRef]
8. Wang, S.; He, G.; Chen, M.; Zuo, T.; Xu, W.; Liu, X. The Role of Antioxidant Enzymes in the Ovaries. *Oxid. Med. Cell. Longev.* **2017**, *2017*, 4371714. [CrossRef]
9. Al-Gubory, K.H.; Garrel, C.; Faure, P.; Sugino, N. Roles of antioxidant enzymes in corpus luteum rescue from reactive oxygen species-induced oxidative stress. *Reprod. Biomed. Online* **2012**, *25*, 551–560. [CrossRef]
10. Lu, J.; Wang, Z.; Cao, J.; Chen, Y.; Dong, Y. A novel and compact review on the role of oxidative stress in female reproduction. *Reprod. Biol. Endocrinol.* **2018**, *16*, 80. [CrossRef]
11. Marcus, A.J.; Woodbury, D. Fetal stem cells from extra-embryonic tissues: Do not discard. *J. Cell. Mol. Med.* **2008**, *12*, 730–742. [CrossRef]
12. Miki, T. Amnion-derived stem cells: In quest of clinical applications. *Stem Cell Res. Ther.* **2011**, *2*, 25. [CrossRef]
13. Li, Z.; Han, Z.C. Introduction of Perinatal Tissue-Derived Stem Cells. In *Perinatal Stem Cells: Biology, Manufacturing and Translational Medicine*; Han, Z.C., Takahashi, T.A., Han, Z., Li, Z., Eds.; Springer: Singapore, 2019; pp. 1–7. [CrossRef]
14. Huang, B.; Ding, C.; Zou, Q.; Lu, J.; Wang, W.; Li, H. Human Amniotic Fluid Mesenchymal Stem Cells Improve Ovarian Function During Physiological Aging by Resisting DNA Damage. *Front. Pharmacol.* **2020**, *11*, 272. [CrossRef]
15. Li, J.; Mao, Q.; He, J.; She, H.; Zhang, Z.; Yin, C. Human umbilical cord mesenchymal stem cells improve the reserve function of perimenopausal ovary via a paracrine mechanism. *Stem Cell Res. Ther.* **2017**, *8*, 55. [CrossRef] [PubMed]
16. Ding, C.; Zou, Q.; Wang, F.; Wu, H.; Chen, R.; Lv, J.; Ling, M.; Sun, J.; Wang, W.; Li, H.; et al. Human amniotic mesenchymal stem cells improve ovarian function in natural aging through secreting hepatocyte growth factor and epidermal growth factor. *Stem Cell Res. Ther.* **2018**, *9*, 55. [CrossRef] [PubMed]
17. Ra, K.; Oh, H.J.; Kim, E.Y.; Kang, S.K.; Ra, J.C.; Kim, E.H.; Lee, B.C. Anti-Oxidative Effects of Human Adipose Stem Cell Conditioned Medium with Different Basal Medium during Mouse Embryo In Vitro Culture. *Animals* **2020**, *10*, 1414. [CrossRef] [PubMed]
18. Ra, K.; Oh, H.J.; Kim, E.Y.; Kang, S.K.; Ra, J.C.; Kim, E.H.; Park, S.C.; Lee, B.C. Comparison of Anti-Oxidative Effect of Human Adipose- and Amniotic Membrane-Derived Mesenchymal Stem Cell Conditioned Medium on Mouse Preimplantation Embryo Development. *Antioxidants* **2021**, *10*, 268. [CrossRef]
19. Dobin, A.; Davis, C.A.; Schlesinger, F.; Drenkow, J.; Zaleski, C.; Jha, S.; Batut, P.; Chaisson, M.; Gingeras, T.R. STAR: Ultrafast universal RNA-seq aligner. *Bioinformatics* **2013**, *29*, 15–21. [CrossRef]
20. Trapnell, C.; Williams, B.A.; Pertea, G.; Mortazavi, A.; Kwan, G.; van Baren, M.J.; Salzberg, S.L.; Wold, B.J.; Pachter, L. Transcript assembly and quantification by RNA-Seq reveals unannotated transcripts and isoform switching during cell differentiation. *Nat. Biotechnol.* **2010**, *28*, 511–515. [CrossRef]
21. Trapnell, C.; Hendrickson, D.G.; Sauvageau, M.; Goff, L.; Rinn, J.L.; Pachter, L. Differential analysis of gene regulation at transcript resolution with RNA-seq. *Nat. Biotechnol.* **2013**, *31*, 46–53. [CrossRef]
22. Reimand, J.; Arak, T.; Adler, P.; Kolberg, L.; Reisberg, S.; Peterson, H.; Vilo, J. g:Profiler—A web server for functional interpretation of gene lists (2016 update). *Nucleic Acids Res.* **2016**, *44*, W83–W89. [CrossRef]
23. Ghiselli, A.; Serafini, M.; Natella, F.; Scaccini, C. Total antioxidant capacity as a tool to assess redox status: Critical view and experimental data. *Free Radic. Biol. Med.* **2000**, *29*, 1106–1114. [CrossRef] [PubMed]
24. Jomova, K.; Raptova, R.; Alomar, S.Y.; Alwasel, S.H.; Nepovimova, E.; Kuca, K.; Valko, M. Reactive oxygen species, toxicity, oxidative stress, and antioxidants: Chronic diseases and aging. *Arch. Toxicol.* **2023**, *97*, 2499–2574. [CrossRef]
25. Ra, K.; Oh, H.J.; Kim, G.A.; Kang, S.K.; Ra, J.C.; Lee, B.C. High Frequency of Intravenous Injection of Human Adipose Stem Cell Conditioned Medium Improved Embryo Development of Mice in Advanced Maternal Age through Antioxidant Effects. *Animals* **2020**, *10*, 978. [CrossRef]

26. Azami, S.H.; Nazarian, H.; Abdollahifar, M.A.; Eini, F.; Farsani, M.A.; Novin, M.G. The antioxidant curcumin postpones ovarian aging in young and middle-aged mice. *Reprod. Fertil. Dev.* **2020**, *32*, 292–303. [CrossRef] [PubMed]
27. Finch, C.E. The menopause and aging, a comparative perspective. *J. Steroid Biochem. Mol. Biol.* **2014**, *142*, 132–141. [CrossRef] [PubMed]
28. Freeman, E.W.; Sammel, M.D.; Lin, H.; Gracia, C.R. Anti-mullerian hormone as a predictor of time to menopause in late reproductive age women. *J. Clin. Endocrinol. Metab.* **2012**, *97*, 1673–1680. [CrossRef]
29. Xu, Y.; Sun, Z. Molecular basis of Klotho: From gene to function in aging. *Endocr. Rev.* **2015**, *36*, 174–193. [CrossRef]
30. Donate-Correa, J.; Martin-Carro, B.; Cannata-Andia, J.B.; Mora-Fernandez, C.; Navarro-Gonzalez, J.F. Klotho, Oxidative Stress, and Mitochondrial Damage in Kidney Disease. *Antioxidants* **2023**, *12*, 239. [CrossRef]
31. Olejnik, A.; Banaszkiewicz, M.; Krzywonos-Zawadzka, A.; Bil-Lula, I. The Klotho protein supports redox balance and metabolic functions of cardiomyocytes during ischemia/reperfusion injury. *Cardiol. J.* **2022**, *29*, 836–849. [CrossRef]
32. Meroni, M.; Dongiovanni, P.; Tiano, F.; Piciotti, R.; Alisi, A.; Panera, N. beta-Klotho as novel therapeutic target in Metabolic Dysfunction-Associated Steatotic Liver Disease (MASLD): A narrative review. *Biomed. Pharmacother.* **2024**, *180*, 117608. [CrossRef]
33. Kuro, O.M. The Klotho proteins in health and disease. *Nat. Rev. Nephrol.* **2019**, *15*, 27–44. [CrossRef]
34. Idrees, M.; Kumar, V.; Khan, A.M.; Joo, M.D.; Lee, K.W.; Sohn, S.H.; Kong, I.K. Cycloastragenol activation of telomerase improves beta-Klotho protein level and attenuates age-related malfunctioning in ovarian tissues. *Mech. Ageing Dev.* **2023**, *209*, 111756. [CrossRef]
35. Xu, C.; Messina, A.; Somm, E.; Miraoui, H.; Kinnunen, T.; Acierno, J., Jr.; Niederlander, N.J.; Bouilly, J.; Dwyer, A.A.; Sidis, Y.; et al. *KLB*, encoding beta-Klotho, is mutated in patients with congenital hypogonadotropic hypogonadism. *EMBO Mol. Med.* **2017**, *9*, 1379–1397. [CrossRef]
36. Konishi, M.; Haraguchi, G.; Ohigashi, H.; Ishihara, T.; Saito, K.; Nakano, Y.; Isobe, M. Adiponectin protects against doxorubicin-induced cardiomyopathy by anti-apoptotic effects through AMPK up-regulation. *Cardiovasc. Res.* **2011**, *89*, 309–319. [CrossRef]
37. Kim, J.; Yang, G.; Kim, Y.; Kim, J.; Ha, J. AMPK activators: Mechanisms of action and physiological activities. *Exp. Mol. Med.* **2016**, *48*, e224. [CrossRef] [PubMed]
38. Steinberg, G.R.; Carling, D. AMP-activated protein kinase: The current landscape for drug development. *Nat. Rev. Drug Discov.* **2019**, *18*, 527–551. [CrossRef] [PubMed]
39. Guan, G.; Chen, Y.; Dong, Y. Unraveling the AMPK-SIRT1-FOXO Pathway: The In-Depth Analysis and Breakthrough Prospects of Oxidative Stress-Induced Diseases. *Antioxidants* **2025**, *14*, 70. [CrossRef] [PubMed]
40. Yun, H.; Park, S.; Kim, M.J.; Yang, W.K.; Im, D.U.; Yang, K.R.; Hong, J.; Choe, W.; Kang, I.; Kim, S.S.; et al. AMP-activated protein kinase mediates the antioxidant effects of resveratrol through regulation of the transcription factor FoxO1. *FEBS J.* **2014**, *281*, 4421–4438. [CrossRef]
41. Zhao, Y.; Hu, X.; Liu, Y.; Dong, S.; Wen, Z.; He, W.; Zhang, S.; Huang, Q.; Shi, M. ROS signaling under metabolic stress: Cross-talk between AMPK and AKT pathway. *Mol. Cancer* **2017**, *16*, 79. [CrossRef]
42. Azzu, V.; Valencak, T.G. Energy Metabolism and Ageing in the Mouse: A Mini-Review. *Gerontology* **2017**, *63*, 327–336. [CrossRef]
43. Reverchon, M.; Rame, C.; Bertoldo, M.; Dupont, J. Adipokines and the female reproductive tract. *Int. J. Endocrinol.* **2014**, *2014*, 232454. [CrossRef] [PubMed]
44. Zhan, F.; Zeng, W.; Yuan, X.; Li, B.; Li, T.; Zu, Y.; Jiang, M.; Li, Y. Field experiment on the effects of sepiolite and biochar on the remediation of Cd- and Pb-polluted farmlands around a Pb-Zn mine in Yunnan Province, China. *Environ. Sci. Pollut. Res. Int.* **2019**, *26*, 7743–7751. [CrossRef] [PubMed]
45. Tao, X.; Ge, S.Q.; Chen, L.; Cai, L.S.; Hwang, M.F.; Wang, C.L. Relationships between female infertility and female genital infections and pelvic inflammatory disease: A population-based nested controlled study. *Clinics* **2018**, *73*, e364. [CrossRef]
46. Xing, X.; Su, L.; Asare, P.F.; Wang, L.; Li, L.; Liu, E.; Yu, B.; Zhu, Y.; Gao, X.; Fan, G. Danzhi Qing'e (DZQE) activates AMPK pathway and regulates lipid metabolism in a rat model of perimenopausal hyperlipidaemia. *Exp. Physiol.* **2016**, *101*, 1406–1417. [CrossRef] [PubMed]
47. Almada, M.; Domingues, M.R.; Doria, M.L.; Fonseca, B.M.; Teixeira, N.A.; Correia-da-Silva, G. Lipidomic approach towards deciphering anandamide effects in rat decidual cell. *J. Cell. Physiol.* **2015**, *230*, 1549–1557. [CrossRef]
48. Yang, Y.; Wang, L.; Chen, C.; Qi, H.; Baker, P.N.; Liu, X.; Zhang, H.; Han, T.L. Metabolic Changes of Maternal Uterine Fluid, Uterus, and Plasma during the Peri-implantation Period of Early Pregnancy in Mice. *Reprod. Sci.* **2020**, *27*, 488–502. [CrossRef]

cells

MDPI

Review

Spotlight on Proteases: Roles in Ovarian Health and Disease

Bhawna Kushawaha and Emanuele Pelosi *

Department of Biochemistry and Molecular Biology, Indiana University, Indianapolis, IN 46202, USA; bkushaw@iu.edu
* Correspondence: epelosi@iu.edu

Abstract: Proteases play crucial roles in ovarian folliculogenesis, regulating several processes from primordial follicle activation to ovulation and corpus luteum formation. This review synthesizes the current knowledge on the diverse functions of proteases in ovarian physiology and pathology. We discuss the classification and regulation of proteases, highlighting their importance in extracellular matrix remodeling, cell signaling, and apoptosis during ovarian follicular development. We explore the roles of several proteases including matrix metalloproteinases, tissue inhibitors of metalloproteinases, the plasminogen activator system, and cathepsins, and their roles in the critical functions of ovarian biology including follicle dynamics and senescence. Furthermore, we address the involvement of proteases in ovarian pathologies, including cancer, polycystic ovary syndrome, and primary ovarian insufficiency. By integrating recent findings from clinical genomics and animal models, this review provides a comprehensive overview of protease functions in the ovary, emphasizing their potential use for therapeutic interventions in reproductive medicine.

Keywords: proteases; ovary; matrix metalloproteinases; cathepsins; ovarian reserve; ovarian cancer; polycystic ovary syndrome; primary ovarian insufficiency

1. Introduction

Several ovarian processes from follicle activation to the maturation and release of oocytes require the catalytic action of protease enzymes [1–4]. Since first reported in 1905 by P. A. Levene, proteases remain at the forefront of basic and clinical research due to their regulatory activity of fundamental and ubiquitous cellular functions [5]. Recent innovations in large-scale genome sequencing and high-throughput differential screening methods have established that the mouse and human genomes contain 628 and 566 protease-encoding genes, respectively [6–10]. In humans, there are 273 extracellular proteases, 277 intracellular proteases, and 16 integral membrane proteases. The mouse degradome includes at least 628 members with 514 true orthologs of human proteases [11–13]. The specific function of several proteases remains unclear, and roughly 100 of these proteins are believed to be inactive due to the lack of critical catalytic domains [14]. Proteases are classified into two main groups: exopeptidases, which cleave peptide bonds near the amino or carboxy terminus (aminopeptidases and carboxypeptidases), and endopeptidases, which cut inside proteins, away from the ends. Based on the type of catalytic domain, proteases are grouped into six classes [15–17]. Serine, cysteine, and threonine proteases use nucleophilic amino acids for cleavage, whereas aspartic, glutamic, and metalloproteases activate water molecules as nucleophiles for the hydrolysis of peptide bonds (Figure 1) [18–20]. Most proteases are synthesized as inactive precursors called zymogens or proenzymes. These contain inhibitory pro-domains that prevent premature proteolytic activity [17,21]. Proteases lack-

 https://doi.org/10.3390/cells14120921

ing pro-domains typically require cofactor binding or post-translational modifications to achieve an active state (Figure 1) [22–24].

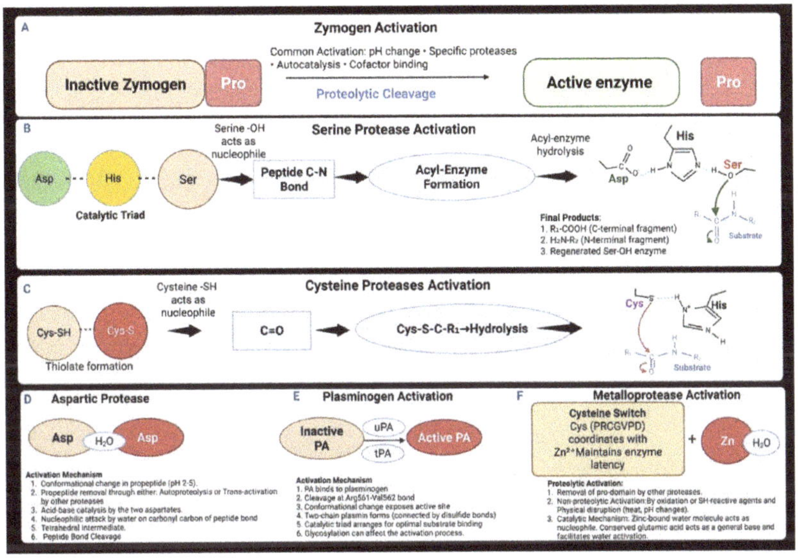

Figure 1. Mechanisms of protease activation: (**A**) zymogen activation through proteolytic cleavage triggered by factors including specific proteases, autocatalysis, or cofactor binding; (**B**) activation of serine proteases featuring the Asp–His–Ser catalytic triad, which facilitates nucleophilic attack on peptide bonds; (**C**) activation of cysteine proteases where the thiolate group serves as the nucleophile; (**D**) activation of aspartic proteases involving dual aspartate residues that use water for peptide bond hydrolysis; (**E**) mechanisms of activation of tissue–type (tPA) and urokinase–type (uPA) plasminogen activators; (**F**) activation of metalloproteases through the cysteine switch and zinc-mediated catalysis.

Proteases modulate growth factor/cytokine bioavailability, control the levels of cell surface/receptor proteins, and are involved in the reorganization of the extracellular matrix (ECM) [25–27]. Therefore, their activity requires fine regulation through several mechanisms including inhibition by specific proteins, degradation by other proteases, and subcellular relocalization [28].

In the ovary, several proteases, including matrix metalloproteinases (MMPs), a disintegrin and metalloproteinase with thrombospondin-like motifs (ADAMTS), cathepsins (CTS), and plasminogen activator (PA)/plasmin system enzymes, have been reported to play critical roles in folliculogenesis, tissue remodeling, and ovulation [1,3,29–33]. These critical ovarian processes require the reorganization of the collagen-rich ECM surrounding the follicles, as well as the basement membrane separating granulosa and theca cells [34–38]. In addition, proteases regulate other fundamental processes including apoptosis and autophagy [39–44]. Ovarian folliculogenesis is tightly regulated by endocrine, paracrine, and autocrine factors, which affect the growth and maturation of oocytes and surrounding granulosa cells [25,45]. MMPs, including MMP-2, MMP-9, MMP-13, and membrane-type MMP-14 and MMP-16, serve as zinc- and calcium-dependent enzymes that collectively degrade proteinaceous ECM components (proteoglycans, laminin, and collagen) [6,46,47]. Their activities are tightly controlled by tissue inhibitors of metalloproteinases (TIMPs) to maintain the critical MMP/TIMP ratio necessary to regulate cell proliferation, differentiation, migration, and survival throughout follicular development [19,35]. The luteinizing hormone surge triggers the transcriptional activation of key proteases, particularly ADAMTS-1 (a disintegrin and metalloproteinase with thrombospondin motifs) and cathepsin L, which play critical roles in follicle rupture [48,49] (Figure 2). ADAMTS-1 facilitates structural remodeling during folliculogenesis by degrading proteoglycans and maintaining basement membrane integrity, whereas its disruption leads to follicle dysgenesis and impaired ovulation [48–52]. Cathepsin B and L, lysosomal cysteine proteases, regulate granulosa cell apoptosis, proliferation, and steroidogenesis while also modulating autophagy. Cathepsin L activity correlates directly with meiotic progression and embryonic development [53,54]. This network of proteases exhibits specific spatio-temporal expression patterns, and their coordinated action enables the cyclical tissue remodeling essential for follicular growth, ovulation, and corpus luteum formation [11]. Despite their ubiquitous function in both germ and somatic cells, specific roles of proteases in ovarian biology and disease are still far from being understood. This review summarizes the current knowledge on this class of proteins and their functions in follicle dynamics, ovulation, and corpus luteum development, as well as their involvement in ovarian disorders.

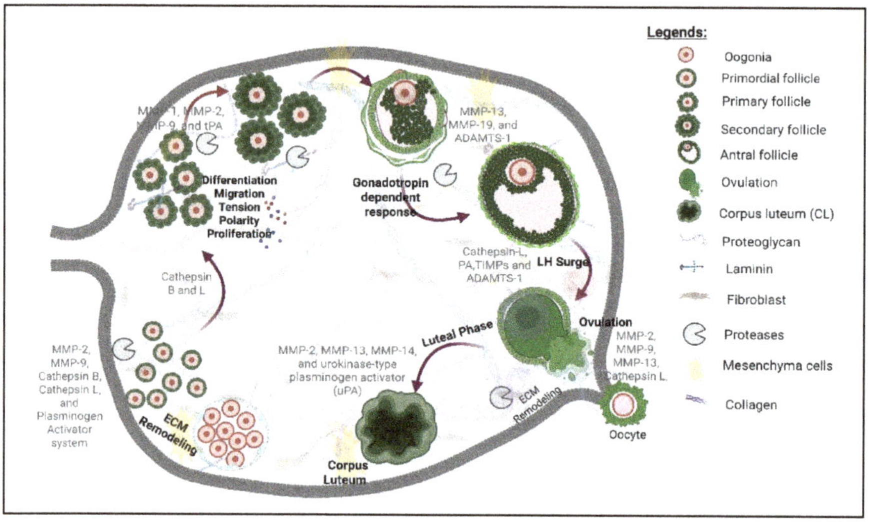

Figure 2. Schematic Overview: Ovarian Follicle Development Through Corpus Luteum Formation and Involvement of Various Protease. Proteases regulate critical processes including follicular activation, differentiation, migration, ECM remodeling, and gonadotropin-dependent responses. Several extracellular matrix components (proteoglycans, laminin, and collagen) and cell types (fibroblasts and mesenchymal cells) interact with follicle cells during the developmental process. Proteases are involved in ECM remodeling, which is critical for follicle growth, oocyte maturation, and luteal phase transition.

2. Materials and Methods

We searched databases including PubMed, Web of Science, and Scopus for the following keywords: "Protease" OR "Proteinase" OR "Peptidase, Ovary" OR "Ovarian, Folliculogenesis" OR "Follicle development, Ovarian pathology" OR "Ovarian disease, Metalloprotease" OR "MMP, Serine protease, Cysteine protease, Aspartic protease, ADAM" OR "ADAMTS, Cathepsin, Ovulation, Atresia, Polycystic ovary syndrome" OR "PCOS, Premature ovarian failure" OR "POF, Ovarian cancer". Inclusion Criteria: Studies focusing on proteases in the ovary, including research on proteases involved in folliculogenesis and ovarian pathologies, and in vivo and in vitro studies in both humans and mammalian models. To ensure relevance while capturing important foundational work, we considered only research articles and reviews published in peer-reviewed journals in the last 30 years.

Exclusion Criteria: Studies focusing solely on non-ovarian tissues or without direct relevance to ovarian function or pathology, articles not peer-reviewed, publications older than 30 years, case reports (unless they provide unique insights not available elsewhere), articles not available in English, and conference abstracts.

3. Roles of Proteases in Ovarian Follicle Development

3.1. Metalloproteinases, Plasminogen Activators, and Their Inhibitors

MMPs, tissue inhibitors of metalloproteinases (TIMPs), and PAs play crucial roles in regulating the remodeling of the ovarian ECM (Table 1) [3,4,6]. MMPs are zinc-dependent endopeptidases that are involved in physiological processes, such as embryonic development, tissue morphogenesis, and angiogenesis, and play critical roles in pathological conditions, including inflammation, cancer, and cardiovascular disease [7–9]. MMPs are classified into several groups based on their substrate specificity and structural features: collagenases (MMP1, 8, 13, and 18), gelatinases (MMP2 and 9), stromelysins (MMP3, 10, and 11), matrilysins (MMP7 and 26), membrane-type MMPs (MMP14, 15, 16, 17, 24, and 25), metalloelastase (MMP12), epilysin (MMP28), and enamelysin (MMP20) [8,10,11,14–16]. MMPs are secreted as zymogens (pro-MMPs) and their function is regulated at multiple levels, including transcription, zymogen activation, and inhibition by TIMPs [6,15,17]. TIMPs are a family of four proteins (TIMP-1, -2, -3, and -4) that bind to the active site of MMPs in a 1:1 stoichiometry, thereby inhibiting their proteolytic activity [7,16,18,19]. In addition, TIMPs have MMP-independent functions and are involved in cell growth and differentiation, angiogenesis, and apoptosis [20–23]. The balance between MMPs and TIMPs is critical for maintaining ECM homeostasis and regulating follicle development in the ovary [2,24]. The activation of primordial follicles involves the breakdown of their basement membrane, allowing the proliferation of granulosa cells [25,27,28]. The inhibition of MMP activity results in a significant reduction in primary and preantral follicle numbers [1,2,34,35]. MMP2 and MMP9 are expressed throughout folliculogenesis in both granulosa cells and oocytes [12,29–31,45]. TIMP-1 and TIMP-2 are also expressed in the granulosa cells and oocytes of primordial and primary follicles in humans [30]. Their expression decreases during the transition from primordial to primary follicles, suggesting that a reduction in TIMP activity may be necessary for follicular activation [30]. TIMP-1-deficient mice show increased MMP activity, particularly MMP2, and MMP9, and display an increased number of primary and preantral follicular compared to wild-type mice [36–38,55] (Table 2).

In addition to MMPs and TIMPs, PAs play an important role in the ovary [56]. PAs convert inactive plasminogen into plasmin, which is involved in the degradation of the ECM proteins, including fibrin, fibronectin, and laminin [57–59]. The PA system consists of a tissue-type plasminogen activator (tPA) and a urokinase-type plasminogen activator (uPA). PAs are regulated by protease inhibitors α2-antiplasmin and α2-macroglobulin, and plasminogen activator inhibitors (PAIs), which belong to the serine proteinase inhibitor (serpin) gene superfamily and include PAI-1, PAI-2, PAI-3, and the protease nexin I [39–43,57,60]. In the ovary, PAs are produced in granulosa cells and oocytes, whereas the majority of PAI-1 is produced within the theca (interstitium) [4,61–63]. uPA and tPA are upregulated during follicle activation and maturation, and their expression is regulated by gonadotropins, growth factors, and cytokines (Table 1) [57,59,64–66]. Li et al. showed that TGFα treatment increases uPA levels and activity in hen granulosa cells across all the follicle stages [66]. Similarly, FSH induces both uPA and tPA, whereas IL-1β suppresses their activity in vitro [67]. In rat ovaries, Hurwitz et al. also reported that IL-1β functions as an inhibitor of the PA system [68].

Other serine proteases that have been reported in the regulation of folliculogenesis include LONP1, FURIN, PAPPA, and matriptase (Table 2). The mitochondrial LONP1 is a multifunctional protease involved in mitochondrial quality control including oxidized protein degradation, protein folding, and mitochondrial DNA copy number homeostasis. The oocyte-specific ablation of *Lonp* using *Gdf9-cre* or *Zp3-cre* results in female infertility due to impaired follicle development, loss of the ovarian reserve, and progressive oocyte death ([69]. *FURIN* encodes a transmembrane serine protease localized in the Golgi apparatus, endosomes, and plasma membrane. The conditional ablation of *Furin* using *Gdf9-cre* or *Zp3-cre* leads to female infertility due to arrested folliculogenesis at the secondary follicle stage [70]. *PAPPA* encodes an extracellular metalloprotease, and *Pappa* knockout females exhibit decreased litter size and ovulatory capacity, which was attributed to the decreased bioavailability of insulin-like growth factor [71–73]. Matriptase, encoded by *Tmprss6*, is a type II transmembrane serine protease that regulates iron homeostasis by cleaving cell surface proteins associated with iron absorption. *Tmprss6*-null females show a severe delay in follicle maturation, likely due to a significant decrease in plasma iron levels [74]. Notably, defective follicle development and female infertility can be reproduced by a low-iron diet [75].

Table 1. Protease functions during ovarian physiology and follicular development.

Protease	Mechanism of Regulation	Specific Stage	Function	Species	Reference
MMP1	Increased expression following hCG administration	Preovulatory	Degradation of collagenous ECM	Rhesus monkey	[18,76]
MMP2 and MMP9	Increased in the granulosa and thecal cells of atretic follicles during proestrus and in corpus luteum during metestrus	Preovulatory follicles	ECM remodeling	Guinea pigs	[45]
MMP2, and MMP9	Localized to the oogonium/oocyte cytoplasm and surface epithelium	Folliculogenesis	ECM remodeling during gonadal development and cell–matrix interactions	Human	[77]
MMP1 and MMP13	Expression increased in response to LH surge	Preovulatory	Degradation of collagenous ECM	Bovine	[78]
MMP1, MMP2, MMP3, MMP9, MMP13	Increased in mRNA expression by gonadotropins	Prehierarchical white (WFs), yellowish (YFs), and preovulatory follicles	Involved in the atresia of the early stage of follicle while not participating in the regulation of advanced stage atresia	Chicken	[24,79]
MMP1, MMP3, and MMP9	Increased MMP1 and MMP3 expression levels in granulosa	Folliculogenesis	MMP9 induced by TGFB1; MMP1and MMP3 stimulated by FSH, LH, P4, and E2	Chicken	[80]
MMP10 and MMP11	Expression patterns changes following hCG administration	Ovulation and luteogenesis	Mmp10 mRNA was increased and MMP11 decreased in granulosa and theca cells during Ovulation	Human and Rats	[81]
MMP13, MMP14, MMP16, ADAMT1	Increased expression in cumulus cells following hCG administration	Ovulation and luteogenesis	Migratory phenotype of the cumulus–oocyte complex at the time of ovulation	Rat	[82]
MMP19	Localized to granulosa and theca-interstitial cells with temporal increases following hCG administration	Preovulatory follicles	ECM remodeling and tissue degradation	Mouse, Rat, Bovineand Human	[76,82,83]

Table 1. *Cont.*

Protease	Mechanism of Regulation	Specific Stage	Function	Species	Reference
MMP2, MMP9, TIMP-1, and TIMP-2	The ratio of MMP-2/TIMP-2 decreased in small antral follicles; the MMP-9/TIMP-1 ratio increased in large-preovulatory follicles	Preovulatory follicles	Tissue reorganization during ovulation	Equine	[29]
TIMP-2 and TIMP-3	Increased transcript abundance of TIMP-2 in yellow atretic follicles; decreased mRNA expression of TIMP-3	Prehierarchical white (WFs), yellowish (YFs), and preovulatory follicles	Involved in the atresia	Chicken	[24,79]
TIMP4	Increased significantly during the luteinization process of granulosa cells	Localized to the theca of antral and preovulatory follicles and adjacent ovarian stroma	Maintenance of luteal function	Mice, Rat	[84,85]
tPA and uPA	Activity increased during the periovulatory period	Granulosa and theca cells	Conversion of plasminogen to plasmin during ovulation	Rat	[66,86,87]
tPA and uPA	TNFα suppressed FSH-stimulated tPA activity but potentiated FSH-induced uPA activity in undifferentiated granulosa cells	Undifferentiated granulosa cells of preantral and antral follicles	Follicular wall remodeling during ovarian follicular development	Rat	[67,88]
PAI-1 and PAI-2	mRNAs upregulated after the gonadotrophin surge	PAI-1 localized to the thecal layer of preovulatory follicles. PAI-2 localized to the granulosa cell	Control plasminogen activator activity associated with ovulation and early corpus luteum formation.	Bovine	[89]
CTSL	Expression increased following hCG administration	Oocyte meiosis, Preovulatory to ovulation	Degradation of the follicular wall	Rat, Rhesus monkey, Bovine	[54,76]
CTSB	Expression increased following hCG administration, Autophagy induction	Preovulatory to ovulation	Regulation of follicular development	Mice, Bovine	[53,90]
CTSB, K, L, and H	Expressed in germinal epithelium throughout the estrous cycle	Oocytes and granulosa cells of primordial, primary follicles and corpus luteum	Degradation of extracellular matrix	Mice	[91,92]
Kallikreins	Response to steroid hormones (androgens and estrogens); various expression patterns with eCG/hCG stimulation	Primordial to ovulation	Proteolytic processing of growth factors and hormones; angiogenesis	Rat	[93,94]

3.2. *Cathepsins*

Cathepsins are classified into three main groups based on their catalytic site residues: cysteine cathepsins (Types B, C, F, H, K, L, O, S, V, W, and X), aspartic cathepsins (Types D and E), and serine cathepsins (Type G) [95,96]. In the ovary, cathepsins are expressed in oocytes, granulosa, and theca cells, and their expression is regulated by gonadotropins and growth factors (Table 1) [92].

i. Cathepsin B (CTSB): CTSB can function both as endo- and exo-(carboxy) peptidase [97,98]. CTSB has been identified as a critical regulator of ovarian reserve maintenance in mice [99]. The inhibition of *Ctsb* by myricetin significantly increased the number of primordial and primary follicles, suggesting a role in follicle activation. This effect seems mediated by the inhibition of autophagy and upregulation of the IGF1R and AKT-mTOR pathways [99]. Similarly, Liang et al. reported that the inhibition of CTSB activity preserved oocyte quality and enhanced developmental competence by mitigating age-related mitochondrial dysfunction and oxidative stress [100]. Chen et al. reported that the silencing of *Ctsb* in mouse granulosa cells decreased apoptosis by downregulating *TNF-α*, *Casp8*, and *Casp3* while upregulating *Bcl2* expression [53]. *Ctsb* knockdown also increased granulosa cell proliferation by activating the *p*-Akt and *p*-ERK pathways [53]. Komatsu et al. reported that Stefin A, an inhibitor of CTSB, blocked the activation of primordial follicles in mouse newborn ovaries *in vitro* [101]. In the follicle fluid of pregnant women undergoing ICSI, Bastu et al. found higher levels of CTSB compared to non-pregnant patients [102].

ii. Cathepsin L (CTSL): *Ctsl* is involved in the activation of primordial follicles adjacent to ovulatory follicles, and its inhibition results in a significant reduction in growing follicle numbers [92]. *Ctsl* expression was detected in large cuboidal cells of small, developing corpora lutea, suggesting possible roles in corpus luteum function [91,103]. Ezz et al. showed that CTSL regulates oocyte meiosis, and its supplementation improves oocyte quality and early embryo development in the bovine [54] (Table 2).
iii. Cathepsin S (CTSS): Song et al. reported that *Ctss* overexpression significantly increased progesterone (P4) and estrogen (E2) production by upregulating *Star* and *Cyp19a1* in rabbit granulosa cells [104,105]. The overexpression of *Ctss* also increased granulosa cell proliferation while decreasing apoptosis by enhancing the expression of *Pcna* and *Bcl2*. Conversely, *Ctss* knockdown significantly decreased the secretion of P4 and E2 while increasing apoptosis [104].

Table 2. Effects of protease gene knockout or protease inhibition on follicle development in model organisms.

Protease	KO/Inhibitor	Effect on Follicular Development	Specific Stage	Molecular Mechanism	Localization	Species	Reference
MMP1, MMP9, MMP10, and MMP19	Inhibitor (GM6001)	Reduced ovulation rate	Preovulatory to ovulation	Degradation of the follicular wall	Granulosa and theca cells	Rhesus monkey	[76]
MMP2	Inhibitor (ZK158252)	Inhibited hCG-induced ovulation and MMP-2 activation	Preovulatory to ovulation	Leukotriene B4-receptor antagonism	Ovarian follicles	Rat	[106]
MMP1, MMP2, and MMP3	Inhibitor (GM6001)	Reduction in CL and E2 with GM6001	Preovulatory to ovulation	Inhibits MMP activity in photostimulated ovaries	-	Hamster	[107,108]
MMP10	Inhibitor AG1478	Up-regulation of Mmp10 by LH.	Ovulation and luteinization.	Suppressed the induction of Mmp10 mRNA	Granulosa cells	Rat	[81]
TIMP-1	KO	Increased number of primary and preantral follicles	Primordial to primary/preantral	Regulation of MMP activity	-	Rodent	[38,109]

Table 2. *Cont.*

Protease	KO/Inhibitor	Effect on Follicular Development	Specific Stage	Molecular Mechanism	Localization	Species	Reference
CTSB	Inhibitor (Myricetin)	Increased oocyte reserve	Primordial to primary	Inhibition of autophagy and upregulation of the IGF1R and AKT-mTOR pathways	Oocytes	Mouse	[99]
CTSL	siRNA	Enhanced fertilization capability and blastocyst formation	Oocytes	Increasing mitochondrial function, reducing accumulated ROS, lowering apoptosis, and recovering lysosome capacity	Oocytes	Mouse	[91]
CTSL	KO	Reduced ovulation rate	Preovulatory to ovulation	Degradation of the follicular wall	Granulosa and theca cells	Mice	[48]
CTSL	rCTSL supplementation	Regulated oocyte meiosis during maturation and early embryo development	Oocyte maturation	Meiotic regulation	Oocytes	Bovine	[54,76]
CTSB	Stefin A	Blocked activation of primordial follicles	Primordial follicles	17β-estradiol increased Stefin A mRNA expression and inhibited follicle development	-	Mouse	[101]
ADAMTS1	KO	Lower numbers of mature follicles and impaired ovulation	Antral to ovulation	Maintenance of follicular basement membrane integrity	Granulosa cells	Mouse	[110,111]
ADAMTS1	KO	Failure of ovulation and fertilization	Preovulatory to ovulation	Expansion of cumulus–oocyte complexes (COCs)	COCs	Mouse	[49,52,112]
ADAMTS9	KO	Ovarian malformation and inability to ovulate	Primordial to ovulation	-	-	Zebrafish	[113]
LONP1	Oocyte-specific KO	Impaired follicular development and progressive oocyte death	Primordial to antral	Regulation of mitochondrial function	Oocytes	Mouse	[69]
FURIN	Oocyte-specific KO	Arrested oogenesis at early secondary follicles	Primary to secondary	-	Oocytes	Mouse	[70]
PAPPA	KO	Decreased litter size and ovulatory capacity	Antral to ovulation	Regulation of IGF bioavailability	-	Mouse	[71–73]

Table 2. *Cont.*

Protease	KO/Inhibitor	Effect on Follicular Development	Specific Stage	Molecular Mechanism	Localization	Species	Reference
TMPRSS6	KO	Retardation in ovarian maturation	Primordial to antral	Regulation of iron homeostasis	-	Mouse	[74]
tPA and uPA	Inhibitor (PAI-1)	Significantly reduced ovulation rate	Preovulatory to ovulation	ECM degradation	-	Rat, Human	[87,114–117]
PA	Protease nexin-1 (SerpinE2)	tPA activity higher in cells from small follicles; SerpinE2 levels higher in large follicles	Antral and basal granulosa cells	SerpinE2 secretion regulated at the transcriptional level	Granulosa cells	Bovine	[41]

4. Role of Proteases in Antral Follicle Development and Ovulation

4.1. Metalloproteinases, Plasminogen Activators, and Their Inhibitors

Ovulation is a complex process involving the rupture of the preovulatory follicle and the release of the oocyte [118,119]. A critical step in this process is the degradation of the basal membrane and ECM surrounding the mature follicle, which are rich in collagen, laminin, and fibronectin [2,57]. MMP2 and MMP9 are expressed in the granulosa and theca cells of preovulatory follicles of both rodents and humans, and their levels increase as the follicles approach ovulation [1,31,34,46,81,118]. Conversely, the inhibition of their activity significantly reduces ovulation rates [31,120–122].

Several studies have analyzed changes in Mmp expression following the administration of human chorionic gonadotropin (hCG), which mimics the luteinizing hormone (LH) surge necessary for ovulation. In rhesus monkeys, MMP1 was found to increase following hCG treatment [18]. *Mmp11* decreased in both humans and rats [81], whereas *Mmp13* increased in bovines [78], but not in humans [83], and *Mmp19* increased in mice [63], rats [123], and humans [83]. ADAMTS1 is a secreted metalloproteinase expressed in the granulosa cell layer of mature follicles in the ovary [51,111,112]. LH stimulates the expression of *Adamts1* in the granulosa cells of the preovulatory follicles and is sustained in a progesterone-dependent manner [48]. *Adamts1*-null female mice display lower numbers of growing follicles and impaired ovulation due to mature oocytes remaining trapped within the antral follicles [110,111]. In *Adamts1* knockout mice, the development of the ovarian medullary vascular network and lymphatic system were severely delayed, and the expansion of cumulus–oocyte complex (COCs) was reduced, causing lower ovulation and fertilization rates [49,52,112]. These findings highlight a role for *Adamts1* in maintaining the structural integrity of follicle basement membranes and supporting lymphangiogenesis, therefore providing new mechanistic insight into ovarian development and disease.

PAs and their inhibitors PAIs are key regulators of the proteolytic cascade involved in ovulation [57,64,124]. During follicle growth, granulosa and theca cells produce tPA and uPA, which facilitate the breakdown of the surrounding basement membrane and the subsequent release of the oocyte [57,59,125–127]. Consistent with these observations, PAI-1 expression was reported to decrease during follicle maturation in the human ovary, allowing higher PA activity [128]. Similarly, in rats and porcine ovaries, tPA and uPA activity increased during the periovulatory period, peaking just before ovulation [86,87,114,124,129]. In addition to tPA and uPA, kallikreins are serine proteases that participate in plasminogen activation and promote vascularization [130–133]. Accumulating evidence suggests that kallikrein expression is regulated by steroid hormones (Table 1). Estrogens stimulate the secretion of KLK1, KLK10, KLK11, and KLK14, whereas androgens trigger the secretion of KLK3 [134,135]. Interestingly, combined stimulation with both androgens and estrogens downregulates KLK3 expression, consistently with the antagonistic effect of estrogens on androgen receptor activity [135–138]. KLK5 and KLK6 were found to be involved in the proteolytic processing of growth factors and hormones essential for follicular development and ovulation [94,139–141].

4.2. Cathepsins

Several studies have reported the involvement of both CTSB and CTSL in regulating ovulation [100,142,143]. In bovine ovaries, Balboula et al. found an increased expression of CTSB in follicle granulosa and theca cells following hCG administration, whereas inhibition of CTSB activity resulted in a significant reduction in ovulation rates [144,145]. The expression of *Ctsl* increases in the granulosa cells of mouse preovulatory follicles following PMSG and hCG administration [48]. In addition, Sriraman et al. found an increased expression of CTSL in the granulosa cells of human preovulatory follicles, with levels peaking just before ovulation [142]. García et al. proposed that the transient expression of progesterone receptor (PR) in human granulosa cells of the preovulatory follicle may play a role in the activation of CTSL [146]. Interestingly, Zhang et al. demonstrated a significant increase in *Ctsl* levels in oocytes of aged mice (8–9 months and 11–12 months), and found that the overexpression of *Ctsl* in the oocytes of young mice (6–8 weeks) substantially diminished their quality, which was restored upon *Ctsl* inhibition [91].

5. Role of Proteases in Corpus Luteum Formation and Function

Following ovulation, the remainder of the ruptured follicle transforms into the corpus luteum (CL). The CL is a transient endocrine gland that is necessary to sustain pregnancy through the production of progesterone, which prepares the uterus for implantation and pregnancy [147–149]. If fertilization does not occur, the CL degenerates leading to the initiation of a new cycle. Dramatic ECM remodeling and angiogenesis are involved in the development, activity, and regression of the CL. MMP2, MMP13, and MMP14 seem to be involved in the regulation of CL dynamics [46,92]. The degradation of the ECM by plasmin allows for the rapid remodeling and vascularization of the CL, which is essential for establishing its function [2,115,135,150]. Wahlberg et al. investigated the formation and function of the CL in plasminogen (*Plg*)-deficient mice, with or without the administration of galardin, a broad-spectrum synthetic MMP inhibitor [151]. The study revealed that CL formation occurred in *Plg*-deficient mice, galardin-treated wild-type mice, and galardin-treated *Plg*-deficient mice, demonstrating that neither the plasminogen activator nor the MMP system is necessary for CL formation. However, serum progesterone levels in *Plg*-deficient mice were reduced by approximately 50%, and galardin treatment did not further decrease progesterone concentrations. These findings suggest that plasmin, but not MMPs, seems to play a significant role in maintaining luteal function, possibly through the proteolytic activation of growth factors and other paracrine factors [151].

6. Role of Proteases in Ovarian Disease

Due to the critical role proteases play in numerous biological processes, including cell proliferation, migration, and programmed cell death, the dysregulation of their function can lead to several ovarian pathologies (Figure 3).

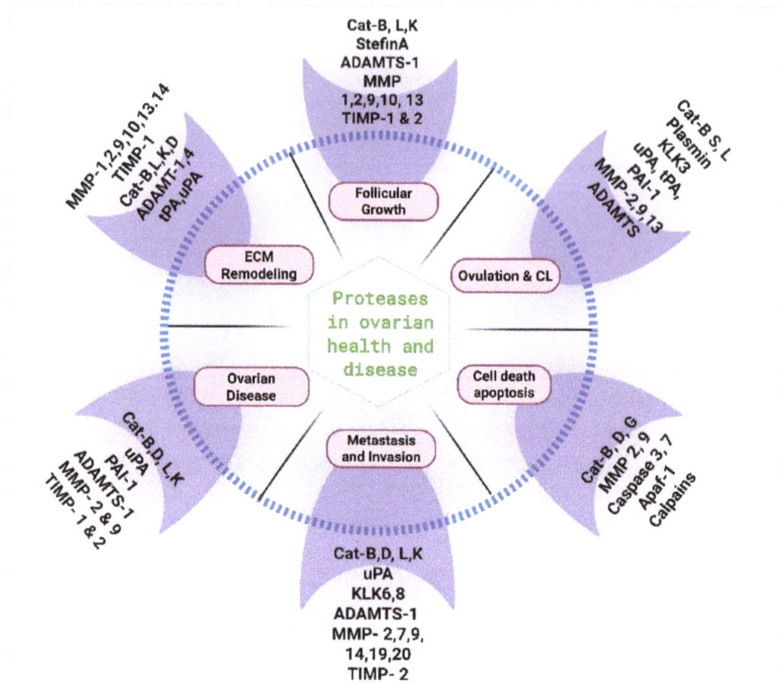

Figure 3. Roles of proteases in ovarian health and disease. Matrix metalloproteinases (MMPs), tissue inhibitors of metalloproteinases (TIMPs), plasminogen activators (PAs), plasminogen activator inhibitors (PAIs), ADAMTS, cathepsins, and kallikreins are involved in critical physiological processes of the ovary, including folliculogenesis, extracellular matrix (ECM) remodeling, apoptosis, ovulation, corpus luteum (CL) development. The dysregulation of proteases can lead to several pathological conditions including cancer, polycystic ovary syndrome (PCOS), primary ovarian insufficiency (POI), and ovulatory irregularities.

6.1. Ovarian Cancer

Ovarian cancer is the fifth leading cause of cancer death in females with long-term survival rates of 20% or less in advanced stages III-IV [152–154]. The dysregulation of protease activity has been linked to ovarian cancer pathogenesis (Table 3) [33,155,156].

i. MMPs and TIMPs: MMPs participate in several processes that are involved in ovarian cancer progression, including the degradation of the ECM, the promotion of angiogenesis, and the induction of epithelial–mesenchymal transition (EMT) [6,35,46,157–159]. Several studies have shown the upregulation of MMPs, such as MMP2 and MMP9, in ovarian cancer tissues compared to normal or benign ovarian tissues, and their expression levels correlate with clinical stage, tumor invasiveness, and metastatic potential [155,157,160,161]. Tumor-derived MMP2 and MMP9 expression has been identified as a negative prognostic indicator in ovarian cancer patients, predicting lower overall survival rates [162–166]. Ovarian cancer cells (Ovcar3) treated with an activator of the PKC pathway, phorbol-12-myristate 13-acetate (PMA), increased *MMP7* and *MMP10* mRNA [167,168]. MMP14 was shown to activate pro-MMP2 to MMP2, playing a role in the development of vasculogenic-like networks and matrix remodeling by aggressive ovarian cancer cells [168–170]. MMP1 activates PAR1, inducing the secretion of angiogenic factors in ovarian carcinoma cells [171]. MMP3 is involved in the estradiol-induced migration and invasion of SKOV3 ovarian cancer cells via the PI3K/Akt/FOXO3 pathway [172]. MMP7 promotes the invasion and metastasis of ovarian cancer cells by activating gelatinases and through the MAPK/ERK and JNK pathways [173,174]. MMP8 upregulates IL-1β, whose expression levels correlate with tumor grade and poor prognosis [175]. The *MMP12* 82A/G polymorphism has been associated with increased susceptibility to ovarian cancer [176,177], and MMP13 in ascitic fluids of ovarian cancer patients has been identified as a potential marker for disease risk and survival outcomes [178]. Taken together, these studies underline the association between the dysregulation of MMP expression and activity and ovarian cancer.

In addition to MMPs, several TIMPs, including TIMP1 and TIMP3, have been found upregulated in ovarian cancer [179,180]. However, Davidson et al. found decreased TIMP levels alongside increased MMP2 in ovarian cancer [181]. These seemingly contradictory results highlight the complex mechanisms involved in ovarian cancer and show how dysregulation of the MMP/TIMP balance may have a more significant impact than the overexpression of a single class of proteins [157,182]. In addition, there are several possible explanations for the conflicting findings of increased levels of both MMPs and their inhibitors: (1) TIMPs regulate processes independent of their protease inhibitory activity, including cell growth, migration, and angiogenesis [183]; (2) the stoichiometric balance between MMPs and TIMPs may be more critical than absolute levels, and elevated TIMP levels may sometimes be insufficient to counteract excessive MMP activity in aggressive cancers [184]; (3) TIMPs have been found to activate MMPs in certain instances [19]; and (4) different tissue compartments may have varying MMP ratios, allowing MMPs to remain active in specific microenvironments despite elevated TIMP levels [185].

ii. The PA and PAI system: In vitro analyses have shown that uPA is highly expressed in several types of cancer cells, including ovarian cancer [186–193]. The overexpression of uPA and PAI-1 was found in more than 75% of primary ovarian carcinomas, and in most metastatic epithelial ovarian cancer (EOC) [194]. Further, Kenny et al. reported that in vitro and in vivo treatments with a uPA receptor (uPAR) antibody inhibited ovarian cancer cell invasion, migration, and adhesion by inhibiting α5-integrin and decreasing the expression of urokinase, uPAR, β3-integrin, and fibroblast growth factor receptor-1 [195]. High levels of PAI-1 have been associated with poor clinical outcomes in ovarian serous carcinoma [187,196]. In ovarian cancer cells, PAI-1 inhibition resulted in cell cycle arrest and decreased proliferation, and, in xenograft models, significantly reduced peritoneal dissemination [196]. Similarly, PAI-1 silencing in SKOV3 cells disrupted the platelet-induced upregulation of the genes involved in proliferation and ECM remodeling [197]. At the molecular level, some reports suggest that PAI-1 inhibits cell adhesion and migration by blocking vitronectin (VN) binding to integrins or by displacing uPAR from VN in the extracellular matrix [198,199]. However, other studies have shown that PAI-1 can enhance cancer cell adhesion [200,201]. As for MMPs/TIMPs, it is unclear why the upregulation of both uPA and PAIs correlates with cancer progression and poor clinical outcomes. Several mechanisms may explain this apparent contradiction: (1) PAI have additional functions beyond uPA inhibition, including activation of pathways that promote tumor growth, angiogenesis, and cell detachment [202–204]; (2) the PAI-1/uPA/uPAR complex can be internalized and recycled, potentially leading to increased uPAR on the cell surface and enhanced invasiveness [205]; (3) PAI-1 can elicit inflammatory responses and immune cell recruitment in the tumor microenvironment, potentially promoting a pro-tumorigenic milieu [206]. Overall, these findings suggest that PAI function is context-dependent and highlight the complex regulation of the PA/PAI system in ovarian cancer.

iii. Cathepsins: Cathepsins and their inhibitors cystatins have also been associated with ovarian cancer. Liu et al. showed that CTSB and its binding proteins AMBP and TSRC1 modulated TNF-induced apoptosis in ovarian cancer cells [207]. Additionally, *Ctsl* knockdown inhibited proliferation, invasion, and tumor growth both in vitro and in vivo, while *Ctsl* overexpression had the opposite effects [208,209]. In malignant serous tumors, cystic fluid levels of CTSB, CTSL, and their inhibitor Cystatin C (Cst3) were significantly elevated compared to benign serous tumors [210]. Gashenko et al. found significantly increased levels of procathepsin B, cystatin B (CstB), and Cst3 in serum and ascite fluids of ovarian cancer patients compared to controls, suggesting their possible use as disease biomarkers [211]. Nishikawa et al. found significantly elevated levels of Cst3, but not CTSB, in ovarian cancer compared to benign samples and healthy controls [212]. Interestingly, invasion assays showed that the inhibition of Cst3 or CTSB suppressed cancer cell invasion in a dose-dependent manner [212]. Once again, some of these findings appear contradictory. Elevated CysC in cancer may represent a compensatory mechanism to control excessive cathepsin activity [213]. In addition, similar to other protease inhibitors, Cst3 may have additional functions including the regulation of immune response and cell signaling [214]. Furthermore, changes in the balance between cathepsins and cystatins may be more important than absolute expression levels of either protein [215]. Finally. Cst3 primarily regulates extracellular cathepsin activity, while pro-tumorigenic effects of CTSB may be, at least in part, intracellular [216].

Xu et al. reported elevated levels of serum CTSK in ovarian cancer patients compared to healthy controls [217]. CTSD was also found associated with ovarian cancer, with epithelial CTSD expression more common (65.1%) in ovarian tumors with low malignant potential (LMP) compared to invasive tumors (43.7%) [218,219]. CTSD was also found to significantly increase the proliferation and migration of human omental microvascular endothelial cells, and the knockdown of *CTSD* was exhibited to be able to suppress migration and invasion in an EOC cell line [220,221]. In addition, the inhibition of CTSS induced apoptosis in cancer cells by downregulating Bcl-2 and c-FLIP [222].

Another serine protease, Hepsin is overexpressed in several cancers including ovarian cancer, but its specific role remains to be elucidated [223–226]. Miao et al. reported that membrane-associated Hepsin localized at desmosomal junctions with its putative proteolytic substrate hepatocyte growth factor (HGF), and showed that Hepsin overexpression promoted ovarian tumor growth in a mouse model [227]. The dysregulation of Hepsin expression could disrupt the integrity and function of epithelial barriers. In addition, the Hepsin-mediated cleavage of substrates including pro-HGF and pro-uPA could promote metastasis [228,229]. Despite a role in mediating cancer formation and progression, sosme reports suggest that high levels of Hepsin could have antitumor effects by reducing oncogenic signaling and increasing autophagy [230]. It is possible that specific functions in cancer depend on the microenvironment or disease stage.

Table 3. Proteases in ovarian cancer.

Protease	Finding in Ovarian Cancer	Localization	Species	Prognostic Value	Role	Reference
MMP1	Activates PAR1	Ovarian carcinoma cells	Human, Epithelial ovary cell lines	Not reported	Induces the secretion of angiogenic factors	[231–234]

Table 3. *Cont.*

Protease	Finding in Ovarian Cancer	Localization	Species	Prognostic Value	Role	Reference
MMP2	Upregulated in ovarian cancer tissues compared to normal/benign ovarian tissues	Ovarian cancer tissue, epithilial, stroma	Human	Negative prognostic indicator with lower overall survival rates	Degrades ECM, promotes angiogenesis, and induces EMT	[161–163,182,235–237]
MMP3	Involved in estradiol-induced migration and invasion	SKOV3 ovarian cancer cells	Human cell line	Not reported	Mediates estrogen-induced cancer progression via PI3K/Akt/FOXO3 pathway	[172]
MMP7	Promotes invasion and metastasis	Ovarian cancer cells	Human cell line	Not reported	Acts through MAPK/ERK and JNK pathways; activates gelatin enzymes	[174]
MMP8	Upregulates IL-1β	Ovarian cancer tissue	Human	Correlates with tumor grade and poor prognosis	Promotes inflammatory microenvironment	[175]
MMP9	Upregulated in ovarian cancer tissues	Ovarian cancer tissue	Human	Correlates with clinical stage, tumor invasiveness, and metastatic potential	Degrades ECM, promotes angiogenesis, and induces EMT; cleaves fibronectin and type IV collagen	[23,165,165,238,239]
MMP10	Increased expression with PKC pathway activation	Ovarian cancer cells (Ovcar3, EOC)	Human cell line	Not reported	Regulated by PKC pathway, Wnt signaling	[167,240]
MMP11	Overexpression in stromal cells	Ovarian carcinomas	Human	Overexpression not correlates with survival.	Tumor progression	[241]
MMP12	82 A/G polymorphism associated with increased susceptibility	Genetic study	Human	Not reported	Genetic predisposition factor	[176,177]
MMP13	Elevated in ascitic fluids	Ascitic fluid	Human	Potential marker for disease risk and survival outcomes	Not reported	[178,179]
MMP14	Activates pro-MMP2 to MMP2	Ovarian cancer cells	Human	Associated with vasculogenic-like networks	Matrix remodeling; activates pro-MMP2	[82,168,242]
TIMP1	Upregulated in ovarian cancer	Ovarian cancer tissue	Human	Not reported	Complex: May have MMP-independent roles in cell growth, migration, and angiogenesis	[180,181]
TIMP3	Upregulated in ovarian cancer	Ovarian cancer tissue	Human	Not reported	Complex regulatory roles beyond MMP inhibition	[243–245]
uPA	Highly expressed in cancer cells; overexpressed in >75% of primary ovarian carcinomas and metastatic EOC samples	Ovarian cancer cells	Human and cell lines	Associated with invasion and metastasis	Promotes invasion, migration, and adhesion	[187–191]

Table 3. *Cont.*

Protease	Finding in Ovarian Cancer	Localization	Species	Prognostic Value	Role	Reference
PAI-1	High levels in ovarian serous carcinoma	Ovarian cancer tissue	Human, cell lines, and xenograft models	Associated with poor clinical outcomes	Complex: Inhibits cell adhesion by blocking vitronectin; disrupts platelet-induced gene upregulation	[189,197,198]
CTSB	Modulates TNF-induced apoptosis; elevated in cystic fluid and serum	Cystic fluid, serum, and cancer cells	Human	Serum procathepsin B significantly elevated compared to healthy controls	Binding proteins AMBP and TSRC1 involved in TNF-induced apoptosis	[210–212]
CTSL	Overexpressed; knockdown inhibits proliferation, invasion, and tumor growth	Cancer cells	Human, cell lines, and mouse models	Associated with paclitaxel resistance	Promotes proliferation and migration; confers chemoresistance	[208,209]
CTSS	Inhibition stimulates TRAIL-induced apoptosis	Cancer cells	Human	Not reported	Downregulation of Bcl-2 and Cbl-mediated c-FLIP by ROS-mediated p53 expression	[222]
Cst3	Elevated in malignant tissues, serum, and cystic fluid	Malignant tissue, Serum, Cystic fluid	Human	Elevated in ovarian cancer compared to benign samples	Complex: May represent failed compensatory mechanism; has additional immune and signaling functions	[210,212–215]
CTSK	Overexpressed in peritoneal metastatic ovarian carcinomas; elevated serum levels	Peritoneal metastases, Serum	Human	Potential biomarker	Associated with peritoneal metastasis	[217,246]
CTSD	Expression more common (65.1%) in tumors with low malignant potential vs. invasive tumors (43.7%); promotes the proliferation and migration of endothelial cells	Epithelial cells, Stromal cells	Human and cell lines	Independent prognostic factor for disease-free survival in invasive ovarian cancer	Pro-angiogenic and pro-metastatic role via ERK1/2 and AKT activation; correlates with microvessel density	[218–221]
Hepsin	Overexpressed in ovarian cancer	Desmosomal junctions	Human and mouse model	Not reported	Cleaves HGF and pro-uPA; localizes with substrate HGF; disrupts epithelial barriers	[223–225,227,229]

6.2. Polycystic Ovary Syndrome (PCOS)

PCOS is an endocrine disorder characterized by ovulatory dysfunction, excess of androgen, and polycystic ovaries, often leading to infertility, insulin resistance, and cardiovascular disease [247–251].

i. MMPs and TIMPs: MMPs and TIMPs have been associated with the pathogenesis of PCOS (Table 4) [252]. It has been reported that MMP2 and MMP9 concentrations are elevated in the follicular fluid of patients with PCOS compared to healthy controls [252,253]. The increased MMP activity was associated with higher levels of androgens, insulin resistance, disrupted follicular development, and ovulatory dysfunction [6,253,254]. Consistent with these observations, it was found that the granulosa cells of women with PCOS express fewer MMP inhibitors TIMP-1 and TIMP-2 compared to healthy controls [252,255]. Recently, Butler et al. reported that women with PCOS showed significantly elevated MMP9 [254]. Interestingly, the ratios of MMP9 to all TIMPs were significantly higher in the PCOS group, while MMP17/TIMP-1 and MMP17/TIMP-2 were lower. Higher expression of Mmp2/9 was also observed in antral follicles compared to the preantral follicle and primordial follicle of a Letrozole-induced PCOS rat model [256].

ii. The PA and PAI system: Elevated PAI-1 levels in plasma have been reported in patients with PCOS compared to controls (Table 4) [257–262]. However, findings regarding PAI-1 distribution within ovarian tissue have been inconsistent. Devin et al. reported increased PAI-1 in granulosa cells of cystic and atretic follicles in mouse models of PCOS [263]. Atiomo et al. detected PAI-1 in granulosa and theca cells without significant differences between PCOS and control ovaries, whereas other authors reported increased PAI-1 expression in follicular fluid from patients with PCOS [257,264,265]. Genetic predisposition seems to contribute to PAI-1 dysregulation in PCOS, which has been reported associated with the 4G/4G and 4G/5G genotypic subtypes in the PAI-1 promoter region, leading to increased protein levels [266]. Kelly et al. observed increased tPA antigen levels inversely correlating with insulin resistance, whereas Tarkun et al. found a direct correlation of PAI-1 levels, even in lean PCOS women [197,259]. Orio et al. reported elevated PAI-1 activity independent of obesity, while Sahay et al. found a correlation with both insulin resistance and obesity [260,267]. Ma et al. provided mechanistic insight through a mouse model demonstrating that PAI-1 deficiency prevented diet-induced obesity and insulin resistance [268,269]. Finally, Ibrahim et al. reported the presence of KLK2 in the serum of women with PCOS in association with hirsutism, but the nature of this relationship remains unclear [270].

iii. Cathepsins: The downregulation of CTSD has been reported in the ovaries of patients with PCOS [271]. CTSD downregulation may contribute to the abnormal follicle development associated with PCOS, leading to ovulatory dysfunction and infertility. Dawood et al. found significantly increased levels of CTSS, among patients with PCOS compared to healthy females [272]. Additionally, genetics may also play a role as CTSB polymorphisms have recently been associated with PCOS risks [273].

Table 4. Protease dysregulation in PCOS.

Protease	Finding in PCOS	Localization	Species	Proposed Pathogenic Role	Reference
MMP2 And MMP9	Elevated in follicular fluid of patients with PCOS compared to healthy controls	Follicular fluid and serum	Human	Associated with higher levels of androgens, insulin resistance, disrupted follicular development, and ovulatory dysfunction. Associated with higher MMP9/TIMP ratios, ECM remodeling, and follicular development	[252,253,256]
MMP2/9	Higher expression in antral follicles compared to preantral and primordial follicles	Ovarian follicles	Rat (Letrozole-induced PCOS model)	ECM remodeling in PCOS ovaries	[256]

Table 4. *Cont.*

Protease	Finding in PCOS	Localization	Species	Proposed Pathogenic Role	Reference
MMP17	Lower MMP17/TIMP-1 and MMP17/TIMP-2 ratios in PCOS	Serum	Human	ECM remodeling	[254]
TIMP-1 and TIMP-2	Decreased expression in granulosa cells of women with PCOS	Granulosa cells	Human	Excessive ECM degradation	[255,274]
PAI-1	Elevated in plasma, granulosa cells, and follicular fluid; homogeneous distribution throughout PCOS ovaries	Plasma, granulosa cells, follicular fluid, and theca cells	Human and mouse (PCOS model)	Associated with insulin resistance; higher expression in 4G/4G and 4G/5G genotypes	[260,262–264,266–269]
Plasminogen	Uniquely present in small follicles of PCOS ovaries	Small follicles	Human	Altered proteolytic activity in early follicular development	[258,259,261]
tPA	Increased antigen levels inversely correlating with insulin resistance	Plasma	Human	Associated with insulin resistance	[275]
KLK2/3	Present in the serum of women with PCOS in association with hirsutism	Serum	Human	Associated with androgsen excess and hirsutism	[270]
CTSB	CTSB polymorphisms contribute to PCOS pathogenesis	Blood	Human	rs12898, rs8898, and rs3779659 variants associated with PCOS risk	[273]
CTSD	Downregulated in ovaries of patients with PCOS	Cytoplasm and cell membrane of stromal and granulosa cells	Human	Abnormal follicle development	[271]
CTSS	Significantly increased levels in patients with PCOS	Serum	Human	Inflammation associated with PCOS	[272]

6.3. Primary Ovarian Insufficiency (POI)

Also known as premature ovarian failure (POF), POI is a condition characterized by the loss of ovarian function before the age of 40 [276,277]. The ovaries of mice with chemotherapy-induced POI showed increased expression of MMP2 and MMP9, and downregulation of TIMP-1 and TIMP-2 compared to control mice [278]. The greater activity of MMPs was associated with the increased apoptosis of the granulosa cells and abnormal folliculogenesis in the POI mice [278]. An et al. reported that the *TIMP2*G > C (rs8179090) and *TIMP2*G > A (rs2277698) genotypes were strongly associated with POI [279]. Pathogenic variants of serine protease *LONP1* have been found associated with POI [69]. Affected patients lack large antral follicles and are infertile, but the molecular mechanisms remain unknown, and functional analyses have not been performed [69]. PAs and PAIs are also associated with POI, particularly regarding the accelerated depletion of the ovarian reserve. Similarly to MMPs, tPA activity was higher in the ovaries of mice with chemotherapy-induced POI and associated with the increased apoptosis of granulosa cells and accelerated depletion of the ovarian reserve [280].

7. Conclusions

Proteases play crucial roles in several processes of ovarian function, including folliculogenesis, ovulation, and corpus luteum development. The balance between proteases and their inhibitors is essential for maintaining normal ovarian function. MMPs, PAs, and cathepsins contribute to the degradation of the follicular basement membrane and the extracellular matrix, enabling follicle activation, maturation, and oocyte release. The dysregulation of protease activity has been associated with several ovarian disorders, including cancer, PCOS, and POI. However, despite extensive evidence of the important role proteases play in ovarian functions, specific molecular mechanisms remain elusive. As proteases represent promising targets for therapeutic applications, future investigations should focus on identifying mechanisms of action and relationships with critical signaling pathways of the ovary.

Author Contributions: Conceptualization—E.P. and B.K.; data curation—B.K.; writing of MS—E.P. and B.K., original draft preparation—B.K.; review and editing—E.P. and B.K.; supervision—E.P. All authors have read and agreed to the published version of the manuscript.

Funding: This research received no external funding.

Institutional Review Board Statement: Not applicable.

Informed Consent Statement: Not applicable. All the figures were created using BioRender (Kushawaha, B. 2025 https://BioRender.com/3htq4d3 accessed on 25 March 2025).

Conflicts of Interest: The authors declare no conflicts of interest.

Abbreviations

The following abbreviations are used in this manuscript:

Abbreviation	Full Form
ADAMTS	A Disintegrin and Metalloproteinase with Thrombospondin-like Motifs
AIFM1	Apoptosis-Inducing Factor Mitochondrion-associated 1
AKT	Protein Kinase B
AMBP	Alpha-1-Microglobulin/Bikunin Precursor
ATG5	Autophagy Related 5
BCL2	B-Cell Lymphoma 2
CASP3/8	Caspase 3/8
CL	Corpus Luteum
COC	Cumulus–Oocyte Complex
CTSB	Cathepsin B
CTSD	Cathepsin D
CTSE	Cathepsin E
CTSG	Cathepsin G
CTSK	Cathepsin K
CTSL	Cathepsin L
CTSS	Cathepsin S
CstB	Cystatin B
Cst3	Cystatin C
CYP17A1	Cytochrome P450 Family 17 Subfamily A Member 1
CYP19A1	Cytochrome P450 Family 19 Subfamily A Member 1
DNA	Deoxyribonucleic Acid
Dpc	Days Post Coitus
Dpp	Days Postpartum
E2	Estradiol
ECM	Extracellular Matrix
EOC	Epithelial Ovarian Cancer
ERK	Extracellular Signal-Regulated Kinase
FAK	Focal Adhesion Kinase
FF	Follicular Fluid
FGF1	Fibroblast Growth Factor 1
FSH	Follicle-Stimulating Hormone

FURIN	Paired Basic Amino Acid Cleaving Enzyme
GCNA	Germ Cell Nuclear Antigen
GDF9	Growth Differentiation Factor 9
hCG	Human Chorionic Gonadotropin
HGF	Hepatocyte Growth Factor
ICSI	Intracytoplasmic Sperm Injection
IGF	Insulin-like Growth Factor
IGF1R	Insulin-like Growth Factor 1 Receptor
IL-1β	Interleukin 1 Beta
JNK	c-Jun N-terminal Kinase
KLK	Kallikrein
LC3-I	Microtubule-associated Protein 1A/1B-Light Chain 3
LH	Luteinizing Hormone
LONP1	Lon Peptidase 1
LMP	Low Malignant Potential
MAPK	Mitogen-Activated Protein Kinase
MMP	Matrix Metalloproteinase
mRNA	Messenger Ribonucleic Acid
mTOR	Mammalian Target of Rapamycin
MYC	Myelocytomatosis Oncogene
NFkB	Nuclear Factor Kappa B
P4	Progesterone
PA	Plasminogen Activator
PAI	Plasminogen Activator Inhibitor
PAPPA	Pregnancy-Associated Plasma Protein A
PAR1	Protease-Activated Receptor 1
PCNA	Proliferating Cell Nuclear Antigen
PCOS	Polycystic Ovary Syndrome
PI3K	Phosphoinositide 3-Kinase
PKC	Protein Kinase C
PMA	Phorbol-12-myristate 13-acetate
PMSG	Pregnant Mare Serum Gonadotropin
POF/POI	Premature Ovarian Failure/Primary Ovarian Insufficiency
PR	Progesterone Receptor
RGD	Arg-Gly-Asp (Arginine–Glycine–Aspartic acid)
ROS	Reactive Oxygen Species
SEC	Securities and Exchange Commission
siRNA	Small Interfering RNA
SMAD	Small Mothers Against Decapentaplegic
SPINK1	Serine Protease Inhibitor Kazal Type 1
STAR	Steroidogenic Acute Regulatory Protein
TGFα	Transforming Growth Factor Alpha
TGF-β	Transforming Growth Factor Beta
TIMP	Tissue Inhibitor of Metalloproteinases
TMPRSS6	Transmembrane Serine Protease 6 (Matriptase-2)
TNF-α	Tumor Necrosis Factor Alpha
tPA	Tissue-type Plasminogen Activator
TRAIL	TNF-Related Apoptosis-Inducing Ligand
TSRC1	Thrombospondin and Calcium-binding domains 1
uPA	Urokinase-type Plasminogen Activator
uPAR	Urokinase-type Plasminogen Activator Receptor
VEGF	Vascular Endothelial Growth Factor
VN	Vitronectin
ZMP-2	Zinc Metalloproteinase-2

ZP3 Zona Pellucida Glycoprotein 3

References

1. Curry, T.E., Jr.; Osteen, K.G. Cyclic Changes in the Matrix Metalloproteinase System in the Ovary and Uterus. *Biol. Reprod.* **2001**, *64*, 1285–1296. [CrossRef]
2. Curry, T.E., Jr.; Osteen, K.G. The Matrix Metalloproteinase System: Changes, Regulation, and Impact throughout the Ovarian and Uterine Reproductive Cycle. *Endocr. Rev.* **2003**, *24*, 428–465. [CrossRef] [PubMed]
3. Rodgers, R.J.; Irving-Rodgers, H.F.; Russell, D.L. Extracellular matrix of the developing ovarian follicle. *Reproduction* **2003**, *126*, 415–424. [CrossRef] [PubMed]
4. Liu, Y.-X. Plasminogen activator/plasminogen activator inhibitors in ovarian physiology. *Front. Biosci. J. Virtual Libr.* **2004**, *9*, 3356–3373. [CrossRef]
5. Levene, P.A. The Cleavage Products of Proteoses. *J. Biol. Chem.* **1905**, *1*, 45–58. [CrossRef]
6. Goldman, S. MMPS and TIMPS in ovarian physiology and pathophysiology. *Front. Biosci.* **2004**, *9*, 2474. [CrossRef] [PubMed]
7. Vu, T.H.; Werb, Z. Matrix metalloproteinases: Effectors of development and normal physiology. *Genes Dev.* **2000**, *14*, 2123–2133. [CrossRef]
8. Nagase, H.; Visse, R.; Murphy, G. Structure and function of matrix metalloproteinases and TIMPs. *Cardiovasc. Res.* **2006**, *69*, 562–573. [CrossRef]
9. Hey, S.; Linder, S. Matrix metalloproteinases at a glance. *J. Cell Sci.* **2024**, *137*, jcs261898. [CrossRef]
10. Page, M.J.; Di Cera, E. Serine peptidases: Classification, structure and function. *Cell. Mol. Life Sci.* **2008**, *65*, 1220–1236. [CrossRef]
11. Levin, G.; Coelho, T.M.; Nóbrega, N.G.; Trombetta-Lima, M.; Sogayar, M.C.; Carreira, A.C.O. Spatio-temporal expression profile of matrix metalloproteinase (Mmp) modulators Reck and Sparc during the rat ovarian dynamics. *Reprod. Biol. Endocrinol. RBE* **2018**, *16*, 116. [CrossRef] [PubMed]
12. Robinson, L.L.L.; Sznajder, N.A.; Riley, S.C.; Anderson, R.A. Matrix metalloproteinases and tissue inhibitors of metalloproteinases in human fetal testis and ovary. *Mol. Hum. Reprod.* **2001**, *7*, 641–648. [CrossRef]
13. Miyakoshi, K.; Murphy, M.J.; Yeoman, R.R.; Mitra, S.; Dubay, C.J.; Hennebold, J.D. The Identification of Novel Ovarian Proteases Through the Use of Genomic and Bioinformatic Methodologies. *Biol. Reprod.* **2006**, *75*, 823–835. [CrossRef]
14. Murphy, G.; Nagase, H. Progress in matrix metalloproteinase research. *Mol. Aspects Med.* **2008**, *29*, 290–308. [CrossRef]
15. Almutairi, S.; Kalloush, H.M.; Manoon, N.A.; Bardaweel, S.K. Matrix Metalloproteinases Inhibitors in Cancer Treatment: An Updated Review (2013–2023). *Molecules* **2023**, *28*, 5567. [CrossRef]
16. Verma, R.P.; Hansch, C. Matrix metalloproteinases (MMPs): Chemical–biological functions and (Q)SARs. *Bioorg. Med. Chem.* **2007**, *15*, 2223–2268. [CrossRef] [PubMed]
17. Donepudi, M.; Grütter, M.G. Structure and zymogen activation of caspases. *Biophys. Chem.* **2002**, *101–102*, 145–153. [CrossRef] [PubMed]
18. Chaffin, C.L.; Stouffer, R.L. Expression of Matrix Metalloproteinases and Their Tissue Inhibitor Messenger Ribonucleic Acids in Macaque Periovulatory Granulosa Cells: Time Course and Steroid Regulation1. *Biol. Reprod.* **1999**, *61*, 14–21. [CrossRef]
19. Lambert, E.; Dassé, E.; Haye, B.; Petitfrère, E. TIMPs as multifacial proteins. *Crit. Rev. Oncol. Hematol.* **2004**, *49*, 187–198. [CrossRef]
20. Moracho, N.; Learte, A.I.R.; Muñoz-Sáez, E.; Marchena, M.A.; Cid, M.A.; Arroyo, A.G.; Sánchez-Camacho, C. Emerging roles of MT-MMPs in embryonic development. *Dev. Dyn.* **2022**, *251*, 240–275. [CrossRef]
21. Moore, C.S.; Crocker, S.J. An Alternate Perspective on the Roles of TIMPs and MMPs in Pathology. *Am. J. Pathol.* **2012**, *180*, 12–16. [CrossRef]
22. Rundhaug, J.E. Matrix metalloproteinases and angiogenesis. *J. Cell. Mol. Med.* **2005**, *9*, 267–285. [CrossRef] [PubMed]
23. Belotti, D.; Paganoni, P.; Manenti, L.; Garofalo, A.; Marchini, S.; Taraboletti, G.; Giavazzi, R. Matrix Metalloproteinases (MMP9 and MMP2) Induce the Release of Vascular Endothelial Growth Factor (VEGF) by Ovarian Carcinoma Cells: Implications for Ascites Formation1. *Cancer Res.* **2003**, *63*, 5224–5229. [PubMed]
24. Wolak, D.; Hrabia, A. Alternations in the expression of selected matrix metalloproteinases (MMP-2, -9, -10, and -13) and their tissue inhibitors (TIMP-2 and -3) and MMP-2 and -9 activity in the chicken ovary during pause in laying induced by fasting. *Theriogenology* **2021**, *161*, 176–186. [CrossRef] [PubMed]
25. Nilsson, E.; Skinner, M.K. Cellular Interactions That Control Primordial Follicle Development and Folliculogenesis. *J. Soc. Gynecol. Investig. JSGI* **2001**, *8*, S17–S20. [CrossRef]
26. Jabłońska-Trypuć, A.; Matejczyk, M.; Rosochacki, S. Matrix metalloproteinases (MMPs), the main extracellular matrix (ECM) enzymes in collagen degradation, as a target for anticancer drugs. *J. Enzyme Inhib. Med. Chem.* **2016**, *31*, 177–183. [CrossRef]
27. Pepling, M.E. From primordial germ cell to primordial follicle: Mammalian female germ cell development. *Genesis* **2006**, *44*, 622–632. [CrossRef]

28. Mazaud, S.; Guyot, R.; Guigon, C.J.; Coudouel, N.; Le Magueresse-Battistoni, B.; Magre, S. Basal membrane remodeling during follicle histogenesis in the rat ovary: Contribution of proteinases of the MMP and PA families. *Dev. Biol.* **2005**, *277*, 403–416. [CrossRef]
29. Sessions, D.R.; Vick, M.M.; Fitzgerald, B.P. Characterization of matrix metalloproteinase-2 and matrix metalloproteinase-9 and their inhibitors in equine granulosa cells in vivo and in vitro. *J. Anim. Sci.* **2009**, *87*, 3955–3966. [CrossRef]
30. Duncan, W.C.; McNeilly, A.S.; Illingworth, P.J. The Effect of Luteal "Rescue" on the Expression and Localization of Matrix Metalloproteinases and Their Tissue Inhibitors in the Human Corpus Luteum. *J. Clin. Endocrinol. Metab.* **1998**, *83*, 2470–2478. [CrossRef]
31. Bagavandoss, P. Differential distribution of gelatinases and tissue inhibitor of metalloproteinase-1 in the rat ovary. *J. Endocrinol.* **1998**, *158*, 221–228. [CrossRef] [PubMed]
32. Robinson, R.S.; Woad, K.J.; Hammond, A.J.; Laird, M.; Hunter, M.G.; Mann, G.E. Angiogenesis and vascular function in the ovary. *Reproduction* **2009**, *138*, 869–881. [CrossRef] [PubMed]
33. Downs, L.S., Jr.; Lima, P.H.; Bliss, R.L.; Blomquist, C.H. Cathepsins B and D Activity and Activity Ratios in Normal Ovaries, Benign Ovarian Neoplasms, and Epithelial Ovarian Cancer. *J. Soc. Gynecol. Investig.* **2005**, *12*, 539–544. [CrossRef]
34. Smith, M.F.; McIntush, E.W.; Ricke, W.A.; Kojima, F.N.; Smith, G.W. Regulation of ovarian extracellular matrix remodelling by metalloproteinases and their tissue inhibitors: Effects on follicular development, ovulation and luteal function. *J. Reprod. Fertil. Suppl.* **1999**, *54*, 367–381. [CrossRef] [PubMed]
35. Smith, M.F.; Ricke, W.A.; Bakke, L.J.; Dow, M.P.D.; Smith, G.W. Ovarian tissue remodeling: Role of matrix metalloproteinases and their inhibitors. *Mol. Cell. Endocrinol.* **2002**, *191*, 45–56. [CrossRef]
36. Nothnick, W.B. Reduction in reproductive lifespan of tissue inhibitor of metalloproteinase 1 (TIMP-1)-deficient female mice. *Reproduction* **2001**, *122*, 923–927. [CrossRef]
37. Nothnick, W.B. Tissue Inhibitor of Metalloproteinase-1 (TIMP-1) Deficient Mice Display Reduced Serum Progesterone Levels during Corpus Luteum Development. *Endocrinology* **2003**, *144*, 5–8. [CrossRef]
38. Stilley, J.A.W.; Birt, J.A.; Nagel, S.C.; Sutovsky, M.; Sutovsky, P.; Sharpe-Timms, K.L. Neutralizing TIMP1 Restores Fecundity in a Rat Model of Endometriosis and Treating Control Rats with TIMP1 Causes Anomalies in Ovarian Function and Embryo Development1. *Biol. Reprod.* **2010**, *83*, 185–194. [CrossRef]
39. Curry, T.E.; Dean, D.D.; Sanders, S.L.; Pedigo, N.G.; Jones, P.B.C. The role of ovarian proteases and their inhibitors in ovulation. *Steroids* **1989**, *54*, 501–521. [CrossRef]
40. Bédard, J.; Brûlé, S.; Price, C.A.; Silversides, D.W.; Lussier, J.G. Serine protease inhibitor-E2 (SERPINE2) is differentially expressed in granulosa cells of dominant follicle in cattle. *Mol. Reprod. Dev.* **2003**, *64*, 152–165. [CrossRef]
41. Cao, M.; Sahmi, M.; Lussier, J.G.; Price, C.A. Plasminogen activator and serine protease inhibitor-E2 (protease nexin-1) expression by bovine granulosa cells in vitro. *Biol. Reprod.* **2004**, *71*, 887–893. [CrossRef] [PubMed]
42. Oh, H.S.; Kim, T.; Gu, D.-H.; Lee, T.S.; Kim, T.H.; Shin, S.; Shin, B.S. Pharmacokinetics of Nafamostat, a Potent Serine Protease Inhibitor, by a Novel LC-MS/MS Analysis. *Molecules* **2022**, *27*, 1881. [CrossRef] [PubMed]
43. Murdoch, W.J.; Gottsch, M.L. Proteolytic Mechanisms in the Ovulatory Folliculo-Luteal Transformation. *Connect. Tissue Res.* **2003**, *44*, 50–57. [CrossRef]
44. Eykelbosh, A.J.; Van Der Kraak, G. A role for the lysosomal protease cathepsin B in zebrafish follicular apoptosis. *Comp. Biochem. Physiol. A. Mol. Integr. Physiol.* **2010**, *156*, 218–223. [CrossRef]
45. Li, J.R.; Shen, T. Expression characteristics of MMP-2 and MMP-9 in guinea pig ovaries during the estrous cycle. *Genet. Mol. Res.* **2015**, *14*, 17329–17340. [CrossRef]
46. Vos, M.C.; van der Wurff, A.A.; Last, J.T.; de Boed, E.A.; Smeenk, J.M.; van Kuppevelt, T.H.; Massuger, L.F. Immunohistochemical expression of MMP-14 and MMP-2, and MMP-2 activity during human ovarian follicular development. *Reprod. Biol. Endocrinol.* **2014**, *12*, 12. [CrossRef] [PubMed]
47. Kato, N.; Motoyama, T. Relation Between Laminin-5 γ2 Chain and Cell Surface Metalloproteinase MT1-MMP in Clear Cell Carcinoma of the Ovary. *Int. J. Gynecol. Pathol.* **2009**, *28*, 49. [CrossRef]
48. Robker, R.L.; Russell, D.L.; Espey, L.L.; Lydon, J.P.; O'Malley, B.W.; Richards, J.S. Progesterone-regulated genes in the ovulation process: ADAMTS-1 and cathepsin L proteases. *Proc. Natl. Acad. Sci. USA* **2000**, *97*, 4689–4694. [CrossRef]
49. Brown, H.M.; Dunning, K.R.; Robker, R.L.; Pritchard, M.; Russell, D.L. Requirement for ADAMTS-1 in extracellular matrix remodeling during ovarian folliculogenesis and lymphangiogenesis. *Dev. Biol.* **2006**, *300*, 699–709. [CrossRef]
50. Russell, D.L.; Doyle, K.M.H.; Ochsner, S.A.; Sandy, J.D.; Richards, J.S. Processing and localization of ADAMTS-1 and proteolytic cleavage of versican during cumulus matrix expansion and ovulation. *J. Biol. Chem.* **2003**, *278*, 42330–42339. [CrossRef]
51. Russell, D.L.; Brown, H.M.; Dunning, K.R. ADAMTS proteases in fertility. *Matrix Biol.* **2015**, *44–46*, 54–63. [CrossRef] [PubMed]
52. Brown, H.M.; Dunning, K.R.; Robker, R.L.; Boerboom, D.; Pritchard, M.; Lane, M.; Russell, D.L. ADAMTS1 Cleavage of Versican Mediates Essential Structural Remodeling of the Ovarian Follicle and Cumulus-Oocyte Matrix During Ovulation in Mice1. *Biol. Reprod.* **2010**, *83*, 549–557. [CrossRef] [PubMed]

53. Chen, C.; Ahmad, M.J.; Ye, T.; Du, C.; Zhang, X.; Liang, A.; Yang, L. Cathepsin B Regulates Mice Granulosa Cells' Apoptosis and Proliferation In Vitro. *Int. J. Mol. Sci.* **2021**, *22*, 11827. [CrossRef]
54. Ezz, M.A.; Takahashi, M.; Rivera, R.M.; Balboula, A.Z. Cathepsin L regulates oocyte meiosis and preimplantation embryo development. *Cell Prolif.* **2023**, *57*, e13526. [CrossRef]
55. Nothnick, W.B.; Soloway, P.; Curry, T.E., Jr. Assessment of the Role of Tissue Inhibitor of Metalloproteinase-1 (TIMP-1) during the Periovulatory Period in Female Mice Lacking a Functional TIMP-1 Gene1. *Biol. Reprod.* **1997**, *56*, 1181–1188. [CrossRef] [PubMed]
56. Fata, J.E.; Ho, A.T.-V.; Leco, K.J.; Moorehead, R.A.; Khokha*, R. Cellular turnover and extracellular matrix remodeling in female reproductive tissues: Functions of metalloproteinases and their inhibitors. *Cell. Mol. Life Sci. CMLS* **2000**, *57*, 77–95. [CrossRef]
57. Ny, T.; Wahlberg, P.; Brändström, I.J.M. Matrix remodeling in the ovary: Regulation and functional role of the plasminogen activator and matrix metalloproteinase systems. *Mol. Cell. Endocrinol.* **2002**, *187*, 29–38. [CrossRef]
58. Bodén, I. The Roles of the Plasminogen Activator and Matrix Metalloproteinase Systems in Ovulation and Corpus Luteum Formation. Ph.D. Thesis, Umeå University, Umeå, Sweden, 2004.
59. Epifano, O.; Riminucci, M.; Manna, C.; Apa, R.; Greco, E.; Lanzone, A.; Stefanini, M.; Canipari, R. In vitro production of plasminogen activator by human granulosa cells. *J. Clin. Endocrinol. Metab.* **1994**, *78*, 174–179. [CrossRef]
60. Zhu, C.; Frederick Woessner, J., Jr. A Tissue Inhibitor of Metalloproteinases and α-Macroglobulins in the Ovulating Rat Ovary: Possible Regulators of Collagen Matrix Breakdown1. *Biol. Reprod.* **1991**, *45*, 334–342. [CrossRef]
61. Peng, X.-R.; Hsueh, A.J.W.; Ny, T. Transient and cell-specific expression of tissue-type plasminogen activator and plasminogen-activator-inhibitor type 1 results in controlled and directed proteolysis during gonadotropin-induced ovulation. *Eur. J. Biochem.* **1993**, *214*, 147–156. [CrossRef]
62. Hägglund, A.-C.; Ny, A.; Leonardsson, G.; Ny, T. Regulation and Localization of Matrix Metalloproteinases and Tissue Inhibitors of Metalloproteinases in the Mouse Ovary during Gonadotropin-Induced Ovulation. *Endocrinology* **1999**, *140*, 4351–4358. [CrossRef] [PubMed]
63. Hägglund, A.-C.; Basset, P.; Ny, T. Stromelysin-3 Is Induced in Mouse Ovarian Follicles Undergoing Hormonally Controlled Apoptosis, but This Metalloproteinase Is Not Required for Follicular Atresia. *Biol. Reprod.* **2001**, *64*, 457–463. [CrossRef]
64. Ogiwara, K.; Hagiwara, A.; Rajapakse, S.; Takahashi, T. The Role of Urokinase Plasminogen Activator and Plasminogen Activator Inhibitor-1 in Follicle Rupture During Ovulation in the Teleost Medaka. *Biol. Reprod.* **2015**, *92*, 1–17. [CrossRef] [PubMed]
65. O'Connell, M.L.; Canipari, R.; Strickland, S. Hormonal regulation of tissue plasminogen activator secretion and mRNA levels in rat granulosa cells. *J. Biol. Chem.* **1987**, *262*, 2339–2344. [CrossRef]
66. Li, M.; Karakji, E.G.; Xing, R.; Fryer, J.N.; Carnegie, J.A.; Rabbani, S.A.; Tsang, B.K. Expression of Urokinase-Type Plasminogen Activator and Its Receptor during Ovarian Follicular Development. *Endocrinology* **1997**, *138*, 2790–2799. [CrossRef]
67. Karakji, E.G.; Tsang, B.K. Regulation of Rat Granulosa Cell Plasminogen Activator System: Influence of Interleukin-1β and Ovarian Follicular Development. *Biol. Reprod.* **1995**, *53*, 1302–1310. [CrossRef] [PubMed]
68. Hurwitz, A.; Finci-Yeheskel, Z.; Dushnik, M.; Milwidsky, A.; Ben-Chetrit, A.; Yagel, S.; Adashi, E.Y.; Mayer, M. Cytokine-mediated regulation of rat ovarian function: Interleukin-1 inhibits plasminogen activator activity through the induction of plasminogen activator inhibitor-1 (PAI-1). *Mol. Cell. Endocrinol.* **1994**, *101*, 307–314. [CrossRef]
69. Sheng, X.; Liu, C.; Yan, G.; Li, G.; Liu, J.; Yang, Y.; Li, S.; Li, Z.; Zhou, J.; Zhen, X.; et al. The mitochondrial protease LONP1 maintains oocyte development and survival by suppressing nuclear translocation of AIFM1 in mammals. *eBioMedicine* **2022**, *75*, 103790. [CrossRef]
70. Meng, T.-G.; Hu, M.-W.; Ma, X.-S.; Huang, L.; Liang, Q.-X.; Yuan, Y.; Hou, Y.; Wang, H.; Schatten, H.; Wang, Z.-B.; et al. Oocyte-specific deletion of furin leads to female infertility by causing early secondary follicle arrest in mice. *Cell Death Dis.* **2017**, *8*, e2846. [CrossRef]
71. Nyegaard, M.; Overgaard, M.T.; Su, Y.-Q.; Hamilton, A.E.; Kwintkiewicz, J.; Hsieh, M.; Nayak, N.R.; Conti, M.; Conover, C.A.; Giudice, L.C. Lack of Functional Pregnancy-Associated Plasma Protein-A (PAPPA) Compromises Mouse Ovarian Steroidogenesis and Female Fertility. *Biol. Reprod.* **2010**, *82*, 1129–1138. [CrossRef]
72. Oxvig, C. The role of PAPP-A in the IGF system: Location, location, location. *J. Cell Commun. Signal.* **2015**, *9*, 177–187. [CrossRef] [PubMed]
73. Mason, E.J.; Grell, J.A.; Wan, J.; Cohen, P.; Conover, C.A. Insulin-like growth factor (IGF)-I and IGF-II contribute differentially to the phenotype of pregnancy associated plasma protein-A knock-out mice. *Growth Horm. IGF Res.* **2011**, *21*, 243–247. [CrossRef]
74. Folgueras, A.R.; de Lara, F.M.; Pendás, A.M.; Garabaya, C.; Rodríguez, F.; Astudillo, A.; Bernal, T.; Cabanillas, R.; López-Otín, C.; Velasco, G. Membrane-bound serine protease matriptase-2 (Tmprss6) is an essential regulator of iron homeostasis. *Blood* **2008**, *112*, 2539–2545. [CrossRef]
75. Tonai, S.; Kawabata, A.; Nakanishi, T.; Lee, J.Y.; Okamoto, A.; Shimada, M.; Yamashita, Y. Iron deficiency induces female infertile in order to failure of follicular development in mice. *J. Reprod. Dev.* **2020**, *66*, 475–483. [CrossRef]
76. Turk, V.; Turk, B.; Turk, D. Lysosomal cysteine proteases: Facts and opportunities. *EMBO J.* **2001**, *20*, 4629–4633. [CrossRef] [PubMed]

77. Turk, V.; Stoka, V.; Vasiljeva, O.; Renko, M.; Sun, T.; Turk, B.; Turk, D. Cysteine cathepsins: From structure, function and regulation to new frontiers. *Biochim. Biophys. Acta Proteins Proteom.* **2012**, *1824*, 68–88. [CrossRef]
78. Oksjoki, S.; Söderström, M.; Vuorio, E.; Anttila, L. Differential expression patterns of cathepsins B, H, K, L and S in the mouse ovary. *Mol. Hum. Reprod.* **2001**, *7*, 27–34. [CrossRef] [PubMed]
79. Peluffo, M.C.; Murphy, M.J.; Talcott Baughman, S.; Stouffer, R.L.; Hennebold, J.D. Systematic Analysis of Protease Gene Expression in the Rhesus Macaque Ovulatory Follicle: Metalloproteinase Involvement in Follicle Rupture. *Endocrinology* **2011**, *152*, 3963–3974. [CrossRef]
80. Lind, A.-K.; Dahm-Kähler, P.; Weijdegård, B.; Sundfeldt, K.; Brännström, M. Gelatinases and their tissue inhibitors during human ovulation: Increased expression of tissue inhibitor of matrix metalloproteinase-1. *Mol. Hum. Reprod.* **2006**, *12*, 725–736. [CrossRef]
81. Bakke, L.J.; Li, Q.; Cassar, C.A.; Dow, M.P.D.; Pursley, J.R.; Smith, G.W. Gonadotropin Surge-Induced Differential Upregulation of Collagenase-1 (MMP-1) and Collagenase-3 (MMP-13) mRNA and Protein in Bovine Preovulatory Follicles. *Biol. Reprod.* **2004**, *71*, 605–612. [CrossRef]
82. Hrabia, A.; Wolak, D.; Kwaśniewska, M.; Kieronska, A.; Socha, J.K.; Sechman, A. Expression of gelatinases (MMP-2 and MMP-9) and tissue inhibitors of metalloproteinases (TIMP-2 and TIMP-3) in the chicken ovary in relation to follicle development and atresia. *Theriogenology* **2019**, *125*, 268–276. [CrossRef] [PubMed]
83. Zhu, G.; Kang, L.; Wei, Q.; Cui, X.; Wang, S.; Chen, Y.; Jiang, Y. Expression and Regulation of MMP1, MMP3, and MMP9 in the Chicken Ovary in Response to Gonadotropins, Sex Hormones, and TGFB1. *Biol. Reprod.* **2014**, *90*, 1–11. [CrossRef]
84. McCord, L.A.; Li, F.; Rosewell, K.L.; Brännström, M.; Curry, T.E. Ovarian Expression and Regulation of the Stromelysins During the Periovulatory Period in the Human and the Rat. *Biol. Reprod.* **2012**, *86*, 78. [CrossRef]
85. Shrestha, K.; Puttabyatappa, M.; Wynn, M.A.; Hannon, P.R.; Al-Alem, L.F.; Rosewell, K.L.; Akin, J.; Curry, T.E., Jr. Protease expression in the human and rat cumulus–oocyte complex during the periovulatory period: A role in cumulus–oocyte complex migration†. *Biol. Reprod.* **2024**, *111*, 845–855. [CrossRef] [PubMed]
86. Rosewell, K.L.; Al-Alem, L.; Zakerkish, F.; McCord, L.; Akin, J.W.; Chaffin, C.L.; Brännström, M.; Curry, T.E. Induction of proteinases in the human preovulatory follicle of the menstrual cycle by human chorionic gonadotropin. *Fertil. Steril.* **2015**, *103*, 826–833. [CrossRef]
87. Bu, S.; Cao, C.; Yang, Y.; Miao, C.; Hu, Z.; Cao, Y.; Sang, Q.A.; Duan, E. Localization and temporal regulation of tissue inhibitor of metalloproteinases-4 in mouse ovary. *Reproduction* **2006**, *131*, 1099–1107. [CrossRef]
88. Simpson, K.S.; Komar, C.M.; Curry, T.E., Jr. Localization and Expression of Tissue Inhibitor of Metalloproteinase-4 in the Immature Gonadotropin-Stimulated and Adult Rat Ovary. *Biol. Reprod.* **2003**, *68*, 214–221. [CrossRef] [PubMed]
89. Liu, Y.-X.; Cajander, S.B.; Ny, T.; Kristensen, P.; Hsueh, A.J.W. Gonadotropin regulation of tissue-type and urokinase-type plasminogen activators in rat granulosa and theca-interstitial cells during the periovulatory period. *Mol. Cell. Endocrinol.* **1987**, *54*, 221–229. [CrossRef]
90. Macchione, E.; Epifano, O.; Stefanini, M.; Belin, D.; Canipari, R. Urokinase Redistribution from the Secreted to the Cell-Bound Fraction in Granulosa Cells of Rat Preovulatory Follicles1. *Biol. Reprod.* **2000**, *62*, 895–903. [CrossRef]
91. Karakji, E.G.; Tsang, B.K. Tumor Necrosis Factor Alpha Inhibits Rat Granulosa Cell Plasminogen Activator Activity in Vitro during Follicular Development. *Biol. Reprod.* **1995**, *52*, 745–752. [CrossRef]
92. Dow, M.P.D.; Bakke, L.J.; Cassar, C.A.; Peters, M.W.; Pursley, J.R.; Smith, G.W. Gonadotropin Surge-Induced Up-Regulation of the Plasminogen Activators (Tissue Plasminogen Activator and Urokinase Plasminogen Activator) and the Urokinase Plasminogen Activator Receptor Within Bovine Periovulatory Follicular and Luteal Tissue. *Biol. Reprod.* **2002**, *66*, 1413–1421. [CrossRef] [PubMed]
93. Li, J.; Balboula, A.Z.; Aboelenain, M.; Fujii, T.; Moriyasu, S.; Bai, H.; Kawahara, M.; Takahashi, M. Effect of autophagy induction and cathepsin B inhibition on developmental competence of poor quality bovine oocytes. *J. Reprod. Dev.* **2020**, *66*, 83–91. [CrossRef]
94. Zhang, K.; Xu, R.; Zheng, L.; Zhang, H.; Qian, Z.; Li, C.; Xue, M.; He, Z.; Ma, J.; Li, Z.; et al. Elevated N-glycosylated cathepsin L impairs oocyte function and contributes to oocyte senescence during reproductive aging. *Aging Cell* **2025**, *24*, e14397. [CrossRef]
95. Holland, A.; Findlay, J.; Clements, J. Kallikrein gene expression in the gonadotrophin-stimulated rat ovary. *J. Endocrinol.* **2001**, *170*, 243–250. [CrossRef] [PubMed]
96. Clements, J.A.; Mukhtar, A.; Holland, A.M.; Ehrlich, A.R.; Fuller, P.J. Kallikrein gene family expression in the rat ovary: Localization to the granulosa cell. *Endocrinology* **1995**, *136*, 1137–1144. [CrossRef]
97. Musil, D.; Zucic, D.; Turk, D.; Engh, R.A.; Mayr, I.; Huber, R.; Popovic, T.; Turk, V.; Towatari, T.; Katunuma, N. The refined 2.15 A X-ray crystal structure of human liver cathepsin B: The structural basis for its specificity. *EMBO J.* **1991**, *10*, 2321–2330. [CrossRef] [PubMed]
98. Cavallo-Medved, D.; Moin, K.; Sloane, B. Cathepsin B: Basis sequence: Mouse. *AFCS-Nat. Mol. Pages* **2011**, *2011*, A000508.
99. Mohanty, A.; Kumari, A.; Kumar, S.L.; Kumar, A.; Birajdar, P.; Beniwal, R.; Athar, M.; Kumar P, K.; Rao, H.B.D.P. Cathepsin B regulates ovarian reserve quality and quantity via mitophagy by modulating IGF1R turnover. *Aging Cell* **2024**, e70066. [CrossRef]

100. Liang, S.; Jiang, H.; Shen, X.-H.; Zhang, J.-B.; Kim, N.-H. Inhibition of cathepsin B activity prevents deterioration in the quality of in vitro aged porcine oocytes. *Theriogenology* **2018**, *116*, 103–111. [CrossRef]
101. Komatsu, K.; Wei, W.; Murase, T.; Masubuchi, S. 17β-Estradiol and cathepsins control primordial follicle growth in mouse ovaries. *Reproduction* **2021**, *162*, 277–287. [CrossRef]
102. Bastu, E.; Gokulu, S.G.; Dural, O.; Yasa, C.; Bulgurcuoglu, S.; Karamustafaoglu Balci, B.; Celik, C.; Buyru, F. The association between follicular fluid levels of cathepsin B, relaxin or AMH with clinical pregnancy rates in infertile patients. *Eur. J. Obstet. Gynecol. Reprod. Biol.* **2015**, *187*, 30–34. [CrossRef]
103. Kondo, Y.; Rajapakse, S.; Ogiwara, K. Involvement of cathepsin L in the degradation and degeneration of postovulatory follicle of the medaka ovary†. *Biol. Reprod.* **2023**, *109*, 904–917. [CrossRef] [PubMed]
104. Matousek, M.; Mitsube, K.; Mikuni, M.; Brännström, M. Inhibition of ovulation in the rat by a leukotriene B4 receptor antagonist. *Mol. Hum. Reprod.* **2001**, *7*, 35–42. [CrossRef]
105. Shahed, A.; Simmons, J.J.; Featherstone, S.L.; Young, K.A. Matrix metalloproteinase inhibition influences aspects of photoperiod stimulated ovarian recrudescence in Siberian hamsters. *Gen. Comp. Endocrinol.* **2015**, *216*, 46–53. [CrossRef] [PubMed]
106. Whited, J.; Shahed, A.; McMichael, C.F.; Young, K.A. Inhibition of Matrix Metalloproteinases (MMPs) in Siberian Hamsters Impedes Photostimulated Recrudescence of Ovaries. *Reproduction* **2010**, *140*, 875–883. [CrossRef]
107. Oksjoki, S.; Sallinen, S.; Vuorio, E.; Anttila, L. Cyclic expression of mRNA transcripts for connective tissue components in the mouse ovary. *Mol. Hum. Reprod.* **1999**, *5*, 803–808. [CrossRef] [PubMed]
108. Shindo, T.; Kurihara, H.; Kuno, K.; Yokoyama, H.; Wada, T.; Kurihara, Y.; Imai, T.; Wang, Y.; Ogata, M.; Nishimatsu, H.; et al. ADAMTS-1: A metalloproteinase-disintegrin essential for normal growth, fertility, and organ morphology and function. *J. Clin. Investig.* **2000**, *105*, 1345–1352. [CrossRef]
109. Mittaz, L.; Russell, D.L.; Wilson, T.; Brasted, M.; Tkalcevic, J.; Salamonsen, L.A.; Hertzog, P.J.; Pritchard, M.A. Adamts-1 Is Essential for the Development and Function of the Urogenital System. *Biol. Reprod.* **2004**, *70*, 1096–1105. [CrossRef]
110. Shozu, M.; Minami, N.; Yokoyama, H.; Inoue, M.; Kurihara, H.; Matsushima, K.; Kuno, K. ADAMTS-1 is involved in normal follicular development, ovulatory process and organization of the medullary vascular network in the ovary. *J. Mol. Endocrinol.* **2005**, *35*, 343–355. [CrossRef]
111. Carter, N.J.; Roach, Z.A.; Byrnes, M.M.; Zhu, Y. Adamts9 is necessary for ovarian development in zebrafish. *Gen. Comp. Endocrinol.* **2019**, *277*, 130–140. [CrossRef]
112. Liu, K.; Brändström, A.; Liu, Y.X.; Ny, T.; Selstam, G. Coordinated expression of tissue-type plasminogen activator and plasminogen activator inhibitor type 1 during corpus luteum formation and luteolysis in the adult pseudopregnant rat. *Endocrinology* **1996**, *137*, 2126–2132. [CrossRef] [PubMed]
113. Bacharach, E.; Itin, A.; Keshet, E. In vivo patterns of expression of urokinase and its inhibitor PAI-1 suggest a concerted role in regulating physiological angiogenesis. *Proc. Natl. Acad. Sci. USA* **1992**, *89*, 10686–10690. [CrossRef]
114. Jones, P.B.; Muse, K.N.; Wilson, E.A.; Curry, T.E. Expression of plasminogen activator (PA) and a PA inhibitor in human granulosa cells from preovulatory follicles. *J. Clin. Endocrinol. Metab.* **1988**, *67*, 857–860. [CrossRef]
115. Chen, B.; Chang, H.-M.; Zhang, Z.; Cao, Y.; Leung, P.C.K. ALK4-SMAD3/4 mediates the effects of activin A on the upregulation of PAI-1 in human granulosa lutein cells. *Mol. Cell. Endocrinol.* **2020**, *505*, 110731. [CrossRef] [PubMed]
116. Song, G.; Jiang, Y.; Wang, Y.; Song, M.; Niu, X.; Xu, H.; Li, M. Modulation of Cathepsin S (CTSS) Regulates the Secretion of Progesterone and Estradiol, Proliferation, and Apoptosis of Ovarian Granulosa Cells in Rabbits. *Animals* **2021**, *11*, 1770. [CrossRef] [PubMed]
117. Song, H.; Qin, Q.; Yuan, C.; Li, H.; Zhang, F.; Fan, L. Metabolomic Profiling of Poor Ovarian Response Identifies Potential Predictive Biomarkers. *Front. Endocrinol.* **2021**, *12*, 774667. [CrossRef]
118. Espey, L.L. Ovarian Proteolytic Enzymes and Ovulation. *Biol. Reprod.* **1974**, *10*, 216–235. [CrossRef]
119. Espey, L.L.; Richards, J.; Neill, J. *Physiology of Reproduction*; Elsevier: Amsterdam, The Netherlands, 2006; Volume 2006, pp. 425–474.
120. Basini, G.; Bussolati, S.; Baioni, L.; Grasselli, F. Gelatinases (MMP2 and MMP9) in swine antral follicle. *BioFactors* **2011**, *37*, 117–120. [CrossRef]
121. Deady, L.D.; Shen, W.; Mosure, S.A.; Spradling, A.C.; Sun, J. Matrix Metalloproteinase 2 Is Required for Ovulation and Corpus Luteum Formation in Drosophila. *PLoS Genet.* **2015**, *11*, e1004989. [CrossRef]
122. Fujihara, M.; Yamamizu, K.; Wildt, D.; Songsasen, N. Expression pattern of matrix metalloproteinases changes during folliculogenesis in the cat ovary. *Reprod. Domest. Anim.* **2016**, *51*, 717–725. [CrossRef]
123. Jo, M.; Curry, T.E., Jr. Regulation of Matrix Metalloproteinase-19 Messenger RNA Expression in the Rat Ovary. *Biol. Reprod.* **2004**, *71*, 1796–1806. [CrossRef] [PubMed]
124. Politis, I.; Srikandakumar, A.; Turner, J.D.; Tsang, B.K.; Ainsworth, L.; Downey, B.R. Changes in and Partial Identification of the Plasminogen Activator and Plasminogen Activator Inhibitor Systems During Ovarian Follicular Maturation in the Pig. *Biol. Reprod.* **1990**, *43*, 636–642. [CrossRef] [PubMed]

125. Zupanič, N.; Počič, J.; Leonardi, A.; Šribar, J.; Kordiš, D.; Križaj, I. Serine pseudoproteases in physiology and disease. *FEBS J.* **2023**, *290*, 2263–2278. [CrossRef] [PubMed]
126. Canipari, R.; Strickland, S. Plasminogen activator in the rat ovary. Production and gonadotropin regulation of the enzyme in granulosa and thecal cells. *J. Biol. Chem.* **1985**, *260*, 5121–5125. [CrossRef]
127. Canipari, R.; O'Connell, M.L.; Meyer, G.; Strickland, S. Mouse ovarian granulosa cells produce urokinase-type plasminogen activator, whereas the corresponding rat cells produce tissue-type plasminogen activator. *J. Cell Biol.* **1987**, *105*, 977–981. [CrossRef]
128. Shen, X.; Minoura, H.; Yoshida, T.; Toyoda, N. Changes in Ovarian Expression of Tissue-Type Plasminogen Activator and Plasminogen Activator Inhibitor Type-1 Messenger Ribonucleic Acids during Ovulation in Rat. *Endocr. J.* **1997**, *44*, 341–348. [CrossRef]
129. Liu, Y.-X. Regulation of the Plasminogen Activator System in the Ovary. *Biol. Signals Recept.* **1999**, *8*, 160–177. [CrossRef]
130. Plendl, J.; Snyman, C.; Naidoo, S.; Sawant, S.; Mahabeer, R.; Bhoola, K.D. Expression of Tissue Kallikrein and Kinin Receptors in Angiogenic Microvascular Endothelial Cells. *Biol. Chem.* **2000**, *381*, 1103–1115. [CrossRef]
131. Hazzard, T.M.; Stouffer, R.L. Angiogenesis in ovarian follicular and luteal development. *Best Pract. Res. Clin. Obstet. Gynaecol.* **2000**, *14*, 883–900. Available online: https://www.sciencedirect.com/science/article/pii/S1521693400901330 (accessed on 21 April 2025). [CrossRef]
132. Tamanini, C.; De Ambrogi, M. Angiogenesis in Developing Follicle and Corpus Luteum. *Reprod. Domest. Anim.* **2004**, *39*, 206–216. [CrossRef]
133. Kobayashi, M.; Yoshino, O.; Nakashima, A.; Ito, M.; Nishio, K.; Ono, Y.; Kusabiraki, T.; Kunitomi, C.; Takahashi, N.; Harada, M.; et al. Inhibition of autophagy in theca cells induces CYP17A1 and PAI-1 expression via ROS/p38 and JNK signalling during the development of polycystic ovary syndrome. *Mol. Cell. Endocrinol.* **2020**, *508*, 110792. [CrossRef] [PubMed]
134. Paliouras, M.; Diamandis, E.P. Intracellular signaling pathways regulate hormone-dependent kallikrein gene expression. *Tumour Biol.* **2008**, *29*, 63–75. [CrossRef]
135. Iwasaki, T.; Tokumori, M.; Matsubara, M.; Ojima, F.; Kamigochi, K.; Aizawa, S.; Ogoshi, M.; Kimura, A.P.; Takeuchi, S.; Takahashi, S. A regulatory mechanism of mouse kallikrein 1 gene expression by estrogen. *Mol. Cell. Endocrinol.* **2023**, *577*, 112044. [CrossRef] [PubMed]
136. Lorenzetti, S.; Marcoccia, D.; Narciso, L.; Mantovani, A. Cell viability and PSA secretion assays in LNCaP cells: A tiered in vitro approach to screen chemicals with a prostate-mediated effect on male reproduction within the ReProTect project. *Reprod. Toxicol.* **2010**, *30*, 25–35. [CrossRef]
137. Raimondo, S.; Gentile, M.; Esposito, G.; Gentile, T.; Ferrara, I.; Crescenzo, C.; Palmieri, M.; Cuomo, F.; De Filippo, S.; Lettieri, G.; et al. Could Kallikrein-Related Serine Peptidase 3 Be an Early Biomarker of Environmental Exposure in Young Women? *Int. J. Environ. Res. Public Health* **2021**, *18*, 8833. [CrossRef] [PubMed]
138. Lawrence, M.G.; Lai, J.; Clements, J.A. Kallikreins on Steroids: Structure, Function, and Hormonal Regulation of Prostate-Specific Antigen and the Extended Kallikrein Locus. *Endocr. Rev.* **2010**, *31*, 407–446. [CrossRef] [PubMed]
139. Borgoño, C.A.; Diamandis, E.P. The emerging roles of human tissue kallikreins in cancer. *Nat. Rev. Cancer* **2004**, *4*, 876–890. [CrossRef]
140. Kurlender, L.; Yousef, G.M.; Memari, N.; Robb, J.-D.; Michael, I.P.; Borgoño, C.; Katsaros, D.; Stephan, C.; Jung, K.; Diamandis, E.P. Differential Expression of a Human Kallikrein 5 (KLK5) Splice Variant in Ovarian and Prostate Cancer. *Tumor Biol.* **2004**, *25*, 149–156. [CrossRef]
141. Ahmed, N.; Dorn, J.; Napieralski, R.; Drecoll, E.; Kotzsch, M.; Goettig, P.; Zein, E.; Avril, S.; Kiechle, M.; Diamandis, E.P.; et al. Clinical relevance of kallikrein-related peptidase 6 (KLK6) and 8 (KLK8) mRNA expression in advanced serous ovarian cancer. *Biol. Chem.* **2016**, *397*, 1265–1276. [CrossRef]
142. Sriraman, V.; Richards, J.S. Cathepsin L Gene Expression and Promoter Activation in Rodent Granulosa Cells. *Endocrinology* **2004**, *145*, 582–591. [CrossRef] [PubMed]
143. Carnevali, O.; Cionna, C.; Tosti, L.; Lubzens, E.; Maradonna, F. Role of cathepsins in ovarian follicle growth and maturation. *Gen. Comp. Endocrinol.* **2006**, *146*, 195–203. [CrossRef] [PubMed]
144. Balboula, A.Z.; Yamanaka, K.; Sakatani, M.; Hegab, A.O.; Zaabel, S.M.; Takahashi, M. Cathepsin B activity is related to the quality of bovine cumulus oocyte complexes and its inhibition can improve their developmental competence. *Mol. Reprod. Dev.* **2010**, *77*, 439–448. [CrossRef] [PubMed]
145. Balboula, A.Z.; Yamanaka, K.; Sakatani, M.; Kawahara, M.; Hegab, A.O.; Zaabel, S.M.; Takahashi, M. Cathepsin B activity has a crucial role in the developmental competence of bovine cumulus–oocyte complexes exposed to heat shock during in vitro maturation. *Reproduction* **2013**, *146*, 407–417. [CrossRef]
146. García, V.; Kohen, P.; Maldonado, C.; Sierralta, W.; Muñoz, A.; Villarroel, C.; Strauss, J.F.; Devoto, L. Transient expression of progesterone receptor and cathepsin-l in human granulosa cells during the periovulatory period. *Fertil. Steril.* **2012**, *97*, 707–713.e1. [CrossRef]

147. Arosh, J.A.; Banu, S.K.; McCracken, J.A. Novel concepts on the role of prostaglandins on luteal maintenance and maternal recognition and establishment of pregnancy in ruminants1. *J. Dairy Sci.* **2016**, *99*, 5926–5940. [CrossRef] [PubMed]
148. Wiltbank, M.C.; Mezera, M.A.; Toledo, M.Z.; Drum, J.N.; Baez, G.M.; García-Guerra, A.; Sartori, R. Physiological mechanisms involved in maintaining the corpus luteum during the first two months of pregnancy. *Anim. Reprod.* **2018**, *15* (Suppl. 1), 805–821. [CrossRef]
149. Cha, J.; Sun, X.; Dey, S.K. Mechanisms of implantation: Strategies for successful pregnancy. *Nat. Med.* **2012**, *18*, 1754–1767. [CrossRef]
150. Ferrara, N.; Chen, H.; Davis-Smyth, T.; Gerber, H.-P.; Nguyen, T.-N.; Peers, D.; Chisholm, V.; Hillan, K.J.; Schwall, R.H. Vascular endothelial growth factor is essential for corpus luteum angiogenesis. *Nat. Med.* **1998**, *4*, 336–340. [CrossRef]
151. Wahlberg, P.; Nylander, Å.; Ahlskog, N.; Liu, K.; Ny, T. Expression and Localization of the Serine Proteases High-Temperature Requirement Factor A1, Serine Protease 23, and Serine Protease 35 in the Mouse Ovary. *Endocrinology* **2008**, *149*, 5070–5077. [CrossRef]
152. Siegel, R.L.; Miller, K.D.; Fuchs, H.E.; Jemal, A. Cancer Statistics, 2021. *CA Cancer J. Clin.* **2021**, *71*, 7–33. [CrossRef]
153. Elias, K.M.; Guo, J.; Bast, R.C. Early Detection of Ovarian Cancer. *Hematol. Oncol. Clin. N. Am.* **2018**, *32*, 903–914. [CrossRef] [PubMed]
154. Tingulstad, S.; Skjeldestad, F.E.; Halvorsen, T.B.; Hagen, B. jørn Survival and prognostic factors in patients with ovarian cancer. *Obstet. Gynecol.* **2003**, *101*, 885–891. [CrossRef] [PubMed]
155. Carey, P.; Low, E.; Harper, E.; Stack, M.S. Metalloproteinases in Ovarian Cancer. *Int. J. Mol. Sci.* **2021**, *22*, 3403. [CrossRef] [PubMed]
156. Scorilas, A.; Borgoño, C.A.; Harbeck, N.; Dorn, J.; Schmalfeldt, B.; Schmitt, M.; Diamandis, E.P. Human Kallikrein 13 Protein in Ovarian Cancer Cytosols: A New Favorable Prognostic Marker. *J. Clin. Oncol.* **2004**, *22*, 678–685. [CrossRef]
157. Zhang, Y.; Chen, Q. Relationship between matrix metalloproteinases and the occurrence and development of ovarian cancer. *Braz. J. Med. Biol. Res.* **2017**, *50*, e6104. [CrossRef]
158. Goldman, S.; Shalev, E. The role of the matrix metalloproteinases in human endometrial and ovarian cycles. *Eur. J. Obstet. Gynecol. Reprod. Biol.* **2003**, *111*, 109–121. [CrossRef]
159. Irving-Rodgers, H.F.; Rodgers, R.J. Extracellular matrix in ovarian follicular development and disease. *Cell Tissue Res.* **2005**, *322*, 89–98. [CrossRef]
160. Kenny, H.A.; Kaur, S.; Coussens, L.M.; Lengyel, E. The initial steps of ovarian cancer cell metastasis are mediated by MMP-2 cleavage of vitronectin and fibronectin. *J. Clin. Investig.* **2008**, *118*, 1367–1379. [CrossRef]
161. Sakata, K.; Shigemasa, K.; Nagai, N.; Ohama, K. Expression of matrix metalloproteinases (MMP-2, MMP-9, MT1-MMP) and their inhibitors (TIMP-1, TIMP-2) in common epithelial tumors of the ovary. *Int. J. Oncol.* **2000**, *17*, 673–754. [CrossRef]
162. Davidson, B.; Goldberg, I.; Gotlieb, W.H.; Kopolovic, J.; Ben-Baruch, G.; Nesland, J.M.; Berner, A.; Bryne, M.; Reich, R. High levels of MMP-2, MMP-9, MT1-MMP and TIMP-2 mRNA correlate with poor survival in ovarian carcinoma. *Clin. Exp. Metastasis* **1999**, *17*, 799–808. [CrossRef]
163. Fu, Z.; Xu, S.; Xu, Y.; Ma, J.; Li, J.; Xu, P. The Expression of Tumor-Derived and Stromal-Derived Matrix Metalloproteinase 2 Predicted Prognosis of Ovarian Cancer. *Int. J. Gynecol. Cancer* **2015**, *25*, 356–362. [CrossRef] [PubMed]
164. Davidson, B.; Goldberg, I.; Gotlieb, W.H.; Kopolovic, J.; Ben-Baruch, G.; Nesland, J.M.; Reich, R. The prognostic value of metalloproteinases and angiogenic factors in ovarian carcinoma. *Mol. Cell. Endocrinol.* **2002**, *187*, 39–45. [CrossRef] [PubMed]
165. Sillanpää, S.; Anttila, M.; Voutilainen, K.; Ropponen, K.; Turpeenniemi-Hujanen, T.; Puistola, U.; Tammi, R.; Tammi, M.; Sironen, R.; Saarikoski, S.; et al. Prognostic significance of matrix metalloproteinase-9 (MMP-9) in epithelial ovarian cancer. *Gynecol. Oncol.* **2007**, *104*, 296–303. [CrossRef] [PubMed]
166. Gong, W.; Liu, Y.; Diamandis, E.P.; Kiechle, M.; Bronger, H.; Dorn, J.; Dreyer, T.; Magdolen, V. Prognostic value of kallikrein-related peptidase 7 (KLK7) mRNA expression in advanced high-grade serous ovarian cancer. *J. Ovarian Res.* **2020**, *13*, 125. [CrossRef]
167. Al-Alem, L.F.; McCord, L.A.; Southard, R.C.; Kilgore, M.W.; Curry, T.E., Jr. Activation of the PKC Pathway Stimulates Ovarian Cancer Cell Proliferation, Migration, and Expression of MMP7 and MMP101. *Biol. Reprod.* **2013**, *89*, 1–7. [CrossRef]
168. Al-Alem, L.; Curry, T.E. Ovarian cancer: Involvement of the matrix metalloproteinases. *Reproduction* **2015**, *150*, R55–R64. [CrossRef]
169. Kamat, A.A.; Fletcher, M.; Gruman, L.M.; Mueller, P.; Lopez, A.; Landen, C.N., Jr.; Han, L.; Gershenson, D.M.; Sood, A.K. The Clinical Relevance of Stromal Matrix Metalloproteinase Expression in Ovarian Cancer. *Clin. Cancer Res.* **2006**, *12*, 1707–1714. [CrossRef]
170. Sood, A.K.; Fletcher, M.S.; Coffin, J.E.; Yang, M.; Seftor, E.A.; Gruman, L.M.; Gershenson, D.M.; Hendrix, M.J.C. Functional role of matrix metalloproteinases in ovarian tumor cell plasticity. *Am. J. Obstet. Gynecol.* **2004**, *190*, 899–909. [CrossRef]
171. Agrawal, A.; Ambad, R.; Lahoti, R.; Muley, P.; Pande, P.S. Role of Artificial Intelligence in PCOS Detection. *J. Datta Meghe Inst. Med. Sci. Univ.* **2022**, *17*, 491. [CrossRef]

172. Gao, X.-W.; Su, X.-T.; Lu, Z.-H.; Ou, J. 17β-Estradiol Prevents Extracellular Matrix Degradation by Downregulating MMP3 Expression via PI3K/Akt/FOXO3 Pathway. *Spine* **2020**, *45*, 292. [CrossRef]
173. Wang, F.-Q.; So, J.; Reierstad, S.; Fishman, D.A. Matrilysin (MMP-7) promotes invasion of ovarian cancer cells by activation of progelatinase. *Int. J. Cancer* **2005**, *114*, 19–31. [CrossRef] [PubMed]
174. Chang, M.-C.; Chen, C.-A.; Chen, P.-J.; Chiang, Y.-C.; Chen, Y.-L.; Mao, T.-L.; Lin, H.-W.; Lin Chiang, W.-H.; Cheng, W.-F. Mesothelin enhances invasion of ovarian cancer by inducing MMP-7 through MAPK/ERK and JNK pathways. *Biochem. J.* **2012**, *442*, 293–302. [CrossRef] [PubMed]
175. Stadlmann, S.; Pollheimer, J.; Moser, P.L.; Raggi, A.; Amberger, A.; Margreiter, R.; Offner, F.A.; Mikuz, G.; Dirnhofer, S.; Moch, H. Cytokine-regulated expression of collagenase-2 (MMP-8) is involved in the progression of ovarian cancer. *Eur. J. Cancer* **2003**, *39*, 2499–2505. [CrossRef]
176. Chen, S.-S.; Song, J.; Tu, X.-Y.; Zhao, J.-H.; Ye, X.-Q. The association between MMP-12 82 A/G polymorphism and susceptibility to various malignant tumors: A meta-analysis. *Int. J. Clin. Exp. Med.* **2015**, *8*, 10845–10854.
177. Liu, L.; Sun, J.; Li, G.; Gu, B.; Wang, X.; Chi, H.; Guo, F. Association between MMP-12-82A/G polymorphism and cancer risk: A meta-analysis. *Int. J. Clin. Exp. Med.* **2015**, *8*, 11896–11904. Available online: https://www.ncbi.nlm.nih.gov/pmc/articles/PMC4612787/ (accessed on 21 April 2025).
178. Hantke, B.; Harbeck, N.; Schmalfeldt, B.; Claes, I.; Hiller, O.; Luther, M.-O.; Welk, A.; Kuhn, W.; Schmitt, M.; Tschesche, H.; et al. Clinical Relevance of Matrix Metalloproteinase-13 Determined with a New Highly Specific and Sensitive ELISA in Ascitic Fluid of Advanced Ovarian Carcinoma Patients. *Biol. Chem.* **2003**, *384*, 1247–1251. [CrossRef] [PubMed]
179. Escalona, R.M.; Kannourakis, G.; Findlay, J.K.; Ahmed, N. Expression of TIMPs and MMPs in Ovarian Tumors, Ascites, Ascites-Derived Cells, and Cancer Cell Lines: Characteristic Modulatory Response Before and After Chemotherapy Treatment. *Front. Oncol.* **2022**, *11*, 796588. [CrossRef]
180. Escalona, R.M.; Bilandzic, M.; Western, P.; Kadife, E.; Kannourakis, G.; Findlay, J.K.; Ahmed, N. TIMP-2 regulates proliferation, invasion and STAT3-mediated cancer stem cell-dependent chemoresistance in ovarian cancer cells. *BMC Cancer* **2020**, *20*, 960. [CrossRef]
181. Davidson, B.; Reich, R.; Berner, A.; Givant-Horwitz, V.; Goldberg, I.; Risberg, B.; Kristensen, G.B.; Trope, C.G.; Bryne, M.; Kopolovic, J.; et al. Ovarian carcinoma cells in serous effusions show altered *MMP-2* and *TIMP-2* mRNA levels. *Eur. J. Cancer* **2001**, *37*, 2040–2049. [CrossRef]
182. Będkowska, G.E.; Piskór, B.; Gacuta, E.; Zajkowska, M.; Osada, J.; Szmitkowski, M.; Dąbrowska, M.; Ławicki, S. Diagnostic Power of Selected Cytokines, MMPs and TIMPs in Ovarian Cancer Patients–ROC Analysis. *Anticancer Res.* **2019**, *39*, 2575–2582. [CrossRef]
183. Stetler-Stevenson, W.G. Tissue Inhibitors of Metalloproteinases in Cell Signaling: Metalloproteinase-Independent Biological Activities. *Sci. Signal.* **2008**, *1*, re6. [CrossRef] [PubMed]
184. Brew, K.; Nagase, H. The tissue inhibitors of metalloproteinases (TIMPs): An ancient family with structural and functional diversity. *Biochim. Biophys. Acta BBA-Mol. Cell Res.* **2010**, *1803*, 55–71. [CrossRef]
185. Cruz-Munoz, W.; Khokha, R. The Role of Tissue Inhibitors of Metalloproteinases in Tumorigenesis and Metastasis. *Crit. Rev. Clin. Lab. Sci.* **2008**, *45*, 291–338. [CrossRef] [PubMed]
186. ASTEDT, B. Immunological identity of urokinase and ovarian carcinoma plasminogen activator released in tissue culture. *Eur. J Cancer* **1981**, *17*, 239–241. [CrossRef]
187. Kuhn, W.; Schmalfeldt, B.; Reuning, U.; Pache, L.; Berger, U.; Ulm, K.; Harbeck, N.; Späthe, K.; Dettmar, P.; Höfler, H.; et al. Prognostic significance of urokinase (uPA) and its inhibitor PAI-1 for survival in advanced ovarian carcinoma stage FIGO IIIc. *Br. J. Cancer* **1999**, *79*, 1746–1751. [CrossRef]
188. Cai, S.; Zhang, P.; Dong, S.; Li, L.; Cai, J.; Xu, M. Downregulation of SPINK13 Promotes Metastasis by Regulating uPA in Ovarian Cancer Cells. *Cell. Physiol. Biochem.* **2018**, *45*, 1061–1071. [CrossRef]
189. Chambers, S.K.; Ivins, C.M.; Carcangiu, M.L. Plasminogen activator inhibitor-1 is an independent poor prognostic factor for survival in advanced stage epithelial ovarian cancer patients. *Int. J. Cancer* **1998**, *79*, 449–454. [CrossRef]
190. Alberti, C.; Pinciroli, P.; Valeri, B.; Ferri, R.; Ditto, A.; Umezawa, K.; Sensi, M.; Canevari, S.; Tomassetti, A. Ligand-dependent EGFR activation induces the co-expression of IL-6 and PAI-1 via the NFkB pathway in advanced-stage epithelial ovarian cancer. *Oncogene* **2012**, *31*, 4139–4149. [CrossRef] [PubMed]
191. Dorn, J.; Harbeck, N.; Kates, R.; Gkazepis, A.; Scorilas, A.; Soosaipillai, A.; Diamandis, E.; Kiechle, M.; Schmalfeldt, B.; Schmitt, M. Impact of expression differences of kallikrein-related peptidases and of uPA and PAI-1 between primary tumor and omentum metastasis in advanced ovarian cancer. *Ann. Oncol.* **2011**, *22*, 877–883. [CrossRef]
192. Wang, L.; Madigan, M.C.; Chen, H.; Liu, F.; Patterson, K.I.; Beretov, J.; O'Brien, P.M.; Li, Y. Expression of urokinase plasminogen activator and its receptor in advanced epithelial ovarian cancer patients. *Gynecol. Oncol.* **2009**, *114*, 265–272. [CrossRef]
193. Harbeck, N.; Krüger, A.; Sinz, S.; Kates, R.E.; Thomssen, C.; Schmitt, M.; Jänicke, F. Clinical Relevance of the Plasminogen Activator Inhibitor Type1—A Multifaceted Proteolytic Factor. *Onkologie* **2001**, *24*, 238–244. [CrossRef]

194. van Dam, P.A.; Coelho, A.; Rolfo, C. Is there a role for urokinase-type plasminogen activator inhibitors as maintenance therapy in patients with ovarian cancer? *Eur. J. Surg. Oncol. EJSO* **2017**, *43*, 252–257. [CrossRef] [PubMed]
195. Kenny, H.A.; Leonhardt, P.; Ladanyi, A.; Yamada, S.D.; Montag, A.; Im, H.K.; Jagadeeswaran, S.; Shaw, D.E.; Mazar, A.P.; Lengyel, E. Targeting the Urokinase Plasminogen Activator Receptor Inhibits Ovarian Cancer Metastasis. *Clin. Cancer Res.* **2011**, *17*, 459–471. [CrossRef]
196. Nakatsuka, E.; Sawada, K.; Nakamura, K.; Yoshimura, A.; Kinose, Y.; Kodama, M.; Hashimoto, K.; Mabuchi, S.; Makino, H.; Morii, E.; et al. Plasminogen activator inhibitor-1 is an independent prognostic factor of ovarian cancer and IMD-4482, a novel plasminogen activator inhibitor-1 inhibitor, inhibits ovarian cancer peritoneal dissemination. *Oncotarget* **2017**, *8*, 89887–89902. [CrossRef]
197. Kelly, T.E.; Spillane, C.L.; Ward, M.P.; Hokamp, K.; Huang, Y.; Tewari, P.; Martin, C.M.; Norris, L.A.; Mohamed, B.M.; Bates, M.; et al. Plasminogen activator inhibitor 1 is associated with high-grade serous ovarian cancer metastasis and is reduced in patients who have received neoadjuvant chemotherapy. *Front. Cell Dev. Biol.* **2023**, *11*, 1150991. [CrossRef] [PubMed]
198. Seiffert, D.; Loskutoff, D.J. Type 1 plasminogen activator inhibitor induces multimerization of plasma vitronectin: A suggested mechanism for the generation of the tissue form of vitronectin in vivo. *J. Biol. Chem.* **1996**, *271*, 29644–29651. [CrossRef]
199. Seiffert, D.; Ciambrone, G.; Wagner, N.V.; Binder, B.R.; Loskutoff, D.J. The somatomedin B domain of vitronectin. Structural requirements for the binding and stabilization of active type 1 plasminogen activator inhibitor. *J. Biol. Chem.* **1994**, *269*, 2659–2666. [CrossRef]
200. Planus, E.; Barlovatz-Meimon, G.; Rogers, R.A.; Bonavaud, S.; Ingber, D.E.; Wang, N. Binding of urokinase to plasminogen activator inhibitor type-1 mediates cell adhesion and spreading. *J. Cell Sci.* **1997**, *110*, 1091–1098. [CrossRef] [PubMed]
201. De Lorenzi, V.; Sarra Ferraris, G.M.; Madsen, J.B.; Lupia, M.; Andreasen, P.A.; Sidenius, N. Urokinase links plasminogen activation and cell adhesion by cleavage of the RGD motif in vitronectin. *EMBO Rep.* **2016**, *17*, 982–998. [CrossRef]
202. Czekay, R.-P.; Wilkins-Port, C.E.; Higgins, S.P.; Freytag, J.; Overstreet, J.M.; Klein, R.M.; Higgins, C.E.; Samarakoon, R.; Higgins, P.J. PAI-1: An Integrator of Cell Signaling and Migration. *Int. J. Cell Biol.* **2011**, *2011*, 562481. [CrossRef]
203. McMahon, G.A.; Petitclerc, E.; Stefansson, S.; Smith, E.; Wong, M.K.K.; Westrick, R.J.; Ginsburg, D.; Brooks, P.C.; Lawrence, D.A. Plasminogen Activator Inhibitor-1 Regulates Tumor Growth and Angiogenesis. *J. Biol. Chem.* **2001**, *276*, 33964–33968. [CrossRef]
204. Andreasen, P.A.; Egelund, R.; Petersen, H.H. The plasminogen activation system in tumor growth, invasion, and metastasis. *Cell. Mol. Life Sci. CMLS* **2000**, *57*, 25–40. [CrossRef]
205. Cortese, K.; Sahores, M.; Madsen, C.D.; Tacchetti, C.; Blasi, F. Clathrin and LRP-1-Independent Constitutive Endocytosis and Recycling of uPAR. *PLoS ONE* **2008**, *3*, e3730. [CrossRef]
206. Placencio, V.R.; DeClerck, Y.A. Plasminogen Activator Inhibitor-1 in Cancer: Rationale and Insight for Future Therapeutic Testing. *Cancer Res.* **2015**, *75*, 2969–2974. [CrossRef]
207. Liu, J.; Guo, Q.; Chen, B.; Yu, Y.; Lu, H.; Li, Y.-Y. Cathepsin B and its interacting proteins, bikunin and TSRC1, correlate with TNF-induced apoptosis of ovarian cancer cells OV-90. *FEBS Lett.* **2006**, *580*, 245–250. [CrossRef]
208. Zhang, L.; Wei, L.; Shen, G.; He, B.; Gong, W.; Min, N.; Zhang, L.; Duan, Y.; Xie, J.; Luo, H.; et al. Cathepsin L is involved in proliferation and invasion of ovarian cancer cells. *Mol. Med. Rep.* **2015**, *11*, 468–474. [CrossRef]
209. Zhang, W.; Wang, S.; Wang, Q.; Yang, Z.; Pan, Z.; Li, L. Overexpression of cysteine cathepsin L is a marker of invasion and metastasis in ovarian cancer. *Oncol. Rep.* **2014**, *31*, 1334–1342. [CrossRef]
210. Kolwijck, E.; Massuger, L.F.A.G.; Thomas, C.M.G.; Span, P.N.; Krasovec, M.; Kos, J.; Sweep, F.C.G.J. Cathepsins B, L and cystatin C in cyst fluid of ovarian tumors. *J. Cancer Res. Clin. Oncol.* **2010**, *136*, 771–778. [CrossRef]
211. Gashenko, E.A.; Lebedeva, V.A.; Brak, I.V.; Tsykalenko, E.A.; Vinokurova, G.V.; Korolenko, T.A. Evaluation of serum procathepsin B, cystatin B and cystatin C as possible biomarkers of ovarian cancer. *Int. J. Circumpolar Health* **2013**, *72*, 21215. [CrossRef]
212. Nishikawa, H.; Ozaki, Y.; Nakanishi, T.; Blomgren, K.; Tada, T.; Arakawa, A.; Suzumori, K. The role of cathepsin B and cystatin C in the mechanisms of invasion by ovarian cancer. *Gynecol. Oncol.* **2004**, *92*, 881–886. [CrossRef]
213. Kos, J.; Werle, B.; Lah, T.; Brunner, N. Cysteine proteinases and their inhibitors in extracellular fluids: Markers for diagnosis and prognosis in cancer. *Int. J. Biol. Markers* **2000**, *15*, 84–89. [CrossRef]
214. Magister, S.; Kos, J. Cystatins in immune system. *J. Cancer* **2013**, *4*, 45–56. [CrossRef]
215. Jedeszko, C.; Sloane, B.F. Cysteine cathepsins in human cancer. *Biol. Chem.* **2004**, *385*, 1017–1027. [CrossRef]
216. Wallin, H.; Bjarnadottir, M.; Vogel, L.K.; Wassélius, J.; Ekström, U.; Abrahamson, M. Cystatins–Extra- and intracellular cysteine protease inhibitors: High-level secretion and uptake of cystatin C in human neuroblastoma cells. *Biochimie* **2010**, *92*, 1625–1634. [CrossRef]
217. Xu, H.; Ma, Y.; Zhang, Y.; Pan, Z.; Lu, Y.; Liu, P.; Lu, B. Identification of Cathepsin K in the Peritoneal Metastasis of Ovarian Carcinoma Using In-silico, Gene Expression Analysis. *J. Cancer* **2016**, *7*, 722–729. [CrossRef]
218. Lösch, A.; Schindl, M.; Kohlberger, P.; Lahodny, J.; Breitenecker, G.; Horvat, R.; Birner, P. Cathepsin D in ovarian cancer: Prognostic value and correlation with p53 expression and microvessel density. *Gynecol. Oncol.* **2004**, *92*, 545–552. [CrossRef]

219. Scambia, G.; Benedetti, P.; Ferrandina, G.; Battaglia, F.; Baiocchi, G.; Mancuso, S. Cathepsin D assay in ovarian cancer: Correlation with pathological features and receptors for oestrogen, progesterone and epidermal growth factor. *Br. J. Cancer* **1991**, *64*, 182–184. [CrossRef]
220. Pranjol, M.Z.I.; Gutowski, N.; Hannemann, M.; Whatmore, J. The Potential Role of the Proteases Cathepsin D and Cathepsin L in the Progression and Metastasis of Epithelial Ovarian Cancer. *Biomolecules* **2015**, *5*, 3260–3279. [CrossRef]
221. Pranjol, M.Z.I.; Gutowski, N.J.; Hannemann, M.; Whatmore, J.L. Cathepsin D non-proteolytically induces proliferation and migration in human omental microvascular endothelial cells via activation of the ERK1/2 and PI3K/AKT pathways. *Biochim. Biophys. Acta BBA-Mol. Cell Res.* **2018**, *1865*, 25–33. [CrossRef]
222. Seo, B.R.; Min, K.; Woo, S.M.; Choe, M.; Choi, K.S.; Lee, Y.-K.; Yoon, G.; Kwon, T.K. Inhibition of Cathepsin S Induces Mitochondrial ROS That Sensitizes TRAIL-Mediated Apoptosis Through p53-Mediated Downregulation of Bcl-2 and c-FLIP. *Antioxid. Redox Signal.* **2017**, *27*, 215–233. [CrossRef]
223. Tripathi, M.; Nandana, S.; Yamashita, H.; Ganesan, R.; Kirchhofer, D.; Quaranta, V. Laminin-332 is a substrate for hepsin, a protease associated with prostate cancer progression. *J. Biol. Chem.* **2008**, *283*, 30576–30584. [CrossRef] [PubMed]
224. Tanimoto, H.; Yan, Y.; Clarke, J.; Korourian, S.; Shigemasa, K.; Parmley, T.H.; Parham, G.P.; O'Brien, T.J. Hepsin, a Cell Surface Serine Protease Identified in Hepatoma Cells, Is Overexpressed in Ovarian Cancer. *Cancer Res.* **1997**, *57*, 2884–2887. [PubMed]
225. Rosewell, K.; Al-Alem, L.; Li, F.; Kelty, B.; Curry, T.E., Jr. Identification of Hepsin and Protein Disulfide Isomerase A3 as Targets of Gelatinolytic Action in Rat Ovarian Granulosa Cells During the Periovulatory Period[1]. *Biol. Reprod.* **2011**, *85*, 858–866. [CrossRef]
226. Duru, N.; Pawar, N.R.; Martin, E.W.; Buzza, M.S.; Conway, G.D.; Lapidus, R.G.; Liu, S.; Reader, J.; Rao, G.G.; Roque, D.M.; et al. Selective targeting of metastatic ovarian cancer using an engineered anthrax prodrug activated by membrane-anchored serine proteases. *Proc. Natl. Acad. Sci. USA* **2022**, *119*, e2201423119. [CrossRef]
227. Miao, J.; Mu, D.; Ergel, B.; Singavarapu, R.; Duan, Z.; Powers, S.; Oliva, E.; Orsulic, S. Hepsin colocalizes with desmosomes and induces progression of ovarian cancer in a mouse model. *Int. J. Cancer J. Int. Cancer* **2008**, *123*, 2041–2047. [CrossRef]
228. Lu, L.; Cole, A.; Huang, D.; Wang, Q.; Guo, Z.; Yang, W.; Lu, J. Clinical Significance of Hepsin and Underlying Signaling Pathways in Prostate Cancer. *Biomolecules* **2022**, *12*, 203. [CrossRef] [PubMed]
229. Herter, S.; Piper, D.E.; Aaron, W.; Gabriele, T.; Cutler, G.; Cao, P.; Bhatt, A.S.; Choe, Y.; Craik, C.S.; Walker, N.; et al. Hepatocyte growth factor is a preferred in vitro substrate for human hepsin, a membrane-anchored serine protease implicated in prostate and ovarian cancers. *Biochem. J.* **2005**, *390*, 125–136. [CrossRef]
230. Willbold, R.; Wirth, K.; Martini, T.; Sültmann, H.; Bolenz, C.; Wittig, R. Correction: Excess hepsin proteolytic activity limits oncogenic signaling and induces ER stress and autophagy in prostate cancer cells. *Cell Death Dis.* **2020**, *11*, 120. [CrossRef]
231. Kenny, H.A.; Lengyel, E. MMP-2 functions as an early response protein in ovarian cancer metastasis. *Cell Cycle* **2009**, *8*, 683–688. [CrossRef]
232. Morales-Vásquez, F.; Castillo-Sánchez, R.; Gómora, M.J.; Almaraz, M.Á.; Pedernera, E.; Pérez-Montiel, D.; Rendón, E.; López-Basave, H.N.; Román-Basaure, E.; Cuevas-Covarrubias, S.; et al. Expression of metalloproteinases MMP-2 and MMP-9 is associated to the presence of androgen receptor in epithelial ovarian tumors. *J. Ovarian Res.* **2020**, *13*, 86. [CrossRef]
233. Wu, X.; Li, H.; Kang, L.; Li, L.; Wang, W.; Shan, B. Activated Matrix Metalloproteinase-2—A Potential Marker of Prognosis for Epithelial Ovarian Cancer. *Gynecol. Oncol.* **2002**, *84*, 126–134. [CrossRef]
234. Zeng, L.; Qian, J.; Zhu, F.; Wu, F.; Zhao, H.; Zhu, H. The prognostic values of matrix metalloproteinases in ovarian cancer. *J. Int. Med. Res.* **2020**, *48*, 0300060519825983. [CrossRef]
235. Li, L.-N.; Zhou, X.; Gu, Y.; Yan, J. Prognostic Value of MMP-9 in Ovarian Cancer: A Meta-analysis. *Asian Pac. J. Cancer Prev.* **2013**, *14*, 4107–4113. [CrossRef]
236. Vos, M.C.; van der Wurff, A.A.M.; van Kuppevelt, T.H.; Massuger, L.F.A.G. The role of MMP-14 in ovarian cancer: A systematic review. *J. Ovarian Res.* **2021**, *14*, 101. [CrossRef]
237. Agarwal, A.; Tressel, S.L.; Kaimal, R.; Balla, M.; Lam, F.H.; Covic, L.; Kuliopulos, A. Identification of a Metalloprotease-Chemokine Signaling System in the Ovarian Cancer Microenvironment: Implications for Antiangiogenic Therapy. *Cancer Res.* **2010**, *70*, 5880–5890. [CrossRef]
238. Wang, F.; Fisher, J.; Fishman, D.A. MMP-1-PAR1 axis mediates LPA-induced epithelial ovarian cancer (EOC) invasion. *Gynecol. Oncol.* **2011**, *120*, 247–255. [CrossRef]
239. Wang, L.; Kong, B. Analysis of the Association of Matrix Metalloproteinase-1 Gene Promoter (rs1799750) Polymorphism and Risk of Ovarian Cancer. *Int. J. Gynecol. Cancer* **2015**, *25*, 961–967. [CrossRef]
240. Hobbs, C.; Huang, Z.; Murphy, S.; Berchuck, A. The function of matrix metalloproteinase 1 in ovarian cancer. *Gynecol. Oncol.* **2019**, *154*, 80. [CrossRef]
241. Mariya, T.; Hirohashi, Y.; Torigoe, T.; Tabuchi, Y.; Asano, T.; Saijo, H.; Kuroda, T.; Yasuda, K.; Mizuuchi, M.; Saito, T.; et al. Matrix metalloproteinase-10 regulates stemness of ovarian cancer stem-like cells by activation of canonical Wnt signaling and can be a target of chemotherapy-resistant ovarian cancer. *Oncotarget* **2016**, *7*, 26806–26822. [CrossRef]

242. Périgny, M.; Bairati, I.; Harvey, I.; Beauchemin, M.; Harel, F.; Plante, M.; Têtu, B. Role of Immunohistochemical Overexpression of Matrix Metalloproteinases MMP-2 and MMP-11 in the Prognosis of Death by Ovarian Cancer. *Am. J. Clin. Pathol.* **2008**, *129*, 226–231. [CrossRef]
243. Hakamy, S.; Assidi, M.; Jafri, M.A.; Nedjadi, T.; Alkhatabi, H.; Al-Qahtani, A.; Al-Maghrabi, J.; Sait, K.; Al-Qahtani, M.; Buhmeida, A.; et al. Assessment of prognostic value of tissue inhibitors of metalloproteinase 3 (TIMP3) protein in ovarian cancer. *Libyan J. Med.* **2021**, *16*, 1937866. [CrossRef]
244. Liu, M.C.P.; Choong, D.Y.H.; Hooi, C.S.F.; Williams, L.H.; Campbell, I.G. Genetic and epigenetic analysis of the *TIMP-3* gene in ovarian cancer. *Cancer Lett.* **2007**, *247*, 91–97. [CrossRef]
245. Cymbaluk-Płoska, A.; Chudecka-Głaz, A.; Pius-Sadowska, E.; Machaliński, B.; Menkiszak, J.; Sompolska-Rzechuła, A. Suitability assessment of baseline concentration of MMP3, TIMP3, HE4 and CA125 in the serum of patients with ovarian cancer. *J. Ovarian Res.* **2018**, *11*, 1. [CrossRef]
246. Fan, X.; Wang, C.; Song, X.; Liu, H.; Li, X.; Zhang, Y. Elevated Cathepsin K potentiates metastasis of epithelial ovarian cancer. *Histol. Histopathol.* **2018**, *33*, 673–680. [CrossRef]
247. Norman, R.J.; Dewailly, D.; Legro, R.S.; Hickey, T.E. Polycystic ovary syndrome. *Lancet* **2007**, *370*, 685–697. [CrossRef]
248. Pasquali, R.; Casimirri, F.; Vicennati, V. Weight control and its beneficial effect on fertility in women with obesity and polycystic ovary syndrome. *Hum. Reprod.* **1997**, *12* (Suppl. 1), 82–87. [CrossRef]
249. Carmassi, F.; De Negri, F.; Fioriti, R.; De Giorgi, A.; Giannarelli, C.; Fruzzetti, F.; Pedrinelli, R.; Dell'Omo, G.; Bersi, C. Insulin resistance causes impaired vasodilation and hypofibrinolysis in young women with polycystic ovary syndrome. *Thromb. Res.* **2005**, *116*, 207–214. [CrossRef]
250. Mannerås-Holm, L.; Baghaei, F.; Holm, G.; Janson, P.O.; Ohlsson, C.; Lönn, M.; Stener-Victorin, E. Coagulation and fibrinolytic disturbances in women with polycystic ovary syndrome. *J. Clin. Endocrinol. Metab.* **2011**, *96*, 1068–1076. [CrossRef]
251. Burchall, G.F.; Piva, T.J.; Linden, M.D.; Gibson-Helm, M.E.; Ranasinha, S.; Teede, H.J. Comprehensive Assessment of the Hemostatic System in Polycystic Ovarian Syndrome. *Semin. Thromb. Hemost.* **2015**, *42*, 55–62. [CrossRef]
252. Ranjbaran, J.; Farimani, M.; Tavilani, H.; Ghorbani, M.; Karimi, J.; Poormonsefi, F.; Khodadadi, I. Matrix metalloproteinases 2 and 9 and MMP9/NGAL complex activity in women with PCOS. *Reproduction* **2016**, *151*, 305–311. [CrossRef]
253. Lewandowski, K.C.; Komorowski, J.; O'Callaghan, C.J.; Tan, B.K.; Chen, J.; Prelevic, G.M.; Randeva, H.S. Increased Circulating Levels of Matrix Metalloproteinase-2 and -9 in Women with the Polycystic Ovary Syndrome. *J. Clin. Endocrinol. Metab.* **2006**, *91*, 1173–1177. [CrossRef] [PubMed]
254. Butler, A.E.; Nandakumar, M.; Sathyapalan, T.; Brennan, E.; Atkin, S.L. Matrix Metalloproteinases, Tissue Inhibitors of Metalloproteinases, and Their Ratios in Women with Polycystic Ovary Syndrome and Healthy Controls. *Int. J. Mol. Sci.* **2025**, *26*, 321. [CrossRef] [PubMed]
255. Lahav-Baratz, S.; Kraiem, Z.; Shiloh, H.; Koifman, M.; Ishai, D.; Dirnfeld, M. Decreased expression of tissue inhibitor of matrix metalloproteinases in follicular fluid from women with polycystic ovaries compared with normally ovulating patients undergoing in vitro fertilization. *Fertil. Steril.* **2003**, *79*, 567–571. [CrossRef] [PubMed]
256. Li, W.; Zhang, Y.; Li, F.; Shi, Y.; Wang, Y. Expression of Vascular Endothelial Growth Factor, Matrix Metalloproteinase-2 and Matrix Metalloproteinase-9 in Polycystic Ovarian Syndrome Rats and Its Implication. *J. Biomater. Tissue Eng.* **2021**, *11*, 841–846. Available online: https://www.ingentaconnect.com/contentone/asp/jbte/2021/00000011/00000005/art00006 (accessed on 22 April 2025). [CrossRef]
257. Atiomo, W.U.; Hilton, D.; Fox, R.; Lee, D.; Shaw, S.; Friend, J.; Wilkin, T.J.; Prentice, A.G. Immunohistochemical detection of plasminogen activator inhibitor-1 in polycystic ovaries. *Gynecol. Endocrinol.* **2000**, *14*, 162–168. [CrossRef]
258. Atiomo, W.U.; Bates, S.A.; Condon, J.E.; Shaw, S.; West, J.H.; Prentice, A.G. The plasminogen activator system in women with polycystic ovary syndrome. *Fertil. Steril.* **1998**, *69*, 236–241. [CrossRef]
259. Tarkun, I.; Cantürk, Z.; Arslan, B.C.; Türemen, E.; Tarkun, P. The plasminogen activator system in young and lean women with polycystic ovary syndrome. *Endocr. J.* **2004**, *51*, 467–472. [CrossRef]
260. Sahay, S.; Jain, M.; Dash, D.; Choubey, L.; Jain, S.; Singh, T.B. Role of plasminogen activator inhibitor type 1 (PAI-1) in PCOS patient. *Int. J. Reprod. Contracept. Obstet. Gynecol.* **2017**, *6*, 4052–4058. [CrossRef]
261. Burchall, G.F.; Pouniotis, D.S.; Teede, H.J.; Ranasinha, S.; Walters, K.A.; Piva, T.J. Expression of the plasminogen system in the physiological mouse ovary and in the pathological polycystic ovary syndrome (PCOS) state. *Reprod. Biol. Endocrinol. RBE* **2019**, *17*, 33. [CrossRef]
262. Ibáñez, L.; Aulesa, C.; Potau, N.; Ong, K.; Dunger, D.B.; De Zegher, F. Plasminogen Activator Inhibitor-1 in Girls with Precocious Pubarche: A Premenarcheal Marker for Polycystic Ovary Syndrome? *Pediatr. Res.* **2002**, *51*, 244–248. [CrossRef]
263. Devin, J.K.; Johnson, J.E.; Eren, M.; Gleaves, L.A.; Bradham, W.S.; Bloodworth, J.R.; Vaughan, D.E. Transgenic overexpression of plasminogen activator inhibitor-1 promotes the development of polycystic ovarian changes in female mice. *J. Mol. Endocrinol.* **2007**, *39*, 9–16. [CrossRef]

264. Shah, A.K.; Yadav, B.K.; Shah, A.K.; Suri, A.; Deo, S.K. Cardiovascular Risk Predictors High Sensitivity C-Reactive Protein and Plasminogen Activator Inhibitor-1 in Women with Lean Phenotype of Polycystic Ovarian Syndrome: A Prospective Case-Control Study. *J. Lab. Physicians* **2022**, *15*, 31–37. [CrossRef] [PubMed]
265. Ambekar, A.S.; Kelkar, D.S.; Pinto, S.M.; Sharma, R.; Hinduja, I.; Zaveri, K.; Pandey, A.; Prasad, T.S.K.; Gowda, H.; Mukherjee, S. Proteomics of Follicular Fluid From Women With Polycystic Ovary Syndrome Suggests Molecular Defects in Follicular Development. *J. Clin. Endocrinol. Metab.* **2015**, *100*, 744–753. [CrossRef] [PubMed]
266. Diamanti-Kandarakis, E.; Palioniko, G.; Alexandraki, K.; Bergiele, A.; Koutsouba, T.; Bartzis, M. The prevalence of 4G5G polymorphism of plasminogen activator inhibitor-1 (PAI-1) gene in polycystic ovarian syndrome and its association with plasma PAI-1 levels. *Eur. J. Endocrinol.* **2004**, *150*, 793–798. [CrossRef]
267. Orio, F.; Palomba, S.; Cascella, T.; Tauchmanová, L.; Nardo, L.G.; Di Biase, S.; Labella, D.; Russo, T.; Tolino, A.; Zullo, F.; et al. Is plasminogen activator inhibitor-1 a cardiovascular risk factor in young women with polycystic ovary syndrome? *Reprod. Biomed. Online* **2004**, *9*, 505–510. [CrossRef]
268. Ma, L.-J.; Mao, S.-L.; Taylor, K.L.; Kanjanabuch, T.; Guan, Y.; Zhang, Y.; Brown, N.J.; Swift, L.L.; McGuinness, O.P.; Wasserman, D.H. Prevention of obesity and insulin resistance in mice lacking plasminogen activator inhibitor 1. *Diabetes* **2004**, *53*, 336–346. Available online: https://diabetesjournals.org/diabetes/article-abstract/53/2/336/11495 (accessed on 22 April 2025). [CrossRef] [PubMed]
269. Wang, L.; Chen, L.; Liu, Z.; Liu, Y.; Luo, M.; Chen, N.; Deng, X.; Luo, Y.; He, J.; Zhang, L.; et al. PAI-1 Exacerbates White Adipose Tissue Dysfunction and Metabolic Dysregulation in High Fat Diet-Induced Obesity. *Front. Pharmacol.* **2018**, *9*, 1087. [CrossRef]
270. Obiezu, C.V.; Scorilas, A.; Magklara, A.; Thornton, M.H.; Wang, C.Y.; Stanczyk, F.Z.; Diamandis, E.P. Prostate-Specific Antigen and Human Glandular Kallikrein 2 Are Markedly Elevated in Urine of Patients with Polycystic Ovary Syndrome. *J. Clin. Endocrinol. Metab.* **2001**, *86*, 1558–1561. [CrossRef]
271. Jin, M.; Cai, J.; Hu, Y.; Lu, X.; Huang, H. Cathepsin D expression in ovaries from polycystic ovarian syndrome patients. *Zhejiang Xue Xue Bao Yi Xue Ban J. Zhejiang Univ. Med. Sci.* **2007**, *36*, 429–432. [CrossRef]
272. Dawood Salman, Z.; Ismaeel Jassim, W. Role of Cathepsin S and Insulin Resistance in Iraqi Women with Polycystic Ovary Syndrome. *J. Obstet. Gynecol. Cancer Res.* **2024**, *9*, 379–384. [CrossRef]
273. Sabeti Akbar-Abad, M.; Majidpour, M.; Sargazi, S.; Ghasemi, M.; Saravani, R. Unraveling the Role of Cathepsin B Variants in Polycystic Ovary Syndrome: Insights from a Case-Control Study and Computational Analyses. *Reprod. Sci. Thousand Oaks Calif* **2025**, *32*, 1166–1179. [CrossRef] [PubMed]
274. Shalev, E.; Goldman, S.; Ben-Shlomo, I. The balance between MMP-9 and MMP-2 and their tissue inhibitor (TIMP)-1 in luteinized granulosa cells: Comparison between women with PCOS and normal ovulatory women. *Mol. Hum. Reprod.* **2001**, *7*, 325–331. [CrossRef] [PubMed]
275. Kelly, C.J.G.; Lyall, H.; Petrie, J.R.; Gould, G.W.; Connell, J.M.C.; Rumley, A.; Lowe, G.D.O.; Sattar, N. A Specific Elevation in Tissue Plasminogen Activator Antigen in Women with Polycystic Ovarian Syndrome. *J. Clin. Endocrinol. Metab.* **2002**, *87*, 3287–3290. [CrossRef] [PubMed]
276. Pelosi, E.; Simonsick, E.; Forabosco, A.; Garcia-Ortiz, J.E.; Schlessinger, D. Dynamics of the Ovarian Reserve and Impact of Genetic and Epidemiological Factors on Age of Menopause. *Biol. Reprod.* **2015**, *92*, 130. [CrossRef]
277. Beck-Peccoz, P.; Persani, L. Premature ovarian failure. *Orphanet J. Rare Dis.* **2006**, *1*, 9. [CrossRef]
278. Yin, R.; Zhang, C.; Gen, A.; Li, Y.; Yang, H.; Tian, X.; Chi, Y. Propofol affects the biological behavior of ovarian cancer SKOV3 cells via ERK1/2-MMP-2/9 signaling pathway. *Trop. J. Pharm. Res.* **2020**, *19*, 233–238. [CrossRef]
279. An, H.J.; Ahn, E.H.; Kim, J.O.; Park, H.S.; Ryu, C.S.; Cho, S.H.; Kim, J.H.; Lee, W.S.; Kim, N.K. Association between tissue inhibitor of metalloproteinase (*TIMP*) genetic polymorphisms and primary ovarian insufficiency (POI). *Maturitas* **2019**, *120*, 77–82. [CrossRef]
280. Hsueh, A.J.; Billig, H.; Tsafriri, A. Ovarian follicle atresia: A hormonally controlled apoptotic process. *Endocr. Rev.* **1994**, *15*, 707–724. [CrossRef]

cells

MDPI

Review

Oxidative Stress and the NLRP3 Inflammasome: Focus on Female Fertility and Reproductive Health

Efthalia Moustakli [1,*], Sofoklis Stavros [2], Periklis Katopodis [1], Charikleia Skentou [3], Anastasios Potiris [2], Periklis Panagopoulos [2], Ekaterini Domali [4], Ioannis Arkoulis [2], Theodoros Karampitsakos [2], Eleftheria Sarafi [5], Theologos M. Michaelidis [5], Athanasios Zachariou [6] and Athanasios Zikopoulos [2]

1 Laboratory of Medical Genetics, Faculty of Medicine, School of Health Sciences, University of Ioannina, 45110 Ioannina, Greece; katopodisper@gmail.com
2 Third Department of Obstetrics and Gynecology, University General Hospital "ATTIKON", Medical School, National and Kapodistrian University of Athens, 12462 Athens, Greece; sfstavrou@med.uoa.gr (S.S.); apotiris@med.uoa.gr (A.P.); perpanag@med.uoa.gr (P.P.); garkoylis@hotmail.com (I.A.); theokarampitsakos@hotmail.com (T.K.); thanzik92@gmail.com (A.Z.)
3 Department of Obstetrics and Gynecology, Medical School of Ioannina, University General Hospital, 45110 Ioannina, Greece; haraskentou@gmail.com
4 First Department of Obstetrics and Gynecology, Alexandra Hospital, Medical School, National and Kapodistrian University of Athens, 11528 Athens, Greece; kdomali@yahoo.fr
5 Department of Biological Applications & Technology, School of Health Sciences, University of Ioannina, 45110 Ioannina, Greece; sarafi.elef@gmail.com (E.S.); tmichael@uoi.gr (T.M.M.)
6 Department of Urology, School of Medicine, University of Ioannina, 45110 Ioannina, Greece; zahariou@otenet.gr
* Correspondence: thaleia.moustakli@gmail.com

Abstract: Chronic inflammation is increasingly recognized as a critical factor in female reproductive health; influencing natural conception and the outcomes of assisted reproductive technologies such as in vitro fertilization (IVF). An essential component of innate immunity, the NLR family pyrin domain-containing 3 (NLRP3) inflammasome is one of the major mediators of inflammatory responses, and its activation is closely linked to oxidative stress. This interaction contributes to a decline in oocyte quality, reduced fertilization potential, and impaired embryo development. In the ovarian milieu, oxidative stress and NLRP3 inflammasome activation interact intricately, and their combined effects on oocyte competence and reproductive outcomes are significant. The aims of this review are to examine these molecular mechanisms and to explore therapeutic strategies targeting oxidative stress and NLRP3 inflammasome activity, with the goal of enhancing female fertility and improving clinical outcomes in reproductive health.

Keywords: inflammation pathways; fertilization potential; redox balance; ovarian aging; mitochondria dysfunction; cytokine signaling

1. Introduction

Female fertility is a complicated biological process that is controlled by the ovarian microenvironment, hormones, and cellular communication. A number of important activities, including follicle formation, oocyte maturation, and conception, depend on the ovarian microenvironment [1]. During the ovulation and luteinization stages, stress responses and cellular signaling must be properly coordinated since a balance between these elements is necessary for the best possible reproductive health [2].

Conversely, fertility may be influenced by illnesses that upset this equilibrium. Worldwide, 10–15% of couples experience infertility, with female infertility accounting for about

50% of instances. Polycystic ovarian syndrome (PCOS), endometriosis, age-related infertility, and infertility that cannot be explained are the main causes [3].

According to new research, oxidative stress and peripheral inflammatory responses may play a role in the emergence of these reproductive diseases and the ensuing infertility in women. More medical studies are necessary since the pathobiology of these processes is intricate and linked [4]. Fertility is greatly impacted by inflammation, a physiological process necessary for ovulation, implantation, and reproduction. For instance, a conditioned inflammatory response promotes cell division and muscle regeneration in embryos. However, there is a clear correlation between uterine malfunction and infertility and chronic low-grade inflammation, which is characterized by persistent immunological activation [5]. Premature ovarian insufficiency (POI), endometriosis, and PCOS are among the conditions that showcase the negative consequences of persistent inflammation [6].

Systemic inflammation in PCOS inhibits folliculogenesis and ovulation by exacerbating insulin resistance and hyperandrogenism. Ovarian reserves are depleted and ovarian aging is accelerated by autoimmune mechanisms and increased pro-inflammatory cytokines in POI, whereas chronic inflammatory lesions linked to endometriosis impair uterine function [7]. Oocyte quality, fertility, and the results of in vitro fertilization (IVF) and other assisted reproductive technologies (ART) are all adversely impacted by these inflammatory alterations. Targeted treatments are necessary to enhance female reproductive health and ART success rates since chronic inflammation and infertility are intertwined [8,9].

One important regulator of inflammation and innate immune responses is the pyrin domain-containing 3 (NLRP3) inflammasome, which belongs to the NLR family. Pro-inflammatory cytokines including interleukin-18 (IL-18) and interleukin-1β (IL-1β) are produced and released when caspase-1 activates this multiprotein complex [10,11]. Although it is useful for identifying tissue injury or inflammation, the NLRP3 inflammasome is also linked to persistent inflammation in the reproductive system.

The activation of the NLRP3 inflammasome is primarily triggered by oxidative stress. An imbalance between the generation of reactive oxygen species (ROS) and the ability of antioxidant mechanisms to neutralize them is the cause of this disorder [12]. ROS serve as signaling molecules in the ovarian microenvironment that promote oocyte maturation during follicular development and ovulation. DNA fragmentation, mitochondrial malfunction, and cellular damage are the results of high ROS generation [13].

Various variables, including age, environmental pollutants, and metabolic changes, can exacerbate these negative effects. In addition to its direct effect on oocyte quality, oxidative stress has been demonstrated to trigger an inflammatory cascade that lowers egg viability and impairs ovarian function by activating the NLRP3 inflammasome in granulosa and theca cells. Poor results may result from this feedback loop between oxidative stress and activation of the inflammasome during IVF [14,15].

This review highlights the combined effects of oxidative stress and NLRP3 inflammasome activation on female fertility by offering a thorough summary of recent studies in this area. To improve women's reproductive health and optimize perinatal outcomes, it is important to highlight the possibility of new treatment targets that could lessen these problems.

2. Oxidative Stress and Female Reproductive Health

2.1. *Definition and Biological Mechanisms of Oxidative Stress*

The body experiences oxidative stress when the generation of ROS surpasses the antioxidant defenses' ability to neutralize them. The majority of ROS are produced naturally as metabolic byproducts by the mitochondria's oxidative phosphorylation mechanism.

These comprise free radicals such as superoxide ions (O_2^-) and hydroxyl radicals (OH·), as well as non-radical species such as hydrogen peroxide (H_2O_2) [16].

Oxidative stress refers to an imbalance between the production of ROS and the body's ability to neutralize them with antioxidants, which can lead to cellular damage and contribute to a variety of diseases, including those affecting the reproductive system. Several important biomarkers that can have an adverse influence on fertility and reproductive outcomes are linked to oxidative stress in the context of female reproductive health. A crucial indicator is nitrosative stress, which is brought on by the generation of reactive nitrogen species (RNS). The signaling molecule nitric oxide (NO) can cause RNS, particularly peroxynitrite, which can harm cellular components and obstruct reproductive functions. It has been demonstrated that excessive NO radicals degrade egg quality, hinder ovarian function, and have a detrimental effect on fertilization and embryo development [17].

Lipid peroxidation is another crucial indicator of oxidative stress that occurs when ROS target the polyunsaturated fatty acids in cellular membranes, producing compounds such as malondialdehyde (MDA) and 4-hydroxy-2-nonenal (HNE). These lipid peroxidation byproducts have the ability to negatively impact biological components, including oocyte membranes, which lowers the viability and fertilization potential of the cells [18]. Overall fertility, embryo development, and oocyte quality have all been found to decline with lipid peroxidation. Similarly, the oxidation of proteins is another hallmark of oxidative stress. Vital biological functions may be disrupted by oxidized proteins, which arise from ROS attacking cellular proteins and causing structural and functional alterations. Infertility can be caused by oxidized proteins in the reproductive system, which can disrupt hormone communication, change the maturation of eggs, and obstruct the implantation of embryos [4].

Antioxidant enzymes including glutathione peroxidase (GPx), catalase (CAT), and superoxide dismutase (SOD) are essential to the body's natural defense against oxidative stress. By neutralizing ROS, these enzymes preserve homeostasis and guard against cellular harm. However, oxidative stress can overwhelm the reproductive system and cause malfunction if ROS levels are too high or antioxidant defense is insufficient [19]. Oxidative stress places granulosa cells and oocytes in the follicular milieu at particular risk. Their low ability to repair DNA and high metabolic activity make them especially vulnerable to damage by ROS. Their general reproductive health, as well as their function, is compromised by this oxidative damage [20] (Table 1). Fertility and pregnancy outcomes can be significantly impacted by a decrease in antioxidant enzyme activity or an excess of ROS.

Table 1. The table describes the mechanisms and effects on reproductive health of the extrinsic and intrinsic variables that lead to oxidative stress in the ovaries.

Factor	Description	Impact
Mitochondrial dysfunction	Ovarian cells' mitochondria, including granulosa cells and oocytes, are the main generators of ROS. Dysfunction inhibits ATP synthesis and increases ROS generation.	Induces mitochondrial deficiencies, oxidative stress buildup, and decreased oocyte quality.
Environmental toxins	Pollutant exposure affects the redox balance of cells, including pesticides, heavy metals, and substances that disrupt hormones.	Reduces antioxidants or directly produces ROS, which increases oxidative stress in ovarian tissues and jeopardizes reproductive capacity.

Table 1. *Cont.*

Factor	Description	Impact
Inflammatory cytokines	ROS generation is stimulated by cytokines such as TNF-α and IL-6, which are elevated in conditions such as PCOS and endometriosis.	Contributes to follicular dysfunction and decreased oocyte competence by raising oxidative stress in ovarian cells and follicular fluid.
Aging	Advanced maternal age increases the generation of ROS in oocytes while decreasing antioxidant efficiency (e.g., SOD and GPx).	Oocyte quality and ovarian reserves are decreased, which leads to infertility and poor ART results.
Lifestyle factors	High-fat diets, heavy alcohol use, and smoking all weaken antioxidant defenses or release exogenous ROS.	Increases oxidative stress and impairs ovarian tissues, especially in vulnerable ovarian cells.

2.2. *Effects of Oxidative Stress on Fertility*

Oxidative stress has a major impact on many areas of female reproductive health. Oocyte quality is one crucial region impacted because oxidative damage hinders both nuclear and cytoplasmic maturation. Damage from ROS can cause double-strand breaks and point mutations in DNA, which lowers the chance of pregnancy [4]. Oocyte metabolism and developmental potential are further jeopardized by mitochondrial dysfunction, which is typified by reduced ATP production and mitochondrial membrane potential.

Additionally, follicular fluid, which is essential for promoting oocyte development and maturation, changes in composition due to oxidative stress [19]. Poorer oocyte retrieval during IVF cycles, higher lipid peroxidation, and decreased antioxidant capacity are all linked to elevated ROS levels in this microenvironment. Granulosa cells, which are crucial for oocyte development, are especially vulnerable to oxidative stress [20,21]. Important growth factors that are essential for follicular development, such as vascular endothelial growth factor (VEGF) and insulin-like growth factor (IGF-1), are not secreted by dysfunctional granulosa cells [22].

Oxidative stress interferes with implantation and early embryonic development in addition to oocyte maturation. Damaged gametes produce zygotes with poor developmental potential, and ROS-induced inflammation in the uterine lining lowers implantation success rates. Oxidative stress also contributes significantly to ovarian aging and premature ovarian insufficiency (POI) [4,17]. Chronic oxidative stress speeds up ovarian aging by causing oocytes to undergo apoptosis and decreasing the follicular pool. Ovarian reserve decreases early in POI due to oxidative stress, which is frequently associated with autoimmune and mitochondrial abnormalities [13,14] (Table 2).

ROS may impair cells, resulting in issues such as inadequate follicular fluid, decreased egg quality, impaired cell function, and difficulty with embryo implantation. Primary ovarian insufficiency (POI) and other disorders can be exacerbated by oxidative stress, which can also accelerate ovarian aging. Inflammation, issues with cell energy production, and DNA damage are the main causes of these effects.

Table 2. This table illustrates several facets of reproductive health that are impacted by oxidative stress.

Aspect of Reproductive Health	Impact of Oxidative Stress	Mechanism
Oocyte quality	Compromised nuclear and cytoplasmic maturation, leading to reduced fertilization potential.	DNA damage (double-strand breaks, mutations) and mitochondrial dysfunction (reduced ATP production and membrane potential).
Follicular fluid composition	Poorer oocyte retrieval during IVF cycles and reduced support for oocyte maturation.	Elevated ROS levels, reduced antioxidant capacity, and increased lipid peroxidation.
Granulosa cell function	Impaired secretion of growth factors critical for follicle development.	Damage to granulosa cells decreases VEGF and IGF-1 secretion, reducing metabolic and structural support for oocytes.
Implantation and embryonic development	Reduced chances of implantation and compromised embryonic potential.	Oxidative damage to gametes and ROS-mediated inflammation in the uterine lining impair zygote development and implantation success.
Ovarian Aging and POI	Accelerated ovarian aging, follicle pool depletion, and early ovarian reserve loss in POI.	Chronic oxidative damage promotes apoptosis in oocytes. Linked to autoimmunity and mitochondrial dysfunction in POI cases.

2.3. Clinical Proof of the Association Between Female Infertility and Oxidative Stress

Numerous studies have demonstrated the negative impact of oxidative stress on female fertility. In women undergoing in vitro fertilization (IVF), lower rates of oocyte retrieval, fertilization, and implantation are linked to elevated levels of oxidative stress indicators in serum or follicular fluid. Furthermore, high concentrations of NO radicals have been linked to impaired ovarian function, reduced egg quality, and poor fertility outcomes. In patients with polycystic ovarian syndrome (PCOS), elevated ROS are also associated with insulin resistance, hyperandrogenism, and persistent anovulation, all of which have a detrimental effect on fertility [23]. Similarly, in endometriosis patients, decreased implantation success rates and inferior oocyte quality are associated with higher levels of oxidative stress in the peritoneal fluid. Clinical assessments of oxidative stress markers, including lipid peroxidation levels and antioxidant enzyme activities, are increasingly used as diagnostic tools to evaluate reproductive health. These biomarkers are valuable in understanding the underlying causes of infertility and may provide insights into potential therapeutic interventions. According to these results, oxidative stress actively contributes to the pathophysiology of infertility rather than just acting as a passive spectator [21,24].

3. The NLRP3 Inflammasome in Reproductive Health

3.1. Overview of the NLRP3 Inflammasome

The NLRP3 inflammasome is one protein that belongs to the Nod-like receptor (NLR) family. It is vital for the innate immune response because it is involved in reacting to damage-associated molecular patterns (DAMPs) or pathogen-associated molecular patterns (PAMPs) under cellular stress [25].

Several crucial procedures are involved in the assembly of the NLRP3 inflammasome. As a sensor in this process, activation of the NLRP3 protein encourages the creation of inflammasomes. NLRP3 is connected to downstream signaling molecules by the adaptor protein ASC (apoptosis-associated speck-like protein containing a CARD) [11]. One important effector, caspase-1, promotes the proteolytic activation of pro-inflammatory cytokines such as interleukin-1β (IL-1β) and interleukin-18 (IL-18), and it causes pyroptosis, a type of programmed cell death [26]. The NLRP3 inflammasome increases the inflammatory response by promoting the maturation of IL-1β and IL-18. Chronic activation or dysregulation of the inflammasome can result in disorders linked to inflammation, including those that affect reproductive health. However, since it encourages tissue healing and offers infection prevention, regulated activation is beneficial [27].

3.2. Mechanisms of NLRP3 Inflammasome Activation

The NLRP3 inflammasome is activated by a variety of signals that fall into a few broad categories. Bacterial toxins, viral RNA, and lipopolysaccharides (LPS) are examples of microbial stimuli that activate Toll-like receptor (TLR)-mediated pathways. Inflammasomes may be activated by increasing advanced glycation end products (AGEs), cholesterol crystals, and excessive glucose levels, all of which are factors in metabolic stress [11,28].

Oxidative stress, which is defined by an excess of ROS, mitochondrial damage, and disruptions in cellular homeostasis, directly triggers the NLRP3 inflammasome. Moreover, endocrine disruptors and other environmental contaminants are linked to elevated inflammasome activity [29]. Prior to NLRP3 inflammasome activation, critical events include lysosomal instability, mitochondrial malfunction, and ionic imbalances. NLRP3 inflammasome activation has been thought to be frequently triggered by a decrease in intracellular K^+. Furthermore, K^+ efflux is not necessary for the alternative NLRP3 inflammasome pathway, but it is necessary for NLRP3 activation in the caspsase-11-mediated non-canonical inflammasome pathway. Additionally, 1L-1β maturation is promoted by the activation of Ca^{2+}-independent phospholipase A2 brought on by potassium efflux [11]. Heavy metals and air pollution play a major role in inflammasome activation in the oocyte microenvironment. These factors often coexist with oxidative stress, perpetuating the inflammatory cycle [30] (Table 3).

Table 3. This table summarizes the various stimuli that can cause the NLRP3 inflammasome to become active.

Stimuli	Description
Microbial stimuli	Lipopolysaccharides (LPS), bacterial toxins, and viral RNA activate NLRP3 through TLR-mediated pathways.
Metabolic stress	Elevated glucose, cholesterol crystals, and advanced glycation end products (AGEs) trigger inflammasome activation.
Oxidative stress	Excess ROS, mitochondrial damage, and disrupted cellular homeostasis directly activate NLRP3.
Environmental toxins	Heavy metals, air pollutants, and endocrine disruptors are linked to increased inflammasome activity

NLRP3 is activated by microbial stimuli, such as viruses and bacteria, via immunological pathways. Environmental contaminants, metabolic stress, and oxidative stress from excessive ROS and mitochondrial damage all activate NLRP3 and are associated with increased inflammasome activity.

3.3. Expression and Role of the NLRP3 Inflammasome in the Reproductive System

There is growing recognition of the role of the NLRP3 inflammasome in controlling female fertility and ovarian function. The NLRP3 inflammasome is expressed by important ovarian cell types, such as granulosa cells, theca cells, and oocytes, which cause inflammatory reactions in both healthy and diseased circumstances [12].

Normal reproductive functions, including uterine remodeling, cytokine release, and local immune cell activation, depend on the NLRP3 inflammasome. A viral outbreak-like phenomenon is frequently used to describe its activation. The NLRP3 inflammasome in granulosa cells temporarily activates cytokines such as interleukin-1β (IL-1β), which are necessary for follicle rupture and oocyte release [25].

Numerous congenital conditions have been connected to inflammasome malfunction at infection sites. Chronic activation in polycystic ovarian syndrome (PCOS) is linked to increased granulosa cell production of NLRP3 inflammasome components. In addition to impairing follicle formation and ovarian maturation, this aberrant inflammasome activation in ovarian and endometrial tissue exacerbates hyperandrogenism. In addition, it results in inflammation and oxidative damage to the endometrium, which lowers oocyte quality and reserves [31].

Additionally, in primary ovarian insufficiency (POI), chronic activation of the NLRP3 inflammasome speeds up follicular atresia and apoptosis, resulting in early ovarian aging. According to these findings, the NLRP3 inflammasome may be a novel therapeutic target for the management of infertility and may be linked to the age-dependent decline in female fertility [32].

3.4. Interplay Between the NLRP3 Inflammasome and Expression of Oxidative Stress and the Role of the NLRP3 Inflammasome in the Reproductive System

Activation of the NLRP3 inflammasome and oxidative stress are closely related feedback loops that worsen ovarian inflammation. NLRP3 is a major factor in the buildup of inflammasomes because ROS harm mitochondria and upset the ionic balance in cells. ROS quickly expose inflammasome components by activating upstream pathways such as NF-κB [11].

Once engaged, the NLRP3 inflammasome increases ROS production, damages mitochondrial function, and erodes the immune system. The structure and function of the placenta and its supporting cells are compromised by this exposure cycle, which exacerbates uterine inflammation [33]. In circumstances such as in vitro fertilization (IVF), when ovarian hormone suppression raises oxidative stress levels, this interaction is especially important [34] (Table 4).

Table 4. Reproductive health is impacted by chronic NLRP3 activation, as this table illustrates.

Aspect of Reproductive Health	Impact of Chronic NLRP3 Activation
Oocyte Quality and Competence	Persistent inflammasome activation damages oocyte mitochondria, DNA integrity, and cytoplasmic maturation, leading to reduced fertilization potential. Elevated levels of IL-1β and IL-18 in follicular fluid have been correlated with lower IVF success rates in patients with poor ovarian response.
Follicular Development	Cytokine release mediated by the inflammasome disrupts granulosa cell function, impairing hormone synthesis and normal follicular development. Excessive inflammation accelerates follicular atresia and reduces ovarian reserve.
Embryo Development and Implantation	Oocytes from inflammasome-dysregulated environments tend to produce embryos with compromised developmental competence. Additionally, uterine inflammation driven by NLRP3 activity can hinder endometrial receptivity, lowering implantation rates in IVF.

Lower ovarian reserves result from the disruption of follicular development. Prolonged inflammation can damage the uterus and oocyte quality, which can impair embryo development and implantation.

3.5. Therapeutic Targeting of the NLRP3 Inflammasome

The important function of the NLRP3 inflammasome in reproductive disorders makes it a promising therapeutic target to increase female fertility. Potential intervention techniques include cytokine blocking, lifestyle changes, anti-inflammatory medications, and inflammasome inhibitors [32]. Due to their encouraging results in preclinical investigations of inflammatory illnesses, small compounds such as MCC950, which selectively suppress NLRP3 activation, are being studied for their potential applications in reproductive health [35]. It has been determined that NLRP3 interacts with melatonin, CoQ10, and N-acetylcysteine (NAC).

Antioxidants assist in lowering oxidative stress by blocking these interactions [36]. Furthermore, anti-IL-1β and anti-IL-18 antibodies may lessen the negative effects on inflammasome function, which could enhance the results of assisted reproductive technologies such as IVF and restore ovarian function. Inflammasome overactivity can be decreased by lowering oxidative stress and systemic inflammation by lifestyle modifications such as regular exercise, a nutritious diet, and quitting smoking [37].

4. Intercoupling of Oxidative Stress and the NLRP3 Inflammasome

The oocyte microenvironment is greatly impacted by the intricate and reciprocal interaction that exists between oxidative stress and the NLRP3 inflammasome [36]. NLRP3 inflammasome activation is largely triggered by oxidative stress, and inflammasome activity also affects mitochondrial function. Through the increased generation of ROS, this cyclic interaction raises oxidative stress, encourages chronic inflammation, and causes cell malfunction, which ultimately results in issues with reproductive health [36,38].

Excess ROS produced by oxidative stress triggers the NLRP3 inflammasome, leading to ionic imbalances, lysosomal instability, and mitochondrial damage. The inflammasome interprets these situations as warning signs of danger. NLRP3, pro-caspase-1, pro-IL-1β, and pro-IL-18 are among the inflammasome components that are expressed as a result of ROS-stimulating redox-sensitive pathways such as NF-κB [12]. Oxidative stress is a key trigger in the activation of the inflammasome, and antioxidants have been demonstrated to prevent this process by lowering ROS levels.

Conversely, NLRP3 inflammasome activation on its own causes mitochondrial malfunction, which raises ROS production even more and prolongs oxidative stress [38]. Pyroptosis is a type of programmed cell death brought on by caspase-1 activity and inflammasome accumulation. ATP and mitochondrial DNA are two more danger signals produced by this process, which worsen oxidative stress and inflammation in nearby cells [39]. This interaction is especially harmful in the context of oocytes, where inflammatory and redox activities are vital and need to be balanced [40].

NLRP3 inflammasome activation and oxidative stress are coupled through a number of overlapping molecular pathways. A major contributing factor is mitochondrial malfunction, since damaged mitochondria emit metabolites such as ROS, mitochondrial DNA, and cardiolipin [41]. The NLRP3 inflammasome is directly stimulated by these compounds. Mitochondrial damage is therefore further exacerbated by decreased oxidative phosphorylation and elevated mitochondrial permeability [42].

This connection is also mediated by redox-sensitive pathways, specifically the NF-κB signaling cascade. Increased ROS levels cause IκB kinase to become active, which phosphorylates and degrades IκB and encourages NF-κB nuclear translocation [43]. The transcription of genes linked to inflammasomes is improved by this process. Important triggers for NLRP3 activation also include oxidative stress-induced disruptions in ion flux, including potassium efflux, calcium influx, and sodium imbalance. Ion channels are immediately impacted by ROS, which intensifies these ion imbalances [44]. Under oxidative stress, thioredoxin-interacting protein (TXNIP) separates from thioredoxin, forming another important mechanism. This connection is further enhanced by the binding and activation of NLRP3 by free TXNIP [45].

These closely related processes show how vulnerable the oxidative stress–NLRP3 inflammasome axis is, especially in the ovarian microenvironment, where disturbed inflammatory and redox balance might have serious consequences for the reproductive health of females [13].

4.1. Role of the Oxidative Stress–NLRP3 Axis in Ovarian Dysfunction

The interaction between oxidative stress and NLRP3 inflammasome activation has significant pathogenic implications in various ovarian diseases [46]. In polycystic ovary syndrome (PCOS), elevated levels of ROS in ovarian tissue and increased expression of NLRP3 components lead to chronic low-grade inflammation. This inflammation contributes to reproductive complications in women with PCOS, as it inhibits oocyte maturation and steroid hormone production [47]. In endometriosis, increased oxidative stress and NLRP3 inflammasome activation aggravate inflammatory lesions, resulting in tissue damage, adhesions, and infertility [48].

Follicle apoptosis is increased by oxidative stress and inflammasome activation, which has an effect on fetal aging as well [14]. Mitochondrial dysfunction is especially risky for older women since it exacerbates oxidative damage and inflammasome activation [49]. Ultimately, oxidative stress and NLRP3 activation have a significant impact on the results of in vitro fertilization (IVF), since they are associated with poor oocyte quality, decreased pregnancy success, and delayed embryonic development [50] (Table 5).

Table 5. This table illustrates the effects of NLRP3 activation and oxidative stress on different reproductive situations.

Condition	Role of Oxidative Stress–NLRP3 Interplay
Polycystic Ovary Syndrome	Elevated ROS and NLRP3 activation lead to chronic inflammation, disrupted steroidogenesis, and impaired oocyte maturation.
Endometriosis	Heightened oxidative stress and NLRP3 activation perpetuate tissue damage, adhesions, and infertility.
Age-related ovarian aging	Accelerates apoptosis and depletion of ovarian follicles, with mitochondrial dysfunction worsening oxidative damage.
In Vitro Fertilization (IVF)	Increased oxidative stress and NLRP3 in follicular fluid correlate with poor oocyte quality and reduced embryo competence.

In polycystic ovarian syndrome, they lead to oocyte maturation impairment, hormonal imbalance, and inflammation. Oxidative damage speeds up cell death and decreases ovarian follicles in age-related ovarian aging.

4.2. Therapeutic Implications in Targeting the Oxidative Stress–NLRP3 Axis

Reproductive results may be enhanced by addressing the connection between oxidative stress and the NLRP3 inflammasome. Antioxidants, lifestyle changes, combination therapy, and direct inflammasome inhibitors are examples of potential treatment approaches [51].

Antioxidants are at the forefront of such strategies, offering dual benefits of reducing oxidative damage and mitigating inflammasome activation. Melatonin, a powerful endogenous antioxidant, stabilizes mitochondrial function, reduces ROS generation, and directly inhibits NLRP3 activation. Melatonin reduces ROS formation and inhibits NLRP3 activation during IVF cycles, improving the quality of the oocytes. It stabilizes the mitochondria and has strong antioxidant properties [52]. Clinical studies indicate that melatonin supplementation during IVF cycles enhances oocyte quality, fertilization rates, and embryo development. Similarly, coenzyme Q10 (CoQ10) enhances mitochondrial efficiency, lowers oxidative damage, and indirectly attenuates NLRP3 activity.

N-acetylcysteine (NAC), a precursor of glutathione, further strengthens cellular defenses against ROS while reducing inflammatory cytokines within the follicular fluid. A healthier ovarian microenvironment is promoted by this dual effect, which raises IVF success rates overall and improves oocyte quality. In addition to their beneficial impact on ROS, vitamins C and E also prevent oxidative damage to biological components, opening up new treatment options.

As promising treatments for reproductive health, direct inflammasome inhibitors such as MCC950 have demonstrated effectiveness in preclinical models of inflammation by directly blocking NLRP3 activation [53]. Changes in lifestyle, such as consistent exercise, antioxidant-rich foods, and stress reduction, can lower inflammation and systemic oxidative stress, which in turn lowers NLRP3 activation and supports reproductive health [54].

Combination treatments that simultaneously target oxidative stress and inflammatory pathways may have further advantages. Given that both systems are less effective in treating diseases such as endometriosis and PCOS, this method may be very useful in treating these conditions and offering a holistic therapeutic approach [5,55].

5. Impact on In Vitro Fertilization (IVF) Outcomes

5.1. Overview of IVF and the Role of the Ovarian Microenvironment

IVF is a crucial component of assisted reproductive technology (ART) that involves controlled sperm stimulation, egg harvesting, insemination, and embryo transfer. However, a number of obstacles, such as problems with oocyte quality, oocyte retrieval, and embryo development, restrict the success rates [4,21,50].

The oocyte microenvironment has a significant impact on the quality of oocytes recovered during an IVF cycle. A delicate balance between inflammation and oxidative stress is necessary for the best follicle growth and oocyte maturation. IVF results can be greatly impacted by dysregulation of this balance, which is typified by elevated oxidative stress and persistent inflammation brought on by NLRP3 inflammasome activation [21].

Controlled ovarian hyperstimulation (COH) is one of the IVF techniques that naturally increases oxidative stress in the oocyte microenvironment [21]. There are multiple reasons behind this rise in oxidative stress. Gonadotropin doses given during COH stimulate hormones, which causes the ovaries' ROS levels to rise noticeably [4]. Oxidative stress is further exacerbated by the stimulated ovarian follicular fluid's frequent high concentrations of ROS and inflammatory cytokines. Additionally, prolonged in vitro culture conditions might harm cells oxidatively, reducing their capacity to proliferate [50].

There are a number of adverse consequences linked to elevated oxidative stress during IVF cycles [21]. Lower oocyte quality results from ROS-induced damage to proteins, lipids, and mitochondrial DNA (mtDNA), which hinders oocyte development and survival. Pregnancy rates are decreased as a result of this oxidative damage, which also impacts the fragile cellular machinery required for effective reproduction. Furthermore, chromosomal abnormalities and significant developmental delays are more likely to occur in embryos formed from oxidatively stressed oocytes, which lowers the likelihood of a successful pregnancy and degrades the quality of the embryo overall [50].

5.2. NLRP3 Inflammasome and IVF Outcomes

Chronic NLRP3 inflammasome activation during the IVF cycle worsens inflammation and oxidative damage in the oocyte microenvironment, which can have negative effects. The reproductive system's essential components are disturbed by this overactivation [21].

First, it has a detrimental effect on follicle growth and oocyte quality. Granulosa cells are unable to sustain healthy follicle growth when the inflammasome is overactivated. Interleukin-1β (IL-1β) and interleukin-18 (IL-18), two inflammatory cytokines, are released into the follicular fluid and are linked to low oocyte quality and a decreased rate of oocyte maturation [4,50]. Oocyte apoptosis is caused by prolonged inflammasome activation, which exacerbates this problem. It further reduces the capacity for oocyte development by inducing mitochondrial malfunction [50].

Additionally, there are negative impacts on the development of the embryo and pregnancy. Sperm abnormalities result from oocyte fertilization that fails to acquire appropriate developmental competence when regulated inflammasome activity is present. In addition to causing embryonic fragmentation and reducing blastocyst development, elevated levels of inflammatory cytokines and ROS in the follicular fluid also degrade sperm quality [56,57].

Lastly, there are significant effects on implantation and embryo retrieval. The uterus's capacity to sustain implantation is compromised by inflammation mediated by inflammasomes, raising the possibility of implantation failure. Higher intraovarian IL-1β levels are significantly associated with lower implantation success during ART cycles, which further limits the total effectiveness of IVF procedures [21,50].

5.3. Clinical Data Connecting the NLRP3 Inflammasome and Oxidative Stress to IVF Results

Recent clinical research shows that oxidative stress and inflammasome activation have a major effect on IVF results. Poor oocyte and ovarian function are linked to elevated levels of ROS in the follicular fluid as well as pro-inflammatory cytokines such as interleukin-1β (IL-1β) and interleukin-18 (IL-18) [58]. The success of IVF may be predicted by these biomarkers. Additionally, granulosa cells from individuals with insufficient or recurrent ovarian reserve failure have been found to express higher NLRP3 inflammasome components, indicating a clear correlation between inflammasome activity and poor reproductive outcomes [59].

Age is also a significant factor since older IVF patients have greater levels of oxidative stress and inflammasome activation, which results in age-related declines in fertility and poorer ART success rates. These findings underscore the particular challenges faced by high-risk or elderly patient populations undergoing IVF [60]. This demonstrates how oxidative stress and inflammasome function can be monitored and altered as therapeutic approaches. IVF success rates may be increased by using elevated oxidative stress and inflammasome activity as indicators and therapeutic targets, allowing for more individualized and efficient care for these patients [50,61].

5.4. Therapeutic Approaches to Mitigate the Impact

Since oxidative stress and NLRP3 inflammasome activation have been demonstrated to adversely impact IVF outcomes, a number of treatments have been studied to mitigate these factors [12]. Antioxidants are a prominent approach to improving success rates. For instance, melatonin, known for its mitochondrial protective properties, has been demonstrated to enhance oocyte quality and lessen oxidative damage to follicular fluid during IVF cycles [62]. Similarly, coenzyme Q10 improves oocyte responsiveness and sperm morphology by enhancing mitochondrial function in mature oocytes. Vitamins E and C scavenge ROS within the oocyte microenvironment, preventing oxidative damage during follicle development [63].

NLRP3 inflammasome inhibitors are a novel and developing strategy. By directly suppressing the NLRP3 inflammasome, compounds such as MCC950 have demonstrated potential in preclinical models. However, their application in reproductive treatment is still being investigated. Because they can lower inflammasome activity, these substances may help with inflammatory IVF failures [64].

Changes in lifestyle can also have a significant impact on lowering systemic oxidative stress and enhancing reproductive results. Oxidative stress can be reduced and overall fertility can be improved by reducing stress, stopping smoking, and eating a diet high in antioxidants [65].

Optimizing IVF protocols during controlled ovarian hyperstimulation (COH) offers another avenue to address oxidative stress. Oxidative damage may be reduced by lowering gonadotropin dosages and customizing stimulation regimens for each patient [66]. Additionally, growth-promoting conditions, such as low-oxygen environments, can shield developing cells from oxidative stress [67].

Lastly, anti-inflammatory medications that target inflammation mediated by inflammasomes represent a complementary approach [68]. Patients with inflammation-related IVF difficulties may benefit from cytokine-neutralizing antibodies or IL-1β blockers, which may enhance embryo receptivity and boost fertilization success [69]. Both oxidative stress and inflammation are addressed by these systems, which combine to improve IVF results in a number of ways.

5.5. Future Directions in Research and Clinical Practice

Even though we now know more about the relationship between oxidative stress, NLRP3 inflammasome activation, and IVF outcomes, further study is still required to improve clinical applicability and success rates. The effects of oxidative stress on follicular fluid and granulosa cells, as well as the control of inflammasome activity, are of significant interest. To show their potential in lowering inflammasome-induced IVF failure, more research is required to assess the safety and effectiveness of inflammasome inhibitors in ART settings in addition to creating trustworthy biomarkers that may be used as indicators of IVF success [14].

Targeting the distinct oxidative and inflammatory profiles of each patient through customized treatment plans is another important field of research. These programs can enhance outcomes by customizing interventions to meet the needs of each individual. However, it is crucial to investigate the long-term impacts of these methods on pregnancy outcomes and the health of the children to guarantee that they are safe and effective in the larger context of reproductive and developmental health [70].

6. Therapeutic Interventions and Future Directions

6.1. NLRP3 Inflammasome Inhibitors

Targeting the NLRP3 inflammasome directly is one method that appears to have potential for lowering intrauterine inflammation and maybe improving IVF outcomes. One important strategy is the application of MCC950, a small-molecule inhibitor that preserves upstream priming pathways while specifically inhibiting the activity of the NLRP3 inflammasome [36]. Preclinical investigations have demonstrated that MCC950 effectively inhibits the release of interleukin-1β (IL-1β) and minimizes oxidative damage in inflammatory situations, making it a promising candidate for fertility treatment [71].

Another intriguing option is the use of IL-1β inhibitors, such as neutralizing antibodies and vaccinations. By activating IL-1β, the NLRP3 inflammasome causes inflammation. IL-1β inhibitors have the potential to improve uterine receptivity and decrease chronic inflammation in IVF patients, which would raise the likelihood of a successful implantation. These inhibitors have the potential to enhance IVF results, despite their primary research being in autoimmune illnesses [72].

Additionally, ROS-modulating therapies offer an indirect way to prevent the activation of the NLRP3 inflammasome. MitoQ and other mitochondria-targeted antioxidants lower ROS to help stop oxidative stress and the activation of the inflammasome. Combining these processes could lead to new developments in fertility treatments by preventing reproductive problems and improving assisted reproductive technology [36] (Table 6).

Table 6. This table illustrates how lifestyle changes can reduce oxidative stress.

Lifestyle Changes	Description
Dietary modifications	Systemic oxidative stress is reduced when antioxidant-rich foods, such as fruits, vegetables, nuts, and seeds, are consumed. NLRP3 activity is modulated by polyphenols such as curcumin and resveratrol.
Exercise	Inflammation is decreased and mitochondrial efficiency is increased with regular moderate exercise. Excessive activity can exacerbate oxidative stress, highlighting the need for balance.
Stress reduction	Inflammation and oxidative damage are exacerbated by ongoing psychological stress. Improved IVF results have been linked to practices such as yoga and mindfulness.

Consuming foods rich in polyphenols and antioxidants helps modulate NLRP3 and lower stress. Regular exercise enhances vitality and reduces inflammation, though excessive exercise may increase oxidative or physical stress.

6.2. Innovative Strategies and Future Directions in Reproductive Health

Promising opportunities to enhance IVF results and reproductive health are presented by novel treatments that target the oxidative stress–NLRP3 inflammasome axis [36]. NLRP3 or related inflammasome genes in reproductive organs can be silenced by gene therapy and RNA interference using instruments such as CRISPR-Cas9 [73]. Additionally, inflammasome components can be targeted in reproductive therapy using antisense oligonucleotides and small interfering RNAs (siRNAs), which have already been produced for other disorders [74]. Further gaining popularity is mitochondrial replacement treatment, which involves putting mitochondria into healthy embryos or oocytes. For patients with mitochondrial dysfunction, this strategy shows potential. More sophisticated treatments that target the mitochondria, including SS-31 peptides, are being researched to improve the quality of oocytes and embryos, lessen oxidative damage, and restore mitochondrial function [75].

Another developing subject is microbial regulation, since studies indicate that vaginal and gut bacteria affect inflammation and systemic oxidative stress. According to reproductive health, probiotics and prebiotics may offer novel ways to control the oxidative–inflammatory axis, making fertility higher. Antioxidants and ROS-scavenging substances are also being added to IVF culture media to lessen oxidative stress during in vitro embryo development. This will increase embryo viability, development efficiency, and IVF success rates [76].

Many questions still surround the oxidative stress–NLRP3 inflammasome axis in reproductive health, despite tremendous advancements in our understanding of this relationship. In follicular fluid, granulosa cells, and uterine tissue, future studies should concentrate on creating trustworthy indicators of oxidative stress and inflammasome activity [14,21,32].

Additionally, non-invasive indicators in blood or urine may improve the tracking of inflammatory and systemic oxidative diseases in IVF patients, allowing for more individualized therapy planning. Tailored treatment customized based on individual oxidative stress and inflammation profiles is key [77]. Machine learning algorithms and other AI-driven tools can evaluate these profiles to predict disease risk and enhance individualized treatment regimens, which helps to improve disease prevention and control. To assess oxidative stress levels, for instance, AI-based predictive models examine biomarkers, and machine learning tailors interventions by combining clinical and molecular data. By facilitating early diagnosis, customized therapy, and improved monitoring in reproductive health, these strategies enhance therapeutic outcomes [78]. More precise and successful interventions may result from the use of AI-driven analytics to integrate genetic, metabolic, and lifestyle aspects.

Examining these treatments' long-term impacts is also crucial, especially the potential effects of oxidative stress and inflammasome modification during IVF on long-term health [79]. Lastly, the combination of transcriptomics, proteomics, metabolomics, and genomics in multi-omics techniques holds great promise for the discovery of new therapeutic targets. These methods increase our comprehension of fertility procedures and results, which eventually results in more potent therapies [80].

7. Conclusions

Oxidative stress and inflammation have a significant effect on female reproductive health, particularly when the NLRP3 inflammasome is activated. One aspect of ovarian activity is oxidative stress, which impairs embryonic growth, fertilization potential, and egg quality. Chronic activation of NLRP3 causes inflammation, which adversely affects the outcome of IVF and biological reproduction. Follicle production, oocyte maturation, and embryo implantation are all impacted by the vicious cycles of ovarian and uterine cell damage caused by oxidative stress and inflammasome activation. Since infertile patients often have elevated levels of oxidative stress markers and cytokines associated with the inflammasome, targeted treatments are crucial. Antioxidants that reduce oxidative damage, such as melatonin, coenzyme Q10, and N-acetylcysteine, can enhance IVF outcomes. Among the possible anti-inflammatory medications are interleukin-1β inhibitors and MCC950. Additionally crucial are lifestyle changes and tailored treatment plans based on each person's particular oxidative and inflammatory profile. Further research is needed to identify biomarkers and develop targeted treatments. Microbiological research, drugs that target the mitochondria, and the development of gene therapy could all significantly improve reproductive results. Improving women's fertility and IVF success requires addressing these variables, which will pave the way for more suitable and efficient reproductive care.

Author Contributions: Conceptualization, E.M. and A.Z. (Athanasios Zikopoulos).; methodology, S.S. and P.K.; validation, A.P., P.K., E.S. and A.Z. (Athanasios Zachariou).; Investigation, I.A. and T.K.; writing—original draft preparation, E.M. and P.K.; writing—review and editing, C.S., A.P., S.S., P.P., E.D., I.A., T.K., E.S., T.M.M., A.Z. (Athanasios Zachariou) and A.Z. (Athanasios Zikopoulos); visualization, E.M. and T.M.M.; project administration, A.Z. (Athanasios Zikopoulos); supervision, A.Z. (Athanasios Zikopoulos). All authors have read and agreed to the published version of the manuscript.

Funding: This research received no external funding.

Data Availability Statement: Not applicable.

Conflicts of Interest: The authors declare no conflicts of interest.

References

1. Hu, B.; Wang, R.; Wu, D.; Long, R.; Ruan, J.; Jin, L.; Ma, D.; Sun, C.; Liao, S. Prospects for fertility preservation: The ovarian organ function reconstruction techniques for oogenesis, growth and maturation in vitro. *Front. Physiol.* **2023**, *14*, 1177443. [CrossRef] [PubMed]
2. Shea, L.D.; Woodruff, T.K.; Shikanov, A. Bioengineering the ovarian follicle microenvironment. *Annu. Rev. Biomed. Eng.* **2014**, *16*, 29–52. [CrossRef]
3. Su, Z.; Ding, P.; Su, W.; Li, X.; Li, Y.; Li, X.; Lao, K.; Wang, Y. Association between oxidative balance score and female infertility from the national health and nutrition examination survey 2013–2018. *Front. Endocrinol.* **2024**, *15*, 1386021. [CrossRef] [PubMed]
4. Agarwal, A.; Aponte-Mellado, A.; Premkumar, B.J.; Shaman, A.; Gupta, S. The effects of oxidative stress on female reproduction: A review. *Reprod. Biol. Endocrinol.* **2012**, *10*, 49. [CrossRef] [PubMed]
5. Kicińska, A.M.; Maksym, R.B.; Zabielska-Kaczorowska, M.A.; Stachowska, A.; Babińska, A. Immunological and Metabolic Causes of Infertility in Polycystic Ovary Syndrome. *Biomedicines* **2023**, *11*, 1567. [CrossRef]
6. Kirshenbaum, M.; Orvieto, R. Premature ovarian insufficiency (POI) and autoimmunity-an update appraisal. *J. Assist. Reprod. Genet.* **2019**, *36*, 2207–2215. [CrossRef] [PubMed]
7. Yang, Z.; Tang, Z.; Cao, X.; Xie, Q.; Hu, C.; Zhong, Z.; Tan, J.; Zheng, Y. Controlling chronic low-grade inflammation to improve follicle development and survival. *Am. J. Reprod. Immunol.* **2020**, *84*, e13265. [CrossRef] [PubMed]
8. Macer, M.L.; Taylor, H.S. Endometriosis and Infertility. *Obstet. Gynecol. Clin. North. Am.* **2012**, *39*, 535–549. [CrossRef]
9. Potiris, A.; Perros, P.; Drakaki, E.; Mavrogianni, D.; Machairiotis, N.; Sfakianakis, A.; Karampitsakos, T.; Vrachnis, D.; Antonakopoulos, N.; Panagopoulos, P.; et al. Investigating the Association of Assisted Reproduction Techniques and Adverse Perinatal Outcomes. *J. Clin. Med.* **2024**, *13*, 328. [CrossRef]
10. Pirzada, R.H.; Javaid, N.; Choi, S. The Roles of the NLRP3 Inflammasome in Neurodegenerative and Metabolic Diseases and in Relevant Advanced Therapeutic Interventions. *Genes* **2020**, *11*, 131. [CrossRef] [PubMed]
11. Kelley, N.; Jeltema, D.; Duan, Y.; He, Y. The NLRP3 Inflammasome: An Overview of Mechanisms of Activation and Regulation. *Int. J. Mol. Sci.* **2019**, *20*, 3328. [CrossRef]
12. Yao, J.; Sterling, K.; Wang, Z.; Zhang, Y.; Song, W. The role of inflammasomes in human diseases and their potential as therapeutic targets. *Signal Transduct. Target. Ther.* **2024**, *9*, 10. [CrossRef]
13. Shi, Y.-Q.; Zhu, X.-T.; Zhang, S.-N.; Ma, Y.-F.; Han, Y.-H.; Jiang, Y.; Zhang, Y.-H. Premature ovarian insufficiency: A review on the role of oxidative stress and the application of antioxidants. *Front. Endocrinol.* **2023**, *14*, 1172481. [CrossRef] [PubMed]
14. Yan, F.; Zhao, Q.; Li, Y.; Zheng, Z.; Kong, X.; Shu, C.; Liu, Y.; Shi, Y. The role of oxidative stress in ovarian aging: A review. *J. Ovarian Res.* **2022**, *15*, 100. [CrossRef] [PubMed]
15. Yang, L.; Chen, Y.; Liu, Y.; Xing, Y.; Miao, C.; Zhao, Y.; Chang, X.; Zhang, Q. The Role of Oxidative Stress and Natural Antioxidants in Ovarian Aging. *Front. Pharmacol.* **2020**, *11*, 617843. [CrossRef] [PubMed]
16. Jomova, K.; Raptova, R.; Alomar, S.Y.; Alwasel, S.H.; Nepovimova, E.; Kuca, K.; Valko, M. Reactive oxygen species, toxicity, oxidative stress, and antioxidants: Chronic diseases and aging. *Arch. Toxicol.* **2023**, *97*, 2499–2574. [CrossRef]
17. Agarwal, A.; Gupta, S.; Sharma, R.K. Role of oxidative stress in female reproduction. *Reprod. Biol. Endocrinol. RBE* **2005**, *3*, 28. [CrossRef] [PubMed]
18. Kaltsas, A.; Zikopoulos, A.; Moustakli, E.; Zachariou, A.; Tsirka, G.; Tsiampali, C.; Dimitriadis, F. The Silent Threat to Women's Fertility: Uncovering the Devastating Effects of Oxidative Stress. *Antioxidants* **2023**, *12*, 1490. [CrossRef]
19. Baldini, G.M.; Lot, D.; Malvasi, A.; Laganà, A.S.; Vimercati, A.; Dellino, M.; Cicinelli, E.; Baldini, D.; Trojano, G. Abnormalities of Oocyte Maturation: Mechanisms and Implications. *Int. J. Mol. Sci.* **2024**, *25*, 12197. [CrossRef]
20. Agarwal, A.; Durairajanayagam, D.; Du Plessis, S.S. Utility of antioxidants during assisted reproductive techniques: An evidence based review. *Reprod. Biol. Endocrinol.* **2014**, *12*, 112. [CrossRef]
21. Chen, Y.; Yang, J.; Zhang, L. The Impact of Follicular Fluid Oxidative Stress Levels on the Outcomes of Assisted Reproductive Therapy. *Antioxidants* **2023**, *12*, 2117. [CrossRef]
22. Evron, A.; Blumenfeld, Z.; Adashi, E.Y.; Kol, S. The Role of Growth Factors in Ovarian Function and Development. *Glob. Libr. Women's Med.* **2015**. Available online: https://www.glowm.com/section-view/item/287 (accessed on 26 December 2024). [CrossRef]
23. Moustakli, E.; Zikopoulos, A.; Skentou, C.; Katopodis, P.; Domali, E.; Potiris, A.; Stavros, S.; Zachariou, A. Impact of Reductive Stress on Human Infertility: Underlying Mechanisms and Perspectives. *Int. J. Mol. Sci.* **2024**, *25*, 11802. [CrossRef] [PubMed]
24. Ruder, E.H.; Hartman, T.J.; Blumberg, J.; Goldman, M.B. Oxidative stress and antioxidants: Exposure and impact on female fertility. *Hum. Reprod. Update* **2008**, *14*, 345–357. [CrossRef] [PubMed]
25. Chen, Y.; Ye, X.; Escames, G.; Lei, W.; Zhang, X.; Li, M.; Jing, T.; Yao, Y.; Qiu, Z.; Wang, Z.; et al. The NLRP3 inflammasome: Contributions to inflammation-related diseases. *Cell Mol. Biol. Lett.* **2023**, *28*, 51. [CrossRef] [PubMed]

26. Molla, M.D.; Akalu, Y.; Geto, Z.; Dagnew, B.; Ayelign, B.; Shibabaw, T. Role of Caspase-1 in the Pathogenesis of Inflammatory-Associated Chronic Noncommunicable Diseases. *J. Inflamm. Res.* **2020**, *13*, 749–764. [CrossRef]
27. Blevins, H.M.; Xu, Y.; Biby, S.; Zhang, S. The NLRP3 Inflammasome Pathway: A Review of Mechanisms and Inhibitors for the Treatment of Inflammatory Diseases. *Front. Aging Neurosci.* **2022**, *14*, 879021. [CrossRef]
28. Yue, Z.; Zhang, X.; Gu, Y.; Liu, Y.; Lan, L.-M.; Liu, Y.; Li, Y.; Yang, G.; Wan, P.; Chen, X. Regulation and functions of the NLRP3 inflammasome in RNA virus infection. *Front. Cell Infect. Microbiol.* **2024**, *13*, 1309128. [CrossRef]
29. Kowalczyk, P.; Sulejczak, D.; Kleczkowska, P.; Bukowska-Ośko, I.; Kucia, M.; Popiel, M.; Wietrak, E.; Kramkowski, K.; Wrzosek, K.; Kaczyńska, K. Mitochondrial Oxidative Stress-A Causative Factor and Therapeutic Target in Many Diseases. *Int. J. Mol. Sci.* **2021**, *22*, 13384. [CrossRef] [PubMed]
30. Bachmann, M.C.; Bellalta, S.; Basoalto, R.; Gómez-Valenzuela, F.; Jalil, Y.; Lépez, M.; Matamoros, A.; von Bernhardi, R. The Challenge by Multiple Environmental and Biological Factors Induce Inflammation in Aging: Their Role in the Promotion of Chronic Disease. *Front. Immunol.* **2020**, *11*, 570083. [CrossRef] [PubMed]
31. Banerjee, S.; Cooney, L.G.; Stanic, A.K. Immune Dysfunction in Polycystic Ovary Syndrome. *ImmunoHorizons* **2023**, *7*, 323–332. [CrossRef]
32. Navarro-Pando, J.M.; Alcocer-Gómez, E.; Castejón-Vega, B.; Navarro-Villarán, E.; Condés-Hervás, M.; Mundi-Roldan, M.; Cordero, M.D. Inhibition of the NLRP3 inflammasome prevents ovarian aging. *Sci. Adv.* **2021**, *7*, eabc7409. [CrossRef] [PubMed]
33. Heid, M.E.; Keyel, P.A.; Kamga, C.; Shiva, S.; Watkins, S.C.; Salter, R.D. Mitochondrial reactive oxygen species induces NLRP3-dependent lysosomal damage and inflammasome activation. *J. Immunol. Baltim. Md.* **2013**, *191*, 5230–5238. [CrossRef] [PubMed]
34. Ra, K.; Park, S.C.; Lee, B.C. Female Reproductive Aging and Oxidative Stress: Mesenchymal Stem Cell Conditioned Medium as a Promising Antioxidant. *Int. J. Mol. Sci.* **2023**, *24*, 5053. [CrossRef] [PubMed]
35. Li, H.; Guan, Y.; Liang, B.; Ding, P.; Hou, X.; Wei, W.; Ma, Y. Therapeutic potential of MCC950, a specific inhibitor of NLRP3 inflammasome. *Eur. J. Pharmacol.* **2022**, *928*, 175091. [CrossRef] [PubMed]
36. Zhao, S.; Chen, F.; Yin, Q.; Wang, D.; Han, W.; Zhang, Y. Reactive Oxygen Species Interact With NLRP3 Inflammasomes and Are Involved in the Inflammation of Sepsis: From Mechanism to Treatment of Progression. *Front. Physiol.* **2020**, *11*, 571810. [CrossRef]
37. Artimovič, P.; Badovská, Z.; Toporcerová, S.; Špaková, I.; Smolko, L.; Sabolová, G.; Rabajdová, M. Oxidative Stress and the Nrf2/PPARγ Axis in the Endometrium: Insights into Female Fertility. *Cells* **2024**, *13*, 1081. [CrossRef]
38. Abais, J.M.; Xia, M.; Zhang, Y.; Boini, K.M.; Li, P.L. Redox Regulation of NLRP3 Inflammasomes: ROS as Trigger or Effector? *Antioxid Redox Signal* **2015**, *22*, 1111–1129. [CrossRef]
39. Wei, Y.; Yang, L.; Pandeya, A.; Cui, J.; Zhang, Y.; Li, Z. Pyroptosis-Induced Inflammation and Tissue Damage. *J. Mol. Biol.* **2022**, *434*, 167301. [CrossRef] [PubMed]
40. Combelles, C.M.; Gupta, S.; Agarwal, A. Could oxidative stress influence the in-vitro maturation of oocytes? *Reprod. Biomed. Online* **2009**, *18*, 864–880. [CrossRef]
41. Zhao, J.; Li, J.; Li, G.; Chen, M. The role of mitochondria-associated membranes mediated ROS on NLRP3 inflammasome in cardiovascular diseases. *Front. Cardiovasc. Med.* **2022**, *9*, 1059576. [CrossRef]
42. Swanson, K.V.; Deng, M.; Ting, J.P.Y. The NLRP3 inflammasome: Molecular activation and regulation to therapeutics. *Nat. Rev. Immunol.* **2019**, *19*, 477–489. [CrossRef]
43. Dagah, O.M.A.; Silaa, B.B.; Zhu, M.; Pan, Q.; Qi, L.; Liu, X.; Liu, Y.; Peng, W.; Ullah, Z.; Yudas, A.F.; et al. Exploring Immune Redox Modulation in Bacterial Infections: Insights into Thioredoxin-Mediated Interactions and Implications for Understanding Host–Pathogen Dynamics. *Antioxidants* **2024**, *13*, 545. [CrossRef] [PubMed]
44. Liao, X.; Liu, Y.; Zheng, J.; Zhao, X.; Cui, L.; Hu, S.; Xia, T.; Si, S. Diverse Pathways of Engineered Nanoparticle-Induced NLRP3 Inflammasome Activation. *Nanomaterials* **2022**, *12*, 3908. [CrossRef] [PubMed]
45. Dagdeviren, S.; Lee, R.T.; Wu, N. Physiological and Pathophysiological Roles of Thioredoxin Interacting Protein: A Perspective on Redox Inflammation and Metabolism. *Antioxid. Redox Signal* **2023**, *38*, 442–460. [CrossRef]
46. Zhou, F.; Li, C.; Zhang, S.Y. NLRP3 inflammasome: A new therapeutic target for high-risk reproductive disorders? *Chin. Med. J.* **2020**, *134*, 20–27. [CrossRef] [PubMed]
47. Szukiewicz, D.; Trojanowski, S.; Kociszewska, A.; Szewczyk, G. Modulation of the Inflammatory Response in Polycystic Ovary Syndrome (PCOS)-Searching for Epigenetic Factors. *Int. J. Mol. Sci.* **2022**, *23*, 14663. [CrossRef] [PubMed]
48. Park, W.; Lim, W.; Kim, M.; Jang, H.; Park, S.J.; Song, G.; Park, S. Female reproductive disease, endometriosis: From inflammation to infertility. *Mol. Cells* **2024**, 100164. [CrossRef] [PubMed]
49. Lane, R.K.; Hilsabeck, T.; Rea, S.L. The role of mitochondrial dysfunction in age-related diseases. *Biochim. Biophys. Acta* **2015**, *1847*, 1387–1400. [CrossRef]
50. Huang, Y.; Cheng, Y.; Zhang, M.; Xia, Y.; Chen, X.; Xian, Y.; Lin, D.; Xie, S.; Guo, X. Oxidative stress and inflammatory markers in ovarian follicular fluid of women with diminished ovarian reserve during in vitro fertilization. *J. Ovarian Res.* **2023**, *16*, 206. [CrossRef] [PubMed]

51. Zong, Y.; Li, H.; Liao, P.; Chen, L.; Pan, Y.; Zheng, Y.; Zhang, C.; Liu, D.; Zheng, M.; Gao, J. Mitochondrial dysfunction: Mechanisms and advances in therapy. *Signal Transduct. Target. Ther.* **2024**, *9*, 124. [CrossRef] [PubMed]
52. Ferlazzo, N.; Andolina, G.; Cannata, A.; Costanzo, M.G.; Rizzo, V.; Currò, M.; Caccamo, D. Is Melatonin the Cornucopia of the 21st Century? *Antioxidants* **2020**, *9*, 1088. [CrossRef] [PubMed]
53. Li, S.; Withaar, C.; Rodrigues, P.G.; Zijlstra, S.N.; de Boer, R.A.; Silljé, H.H.; Meems, L.M. The NLRP3-inflammasome inhibitor MCC950 improves cardiac function in a HFpEF mouse model. *Biomed. Pharmacother.* **2024**, *181*, 117711. [CrossRef]
54. Sharifi-Rad, M.; Anil Kumar, N.V.; Zucca, P.; Varoni, E.M.; Dini, L.; Panzarini, E.; Rajkovic, J.; Tsouh Fokou, P.V.; Azzini, E.; Peluso, I.; et al. Lifestyle, Oxidative Stress, and Antioxidants: Back and Forth in the Pathophysiology of Chronic Diseases. *Front. Physiol.* **2020**, *11*, 694. [CrossRef] [PubMed]
55. Zeber-Lubecka, N.; Ciebiera, M.; Hennig, E.E. Polycystic Ovary Syndrome and Oxidative Stress—From Bench to Bedside. *Int. J. Mol. Sci.* **2023**, *24*, 14126. [CrossRef] [PubMed]
56. Di Renzo, L.; De Lorenzo, A.; Fontanari, M.; Gualtieri, P.; Monsignore, D.; Schifano, G.; Alfano, V.; Marchetti, M.; Sierr, O.B.O. Immunonutrients involved in the regulation of the inflammatory and oxidative processes: Implication for gamete competence. *J. Assist. Reprod. Genet.* **2022**, *39*, 817–846. [CrossRef] [PubMed]
57. Pan, Y.; Pan, C.; Zhang, C. Unraveling the complexity of follicular fluid: Insights into its composition, function, and clinical implications. *J. Ovarian Res.* **2024**, *17*, 237. [CrossRef] [PubMed]
58. Liu, S.; Jia, Y.; Meng, S.; Luo, Y.; Yang, Q.; Pan, Z. Mechanisms of and Potential Medications for Oxidative Stress in Ovarian Granulosa Cells: A Review. *Int. J. Mol. Sci.* **2023**, *24*, 9205. [CrossRef] [PubMed]
59. Sampaio, O.G.M.; Santos, S.A.A.R.; Damasceno, M.d.B.M.V.; Joventino, L.B.; Schneider, A.; Masternak, M.M.; Campos, A.R.; Cavalcante, M.B. Impact of repeated ovarian hyperstimulation on the reproductive function. *J. Reprod. Immunol.* **2024**, *164*, 104277. [CrossRef] [PubMed]
60. Kesharwani, D.K.; Mohammad, S.; Acharya, N.; Joshi, K.S. Fertility With Early Reduction of Ovarian Reserve. *Cureus* **2022**, *14*, e30326. [CrossRef]
61. Mauchart, P.; Vass, R.A.; Nagy, B.; Sulyok, E.; Bódis, J.; Kovács, K. Oxidative Stress in Assisted Reproductive Techniques, with a Focus on an Underestimated Risk Factor. *Curr. Issues Mol. Biol.* **2023**, *45*, 1272–1286. [CrossRef] [PubMed]
62. Rodríguez-Varela, C.; Labarta, E. Clinical Application of Antioxidants to Improve Human Oocyte Mitochondrial Function: A Review. *Antioxidants* **2020**, *9*, 1197. [CrossRef] [PubMed]
63. Romero, S.; Pella, R.; Zorrilla, I.; Berrío, P.; Escudero, F.; Pérez, Y.; García, M.; Gonzalez, C.; Orihuela, P. Coenzyme Q10 improves the in vitro maturation of oocytes exposed to the intrafollicular environment of patients on fertility treatment. *Jbra Assist. Reprod.* **2020**, *24*, 283–288. [CrossRef]
64. Zahid, A.; Li, B.; Kombe, A.J.K.; Jin, T.; Tao, J. Pharmacological Inhibitors of the NLRP3 Inflammasome. *Front. Immunol.* **2019**, *10*, 2538. [CrossRef]
65. Walke, G.; Gaurkar, S.S.; Prasad, R.; Lohakare, T.; Wanjari, M. The Impact of Oxidative Stress on Male Reproductive Function: Exploring the Role of Antioxidant Supplementation. *Cureus* **2023**, *15*, e42583. [CrossRef]
66. Galatis, D.; Kalopita, K.; Grypiotis, I.; Flessas, I.; Kiriakopoulos, N.; Micha, G. Researching the Phenomenon of Poor Ovarian Responders and Management Strategies in IVF: A Narrative Review. *Acta Medica Acad.* **2022**, *51*, 108–122. [CrossRef]
67. Moustakli, E.; Zikopoulos, A.; Skentou, C.; Bouba, I.; Tsirka, G.; Stavros, S.; Vrachnis, D.; Vrachnis, N.; Potiris, A.; Georgiou, I.; et al. Sperm Mitochondrial Content and Mitochondrial DNA to Nuclear DNA Ratio Are Associated with Body Mass Index and Progressive Motility. *Biomedicines* **2023**, *11*, 3014. [CrossRef]
68. Stavros, S.; Potiris, A.; Molopodi, E.; Mavrogianni, D.; Zikopoulos, A.; Louis, K.; Karampitsakos, T.; Nazou, E.; Sioutis, D.; Christodoulaki, C.; et al. Sperm DNA Fragmentation: Unraveling Its Imperative Impact on Male Infertility Based on Recent Evidence. *Int. J. Mol. Sci.* **2024**, *25*, 10167. [CrossRef] [PubMed]
69. Fan, L.; Guan, F.; Ma, Y.; Zhang, Y.; Li, L.; Sun, Y.; Cao, C.; Du, H.; He, M. Acetylcysteine improves oocyte quality through modulating the Nrf2 signaling pathway to ameliorate oxidative stress caused by repeated controlled ovarian hyperstimulation. *Reprod. Fertil. Dev.* **2022**, *34*, 736–750. [CrossRef] [PubMed]
70. Agarwal, A.; Majzoub, A. Role of Antioxidants in Assisted Reproductive Techniques. *World J. Mens. Health* **2017**, *35*, 77. [CrossRef] [PubMed]
71. Das, B.; Sarkar, C.; Rawat, V.S.; Kalita, D.; Deka, S.; Agnihotri, A. Promise of the NLRP3 Inflammasome Inhibitors in In Vivo Disease Models. *Molecules* **2021**, *26*, 4996. [CrossRef] [PubMed]
72. Chen, Y.; Miao, C.; Zhao, Y.; Yang, L.; Wang, R.; Shen, D.; Ren, N.; Zhang, Q. Inflammasomes in human reproductive diseases. *Mol. Hum. Reprod.* **2023**, *29*, gaad035. [CrossRef] [PubMed]
73. Xu, C.; Lu, Z.; Luo, Y.; Liu, Y.; Cao, Z.; Shen, S.; Li, H.; Liu, J.; Chen, K.; Chen, Z.; et al. Targeting of NLRP3 inflammasome with gene editing for the amelioration of inflammatory diseases. *Nat. Commun.* **2018**, *9*, 4092. [CrossRef]
74. Barresi, V.; Musmeci, C.; Rinaldi, A.; Condorelli, D.F. Transcript-Targeted Therapy Based on RNA Interference and Antisense Oligonucleotides: Current Applications and Novel Molecular Targets. *Int. J. Mol. Sci.* **2022**, *23*, 8875. [CrossRef]

75. Wang, Z.-H.; Wang, Z.-J.; Liu, H.-C.; Wang, C.-Y.; Wang, Y.-Q.; Yue, Y.; Zhao, C.; Wang, G.; Wan, J.-P. Targeting mitochondria for ovarian aging: New insights into mechanisms and therapeutic potential. *Front. Endocrinol.* **2024**, *15*, 1417007. [CrossRef] [PubMed]
76. Clemente-Suárez, V.J.; Redondo-Flórez, L.; Rubio-Zarapuz, A.; Martín-Rodríguez, A.; Tornero-Aguilera, J.F. Microbiota Implications in Endocrine-Related Diseases: From Development to Novel Therapeutic Approaches. *Biomedicines* **2024**, *12*, 221. [CrossRef] [PubMed]
77. Rosen, E.M.; Mínguez-Alarcón, L.; Meeker, J.D.; Williams, P.L.; Milne, G.L.; Hauser, R.; Ferguson, K.K. Urinary oxidative stress biomarker levels and reproductive outcomes among couples undergoing fertility treatments. *Hum. Reprod. Oxf. Engl.* **2019**, *34*, 2399–2409. [CrossRef] [PubMed]
78. Moingeon, P. Artificial intelligence-driven drug development against autoimmune diseases. *Trends Pharmacol. Sci.* **2023**, *44*, 411–424. [CrossRef] [PubMed]
79. Pantic, I.; Paunovic, J.; Pejic, S.; Drakulic, D.; Todorovic, A.; Stankovic, S.; Vucevic, D.; Cumic, J.; Radosavljevic, T. Artificial intelligence approaches to the biochemistry of oxidative stress: Current state of the art. *Chem. Biol. Interact.* **2022**, *358*, 109888. [CrossRef] [PubMed]
80. Horgan, R.P.; Kenny, L.C. 'Omic' technologies: Genomics, transcriptomics, proteomics and metabolomics. *Obstet. Gynaecol.* **2011**, *13*, 189–195. [CrossRef]

cells

MDPI

Review

The Role of Hypoxia-Inducible Factor-1α (HIF-1α) in the Progression of Ovarian Cancer: Perspectives on Female Infertility

Md Ataur Rahman [1], Maroua Jalouli [2], Sujay Kumar Bhajan [3], Mohammed Al-Zharani [2] and Abdel Halim Harrath [4,*]

1 Department of Oncology, Karmanos Cancer Institute, Wayne State University, Detroit, MI 48201, USA; ataur1981rahman@hotmail.com
2 Department of Biology, College of Science, Imam Mohammad Ibn Saud Islamic University (IMSIU), Riyadh 11623, Saudi Arabia; mejalouli@imamu.edu.sa (M.J.); mmylzahrani@imamu.edu.sa (M.A.-Z.)
3 Department of Biotechnology and Genetic Engineering, Faculty of Life Sciences, Bangabandhu Sheikh Mujibur Rahman Science and Technology University, Gopalganj 8100, Bangladesh; sujaybge@gmail.com
4 Zoology Department, College of Science, King Saud University, Riyadh 11451, Saudi Arabia
* Correspondence: hharrath@ksu.edu.sa

Abstract: Hypoxia-Inducible Factor-1α (HIF-1α) is crucial in the progression of ovarian cancer, especially in influencing its tumor microenvironment and promoting pathogenic pathways that worsen female infertility. In hypoxic settings, HIF-1α is stabilized and activates the transcription of genes associated with angiogenesis, metabolic reprogramming, epithelial-to-mesenchymal transition, and therapeutic resistance. Angiogenesis and glycolytic reprogramming mediated by HIF-1 tumor proliferation, survival, and metastasis. Its dysfunction concurrently impairs ovarian homeostasis, undermining follicular growth, hormone synthesis, and the ovarian vascular network, consequently contributing to infertility. Moreover, HIF-1α induces persistent inflammation and oxidative stress, promoting an environment damaging to reproductive health. Due to its dual function in ovarian cancer growth and infertility, HIF-1α is a potential therapeutic target. Strategies including small molecule inhibitors and nanoparticle-mediated delivery of drugs possess the potential to reduce HIF-1α activity, hence reducing cancer progression while protecting fertility. This review seeks to clarify the molecular basis of HIF-1α in ovarian cancer and its effects on female infertility, providing insights into novel treatment approaches that target both controlling the disease and preserving fertility.

Keywords: Hypoxia-Inducible Factor-1α (HIF-1α); ovarian cancer; angiogenesis; female infertility; tumor microenvironment

1. Introduction

HIF-1α functions as a transcription factor that modulates the expression of various genes essential for tumor growth, metastasis, and drug resistance [1]. In normoxic conditions, HIF-1α is subject to proteasomal degradation; however, in a hypoxic tumor microenvironment, it becomes stabilized and translocate to the nucleus, where it activates genes associated with angiogenesis, glycolysis, and epithelial-to-mesenchymal transition (EMT) [2]. These processes collectively allow ovarian cancer cells to survive in adverse conditions, avoid immune detection, and withstand conventional treatments [3]. In addition to its involvement in tumor progression, HIF-1α significantly impacts female reproductive health [4]. Ovarian function is significantly influenced by oxygen homeostasis. Disruptions in hypoxic signaling adversely affect follicular development, ovulation, and hormone production, which can lead to infertility [5]. In ovarian cancer, elevated HIF-1α levels can modify the ovarian vascular network, resulting in compromised blood supply to follicles and heightened oxidative stress, ultimately diminishing ovarian reserve and oocyte quality [6]. HIF-1α is essential for cellular adaptation to hypoxia and is crucial in the early stages of placental development. The regulation of angiogenesis, metabolism, and trophoblast function is essential for proper implantation and fetal nourishment [7]. Alterations in HIF-1α function are associated with pregnancy complications, including preeclampsia, intrauterine growth restriction (IUGR), and placental insufficiency, underscoring its critical role in fetal-maternal health (Figure 1) [8]. Additionally, HIF-1α-mediated inflammation intensifies damage to ovarian tissues, thereby reducing reproductive potential.

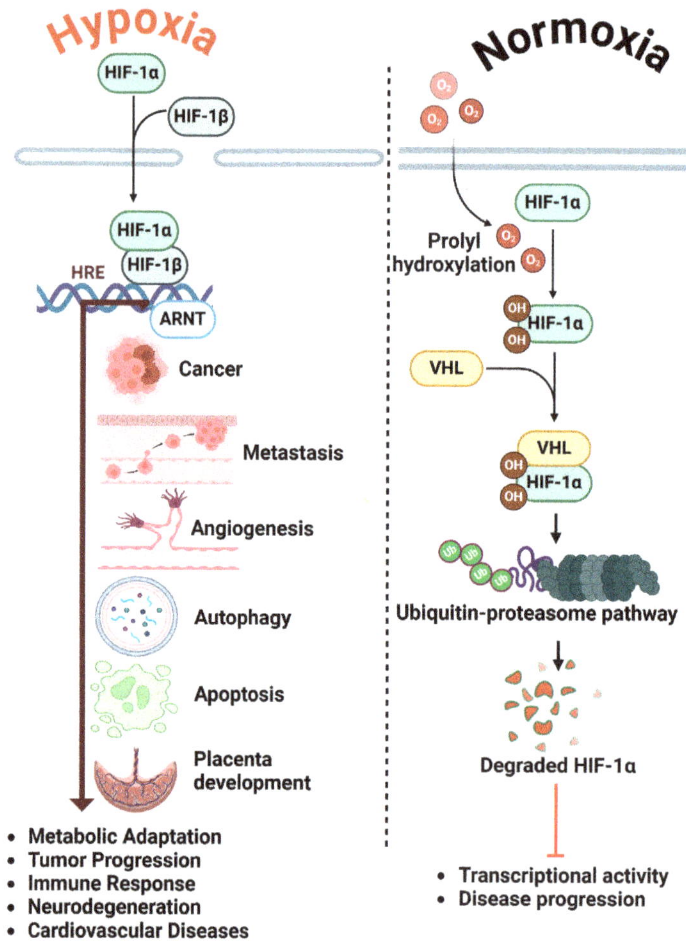

Figure 1. Hypoxia and Normoxia in HIF Signaling. During hypoxia (reduced oxygen levels), prolyl hydroxylase (PHD) activity is suppressed, leading to the stabilization of HIF-α. This stabilized HIF-α translocates to the nucleus and activates target genes associated with tumor progression, angiogenesis, metastasis, autophagy, apoptosis, metabolic adaptation, immune response, neurodegeneration, and cardiovascular diseases. Under normoxic conditions, HIF-α is hydroxylated by PHDs, resulting in its ubiquitination by the von Hippel-Lindau (VHL) complex and subsequent proteasomal degradation, hence inhibiting its transcriptional activity and influencing cellular adaptability and disease progression. The figure was created and modified using the BioRender online commercial platform.

Ovarian cancer represents a significant threat among gynecological malignancies globally, frequently identified at advanced stages owing to its asymptomatic nature and the absence of effective early detection markers [9,10]. Ovarian cancer continues to exhibit high recurrence rates and poor prognosis, despite advancements in treatment modalities such as surgery and chemotherapy [11]. The tumor microenvironment, especially hypoxia, is a critical factor in the progression of ovarian cancer and its resistance to therapy [12]. Hypoxia, characterized by diminished oxygen levels, is a defining feature of solid tumors and significantly influences cancer progression through the induction of cellular adaptations that enhance survival, angiogenesis, metabolic reprogramming, and invasion (Figure 1) [13]. HIF-1α serves as a principal regulator of cellular responses to hypoxic conditions and has been thoroughly investigated for its role in cancer biology [14].

The relationship between ovarian cancer progression and infertility via HIF-1α signaling underscores the necessity for targeted therapeutic interventions. Ongoing research focuses on small-molecule inhibitors of HIF-1α, nanoparticle-mediated drug delivery systems for targeted inhibition, and metabolic reprogramming strategies to mitigate hypoxia-induced tumor aggressiveness [15]. Lifestyle and environmental factors that influence HIF-1α activity, including oxidative stress and dietary interventions, may provide supplementary strategies to enhance reproductive outcomes while addressing cancer progression [16]. Despite increasing evidence connecting HIF-1α to ovarian cancer and infertility, notable research gaps persist in elucidating its specific molecular mechanisms and potential therapeutic interventions. This review provides an analysis of HIF-1α's role in ovarian cancer progression and its negative impact on female fertility. Elucidating these mechanisms may facilitate the development of novel therapeutic strategies that enhance cancer prognosis while preserving reproductive health in affected women.

2. HIF-1α in Ovarian Cancer Progression

The HIF-1α signaling pathway promotes ovarian cancer growth through the augmentation of angiogenesis, metabolic reprogramming, epithelial–mesenchymal transition (EMT), chemoresistance, and immune evasion (Figure 2) [17]. Targeting HIF-1α and its downstream pathways constitutes a viable approach for enhancing treatment outcomes in ovarian cancer. Thus playing a significant role in the aggressive characteristics of ovarian cancer [18].

2.1. *Angiogenesis and Metabolic Reprogramming*

Angiogenesis, the development of new blood vessels, is essential for tumor viability under hypoxic conditions. HIF-1α enhances angiogenesis mainly through the upregulation of vascular endothelial growth factors (VEGF), which is a significant angiogenic factor that encourages endothelial cell proliferation and the formation of blood vessels [19]. Elevated VEGF expression promotes the formation of an aberrant vascular network in ovarian tumors, thereby improving nutrient and oxygen supply to cancer cells [20]. Newly formed blood vessels frequently exhibit structural defects, resulting in irregular perfusion and ongoing tumor hypoxia, which in turn stabilizes HIF-1α and promotes tumor progression (Figure 2) [21]. Besides angiogenesis, HIF-1α is crucial for the metabolic reprogramming of ovarian cancer cells. In hypoxic environments, cancer cells transition from oxidative phosphorylation to aerobic glycolysis, a phenomenon referred to as the Warburg effect [22]. HIF-1α enhances the expression of glycolytic enzymes, including hexokinase (HK), lactate dehydrogenase A (LDHA), and glucose transporters (GLUT1), thereby promoting glucose uptake and lactate production [23]. This metabolic adaptation confers a survival advantage to ovarian cancer cells by sustaining ATP production and establishing an acidic microenvironment that facilitates immune evasion and enhances invasiveness (Figure 2).

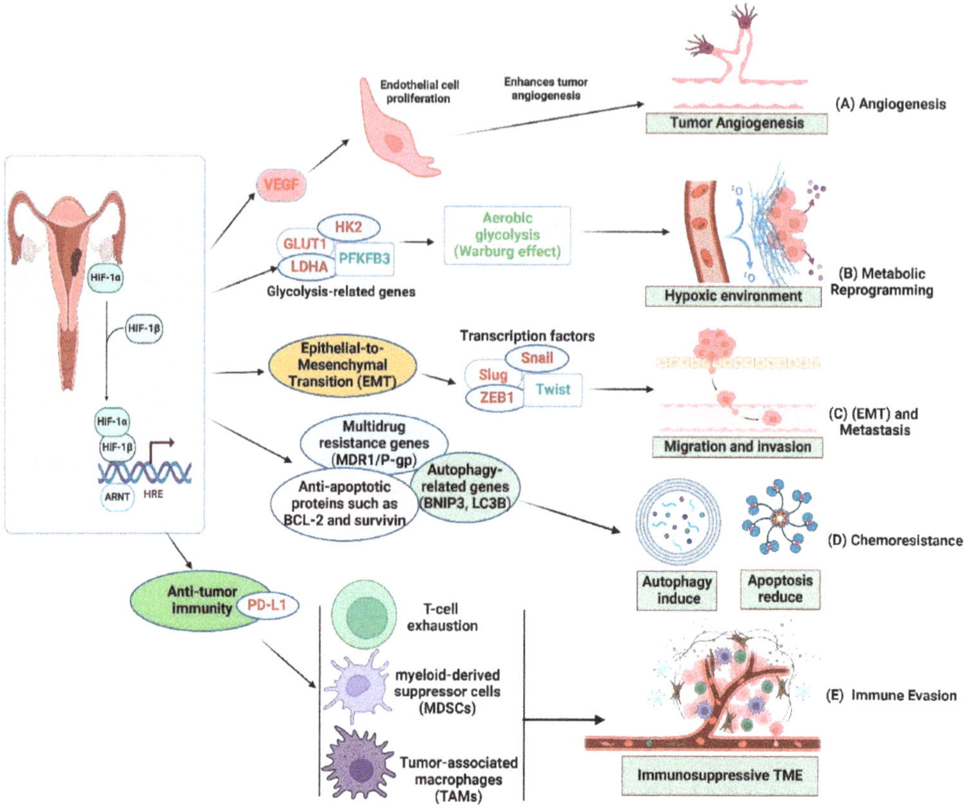

Figure 2. Molecular mechanism of HIF-1α signaling in ovarian cancer progression. HIF-1α migrates to the nucleus, where it dimerizes with HIF-1β and is associated with hypoxia-response elements (HREs) in the promoter regions of target genes. (**A**) Angiogenesis: HIF-1α upregulates VEGF, promoting endothelial cell proliferation and tumor angiogenesis, enhancing oxygen and nutrient supply. (**B**) Metabolic Reprogramming: HIF-1α induces glycolysis-related genes (GLUT1, HK2, PFKFB3, LDHA), shifting metabolism towards aerobic glycolysis (Warburg effect) for survival in hypoxia. (**C**) Epithelial-to-Mesenchymal Transition (EMT) and Metastasis: HIF-1α upregulates EMT regulators (Snail, Slug, Twist, ZEB1), downregulating E-cadherin and upregulating N-cadherin and vimentin, enhancing migration, invasion, and peritoneal dissemination. (**D**) Chemoresistance: HIF-1α promotes MDR1/P-gp, autophagy genes (BNIP3, LC3B), and anti-apoptotic proteins (BCL-2, survivin), reducing cisplatin-induced apoptosis. (**E**) Immune Evasion: HIF-1α induces PD-L1 expression, leading to T-cell exhaustion, and promotes MDSCs and TAMs, fostering an immunosuppressive tumor microenvironment (TME). The figure was created and modified using the BioRender online commercial platform.

2.2. Epithelial-to-Mesenchymal Transition (EMT)

Epithelial-to-Mesenchymal Transition (EMT) refers to the biological process whereby epithelial cells acquire mesenchymal characteristics, leading to increased motility and invasiveness [24]. This transition is crucial in various physiological and pathological contexts, including embryogenesis, wound healing, and cancer metastasis [25]. EMT is a vital mechanism in the metastasis of ovarian cancer, allowing epithelial cancer cells to adopt mesenchymal traits that facilitate increased motility and invasion. HIF-1α activates essential EMT transcription factors, including Snail, Slug, and Twist, leading to the suppression of epithelial markers, such as E-cadherin, and the induction of mesenchymal markers like

N-cadherin and vimentin (Figure 2) [26]. This transition enhances the metastatic capability of ovarian cancer cells, facilitating their detachment from the primary tumor, invasion of adjacent tissues, and spread to remote organs. Additionally, HIF-1α-mediated EMT increases resistance to anoikis, a type of programmed cell death that takes place when epithelial cells detach from the extracellular matrix [27].

2.3. Resistance to Therapy

The development of resistance to chemotherapy and targeted therapies represents a significant challenge in the treatment of ovarian cancer. HIF-1α plays a role in therapy resistance via various mechanisms, notably by activating multidrug resistance (MDR) proteins, such as P-glycoprotein (P-gp) and breast cancer resistance protein (BCRP), which facilitate the efflux of chemotherapeutic agents from cancer cells [28]. HIF-1α also enhances DNA repair mechanisms and suppresses apoptosis through the upregulation of anti-apoptotic proteins, such as Bcl-2, thereby diminishing the effectiveness of conventional treatments [1]. These mechanisms underscore the essential function of HIF-1α in the progression of ovarian cancer. Targeting HIF-1α and its downstream pathways offers a promising therapeutic strategy to address tumor growth, metastasis, and drug resistance, thereby enhancing patient outcomes [29].

3. Impact of HIF-1α on Female Infertility

Ovarian cancer has a substantial effect on female reproductive health, with HIF-1α serving a critical function in the mechanisms associated with infertility [30]. HIF-1α serves as a key regulator of the hypoxic response, influencing tumor progression and impacting ovarian function through mechanisms such as angiogenesis, metabolic dysregulation, oxidative stress, inflammation, and hormonal imbalances [31]. These disruptions impair follicular development, oocyte quality, and overall reproductive capacity (Figure 3). Examining the role of HIF-1α in infertility is crucial for formulating targeted interventions to address reproductive dysfunction in women with ovarian cancer.

3.1. Disruption of Ovarian Function and Follicular Development

Ovarian function relies on a precise equilibrium of oxygen supply, hormonal signaling, and vascular integrity, all of which are governed by hypoxic responses [32]. Mild hypoxia contributes to follicular development under normal physiological conditions by stimulating angiogenesis and ensuring an adequate nutrient supply to growing follicles [33]. Excessive hypoxia resulting from HIF-1α overexpression in ovarian cancer disrupts the balance, impairing folliculogenesis and reducing ovarian reserve. HIF-1α modifies the ovarian microenvironment by disrupting the expression of essential growth factors for follicular maturation, such as vascular endothelial growth factor (VEGF), transforming growth factor-beta (TGF-β), and insulin-like growth factors (IGFs) [34,35]. Excessive VEGF production results in atypical blood vessel development in the ovaries, which disrupts follicular blood flow and induces localized hypoxia [36]. This leads to impaired folliculogenesis, reduced survival of granulosa cells, and heightened rates of follicular atresia. Excessive activation of HIF-1α also promotes granulosa cell apoptosis by increasing pro-apoptotic factors like Bcl-2-associated X protein (BAX) and decreasing anti-apoptotic protein expression, which accelerates follicular depletion [37]. Inadequate follicular development adversely impacts oocyte quality, resulting in a heightened risk of aneuploidy, diminished fertilization potential, and early embryonic arrest [38]. Moreover, changes in mitochondrial function caused by hypoxia lead to impaired ATP production, which elevates oxidative stress in oocytes and diminishes their developmental competence [39]. Consequently, women diagnosed with ovarian cancer frequently exhibit reduced ovarian reserve, decreased fertilization rates, and an increased risk of infertility [40].

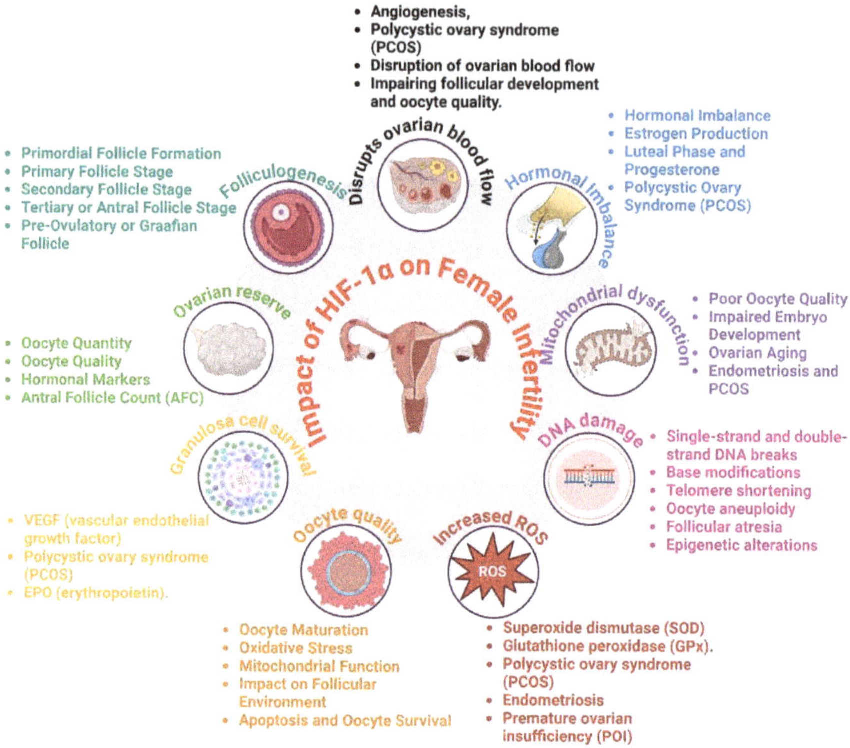

Figure 3. Impact of HIF-1α on Female Infertility. Dysregulated folliculogenesis, reduced ovarian reserve, granulosa cell apoptosis, and impaired oocyte quality. Elevated ROS, DNA damage, mitochondrial dysfunction, and increased pro-inflammatory cytokines (TNF-α, IL-6, IL-1β) accelerate ovarian aging and fibrosis. Suppressed estrogen synthesis (via aromatase inhibition), reduced AMH levels, impaired ovarian blood flow, and hypoxia-induced infertility. The figure was created and modified using the BioRender online commercial platform.

3.2. *Oxidative Stress and Chronic Inflammation*

HIF-1α is crucial in the regulation of oxidative stress responses, with its overactivation in ovarian cancer resulting in an accumulation of excessive reactive oxygen species (ROS) [41]. ROS are produced during cellular metabolism and contribute to physiological signaling; however, their overproduction in hypoxic conditions can harm ovarian tissues, increasing the risk of infertility. Elevated levels of ROS disrupt folliculogenesis through the induction of DNA damage, lipid peroxidation, and mitochondrial dysfunction in ovarian cells [42]. Oxidative stress impacts granulosa and theca cells, resulting in premature ovarian aging and heightened follicular atresia [43]. Oxidative stress also disturbs the equilibrium of antioxidants like superoxide dismutase (SOD) and glutathione (GSH), increasing the susceptibility of ovarian cells to apoptosis [44].

Inflammation represents a significant factor modulated by HIF-1α in ovarian cancer. Chronic inflammation in the ovarian microenvironment leads to tissue fibrosis, impairing normal ovarian function [33]. HIF-1α increases the synthesis of pro-inflammatory cytokines, including tumor necrosis factor-alpha (TNF-α), interleukin-6 (IL-6), and interleukin-1 beta (IL-1β), thereby facilitating immune cell infiltration and oxidative stress [45]. The inflammatory environment hastens ovarian dysfunction, diminishes ovarian reserve, and adversely affects fertility. Chronic inflammation mediated by HIF-1α pathways is associated with endometriosis-like conditions in patients with ovarian cancer [46]. Endometriosis is defined by the growth of ectopic endometrial tissue, which is aggravated by hypoxic conditions, angiogenesis, and the production of cytokines [47]. An inflammatory and hypoxic ovarian microenvironment contributes to increased fibrosis and scarring, which further compromises ovarian function and results in reduced fertility rates.

3.3. Hormonal Imbalance and Vascular Abnormalities

Dysregulation of HIF-1α in ovarian cancer disrupts hormonal homeostasis, resulting in reproductive dysfunction. The hypothalamic–pituitary–ovarian (HPO) axis regulates reproductive hormones, such as follicle-stimulating hormone (FSH), luteinizing hormone (LH), estrogen, and progesterone [48]. HIF-1α disrupts this regulation by modifying hormone synthesis and receptor sensitivity in ovarian tissues [4]. The overexpression of HIF-1α significantly suppresses estrogen synthesis. HIF-1α suppresses aromatase, the enzyme that converts androgens to estrogen, resulting in reduced estrogen levels [49]. The hormonal imbalance adversely impacts endometrial receptivity and oocyte maturation, thereby diminishing the likelihood of successful implantation and pregnancy [50]. Furthermore, diminished estrogen levels are associated with ovarian insufficiency, thereby elevating the risk of premature ovarian failure in patients with ovarian cancer. HIF-1α influences the secretion of anti-Müllerian hormone (AMH), which serves as a marker of ovarian reserve [51]. Increased levels of HIF-1α are associated with diminished AMH expression, suggesting a decline in ovarian follicles and a shortened reproductive lifespan [52]. Vascular abnormalities caused by HIF-1α further contribute to infertility by impairing ovarian blood flow [53]. An irregular and dysfunctional vascular network hinders oxygen and nutrient delivery to follicles, resulting in hypoxic stress and heightened follicular atresia. Defective vasculature contributes to ovarian hyperstimulation syndrome (OHSS) in women undergoing assisted reproductive technologies (ART), complicating fertility treatments [54].

4. Clinical Implications and Potential Therapeutic Approaches

Ovarian cancer presents a considerable challenge owing to its late-stage diagnosis, aggressive progression, and elevated recurrence rates [55]. HIF-1α is crucial in the pathogenesis of ovarian cancer, facilitating angiogenesis, metabolic reprogramming, EMT, and resistance to therapy [4]. The dysregulation of this system has significant implications for female reproductive health, resulting in infertility due to oxidative stress, chronic inflammation, hormonal imbalances, and vascular abnormalities. Comprehending the clinical implications of HIF-1α establishes a basis for targeted therapeutic strategies that can enhance cancer prognosis while defending fertility. Various strategies, such as pharmacological inhibitors, antioxidant therapy, anti-inflammatory agents, nanoparticle-based drug delivery, and lifestyle modifications, have been identified as potential interventions to reduce HIF-1α-driven ovarian cancer progression and infertility.

4.1. Inhibitors of HIF-1α

Inhibiting HIF-1α with targeted agents offers a dual therapeutic strategy that effectively reduces ovarian cancer proliferation while safeguarding reproductive health.

Numerous HIF-1α inhibitors have been investigated for their anti-cancer effects, with many demonstrating potential advantages in maintaining ovarian function (Table 1). Inhibition of HIF-1α represents a significant approach in cancer treatment, given its involvement in tumor growth, metastasis, and drug resistance. Multiple compounds that target HIF-1α via distinct mechanisms have been identified, providing therapeutic advantages in ovarian cancer.

PX-478 is a small molecule that inhibits HIF-1α transcription and protein accumulation. PX-478 inhibits angiogenesis and therapy resistance by reducing HIF-1α levels, thereby effectively suppressing the growth of ovarian cancer. Furthermore, PX-478 enhances ovarian vascular function, thereby preventing follicular atresia, which is a significant factor in female infertility [56]. Acriflavine inhibits the binding of HIF-1α to HIF-1β, thereby obstructing its transcriptional function. This disruption results in decreased cancer metastasis and angiogenesis, establishing it as an effective anti-tumor agent. Acriflavine also diminishes ovarian hypoxia, a significant factor contributing to follicular degeneration and suboptimal oocyte quality in women experiencing infertility [57]. Echinomycin disrupts HIF-1 DNA binding, inhibiting its ability to activate genes associated with tumor survival and metastasis. This leads to reduced aggressiveness of tumors in ovarian cancer. Echinomycin additionally safeguards ovarian follicles against oxidative stress, thereby promoting improved reproductive function [58]. Digoxin, commonly employed in the treatment of cardiac conditions, has been shown to reduce HIF-1α protein levels. It restricts ovarian tumor progression by inhibiting angiogenesis and cellular proliferation. Additionally, it enhances ovarian microcirculation, stabilizes hormonal balance, and mitigates reproductive dysfunction [59]. 2-Methoxyestradiol (2-ME) induces destabilization of the HIF-1α protein, thereby inhibiting its transcriptional influence on tumor growth. It has demonstrated potential in inhibiting cancer progression while concurrently restoring hormonal equilibrium and ovarian functionality in women experiencing infertility challenges [60]. YC-1 suppresses the synthesis of HIF-1α and the signaling of VEGF, both of which are critical in tumor vascularization and metastasis. The capacity to mitigate inflammatory damage in ovarian tissues positions it as a viable option for fertility preservation during ovarian cancer treatment [61]. LW6 facilitates the degradation of HIF-1α through the proteasome pathway, consequently restricting tumor adaptation to hypoxic conditions. Furthermore, it safeguards ovarian follicles from stress-induced apoptosis, thereby enhancing ovarian function and decreasing the likelihood of premature ovarian failure [62]. PX-12 inhibits the stabilization of HIF-1α under hypoxic conditions by targeting the thioredoxin system, thereby diminishing tumor resistance to chemotherapy. It restores oxidative balance in ovarian tissues, which is essential for preserving follicular integrity and reproductive potential [63]. KRIBB11 inhibits HIF-1α's capacity to initiate gene transcription, thereby diminishing cancer cell invasion and resistance to therapy. It enhances ovarian follicular survival by mitigating hypoxia-induced stress, thus providing advantages for women facing reproductive issues [17]. Bortezomib facilitates the degradation of HIF-1α, thereby diminishing tumor adaptation to hypoxic conditions and improving the effectiveness of chemotherapy. Furthermore, it enhances ovarian cell viability, thereby decreasing the likelihood of infertility caused by chemotherapy [64].

Table 1. Inhibitors of HIF-1α present potential therapeutic strategies for targeting HIF-1α in the management of ovarian cancer and female infertility.

HIF-1α Inhibitor	Mechanism of Action	Potential Benefits for Ovarian Cancer	Potential Benefits for Female Infertility	Commercial Availability	Ref.
PX-478	Directly inhibits HIF-1α transcription and protein accumulation	Suppresses tumor angiogenesis, reduces therapy resistance	Improves ovarian vascular function, prevents follicular atresia	Yes	[56]
Acriflavine	Disrupts HIF-1α dimerization, blocking its transcriptional activity	Inhibits cancer progression and metastasis	Reduces hypoxia-induced ovarian dysfunction	Yes (Clinical use for infections)	[57]
Echinomycin	Binds to HIF-1 DNA binding sites, preventing target gene expression	Reduces hypoxia-driven tumor growth and survival	Protect ovarian follicles from oxidative damage	No (Investigational)	[58]
Digoxin	Downregulates HIF-1α protein levels via inhibition of its synthesis	Decreases cancer cell proliferation and vascular abnormalities	Improves ovarian microcirculation, enhances hormonal balance	Yes (FDA-approved cardiac drug)	[59]
2-Methoxyestradiol (2-ME)	Destabilizes HIF-1α protein and inhibits its transcription	Reduces angiogenesis and tumor progression	Restores hormonal equilibrium and ovarian function	No (Investigational)	[60]
YC-1	Inhibits HIF-1α synthesis and blocks downstream VEGF signaling	Suppresses tumor growth and hypoxia adaptation	Reduces inflammatory damage to ovarian tissues	No (Investigational)	[61]
LW6	Promotes HIF-1α degradation via the proteasome pathway	Decreases in hypoxia-induced drug resistance in ovarian cancer	Protect ovarian cells from stress-induced apoptosis	No (Investigational)	[62]
PX-12	Inhibits thioredoxin, preventing HIF-1α stabilization under hypoxia	Increases chemotherapy sensitivity, prevents metastasis	Restores oxidative balance in ovarian tissues	No (Investigational)	[63]
KRIBB11	Blocks HIF-1α transcriptional activation and nuclear localization	Reduces cancer cell invasion and therapy resistance	Improves ovarian follicular survival by limiting hypoxic stress	No (Investigational)	[17]
Bortezomib	Inhibits the proteasome, leading to degradation of HIF-1α	Prevents tumor adaptation to hypoxia, enhances drug efficacy	Enhances ovarian cell viability and prevents premature ovarian failure	Yes (FDA-approved for multiple myeloma)	[64]

The incorporation of HIF-1α inhibitors into clinical practice offers promising prospects for enhancing ovarian cancer outcomes while protecting fertility. Several critical areas necessitate additional investigation. Numerous HIF-1α inhibitors exhibit low bioavailability, necessitating nanoparticle-based delivery systems to improve stability and tissue penetration [65]. Additionally, the development of liposomal formulations, polymeric nanoparticles, and targeted drug carriers is essential for improved therapeutic outcomes [66]. Although numerous HIF-1α inhibitors have demonstrated potential in preclinical studies, additional clinical trials are required to evaluate their long-term impacts on ovarian function and fertility [4]. It is essential to ensure low toxicity, minimal off-target effects, and reproductive safety prior to widespread clinical application.

4.2. Antioxidant Therapy

Oxidative stress significantly results from HIF-1α overexpression in ovarian cancer and is essential in ovarian dysfunction. Accumulation of ROS leads to damage in ovarian follicles, diminishes oocyte quality, and accelerates premature ovarian aging [67]. Antioxidant therapy has emerged as a potential strategy to mitigate these effects, providing both anti-cancer and fertility-preserving advantages. Resveratrol is a polyphenolic compound present in grapes and berries, exhibiting significant antioxidant and anti-inflammatory properties. It directly inhibits the expression of HIF-1α and mitigates oxidative stress-induced DNA damage in ovarian tissues [68]. Preclinical studies indicate that resveratrol supplementation may enhance ovarian reserve and follicular health in cancer patients receiving chemotherapy. N-Acetylcysteine (NAC) serves as a precursor to glutathione, an essential cellular antioxidant that mitigates ROS and safeguards against oxidative damage. NAC has demonstrated the ability to suppress HIF-1α activity, inhibit tumor growth, and enhance chemosensitivity in ovarian cancer [69]. The capacity to restore redox balance in ovarian cells indicates a possible role in enhancing fertility outcomes. Melatonin functions as an effective free radical scavenger and has shown protective effects against oxidative stress in the ovaries. It regulates HIF-1α expression, diminishes VEGF-induced aberrant angiogenesis, and enhances mitochondrial function in oocytes [70]. Clinical studies indicate that melatonin supplementation improves oocyte quality and enhances reproductive success in women with ovarian dysfunction [71]. Antioxidant therapy shows potential in reducing the negative impacts of HIF-1α-induced oxidative stress. Further investigation is required to determine optimal dosing, timing, and combination strategies with existing therapies.

4.3. Anti-Inflammatory Agents

Inflammation significantly contributes to the progression of ovarian cancer by enhancing tumor survival, facilitating angiogenesis, and increasing resistance to therapy [72]. Inhibiting HIF-1α to target inflammation presents a promising strategy for enhancing treatment efficacy and decreasing tumor aggressiveness [73]. Chronic inflammation significantly contributes to the progression of ovarian cancer and infertility, with HIF-1α serving a pivotal role in the enhancement of pro-inflammatory cytokine production [45]. Targeting inflammatory pathways may reduce tumor aggressiveness and preserve ovarian function, playing a dual role in managing ovarian cancer progression through the inhibition of HIF-1α-driven inflammation and in the treatment of female infertility by safeguarding ovarian function and minimizing reproductive tissue damage (Table 2).

Curcumin, a polyphenolic compound derived from turmeric, inhibits the NF-κB and HIF-1α signaling pathways, leading to a reduction in inflammatory cytokines and oxidative stress. Curcumin inhibits tumor proliferation, diminishes angiogenesis, and improves chemotherapy efficacy in ovarian cancer. Curcumin mitigates inflammation, thereby preventing ovarian tissue damage and enhancing oocyte quality and reproductive outcomes [74]. Resveratrol, present in red grapes, inhibits HIF-1α and cytokine production, thereby decreasing tumor hypoxia and inflammation. It suppresses metastasis and chemotherapy resistance, positioning it as a promising candidate for ovarian cancer treatment. Resveratrol also maintains ovarian function by postponing ovarian aging and mitigating oxidative stress, thereby enhancing fertility potential [75]. Aspirin inhibits COX-2 and prostaglandin synthesis, thereby mitigating chronic inflammation linked to the progression of ovarian cancer. Research indicates that aspirin diminishes cancer cell proliferation and metastasis, establishing its role as an important chemopreventive agent. This also improves ovarian blood flow, mitigating infertility associated with endometriosis and promoting reproductive health [76]. Melatonin modulates the levels of HIF-1α, IL-6, and TNF-α, thereby mitigating inflammation induced by tumors. Melatonin inhibits tumor progression, enhances chemotherapy response, and decreases oxidative stress in ovarian cancer. Melatonin is essential for ovarian health, as it improves oocyte quality, safeguards ovarian follicles, and inhibits premature ovarian aging [77]. Sulforaphane, present in cruciferous vegetables, inhibits the activity of NF-κB and HIF-1α, leading to a reduction in the production of inflammatory cytokines. In ovarian cancer, it inhibits tumor growth, increases sensitivity to chemotherapy, and prevents metastasis. Sulforaphane enhances the stability of the ovarian microenvironment, mitigates oxidative stress, and improves fertility outcomes [78]. Omega-3 fatty acids influence pro-inflammatory cytokines, thereby decreasing oxidative stress in ovarian tissues. In ovarian cancer, tumor growth is suppressed, and immune response is enhanced. Omega-3 fatty acids improve ovarian response, promote follicular development, and mitigate inflammation-associated infertility [79]. Quercetin, a flavonoid derived from plants, inhibits HIF-1α, TNF-α, and NF-κB, thereby obstructing inflammation induced by hypoxia. In ovarian cancer, it diminishes tumor progression, inhibits epithelial–mesenchymal transition (EMT), and improves the effectiveness of chemotherapy. Quercetin protects ovarian follicles from inflammatory damage, thereby preserving ovarian function and fertility potential [80]. NAC enhances glutathione concentration, thereby diminishing oxidative stress and inflammation in ovarian tissues. In ovarian cancer, it improves the response to chemotherapy and inhibits tumor resistance. NAC prevents ovarian fibrosis, restores hormonal balance, and enhances overall reproductive health [81]. Boswellia Serrata inhibits 5-LOX and NF-κB, thereby reducing inflammation in ovarian cancer and reproductive tissues. It reduces tumor-associated inflammation, inhibits metastasis, and enhances the efficacy of chemotherapy. Boswellia enhances ovarian function and mitigates inflammatory damage in cases of infertility, thereby improving fertility outcomes [82]. Gingerol inhibits COX-2, TNF-α, and IL-6, thereby decreasing systemic inflammation. Gingerol inhibits tumor progression, enhances chemotherapy response, and reduces drug resistance in ovarian cancer. It also mitigates inflammation-related ovarian dysfunction, thereby enhancing reproductive health and fertility [83].

Additionally, metformin, initially utilized as an anti-diabetic medication, has garnered interest in its anti-inflammatory and anti-cancer effects. It inhibits the stabilization of HIF-1α through the activation of AMP-activated protein kinase (AMPK), which results in reduced tumor growth and improved efficacy of chemotherapy [84]. Metformin enhances insulin sensitivity, diminishes oxidative stress, and safeguards ovarian function in patients with polycystic ovary syndrome (PCOS) and ovarian cancer [85]. Interleukin-6 (IL-6) inhibitors play a significant role as IL-6 is a crucial inflammatory cytokine that is upregulated by HIF-1α in ovarian tumors. Monoclonal antibodies targeting IL-6, including tocilizumab, demonstrate potential in mitigating tumor-associated inflammation and enhancing immune responses. IL-6 inhibitors may enhance fertility outcomes and preserve ovarian function by reducing inflammatory stress in the ovarian microenvironment. Non-Steroidal Anti-Inflammatory Drugs (NSAIDs), including aspirin and ibuprofen, have been investigated for their potential to lower ovarian cancer risk through the inhibition of cyclooxygenase-2 (COX-2), an important mediator of inflammation [86]. Research indicates that NSAIDs may enhance ovarian blood flow, decrease fibrosis, and improve follicular viability in women at risk for ovarian failure. Anti-inflammatory therapies demonstrate significant potential in addressing oncological and reproductive issues [87]. Long-term effects on ovarian physiology necessitate additional research to confirm safety and efficacy. Future research should prioritize the optimization of bioavailability, formulation, and clinical applications to achieve dual benefits in oncology and reproductive medicine.

Table 2. Anti-inflammatory agents may serve as a potential therapeutic approach for targeting HIF-1α in the control of ovarian cancer and the management of female infertility.

Anti-Inflammatory Agent	Mechanism of Action	Potential Benefits for Ovarian Cancer	Potential Benefits for Female Infertility	Limitations	Ref.
Curcumin	Inhibits NF-κB and HIF-1α signaling, reducing inflammation and oxidative stress	Suppresses tumor growth, reduces angiogenesis	Protects ovarian follicles, improves oocyte quality	Poor Stability, Toxicity and Biocompatibility	[74]
Resveratrol	Downregulates HIF-1α and cytokine production, reducing hypoxia-driven inflammation	Inhibits metastasis, enhances chemotherapy sensitivity	Preserves ovarian function, delays ovarian aging	Stability and Aggregation, Blood–Brain Barrier (BBB) Penetration	[75]
Aspirin	Blocks COX-2 and prostaglandins, reducing chronic inflammation	Decreases cancer cell proliferation, lowers risk of metastasis	Improves ovarian blood flow, reduces endometriosis-related infertility	Rapid Clearance, Limited Targeting Efficiency, Poor Stability,	[76]
Melatonin	Acts as an antioxidant and anti-inflammatory agent, reducing IL-6 and TNF-α levels	Inhibits HIF-1α-mediated tumor progression	Enhances ovarian reserve, protects against oxidative stress	Limited Targeting Efficiency, Regulatory Challenges, Rapid Clearance and Short Half-Life	[77]
Sulforaphane	NF-κB and HIF-1α, reducing inflammatory cytokines	Prevents tumor progression and therapy resistance	Improves ovarian microenvironment, enhances fertility potential	Stability and Aggregation, Cost of Production, Toxicity and Biocompatibility	[78]
Omega-3 Fatty Acids	Reduces pro-inflammatory cytokines and oxidative stress	Suppresses tumor growth, improves immune response	Enhances ovarian response, reduces inflammation-induced infertility	Poor Targeting Efficiency, Short Circulation Half-Life	[79]
Quercetin	Inhibits HIF-1α and TNF-α, preventing hypoxia-driven inflammation	Reduces cancer progression and EMT	Protects ovarian follicles from inflammatory damage	Stability and Aggregation, Blood–Brain Barrier (BBB) Penetration, Limited Targeting Efficiency	[80]
N-Acetylcysteine (NAC)	Boosts glutathione levels, reduces oxidative stress and inflammation	Increases chemotherapy efficacy, protects normal cells	Prevents ovarian fibrosis, restores hormonal balance	Toxicity Concerns, Biodegradability and Clearance, Scalability and Cost	[81]
Boswellia Serrata (Frankincense)	Inhibits 5-LOX and NF-κB, reducing inflammatory responses	Lowers tumor-related inflammation, prevents metastasis	Supports ovarian function, reducing inflammatory stress on reproductive tissues	Manufacturing Challenges, Biocompatibility, Limited Targeting	[82]
Gingerol (from Ginger)	Suppresses COX-2, TNF-α, and IL-6, reducing systemic inflammation	Inhibits ovarian cancer progression, enhances chemotherapy response	Reduces inflammation-related ovarian dysfunction, improves reproductive outcomes	Limited Tissue Penetration, Possible Drug Resistance, Toxicity, Regulatory Challenge	[83]

4.4. Nanoparticle-Based Drug Delivery

Nanoparticle-based drug delivery systems present an effective approach for targeting HIF-1α, facilitating accurate and efficient drug administration while reducing systemic toxicity. Nanocarriers improve drug stability, bioavailability, and penetration into hypoxic tumor regions, presenting a promising strategy for ovarian cancer treatment. Furthermore, specific nanoparticles have demonstrated the potential to safeguard ovarian function, thereby contributing to both cancer management and fertility preservation [12]. Nanoparticle-based drugs enhance specificity and efficacy in targeting HIF-1α, offering a dual benefit in the management of ovarian cancer progression while protecting female fertility (Table 3).

Lipid-based nanoparticles that encapsulate HIF-1α siRNA effectively silence the gene associated with hypoxia adaptation in cancer cells. These nanoparticles inhibit tumor growth and enhance chemotherapy efficacy by decreasing HIF-1α expression. Lower HIF-1α activity mitigates ovarian hypoxia, thereby safeguarding follicular development and maintaining ovarian function from a fertility standpoint [88]. PX-478, a recognized inhibitor of HIF-1α, has been integrated into polymeric nanoparticles to enhance its delivery and bioavailability. These nanoparticles inhibit HIF-1α transcriptional activity, thereby reducing angiogenesis and tumor invasion. Furthermore, they enhance ovarian vascularization, thereby decreasing hypoxic stress that may adversely affect fertility [61]. Acriflavine serves as an effective inhibitor of HIF-1α dimerization, and its formulation in nanoparticles improves drug penetration within tumors. Disruption of HIF-1α function inhibits ovarian cancer metastasis. Additionally, it inhibits HIF-1α-induced follicular atresia, thereby promoting improved ovarian health and reproductive capacity [57]. Polymeric nanoparticles loaded with curcumin demonstrate anti-inflammatory properties and inhibit HIF-1α. Targeting HIF-1α and NF-κB signaling reduces inflammation in tumors and enhances sensitivity to chemotherapy. Curcumin nanoparticles also safeguard ovarian follicles against oxidative stress-related damage, thereby enhancing fertility outcomes [89]. Resveratrol, a natural antioxidant, has been encapsulated in lipid-based nanoparticles to enhance its bioavailability and therapeutic efficacy. It regulates HIF-1α signaling and mitochondrial function, thereby effectively inhibiting cancer progression. Resveratrol enhances oocyte quality, ovarian reserve, and reproductive outcomes, thus serving as a dual-benefit therapy [90]. Melatonin-loaded solid lipid nanoparticles demonstrate antioxidant properties and stabilize mitochondria, leading to a significant reduction in cancer aggressiveness induced by hypoxia. These nanoparticles protect ovarian tissue by decreasing reactive oxygen species (ROS) levels, minimizing apoptosis, and enhancing overall reproductive health [77]. NAC nanoparticles restore redox balance and inhibit HIF-1α signaling, thereby enhancing chemosensitivity in ovarian cancer cells. NAC nanoparticles improve ovarian function by mitigating oxidative stress, a significant contributor to infertility [91]. Liposomal formulations of doxorubicin facilitate targeted drug delivery to hypoxic tumor regions, thereby minimizing off-target toxicity. These nanoparticles effectively eliminate ovarian cancer cells while minimizing ovarian damage, thus reducing the risk of infertility post-chemotherapy [92]. Gold nanoparticles linked to HIF-1α inhibitors improve tumor-specific inhibition, reducing angiogenesis and tumor hypoxia. Furthermore, these nanoparticles enhance ovarian blood circulation, thereby preventing premature ovarian aging and maintaining fertility [17]. Selenium-based nanoparticles demonstrate antioxidant and anti-inflammatory effects, modulating HIF-1α pathways in ovarian cancer. They protect ovarian follicles against oxidative stress, improve ovarian function, and maintain hormonal equilibrium, thereby promoting reproductive health [93].

Table 3. Nanoparticle-based drugs for potential therapeutic approaches targeting HIF-1α in ovarian cancer control and female infertility.

No.	Nanoparticle-Based Drug	Type of Nanoparticle	Mechanism of Action	Potential Benefits for Ovarian Cancer	Potential Benefits for Female Infertility	References
1	HIF-1α siRNA Nanoparticles	Lipid-based	Silences HIF-1α expression, reducing tumor hypoxia	Inhibits tumor progression, enhances therapy response	Protects ovarian function by reducing hypoxia-induced damage	[88]
2	PX-478-Loaded Nanoparticles	Polymeric	HIF-1α inhibitor, blocks transcriptional activity	Suppresses angiogenesis, EMT, and drug resistance	Improves ovarian vascularization, reduces oxidative stress	[61]
3	Acriflavine-Encapsulated Nanoparticles	Liposomal	Disrupts HIF-1α dimerization, blocking its function	Inhibits tumor growth and metastasis	Prevents HIF-1α-induced follicular atresia	[57]
4	Curcumin Nanoparticles	Polymeric	Anti-inflammatory, inhibits HIF-1α and NF-κB signaling	Reduces inflammation, enhances chemotherapy efficacy	Protect ovarian follicles from oxidative stress	[89]
5	Resveratrol-Loaded Nanoparticles	Lipid-based	Antioxidant, modulates HIF-1α and mitochondrial function	Inhibits cancer cell proliferation, reduces therapy resistance	Improves oocyte quality and ovarian reserve	[90]
6	Melatonin-Conjugated Nanoparticles	Solid lipid	Antioxidant, stabilizes mitochondrial function, downregulates HIF-1α	Reduces hypoxia-induced cancer aggressiveness	Protects ovarian tissue from ROS and apoptosis	[77]
7	N-Acetylcysteine (NAC) Nanoparticles	Polymeric	Restores redox balance, inhibits HIF-1α signaling	Increases chemosensitivity, prevents tumor relapse	Enhances ovarian function and fertility outcomes	[91]
8	Doxorubicin-Loaded Nanoparticles	Liposomal	Chemotherapy drug, enhances drug delivery to hypoxic tumors	Targets hypoxic cancer cells more effectively	Reduces off-target toxicity to ovarian tissue	[92]
9	HIF-1α Inhibitor-Conjugated Gold Nanoparticles	Gold-based	Enhances targeted inhibition of HIF-1α in tumors	Suppresses angiogenesis and tumor hypoxia	Prevents premature ovarian aging by improving blood flow	[17]
10	Selenium Nanoparticles	Metal-based	Antioxidant, regulates oxidative stress and HIF-1α pathways	Protects against cancer progression, boosts immune response	Enhances ovarian follicle survival and hormonal balance	[93]

Nanoparticle-based drug delivery offers an innovative and efficient method for targeting HIF-1α in the treatment of ovarian cancer, while also addressing issues related to female infertility. The integration of HIF-1α-targeted nanoparticles with antioxidant therapy and reproductive medicine has the potential to significantly alter treatment paradigms for ovarian cancer as nanomedicine advances. Personalized approaches and advanced drug delivery systems may enable effective control of ovarian cancer while preserving reproductive potential, thereby enhancing quality of life and long-term health outcomes for affected women.

5. Limitations and Future Directions

5.1. Research Gaps and Emerging Trends

Although there is an increasing amount of evidence connecting Hypoxia-Inducible Factor-1α (HIF-1α) to the progression of ovarian cancer and female infertility, significant gaps persist in our comprehension of this relationship. Preclinical studies have established the involvement of HIF-1α in tumor angiogenesis, metabolic reprogramming, EMT, and therapy resistance; however, the translation of these findings into clinical applications poses considerable challenges. A significant limitation is the absence of extensive clinical trials assessing HIF-1α inhibitors in patients with ovarian cancer. Small-molecule inhibitors, such as PX-478 and YC-1, demonstrate potential in preclinical models; however, their long-term efficacy, specificity, and safety in human patients are still uncertain [94,95]. The potential side effects of HIF-1α suppression on normal ovarian physiology and fertility require further investigation.

Recent trends indicate that precision medicine and genetic profiling may improve the efficacy of HIF-1α-targeted therapies. Identifying patient subgroups with unique HIF-1α expression patterns may enhance treatment strategies and reduce adverse effects. Advancements in nanoparticle-based drug delivery present a promising approach to enhance drug bioavailability, minimize toxicity, and improve treatment specificity [96]. A significant research gap pertains to the dual influence of HIF-1α on ovarian cancer and fertility. Targeting HIF-1α may inhibit tumor progression; however, its involvement in normal ovarian function is intricate. Future research should aim to clarify the molecular mechanisms through which HIF-1α dysregulation impacts follicular development, oocyte quality, and ovarian reserve.

5.2. Integrating Cancer Treatment with Fertility Preservation Strategies

A major challenge in ovarian cancer management is the need to balance effective treatment with the preservation of fertility. Conventional cancer treatments, such as chemotherapy and radiation, frequently result in ovarian failure, thereby diminishing the likelihood of natural conception. Assisted reproductive technologies (ART), including ovarian tissue cryopreservation and in vitro maturation (IVM), present potential benefits for cancer survivors [97]; however, their success rates are inconsistent and often remain in the experimental phase across various contexts. Integrating HIF-1α-targeted therapies with fertility preservation strategies is essential to address this gap. Co-administration of antioxidants, such as resveratrol and melatonin, with cancer treatments may protect ovarian function and mitigate oxidative stress-induced damage [98]. Utilizing HIF-1α inhibitors alongside ovarian cryoprotection techniques may provide a more integrated strategy for enhancing reproductive outcomes in cancer patients [17]. Future research should investigate personalized treatment strategies that consider both oncological and reproductive outcomes. The integration of oncology, reproductive medicine, and molecular biology facilitates the development of innovative therapeutic strategies aimed at concurrently addressing ovarian cancer and preserving fertility, thereby enhancing the quality of life for affected women.

6. Conclusions

Hypoxia-Inducible Factor-1α (HIF-1α) is integral to the progression of ovarian cancer, facilitating angiogenesis, metabolic reprogramming, epithelial-to-mesenchymal transition, and resistance to therapy. In addition to its oncogenic effects, dysregulation of HIF-1α adversely affects female fertility by impairing ovarian function, increasing oxidative stress, triggering chronic inflammation, and disrupting hormonal balance. Examining the complex relationship between HIF-1α, cancer progression, and reproductive health is essential for formulating targeted therapeutic strategies that enhance cancer prognosis while safeguarding fertility. Various therapeutic strategies have been developed, such as HIF-1α inhibitors, antioxidant therapy, anti-inflammatory agents, and nanoparticle-based drug delivery systems [99]. Despite the significant potential of these strategies, clinical translation is hindered by a scarcity of human trials, possible side effects, and the intricate nature of ovarian physiology. Future research must prioritize the integration of cancer treatment with fertility preservation strategies, utilizing advancements in precision medicine, nanotechnology, and reproductive biomedicine. A multidisciplinary approach integrating oncology and reproductive medicine is crucial for enhancing treatment outcomes. Tailoring therapies to individual patient needs can effectively combat ovarian cancer while preserving reproductive potential, thereby enhancing both survival rates and quality of life for affected women. Ongoing research and collaboration in this field will facilitate the development of innovative, patient-centered therapeutic solutions.

Author Contributions: Conceptualization, M.A.R. and A.H.H.; writing—original draft, M.A.R., M.J., S.K.B., M.A.-Z. and A.H.H.; writing—review and editing, M.A.R., M.J., S.K.B. and M.A.-Z.; visualization, M.A.R., M.J., S.K.B., M.A.-Z. and A.H.H.; supervision, A.H.H.; funding acquisition, M.J. All authors have read and agreed to the published version of the manuscript.

Funding: This work was supported and funded by the Deanship of Scientific Research at Imam Mohammad Ibn Saud Islamic University (IMSIU) (grant number IMSIU-DDRSP2501).

Institutional Review Board Statement: Not applicable.

Informed Consent Statement: Not applicable.

Data Availability Statement: No new data were created or analyzed in this study.

Conflicts of Interest: The authors declare no conflicts of interest.

References

1. Zhao, Y.; Xing, C.; Deng, Y.; Ye, C.; Peng, H. HIF-1α signaling: Essential roles in tumorigenesis and implications in targeted therapies. *Genes Dis.* **2024**, *11*, 234–251. [CrossRef] [PubMed]
2. Fakhri, S.; Moradi, S.Z.; Faraji, F.; Kooshki, L.; Webber, K.; Bishayee, A. Modulation of hypoxia-inducible factor-1 signaling pathways in cancer angiogenesis, invasion, and metastasis by natural compounds: A comprehensive and critical review. *Cancer Metastasis Rev.* **2024**, *43*, 501–574. [CrossRef] [PubMed]
3. Muttiah, B.; Muhammad Fuad, N.D.; Jaafar, F.; Abdullah, N.A.H. Extracellular Vesicles in Ovarian Cancer: From Chemoresistance Mediators to Therapeutic Vectors. *Biomedicines* **2024**, *12*, 1806. [CrossRef]
4. Dai, W.; Guo, R.; Na, X.; Jiang, S.; Liang, J.; Guo, C.; Fang, Y.; Na, Z.; Li, D. Hypoxia and the endometrium: An indispensable role for HIF-1α as therapeutic strategies. *Redox Biol.* **2024**, *73*, 103205. [CrossRef]
5. Kumari, P.; Bharti, V.K.; Kumar, K.; Sharma, I. Hypobaric hypoxia affects the reproductive physiology of dairy cattle. *Anim. Reprod. Update* **2023**, *3*, 6–17. [CrossRef]
6. Xu, Y.-P.; Fu, J.-C.; Hong, Z.-L.; Zeng, D.-F.; Guo, C.-Q.; Li, P.; Wu, J.-X. Psychological stressors involved in the pathogenesis of premature ovarian insufficiency and potential intervention measures. *Gynecol. Endocrinol.* **2024**, *40*, 2360085. [CrossRef]
7. Zhao, H.; Wong, R.J.; Stevenson, D.K. The impact of hypoxia in early pregnancy on placental cells. *Int. J. Mol. Sci.* **2021**, *22*, 9675. [CrossRef]
8. Chiang, Y.-T.; Seow, K.-M.; Chen, K.-H. The pathophysiological, genetic, and hormonal changes in preeclampsia: A systematic review of the molecular mechanisms. *Int. J. Mol. Sci.* **2024**, *25*, 4532. [CrossRef]
9. Shabir, S.; Gill, P.K. Global scenario on ovarian cancer–Its dynamics, relative survival, treatment, and epidemiology. *Adesh Univ. J. Med. Sci. Res.* **2020**, *2*, 17–25. [CrossRef]
10. Konishi, I.; Abiko, K.; Hayashi, T.; Koshiyama, M.; Matsumura, N.; Baba, T.; Yamanoi, K.; Yamaguchi, K.; Hamanishi, J.; Mandai, M. Ovarian Cancer Screening: Where Do We Stand Now? *Acad. Oncol.* **2025**, *2*. [CrossRef]
11. Tavares, V.; Marques, I.S.; Melo, I.G.d.; Assis, J.; Pereira, D.; Medeiros, R. Paradigm Shift: A Comprehensive Review of Ovarian Cancer Management in an Era of Advancements. *Int. J. Mol. Sci.* **2024**, *25*, 1845. [CrossRef] [PubMed]

12. Liu, Y.; Kong, L.; Yu, Y.; Zang, J.; Zhang, L.; Guo, R.-B.; Li, S.-T.; Cheng, L.; Li, X.-T.; Chen, Y.-Q. Tumor Microenvironment Responsive Key Nanomicelles for Effective Against Invasion and Metastasis in Ovarian Cancer Using Mice Model. *Int. J. Nanomed.* **2025**, *20*, 215–238. [CrossRef] [PubMed]
13. Manian, M.; Taherian, M.; Nickho, H.; Emami Nejad, A.; Shaverdi, S. Hypoxia, Stem Cells and Cancer Stem Cells. In *Cancer Stem Cells and Cancer Therapy*; Springer: Berlin/Heidelberg, Germany, 2025; pp. 29–114.
14. Tirpe, A.A.; Gulei, D.; Ciortea, S.M.; Crivii, C.; Berindan-Neagoe, I. Hypoxia: Overview on hypoxia-mediated mechanisms with a focus on the role of HIF genes. *Int. J. Mol. Sci.* **2019**, *20*, 6140. [CrossRef]
15. Zou, Z.; Luo, T.; Wang, X.; Wang, B.; Li, Q. Exploring the interplay between triple-negative breast cancer stem cells and tumor microenvironment for effective therapeutic strategies. *J. Cell. Physiol.* **2024**, *239*, e31278. [CrossRef]
16. Yu, X.; Xu, J.; Song, B.; Zhu, R.; Liu, J.; Liu, Y.F.; Ma, Y.J. The role of epigenetics in women's reproductive health: The impact of environmental factors. *Front. Endocrinol.* **2024**, *15*, 1399757. [CrossRef]
17. Wang, X.; Du, Z.-w.; Xu, T.-m.; Wang, X.-j.; Li, W.; Gao, J.-l.; Li, J.; Zhu, H. HIF-1α is a rational target for future ovarian cancer therapies. *Front. Oncol.* **2021**, *11*, 785111. [CrossRef]
18. Ma, Z.; Xiang, X.; Li, S.; Xie, P.; Gong, Q.; Goh, B.C.; Wang, L. Targeting hypoxia-inducible factor-1, for cancer treatment: Recent advances in developing small-molecule inhibitors from natural compounds. *Semin. Cancer Biol.* **2022**, *80*, 379–390. [CrossRef]
19. Fang, J.; Yan, L.; Shing, Y.; Moses, M.A. HIF-1α-mediated up-regulation of vascular endothelial growth factor, independent of basic fibroblast growth factor, is important in the switch to the angiogenic phenotype during early tumorigenesis. *Cancer Res.* **2001**, *61*, 5731–5735.
20. Zhou, W.; Zeng, T.; Chen, J.; Tang, X.; Yuan, Y.; Hu, D.; Zhang, Y.; Li, Y.; Zou, J. Aberrant angiogenic signaling pathways: Accomplices in ovarian cancer progression and treatment. *Cell. Signal.* **2024**, *120*, 111240. [CrossRef]
21. Basheeruddin, M.; Qausain, S. Hypoxia-Inducible Factor 1-Alpha (HIF-1α) and Cancer: Mechanisms of Tumor Hypoxia and Therapeutic Targeting. *Cureus* **2024**, *16*, e70700. [CrossRef]
22. Oliveira, R.C.; Cavalcante, G.C.; Soares-Souza, G.B. Exploring Aerobic Energy Metabolism in Breast Cancer: A Mutational Profile of Glycolysis and Oxidative Phosphorylation. *Int. J. Mol. Sci.* **2024**, *25*, 2585. [CrossRef] [PubMed]
23. Shi, Y.; Lin, X.; Wang, J.; Zhou, Z.; Chen, S.; Chen, G. Advances of HIF-1α/glycolysis axis in non-small cell lung cancer. *Oncol. Rep.* **2024**, *51*, 55. [CrossRef]
24. Yao, W.; Wang, Z.; Ma, H.; Lin, Y.; Liu, X.; Li, P.; He, X. Epithelial-mesenchymal plasticity (EMP) in wound healing: Exploring EMT mechanisms, regulatory network, and therapeutic opportunities. *Heliyon* **2024**, *10*, e34269. [CrossRef] [PubMed]
25. Mukerjee, N.; Nag, S.; Bhattacharya, B.; Alexiou, A.; Mirgh, D.; Mukherjee, D.; Adhikari, M.D.; Anand, K.; Muthusamy, R.; Gorai, S. Clinical impact of epithelial–mesenchymal transition for cancer therapy. *Clin. Transl. Discov.* **2024**, *4*, e260. [CrossRef]
26. Tam, S.Y.; Wu, V.W.; Law, H.K. Hypoxia-induced epithelial-mesenchymal transition in cancers: HIF-1α and beyond. *Front. Oncol.* **2020**, *10*, 486. [CrossRef]
27. Ucaryilmaz Metin, C.; Ozcan, G. The HIF-1α as a potent inducer of the hallmarks in gastric cancer. *Cancers* **2022**, *14*, 2711. [CrossRef]
28. Yong, L.; Tang, S.; Yu, H.; Zhang, H.; Zhang, Y.; Wan, Y.; Cai, F. The role of hypoxia-inducible factor-1 alpha in multidrug-resistant breast cancer. *Front. Oncol.* **2022**, *12*, 964934. [CrossRef]
29. Qannita, R.A.; Alalami, A.I.; Harb, A.A.; Aleidi, S.M.; Taneera, J.; Abu-Gharbieh, E.; El-Huneidi, W.; Saleh, M.A.; Alzoubi, K.H.; Semreen, M.H. Targeting hypoxia-inducible Factor-1 (HIF-1) in Cancer: Emerging therapeutic strategies and pathway regulation. *Pharmaceuticals* **2024**, *17*, 195. [CrossRef]
30. Li, W.-N.; Wu, M.-H.; Tsai, S.-J. Hypoxia and reproductive health: The role of hypoxia in the development and progression of endometriosis. *Reproduction* **2021**, *161*, F19–F31. [CrossRef]
31. Pereira, M.; Matuszewska, K.; Jamieson, C.; Petrik, J. Characterizing endocrine status, tumor hypoxia and immunogenicity for therapy success in epithelial ovarian cancer. *Front. Endocrinol.* **2021**, *12*, 772349. [CrossRef]
32. Guo, Y.; Xue, L.; Tang, W.; Xiong, J.; Chen, D.; Dai, Y.; Wu, C.; Wei, S.; Dai, J.; Wu, M.; et al. Ovarian microenvironment: Challenges and opportunities in protecting against chemotherapy-associated ovarian damage. *Hum. Reprod. Update* **2024**, *30*, 614–647. [CrossRef] [PubMed]
33. Thompson, J.G.; Brown, H.M.; Kind, K.L.; Russell, D.L. The Ovarian Antral Follicle: Living on the Edge of Hypoxia or Not? *Biol Reprod.* **2015**, *92*, 153. [CrossRef] [PubMed]
34. Pan, Y.; Pan, C.; Zhang, C. Unraveling the complexity of follicular fluid: Insights into its composition, function, and clinical implications. *J. Ovarian Res.* **2024**, *17*, 237. [CrossRef] [PubMed]
35. Ray, A.; Cleary, M.P. The potential role of leptin in tumor invasion and metastasis. *Cytokine Growth Factor. Rev.* **2017**, *38*, 80–97. [CrossRef]
36. Guo, X.; Yi, H.; Li, T.C.; Wang, Y.; Wang, H.; Chen, X. Role of vascular endothelial growth factor (VEGF) in human embryo implantation: Clinical implications. *Biomolecules* **2021**, *11*, 253. [CrossRef]

37. Shi, Y.-Q.; Zhu, X.-T.; Zhang, S.-N.; Ma, Y.-F.; Han, Y.-H.; Jiang, Y.; Zhang, Y.-H. Premature ovarian insufficiency: A review on the role of oxidative stress and the application of antioxidants. *Front. Endocrinol.* **2023**, *14*, 1172481. [CrossRef]
38. Baldini, G.M.; Ferri, D.; Malvasi, A.; Laganà, A.S.; Vimercati, A.; Dellino, M.; Baldini, D.; Trojano, G. Genetic Abnormalities of Oocyte Maturation: Mechanisms and Clinical Implications. *Int. J. Mol. Sci.* **2024**, *25*, 13002. [CrossRef]
39. Gualtieri, R.; Kalthur, G.; Barbato, V.; Di Nardo, M.; Adiga, S.K.; Talevi, R. Mitochondrial dysfunction and oxidative stress caused by cryopreservation in reproductive cells. *Antioxidants* **2021**, *10*, 337. [CrossRef]
40. Shandley, L.M.; Fothergill, A.; Spencer, J.B.; Mertens, A.C.; Cottrell, H.N.; Howards, P.P. Impact of cancer treatment on risk of infertility and diminished ovarian reserve in women with polycystic ovary syndrome. *Fertil. Steril.* **2018**, *109*, 516–525.e511. [CrossRef]
41. Zhang, J.; Yao, M.; Xia, S.; Zeng, F.; Liu, Q. Systematic and comprehensive insights into HIF-1 stabilization under normoxic conditions: Implications for cellular adaptation and therapeutic strategies in cancer. *Cell Mol Biol Lett.* **2025**, *30*, 2. [CrossRef]
42. Wang, X.; Liu, Y.; Wang, J.; Lu, X.; Guo, Z.; Lv, S.; Sun, Z.; Gao, T.; Gao, F.; Yuan, J. Mitochondrial Quality Control in Ovarian Function: From Mechanisms to Therapeutic Strategies. *Reprod. Sci.* **2024**, 1–15. [CrossRef] [PubMed]
43. Zhou, C.; Li, D.; He, J.; Luo, T.; Liu, Y.; Xue, Y.; Huang, J.; Zheng, L.; Li, J. TRIM28-Mediated Excessive Oxidative Stress Induces Cellular Senescence in Granulosa Cells and Contributes to Premature Ovarian Insufficiency In Vitro and In Vivo. *Antioxidants* **2024**, *13*, 308. [CrossRef] [PubMed]
44. Long, Y.; Shi, H.; Ye, J.; Qi, X. Exploring Strategies to Prevent and Treat Ovarian Cancer in Terms of Oxidative Stress and Antioxidants. *Antioxidants* **2025**, *14*, 114. [CrossRef] [PubMed]
45. Ullah, A.; Chen, Y.; Singla, R.K.; Cao, D.; Shen, B. Pro-inflammatory Cytokines and CXC Chemokines as Game-Changer in Age-Associated Prostate Cancer and Ovarian Cancer: Insights from Preclinical and Clinical Studies' Outcomes. *Pharmacol. Res.* **2024**, *204*, 107213. [CrossRef]
46. Hoffmann-Młodzianowska, M.; Maksym, R.B.; Pucia, K.; Kuciak, M.; Mackiewicz, A.; Kieda, C. Endometriosis development in relation to hypoxia: A murine model study. *Mol. Med.* **2024**, *30*, 195. [CrossRef]
47. Oală, I.E.; Mitranovici, M.-I.; Chiorean, D.M.; Irimia, T.; Crișan, A.I.; Melinte, I.M.; Cotruș, T.; Tudorache, V.; Moraru, L.; Moraru, R. Endometriosis and the role of pro-inflammatory and anti-inflammatory cytokines in pathophysiology: A narrative review of the literature. *Diagnostics* **2024**, *14*, 312. [CrossRef]
48. Goldsammler, M.; Merhi, Z.; Buyuk, E. Role of hormonal and inflammatory alterations in obesity-related reproductive dysfunction at the level of the hypothalamic-pituitary-ovarian axis. *Reprod. Biol. Endocrinol.* **2018**, *16*, 1–10. [CrossRef]
49. Wu, G.; Li, C.; Tao, J.; Liu, Z.; Li, X.; Zang, Z.; Fu, C.; Wei, J.; Yang, Y.; Zhu, Q. FSH mediates estradiol synthesis in hypoxic granulosa cells by activating glycolytic metabolism through the HIF-1α–AMPK–GLUT1 signaling pathway. *J. Biol. Chem.* **2022**, *298*, 101830. [CrossRef]
50. Neykova, K.; Tosto, V.; Giardina, I.; Tsibizova, V.; Vakrilov, G. Endometrial receptivity and pregnancy outcome. *J. Matern.-Fetal Neonatal Med.* **2022**, *35*, 2591–2605. [CrossRef]
51. Shaw, S.; Gidugu, H.; Bhaumik, G.; Reddy, M.P.K.; Panjwani, U.; Ghosh, D. Anti-Mullerian hormone and macrophage migration inhibitory factor determine the reproductive health of Ladakhi women residing at 3,500 m. *High. Alt. Med. Biol.* **2021**, *22*, 317–326. [CrossRef]
52. Fu, X.; Shi, L.; Liu, P.; Jiao, Y.; Guo, S.; Chen, Q.; Zheng, Q.; Chen, X.; Wang, Y. Expression and clinical significance of HIF-1α in follicular fluid and granulosa cells in infertile PCOS patients. *Reprod. Sci.* **2023**, *30*, 2263–2274. [CrossRef] [PubMed]
53. Zhang, L.-Y.; Zhang, K.; Zhao, X.; Tao, H.-P.; Jia, G.-X.; Fang, Y.-G.; Hou, Y.-P.; Yang, Q.-E. Fetal hypoxia exposure induces Hif1a activation and autophagy in adult ovarian granulosa cells. *Biol. Reprod.* **2024**, *111*, 1220–1234. [CrossRef] [PubMed]
54. Leal, C.R.; Zanolla, K.; Spritzer, P.M.; Reis, F.M. Assisted reproductive technology in the presence of polycystic ovary syndrome: Current evidence and knowledge gaps. *Endocr. Pract.* **2024**, *30*, 64–69. [CrossRef]
55. Garlisi, B.; Lauks, S.; Aitken, C.; Ogilvie, L.M.; Lockington, C.; Petrik, D.; Eichhorn, J.S.; Petrik, J. The complex tumor microenvironment in ovarian cancer: Therapeutic challenges and opportunities. *Curr. Oncol.* **2024**, *31*, 3826–3844. [CrossRef]
56. Welsh, S.; Williams, R.; Kirkpatrick, L.; Paine-Murrieta, G.; Powis, G. Antitumor activity and pharmacodynamic properties of PX-478, an inhibitor of hypoxia-inducible factor-1α. *Mol. Cancer Ther.* **2004**, *3*, 233–244. [CrossRef]
57. Piorecka, K.; Kurjata, J.; Stanczyk, W.A. Acriflavine, an Acridine Derivative for Biomedical Application: Current State of the Art. *J. Med. Chem.* **2022**, *65*, 11415–11432. [CrossRef]
58. Kong, D.; Park, E.J.; Stephen, A.G.; Calvani, M.; Cardellina, J.H.; Monks, A.; Fisher, R.J.; Shoemaker, R.H.; Melillo, G. Echinomycin, a small-molecule inhibitor of hypoxia-inducible factor-1 DNA-binding activity. *Cancer Res.* **2005**, *65*, 9047–9055. [CrossRef]
59. Zhang, H.; Qian, D.Z.; Tan, Y.S.; Lee, K.; Gao, P.; Ren, Y.R.; Rey, S.; Hammers, H.; Chang, D.; Pili, R. Digoxin and other cardiac glycosides inhibit HIF-1α synthesis and block tumor growth. *Proc. Natl. Acad. Sci. USA* **2008**, *105*, 19579–19586. [CrossRef]
60. Becker, C.M.; Rohwer, N.; Funakoshi, T.; Cramer, T.; Bernhardt, W.; Birsner, A.; Folkman, J.; D'Amato, R.J. 2-Methoxyestradiol inhibits hypoxia-inducible factor-1α and suppresses growth of lesions in a mouse model of endometriosis. *Am. J. Pathol.* **2008**, *172*, 534–544. [CrossRef]

61. Bui, B.P.; Nguyen, P.L.; Lee, K.; Cho, J. Hypoxia-inducible factor-1: A novel therapeutic target for the management of cancer, drug resistance, and cancer-related pain. *Cancers* **2022**, *14*, 6054. [CrossRef]
62. Lee, K.; Kang, J.E.; Park, S.-K.; Jin, Y.; Chung, K.-S.; Kim, H.-M.; Lee, K.; Kang, M.R.; Lee, M.K.; Song, K.B. LW6, a novel HIF-1 inhibitor, promotes proteasomal degradation of HIF-1α via upregulation of VHL in a colon cancer cell line. *Biochem. Pharmacol.* **2010**, *80*, 982–989. [CrossRef] [PubMed]
63. Kim, Y.H.; Coon, A.; Baker, A.F.; Powis, G. Antitumor agent PX-12 inhibits HIF-1α protein levels through an Nrf2/PMF-1-mediated increase in spermidine/spermine acetyl transferase. *Cancer Chemother. Pharmacol.* **2011**, *68*, 405–413. [CrossRef] [PubMed]
64. Kao, C.; Chao, A.; Tsai, C.; Chuang, W.; Huang, W.; Chen, G.; Lin, C.; Wang, T.; Wang, H.; Lai, C. Bortezomib enhances cancer cell death by blocking the autophagic flux through stimulating ERK phosphorylation. *Cell Death Dis.* **2014**, *5*, e1510. [CrossRef]
65. Sun, S.; Wang, P.; Sun, S.; Liang, X. Applications of micro/nanotechnology in ultrasound-based drug delivery and therapy for tumor. *Curr. Med. Chem.* **2021**, *28*, 525–547. [CrossRef]
66. Begines, B.; Ortiz, T.; Pérez-Aranda, M.; Martínez, G.; Merinero, M.; Argüelles-Arias, F.; Alcudia, A. Polymeric nanoparticles for drug delivery: Recent developments and future prospects. *Nanomaterials* **2020**, *10*, 1403. [CrossRef]
67. Wang, L.; Tang, J.; Wang, L.; Tan, F.; Song, H.; Zhou, J.; Li, F. Oxidative stress in oocyte aging and female reproduction. *J. Cell. Physiol.* **2021**, *236*, 7966–7983. [CrossRef]
68. Mozafaryan, M.J.; Rezaei, P.; Masoudpoor, F.; Bahri, A.; Samini, M.; Farkhondeh, T.; Pourhanifeh, M.H.; Samarghandian, S. Resveratrol in the Treatment of Gynecological Cancer: Mechanisms and Therapeutic Potential. *Curr. Med. Chem.* **2025**. [CrossRef]
69. Lee, E.J.; Park, S.J.; Lee, C.; Yim, G.W.; Kim, J.W.; Kim, H.S. Hypoxia-induced maspin expression affects the prognosis of ovarian clear cell carcinoma. *Vivo* **2022**, *36*, 212–220. [CrossRef]
70. Tamura, H.; Takasaki, A.; Taketani, T.; Tanabe, M.; Kizuka, F.; Lee, L.; Tamura, I.; Maekawa, R.; Asada, H.; Yamagata, Y. Melatonin as a free radical scavenger in the ovarian follicle. *Endocr. J.* **2013**, *60*, 1–13. [CrossRef]
71. Guo, Y.M.; Sun, T.C.; Wang, H.P.; Chen, X. Research progress of melatonin (MT) in improving ovarian function: A review of the current status. *Aging* **2021**, *13*, 17930. [CrossRef]
72. Savant, S.S.; Sriramkumar, S.; O'Hagan, H.M. The role of inflammation and inflammatory mediators in the development, progression, metastasis, and chemoresistance of epithelial ovarian cancer. *Cancers* **2018**, *10*, 251. [CrossRef] [PubMed]
73. DiGiacomo, J.W.; Gilkes, D.M. Tumor hypoxia as an enhancer of inflammation-mediated metastasis: Emerging therapeutic strategies. *Target. Oncol.* **2018**, *13*, 157–173. [CrossRef] [PubMed]
74. Lin, Y.G.; Kunnumakkara, A.B.; Nair, A.; Merritt, W.M.; Han, L.Y.; Armaiz-Pena, G.N.; Kamat, A.A.; Spannuth, W.A.; Gershenson, D.M.; Lutgendorf, S.K. Curcumin inhibits tumor growth and angiogenesis in ovarian carcinoma by targeting the nuclear factor-κB pathway. *Clin. Cancer Res.* **2007**, *13*, 3423–3430. [CrossRef] [PubMed]
75. Guo, K.; Feng, Y.; Zheng, X.; Sun, L.; Wasan, H.S.; Ruan, S.; Shen, M. Resveratrol and its analogs: Potent agents to reverse epithelial-to-mesenchymal transition in tumors. *Front. Oncol.* **2021**, *11*, 644134. [CrossRef]
76. Finetti, F.; Travelli, C.; Ercoli, J.; Colombo, G.; Buoso, E.; Trabalzini, L. Prostaglandin E2 and Cancer: Insight into Tumor Progression and Immunity. *Biology* **2020**, *9*, 434. [CrossRef]
77. Zare, H.; Shafabakhsh, R.; Reiter, R.J.; Asemi, Z. Melatonin is a potential inhibitor of ovarian cancer: Molecular aspects. *J. Ovarian Res.* **2019**, *12*, 1–8. [CrossRef]
78. Pastorek, M.; Simko, V.; Takacova, M.; Barathova, M.; Bartosova, M.; Hunakova, L.; Sedlakova, O.; Hudecova, S.; Krizanova, O.; Dequiedt, F. Sulforaphane reduces molecular response to hypoxia in ovarian tumor cells independently of their resistance to chemotherapy. *Int. J. Oncol.* **2015**, *47*, 51–60. [CrossRef]
79. Freitas, R.D.; Campos, M.M. Protective effects of omega-3 fatty acids in cancer-related complications. *Nutrients* **2019**, *11*, 945. [CrossRef]
80. Vafadar, A.; Shabaninejad, Z.; Movahedpour, A.; Fallahi, F.; Taghavipour, M.; Ghasemi, Y.; Akbari, M.; Shafiee, A.; Hajighadimi, S.; Moradizarmehri, S.; et al. Quercetin and cancer: New insights into its therapeutic effects on ovarian cancer cells. *Cell Biosci.* **2020**, *10*, 32. [CrossRef]
81. Shahveghar Asl, Z.; Parastouei, K.; Eskandari, E. The effects of N-acetylcysteine on ovulation and sex hormones profile in women with polycystic ovary syndrome: A systematic review and meta-analysis. *Br. J. Nutr.* **2023**, *130*, 202–210. [CrossRef]
82. Yadav, V.R.; Prasad, S.; Sung, B.; Gelovani, J.G.; Guha, S.; Krishnan, S.; Aggarwal, B.B. Boswellic acid inhibits growth and metastasis of human colorectal cancer in orthotopic mouse model by downregulating inflammatory, proliferative, invasive and angiogenic biomarkers. *Int. J. Cancer* **2012**, *130*, 2176–2184. [CrossRef] [PubMed]
83. Ayustaningwarno, F.; Anjani, G.; Ayu, A.M.; Fogliano, V. A critical review of Ginger's (*Zingiber officinale*) antioxidant, anti-inflammatory, and immunomodulatory activities. *Front. Nutr.* **2024**, *11*, 1364836. [CrossRef] [PubMed]
84. Rocha, G.Z.; Dias, M.M.; Ropelle, E.R.; Osório-Costa, F.; Rossato, F.A.; Vercesi, A.E.; Saad, M.J.; Carvalheira, J.B. Metformin amplifies chemotherapy-induced AMPK activation and antitumoral growth. *Clin. Cancer Res.* **2011**, *17*, 3993–4005. [CrossRef] [PubMed]

85. Ayhan, S.; Hancerliogullari, N.; Guney, G.; Gozukucuk, M.; Caydere, M.; Guney, S.S.; Tokmak, A.; Ustun, Y. Does the addition of metformin to carboplatin treatment decreases ovarian reserve damage associated with carboplatin usage? *J. Ovarian Res.* **2023**, *16*, 184. [CrossRef]
86. Mills, G.B. Mechanisms underlying chemoprevention of ovarian cancer. *Clin. Cancer Res.* **2002**, *8*, 7–10.
87. Baandrup, L.; Faber, M.T.; Christensen, J.; Jensen, A.; Andersen, K.K.; Friis, S.; Kjaer, S.K. Nonsteroidal anti-inflammatory drugs and risk of ovarian cancer: Systematic review and meta-analysis of observational studies. *Acta Obs. Gynecol. Scand.* **2013**, *92*, 245–255. [CrossRef]
88. Chen, Z.; Zhang, T.; Wu, B.; Zhang, X. Insights into the therapeutic potential of hypoxia-inducible factor-1α small interfering RNA in malignant melanoma delivered via folate-decorated cationic liposomes. *Int. J. Nanomed.* **2016**, *11*, 991–1002.
89. Mundekkad, D.; Cho, W. Applications of curcumin and its nanoforms in the treatment of cancer. *Pharmaceutics* **2023**, *15*, 2223. [CrossRef]
90. Battaglia, R.; Caponnetto, A.; Caringella, A.; Cortone, A.; Ferrara, C.; Smirni, S.; Iannitti, R.; Purrello, M.; D'amato, G.; Fioretti, B. Resveratrol treatment induces Mito-miRNome modification in follicular fluid from aged women with a poor prognosis for in vitro fertilization cycles. *Antioxidants* **2022**, *11*, 1019. [CrossRef]
91. Li, X.; Wang, Z.; Wang, H.; Xu, H.; Sheng, Y.; Lian, F. Role of N-acetylcysteine treatment in women with advanced age undergoing IVF/ICSI cycles: A prospective study. *Front. Med.* **2022**, *9*, 917146. [CrossRef]
92. Perche, F.; Patel, N.R.; Torchilin, V.P. Accumulation and toxicity of antibody-targeted doxorubicin-loaded PEG–PE micelles in ovarian cancer cell spheroid model. *J. Control. Release* **2012**, *164*, 95–102. [CrossRef] [PubMed]
93. Shalaby, O.E.; Ahmed, Y.H.; Mekkawy, A.M.; Mahmoud, M.Y.; Elbargeesy, G. The ameliorative effect of selenium-loaded chitosan nanoparticles against silver nanoparticles-induced ovarian toxicity in female albino rats. *J. Ovarian Res.* **2025**, *18*, 4. [CrossRef] [PubMed]
94. Koh, M.Y.; Spivak-Kroizman, T.; Venturini, S.; Welsh, S.; Williams, R.R.; Kirkpatrick, D.L.; Powis, G. Molecular mechanisms for the activity of PX-478, an antitumor inhibitor of the hypoxia-inducible factor-1α. *Mol. Cancer Ther.* **2008**, *7*, 90–100. [CrossRef]
95. Chau, N.-M.; Rogers, P.; Aherne, W.; Carroll, V.; Collins, I.; McDonald, E.; Workman, P.; Ashcroft, M. Identification of novel small molecule inhibitors of hypoxia-inducible factor-1 that differentially block hypoxia-inducible factor-1 activity and hypoxia-inducible factor-1α induction in response to hypoxic stress and growth factors. *Cancer Res.* **2005**, *65*, 4918–4928. [CrossRef]
96. Elumalai, K.; Srinivasan, S.; Shanmugam, A. Review of the efficacy of nanoparticle-based drug delivery systems for cancer treatment. *Biomed. Technol.* **2024**, *5*, 109–122. [CrossRef]
97. Abir, R.; Ben-Aharon, I.; Garor, R.; Yaniv, I.; Ash, S.; Stemmer, S.; Ben-Haroush, A.; Freud, E.; Kravarusic, D.; Sapir, O. Cryopreservation of in vitro matured oocytes in addition to ovarian tissue freezing for fertility preservation in paediatric female cancer patients before and after cancer therapy. *Hum. Reprod.* **2016**, *31*, 750–762. [CrossRef]
98. Al-Shahat, A.; Hulail, M.A.; Soliman, N.M.; Khamis, T.; Fericean, L.M.; Arisha, A.H.; Moawad, R.S. Melatonin mitigates cisplatin-induced ovarian dysfunction via altering steroidogenesis, inflammation, apoptosis, oxidative stress, and PTEN/PI3K/Akt/mTOR/AMPK signaling pathway in female rats. *Pharmaceutics* **2022**, *14*, 2769. [CrossRef]
99. Luo, J.; Wang, H.; Chen, J.; Wei, X.; Feng, J.; Zhang, Y.; Zhou, Y. The Application of Drugs and Nano-Therapies Targeting Immune Cells in Hypoxic Inflammation. *Int. J. Nanomed.* **2024**, *19*, 3441–3459. [CrossRef]

cells

MDPI

Article

Exploiting Cancer Dormancy Signaling Mechanisms in Epithelial Ovarian Cancer Through Spheroid and Organoid Analysis

Emily J. Tomas [1,2], Yudith Ramos Valdes [1], Jennifer Davis [1], Bart Kolendowski [1], Adrian Buensuceso [1], Gabriel E. DiMattia [1,3,4] and Trevor G. Shepherd [1,2,3,5,*]

1. The Mary and John Knight Translational Ovarian Cancer Research Unit, Verspeeten Family Cancer Centre, London, ON N6A 5W9, Canada
2. Department of Anatomy & Cell Biology, Western University, London, ON N6A 5C1, Canada
3. Department of Oncology, Western University, London, ON N6A 5C1, Canada
4. Department of Biochemistry, Western University, London, ON N6A 5C1, Canada
5. Department of Obstetrics & Gynaecology, Western University, London, ON N6A 5C1, Canada

* Correspondence: tshephe6@uwo.ca; Tel.: +1-519-685-8500 (ext. 56347)

Abstract: Epithelial ovarian cancer (EOC) exhibits a unique mode of metastasis, involving spheroid formation in the peritoneum. Our research on EOC spheroid cell biology has provided valuable insights into the signaling plasticity associated with metastasis. We speculate that EOC cells modify their biology between tumour and spheroid states during cancer dormancy, although the specific mechanisms underlying this transition remain unknown. Here, we present novel findings from direct comparisons between cultured EOC spheroids and organoids. Our results indicated that AMP-activated protein kinase (AMPK) activity was significantly upregulated and protein kinase B (Akt) was downregulated in EOC spheroids compared to organoids, suggesting a clear differential phenotype. Through RNA sequencing analysis, we further supported these phenotypic differences and highlighted the significance of cell cycle regulation in organoids. By inhibiting the G2/M checkpoint via kinase inhibitors, we confirmed that this pathway is essential for organoids. Interestingly, our results suggest that specifically targeting aurora kinase A (AURKA) may represent a promising therapeutic strategy since our cells were equally sensitive to Alisertib treatment as both spheroids and organoids. Our findings emphasize the importance of studying cellular adaptations of EOC cells, as there may be different therapeutic targets depending on the step of EOC disease progression.

Keywords: epithelial ovarian cancer; high-grade serous; spheroids; organoids; cancer dormancy; transcriptome; G2/M checkpoint pathway

1. Introduction

Epithelial ovarian cancer (EOC) is one of the most lethal forms of gynaecological malignancy, with a survival rate of less than 30% for late-stage presentation, and due to the recurrence of chemotherapy-resistant disease [1]. The most common type of EOC is high-grade serous ovarian cancer (HGSOC), which is often diagnosed at an advanced stage [2,3]. HGSOC originates from the fallopian tube epithelium, where cells acquire an inactivating *TP53* mutation, resulting in further genomic instability and increased invasiveness [4,5]. This leads to the formation of serous tubal intraepithelial carcinoma (STIC) lesions that can seed on the ovary for tumour formation [6]. These malignant tumour cells can detach and aggregate into heterogeneous clusters known as spheroids within the peritoneal cavity [7].

As these spheroids transit within the peritoneal fluid, or malignant ascites, they can adhere to the mesothelial surface and form secondary tumours. Numerous studies have shown that cells undergo changes in their pathobiology as they adapt from a tumour state to a spheroid state [8–11]. This process can be described as cancer dormancy, which explains the clinical problem of cancer recurrence from minimal underlying disease that has evaded chemotherapy [12,13]. Nonetheless, it is currently not well understood in the context of EOC metastasis.

Researchers have used three-dimensional (3D) in vitro culture model systems to investigate the specific cellular and molecular biology of EOC cells during disease progression. Spheroids have been extensively studied using suspension cultures, with Ultra-Low Attachment (ULA) plates and hanging droplet methods, to understand the mechanisms of cell survival [14]. This includes the activation of intracellular signaling pathways that induce cancer dormancy, such as cellular quiescence, cell adhesion, epithelial–mesenchymal transition (EMT), stem-like phenotype, bioenergetic stress response, and autophagy [13–21]. As cells transition from an adherent to spheroid culture, AMP-activated protein kinase (AMPK) activity is induced in response to cell bioenergetic stress, and protein kinase B (Akt) activity is reduced to decrease cell proliferation [18,22–25]. The process of cancer dormancy can be inhibited by targeting these specific proteins in HGSOC cells, leading to cell death within spheroids. Thus, researchers focus on discovering novel targeted agents to effectively eradicate dormant residual disease that may be evading chemotherapy-induced killing.

Several advancements have been made in 3D culture model systems, which have been integrated into current EOC studies. One such model is organoids which is very effective in replicating tumour-like structures [26–28]. Using an extracellular matrix (ECM) reagent, cells from any patient-derived sample (i.e., ascites fluid or dissociated tumours) can proliferate into a 3D structure that reproduces the cell–cell and cell–matrix interactions observed in vivo. With respect to EOC, various groups have made significant contributions to the development of patient-derived organoids (PDOs) and validated their ability to reproduce the patient-specific characteristics of original solid tumours [28–32]. As a result, PDOs have been harnessed for high-throughput drug testing of chemotherapy and many targeted agents, as well as for investigating potential mechanisms of platinum and/or Poly ADP-ribose polymerase (PARP) inhibitor resistance in patients [32–37]. Consequently, it is critical to incorporate 3D organoids into current research to comprehend the biology of HGSOC tumours and distinguish it from spheroid biology.

In this study, we compared the spheroid model system to the organoid model system using several new patient ascites-derived HGSOC cell lines to explore the underlying cellular pathobiology during disease progression. Our results indicate that EOC cells have differential expression patterns between spheroids and organoids, which validates our previous studies comparing adherent to spheroid cultures. We observed altered AMPK activity and Akt signaling when comparing adherent, spheroid and organoid cultures. However, spheroid and organoid assessments may provide a more accurate understanding of biological switching behaviours as cells transition from tumours to spheroids and back to tumour structures as metastatic lesions. This was highlighted with a transcriptomic analysis that confirmed a dormancy phenotype in spheroids and the importance of cell cycle checkpoint regulation in organoids. Overall, our results provided potentially new mechanistic discoveries that could be targeted therapeutically for EOC.

2. Materials and Methods

2.1. Cell Lines

The following HGSOC-established cell lines were used for analysis and purchased from ATCC (Cedarlane, Burlington, ON, Canada): OVCAR8, OVCAR4 and OVCAR3

(Table 1). The following HGSOC patient ascites-derived immortalized cell lines were used: iOvCa182, iOvCa195, iOvCa198, iOvCa246, iOvCa256, iOvCa398 and iOvCa411 (Table 1). iOvCa cell lines were derived from the malignant ascites fluid collected during debulking surgery and continuously passaged on T-75 flasks (Sarstedt, Newton, NC, USA) until immortalized [22]. Patient consent for the clinical specimens from which these cell lines were derived was obtained according to our institution's research ethics board-approved protocol (#115904). The cells were maintained at 37 °C and a CO_2 concentration of 5% in adherent cultures on tissue culture-treated polystyrene plates (Sarstedt, Newton, NC, USA) with 10% foetal bovine serum (FBS) (Wisent, Saint-Jean-Baptiste, QC, Canada) in either RPMI-1640 (Gibco, Grand Island, NY, USA) for OVCAR8 and OVCAR3 cell lines, or DMEM/F-12 (Gibco) for OVCAR4 and iOvCa cell lines. All cell lines were confirmed to be HGSOC based on their marker expression of PAX8 and p53 (Figure S1A,B; Table 1). All cell lines were validated by STR analysis (The Centre for Applied Genomics, The Hospital for Sick Children, Toronto, ON, Canada) and tested for mycoplasma routinely using the Universal Mycoplasma Detection Kit (ATCC; Cedarlane, Burlington, ON, Canada).

Table 1. Overview of characteristics for HGSOC-established and patient ascites-derived immortalized cell lines.

Cell Line	Histotype	FIGO Staging	CNV Status	Mutations †	p53 Immunoblot Signal	Chemotherapy Prior to Derivation
OVCAR8	HGSC			TP53 (SA) # KRAS P121H ERBB2 G776V CTNNB1 E26R	Low	N/A
OVCAR4	HGSC			TP53 L130V #	Low	N/A
OVCAR3	HGSC			TP53 R248Q # PIK3R1 c.1746-2A > T (SA)	High	N/A
iOvCa182	HGSC	IIIB	84.74%	TP53 K132R	High	6 cycles of carboplatin and paclitaxel; trebananib, angiopoietin inhibitor + weekly paclitaxel for 6 months
iOvCa195	Mixed HGSC and Endometrioid	IV	77.89%	TP53 E171fs*61 BRCA1 Y978*	NS	NONE
iOvCa198	HGSC	IIIB	80.37%	TP53 Q192* PIK3CA I391M KDR Q472H	NS	6 cycles of carboplatin and paclitaxel
iOvCa246	HGSC	IIIC	81.09%	TP53 c.560-2A > T (SA)	NS	NONE
iOvCa256	HGSC Endometrial	IIIA	55.63%	TP53 Y220C KIT M541L	Low	4 cycles of carboplatin and weekly paclitaxel
iOvCa398	HGSC	IC	70.88%	n.d.	High	NONE
iOvCa411	HGSC	IIIC	67.84%	TP53 Y234H	High	6 cycles of carboplatin and paclitaxel; 5 cycles doxorubicin; weekly taxol for 8 months; topotecan; etoposide

HGSC—high-grade serous carcinoma. The staging was based on The International Federation of Gynecology and Obstetrics (FIGO) staging system. # Leroy et al. 2014. Human Mutation [38]. † Cellosaurus for mutations in OVCAR cell lines; mutations in iOvCa cell lines were determined based on Oncogene Hotspot analysis. SA—Splice Acceptor Mutation; n.d.—Not determined; NS—no signal (refer to Supplementary Figure S1B).

2.2. Three-Dimensional Spheroid Culture

The cells were seeded at 5×10^5 or 1×10^6 cells/well in suspension using 6-well ULA plates (Corning, New York, NY, USA) and supplemented with their corresponding media (RPMI-1640 or DMEM/F12 with 10% FBS). The spheroids were collected after 3 days in culture.

2.3. Three-Dimensional Organoid Culture

The cells were seeded at 2.5×10^5 or 5×10^5 cells/well and resuspended in 50 μL of Cultrex Basement Membrane Extract (BME) PathClear Type 2 (Cedarlane, Burlington, ON, Canada) as droplets on 24-well tissue culture-treated polystyrene plates (Sarstedt, Newton, NC, USA). The droplets were overlaid with EOC organoid-specific media containing Advanced DMEM/F-12 (Invitrogen, Waltham, MA, USA) and supplemented with B-27™ (Invitrogen, Waltham, MA, USA), Forskolin (Cedarlane, Burlington, ON, Canada), GlutaMAX™ (Invitrogen, Waltham, MA, USA), HEPES (Wisent, Saint-Jean-Baptiste, QC, Canada), Human EGF (Peprotech Inc., Cranbury, NJ, USA), Human FGF-10 (Peprotech Inc., Cranbury, NJ, USA), Nicotinamide (MilliporeSigma, Oakville, ON, Canada), N-Acetyl-L-cysteine (MilliporeSigma, Oakville, ON, Canada), Recombinant Human Noggin (R&D Systems, Toronto, ON, Canada) and Y-27632 dihydrochloride (MilliporeSigma, Oakville, ON, Canada) (Formula provided in Table S1). After 21 days in culture, the organoids were released from Cultrex BME by dissolving the matrix dome in 500 μL of Cell Recovery Solution (Corning, New York, NY, USA) incubated on ice for 1.5–2 h.

2.4. Organoid Growth Analysis

The cells were seeded at 5×10^4 cells/well in 50 μL of Cultrex BME on a 24-well tissue culture plate (Corning, New York, NY, USA). For manual analysis, brightfield images were taken at the 21-day endpoint using a Leica DMI4000B inverted microscope (Teaneck, NJ, USA). ImageJ Software v1.54m was used to manually outline each organoid structure to quantify organoid number (count/well or image) and average organoid area (μm^2). For automated analysis, the Incucyte® S3 System (Sartorius, Oakville, ON, Canada) was used to capture brightfield images at 12 h increments over 21 days. The Incucyte® Organoid Analysis Software Module v2023B (Sartorius, Oakville, ON, Canada) was used to automatically outline each organoid structure to quantify the total organoid area per image or well (μm^2), organoid number, organoid darkness and organoid eccentricity.

2.5. Antibodies

The following primary antibodies, from Cell Signaling Technologies (Whitby, ON, Canada), were utilized at 1:1000 dilution: phosphorylated AMPKα at T172 (2535), AMPKα (5832), phosphorylated Akt at S473 (4060), Akt (9272), AURKA (14475), phosphorylated cdc2 at Y15 (CDK1, 4539), cdc2 (CDK1, 9116), phosphorylated cdc25C at S216 (4901), cdc25C (4688), phosphorylated γH2A.X at S139 (9718), cleaved PARP (9541), p53 (2527), PCNA (2586), PLK1 (4513) and WEE1 (13084). The antibodies for vinculin (V9264, 1:40,000) and actin (A2066, 1:20,000) were purchased from MilliporeSigma (Oakville, ON, Canada). The antibody for PAX8 (10336-1-AP, 1:500) was purchased from Proteintech (Rosemont, IL, USA). The HRP-conjugated secondary antibodies used were anti-rabbit IgG (NA934V, 1:10,000), anti-mouse IgG (NA931V, 1:10,000), or anti-rabbit FITC (F9887, 1:300) from MilliporeSigma (Oakville, ON, Canada).

2.6. Protein Isolation

The adherent cells at 80% confluency were washed with PBS and scraped into a modified radioimmunoprecipitation assay (RIPA) lysis buffer solution. Spheroids and organoids were collected into 15 mL or 50 mL tubes, washed with cold PBS, spun down and then lysed using the same lysis buffer. The cells were stored at −80 °C until the next day and then lysed with vortexing every 5 min. The lysates were centrifuged (21,000× *g*, 4 °C, 20 min) and the supernatant was collected for protein analysis.

2.7. Immunoblotting

The protein lysates were prepared at 15–30 μg per sample being mixed with modified RIPA buffer and resolved on 8–12% polyacrylamide SDS gels by electrophoresis at 100 volts and 120 volts each for 1 h (a total of 2 h running time) in running buffer. The proteins were transferred to a PDVF membrane (Bio-Rad, Mississauga, ON, Canada) for 1 h at 100 volts in transfer buffer. The membranes were blocked with 5% skim milk or BSA in Tris-buffered saline-Tween 20 (TBST) for 1 h at room temperature. The protein-specific primary antibodies were diluted in 5% BSA + TBST, except for p53 and cdc2, which used 5% skim milk + TBST and incubated overnight at 4°C with gentle agitation. The membranes were incubated for 1 h at room temperature with a secondary antibody in 5% BSA + TBST. Protein detection was achieved through a chemiluminescence reagent, and imaging was performed using the Chemidoc™ MP 7 System (Bio-Rad, Mississauga, ON, Canada). Densitometric analyses were performed using ImageLab™ 6.1 Software (Bio-Rad, Mississauga, ON, Canada).

2.8. Hematoxylin and Eosin Staining

Intact spheroids and organoids were washed with cold PBS and then fixed with 10% Formalin for 15 min. The pellets were resuspended in 3% agarose in PBS, polymerized at 4 °C overnight, then placed into cassettes and processed through graded alcohols to paraffin. The embedded pellets in paraffin blocks were cut at 5 μm sections onto charged slides (Thermo Fisher Scientific, Mississauga, ON, Canada), and H&E staining was performed. Images of H&E-stained sections were captured using an Aperio ScanScope slide scanner (Leica, Teaneck, NJ, USA) and exported using the Aperio ScanScope Console v101.0.4.413 (Leica, Teaneck, NJ, USA).

2.9. Immunohistochemistry

Previously archived formalin-fixed, paraffin-embedded blocks were sectioned at a thickness of 5 μm. Immunohistochemical staining for Antigen Kiel 67 (Ki67) with hematoxylin counterstaining was completed. Sectioning and staining were performed by the Molecular Pathology Core Facility at the Robarts Research Institute (London, ON, Canada). Images of IHC-stained sections were captured using an Aperio ScanScope slide scanner (Leica, Teaneck, NJ, USA). Images of individual stained sections were exported using the Aperio ScanScope Console v101.0.4.413 (Leica, Teaneck, NJ, USA). Using ImageJ Software, the imaged slides were analyzed for staining intensity. The nuclei were masked using the Trainable Weka Segmentation plugin, and the masked regions were counted using a minimum area of 100 and a maximum area of 1000 pixels^2. Each pixel was graded based on the staining intensity as high positive (0–60), positive (60–120), low positive (120–180) and negative (>180) groups.

2.10. Immunofluorescence

Intact spheroids and organoids were collected into 50 mL conical tubes and allowed to settle passively. Using Cryomatrix (Thermo Fisher Scientific, Mississauga, ON, Canada), the pellets were embedded in a 15 × 15 × 5 mm disposable base mold (Thermo Fisher Scientific, Mississauga, ON, Canada). The blocks were cryosectioned at 7 μm onto charged slides and the tissues were fixed using 4% Neutral Buffered Formalin (Thermo Fisher Scientific, Mississauga, ON, Canada), permeabilized with 0.1% Triton-X100 in PBS and blocked with 5% BSA in 0.1% Triton-X100 PBS. The primary antibodies were diluted in blocking solution and incubated overnight at 4 °C. The secondary antibody was diluted in blocking solution and incubated for 1 h at room temperature. Alexa Fluor Phalloidin (Thermo Fisher Scientific, Mississauga, ON, Canada), at 1:40 dilution, was incubated for 1 h, then slides were mounted using VectaShield (Vector Laboratories, Newark, CA, USA) containing a 1:1000 dilution of DAPI (MilliporeSigma, Oakville, ON, Canada). Fluorescent images were captured using the Olympus AX70 upright microscope and ImagePro Software 10.0.15.

2.11. RNA Isolation

The spheroids and organoids were collected into 15 mL or 50 mL tubes, washed with cold PBS and pelleted. RNEasy Spin Column Kit (Qiagen, Toronto, ON, Canada) was used to isolate RNA, and genomic DNA was removed by RNase-free DNaseI (Qiagen, Toronto, ON, Canada), both using the manufacturer's instructions. RNA concentrations, $A_{260/280}$ and $A_{260/230}$, were measured using the NanoDrop Microvolume Spectrophotometer (Thermo Fisher Scientific, Mississauga, ON, Canada).

2.12. Transcriptomic Analysis

For RNA sequencing analysis, isolated RNA from iOvCa spheroid and organoids (n = 7) were collected in duplicate and sent to the London Regional Genomics Centre at the Robarts Research Institute for library preparation. The RNA quality was first confirmed using 2100 Bioanalyzer Instrument (Agilent Technologies, Mississauga, ON, Canada), and sequencing was performed on an Illumina NextSeq using High Output 75 cycle sequencing kits. Raw sequencing reads were uploaded to the Galaxy platform and tested for quality using FastQC and MultiQC to measure the per base sequence quality, GC content, N content, overrepresented sequences and adapter content. The reads were mapped to the human genome (hg38) using HISAT2. The differential gene expression was determined using DESeq2. Pathway analysis was performed by gene set enrichment analysis (GSEA) using all gene set collections from the Molecular Signatures Database, focusing on the Cancer Hallmark and Curated gene sets.

2.13. Quantitative Reverse Transcription-(qRT)-PCR

Complementary DNA (cDNA) was generated from total RNA using the HighCapacity cDNA Reverse Transcriptase Kit (Thermo Fisher Scientific, Mississauga, ON, Canada) according to the manufacturer's instructions. The reactions were completed using 1 µg of RNA while using the MyCycler thermocycler (Bio-Rad, Mississauga, ON, Canada) with the following cycles: 25 °C for 10 min, 37 °C for 120 min, 85 °C for 5 min and 4 °C until sample retrieval. qRT-PCR was performed using the Brilliant II SYBR Green Master Mix (Agilent Technologies, Mississauga, ON, Canada) according to the manufacturer's instructions. The reactions were performed using a QuantStudio3 RT-PCR system (Thermo Fisher Scientific, Mississauga, ON, Canada), and data analysis was performed using the QuantStudio Design and Analysis Software v1.4.3 with the $2^{-\Delta\Delta CT}$ method for relative fold-change calculations. The primers were purchased from Invitrogen (Waltham, MA, USA) (Table S1).

2.14. Kinase Inhibitor Treatments

Adavosertib (10 mM), Alisertib (20 mM) and Volasertib (10 mM) were purchased from Cedarlane (Burlington, ON, Canada), reconstituted in dimethylsulfoxide (DMSO) and stored at −20 °C until use. The adherent cells were seeded at 2×10^4–6×10^4 cells/well on 96-well standard tissue culture plates and treated 24 h after plating for dose response curves. For spheroids and organoids, the cells were seeded at 5×10^4 cells/well in 96-well round bottom ULA plates and in 10 µL of Cultrex® BME on 96-well standard tissue culture plates with their corresponding media types and left for 3 days. The cells were treated with a 3-fold serial dilution of Adavosertib (0.001–10 µM), Alisertib (0.01–100 µM) or Volasertib (0.001–10 µM) for 72 h. An alamarBlue™ assay (Thermo Fisher Scientific, Mississauga, ON, Canada) was completed for cell viability, and raw data were normalized to DMSO (Adavosertib and Volasertib) or the lowest drug concentration (Alisertib) as the control.

2.15. Statistical Analysis

The graphs were generated using GraphPad Prism 10 (GraphPad Software, San Diego, CA, USA), and most data were expressed as mean ± SEM. Student's t-test, an ordinary one-way ANOVA, or a two-way ANOVA with Tukey's multiple comparisons test was performed, with results being considered significant at $p < 0.05$. Specific details on graphs, statistical tests and significance values are provided in each figure legend.

3. Results

3.1. HGSOC Cell Lines Showed Distinct Spheroid and Organoid Morphologies

To validate our culture systems for model development, three established HGSOC cell lines (i.e., OVCAR3, OVCAR4 and OVCAR8) were grown as spheroids on ULA plates and as organoids in Cultrex© BME with an EOC organoid specific media (Table S1). This was important in developing the appropriate culture conditions for the organoid model system in comparison to our suspension spheroid model. Brightfield imaging, as well as hematoxylin and eosin (H&E) staining on sections was conducted for these OVCAR cells lines for a morphological analysis of individual structures (Figure 1A). We characterized the spheroid morphologies based on the density of cellular aggregation as compact, grape-like or sparse clusters [22,25,39]. As expected, OVCAR8 cells formed large, compact spheroids, whereas OVCAR4 and OVCAR3 cells formed large, grape-like clusters as observed previously when studying HGSOC spheroid formation (Figure 1A). Comparatively, we characterized organoids as dense structures that appeared solid or complex structures that appeared as a disorganized mixture of cystic formations and cell clumps [28,40]. All the OVCAR cell lines formed similarly large, dense organoids. This uniform morphology was unexpected to other studies on EOC PDOs, indicating a lack of heterogeneity within these established cell lines.

Thus, we incorporated our own patient ascites-derived immortalized cell lines (iOvCa) into this analysis. These cell lines were developed from patient ascites samples and passaged on standard tissue culture plates until immortalisation, as previously described [22]. Identical methods were conducted using six iOvCa cell lines after being cultured in these two model systems (Figure 1B). This was important to establish how these particular cells appear, grow and behave within these models, as they have not been widely used in our previous research. Initially, we noted that the iOvCa spheroids and organoids appeared very heterogeneous with varying sizes and morphologies, as seen in the brightfield and H&E-stained images (Figure 1B). Using the same categorisations above, the iOvCa398 cells appeared as compact spheroids, and the iOvCa246 and iOvCa256 cells adopted a grape-like structure. The remaining three cell lines (i.e., iOvCa195, iOvCa198 and iOvCa411) formed sparse clusters with abundant cellular debris, indicating some cell death during spheroid formation [41,42]. Again, the iOvCa organoids varied, where half of the cell lines appeared as dense (i.e., iOvCa198, iOvCa246 and iOvCa411) and the other half appeared complex with a sporadic mix of dense and cystic structures (i.e., iOvCa195, iOvCa256 and iOvCa398). In addition, iOvCa256 and iOvCa398 organoids appeared to have necrotic cores as seen upon H&E staining (Figure 1B). Overall, the spheroid and organoid morphologies largely varied, where the iOvCa cell lines better portrayed HGSOC heterogeneity than the OVCAR cell lines, and the organoids better resembled PDOs, as published. Hence, we focused our analyses on the iOvCa cell lines moving forward.

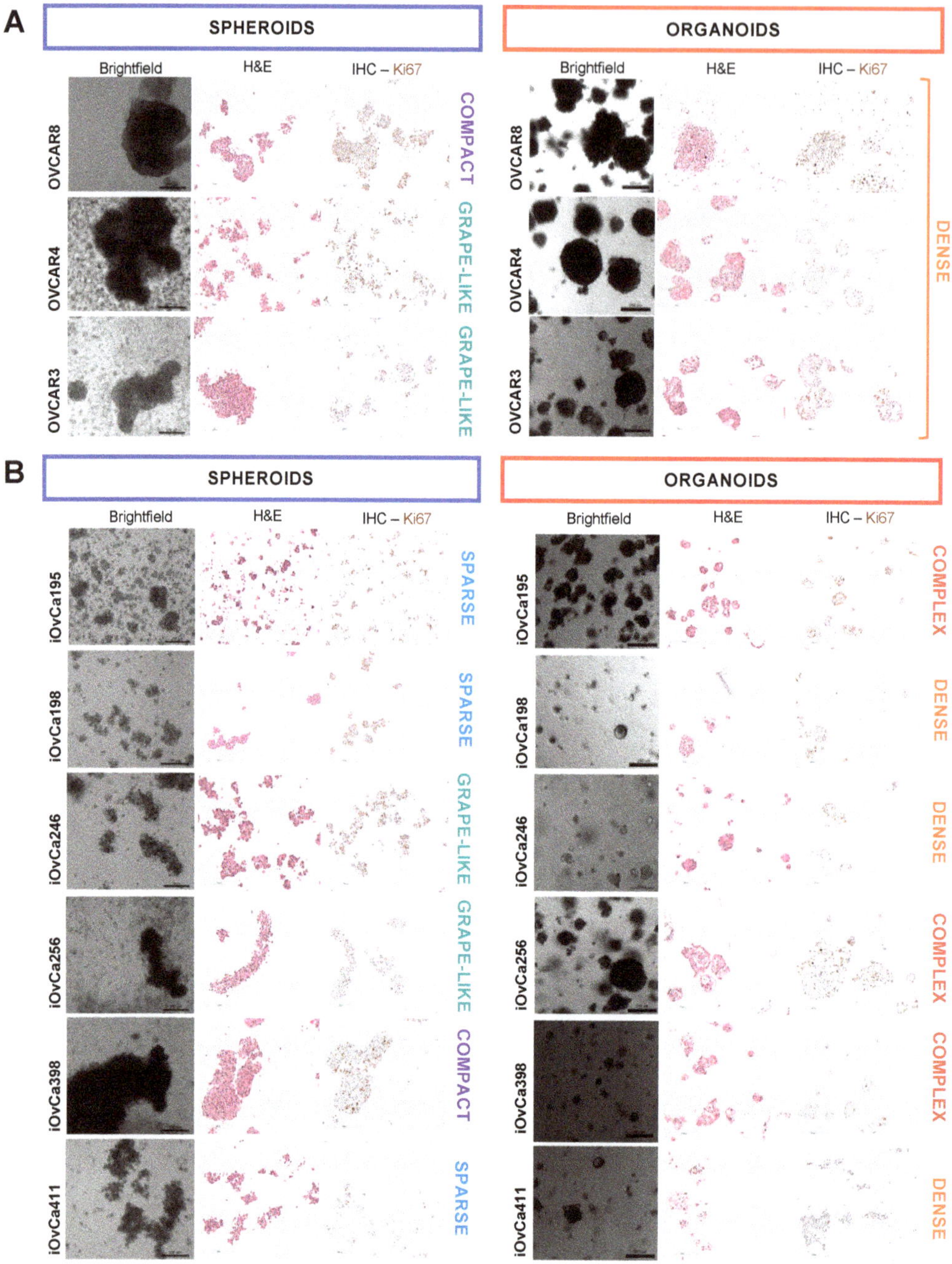

Figure 1. Morphological and histological comparisons of HGSOC cell lines cultured using 3D model systems. (**A**,**B**) Brightfield images of established cells lines and patient ascites-derived immortalized cell lines cultured as spheroids and organoids. The structures were categorized based on appearance as compact, grape-like, or sparse spheroids, and dense or complex organoids, as indicated on the right side of each cell line-associated image. Spheroids and organoids were also collected for formalin-fixed paraffin-embedding (FFPE), sectioning and staining for H&E and IHC with the Ki67 proliferation marker. The scale bar is 200 μm or 250 μm for brightfield, and 200 μm for H&E and IHC images.

3.2. HGSOC Organoid Varying Growth Dynamics Relate to Morphology

Since our previous 3D models using spheroids in suspension have shown a dormancy phenotype in comparison to 2D adherent culture, we wanted to determine whether spheroids differ to organoids in a similar manner. Also, organoids have become a popular model system in cancer research, and it is imperative to understand organoid growth dynamics in the context of EOC, which is not well known. The results showed that the iOvCa cell lines demonstrated great heterogeneity with organoid growth. The iOvCa195 cells formed the most organoids and the iOvCa198 cells formed the least after 3 weeks of culture (Figure S2A). The largest organoids were greater than 20,000 μm^2 in individual organoid area (i.e., iOvCa246), the smallest were less than 5000μm^2 (i.e., iOvCa198, iOvCa398 and iOvCa411), and the rest had an organoid size in between 10,000 and 15,000 μm^2 in area (i.e., iOvCa195 and iOvCa256) (Figure S2B). Although when observing growth over time, the total organoid number often reached a maximum by day 3 in culture, indicating that the capacity to form an organoid was established by that time point (Figure S2E). As predicted, the total organoid area had a steady increase from day 1 for the iOvCa195, iOvCa246 and iOvCa411 cells, yet the other three cell lines plateaued after days 3–6 in culture (Figure S2F).

Furthermore, we calculated the average organoid area based on organoid number and total organoid area measurements, which demonstrated that each cell line had its own unique linear growth rate (Figure 2A; Video S1). Of note, the iOvCa195 and iOvCa411 lines had the greatest increase in average organoid area over time. All iOvCa growth rates, based on a simple linear regression analysis, were significantly different from each other, except for iOvCa246 and iOvCa398, which both had a similar morphology. Moreover, we incorporated organoid darkness and organoid eccentricity characteristics as another way to quantify HGSOC organoid morphology. Organoid darkness gradually increased over time in each cell line, with significantly different endpoint values. Correlating these values to morphology, iOvCa246 organoids had the second highest darkness value (59.9 ± 1.2), which could indicate a dense morphology; in contrast, iOvCa256 organoids had the lowest value (55.6 ± 0.2) in conjunction with a complex morphology and necrotic centre (white arrows; Figure 2B,D). However, the other four cell lines had inconsistent darkness values that did not match their designed morphology. The eccentricity of iOvCa organoids also gradually increased over time for each cell line, indicating that the organoids become more irregularly shaped as they grow (Figure 2C). The cell lines were grouped based on their endpoint organoid eccentricity values to categorize them into two shapes: those with a value greater than 0.6 had an irregular shape, like iOvCa398 organoids with the highest value of 0.714 ± 0.008, and those with a value of less than 0.6, like iOvCa195 organoids with the lowest value of 0.590 ± 0.013 (yellow arrows; Figure 2C,D), were regular. Taken together, there is clear heterogeneity in iOvCa organoid growth characteristics that can easily be quantified through various methods. In contrast, the OVCAR cell lines had the largest organoids based on number and size after 3 weeks of growth (Figure S2A–C). With growth over time, the OVCAR cell lines had no significant differences with organoid number, total organoid area and average organoid area (Figure S3A–C). However, OVCAR4 did have a higher darkness value, which is apparent in the brightfield images (Figure S3D,F). Based on the organoid growth characteristics, this indicates that the OVCAR cells lines are not an accurate representative model for observing the heterogeneity seen in HGSOC.

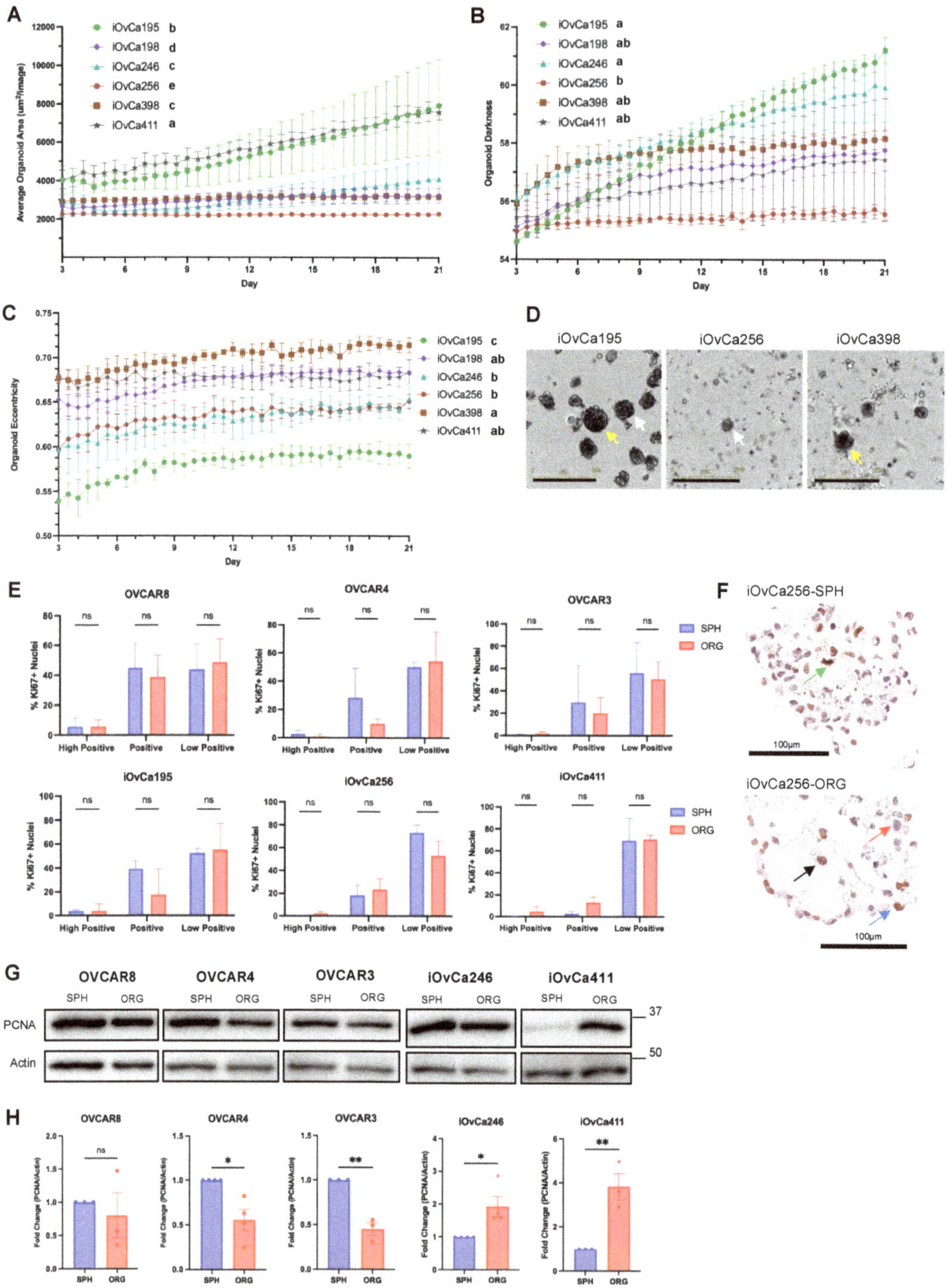

Figure 2. Growth analysis in relation to proliferation status of HGSOC spheroids and organoids. (**A**) Average organoid number (μm^2/image) over time was calculated using organoid number and total organoid area measurements. (**B**) Organoid darkness was measured to observe the density of cells or debris within a structure over time. (**C**) Organoid eccentricity was measured to observe the

difference in shape over time. (**D**) Representative brightfield images of iOvCa195, iOvCa256 and iOvCa398 organoids at day 21, with white arrows for organoid darkness and yellow arrows for organoid eccentricity. The scale bar is 600 μm. (**E**) The IHC sections were used to measure the percentage of Ki67 positive (Ki67+) nuclei, and they were binned based on pixel intensity within ImageJ for each indicated cell line (n = 2–3). (**F**) High magnification images of representative organoids with arrows indicating high positive (green), positive (blue), low positive (black), Ki67+ nuclei and Ki67- nuclei (red). (**G**) Immunoblots of PCNA were completed to confirm the proliferative index of spheroids versus organoids. (**H**) Relative quantification of PCNA in spheroids (SPH) and organoids (ORG). The graphs were generated using GraphPad Prism 10 demonstrating mean ± SD for (**E**) and mean ± SEM for (**A–C**,**H**). Statistical analysis for (**A**) was completed on the simple linear regression curve of each cell line over all timepoints; statistical analysis for (**B**,**C**) was completed on only the Day 21 data, both using an ordinary one-way ANOVA with Tukey's multiple comparisons test and displayed with compact lettering for each cell line, indicated as bold letters; statistical analysis for (**E**,**H**) was a Student's *t* test for each cell line ($n \geq 3$, ns = not significant, * $p < 0.05$, ** $p < 0.01$).

To gain more knowledge on the 3D structure of spheroids and organoids, quantification of Ki67 immunohistochemistry (IHC) was completed to assess cell proliferation. Ki67 staining was grouped based on pixel intensity as high positive (0–60), positive (60–120) and low positive (120–180). Unexpectedly, there were no significant differences in Ki67-positive (Ki67+) nuclei between spheroids and organoids in the different Ki67 staining intensity groups (Figure 2E,F). To further validate these findings, immunoblots were conducted for proliferative cell nuclear antigen (PCNA). The results showed that iOvCa246 and iOvCa411 organoids had higher PCNA protein levels than their spheroids (Figure 2G,H). However, the remaining iOvCa cell lines showed no significant differences in PCNA levels, matching the Ki67 staining data (Figure S2G,H). Interestingly, the OVCAR4 and OVCAR3 spheroids had higher PCNA protein levels than their organoids (Figure 2G,H), reaffirming their deviation from patient-associated HGSOC phenotypes. Overall, these results, using two standard cell proliferation markers, indicate that there are very little differences between spheroids and organoids in their proliferation status.

3.3. Altered AMPK and Akt Signaling in HGSOC Spheroids and Organoids Indicates Biological Switching

Previously, we determined the pathways involved in spheroid formation, including cellular quiescence, bioenergetic stress and autophagy, which indicated a switch to dormancy [13,14]. However, we only have knowledge on the molecular differences between adherent and spheroid models and have not assessed these pathways in organoids. As of now, it is not clear whether organoids may differ from spheroids in terms of proliferation and growth. Thus, we observed AMPK, a known kinase that is highly active during the bioenergetic stress response and is required for cell survival during spheroid formation [18,22]. The iOvCa cell lines showed >4-fold increase in phosphorylated-AMPK (P-AMPK) levels at T172 in spheroids compared to adherent and organoid cultures (Figure 3A,B). We noted that the total AMPK protein levels fluctuated between spheroid and organoid samples, which could exacerbate this effect in some samples. The net result, however, supports the notion of an increased requirement for active AMPK within spheroids for cell survival, as we demonstrated previously [18,22]. The exceptions were the iOvCa198 and iOvCa246 cells, with no significant difference in levels of P-AMPK in all culture conditions, but the iOvCa198 cells seemed to have followed a similar trend with potentially higher levels in spheroids. To address whether differences in culture media composition may contribute to signaling changes, particularly with the additional supplements in the EOC organoid specific media, we cultured spheroids in this medium to assess AMPK protein levels. Indeed, we observed similar P-AMPK protein levels between spheroids grown in each medium, indicating that at least this difference in culture conditions has no effect on AMPK pathway activity (Figure S4). This suggests that AMPK is important during spheroid formation in the majority of iOvCa cell lines, and not necessarily for organoid growth. Conversely, we saw a dramatic increase (>10-fold) in P-AMPK levels in the OVCAR8 and OVCAR4 organoids compared to adherent cultures, while still having high levels in spheroids (Figure S3G,H). Also, the OVCAR3 cells had no significant differences between any culture models, similar to iOvCa198 cells. This reiterates that the iOvCa cell lines better demonstrate HGSOC molecular signaling changes.

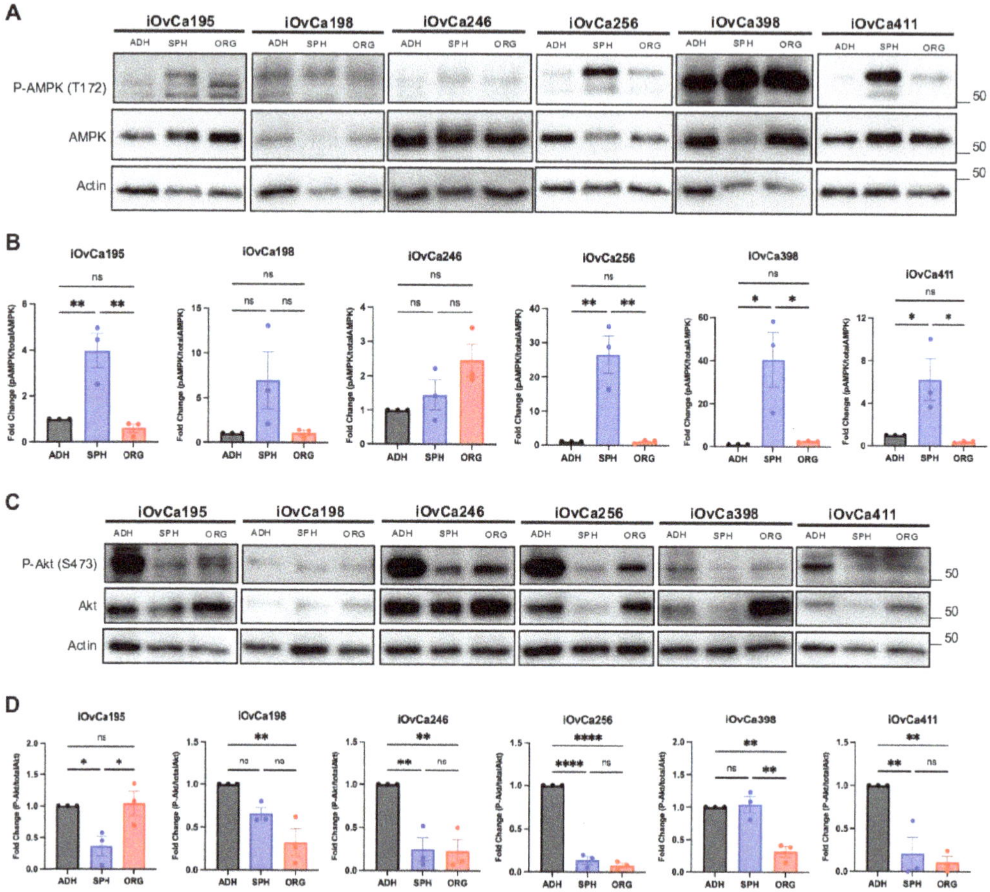

Figure 3. Altered phosphorylation status of AMPK and Akt between culture model systems. (**A**,**B**) Relative protein expression and quantified fold change in phosphorylated AMPK at T172 (P-AMPK) to total AMPK in adherent (ADH), spheroid (SPH) and organoid (ORG) cultures. (**C**,**D**) Relative protein expression and quantified fold change in phosphorylated Akt at S473 (P-Akt) to total Akt in adherent (ADH), spheroid (SPH) and organoid (ORG) cultures. The graphs were generated using GraphPad Prism 10 demonstrating mean ± SEM and a one-way ANOVA with Tukey's multiple comparisons test for statistical analysis of each individual cell line (n = 3, ns = not significant * $p < 0.05$, ** $p < 0.01$, **** $p < 0.0001$).

Next, we interrogated the phosphorylation status of Akt, a prominent kinase down-regulated during spheroid formation, as cells switch from a proliferative phenotype to a quiescent or dormant phenotype [25,43]. As predicted, there was a significant decrease in phosphorylated-Akt (P-Akt) at S473 in most iOvCa cell lines from adherent to spheroid cultures, except in the iOvCa198 and iOvCa398 lines, which had similar levels (Figure 3C,D). Interestingly, the iOvCa195 cells showed no significant difference in P-Akt in organoids compared to adherent cultures, while the rest had decreased levels in organoid cultures. Furthermore, only the iOvCa195 and iOvCa398 cell lines showed significant differences between their spheroid and organoid P-Akt levels, but with opposites trends (Figure 3C,D). Similarly, the OVCAR cell lines all showed decreased P-Akt levels from adherent to spheroid cultures, but varying levels in organoids (Figure S3I,J). The OVCAR8 and OVCAR3 levels showed similar spheroid and organoids levels, while the OVCAR4 organoids had higher P-Akt levels compared to both levels in adherent and spheroid cells. In conclusion, these results display the different molecular signaling activity between our HGSOC spheroids and organoids and demonstrate the potential of HGSOC cells to transition their cell-signaling behaviours depending on their 3D culture state. Based on the above results, there is a lack of consistency in the differences between spheroid and organoid molecular signaling activity in terms of dormant to proliferative switching. Therefore, a comprehensive transcriptomic analysis may better elucidate the underlying biology of HGSOC spheroids and organoids.

3.4. Differential Gene Expression Between HGSOC Spheroids and Organoids

We postulate that there are likely many different intracellular signaling programs altered in spheroids and organoids that could control their growth and morphology beyond those interrogated previously. Hence, we conducted bulk RNA-sequencing analysis on spheroids and organoids using seven iOvCa cell lines. Principal component analysis (PCA) revealed clustering based on cell line rather than culture condition, with only a few single sample outliers seen in iOvCa195-ORG, iOvCa198-SPH and iOvCa411-ORG samples (Figure 4A). This most likely reflects that each cell line recapitulates the original patient source and associated genotype. Differential gene expression analysis was also conducted to identify the top upregulated genes in spheroids and in organoids across all cell lines (Figure 4B). There were a similar number of genes significantly expressed in each model, where 265 were upregulated in organoids, in comparison to 280 upregulated in spheroids. Based on this differential expression, gene set enrichment analysis (GSEA) was completed to identify significantly elevated pathways in the Cancer Hallmarks Collection (Figure 4C). We found fourteen pathways up in spheroids and ten pathways up in organoids ($FDR < 0.05$). Based on our previous work on HGSOC spheroid formation in comparison to adherent cultures, we anticipated seeing similarly regulated pathways in our iOvCa spheroids. This was confirmed with TGFβ signaling and oxidative phosphorylation Cancer Hallmark pathways being in the top 3 elevated pathways in spheroids [44–46].

Since we have copious knowledge on the intercellular signaling changes that occur in spheroids, we pursued the pathways identified in organoids, as our interrogations into pathway analysis is more limited. More specifically, we chose G2/M checkpoint and E2F target pathways because their normalized enrichment scores were the highest, at −2.98 and −2.89, respectively (Figure 4D,E). This was intriguing, as these cancer hallmark pathways are both involved in cell cycle regulation, which we speculate is crucial for actively growing organoids, even though we noted little difference in cell proliferation in previous data [47–49]. To further characterize these pathways, validation of the RNA-sequencing results was conducted with genes relevant to the G2/M checkpoint pathway (i.e., *AURKB*, *HMGCR*, and *SQLE*) and E2F target pathway (i.e., *AURKB* and *IQGAP*), either from the Cancer Hallmark or Curated gene sets associated with those pathways. Based on the transcript expression, there is a clear, significant increase in gene expression from spheroids to organoids for nearly all cell lines ($\log2FC < -1$), indicating that these pathways may be important for cell growth in organoids (Figure 4F).

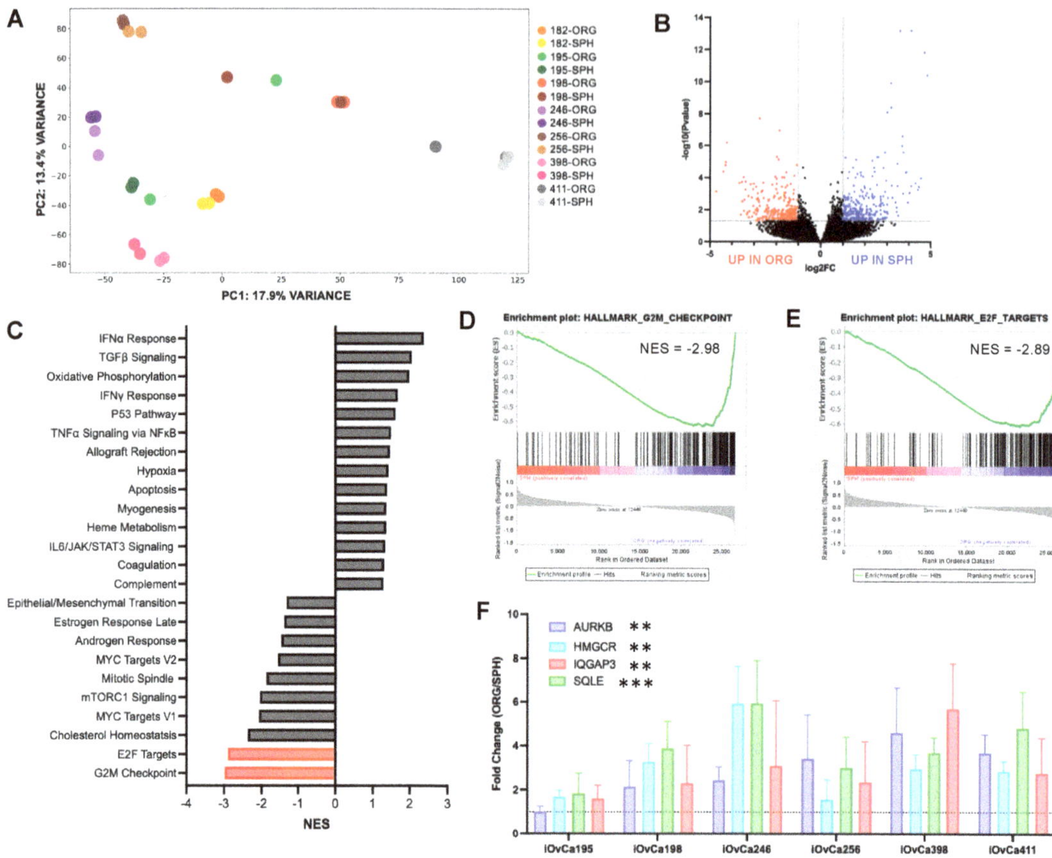

Figure 4. Discovery-based analysis of HGSOC spheroids and organoids based on bulk RNA-sequencing results. (**A**) Principal component analysis plot of all samples, where each colour indicates a cell line and each hue change differentiates spheroids (SPH) from organoids (ORG). (**B**) Differential gene expression between spheroids and organoids with the coloured dots (blue—elevated in organoids, red—elevated in spheroids) indicating significant genes. (**C**) Normalized enrichment scores (NES) of cancer hallmark gene sets in spheroids (NES = positive) and organoids (NES = negative) with the red bars showing two pathways of interest, G2/M checkpoint and E2F Targets (FDR < 0.05). (**D**,**E**) Enrichment plots for G2M checkpoint and E2F target grouped gene sets within the cancer hallmark pathways. (**F**) RT-qPCR validation of selected genes within the G2M checkpoint and E2F target cancer hallmark and curated sets, where any bar above the dotted line (fold change = 1) shows an increase. The graphs were generated using GraphPad Prism 10, demonstrating mean ± SEM with a Student's *t* test for statistical analysis grouping all cell lines to show the difference between spheroids and organoids (n = 3, ** $p < 0.01$, *** $p < 0.001$).

3.5. G2/M Checkpoint Is a Key Regulatory Pathway in HGSOC Organoids

To confirm that the G2/M checkpoint pathway is important for organoid growth rather than spheroid cell survival, we first assessed the activity of key proteins involved in this pathway. At the G2/M checkpoint, the cell division cycle 25C (cdc25C) was phosphorylated at S216 to sequester it, preventing transition into mitosis [50,51]. When the cells are ready to progress into mitosis, active cdc25C dephosphorylates cyclin-dependent kinase 1 (CDK1) at Y15 so that it can bind to Cyclin B1 for the activation of mitosis. Thus, we determined the baseline levels of phosphorylated-CDK1 (P-CDK1) and phosphorylated-cdc25C (P-cdc25C) in the iOvCa cell lines to indicate proper G2/M checkpoint regulation.

The results demonstrated significantly higher levels (>1.25-fold) of P-CDK1 in organoids compared to spheroids, except in the iOvCa256 and iOvCa411 lines (Figure 5A,B). In contrast, only one cell line (i.e., iOvCa398) had increased P-cdc25C levels in organoids compared with spheroids, and the remaining cell lines had a similar trend but without a significant difference (Figure 5A,C). Furthermore, we conducted a preliminary assessment of the expression levels of proteins involved in directly regulating the activity of cdc25C and CDK1, including aurora kinase A (AURKA), polo-like kinase 1 (PLK1) and Mitosis inhibitor protein kinase Wee1 (Wee1). The iOvCa cell lines appear to have differing levels of AURKA and PLK1, but not Wee1, in organoids compared to spheroids (Figure S5). Higher expression of these proteins in organoids suggests elevated G2/M checkpoint cell cycle regulation indicating that functionally targeting these proteins may prove useful.

To determine if the cells within organoids rely on G2/M checkpoint for cell survival more than within spheroids, we chose to selectively inhibit three regulatory proteins in this pathway (i.e., AURKA, PLK1, Wee1). Blocking CDK1 activity was accomplished with the Wee1 inhibitor, Adavosertib, while impeding cdc25C activity was accomplished with either an AURKA inhibitor, Alisertib, or a PLK1 inhibitor, Volasertib [52–54]. All these inhibitors have already shown anti-tumour effects in preclinical experiments and have ongoing Phase I/II clinical trials with ovarian cancer patients, either alone or in combination with chemotherapy [55,56]. We first assessed the IC_{50} values for each of the compounds in our lines in standard adherent culture to evaluate our results as compared with previous publications (Figure S6A). The activity of these inhibitors was validated through the accumulation of P-CDK1 and P-cdc25C with treatment of each inhibitor, compared to the DMSO control (Figure S6B). We also observed an increase in the DNA damage marker gamma H2A histone family member X (γH2A.X) and an increase in the apoptosis marker cleaved PARP under treated conditions with the lowest adherent IC_{50} value (Figure S6B). When comparing the spheroid and organoid dose response curves, we found that there are significantly different sensitivities to these inhibitors between model systems. Spheroids were consistently less sensitive to the Wee1 inhibitor Adavosertib than organoids in all three iOvCa cell lines that we tested (Figure 6A). Additionally, there were no cytotoxic effects on spheroids treated with the PLK1 inhibitor Volasertib as compared to organoids, except for iOvCa246 cells, which showed no difference in sensitivity between spheroids and organoids (Figure 6C). However, when spheroids and organoids were treated with the AURKA inhibitor Alisertib, they either had no difference in sensitivity or showed the opposite trend, with organoids being less sensitive than spheroids as observed with the iOvCa198 line (Figure 6B). To quantify and perform statistical analysis of drug sensitivities between spheroids and organoids, we used the area under the curve (AUC) calculations. The AUC for Adavosertib and Volasertib shows a clear significant difference in sensitivity between organoids and spheroids, where spheroids are less sensitive (Figure 6D). Again, the AUC for Alisertib showed no significant difference in sensitivity for the iOvCa195 and iOvCa246 cells, but the opposite trend for the iOvCa198 cells, where spheroids are more sensitive than organoids (Figure 6D). To conclude, G2/M checkpoint regulation is critical for organoid cell survival rather than spheroid cell survival in EOC. This corresponds to our previous results, which indicate that spheroids are more dormant and our organoids are more proliferative. Targeting this biological switch could pose as a novel therapeutic strategy for HGSOC.

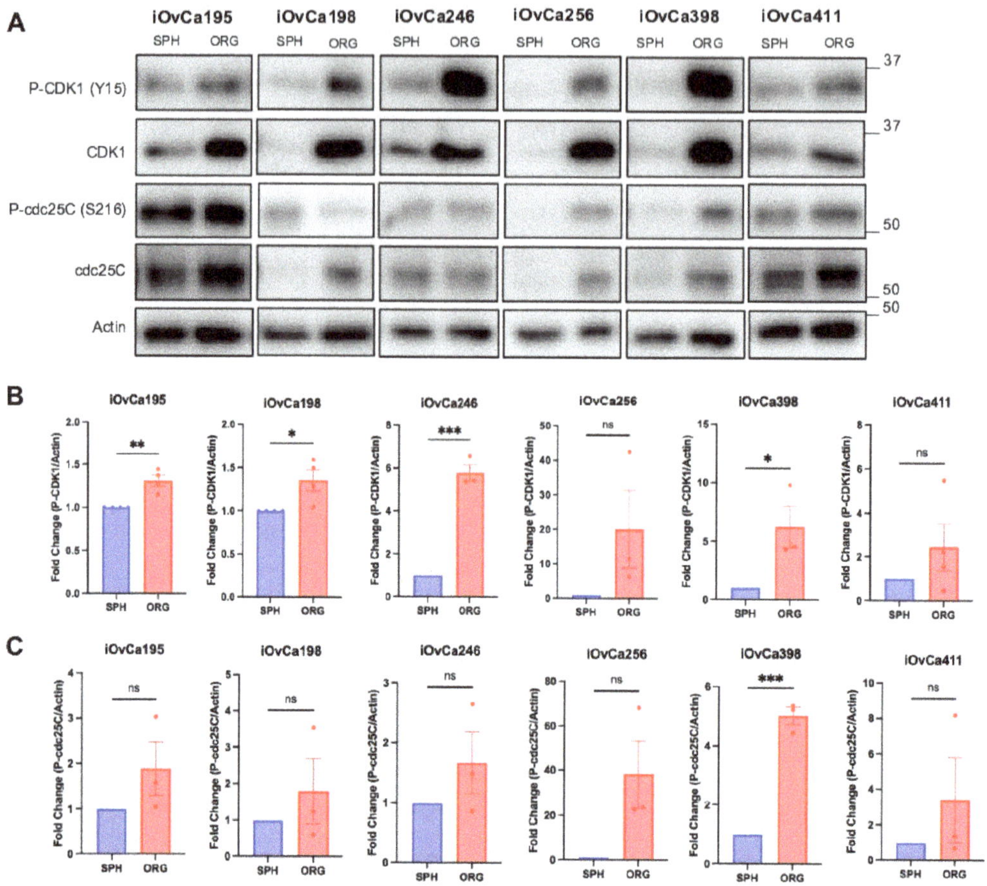

Figure 5. Expression and regulation of G2M checkpoint proteins in HGSOC organoids versus spheroids. (**A**) Relative protein expression of phosphorylated CDK1 at Y15 (P-CDK1), total CDK1, phosphorylated cdc25C at S216 (P-cdc25C) and total cdc25C in spheroid (SPH) and organoid (ORG) cultures. (**B**,**C**) Quantified fold change in P-CDK1 and P-cdc25C to actin housekeeping protein. The graphs were generated using GraphPad Prism 10 demonstrating mean ± SEM a Student's t test for statistical analysis (n ≥ 3, ns = not significant, * $p < 0.05$, ** $p < 0.01$, *** $p < 0.001$).

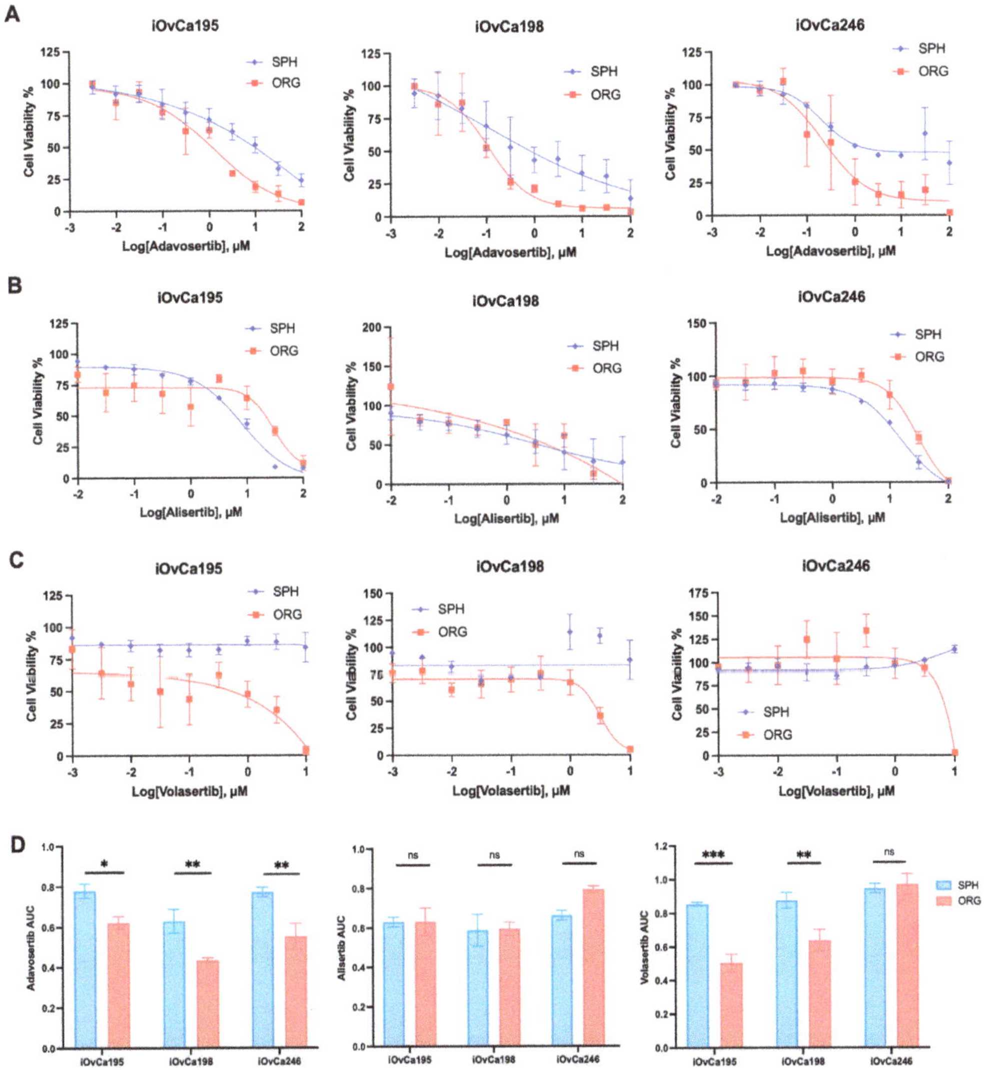

Figure 6. Varying sensitivities to G2M checkpoint small molecule inhibitors on HGSOC spheroid and organoid viability. (**A–C**) Dose response curves of a Adavosertib, Alisertib and Volasertib normalized to either DMSO control (Adavosertib and Volasertib) or the lowest drug concentration (Alisertib) based on the alamarBlue readings in spheroid (SPH) and organoid (ORG) cultures. (**D**) Bar graphs of area under the curve (AUC) for each of the above dose–response curves to demonstrate sensitivity between spheroids and organoids. The graphs were generated using GraphPad Prism 10, demonstrating mean ± SEM with two-way ANOVA for statistical analysis (n = 3–4, ns = not significant, * $p < 0.05$, ** $p < 0.01$, *** $p < 0.001$).

4. Discussion

HGSOC is known to be a very heterogeneous disease that complicates treatment options for patients [57–59]. The unique biological changes that occur during the metastatic process has have been well elucidated in the context of 3D cell–cell interactions. To understand this complex biology, 3D in vitro culture model systems should be incorporated into our research protocols. Here, we employed an EOC organoid model system in conjunction with the spheroid model system. We found that HGSOC heterogeneity is very apparent with our patient ascites-derived immortalized (iOvCa) cell lines. This was expected, as other studies have demonstrated interpatient heterogeneity through single-cell transcriptomics, where HGSOC patient ascites samples cluster by themselves [5,32]. However, there are key proteins and associated signaling pathways that are crucial for cell survival within each model system. This was demonstrated with the bioenergetic stress pathway in terms of increased P-AMPK in spheroids compared to organoids. In addition, our bulk RNA-sequencing analysis provided an extensive list of differentially regulated pathways between spheroids and organoids. One iOvCa cell line (i.e., iOvCa182) was included in our RNA sequencing data but not in our further analysis due to its difficult nature of being maintained in adherent cultures. With our downstream analysis into the G2/M checkpoint pathway, we saw increased sensitivity to G2/M checkpoint-selective inhibitors in organoids compared to spheroids, except with our AURKA inhibitor Alisertib, which could pose as a new therapeutic target for HGSOC. Therefore, these biological differences should be considered when discovering new targeted treatment options for patients, as they may respond differently, depending on their disease progression.

In this study, we employed multiple ways to quantify organoid growth dynamics and correlate them with morphology. Organoid eccentricity is defined as the ability to bud or the irregularity in 3D shape, whereas darkness is defined as the change in organoid brightness over time, which could be associated with an accumulation of cells or debris within 3D structures. The H&E-stained sections best elucidate these organoid characteristics by visualizing their core. The OVCAR cell lines clearly grow more aggressive organoids with increased central cells (i.e., darkness) and active budding for irregularly shaped structures (i.e., eccentricity), whereas the iOvCa cell lines grow more slowly and into round shapes, sometimes with hollow centres, as cystic-like structures. These discrepancies between OVCAR and iOvCa cell lines demonstrates that patient-derived samples better represent the morphology of HGSOC patient tumours, as seen in many other PDO studies [27,28,30]. When comparing organoid to spheroid morphology, the results showed that all iOvCa cell lines had a unique ability to form spheroids and organoids. Moreover, the iOvCa cell lines demonstrated an aggressive morphology, as spheroids do not correlate with an aggressive organoid morphology and vice versa.

Additionally, these 3D morphologies complement the differing AMPK and Akt protein expressions and, more importantly, their activity levels within spheroids and organoids. For example, the iOvCa398 cells formed very small, complex organoids but were able to survive well as spheroids forming large, compact structures. The iOvCa398 compact spheroids correlated with an increase in P-AMPK levels, which we have previously seen, as cells transition into a dormant state for metastasis [18,22,23]. In contrast, the iOvCa195 cells formed medium-sized, complex organoids but formed very small sparse spheroids. We suspected that organoid growth would correspond to high P-Akt levels and lower levels in spheroids, which is displayed with the iOvCa195 line. We believe that these influences in signaling activity may be critical for the growth of organoids and dormancy in spheroids for this cell line. However, it seems that AMPK signaling may be more influential on spheroid formation rather than Akt signaling on organoid growth. Of course, this would need to be investigated further by observing the downstream AMPK and Akt pathway proteins, such as ULK1 for macro-autophagy regulation [17,18,23,25,39,60].

With the development of the organoid model system, we have vastly expanded our ability to study the important cell biology of this disease in a 3D context. Organoids have become a popular research avenue in all cancer fields, as it provides a reproducible way to observe mini tumour-like structures ex vivo [40,61]. For researchers, it is necessary to gain more insight into the 3D interactions within tumours or metastatic lesions to better eradicate HGSOC in patients. Using solely 2D models limits our understanding of cancer cells and their molecular adaptations to survive during disease progression, especially with HGSOC, where it spreads vastly within the peritoneal cavity, as spheroids easily form secondary tumours [62]. This was clearly demonstrated with our immunoblots for P-AMPK and P-Akt, where organoid protein activity levels did not always correlate to adherent culture as speculated. Therefore, this study provides evidence that both the 3D spheroid and organoid model systems should be utilized for EOC research moving forward, as 2D adherent cultures are not accurate to the cellular behaviour in patients.

In conjunction with previously published results, this study confirmed that there are significant changes in protein activity important for the cancer dormancy pathways in HGSOC. Specifically, we provided some insight into the integral signaling pathways for cancer dormancy and biological switching behaviour in EOC. The Ki67 and PCNA quantifications revealed that the spheroids and organoids did not differ in their proliferative capacity in most cell lines. This was an unexpected result but could be due to the fact that Ki67 is expressed in all cell cycles except for G0, indicating an overabundance of Ki67 in all cells within spheroid structures prior to initiating quiescence [63]. Likewise, PCNA functions during DNA replication and repair, making it critical only for the S phase, indicating more than proliferation induction in cells [64]. On the other hand, our transcriptomic analysis demonstrated upregulation in G2/M checkpoint regulation, E2F Targets and MYC targets in organoids compared to spheroids. These results confirmed that there is a difference in proliferative capacity between culture model systems, but with regard to the regulation of cell cycle checkpoints. This indicates that Ki67 and PCNA may not accurately represent the proliferation index of these 3D structures. Since the process cancer dormancy is a complex relationship between decreased proliferation, increased quiescence and resistance to apoptosis, utilizing the appropriate markers is important to understand this interplay during EOC metastasis.

In recent years, researchers have expanded their search for new HGSOC-targeted therapies to include G2/M checkpoint inhibitors. By blocking the G2/M checkpoint, it forces cancer cells into premature mitosis without proper DNA repair; thus, cells accumulate DNA damage, and it triggers cell death via apoptosis [65–67]. Given our RNA sequencing data comparing spheroids and organoids, we speculated that each model system would have a different sensitivity to G2/M inhibition. Here, we were able to provide evidence that targeting the G2/M checkpoint pathway is effective. Adavosertib is a well-known Wee1 inhibitor that has been used to inhibit G2/M transition in many other cancers as well as HGSOC [52,68]. Volasertib is a more recently developed selective PLK1 inhibitor that had antitumour activity in platinum-resistant ovarian cancer [69]. Lastly, Alisertib is an AURKA inhibitor that is being tested on solid tumours in Phase I or II clinical trials that result in mitotic catastrophe [53,70–72]. Adavosertib and Volasertib clearly showed the difference in sensitivity between model systems where spheroids were less sensitive than organoids regardless of each unique cell line morphology. In contrast, Alisertib showed no difference in sensitivity between spheroids and organoids. This could be due to a shared dependency on AURKA activity within spheroids and organoids independent of cell cycle regulation and mitosis. These other roles include enhancing cancer cell stemness, initiating EMT, or inducing glucose metabolism [73–75].

Based on these results, we believe that targeting AURKA could be a new alternative therapy for HGSOC, especially since AURKA is often amplified in ovarian cancer [76]. Alisertib was able to effectively kill cells within dormant spheroids as well as cells within proliferative organoids. We speculate that this may be due to other downstream effects of targeting AURKA in cells or the potential off-target effects of Alisertib. For example, AURKA also regulates centrosome maturation, chromosomal alignment and separation, and assists with the DNA damage response [70,77,78]. Even though Alisertib is highly selective for AURKA, it may affect the activity of Aurora Kinase B, albeit very minimally. Since HGSOC cells often have a large accumulation of genomic instability, inhibiting these processes could lead to mitotic catastrophe, regardless of the biological state. Of course, the full mechanism of action has not been elucidated but could be extrapolated in the future. As an extension of this, we could improve the efficacy of Alisertib in HGSOC by combining it with chemotherapeutics or other targeted agents to overcome chemoresistance or the recurrence of disease.

The few limitations of this study should be considered when interpreting the results. First, the experiments were conducted using cell lines in in vitro systems, which are valuable for controlled mechanistic studies but do not fully recapitulate the complexity of in vivo biological environments. Even though we used recently developed ascites derived patient samples, they were immortalized, which can limit the translation of these findings to physiological conditions [79]. Second, the use of spheroids and organoids introduces additional complexity due to intercellular spatial differences. These three-dimensional models, though more representative of tissue architecture than traditional monolayer cultures, exhibit gradients in oxygen concentration, nutrient availability, and cellular composition across their structure [62,80,81]. We could invoke our new technology of spatial analysis of spheroids (i.e., SPoRTS) and apply it to organoids for a better understanding of these gradients throughout these 3D structures. Therefore, further studies using in vivo models or clinical samples, such as PDOs, are necessary to validate these findings and better understand their relevance in physiological and pathological contexts.

5. Conclusions

In conclusion, utilizing in vitro 3D model systems can be an effective way to better understand the biological complexities of EOC. These cumulative results directly correlate with our previous work that describes the biological changes during metastasis in the context of cancer dormancy. It is clear that there are intercellular signaling changes in cells, rather than strictly morphological, which indicates a dormant phenotype in spheroids compared to a proliferative phenotype in organoids. To determine new targeted therapies and their translational applicability to patients, we provide more evidence by using 3D models with patient-derived samples in current research studies.

Supplementary Materials: The following supporting information can be downloaded at: https://www.mdpi.com/article/10.3390/cells14020133/s1, Figure S1: Confirmation of HGSOC histotype in cultured cell lines; Figure S2: Additional organoid growth measurements with proliferation analysis; Figure S3: Spheroid to organoid comparisons in established OVCAR cell lines; Figure S4: The AMPK protein activity is not affected by the supplements added in EOC organoid specific media; Figure S5: Key proteins in the regulation of the G2/M checkpoint pathway; Figure S6: G2/M checkpoint inhibitor activity levels in adherent cultures; Table S1: EOC organoid specific media formula; Table S2: Primer sets for RT-qPCR validation of RNA-sequencing results; Table S3: Summary of results; Video S1: iOvCa195 organoid growth over time at 12 h increments.

Author Contributions: Conceptualization, E.J.T. and T.G.S.; methodology, E.J.T., Y.R.V., B.K., G.E.D. and T.G.S.; software, B.K.; validation, E.J.T.; formal analysis, E.J.T. and B.K.; investigation, E.J.T., Y.R.V. and J.D.; resources, Y.R.V., G.E.D. and T.G.S.; data curation, E.J.T., J.D., B.K. and A.B.; writing—original draft, E.J.T.; writing—review and editing, E.J.T., A.B. and T.G.S.; visualization, E.J.T.; supervision, T.G.S.; funding acquisition, T.G.S. All authors have read and agreed to the published version of the manuscript.

Funding: We acknowledge the funding support from the Canadian Institutes for Health Research to TGS (Grant #451405) and from the London Health Sciences Foundation through donations to the Mary & John Knight Translational Ovarian Cancer Research Unit. EJT was supported by the Karen Pooley Scholarship in Ovarian Cancer Research with funds from the London Health Sciences Foundation.

Institutional Review Board Statement: The study was conducted according to the guidelines of the Declaration of Helsinki and approved by the Institutional Review Board of Western University (protocol #115904).

Informed Consent Statement: Patient consent for clinical specimens was obtained according to Western University institution research ethics board approved protocol #115904.

Data Availability Statement: The data discussed in this publication have been deposited in NCBI's Gene Expression Omnibus (Edgar et al., 2002) and are accessible through GEO Series accession number GSE284507 [82].

Acknowledgments: We would like to thank all the members of the Mary & John Knight Translational Ovarian Cancer Research Unit for their support and guidance. We are very grateful to the many donors through the London Health Sciences Foundation for additional infrastructure funding, including the Leica DMI 400B inverted microscope, Bio-Rad Chemidoc, Aglient BioTek Synergy H1 multimode reader and IncuCyte S3 live-cell imaging system used in this study.

Conflicts of Interest: The authors declare no conflicts of interest for this study.

References

1. Lheureux, S.; Gourley, C.; Vergote, I.; Oza, A.M. Epithelial Ovarian Cancer. *Lancet* **2019**, *393*, 1240–1253. [CrossRef] [PubMed]
2. Labidi-Galy, S.I.; Papp, E.; Hallberg, D.; Niknafs, N.; Adleff, V.; Noe, M.; Bhattacharya, R.; Novak, M.; Jones, S.; Phallen, J.; et al. High Grade Serous Ovarian Carcinomas Originate in the Fallopian Tube. *Nat. Commun.* **2017**, *8*, 1093. [CrossRef]
3. Karst, A.M.; Levanon, K.; Drapkin, R. Modeling High-Grade Serous Ovarian Carcinogenesis from the Fallopian Tube. *Proc. Natl. Acad. Sci. USA* **2011**, *108*, 7547–7552. [CrossRef] [PubMed]
4. Lee, S.; Zhao, L.; Rojas, C.; Bateman, N.W.; Yao, H.; Lara, O.D.; Celestino, J.; Morgan, M.B.; Nguyen, T.V.; Conrads, K.A.; et al. Molecular Analysis of Clinically Defined Subsets of High-Grade Serous Ovarian Cancer. *Cell Rep.* **2020**, *31*, 107502. [CrossRef] [PubMed]
5. Izar, B.; Tirosh, I.; Stover, E.H.; Wakiro, I.; Cuoco, M.S.; Alter, I.; Rodman, C.; Leeson, R.; Su, M.J.; Shah, P.; et al. A Single-Cell Landscape of High-Grade Serous Ovarian Cancer. *Nat. Med.* **2020**, *26*, 1271–1279. [CrossRef] [PubMed]
6. Lengyel, E. Ovarian Cancer Development and Metastasis. *Am. J. Pathol.* **2010**, *177*, 1053–1064. [CrossRef] [PubMed]
7. Goyeneche, A.; Lisio, M.A.; Fu, L.; Srinivasan, R.; Capuccino, J.V.; Gao, Z.H.; Telleria, C. The Capacity of High-Grade Serous Ovarian Cancer Cells to Form Multicellular Structures Spontaneously along Disease Progression Correlates with Their Orthotopic Tumorigenicity in Immunosuppressed Mice. *Cancers* **2020**, *12*, 699. [CrossRef]
8. Das, T.; Meunier, L.; Barbe, L.; Provencher, D.; Guenat, O.; Gervais, T.; Mes-Masson, A.M. Empirical Chemosensitivity Testing in a Spheroid Model of Ovarian Cancer Using a Microfluidics-Based Multiplex Platform. *Biomicrofluidics* **2013**, *7*, 011805. [CrossRef] [PubMed]
9. Raghavan, S.; Ward, M.R.; Rowley, K.R.; Wold, R.M.; Takayama, S.; Buckanovich, R.J.; Mehta, G. Formation of Stable Small Cell Number Three-Dimensional Ovarian Cancer Spheroids Using Hanging Drop Arrays for Preclinical Drug Sensitivity Assays. *Gynecol. Oncol.* **2015**, *138*, 181–189. [CrossRef] [PubMed]
10. Liao, J.; Qian, F.; Tchabo, N.; Mhawech-Fauceglia, P.; Beck, A.; Qian, Z.; Wang, X.; Huss, W.J.; Lele, S.B.; Morrison, C.D.; et al. Ovarian Cancer Spheroid Cells with Stem Cell-like Properties Contribute to Tumor Generation, Metastasis and Chemotherapy Resistance through Hypoxia-Resistant Metabolism. *PLoS ONE* **2014**, *9*, e84941. [CrossRef]
11. Dean, M.; Jin, V.; Bergsten, T.M.; Austin, J.R.; Lantvit, D.D.; Russo, A.; Burdette, J.E. Loss of PTEN in Fallopian Tube Epithelium Results in Multicellular Tumor Spheroid Formation and Metastasis to the Ovary. *Cancers* **2019**, *11*, 884. [CrossRef]
12. Summers, M.A.; McDonald, M.M.; Croucher, P.I. Cancer Cell Dormancy in Metastasis. *Cold Spring Harb. Perspect. Med.* **2020**, *10*, a037556. [CrossRef]
13. Shepherd, T.G.; Dick, F.A. Principles of Dormancy Evident in High-Grade Serous Ovarian Cancer. *Cell Div.* **2022**, *17*, 2. [CrossRef] [PubMed]
14. Tomas, E.; Shepherd, T.G. Insights into High-Grade Serous Carcinoma Pathobiology Using Three-Dimensional Culture Model Systems. *J. Ovarian Res.* **2023**, *16*, 70. [CrossRef] [PubMed]

15. Sodek, K.L.; Ringuette, M.J.; Brown, T.J. Compact Spheroid Formation by Ovarian Cancer Cells Is Associated with Contractile Behavior and an Invasive Phenotype. *Int. J. Cancer* **2009**, *124*, 2060–2070. [CrossRef] [PubMed]
16. MacDonald, J.; Ramos-Valdes, Y.; Perampalam, P.; Litovchick, L.; DiMattia, G.E.; Dick, F.A. A Systematic Analysis of Negative Growth Control Implicates the DREAM Complex in Cancer Cell Dormancy. *Mol. Cancer Res.* **2017**, *15*, 371–381. [CrossRef] [PubMed]
17. Peart, T.; Ramos Valdes, Y.; Correa, R.J.M.; Fazio, E.; Bertrand, M.; McGee, J.; Préfontaine, M.; Sugimoto, A.; DiMattia, G.E.; Shepherd, T.G. Intact LKB1 Activity Is Required for Survival of Dormant Ovarian Cancer Spheroids. *Oncotarget* **2015**, *6*, 22424–22438. [CrossRef]
18. Laski, J.; Singha, B.; Wang, X.; Ramos Valdes, Y.; Collins, O.; Shepherd, T.G. Activated CAMKKβ-AMPK Signaling Promotes Autophagy in a Spheroid Model of Ovarian Tumour Metastasis. *J. Ovarian Res.* **2020**, *13*, 58. [CrossRef] [PubMed]
19. Iwanicki, M.P.; Davidowitz, R.A.; Ng, M.R.; Besser, A.; Muranen, T.; Merritt, M.; Danuser, G.; Ince, T.; Brugge, J.S. Ovarian Cancer Spheroids Use Myosin-Generated Force to Clear the Mesothelium. *Cancer Discov.* **2011**, *1*, 144–157. [CrossRef] [PubMed]
20. Puiffe, M.L.; Le Page, C.; Filali-Mouhim, A.; Zietarska, M.; Ouellet, V.; Tonin, P.N.; Chevrette, M.; Provencher, D.M.; Mes-Masson, A.-M. Characterization of Ovarian Cancer Ascites on Cell Invasion, Proliferation, Spheroid Formation, and Gene Expression in an in Vitro Model of Epithelial Ovarian Cancer. *Neoplasia* **2007**, *9*, 820–829. [CrossRef] [PubMed]
21. Burston, H.E.; Kent, O.A.; Communal, L.; Udaskin, M.L.; Sun, R.X.; Brown, K.R.; Jung, E.; Francis, K.E.; Rose, J.L.; Lowitz, J.; et al. Inhibition of Relaxin Autocrine Signaling Confers Therapeutic Vulnerability in Ovarian Cancer. *J. Clin. Investig.* **2021**, *131*, e142677. [CrossRef]
22. Buensuceso, A.; Ramos Valdes, Y.; DiMattia, G.E.; Shepherd, T.G. AMPK-Independent LKB1 Activity Is Required for Efficient Epithelial Ovarian Cancer Metastasis. *Mol. Cancer Res.* **2020**, *18*, 488–500. [CrossRef] [PubMed]
23. Willows, R.; Sanders, M.J.; Xiao, B.; Patel, B.R.; Martin, S.R.; Read, J.; Wilson, J.R.; Hubbard, J.; Gamblin, S.J.; Carling, D. Phosphorylation of AMPK by Upstream Kinases Is Required for Activity in Mammalian Cells. *Biochem. J.* **2017**, *474*, 3059–3073. [CrossRef] [PubMed]
24. Endo, H.; Owada, S.; Inagaki, Y.; Shida, Y.; Tatemichi, M. Glucose Starvation Induces LKB1-AMPK-Mediated MMP-9 Expression in Cancer Cells. *Sci. Rep.* **2018**, *8*, 10122. [CrossRef]
25. Correa, R.J.M.; Valdes, Y.R.; Peart, T.M.; Fazio, E.N.; Bertrand, M.; McGee, J.; Préfontaine, M.; Sugimoto, A.; DiMattia, G.E.; Shepherd, T.G. Combination of AKT Inhibition with Autophagy Blockade Effectively Reduces Ascites-Derived Ovarian Cancer Cell Viability. *Carcinogenesis* **2014**, *35*, 1951–1961. [CrossRef] [PubMed]
26. Lõhmussaar, K.; Kopper, O.; Korving, J.; Begthel, H.; Vreuls, C.P.H.; van Es, J.H.; Clevers, H. Assessing the Origin of High-Grade Serous Ovarian Cancer Using CRISPR-Modification of Mouse Organoids. *Nat. Commun.* **2020**, *11*, 2660. [CrossRef]
27. de Witte, C.J.; Espejo Valle-Inclan, J.; Hami, N.; Lõhmussaar, K.; Kopper, O.; Vreuls, C.P.H.; Jonges, G.N.; van Diest, P.; Nguyen, L.; Clevers, H.; et al. Patient-Derived Ovarian Cancer Organoids Mimic Clinical Response and Exhibit Heterogeneous Inter- and Intrapatient Drug Responses. *Cell Rep.* **2020**, *31*, 107762. [CrossRef]
28. Kopper, O.; de Witte, C.J.; Lõhmussaar, K.; Valle-Inclan, J.E.; Hami, N.; Kester, L.; Balgobind, A.V.; Korving, J.; Proost, N.; Begthel, H.; et al. An Organoid Platform for Ovarian Cancer Captures Intra- and Interpatient Heterogeneity. *Nat. Med.* **2019**, *25*, 838–849. [CrossRef]
29. Filippou, P.S.; Aleman, J.; Garreta, E.; Wang, Y.; Yang, J.; Huang, S.; Cheng, S.; Jin, Y.; Zhang, N. Application of Ovarian Cancer Organoids in Precision Medicine: Key Challenges and Current Opportunities. *Front. Cell Dev. Biol.* **2021**, *9*, 701429. [CrossRef]
30. Nero, C.; Vizzielli, G.; Lorusso, D.; Cesari, E.; Daniele, G.; Loverro, M.; Scambia, G.; Sette, C. Patient-Derived Organoids and High Grade Serous Ovarian Cancer: From Disease Modeling to Personalized Medicine. *J. Exp. Clin. Cancer Res.* **2021**, *40*, 116. [CrossRef] [PubMed]
31. Yee, C.; Dickson, K.-A.; Muntasir, M.N.; Ma, Y.; Marsh, D.J. Three-Dimensional Modelling of Ovarian Cancer: From Cell Lines to Organoids for Discovery and Personalized Medicine. *Front. Bioeng. Biotechnol.* **2022**, *10*, 836984. [CrossRef] [PubMed]
32. Senkowski, W.; Gall-Mas, L.; Marin Falco, M.; Li, Y.; Lavikka, K.; Kriegbaum, M.C.; Oikkonen, J.; Bulanova, D.; Pietras, E.J.; Voßgröne, K.; et al. A Platform for Efficient Establishment, Expansion and Drug Response Profiling of High-Grade Serous Ovarian Cancer Organoids. *Dev. Cell* **2023**. [CrossRef]
33. Gorski, J.W.; Zhang, Z.; McCorkle, J.R.; DeJohn, J.M.; Wang, C.; Miller, R.W.; Gallion, H.H.; Dietrich, C.S.; Ueland, F.R.; Kolesar, J.M. Utilizing Patient-Derived Epithelial Ovarian Cancer Tumor Organoids to Predict Carboplatin Resistance. *Biomedicines* **2021**, *9*, 1021. [CrossRef]
34. Tao, M.; Wu, X. The Role of Patient-Derived Ovarian Cancer Organoids in the Study of PARP Inhibitors Sensitivity and Resistance: From Genomic Analysis to Functional Testing. *J. Exp. Clin. Cancer Res.* **2021**, *40*, 338. [CrossRef]
35. Tao, M.; Sun, F.; Wang, J.; Wang, Y.; Zhu, H.; Chen, M.; Liu, L.; Liu, L.; Lin, H.; Wu, X. Developing Patient-Derived Organoids to Predict PARP Inhibitor Response and Explore Resistance Overcoming Strategies in Ovarian Cancer. *Pharmacol. Res.* **2022**, *179*, 106232. [CrossRef] [PubMed]

36. Nanki, Y.; Chiyoda, T.; Hirasawa, A.; Ookubo, A.; Itoh, M.; Ueno, M.; Akahane, T.; Kameyama, K.; Yamagami, W.; Kataoka, F.; et al. Patient-Derived Ovarian Cancer Organoids Capture the Genomic Profiles of Primary Tumours Applicable for Drug Sensitivity and Resistance Testing. *Sci. Rep.* **2020**, *10*, 12581. [CrossRef]
37. Phan, N.; Hong, J.J.; Tofig, B.; Mapua, M.; Elashoff, D.; Moatamed, N.A.; Huang, J.; Memarzadeh, S.; Damoiseaux, R.; Soragni, A. A Simple High-Throughput Approach Identifies Actionable Drug Sensitivities in Patient-Derived Tumor Organoids. *Commun. Biol.* **2019**, *2*, 78. [CrossRef] [PubMed]
38. Leroy, B.; Girard, L.; Hollestelle, A.; Minna, J.D.; Gazdar, A.F.; Soussi, T. Analysis of TP53 Mutation Status in Human Cancer Cell Lines: A Reassessment. *Hum. Mutat.* **2014**, *35*, 756–765. [CrossRef]
39. Singha, B.; Laski, J.; Ramos Valdes, Y.; Liu, E.; DiMattia, G.E.; Shepherd, T.G. Inhibiting ULK1 Kinase Decreases Autophagy and Cell Viability in High-Grade Serous Ovarian Cancer Spheroids. *Am. J. Cancer Res.* **2020**, *10*, 1384–1399.
40. Maenhoudt, N.; Defraye, C.; Boretto, M.; Jan, Z.; Heremans, R.; Boeckx, B.; Hermans, F.; Arijs, I.; Cox, B.; Van Nieuwenhuysen, E.; et al. Developing Organoids from Ovarian Cancer as Experimental and Preclinical Models. *Stem Cell Rep.* **2020**, *14*, 717–729. [CrossRef]
41. Burleson, K.M.; Casey, R.C.; Skubitz, K.M.; Pambuccian, S.E.; Oegema, T.R.; Skubitz, A.P.N. Ovarian Carcinoma Ascites Spheroids Adhere to Extracellular Matrix Components and Mesothelial Cell Monolayers. *Gynecol. Oncol.* **2004**, *93*, 170–181. [CrossRef] [PubMed]
42. Desjardins, M.; Xie, J.; Gurler, H.; Muralidhar, G.G.; Sacks, J.D.; Burdette, J.E.; Barbolina, M.V. Versican Regulates Metastasis of Epithelial Ovarian Carcinoma Cells and Spheroids. *J. Ovarian Res.* **2014**, *7*, 70. [CrossRef] [PubMed]
43. Peart, T.M.; Correa, R.J.M.; Valdes, Y.R.; DiMattia, G.E.; Shepherd, T.G. BMP Signalling Controls the Malignant Potential of Ascites-Derived Human Epithelial Ovarian Cancer Spheroids via AKT Kinase Activation. *Clin. Exp. Metastasis* **2012**, *29*, 293–313. [CrossRef] [PubMed]
44. Rafehi, S.; Valdes, Y.R.; Bertrand, M.; McGee, J.; Préfontaine, M.; Sugimoto, A.; Dimattia, G.E.; Shepherd, T.G. TGFβ Signaling Regulates Epithelial-Mesenchymal Plasticity in Ovarian Cancer Ascites-Derived Spheroids. *Endocr. Relat. Cancer* **2016**, *23*, 147–159. [CrossRef] [PubMed]
45. Frederick, M.I.; Nassef, M.Z.; Borrelli, M.J.; Kuang, S.; Buensuceso, A.; More, T.; Cordes, T.; O'Donoghue, P.; Shepherd, T.G.; Hiller, K.; et al. Metabolic Adaptation in Epithelial Ovarian Cancer Metastasis. *BBA-Mol. Basis Dis.* **2024**, *1870*, 167312. [CrossRef]
46. Kim, Y.S.; Gupta Vallur, P.; Jones, V.M.; Worley, B.L.; Shimko, S.; Shin, D.H.; Crawford, L.T.C.; Chen, C.W.; Aird, K.M.; Abraham, T.; et al. Context-Dependent Activation of SIRT3 Is Necessary for Anchorage-Independent Survival and Metastasis of Ovarian Cancer Cells. *Oncogene* **2020**, *39*, 1619–1633. [CrossRef]
47. Chuang, T.C.; Shao, W.S.; Hsu, S.C.; Lee, S.L.; Kao, M.C.; Wang, V. Baicalein Induces G2/M Cell Cycle Arrest Associated with ROS Generation and CHK2 Activation in Highly Invasive Human Ovarian Cancer Cells. *Molecules* **2023**, *28*, 1039. [CrossRef] [PubMed]
48. Zhan, L.; Zhang, Y.; Wang, W.; Song, E.; Fan, Y.; Wei, B. E2F1: A Promising Regulator in Ovarian Carcinoma. *Tumor Biol.* **2016**, *37*, 2823–2831. [CrossRef] [PubMed]
49. Zhou, Q.; Zhang, F.; He, Z.; Zuo, M.Z. E2F2/5/8 Serve as Potential Prognostic Biomarkers and Targets for Human Ovarian Cancer. *Front. Oncol.* **2019**, *9*, 161. [CrossRef]
50. Lynch, K.N.; Liu, J.F.; Kesten, N.; Chow, K.H.; Shetty, A.; He, R.; Afreen, M.F.; Yuan, L.; Matulonis, U.A.; Growdon, W.B.; et al. Enhanced Efficacy of Aurora Kinase Inhibitors in G2/m Checkpoint Deficient Tp53 Mutant Uterine Carcinomas Is Linked to the Summation of Lkb1–Akt–P53 Interactions. *Cancers* **2021**, *13*, 2195. [CrossRef] [PubMed]
51. Li, J.; Lu, J.; Xu, M.; Yang, S.; Yu, T.; Zheng, C.; Huang, X.; Pan, Y.; Chen, Y.; Long, J.; et al. ODF2L Acts as a Synthetic Lethal Partner with WEE1 Inhibition in Epithelial Ovarian Cancer Models. *J. Clin. Investig.* **2023**, *133*, e161544. [CrossRef] [PubMed]
52. Roering, P.; Siddiqui, A.; Heuser, V.D.; Potdar, S.; Mikkonen, P.; Oikkonen, J.; Li, Y.; Pikkusaari, S.; Wennerberg, K.; Hynninen, J.; et al. Effects of Wee1 Inhibitor Adavosertib on Patient-Derived High-Grade Serous Ovarian Cancer Cells Are Multiple and Independent of Homologous Recombination Status. *Front. Oncol.* **2022**, *12*, 954430. [CrossRef] [PubMed]
53. Ding, Y.H.; Zhou, Z.W.; Ha, C.F.; Zhang, X.Y.; Pan, S.T.; He, Z.X.; Edelman, J.L.; Wang, D.; Yang, Y.X.; Zhang, X.; et al. Alisertib, an Aurora Kinase A Inhibitor, Induces Apoptosis and Autophagy but Inhibits Epithelial to Mesenchymal Transition in Human Epithelial Ovarian Cancer Cells. *Drug Des. Dev. Ther.* **2015**, *9*, 425–464. [CrossRef]
54. Gasimli, K.; Raab, M.; Tahmasbi Rad, M.; Kurunci-Csacsko, E.; Becker, S.; Strebhardt, K.; Sanhaji, M. Sequential Targeting of PLK1 and PARP1 Reverses the Resistance to PARP Inhibitors and Enhances Platin-Based Chemotherapy in BRCA-Deficient High-Grade Serous Ovarian Cancer with KRAS Amplification. *Int. J. Mol. Sci.* **2022**, *23*, 10892. [CrossRef]
55. Borah, N.A.; Reddy, M.M. Aurora Kinase B Inhibition: A Potential Therapeutic Strategy for Cancer. *Molecules* **2021**, *26*, 1981. [CrossRef] [PubMed]
56. Ngoi, N.Y.L.; Sundararajan, V.; Tan, D.S.P. Exploiting Replicative Stress in Gynecological Cancers as a Therapeutic Strategy. *Int. J. Gynecol. Cancer* **2020**, *30*, 1224–1238. [CrossRef] [PubMed]

57. Kroeger, P.T.; Drapkin, R. Pathogenesis and Heterogeneity of Ovarian Cancer. *Curr. Opin. Obstet. Gynecol.* **2017**, *29*, 26–34. [CrossRef] [PubMed]
58. Lawrenson, K.; Sproul, D.; Grun, B.; Notaridou, M.; Benjamin, E.; Jacobs, I.J.; Dafou, D.; Sims, A.H.; Gayther, S.A. Modelling Genetic and Clinical Heterogeneity in Epithelial Ovarian Cancers. *Carcinogenesis* **2011**, *32*, 1540–1549. [CrossRef] [PubMed]
59. Geistlinger, L.; Oh, S.; Ramos, M.; Schiffer, L.; LaRue, R.S.; Henzler, C.M.; Munro, S.A.; Daughters, C.; Nelson, A.C.; Winterhoff, B.J.; et al. Multiomic Analysis of Subtype Evolution and Heterogeneity in High-Grade Serous Ovarian Carcinoma. *Cancer Res.* **2021**, *80*, 4335–4345. [CrossRef] [PubMed]
60. Li, N.; Huang, D.; Lu, N.; Luo, L. Role of the LKB1/AMPK Pathway in Tumor Invasion and Metastasis of Cancer Cells (Review). *Oncol. Rep.* **2015**, *34*, 2821–2826. [CrossRef]
61. Simian, M.; Bissell, M.J. Organoids: A Historical Perspective of Thinking in Three Dimensions. *J. Cell Biol.* **2017**, *216*, 31–40. [CrossRef] [PubMed]
62. Brodeur, M.N.; Simeone, K.; Leclerc-Deslauniers, K.; Fleury, H.; Carmona, E.; Provencher, D.M.; Mes-Masson, A.M. Carboplatin Response in Preclinical Models for Ovarian Cancer: Comparison of 2D Monolayers, Spheroids, Ex Vivo Tumors and in Vivo Models. *Sci. Rep.* **2021**, *11*, 18183. [CrossRef]
63. Khashaba, M.; Fawzy, M.; Abdel-Aziz, A.; Eladawei, G.; Nagib, R. Subtyping of High Grade Serous Ovarian Carcinoma: Histopathological and Immunohistochemical Approach. *J. Egypt. Natl. Cancer Inst.* **2022**, *34*, 6. [CrossRef] [PubMed]
64. Bologna-Molina, R.; Mosqueda-Taylor, A.; Molina-Frechero, N.; Mori-Estevez, A.D.; Sánchez-Acuña, G. Comparison of the Value of PCNA and Ki-67 as Markers of Cell Proliferation in Ameloblastic Tumors. *Med. Oral Patol. Oral Cir. Bucal* **2013**, *18*, e174–e179. [CrossRef]
65. Fernando, M.; Duijf, P.H.G.; Proctor, M.; Stevenson, A.J.; Ehmann, A.; Vora, S.; Skalamera, D.; Adams, M.; Gabrielli, B. Dysregulated G2 Phase Checkpoint Recovery Pathway Reduces DNA Repair Efficiency and Increases Chromosomal Instability in a Wide Range of Tumours. *Oncogenesis* **2021**, *10*, 41. [CrossRef] [PubMed]
66. Zhang, T.; Zheng, S.; Liu, Y.; Li, X.; Wu, J.; Sun, Y.; Liu, G. DNA Damage Response and PD-1/PD-L1 Pathway in Ovarian Cancer. *DNA Repair* **2021**, *102*, 103112. [CrossRef] [PubMed]
67. Barnaba, N.; LaRocque, J.R. Targeting Cell Cycle Regulation via the G2-M Checkpoint for Synthetic Lethality in Melanoma. *Cell Cycle* **2021**, *20*, 1041–1051. [CrossRef] [PubMed]
68. Chen, C.P.; Yeh, C.N.; Pan, Y.R.; Huang, W.K.; Hsiao, Y.T.; Lo, C.H.; Wu, C.E. Wee1 Inhibition by MK1775 Potentiates Gemcitabine through Accumulated Replication Stress Leading to Apoptosis in Biliary Tract Cancer. *Biomed. Pharmacother.* **2023**, *166*, 115389. [CrossRef]
69. Affatato, R.; Carrassa, L.; Chilà, R.; Lupi, M.; Restelli, V.; Damia, G. Identification of PLK1 as a New Therapeutic Target in Mucinous Ovarian Carcinoma. *Cancers* **2020**, *12*, 672. [CrossRef] [PubMed]
70. Pérez-Fidalgo, J.A.; Gambardella, V.; Pineda, B.; Burgues, O.; Piñero, O.; Cervantes, A. Aurora Kinases in Ovarian Cancer. *ESMO Open* **2020**, *5*, e000718. [CrossRef] [PubMed]
71. Ganapathi, R.N.; Norris, E.J.; Sutker, A.P.; Klotz, K.E.; Ganapathi, M.K. Targeting Aurora A Kinase (AAK) in Platinum-Resistant High Grade Serous Ovarian Cancer. *Front. Oncol.* **2020**, *10*, 1354. [CrossRef]
72. Manfredi, M.G.; Ecsedy, J.A.; Chakravarty, A.; Silverman, L.; Zhang, M.; Hoar, K.M.; Stroud, S.G.; Chen, W.; Shinde, V.; Huck, J.J.; et al. Characterization of Alisertib (MLN8237), an Investigational Small-Molecule Inhibitor of Aurora A Kinase Using Novel in Vivo Pharmacodynamic Assays. *Clin. Cancer Res.* **2011**, *17*, 7614–7624. [CrossRef]
73. Sun, H.; Wang, H.; Wang, X.; Aoki, Y.; Wang, X.; Yang, Y.; Cheng, X.; Wang, Z.; Wang, X. Aurora-A/SOX8/FOXK1 Signaling Axis Promotes Chemoresistance via Suppression of Cell Senescence and Induction of Glucose Metabolism in Ovarian Cancer Organoids and Cells. *Theranostics* **2020**, *10*, 6928–6945. [CrossRef] [PubMed]
74. Xia, Z.; Wei, P.; Zhang, H.; Ding, Z.; Yang, L.; Huang, Z.; Zhang, N. AURKA Governs Self-Renewal Capacity in Glioma-Initiating Cells via Stabilization/Activation of β-Catenin/Wnt Signaling. *Mol. Cancer Res.* **2013**, *11*, 1101–1111. [CrossRef] [PubMed]
75. D'Assoro, A.B.; Liu, T.; Quatraro, C.; Amato, A.; Opyrchal, M.; Leontovich, A.; Ikeda, Y.; Ohmine, S.; Lingle, W.; Suman, V.; et al. The Mitotic Kinase Aurora-A Promotes Distant Metastases by Inducing Epithelial-to-Mesenchymal Transition in ER + Breast Cancer Cells. *Oncogene* **2014**, *33*, 599–610. [CrossRef]
76. Zheng, D.; Li, J.; Yan, H.; Zhang, G.; Li, W.; Chu, E.; Wei, N. Emerging Roles of Aurora-A Kinase in Cancer Therapy Resistance. *Acta Pharm. Sin. B* **2023**, *13*, 2826–2843. [CrossRef] [PubMed]
77. Aquino-Acevedo, A.N.; Orengo-Orengo, J.A.; Cruz-Robles, M.E.; Saavedra, H.I. Mitotic Kinases Are Emerging Therapeutic Targets against Metastatic Breast Cancer. *Cell Div.* **2024**, *19*, 21. [CrossRef]
78. Alkhateeb, K.J.; Crane, J.E.; Sak, M.; Jorgensen, C.J.; O'Donnell, J.P.; Zumbar, C.T.; Wozniak, J.A.; Salazar, C.R.; Parwani, A.V.; Lehman, N.L. Aurora-A Kinase Is Differentially Expressed in the Nucleus and Cytoplasm in Normal Müllerian Epithelium and Benign, Borderline and Malignant Serous Ovarian Neoplåsms. *Diagn. Pathol.* **2021**, *16*, 98. [CrossRef]
79. Domcke, S.; Sinha, R.; Levine, D.A.; Sander, C.; Schultz, N. Evaluating Cell Lines as Tumour Models by Comparison of Genomic Profiles. *Nat. Commun.* **2013**, *4*, 2126. [CrossRef]

80. Kim, S.; Choung, S.; Sun, R.X.; Ung, N.; Hashemi, N.; Fong, E.J.; Lau, R.; Spiller, E.; Gasho, J.; Foo, J.; et al. Comparison of Cell and Organoid-Level Analysis of Patient-Derived 3D Organoids to Evaluate Tumor Cell Growth Dynamics and Drug Response. *SLAS Discov.* **2020**, *25*, 744–754. [CrossRef]
81. Cody, N.A.; Zietarska, M.; Filali-Mouhim, A.; Provencher, D.M.; Mes-Masson, A.-M.; Tonin, P.N. Influence of Monolayer, Spheroid, and Tumor Growth Conditions on Chromosome 3 Gene Expression in Tumorigenic Epithelial Ovarian Cancer Cell Lines. *BMC Med. Genom.* **2008**, *1*, 34. [CrossRef] [PubMed]
82. Edgar, R.; Domrachev, M.; Lash, A.E. Gene Expression Omnibus: NCBI Gene Expression and Hybridization Array Data Repository. *Nucleic Acids Res.* **2002**, *30*, 207–210. [CrossRef] [PubMed]

cells

MDPI

Article

Identification of Prognostic Biomarkers of Ovarian High-Grade Serous Carcinoma: A Preliminary Study Using Spatial Transcriptome Analysis and Multispectral Imaging

Haeyoun Kang [1,†], Je-Gun Joung [2,3,4,†], Hyun Park [5], Min Chul Choi [6], Doohyun Koh [2], Ju-Yeon Jeong [3], Jimin Lee [3], Sook-Young Kim [3], Daun Jung [7], Sohyun Hwang [1,3,*] and Hee Jung An [1,3,*]

1 Department of Pathology, CHA Bundang Medical Center, CHA University School of Medicine, Seongnam 13496, Republic of Korea; hykang@cha.ac.kr
2 Department of Biomedical Science, CHA University, Seongnam 13496, Republic of Korea; jgjoung@cha.ac.kr (J.-G.J.); timoplay12@naver.com (D.K.)
3 CHA Future Medicine Research Institute, CHA Bundang Medical Center, CHA University School of Medicine, Seongnam 13496, Republic of Korea; dufakthd@chamc.co.kr (J.-Y.J.); c094002@chamc.co.kr (J.L.); skylove012@chamc.co.kr (S.-Y.K.)
4 Department of Biomedical Informatics, CHA University, Seongnam 13496, Republic of Korea
5 Department of Gynecological Oncology, Comprehensive Gynecologic Cancer Center, CHA Bundang Medical Center, CHA University School of Medicine, Seongnam 13496, Republic of Korea; p06162006@cha.ac.kr
6 Department of Obstetrics and Gynecology, Asan Medical Center, University of Ulsan College of Medicine, Seoul 05505, Republic of Korea; oursk@naver.com
7 CHA Biomedical Research Institute, CHA Bundang Medical Center, Seongnam 13496, Republic of Korea; jhd2800@chamc.co.kr
* Correspondence: blissfulwin@cha.ac.kr (S.H.); hjahn@cha.ac.kr (H.J.A.); Tel.: +82-31-780-6019 (S.H.); +82-31-780-5440 (H.J.A.)
† These authors contributed equally to this work.

Abstract: Ovarian cancer is a lethal malignancy, with most patients initially responding to chemotherapy but frequently experiencing recurrence. Previous studies primarily examined tumor characteristics using limited genetic markers or bulk RNA sequencing. Here, we used spatial transcriptomics via the GeoMx® platform, alongside multispectral immune cell immunofluorescence (IF), to identify biomarkers associated with disease progression following first-line treatment of high-grade serous carcinoma (HGSC). We identified several spatial biomarkers linked to non-recurrence, including elevated *NKG7* expression in CD45+ immune cell regions ($p = 0.0011$) and higher *TFPI2* and *PIGR* expression in tumor areas ($p = 2.09 \times 10^{-6}$), both associated with improved progression-free survival. Multispectral IF revealed significantly higher regulatory T cell (Treg) to CD8+ T cell ratios in the tumor nests and stroma of recurrent patients ($p = 0.016, 0.048$). Tregs were also found closer to cancer cells or macrophages than CD8+ T cells in recurrent tumors ($p = 0.048$), correlating with poor survival. Integrated analysis showed that immune cell density and immune pathway scores in the recurrent group positively correlated with cancer pathway scores, except for NF-κB. This comprehensive analysis revealed clues to interactions between different immune cells and identified biomarkers that may be useful for predicting recurrence of HGSC.

Keywords: ovarian cancer; biomarkers; spatial transcriptomics; cancer recurrence; tumor microenvironment

1. Introduction

Ovarian cancer (OC) is a highly lethal gynecological malignancy; indeed, it is one of the most common causes of cancer-related mortality in women worldwide [1]. Although most OC patients are responsive to chemotherapy, they frequently present with recurrent metastatic lesions that result in poor overall survival [2]. Therefore, precise diagnosis and treatment based on distinctive clinical, molecular, and immunologic characteristics of the tumor in each patient are imperative for strategic treatment planning and accurate prediction of recurrence.

High-grade serous carcinoma (HGSC), the most common and deadliest type of OC, is characterized by its unique clinical, genomic, and immunologic profile and high rates of recurrence. HGSCs present with a complex landscape of cellular and molecular diversity, with varied morphological patterns and tissue architectures within individual tumors [3]. The molecular profile of these cancers is characterized by a high frequency of TP53 tumor suppressor gene mutations coupled with chromosomal instability, resulting in extensive copy number variations [4,5]. The prognosis of HGSC is affected significantly by the immune profile of the tumor microenvironment (TME) [6]. These immune profiles within the TME are associated with survival outcomes and/or tumor recurrence. Crucially, the TME in HGSC patients displays remarkable variability, not only between patients but also among distinct tumor sites within a single patient, and even across different regions within an individual tumor. This huge diversity underscores the need for spatial transcriptomic approaches to comprehensively map and understand the intricate interactions between tumor cells and immune components across various tumor regions, thereby providing a more nuanced understanding of HGSC biology and potentially guiding more effective, personalized treatment strategies.

Recently, spatial transcriptomics approaches to assessing the localization of mRNA molecules within the tissue architecture of histological sections have been developed. These approaches provide valuable insight into the spatial characteristics of gene expression within tissues and are a promising method for discovering novel biomarkers for disease progression and treatment responses. Among the different types of spatial transcriptomics approaches, we utilized the GeoMx® Digital Spatial Profiling (DSP) System, which can visualize formalin-fixed paraffin-embedded (FFPE) tissue sections stained with fluorescently labeled antibodies. Regions of interest (ROI) within the tissue are selected prior to spatial RNA profiling of multiple genes [3]. The capability for a simultaneous high-level multiplex approach to examining tissue areas of interest, selected on the basis of marker expression, makes DSP suitable for analyzing multiple biological markers within specific areas of a tissue sample at the same time.

The aim of this study was to use this spatial transcriptome analysis approach to identify biomarkers associated with the progression of HGSCs after first-line treatment. We investigated diverse gene signatures related to immune infiltration, taking into consideration their spatial enrichment patterns. Furthermore, we used multispectral imaging assessment (i.e., multiplex immunofluorescence (IF) analysis) by the Vectra Polaris™ system (Akoya Bioscience, Marlborough, MA, USA) to computationally identify and quantify different immune cell types within the tumor and stromal regions. This approach also allowed us to analyze the distance between cancer cells and different types of immune cells. By integrating these two methods, we sought to gain detailed insight into the spatial dynamics of gene expression, as well as immune cell interactions within the TME. A comprehensive understanding of these interactions is critical for improving treatment methods and enabling a personalized approach to this type of OC.

2. Materials and Methods

2.1. Patients and Construction of Tissue Microarray (TMA)

This study examined fresh and paraffin-embedded tissues from 12 cases selected from the archives; these patients had HGSC and were treated with standard therapy (radical hysterectomy with bilateral salpingo-oophorectomy, followed by platinum-based chemotherapy) at CHA Bundang Medical Center between 2007 and 2014. This study was approved by the institutional review board of CHA Bundang Medical Center (IRB #CHAMC 2020-03-004). Patient informed consent was received. The biospecimen and data used in this study were provided by Bundang CHA Biobank, Korea Biobank Network (KBN4-A05).

For NanoString GeoMx® DSP (NanoString Technologies, Seattle, WA, USA), a TMA was constructed after two board-certified pathologists (HK and HA) reviewed representative whole slide tumor sections and provided a histologic diagnosis according to WHO classifications (2020) [7]. One representative core (5 mm diameter) was obtained from a representative FFPE block of tissue from each case. Table 1 summarizes the clinicopathologic characteristics of all 12 patients.

Table 1. Clinicopathologic characteristics of 12 patients.

Patient ID	Age Group at Dx	Primary Site	Stage	Tumor Size	PFS > 5y	Recurrence Status	Lymph Node Metastasis	Distant Metastasis
Pat05	70s	Ovary	3c	14	Yes	Non-recurred	Absent	Absent
Pat09	40s	Ovary	3c	5.3	No	Recurred	Absent	Absent
Pat10	50s	Ovary	3c	7	No	Recurred	Present	Absent
Pat11	60s	Ovary	3b	6	Yes	Non-recurred	Absent	Absent
Pat12	50s	Ovary	2c	7.5	Yes	Non-recurred	Absent	Absent
Pat13	30s	Ovary	3c	4	Yes	Non-recurred	Present	Absent
Pat15	50s	Ovary	3c	10	No	Recurred	Present	Absent
Pat16	50s	Ovary	3c	12	No	Recurred	Present	Absent
Pat20	50s	Ovary	3c	10	No	Recurred	Present	Absent
Pat21	50s	Ovary	3c	12	No	Recurred	Present	Absent
Pat42	50s	Ovary	3c	15	No	Recurred	Present	Absent
Pat53	60s	Ovary	3c	12	No	Recurred	Present	Absent

Dx: Diagnosis.

2.2. Spatial Transcriptome Profiling

The NanoString GeoMx® DSP (NanoString Technologies, Seattle, WA, USA) was used to identify and visualize tumor cells, immune cells, and stromal cells. Each of these areas was viewed by immunofluorescent staining of morphology markers: PanCK for tumor tissue, CD45 for immune cells, and SMA for stroma. DAPI was used to visualize cell nuclei. In 11 cases, two rectangular-shaped ROIs per core were annotated by a board-certified pathologist (HK), whereas one ROI was annotated within a smaller piece of tumor tissue from one case (Patient 12). Each rectangular ROI had an area of $660 \times 785\ \mu m^2$. In total, 23 ROIs were annotated, each containing three areas of interest (AOIs) showing expression of the selected morphologic markers (i.e., PanCK, SMA, and CD45). In total, 69 AOIs were generated by automated tissue segmentation conducted using NanoString GeoMx® DSP software and based on expression of the three morphology markers. The data were confirmed by a board-certified pathologist (HK).

After automated tissue segmentation, oligonucleotides were released from the selected ROIs upon localized exposure to UV light. Photocleaved oligonucleotides were collected via microcapillary aspiration and dispensed into a 96-well plate. Localized UV exposure and PC-oligonucleotide collection were conducted for each ROI. Oligonucleotides collected from each AOI were amplified by Polymerase Chain Reaction (PCR). The PCR products were pooled and purified using AMPure XP beads (Beckman Coulter Diagnostics Inc., Brea, CA, USA). Purified libraries were pooled and sequenced separately. Library quantity and quality were measured using a Qubit 4.0 (Thermo Fisher Scientific, Waltham, MA, USA) and TapeStation 4200 (Agilent Technologies, Santa Clara, CA, USA). Sequencing was performed on a NovaSeq 6000 platform, with 38 bp paired-end reads (Illumina, San Diego, CA, USA). The whole transcriptome within each AOI within a selected ROI was analyzed.

2.3. Spatial Transcriptome Data Processing

Sequencing for spatial transcriptome data was conducted using the GeoMx® NGS Pipeline. After sequencing, reads were aligned to identify the unique ID of the probe. Duplicate reads were removed using the unique identifier region, and then the reads were converted to digital counts. The presence or absence of a gene was defined by the limit of quantitation (LOQ), which is the geometric mean of the negative probes multiplied by the geometric standard deviation raised to the second power. Of the 18,677 genes assayed, 16,132 (86.4%) were above the LOQ in 10% of the AOIs. Q3 normalization (75% of counts in each sample) was used for downstream analysis to account for variations in the area of the AOIs.

2.4. Gene Expression Analysis

The SpatialDecon R package v1.6.0, which performs deconvolution of mixed cells from spatial gene expression data based on constrained log-normal regression, was used to estimate cell abundance based on the spatial gene expression dataset derived from each AOI [8]. The SafeTME matrix, a cell profile matrix used to estimate immune and stromal cell types in tumors, was used for deconvolution [5]. DEGs were identified using the edgeR R package [6]. For multiple comparisons, we used the Benjamini–Hochberg *p*-value correction procedure. Gene Ontology analysis was performed using the clusterProfiler R package v4.4.4 [9]. Data were visualized as a pie chart constructed using the ggplot2 R package v3.5.1.

2.5. Multispectral Imaging of Immune Cells Using VECTRA

Multispectral staining was performed using a multiplex immunohistochemistry (IHC) method, followed by the generation of fluorescent images using the fluorophores contained in the Opal 7 kit (Akoya Biosciences, Marlborough, MA, USA). Multispectral staining was conducted using an automated staining apparatus (Bond Rx, Leica Biosystems, Nussloch, Germany) and specific antibodies (see below). After staining, whole slide scanning was performed to obtain images (magnification ×20) using Vectra Polaris™ (Akoya Biosciences, Marlborough, MA, USA). Target cells were identified based on protein expression levels, and autofluorescence was minimized using control tissues. A fluorescence cut-off level was established for certain target proteins, and the optimal cut-off value was derived using multiple analyses of tissues and receiver operating characteristics curve analysis. Quantification of the fluorescent images was performed on a pixel-by-pixel basis using Phenochart v1.1.0 within inForm® software 2.5.1 (Akoya Bioscience, Marlborough, MA, USA). The resulting numeric data facilitated two-dimensional differentiation of tumor nests and non-neoplastic stroma within the TME. Classification of immune cells within the TME was performed using cell type-specific antibodies, including anti-CD8 (for cytotoxic T cells), anti-PD1 (an immune checkpoint inhibitor), anti-CD68 (for macrophages), anti-FOXP3 (for Treg cells), anti-CD56 (for NK/T cells), anti-CK (for carcinoma cells), and DAPI (for cell nuclei). Cell density and the distance between cell types in the tumor nests and stroma were then analyzed.

2.6. Integrative Analysis of Multispectral Immune Cell and Spatial Transcriptome Data

The score for each identified immune and cancer cell pathway was calculated by the GSVA R package v1.46.0 [10] using gene sets derived from the mSigDB hallmark gene sets database v.7.5.1 [11] and WikiPathways v7.5.1 [12]. Pathways related to NF-κB signaling, Warburg effect signaling, RAS signaling, Wnt signaling, and autophagy were identified from the WikiPathways database; all other pathways were identified from the mSigDB hallmark gene sets database. Correlations between gene expression, pathway scores, and immune cell density were calculated by Spearman's correlation analysis. Correlation plots were constructed using corrplot within the R package.

2.7. Statistical Analysis

Survival analysis was performed using a Cox proportional hazards model within the survival R package. The statistical comparison of data from the two groups was performed using the Wilcoxon rank sum test. All plots were drawn in the ggplot2 package in R. p values < 0.05 were considered significant.

3. Results

3.1. Spatial Transcriptome Analysis and Gene Profiling of Selected ROI

To identify gene profiles associated with immunogenetic responses in HGSC patients, we used the NanoString GeoMx® DSP to examine TMA slides from 12 patients (four non-recurring and eight recurring cases) who received standard treatment. The four patients without recurrence were named the "non-recur group" and the eight patients with recurrence were named the "recur group". The tumor tissue, immune cells, and stromal regions were visualized using fluorescently labeled antibodies specific for PanCK (for the tumor area), CD45 (for the immune cell area), and SMA (for the stroma area). In total, 23 ROIs were selected for RNA sequencing (average 1.9 ROIs per patient).

The deconvolution method was used to estimate the proportion (%) of each immune cell type in each area within the selected ROI for each HGSC sample (Figure 1A and Supplementary Table S1). Each area showed distinctive characteristics with respect to cell-type enrichment. In the CD45 immune cell area, we found enrichment of mainly CD8+ T cells and macrophages (29.03% and 37.59%, respectively). In the PanCK tumor area, we found that the percentages of CD4+ T cells, CD8+ T cells, and neutrophils tended to be higher than those of other cell types (22.75%, 19.36%, and 27.24%, respectively). In the stroma area, the percentages of CD8+ T, CD4+ T cells, and macrophages were highest (26.99%, 13.23%, and 14.11%, respectively). The percentage of macrophages across the entire ROI was higher in the non-recur group than in the recur group (28.70% vs. 14.48%, respectively; $p = 0.015$) (Figure 1B). When we compared differences in the percentages of the different cell types within each area between the non-recur and recur groups, we found that the percentage of monocytes in the tumor area was significantly lower in the recur group than in the non-recur group (0.00% vs. 1.23%, respectively; $p = 0.034$). Among the CD45 immune cell population, we found that the percentage of neutrophils was significantly higher in the recur group than in the non-recur group (2.58% vs. 0.71%, respectively; $p = 0.03$).

3.2. Identification of Differentially Expressed Genes (DEGs) Between the Non-Recur and Recur Groups

DEGs ($|\log_2$FC (fold change)$| \geq 1.0$ and $p < 0.001$) in each area within the recur and non-recur groups were identified (Figure 2A,B and Supplementary Table S2). Among the DEGs, *TFPI2* showed the greatest overexpression in the non-recur group ($p = 2.09 \times 10^{-6}$ in the tumor area); indeed, expression of this gene was highest across all three areas in this group. Higher expression of this gene was associated with a better prognosis in the non-recur group (progression-free survival (PFS), $p = 0.021$; median cut-off = 13.86 (immune area); Figure 2C). Expression of *PIGR* was significantly higher in the non-recur group, and was associated with a better prognosis (PFS, $p = 8.8 \times 10^{-3}$; median cut-off = 17.67 (immune areas), Figure 2D). By contrast, the expression of *EYA2* was higher in the immune cells and tumor areas of the recur group. *CDH5*, *ID1*, *H2BC10*, and *ITLN1* were common DEGs in the tumor and stroma areas. Next, we conducted Gene Ontology analysis to identify enriched functions in the tumor and stroma areas (adj. p-value < 0.05; Supplementary Tables S3 and S4). The data revealed that biological functions such as 'protein-DNA complex assembly', 'chromatin remodeling', and 'organ- or tissue-specific immune response' (adj. p-value = 1.69×10^{-7}, 1.67×10^{-5}, and 0.003, respectively) were enriched in the tumor area of the non-recur group (Figure 2E).

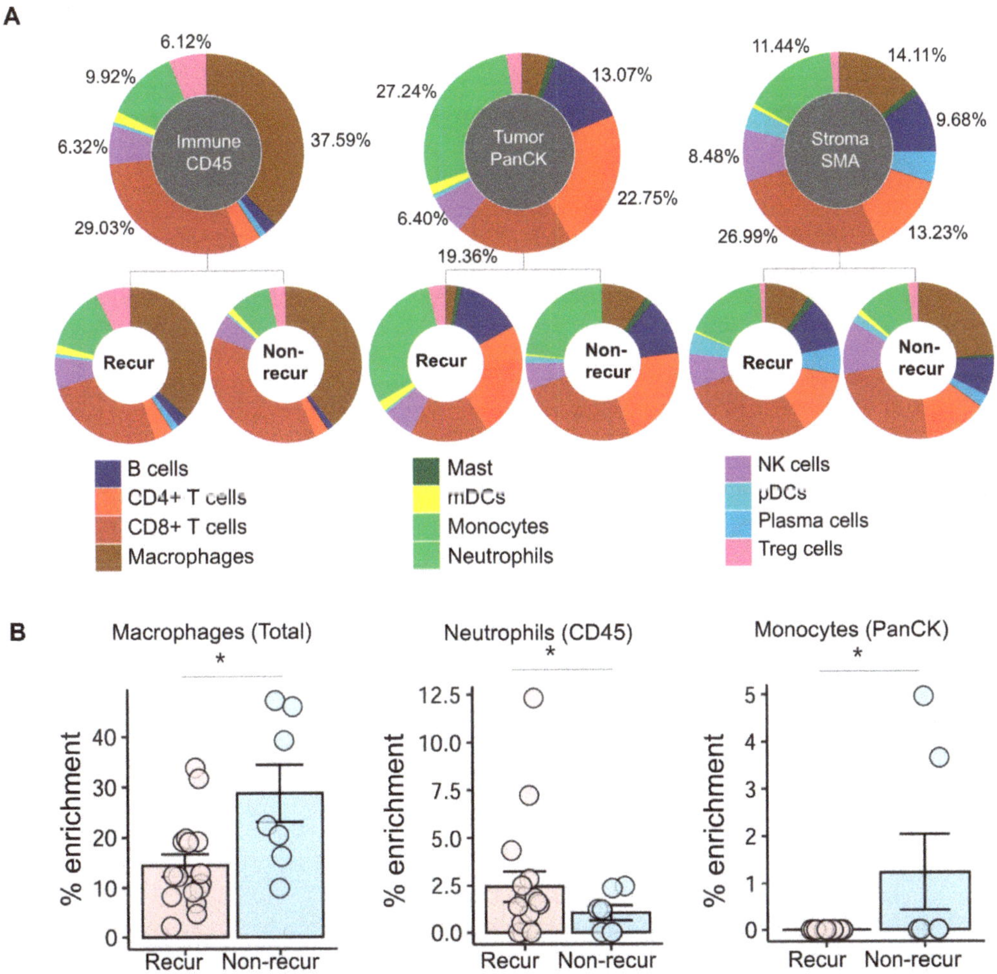

Figure 1. The proportions of cell types within each of the captured areas in ovarian cancer tissue sections. (**A**) The percentage of each cell type in the tumor (PanCK), immune (CD45), and stromal (SMA) areas within the tumor microenvironment, depicted as pie charts. The upper pie charts display combined data from all samples, whereas the bottom charts show data specific to each area in the recur and non-recur groups. (**B**) Comparison of macrophages in the total three areas, neutrophils in the CD45 (immune cell) area, and monocytes in the PanCK (tumor) area. mDCs: myeloid dendritic cells, pDCs: plasmacytoid dendritic cells and *: $p < 0.05$.

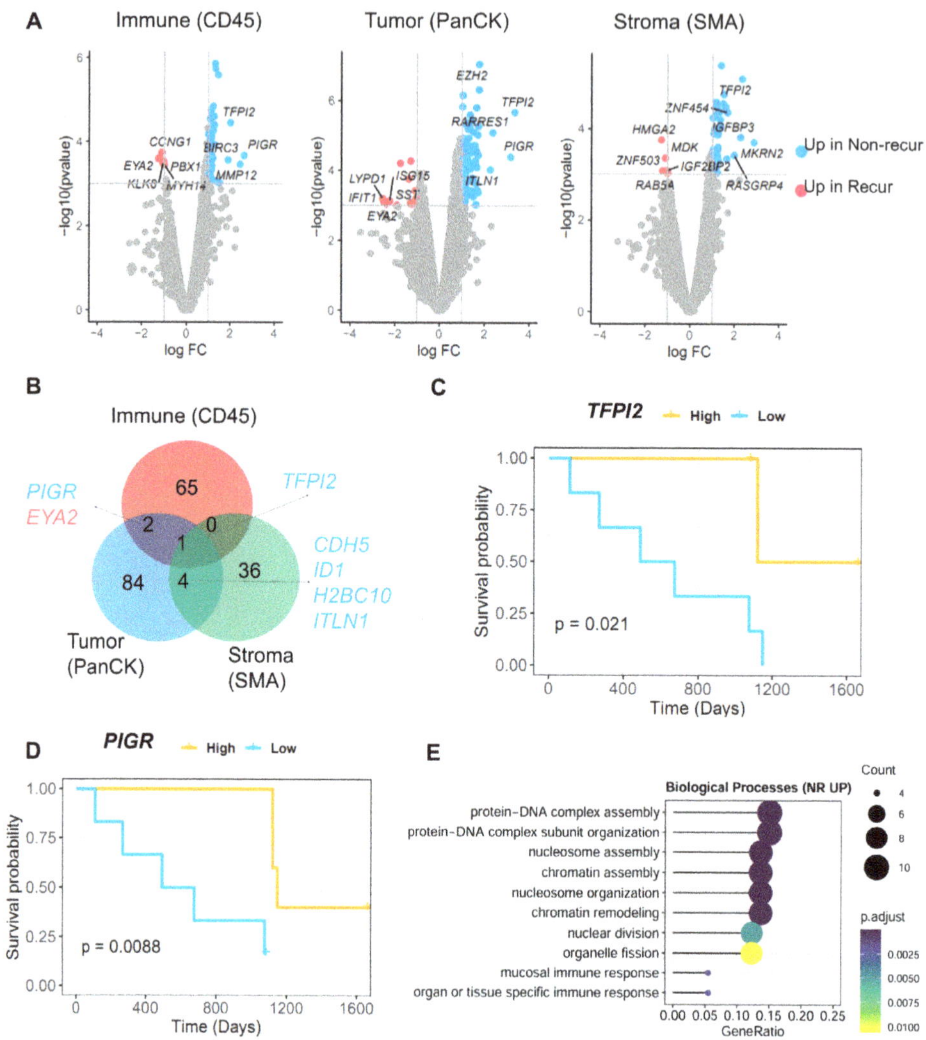

Figure 2. Differences in transcriptomic data between the recur and non-recur groups. (**A**) Volcano plots showing DEGs in each area, and (**B**) Venn diagram identifying overlapped DEGs. (**C**) Progression-free survival (PFS) analysis between high and low groups of *TFPI2* expression, and (**D**) *PIGR* expression. (**E**) Enriched Gene Ontology terms associated with up-regulated genes in the PanCK area of the non-recur (NR) than in that of recur groups. NR UP: Up-regulated in non-recur group and *p*. adjust: adjusted *p*-value.

3.3. Identification of TME-Related Markers

To identify potential therapeutic targets related to tumor recurrence and survival, we examined common genes between DEG genes and the pan-cancer microenvironment gene set suggested by Bagaev et al. [13]. For each immune cell, tumor, and stroma area, we examined statistically significant differences in gene expression between the non-recur and recur groups (Figure 3A). In the immune area, *NKG7* and *CYBB* were significantly overexpressed in the non-recur group ($Log_2FC = 1.32$ and 1.12 at $p = 1.79 \times 10^{-6}$ and 1.77×10^{-5}, respectively). Two genes (*CDH5* and *TRAC*) showed higher expression in the stroma area of the non-recur group ($Log_2FC = 1.28$ and 1.23 at $p = 4.23 \times 10^{-6}$ and

2.60×10^{-4}, respectively). Analysis of PFS related to the genes showing significantly higher or lower expression in the recur and non-recur groups (Supplementary Figure S1) revealed that high expression of *NKG7* in the immune area was associated with a favorable prognosis ($p = 6.0 \times 10^{-3}$, median cut-off = 53.5) (Figure 3B).

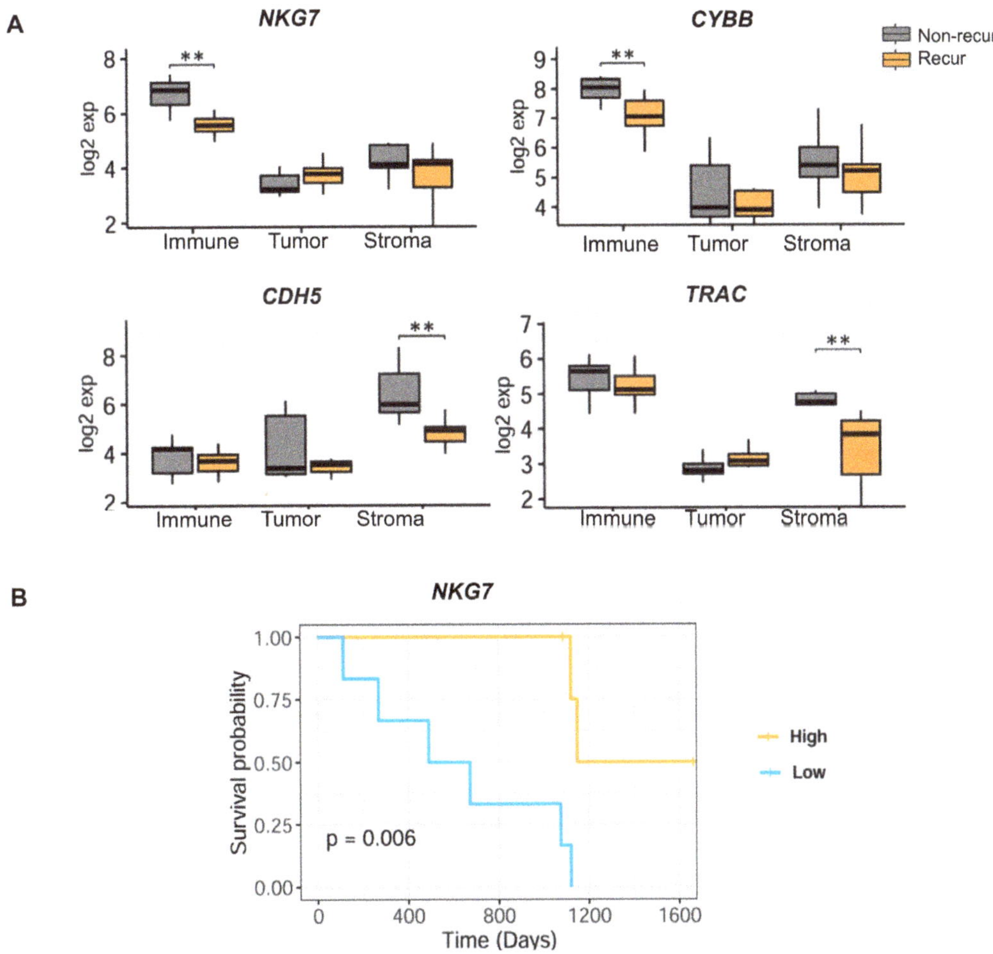

Figure 3. Expression and prognosis related to the expression of TME-associated genes. (**A**) Comparison of the expression of genes associated with TME in the recur and non-recur groups. (**B**) Analysis of PFS related to the expression of *NKG7*. The expression level is classified as low or high based on the median cut-off value. ** $p < 0.01$.

3.4. Multispectral Quantitative Computational Assessment of Immune Cell Density and Distance Between Tumor and Immune Cells Within Tumor Nests and Stroma

To examine and visualize the distribution of immune cells within HGSC tissue, we used multiplex IF markers to identify these cells by detecting expression of the immune checkpoint molecule PD1 expressed by CD8+ T cells as well as other immune cells. The density and spatial distribution (i.e., distance) between CD8+ T cells, which are pivotal for tumor cell cytotoxicity, and CD68+ macrophages, particularly tumor-associated macrophages that play a critical role in the TME, were analyzed. Our earlier analyses of cell percentages in the different areas identified both of these cell types as predominant components. In addition to PD1-CD8+ T cells and PD1+CD8+ T cells, we analyzed regulatory T cells (FOXP3+), which contribute to the suppressive TME through their actions on CD8+ T cells, macrophages, and cancer cells.

The overall mean density of total CD8+ T cells, PD1-CD8+ T cells, PD1+CD8+ T cells, regulatory Treg cells, and macrophages was higher in both tumor nests and stromal areas in the non-recur group than in those of the recur group (Supplementary Table S5). The mean CD8+ T cell density in the tumor nest and stromal areas of the non-recur group was higher than in that of the recur group (289.0 vs. 46.6 and 320.7 vs. 110.0, respectively); however, the difference was not statistically significant. The mean macrophage density in the tumor nest and stromal area of the non-recur group was higher than in the recur group (270.3 vs. 137.2 and 443.2 vs. 332.5, respectively), but again, the difference was not statistically significant. Figure 4A shows a representative multiplex IF image of immune cells from a patient in the non-recur group (P12) and from a patient in the recur group (P53); the image shows high infiltration of CD8+ T cells and macrophages in the former, and scant infiltration by these cells in the latter. Although the difference in the overall mean density of immune cells between the recur and non-recur group was not statistically significant, the Treg/total CD8+ T cell ratio in both the tumor nest and stroma area was significantly higher in the recur group than in the non-recur group, by 6.40-fold ($p = 0.016$) and 2.37-fold ($p = 0.048$), respectively (Figure 4B and Supplementary Table S5). The Treg/PD1-CD8+ T cell and macrophage/PD1-CD8+ T cell ratios were also significantly higher (by 4.1-fold and 2.28-fold, respectively) in the tumor nest of the recur group than in that of the non-recur group (both $p = 0.048$, Supplementary Table S5).

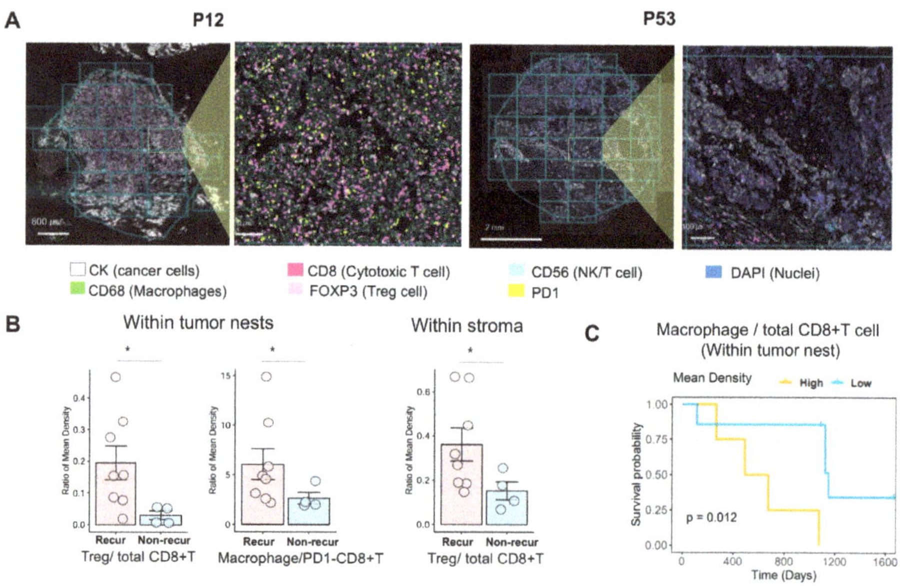

Figure 4. *Cont.*

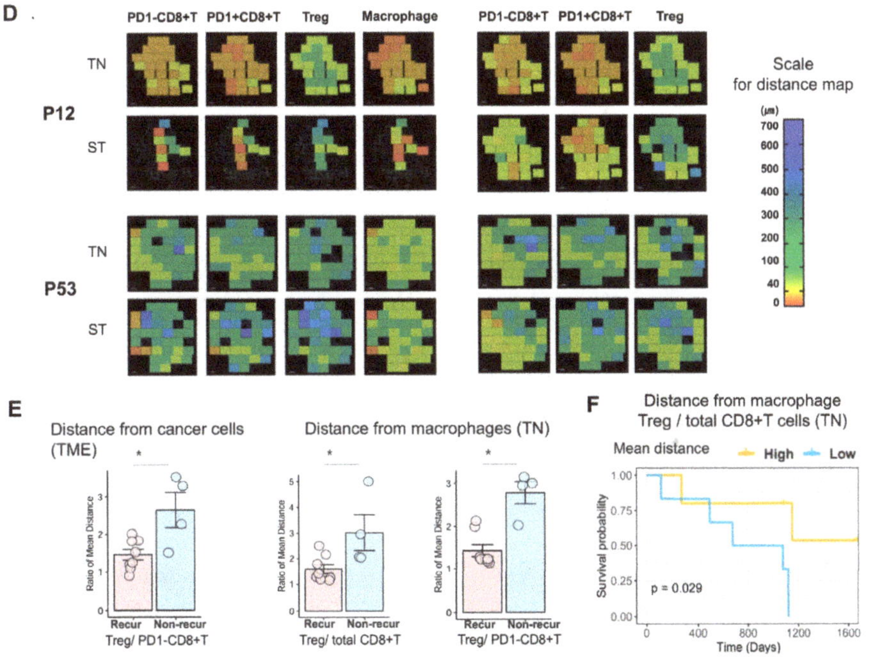

Figure 4. Comparison of mIF immune cell density, and distances between cell types in the non-recur and recur groups. (**A**) mIF immune cell densities of representative patients of recur (P53) and non-recur (P12) groups. Many immune cells, including CD8+ T cells and macrophages, were highly infiltrated in P12, whereas they were scant in P53. (**B**) Comparison of mIF immune cell density between the two groups. The box plots show the Treg/PD1-CD8+ T cell ratio and the macrophage/PD1-CD8+ T cell ratio in tumor nests, and the Treg/total CD8+ T cell ratio in the stroma. (**C**) PFS difference between high and low groups of macrophage/total CD8+ T cell in mIF immune cell density. (**D**) Representative images showing the mean distance from cancer cells to the nearest immune cell. The distances between cancer cells and immune cells, and the distance between macrophages and immune cells in P12 (non-recur group) were much shorter (red color) than in P53 (recur group). (**E**) Comparison of the mIF immune cell mean distance data from the two groups. Data depict the Treg/PD1-CD8+ T ratio in the TME, and Treg/total CD8+ T cell and Treg/PD1-CD8+ T cell ratios in tumor nests. (**F**) PFS difference between high and low groups of Treg/total CD8+ T cell (tumor nest, TN) in mIF immune cell mean distance from macrophages. * $p < 0.05$.

When investigating the correlation between immune cell density and recurrence status, we found that the densities of most immune cells correlated positively with each other; however, the macrophage density in the stroma of the recur group correlated negatively with the density of other immune cells, especially CD8+ T cells and Treg cells (Supplementary Figure S2). Survival analysis in the context of immune cell density revealed that the group with a higher macrophage/total CD8+ T cell density ratio within the tumor cell nest, meaning that the macrophage density was higher than that of total CD8+ T cells, exhibited a significantly poorer prognosis ($p = 0.012$, mean cut-off = 4.35; Figure 4C).

We assumed that actively interacting cells ought to be closer together; therefore, we measured the shortest distance between immune cells and cancer cells. Representative images showing the distance from a cancer cell to the nearest immune cell in the non-recur and recur groups are presented (P12, non-recur; P53, recur) in Figure 4D. We found that the distance between cancer cells and immune cells in the non-recur group was shorter than in the recur group, although the difference was not statistically significant (i.e., recur/non-recur in total CD8+ T cells: 1.40-fold) (Supplementary Table S6). The Treg/PD1-CD8+ T cell distance ratio from cancer cells was significantly smaller for the recur group than for the non-recur group, by 0.55-fold (p = 0.048) (Figure 4E and Supplementary Table S6). In addition, we observed that the Treg/total CD8+ T cell distance from macrophages in the tumor nests was significantly shorter in the recur group than in the non-recur group, by 0.53-fold (p = 0.048) (Figure 4E and Supplementary Table S7). Survival analysis in the context of distance ratios revealed that a lower Treg/total CD8+ T cell distance within the tumor cell nest, which suggests that Treg cells are closer to macrophages than CD8+ T cells, was associated with a poorer prognosis (p = 0.029, mean cut-off = 1.80; Figure 4F).

3.5. *Integrative Analysis of Multispectral Immune Cell Density and Spatial Transcriptome Data Related to the TME*

Finally, we performed an integrative analysis of the spatial transcriptome data and multispectral quantitative immune cell assessment data. Before integrating the two data sets, we first checked whether data derived from GeoMx and multispectral quantitative analysis of immune cells were consistent; for example, CD68 expression by macrophages, and CD8A expression by total CD8+ T cells. Data from multispectral quantitative analysis revealed that the macrophage density in the tumor nest correlated strongly with CD68 expression in the tumor area, as assessed by GeoMx (correlation coefficient = 0.66, p = 0.02), and that total CD8+ T cell density in the tumor nest correlated positively with CD8A expression in the tumor area; however, the correlation was not statistically significant (correlation coefficient = 0.43, p = 0.17; Supplementary Figure S3). Therefore, we integrated these two data sets.

Upon investigating the correlation between immune cell density and the immune and cancer pathway scores derived from transcriptome data, we found that both immune cell density and immune pathway scores correlated positively with cancer pathway scores in the recur group; an exception was the NF-kB signaling pathway. However, in the non-recur group, immune cell density correlated positively with the immune pathway score, and negatively with most cancer pathway scores (including the PI3K/mTOR and Wnt signaling pathways; Figure 5). In addition, the NF-κB pathway score correlated negatively with total CD8+ T cells in the recur group, but positively in the non-recur group.

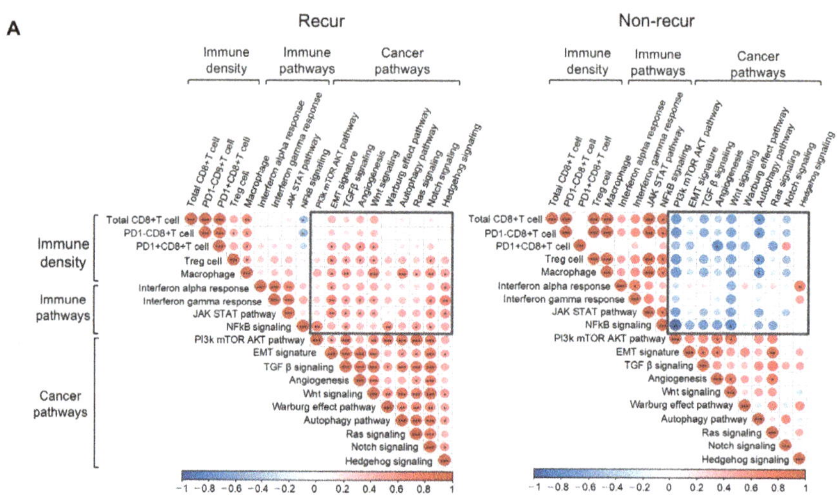

Figure 5. *Cont.*

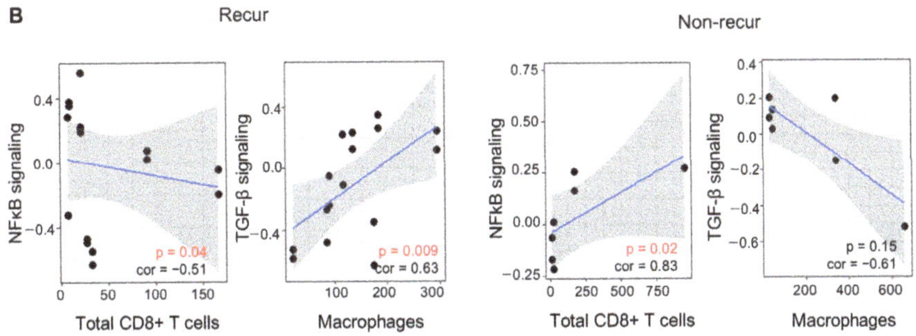

Figure 5. Correlation between the multispectral immune cell density and the immune and cancer pathway scores derived from the transcriptome data, based on recurrence status. (**A**) Correlation between immune cell density and the immune and cancer pathway scores derived from transcriptome data. The most significant differences are shown in black boxes. Colors represent the range of the correlation coefficient, from 1 (red) to −1 (blue). Statistical significance is denoted by an asterisk (* $p < 0.05$, ** $p < 0.01$, and *** $p < 0.001$). (**B**) Dot plots showing representative pathways (NF-κB signaling and TGF-β signaling pathway) and immune cell density (total CD8+ T cells and macrophages). The blue line and gray area represent the trend and the confidence interval of the correlation, respectively.

4. Discussion

The TME plays a crucial role in tumorigenesis; therefore, studying the TME in detail may provide comprehensive knowledge regarding responses to treatment [14]. In this study, we characterized gene expression profiles in the TME and examined their correlation with the recurrence of HGSC. We achieved this by conducting spatial transcriptome analysis of various histologic regions within FFPE slides of HGSC tissue. While bulk RNA sequencing has limitations with respect to representing the spatial diversity of tissue in cancer samples [15], spatial transcriptomics analysis of the tumor and TME provides more precise information that may give clues to the mechanisms underlying the recurrence of HGSC.

We identified several biomarkers that are associated with recurrence status. We found that overexpression of *TFPI2* and *PIGR* was associated with non-recurrence and a better prognosis. *TFPI2* (Tissue Factor Pathway inhibitor 2) encodes a member of the serine proteinase inhibitor family, which plays a multifaceted role in regulating protease activity, tissue remodeling, and cancer progression. *TFPI2* has been identified as a tumor suppressor gene in several types of cancer [16], and is a diagnostic marker for clear cell type OC; however, our results suggest that it could be a prognostic marker for HGSC. PIGR (Polymeric Immunoglobulin Receptor) is a transmembrane protein that transports polymeric IgA across the intestinal epithelium. A recent report shows that IgA-PIGR colonization is associated with improved survival of patients with HGSC [17], and IgA transcytosis through malignant epithelial cells via the IgA-PIGR interaction elicits transcriptional changes that sensitize tumor cells to cytolytic killing by T cells; therefore, PIGR might antagonize the growth of HGSC cells by regulating coordinated tumor cell, T cell, and B cell responses. *EYA2* (Eyes Absent Homolog 2) has been shown to be overexpressed in ovarian cancer cases and is associated with poorer survival outcomes, particularly in advanced-stage disease. Functionally, EYA2 acts as a transcriptional coactivator and has been shown to promote tumor growth in xenograft models [18].

In addition, among TME-related genes, we found that overexpression of *NKG7* in the immune cell area was associated with a favorable prognosis. NKG7 (natural killer cell granule protein-7) is associated with the quantity and trafficking (as well as calcium release) of cytolytic granules within cytotoxic CD8+ T cells, which are essential for effective anti-tumor cytotoxicity. NKG7 is regulated by the PD-1-NFAT signaling pathway; indeed, a previous report shows that patients with increased expression of NKG7 exhibit a complete response to anti-PD-1 therapy [19]. According to our results, increased expression of *NKG7* might also be associated with responses to conventional chemotherapy, as well as to PD1/PD-L1 immune checkpoint inhibitor therapy. In addition, we found that high expression of *NKG7* by immune cells correlated significantly with PFS of patients with HGSC, which, to the best of our knowledge, is the first report of a major role for NKG7 in HGSC.

We also performed multispectral quantitative IF assessments to examine the proximity between immune cells and cancer cells, as well as the density of each type of immune cell within the tumor and TME; this is because cells exhibiting closer proximity might be engaging in more active interactions. Our quantitative assessment revealed that the Treg/total CD8+ T cell ratio was significantly higher in the recur group than in the non-recur group. Treg cells suppress immune responses [20], and their presence often correlates with a poor prognosis because Tregs can promote tumor growth by suppressing the anti-tumor activity of CD8+ T cells [21]. Therefore, the ratio of Tregs to CD8+ T cells, rather than the density of Treg cells alone, could be a predictive marker for treatment responses in HGSC, a finding that agrees with previous reports [22–24]. Regarding the proximity of cells, we observed that the Treg/PD1-CD8+ T cell distance ratio from cancer cells, and that of Treg/total CD8+ T cells from cancer cells or macrophages, was significantly lower in the recur group than in the non-recur group (0.51-fold vs. 0.53-fold, respectively), which means that Treg cells are closer to cancer cells or macrophages than CD8+ T cells, and suggests that Treg cells in the recur group are more effective at suppressing the immune response than PD1-CD8+ or total CD8+ cytotoxic T cells are at killing cancer cells.

Next, we integrated the spatial transcriptomics data with the multispectral IF assessment data. Because multispectral immune cell densities were assessed in whole tissues, while spatial transcriptomics was conducted for only 69 regions within the same TMA, it was necessary to determine whether the assessment of these two different experimental datasets would introduce errors into the results of the integral analysis. To address this, we evaluated the correlation between the multispectral immune cell density and expression of the corresponding immune cell marker genes (e.g., CD68 for macrophages and CD8A for total CD8+ T cells) across the whole tissue area and in the overlapping regions. The results for the whole tissue area were highly consistent with those for the overlapping regions (Supplementary Figures S3 and S4), thereby confirming that our 69 selected regions were representative of the whole tumor tissue.

According to our integrated analysis of these spatial assessments, the immune cell density and immune pathways within the tumor nests of the recur group correlated positively with cancer pathways (with the exception of the NF-κB pathway). This finding implies that tumors with a TME in which the JAK/STAT or TGF-β signaling pathways in immune cells engage in crosstalk and positive interactions with other cancer signaling pathways, such as the PI3K/mTOR or Wnt signaling pathways, are prone to recurrence. By contrast, tumors exhibiting an inverse correlation between immune pathways and cancer pathways are less likely to recur. In addition, the NF-κB pathway correlated negatively with the total CD8+ T cell count in the recur group, but positively in the non-recur group. In general, NF-κB signaling plays a role in activating CD8+ T cells [25]; however, in some specific contexts and cellular environments, it may negatively impact CD8+ T cells. Possible explanations are that in some contexts, including cancer, activation of NF-κB is compromised by failure of the Rela (also known as p65, encoded by *Rela*) subunit in CD8+ T cells [26], or that CD8+ T cell responses can be modulated indirectly by Treg cells, which are also activated by NF-κB signaling [27]. However, we found that the TGF-β signaling pathway correlated positively with macrophages in the recur group, but negatively in the non-recur group, which might be related to the function of the TGF-β signaling pathway in promoting polarization of macrophages to the M2 phenotype, thereby creating an immunosuppressive environment and enhancing tumor progression.

Although this study is limited by the small number of cases, our novel integrated spatial analyses have profound implications not only for our understanding of the molecular tumor immune microenvironment in HGSC but also for the development of novel therapeutic targets for precision medicine approaches. In future studies, we will first investigate the potential of *NKG7* expression as a biomarker for immunotherapy responsiveness and examine the existence of tumor subclones associated with immune evasion. Finally, validation of these findings in a larger cohort is planned to strengthen the robustness and broader applicability of the results.

5. Conclusions

We identified the expression of *NKG7* by immune cells, as well as *TFPI2* and *PIGR* in the tumor area, as potential biomarkers that predict recurrence and survival of patients with HGSC. The Treg/total CD8+ T cell ratio and the distance ratio of Treg/total CD8+ T cells from cancer cells or macrophages are the other markers that predict recurrence and poorer survival. Interestingly, the correlation between immune cells and cancer pathways suggests that an immunosuppressive TME may be responsible for recurrence.

Supplementary Materials: The following supporting information can be downloaded at: https://www.mdpi.com/article/10.3390/cells14100681/s1, Figure S1: Progression-free survival based on expression of genes in the tumor microenvironment, Figure S2: Correlation between multispectral immune cell density and recurrence of HGSCs, Figure S3: Expression of immune cell marker genes in GeoMx spatial transcriptome data is similar to the corresponding mIF immune cell densities, Figure S4: Comparison of immune cell densities between overlapping ROIs and the whole tissue in the tumor nest with ex7pression of immune cell marker genes in GeoMx spatial transcriptome data; Table S1: The proportion (%) of cells in the immune (CD45), tumor (PanCK), and stromal (SMA) areas, Table S2: Differentially expressed genes in each compartment, Table S3: Enriched Gene Ontology terms for up-regulated genes in the tumor (PanCK) area, Table S4: Enriched Gene Ontology terms for up-regulated genes in the stroma (SMA) area, Table S5: Mean immune cell density according to recurrence status (/mm2), Table S6: Mean distance of cancer cells to the nearest immune cells (μm), Table S7: Mean distance of macrophages to the nearest immune cell (μm).

Author Contributions: Conceptualization, H.K., S.H. and H.J.A.; formal analysis, H.K., J.-G.J., D.K. and S.H.; investigation, H.P., M.C.C., J.-Y.J., J.L. and S.-Y.K.; resources, H.P. and M.C.C.; data curation, H.K., J.-G.J., D.J. and S.H.; writing—original draft preparation, H.K., J.-G.J., J.-Y.J., J.L., S.-Y.K., D.J., S.H. and H.J.A.; writing—review and editing, D.J., S.H. and H.J.A.; supervision, H.J.A.; validation, S.H.; project administration, S.H.; funding acquisition, S.H. and H.J.A. All authors have read and agreed to the published version of the manuscript.

Funding: This research was supported by grants (HI16C1559 and HI21C0977) from the Korea Health Technology R&D Project through The Korea Health Industry Development Institute (KHIDI), funded by the Ministry of Health and Welfare; and by grants (NRF-2019R1A6A1A03032888, NRF-2022R1F1A1071285 and NRF-2022R1A2C109179712) funded by the Ministry of Education and the Ministry of Science and ICT.

Institutional Review Board Statement: This study was approved by the institutional review board of CHA Bundang Medical Center (IRB #CHAMC 2020-03-004).

Informed Consent Statement: Informed consent was obtained from all subjects involved in the study.

Data Availability Statement: All expression data are available at the GEO Database (https://www.ncbi.nlm.nih.gov/geo/) (accessed on 3 May 2025) with accession number GSE279969.

Acknowledgments: We thank Jae-Hwan Kim of the DAAN Cancer Laboratory, Yonsei University College of Medicine, for technical support with multispectral staining analysis. We also thank Stefan Phelan, Liuliu Pan, and Yan Liang for technical support with NanoString GeoMx® Digital Spatial Profiler. The biospecimen and data used in this study were provided by Bundang CHA Biobank, Korea Biobank Network (KBN4-A05).

Conflicts of Interest: The authors declare no conflicts of interest.

Abbreviations

The following abbreviations are used in this manuscript.

adj.	adjusted
AOI	areas of interest
CDH5	cadherin 5
CYBB	cytochrome b-245 beta chain
DEGs	differentially expressed genes
DSP	Digital Spatial Profiling
Dx	diagnosis
FC	fold change
FFPE	formalin-fixed paraffin-embedded
HGSC	high-grade serous carcinomas
IF	immunofluorescence
LOQ	limit of quantitation
NKG7	natural killer cell granule protein 7
OC	ovarian cancer
PCR	Polymerase Chain Reaction
PFS	progression-free survival
PIGR	polymeric immunoglobulin receptor
ROI	regions of interest
TFPI2	tissue factor pathway inhibitor 2
TMA	tissue microarray
TME	tumor microenvironment
TN	tumor nest
TRAC	T cell receptor alpha constant
Tregs	regulatory T cells

References

1. Bray, F.; Laversanne, M.; Sung, H.; Ferlay, J.; Siegel, R.L.; Soerjomataram, I.; Jemal, A. Global cancer statistics 2022: GLOBOCAN estimates of incidence and mortality worldwide for 36 cancers in 185 countries. *CA Cancer J. Clin.* **2024**, *74*, 229–263. [CrossRef] [PubMed]
2. Pogge von Strandmann, E.; Reinartz, S.; Wager, U.; Muller, R. Tumor-Host Cell Interactions in Ovarian Cancer: Pathways to Therapy Failure. *Trends Cancer* **2017**, *3*, 137–148. [CrossRef] [PubMed]
3. Zollinger, D.R.; Lingle, S.E.; Sorg, K.; Beechem, J.M.; Merritt, C.R. GeoMx RNA Assay: High Multiplex, Digital, Spatial Analysis of RNA in FFPE Tissue. *Methods Mol. Biol.* **2020**, *2148*, 331–345. [CrossRef] [PubMed]
4. Hohn, A.K.; Brambs, C.E.; Hiller, G.G.R.; May, D.; Schmoeckel, E.; Horn, L.C. 2020 WHO Classification of Female Genital Tumors. *Geburtshilfe Frauenheilkd* **2021**, *81*, 1145–1153. [CrossRef]
5. Danaher, P.; Kim, Y.; Nelson, B.; Griswold, M.; Yang, Z.; Piazza, E.; Beechem, J.M. Advances in mixed cell deconvolution enable quantification of cell types in spatial transcriptomic data. *Nat. Commun.* **2022**, *13*, 385. [CrossRef]
6. Robinson, M.D.; McCarthy, D.J.; Smyth, G.K. edgeR: A Bioconductor package for differential expression analysis of digital gene expression data. *Bioinformatics* **2010**, *26*, 139–140. [CrossRef]
7. Herrington, C.S. (Ed.) *WHO Classification of Tumours Female Genital Tumours*, 5th ed.; International Agency for Research on Cancer: Lyon, France, 2020.

8. Griswold, M.; Danaher, P. SpatialDecon: Deconvolution of Mixed Cells from Spatial and/or Bulk Gene Expression Data. *R package*, Version 1.6.0. 2022.
9. Yu, G.; Wang, L.G.; Han, Y.; He, Q.Y. clusterProfiler: An R package for comparing biological themes among gene clusters. *OMICS* **2012**, *16*, 284–287. [CrossRef]
10. Hanzelmann, S.; Castelo, R.; Guinney, J. GSVA: Gene set variation analysis for microarray and RNA-seq data. *BMC Bioinform.* **2013**, *14*, 7. [CrossRef]
11. Liberzon, A.; Birger, C.; Thorvaldsdottir, H.; Ghandi, M.; Mesirov, J.P.; Tamayo, P. The Molecular Signatures Database (MSigDB) hallmark gene set collection. *Cell Syst.* **2015**, *1*, 417–425. [CrossRef]
12. Agrawal, A.; Balci, H.; Hanspers, K.; Coort, S.L.; Martens, M.; Slenter, D.N.; Ehrhart, F.; Digles, D.; Waagmeester, A.; Wassink, I.; et al. WikiPathways 2024: Next generation pathway database. *Nucleic Acids Res.* **2024**, *52*, D679–D689. [CrossRef]
13. Bagaev, A.; Kotlov, N.; Nomie, K.; Svekolkin, V.; Gafurov, A.; Isaeva, O.; Osokin, N.; Kozlov, I.; Frenkel, F.; Gancharova, O.; et al. Conserved pan-cancer microenvironment subtypes predict response to immunotherapy. *Cancer Cell* **2021**, *39*, 845–865.e7. [CrossRef] [PubMed]
14. Yang, Y.; Yang, J.; Zhao, X.; Wei, X. Tumor Microenvironment in Ovarian Cancer: Function and Therapeutic Strategy. *Front. Cell Dev. Biol.* **2020**, *8*, 758. [CrossRef] [PubMed]
15. Stur, E.; Corvigno, S.; Xu, M.; Chen, K.; Tan, Y.; Lee, S.; Liu, J.; Ricco, E.; Kraushaar, D.; Castro, P.; et al. Spatially resolved transcriptomics of high-grade serous ovarian carcinoma. *iScience* **2022**, *25*, 103923. [CrossRef] [PubMed]
16. Kobayashi, H.; Imanaka, S.; Matsubara, S.; Shigetomi, H.; Yoshimoto, C. Dual Role of Tissue Factor Pathway Inhibitor 2—A Novel Serodiagnostic Marker for Ovarian Cancer—In Human Cancers. *Int. J. Transl. Med.* **2024**, *4*, 419–438. [CrossRef]
17. Berntsson, J.; Lundgren, S.; Nodin, B.; Uhlen, M.; Gaber, A.; Jirstrom, K. Expression and prognostic significance of the polymeric immunoglobulin receptor in epithelial ovarian cancer. *J. Ovarian Res.* **2014**, *7*, 26. [CrossRef]
18. Zhang, L.; Yang, N.; Huang, J.; Buckanovich, R.J.; Liang, S.; Barchetti, A.; Vezzani, C.; O'Brien-Jenkins, A.; Wang, J.; Ward, M.R.; et al. Transcriptional coactivator Drosophila eyes absent homologue 2 is up-regulated in epithelial ovarian cancer and promotes tumor growth. *Cancer Res.* **2005**, *65*, 925–932. [CrossRef]
19. Wen, T.; Barham, W.; Li, Y.; Zhang, H.; Gicobi, J.K.; Hirdler, J.B.; Liu, X.; Ham, H.; Peterson Martinez, K.E.; Lucien, F.; et al. NKG7 Is a T-cell-Intrinsic Therapeutic Target for Improving Antitumor Cytotoxicity and Cancer Immunotherapy. *Cancer Immunol. Res.* **2022**, *10*, 162–181. [CrossRef]
20. Noyes, D.; Bag, A.; Oseni, S.; Semidey-Hurtado, J.; Cen, L.; Sarnaik, A.A.; Sondak, V.K.; Adeegbe, D. Tumor-associated Tregs obstruct antitumor immunity by promoting T cell dysfunction and restricting clonal diversity in tumor-infiltrating CD8+ T cells. *J. Immunother. Cancer* **2022**, *10*, e004605. [CrossRef]
21. Li, C.; Jiang, P.; Wei, S.; Xu, X.; Wang, J. Regulatory T cells in tumor microenvironment: New mechanisms, potential therapeutic strategies and future prospects. *Mol. Cancer* **2020**, *19*, 116. [CrossRef]
22. Baras, A.S.; Drake, C.; Liu, J.J.; Gandhi, N.; Kates, M.; Hoque, M.O.; Meeker, A.; Hahn, N.; Taube, J.M.; Schoenberg, M.P.; et al. The ratio of CD8 to Treg tumor-infiltrating lymphocytes is associated with response to cisplatin-based neoadjuvant chemotherapy in patients with muscle invasive urothelial carcinoma of the bladder. *Oncoimmunology* **2016**, *5*, e1134412. [CrossRef]
23. Hayashi, Y.; Ueyama, A.; Funaki, S.; Jinushi, K.; Higuchi, N.; Morihara, H.; Hirata, M.; Nagira, Y.; Saito, T.; Kawashima, A.; et al. In situ analysis of $CCR8^+$ regulatory T cells in lung cancer: Suppression of $GzmB^+$ $CD8^+$ T cells and prognostic marker implications. *BMC Cancer* **2024**, *24*, 627. [CrossRef] [PubMed]
24. Preston, C.C.; Maurer, M.J.; Oberg, A.L.; Visscher, D.W.; Kalli, K.R.; Hartmann, L.C.; Goode, E.L.; Knutson, K.L. The ratios of CD8+ T cells to CD4+CD25+ FOXP3+ and FOXP3- T cells correlate with poor clinical outcome in human serous ovarian cancer. *PLoS ONE* **2013**, *8*, e80063. [CrossRef] [PubMed]
25. Yatim, N.; Jusforgues-Saklani, H.; Orozco, S.; Schulz, O.; Barreira da Silva, R.; Reis e Sousa, C.; Green, D.R.; Oberst, A.; Albert, M.L. RIPK1 and NF-kappaB signaling in dying cells determines cross-priming of $CD8^+$ T cells. *Science* **2015**, *350*, 328–334. [CrossRef] [PubMed]
26. Ling, W.; Rayman, P.; Uzzo, R.; Clark, P.; Kim, H.J.; Tubbs, R.; Novick, A.; Bukowski, R.; Hamilton, T.; Finke, J. Impaired activation of NFkappaB in T cells from a subset of renal cell carcinoma patients is mediated by inhibition of phosphorylation and degradation of the inhibitor, IkappaBalpha. *Blood* **1998**, *92*, 1334–1341. [CrossRef]
27. Hovelmeyer, N.; Schmidt-Supprian, M.; Ohnmacht, C. NF-kappaB in control of regulatory T cell development, identity, and function. *J. Mol. Med.* **2022**, *100*, 985–995. [CrossRef]

cells

MDPI

Review

Biological Aging and Uterine Fibrosis in Cattle: Reproductive Trade-Offs from Enhanced Productivity

Yuta Matsuno *,† and Kazuhiko Imakawa

Research Institute of Agriculture, Tokai University, Kumamoto 862-8652, Japan; imakawa.kazuhiko.s@tokai.ac.jp
* Correspondence: yuta.matsuno@tokai.ac.jp
† Current address: Department of Animal Science, School of Agriculture, Tokai University, Kumamoto 861-2205, Japan.

Abstract: Reproductive efficiency in cattle remains sub-optimal, with pregnancy rates often below 50%, despite fertilization rates approaching 100%, indicating that implantation failure and/or early embryonic loss are major limiting factors. This disparity highlights the need to understand the biological and physiological mechanisms underlying implantation failure. This review elucidates the cellular and molecular mechanisms underlying reduced pregnancy rates, with a particular focus on biological aging and fibrosis in the reproductive organs as emerging contributors to uterine dysfunction. Accumulated evidence suggests that metabolic demands associated with intensive breeding strategies aimed at maximizing meat and milk productivity may induce multiple forms of stress, including oxidative stress, metabolic stress, and inflammation, which accelerate biological aging and fibrosis in the female reproductive tract. However, the direct molecular mechanisms remain poorly characterized. We hypothesize that biological aging and fibrosis are interconnected mechanisms contributing to impaired uterine function, resulting in reduced implantation rates. By summarizing recent findings and adopting a comparative perspective, this review explores the extent to which insights from human and mouse models can be applied to cattle, considering species-specific reproductive physiology and metabolic adaptations. It explores their relevance to reproductive inefficiencies and discusses potential strategies to enhance fertility and extend bovine reproductive longevity.

Keywords: reproductive tract; uterus; senescence; senescence-associated secretory phenotype (SASP); biological aging; fibrosis; ruminants; cattle

1. Introduction

Pregnancy represents a profound physiological transformation in eutherian mammals, requiring adaptive processes across the metabolic, endocrine, and immune systems. Key modifications, such as the structural remodeling of reproductive tissues, hormonal changes, and immunological adaptations, are intricately coordinated and essential for successful implantation, placental development, and maintaining maternal health and fetal growth. However, disruptions in uterine remodeling, hormonal balance, or immune regulation can result in implantation failure, fetal growth restriction, gestational complications, or delays to establishing pregnancy.

Despite extensive research in cattle reproduction, pregnancy rates remain below 50%, despite fertilization rates approaching 100%. Research on cattle reproduction consistently reports fertilization rates approaching 90–100% for heifers, beef, and moderate-yielding dairy cows, though rates may be lower in high-producing dairy cows [1–3]. However, pregnancy rates often fall below 50% due to significant embryonic losses. Early embryonic

 https://doi.org/10.3390/cells14130955

mortality, occurring within the first 16 days post breeding, is the primary factor contributing to pregnancy failure [1,4]. A meta-analysis of beef cattle reproduction found that by day 32 of gestation, 47.9% of cows inseminated were not pregnant [5]. This disparity highlights the need to investigate biological events such as uterine receptivity, embryo–maternal interaction, and metabolic regulation, all of which may contribute to early pregnancy failure and sub-optimal reproductive outcomes.

Accumulated evidence suggests that the cumulative effects of these physiological changes contribute to biological aging of female reproductive organs. Biological aging is associated with decreased fertility and increased pregnancy complications, accompanied by cellular senescence [6], chronic inflammation [7], and altered hormonal balance [8], often leading to fibrosis, particularly in the ovaries and the uterus. Fibrosis, characterized by excessive accumulation of extracellular matrix (ECM) proteins, disrupts normal reproductive function and contributes to infertility in humans and mice [9–12]. While studies using mouse and human models have provided valuable insights, it remains to be determined whether these findings can be directly applied to cattle, particularly given their ruminant physiology and production context. Furthermore, the applicability of these insights extends beyond fundamental research; whether current therapeutic strategies developed for humans, primates, and mice can be effectively translated to cattle remains inconclusive.

Excessive energy intake disrupts metabolic homeostasis, leading to oxidative stress and chronic inflammation, both of which are key drivers of biological aging [13,14]. In cattle, metabolic stress induced by high-energy feeds and intensive selective breeding for increased milk or meat production has been implicated in accelerating biological aging [15]. Similarly, high-yield selection and intensive breeding practices prioritize productivity over physiological balance, which can impair ECM remodeling in reproductive organs. Impaired ECM remodeling promotes excessive collagen deposition, reducing tissue elasticity and predisposing the uterus to fibrosis. These disruptions may increase metabolic strain and exacerbate tissue fibrosis. Increased dairy milk yield per cow has been associated with chronic nutrient depletion and hormonal imbalances, which may influence epigenetic modifications and have been implicated in the development of uterine fibrosis, potentially affecting reproductive efficiency.

Despite recent fertility-focused breeding indices and protocols, such as the Double Ovsynch, that have improved estrous synchronization and fertilization rates [16–19], overall reproductive success remains suboptimal, with pregnancy rates in dairy cows still below 50% [20]. This suggests that existing reproductive management protocols may enhance biological aging and compromised uterine function under intensive production practices.

While postpartum uterine infections and hormonal imbalances are well-recognized factors contributing to prolonged calving intervals in dairy and beef cows, an increasing number of cows exhibit extended calving intervals in the absence of apparent infections or endocrine dysfunction. This observation suggests that subclinical processes, such as biological aging and progressive uterine fibrosis, may play a hidden role in compromising uterine recovery and function, particularly under intensive production systems.

In this review, we explore the role of biological aging and uterine fibrosis as underrecognized, yet potentially critical contributors to declining fecundity in cattle. By adopting a comparative perspective from livestock and biomedical research, we aim to provide a comprehensive understanding of the interplay between biological aging and fibrosis in the bovine uterus. Furthermore, we discuss how these biological processes affect the balance between production efficiency and fecundity, highlighting key knowledge gaps and future research directions for improving fertility and sustaining productivity in cattle.

2. Biological Aging

2.1. Biological Aging and Uterine Function and Infertility

Biological aging is characterized by progressive dysfunction in cells, tissues, and organs [21]. In 2023, twelve molecular, cellular, and systemic hallmarks of biological aging were consolidated based on accumulating evidence: genomic instability, telomere attrition, epigenetic alterations, loss of proteostasis, disabled macroautophagy, deregulated nutrient-sensing, mitochondrial dysfunction, cellular senescence, stem cell exhaustion, altered intercellular communication, chronic inflammation, and dysbiosis. These hallmarks fall into three categories: primary, antagonistic, and integrative [22–24] (Figure 1). Primary hallmarks are the initial triggers of cellular damage, including genomic instability, telomere attrition, and epigenetic alterations. Antagonistic hallmarks represent responses to these kinds of damage, such as cellular senescence and mitochondrial dysfunction, that are initially protective but become deleterious if persistent. Integrative hallmarks are the consequences of accumulated damage and compensatory failure, such as stem cell exhaustion and altered intercellular communication, which directly impact tissue and organ function. This hierarchical model provides a conceptual framework to understand how biological aging-related processes, once triggered, can converge to impair tissue homeostasis, including uterine function.

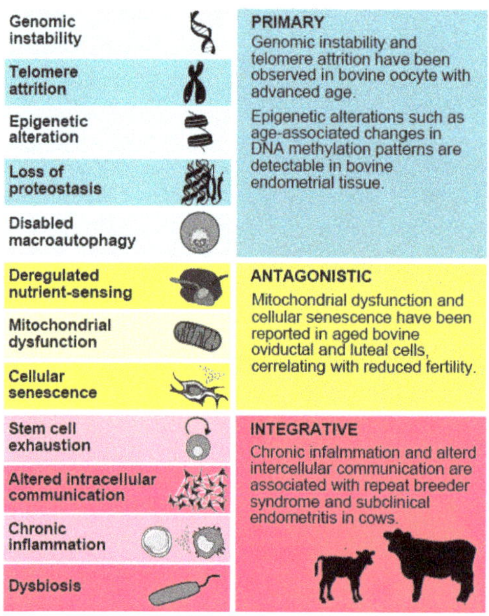

Figure 1. The hallmarks of biological aging categorized into primary, antagonistic, and integrative groups, as proposed by López-Otín et al. (2023) [24]. The 12 key hallmarks of biological aging: genomic instability, telomere shortening, epigenetic modifications, impaired proteostasis, dysfunctional macroautophagy, disrupted nutrient sensing, mitochondrial impairment, cellular senescence, stem cell depletion, altered intercellular signaling, chronic inflammation, and microbiome imbalance. Examples of cattle-specific relevance are noted where available: Genomic instability and telomere attrition have been observed in bovine oocytes with advancing age. Epigenetic alterations such as age-associated changes in DNA methylation patterns are detectable in bovine endometrial tissue. Mitochondrial dysfunction and cellular senescence have been reported in aged bovine oviductal and luteal cells, correlating with reduced fertility. Chronic inflammation and altered intercellular communication are associated with repeat-breeder syndrome and subclinical endometritis in cows. Source: adapted from [24].

These hallmarks play critical roles in reproductive organs, where age-related changes can directly influence fertility. Researchers have increasingly used mouse models to elucidate their roles [25]. Uterine biological aging affects various components, including the myometrium, neurofibers, and endometrium, leading to poorer reproductive performance in mice, rats, and women [23,26]. Recent findings from mouse models suggest that these changes are mediated by altered genetic and epigenetic mechanisms, metabolic conditions, immune system dynamics, and other components of the female reproductive tract [23].

In cattle, several studies have shown that reproductive efficiency decreases with age. The productive lifespan of dairy cattle typically peaks between 3.5 and 5 years, after which fertility begins to decline. Cows aged 7 years or older tend to exhibit significantly reduced reproductive efficiency [27]. A study on Friesian Holstein dairy cattle in Indonesia (2017–2018) reported that, as cows aged from 4 to 5 years, the number of services per conception increased from 3.2 ± 1.8 to 4.3 ± 1.9 ($p < 0.05$), indicating reduced reproductive efficiency [27]. Additionally, Osoro and Wright (1992) found that pregnancy rates declined from 90% in younger Blue-Gray cows to 83% in Hereford × Friesian cows older than 7 years ($p < 0.05$) [28]. This decline is compounded by parity effects, as fertility tends to decrease with each successive calving until dairy cattle reach physiological adaptation to repeated calving rather than chronological age [29]. Biological aging is also associated with the reduced developmental competence of oocyte in crossbred Hereford cows [30]. Older crossbred Hereford cows exhibit elevated circulating follicle-stimulating hormone (FSH) levels, reduced follicular recruitment, smaller ovulatory follicles, increased plasma estradiol levels, and a tendency toward lower luteal-phase progesterone levels [31].

2.2. *Cellular Senescence in Uterine Biological Aging*

Recent studies highlight the role of cellular senescence in uterine biological aging and related infertility in human and mouse models. Cellular senescence is a hallmark of biological aging and is characterized by a state of stable and generally irreversible growth arrest, in which cells permanently cease to divide but remain metabolically active [32,33]. This arrest is triggered by various stresses, such as DNA damage, oxidative stress, and telomere damage [32,33] (Figure 2). Senescent cells exhibit distinct morphological, biochemical, and functional changes, including the senescence-associated secretory phenotype (SASP), which disrupt normal tissue homeostasis and contribute to age-related pathologies [34].

The role of senescent cells in reproductive organs is complex; while senescent cells contribute to ovarian follicle decline and pregnancy complications in human and mouse models [6], they also support embryo/fetal and placental development and parturition through the SASP under normal physiological conditions [35,36]. However, the excessive accumulation of senescent cells is linked to age-related pathologies [37]. For instance, increased senescence and decreased stemness in human endometrial stromal cells during the proliferative phase are associated with implantation failure [38]. Senescent cells in murine ovaries may accelerate reproductive biological aging via SASP [39]. While these findings highlight that cellular senescence has both positive and negative effects on female reproductive biological aging and fertility, the applicability of these observations to cattle remains uncertain and requires careful validation. Cattle exhibit distinct reproductive physiology compared to rodents and primates, such as a 21-day estrous cycle, non-invasive epitheliochorial placentation, and multiple follicular waves per cycle. Therefore, further studies are needed to determine whether senescent cell functions observed in human and mouse models apply similarly to bovine reproductive aging.

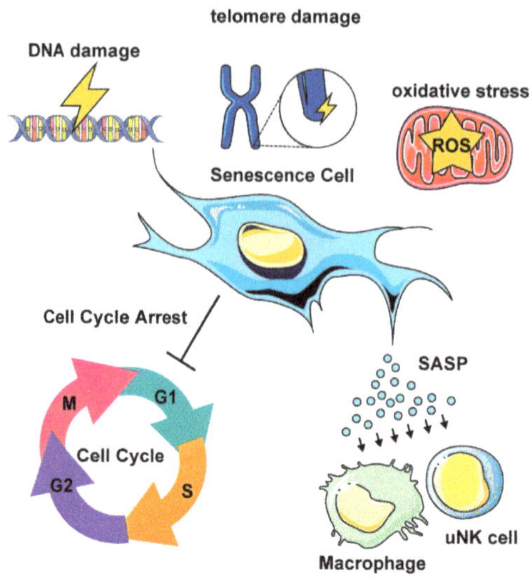

Figure 2. Cellular senescence in the uterus. DNA damage, telomere damage, and oxidative stress drive cellular senescence. Senescent cells show cell cycle arrest and senescent-associated secretory phenotype (SASP). The SASP recruits immune cells, such as macrophages and uterine natural killer (uNK) cells.

Human and mouse models have also revealed that cellular senescence is involved in endometriosis and related infertility. Human endometriotic lesions (consisting of ectopic endometrium), often associated with infertility, exhibit increased expression of senescence markers like $p16^{Ink4a}$ and decreased lamin B1 (*LMNB1*) [40,41]. Oxidative stress in human endometriotic lesions upregulates senescence-associated proteins, contributing to inflammation and infertility [42,43]. The upregulation of Sirtuin 1 (*SIRT1*), a nicotinamide adenine dinucleotide (NAD)-dependent protein deacetylase, promotes epithelial–mesenchymal transition by inducing senescence escape in murine endometriosis [44]. Premature immunosenescence may impair immune surveillance, allowing human endometriotic stem cell migration [45]. During the peri-implantation period, pro-senescent decidual responses have been implicated in recurrent pregnancy loss, further highlighting the interplay between senescence and immune regulation in the human endometrium [46]. Endometrial stromal cellular senescence may contribute to human endometrium dysfunction and implantation failure [47]. Thus, cellular senescence has emerged as a key factor in uterine function and pathology, requiring careful balance to maintain reproductive health.

In cattle, tissue-specific senescence effects have been identified in reproductive organs. Aged oviductal epithelial cells show the upregulation of proinflammatory genes and downregulation of ECM-related genes, accompanied by decreased proliferation rates and lower ciliary activity [48–50]. Aged endometrial cells show the activation of inflammation-related and interferon-signaling pathways, accompanied by lower cell viability and proliferation rates [51]. Aged corpus luteum shows decreased progesterone production and secretion, particularly during the early luteal phase [52]. These alterations in bovine reproductive tissues contribute to the overall decline in fertility observed with advanced maternal age in cows.

2.3. Molecular Drivers of Biological Aging in the Uterus

Recent studies have demonstrated that several molecular mechanisms are related to biological aging [24]. DNA damage, particularly the accumulation of G-quadruplexes, plays a central role in biological aging by altering gene expression and chromatin structure [53,54]. While reactive oxygen species (ROS) have long been implicated in biological aging, the mammalian target of rapamycin (mTOR) signaling pathway has also emerged as a key regulator of biological aging [55,56]. Non-enzymatic molecular damage, such as advanced glycation end products (AGEs), contributes to biological aging through various spontaneous chemical reactions [57,58]. Telomere shortening and the overall loss of molecular fidelity are also key factors for biological aging and genomic stability [59,60].

In humans, weight gain, obesity, and metabolic stress are associated with accelerated biological aging [61]. Studies have shown that weight gain from early to middle adulthood is linked to higher risks of chronic diseases and decreased odds of healthy biological aging [62]. Obesity and weight gain are inversely associated with telomere length [63]. Metabolic syndrome and associated oxidative stress can contribute to accelerated biological aging [64].

In cattle, inflammaging, which refers to chronic inflammation associated with biological aging, has been implicated in the deterioration of bovine uterine function and age-related reproductive inefficiency [65]. Additionally, while genetic selection has significantly improved both meat and milk yield [66], these achievements have come with unintended consequences, including increased metabolic demands, associated health issues and reduced welfare [67]. This stress includes: altered nutrient metabolism, dysfunctional inflammatory responses, and oxidative stress [68]. Due to the immense metabolic demands of high milk production, high-yielding dairy cows experience a negative energy balance characterized by elevated blood beta-hydroxybutyrate and urea levels [69,70].

How genetic selection interacts with intensive milk production reveals broader biological trade-offs. Intensive breeding practices not only amplify metabolic stress but also accelerate epigenetic aging in dairy cows [15]. This epigenetic aging process impairs bovine immunity, increasing susceptibility to infections such as metritis and endometritis in dairy cows [71]. The demands of milk production also induce endoplasmic reticulum stress in the liver and mammary epithelial cells, compounding health issues and affecting milk yield [72]. Breeding programs focused on milk production have also led to reproduction insufficiencies, such as lower fertility rates and increased calving intervals [73]. The combined effects of genetic selection and intensive milk production highlight the need for sustainable breeding practices.

2.4. Epigenetic Aging

Epigenetic aging is defined as age-related changes in DNA methylation, histone modifications, and non-coding RNA (ncRNA) expression, affecting gene regulation and genomic stability, referring to the integrity and maintenance of the DNA structure, without altering the DNA sequence [74,75]. In reproductive organs and tissues, epigenetic aging has profound implications, as it is closely linked to reduced fertility and an increased risk of reproductive cancers in both males and females in human studies [76,77].

In females, oocyte quality declines with age, showing reduced DNA methylation, decreased DNA methyltransferases and histone deacetylase expression, and the loss of heterochromatin marks in human and mouse models [78,79]. These changes contribute to genomic instability, increased DNA damage, and elevated retrotransposon activity in aged oocytes [79].

Epigenetic clocks, which assess biological age based on DNA methylation patterns, provide a robust tool for evaluating reproductive biological aging [80–83]. Several models, including Horvath's clock, Hannum's clock, PhenoAge, and GrimAge [84], have been adapted for ruminants, such as cattle, sheep, and goats to estimate the biological age and assess their reproductive potential [85].

These clocks are valuable for identifying accelerated biological aging linked to health outcomes. Additionally, they can be used to evaluate anti-biological aging interventions and mechanisms [84], although computational refinements are required to enhance their accuracy and utility [86]. Epigenetic aging also relates to nutrient sensing, mitochondrial activity, and stem cell composition, highlighting its complex interactions with metabolism and cellular function [87]. Epigenetic changes related to metabolism, immunity, and autophagy have significant implications for fertility, particularly in bovine reproductive management [88,89].

Emerging tools such as cell-free DNA (cfDNA) in the blood can be used as a minimally invasive biomarker to detect age-associated epigenomic changes [90]. The use of cfDNA could provide insights into inferring age-related epigenetic alterations in the female reproductive tract of cows without the need for direct tissue sampling, offering practical and early interventions, improved breeding strategies, and optimized reproductive management, emphasizing the importance of integrating epigenetic insights into bovine production systems.

2.5. Epigenetic Aging in Pregnancy

Recent studies suggest an association between pregnancy and accelerated epigenetic aging in women and mice. For instance, multiple investigations report faster epigenetic aging with an increasing number of giving birth, as measured by epigenetic clocks [91–93]. Biomarkers such as telomere shortening and alteration in DNA methylation patterns further support this association [91,93]. Women with preeclampsia exhibit more pronounced epigenetic aging and increased cellular senescence markers compared to normotensive pregnancies [94]. Maternal epigenetic aging during pregnancy is associated with shorter gestation and lower birthweight [95].

In cattle, DNA methylation, among other epigenetic mechanisms, plays a significant role in regulating gene expression in uterine tissues. Analysis of endometrial DNA methylation patterns has identified 1958 differentially methylated CpG sites between pregnant and nonpregnant cows [96]. Additionally, over 1000 significant correlations between endometrial DNA methylation and gene expression have been identified, with 52% showing a negative relationship [97]. Temporal regulation is also observed; for example, 42% of genes exhibiting significant methylation-expression correlations are differentially expressed between pregnant and cycling cows [97]. Furthermore, methionine supplementation alters 3430 CpG sites, potentially influencing metabolic and reproductive pathways via epigenetic regulation [96]. These results suggest that nutritional interventions may modulate epigenetic regulation of endometrial gene networks, potentially influencing reproductive success. These findings have significant implications for bovine breeding, as epigenetic modifications may contribute to biological aging, reduced fertility, and fecundity.

3. Fibrosis

3.1. Fibrosis in Female Reproductive Organs

Fibrosis, defined as excessive ECM accumulation, leads to tissue scarring and organ dysfunction, significantly affecting reproductive health (Figure 3A). It commonly occurs in chronic inflammatory diseases and has profound implications for ovarian and uterine physiology. In ovaries, biological aging is characterized by increased fibrosis, which disrupts ovulation, reduces fertility, and increases the risk of ovarian cancer [98–100]. This progression involves a combination of chronic inflammation, mitochondrial dysfunction, and altered immune cell populations, contributing to a proinflammatory cellular microenvironment [100,101]. Ovarian fibrosis impairs ovulation events including oocyte release from follicles, follicle rupturing, corpus luteum formation, and post-ovulatory tissue remodeling, critical for maintaining ovarian function in aged mice [102]. Notably, the cytokine signature in human follicular fluid from aged ovaries, characterized by elevated fibroinflammatory markers, correlates inversely with anti-mullerian hormone (AMH) levels, a biomarker of ovarian reserve [103].

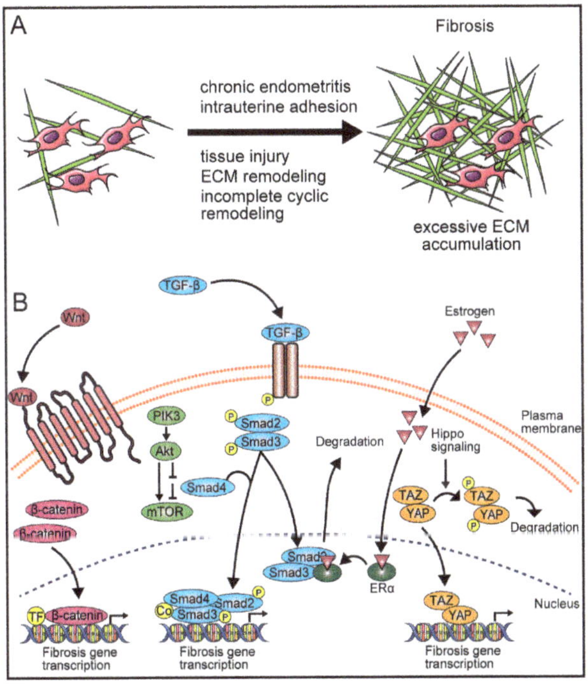

Figure 3. Molecular mechanisms lead to uterine fibrosis. (**A**) Schematic representation of fibrosis progression. This illustration outlines the key cellular and molecular events driving fibrosis development, including excessive ECM accumulation, fibroblast activation, and inflammatory signaling. (**B**) Integrated signaling pathways involved in fibrosis modulation. These include: the wingless-type MMTV integration site family (Wnt)/β-catenin pathway which promotes fibroblast activation and ECM production. Evidence is currently limited to human and rodent models; bovine-specific data are lacking. The transforming growth factor beta (TGF-β) signaling pathway, primarily via the suppressor of mother against decapentaplegic (Smad) cascade, plays a central role in fibrosis by inducing myofibroblast differentiation and collagen synthesis. Validated in bovine and equine endometrium. The phosphatidylinositol-3 kinase (PI3K)-Akt-mammalian target of rapamycin (mTOR) pathway contributes to cell survival, proliferation, and metabolic adaptations that exacerbate fibrosis. There has been no direct validation in cattle uterine tissue, though pathway components are conserved. Estrogen modulates fibrosis by inhibiting the TGF-β/Smad signaling pathway, leading to Smad complex degradation and reduced fibrotic response. This anti-fibrotic role is supported by in vitro studies using bovine endometrial epithelial cells. The Hippo-Yes-associated protein (YAP)/transcriptional coactivator with PDZ-binding motif (TAZ) pathway regulates cellular mechanotransduction and fibrosis-related gene expression, influencing fibroblast activation and tissue remodeling. Established in human uterine fibroid models; not yet validated in bovine uterus. These pathways collectively determine the balance between tissue repair and fibrotic progression. Species-specific differences should be considered when interpreting these molecular interactions in the context of cattle reproduction.

Uterine fibrosis develops through multiple interconnected mechanisms. Chronic endometritis has been identified as a key driver, inducing persistent inflammatory responses that promote fibrotic tissue deposition [10]. In humans, ECM remodeling plays a critical role, influencing mechanotransduction pathways that facilitate fibroid growth and uterine structural changes [104,105]. For example, studies of intrauterine adhesions in clinical cases have revealed heightened levels of fibrotic markers and the aberrant activation of stem cell pathways, which lead to infertility and menstrual dysfunction [9–11]. In murine models, the repeated ECM remodeling of the reproductive tract, particularly the uterus, oviduct, and vagina by estrous cycles promotes fibrosis accumulation and inflammatory changes due to incomplete resolution of collagen deposition [12]. While uterine fibrosis is linked to surgical interventions in both humans and mice, its etiology also encompasses chronic inflammation, incomplete cyclic remodeling, and tissue injuries [10,12].

In cattle, endometritis can result in uterine fibrosis [106–108]. A recent study on first-parity Holstein cows reports that those with early postpartum metritis show increased adenomyosis and uterine fibrosis later in life, suggesting that acute inflammatory events may contribute to persistent pro-fibrosis [109]. Additionally, the lower expression of colony-stimulating factor 2, an embryokine with roles in early pregnancy, has been observed in the endometrial epithelium of older cows, implying the impaired regenerative capacity of the biological aging uterus [110]. Molecular pathways involving transforming growth factor beta (TGF-β)1 signaling via transforming growth factor beta receptor 3 (TGFBR3)/Suppressor of mother against decapentaplegic (Smad) 2/3 have been shown to promote fibrosis in bovine endometrial epithelial cells, although 17β-estradiol may mitigate these effects through G protein-coupled estrogen receptor 1 (GPER1)-mediated mechanisms [108]. Furthermore, collagen-specific chaperone heat shock protein 47 (HSP47), a marker of active collagen synthesis, has been found in higher abundance in the cervix, uterus, and oviducts of aged cows, coinciding with the accumulation of denatured collagen and fibrotic matrix [111,112]. AMH and its receptor (AMHR2) are differentially expressed in reproductive tissues of older cows compared to younger or different breeds, and AMH has been shown to upregulate HSP47 expression in uterine epithelial cells [113,114]. These findings suggest that hormonal signaling pathways may influence the onset or progression of uterine fibrosis. Finally, histopathological studies have highlighted the co-existence of chronic inflammation and fibrosis within the endometrial tissues of repeat breeder cows, further supporting the inflammatory–fibrotic axis in bovine reproductive biological aging [115].

Endometrial biopsies in cattle with reproductive disorders commonly reveal chronic endometritis as the predominant histopathological finding [116–118]. Key features include mononuclear leukocyte infiltration, dilated uterine glands, and periglandular fibrosis [117,118]. Subclinical endometritis is frequently observed in repeat breeder cows, with inflammatory cell infiltration, epithelial degeneration, and desquamation [118]. Histopathological lesions can be present in cows with or without clinical signs of reproductive disorders [116]. Epithelial height, segmented cell counts, and inflammatory changes show correlations between uterine horns and across time points [119]. Gland number, dilation, and fibrosis are interrelated, with a negative correlation between glandular characteristics and inflammatory changes [119]. Endometrial biopsy is considered a valuable diagnostic tool for assessing fertility potential in dairy cows [116,118].

Equine endometrial fibrosis is a well-characterized condition that serves as a valuable comparative model for studying uterine fibrosis in cattle. It is defined by progressive endometrial and periglandular fibrosis, a reduction in the uterine glands, and is strongly associated with biological aging [120]. Histologically, it features myofibroblast differentiation and aberrant deposition of ECM proteins, including collagen accumulation outside the basement membrane [121]. Additionally, fibrotic remodeling is not limited to the endometrium, studies have also identified fibrosis and myometrial atrophy in the uterine smooth muscle layer of mares [120], suggesting broader uterine involvement. Proposed mechanisms include vascular compromise, lymphatic edema, and chronic inflammation.

Neutrophil extracellular traps, while primarily involved in pathogen defense, have been implicated in the upregulation of fibrosis-associated genes [122]. Pro-fibrotic cytokines, interleukins, and prostaglandins also contribute to fibrogenesis [122]. Understanding the signaling pathways involved in myofibroblast proliferation and ECM remodeling in mares provides mechanistic insights that are translatable to bovine uterine fibrosis.

3.2. Mechanisms of Fibrosis

Key signaling pathways, including TGF-β, wingless-type MMTV integration site family (Wnt)/β-catenin, and Hippo, regulate fibrosis development through ECM deposition and myofibroblast differentiation (Figure 3B) [123–125]. TGF-β signaling is initiated when ligands bind to type II and type I serine/threonine kinase receptors on the cell surface. This activates the phosphorylation of the receptor-regulated Smad proteins, which form complexes with Smad4 and translocate to the nucleus to regulate gene expression [126]. When the Hippo pathway is activated, mammalian Ste20-like kinase 1/2 (MST1/2) kinases phosphorylate and activate large tumor suppressor 1/2 (LATS1/2), which in turn phosphorylate the Yes-associated protein (YAP) and transcriptional coactivator with PDZ-binding motif (TAZ), leading to their cytoplasmic sequestration or degradation, which prevent its transcriptional activation. When the Hippo pathway is inactive, unphosphorylated YAP/TAZ translocate into the nucleus, where they bind to TEA domain family member (TEAD) transcription factors, promoting the expression of downstream genes [127]. The activation of the Wnt/β-catenin pathway leads to the stabilization and nuclear translocation of β-catenin, promoting the transcription of fibrosis-related genes [128].

These pathways interact extensively, forming a cross regulation network that amplifies fibrotic responses. For instance, TGF-β signaling enhances β-catenin stabilization, thereby upregulating the transcriptional regulation of ECM components in human fibroid and myometrial cells from hysterectomy surgeries [129]. The Hippo pathway, through YAP/TAZ, further contributes to human fibrosis development [124]. Such interactions illustrate the intricate and dynamic interplay between these pathways in uterine fibrosis development. Additionally, the dual specificity protein phosphatase 4 (DUSP4)/glycogen synthase kinase-3 beta (GSK3β)/snail family transcriptional repressor 1 (SNAI1) pathway, implicated in the human endometrial fibrosis of patients with intrauterine adhesions [130], appears to function downstream of Wnt/β-catenin signaling, suggesting its role as a secondary amplifier of fibrotic gene expression.

Hormonal factors, particularly estrogens, are involved in this cross-regulation. Estrogens modulate ECM dynamics by promoting hypercoagulability and altering fibrin network formation, particularly during the menstrual cycle in women [131,132]. Estrogen signaling plays a crucial role in human fibroid pathobiology, involving both genomic and non-genomic pathways such as mitogen-activated protein kinase (MAPK) and phosphatidylinositol-3 kinase (PI3K)-Akt-mTOR [133]. Moreover, a human breast cancer cell study shows that estrogens can bind to estrogen receptor alpha (ERα), interacting with the Smad complex to facilitate its degradation resulting in the inhibition of TGF-β signaling and thereby exerting anti-fibrotic effects [134]. While most mechanistic insights into estrogen's anti-fibrotic roles derive from human studies, evidence in cattle has also emerged. Specifically, 17β-Estradiol attenuates TGF-β1-induced fibrosis in bovine endometrial epithelial cells via the GPER1-mediated inhibition of TGFBR3/Smad2/3 signaling [108]. These findings indicate that estrogens may play conserved but potentially species-specific roles in modulating uterine fibrosis. These multifaced actions of estradiol highlight its complex contribution to uterine homeostasis. However, the extent to which this inhibition counterbalances other pro-fibrotic signals, such as those mediated by YAP/TAZ or neurogenic locus notch homolog protein 1 (Notch1), remains unclear.

Although Notch1 signaling upregulates collagen expression, promoting fibrosis in mouse uterus [135], the applicability of this finding to cattle remains uncertain. Cross-species differences in reproductive physiology and gene regulation necessitate further studies to validate the role of Notch1 signaling in bovine uterine fibrosis. Activin A, a cytokine belonging to the TGF-β superfamily, plays a role in inflammation, tissue repair, and fibrosis, suggesting its involvement in the complexity of signaling crosstalk associated with uterine fibroid development in reproductive-age women [136]. Understanding the molecular mechanisms underlying uterine fibrosis across species will help researchers develop targeted therapeutic interventions to restore uterine function and enhance fertility.

Recent research highlights fibroblasts as active regulators of inflammation and fibrosis in various organs. In the context of endometrial fibrosis, fibroblasts contribute to pathological tissue remodeling by interacting with immune cells, such as macrophages and T cells, and releasing pro-fibrotic mediators, including TGF-β, interleukin 6 (IL6), and vascular endothelial growth factor (VEGF). This role is particularly pronounced in conditions like endometriosis and intrauterine adhesions [137–139]. Myofibroblasts, derived from precursors like endothelial cells via endothelial-to-mesenchymal transition, are key drivers of fibrosis [137]. Single-cell RNA sequencing reveals increased proportions of myofibroblasts, pericytes, and macrophages in human ectopic endometrium, suggesting a coordinated response that contributes to tissue remodeling and fibrosis [140]. Myometrial cells can transition to myofibroblast-like phenotypes, forming collagen- and fibronectin-rich nodules, which are hallmarks of fibrotic progression in uterine tissues [141]. Human endometrial mesenchymal stem cells can differentiate into stromal fibroblasts, with those from endometriosis patients showing progesterone resistance and an inflammatory phenotype [142]. These findings highlight the interplay between immune dysregulation and stromal cell dysfunction in fibrosis progression. Recent studies indicate that immunoglobulin G (IgG) accumulates across aged tissues, such as in the brain, liver, and spleen, in both humans and mice [143], while IgG accumulation has been shown to activate macrophages and induce fibrosis and inflammation in adipose tissue [144]. Understanding these mechanisms is essential for developing targeted therapies for endometrial fibrosis.

The coagulation system, particularly fibrin formation, plays an important role in fibrosis development [145]. However, the relationship between fibrin and fibrosis is complex, involving both protective and pathogenic roles. For instance, fibrin accumulation serves as a scaffold for fibroblast recruitment and ECM deposition, while excessive fibrin degradation can exacerbate inflammation, indirectly promoting fibrosis. While some studies using murine models suggest that fibrin removal is not sufficient to prevent fibrosis [146,147], others indicate that thrombin inhibition can reduce fibrosis in pulmonary and hepatic tissues [148]. The plasminogen activation system also contributes to fibrosis, with plasminogen activator inhibitor-1 (PAI-1) promoting fibrosis by inhibiting fibrinolysis and enhancing ECM stability [149]. Notably, coagulation and fibrinolysis factors are dysregulated in fibrotic diseases such as lung and renal fibrosis in humans, where protease-activated receptor (PAR) signaling and growth factor activation (e.g., TGF-β) play a central role [150,151].

Studies in rat models show that excessive hormone treatment, such as dehydroepiandrosterone (DHEA), can induce uterine fibrosis by promoting coagulation activity and ECM deposition, leading to fibrotic remodeling [152]. However, species-specific differences between rodents and cattle necessitate careful interpretation of these findings in the context of bovine reproductive health.

Notably, coagulation- and fibrinolysis-related factors are highly expressed at the trophoblast–endometrium interface during the implantation period in cattle and sheep, which is essential for ECM remodeling and successful implantation [153–156]. Disruptions in the balance between coagulation and fibrinolysis may contribute to persistent inflammation and excessive ECM deposition, resulting in uterine fibrosis. Maintaining this balance is important not only for successful implantation but also in preventing pathological fibrosis that could affect reproductive efficiency.

4. Future Direction

4.1. Targeting Senescent Cells in the Bovine Uterus: Promise and Limitations

Pharmacological approaches, including stem cell therapy, anti-biological aging drugs, and herbal medicines, have been investigated for their ability to influence uterine biological aging. For instance, mesenchymal stem cells have demonstrated the ability to enhance endometrial regeneration through paracrine signaling, while compounds like metformin and resveratrol exhibit anti-senescence effects by decreasing the expression of senescence-related markers, such as p21 and RB transcriptional corepressor 1 (RB1) hypophosphorylation in rodent and human models [22,23].

Senolytic drugs selectively induce apoptosis in senescent cells, which influences female reproductive biological aging and pregnancy outcomes. For example, the Dasatinib and Quercetin combination (D+Q) has shown efficacy in reducing senescent cell burden in murine preclinical models, thereby enhancing tissue function. In mice, F4/80+ macrophages play a role in clearing senescent cells from the postpartum uterus, preserving uterine function [157]. Recent studies in rodents and primates have explored senotherapeutics as a strategy to counteract reproductive biological aging [158]. The development of senolytic treatments could lead to breakthroughs in managing age-related infertility and improving uterine receptivity during the implantation period. However, the translation of senolytic therapies to cattle presents major challenges. Practical considerations include differences in pharmacokinetics due to ruminant digestion, the route and frequency of drug administration in large animals, tissue distribution barriers, and safety concerns. In addition, regulatory hurdles for veterinary use of senolytics in food-producing animals remain substantial. Thus, while these compounds are promising in preclinical models, their application to bovine reproductive medicine requires further validation and adaptation.

Research on eliminating senescent cells in rodents and primates has shown promise. Several strategies target and eliminate these cells. For example, natural killer group 2 member D chimeric antigen receptor (NKG2D-CAR) T cells effectively remove senescent cells in aged mice and nonhuman primates, highlighting the potential of immunotherapy in managing senescence [159]. Similarly, mitochondria-targeted tamoxifen (MitoTam) exploits low adenine nucleotide translocase-2 expression to selectively eliminate senescent cells in mice [160]. These strategies highlight the diversity of potential therapeutic targets, ranging from mitochondrial dysfunction to anti-apoptotic signaling pathways. The inhibition of the anti-apoptotic proteins B-cell lymphoma-w (BCL-W) and B-cell lymphoma-extra-large (BCL-XL) has been shown to induce apoptosis in senescent cells in the lungs of ionizing radiation-exposed mice [161]. However, the potential off-target effects and long-term safety of these interventions necessitate careful evaluation before implementation.

Sodium–Glucose Transport Protein 2 (SGLT2) inhibitors, initially developed for diabetic treatment, exhibit senolytic properties. Canagliflozin, an SGLT2 inhibitor, reduces senescent cells in visceral adipose tissue, extends lifespan in obese and aging mice [162], and enhances immune clearance by downregulating programmed cell death ligand 1 (PD-L1) expression [163,164]. This senolytic activity may stem from its ability to modulate oxidative stress and inflammation through nuclear factor, erythroid derived 2, like 2 (NFE2L2) activation [165]. However, the applicability of SGLT2 inhibitors to cattle remains uncertain due to fundamental differences in glucose metabolism and renal physiology between humans and ruminants. In cattle, glucose homeostasis is primarily regulated through hepatic gluconeogenesis [166], and renal glucose reabsorption mechanisms may differ functionally, potentially affecting both drug efficacy and pharmacokinetics. Further research is necessary to evaluate whether SGLT2 inhibitors exert similar senolytic or anti-inflammatory effects in bovine tissues, and whether their administration would be feasible and safe in a veterinary reproductive context.

Senolytic drugs show promise in treating age-related conditions in human and mouse models [167], but their effects on female reproductive biological aging remain unclear. Treatment with D+Q reduced uterine fibrosis and increased p53 expression in aged mice [168], indicating potential benefits for uterine function. However, these treatments failed to improve ovarian reserve or fertility [39]. In human endometrial stromal cells, D+Q enhanced decidualization and reduced senescence markers, suggesting their role in improving uterine cellular function [169].

Sodium tanshinone IIA sulfonate, a natural compound from a traditional Chinese herbal medicine, reduces lesion development by inducing cellular senescence in a mouse model of deep endometriosis [170]. This finding highlights the context-dependent effects of senescence, where its induction in pathological contexts may help limit disease progression, in contrast to senescent cell clearance strategies used in other conditions. Although senolytic therapies aim to mitigate uterine dysfunction, their impact on extending reproductive longevity remains uncertain [158].

Recent studies have shown that senescent cells form a heterogeneous population [171,172], highlighting the need for precise elimination strategies targeting senescent cells that contribute to chronic inflammation and hormonal dysregulation, ultimately impairing uterine function and fertility. Emerging approaches such as single-cell transcriptomics and proteomics may aid in identifying unique markers of senescent cell subtypes in reproductive tissues, facilitating the development of precise senolytic therapies. While single-cell transcriptomic approaches are not yet routine in livestock research, they hold promise for identifying bovine-specific senescence markers and tailoring selective interventions. However, current senolytic therapies remain largely experimental in cattle, and future studies must address pharmacokinetics, tissue targeting, and safety under field conditions.

4.2. Preventing Fibrosis Development in Female Reproductive Tissues: From Rodents to Ruminants

Fibrosis is a wound-healing response that can become pathogenic if it continues unchecked. Key mechanisms for fibrosis resolution include degradation of ECM, elimination of fibrogenic myofibroblasts through apoptosis, senescence, or reprogramming, and the modulation of inflammatory responses [173,174]. In the liver, hepatic stellate cell (HSC) apoptosis and reduced expression of metalloproteinase inhibitors play important roles in spontaneous fibrosis resolution [175]. Additionally, macrophages contribute through phenotypic switching, which modulates inflammatory responses and promotes tissue remodeling [176]. Targeting HSC activation, inducing HSC apoptosis, or modulating fibrosis-related signaling pathways are promising strategies for promoting liver fibrosis regression [177].

Recent studies highlight the potential of antifibrotic therapies to reverse fibrosis in female reproductive organs. However, direct evidence linking these therapies to improved fertility and reproductive longevity remains limited in mouse models [12,99]. In vitro studies using bovine endometrial cell cultures have provided valuable insights into fibrosis regulation in cattle. Ginsenoside Rg1 alleviates lipopolysaccharide-induced fibrosis of bovine endometrial epithelial cells by inhibiting reactive oxygen species-activated NLR family pyrin domain containing 3 (NLRP3) [107]. Similarly, melatonin mitigates endometrial fibrosis in bovine endometritis by regulating TGF-β/Smad and MAPK signaling pathways via MT2 receptor [106].

Epigenetic aging, associated with altered coagulation and fibrotic remodeling, is thought to play a role in exacerbating fibrosis, yet the precise molecular mechanisms remain under investigation [178–180]. In vitro studies have demonstrated that histone deacetylase inhibitors, such as valproic acid and trichostatin A, can suppress fibrotic gene expression and epithelial-to-mesenchymal transition in uterine and hepatic fibroblasts [178,180]. In vivo animal models further support the anti-fibrotic effects of DNA methyltransferase inhibitors and SIRT1 activators in reversing fibrosis-associated gene expression and improving reproductive outcomes [179]. Despite these promising findings, the in vivo efficacy of these interventions in cattle remains unproven. Challenges include: A lack of long-term studies on reproductive outcomes (e.g., pregnancy rate, uterine involution). Limited pharmacokinetic and safety data for repeated use in livestock. Regulatory barriers to implementing molecular therapies in food-producing animals. Thus, while epigenetic and molecular antifibrotic strategies offer mechanistic promise, their applicability in bovine reproductive practice requires careful evaluation through well-designed translational studies, especially those incorporating fertility endpoints relevant to dairy and beef production systems.

4.3. Microbiome-Based Strategies to Modulate Bovine Uterine Immunity

Unlike primates and rodents, cattle present unique challenges in applying senolytic drugs directly to modulate uterine biological aging. The rumen microbiome plays a role in nutrient metabolism, immune modulation, and systemic inflammation, offering a potential target for interventions aimed at improving reproductive function. Specifically, short chain fatty acids such as butyrate, produced by fiber-fermenting bacteria, have been shown to exert anti-inflammatory effects by inhibiting NF-κB signaling and enhancing regulatory T-cell responses, which may indirectly influence uterine immune tone [181,182]. Moreover, certain microbial metabolites can modulate endocrine signaling, such as estrogen metabolism, via enterohepatic circulation, thereby affecting reproductive tissues including the uterus.

Instead of relying solely on direct senolytic treatments, modifying the rumen microbiome or developing targeted delivery strategies may provide a more viable approach to optimizing uterine health in cattle.

Probiotic interventions, particularly intravaginal administration of lactic acid bacteria (LAB), have demonstrated the ability to influence the bovine uterine immune environment through multiple mechanisms [183]. LAB modulate the mucosal immune system by stimulating pattern recognition receptors, including Toll-like receptors, on epithelial and immune cells, leading to altered cytokine production. Specifically, LAB administration reduces local proinflammatory cytokines such as IL-1β, IL-6, and IL-8, while increasing anti-inflammatory cytokines like IL-10 and enhancing secretory immunoglobulin A production [184–186]. These immune shifts correlate with lower incidences of uterine infections, such as metritis and Escherichia coli challenge, as well as improvements in reproductive parameters, including earlier estrus expression and increased conception rates [186].

In addition, a microbiome-targeted phytochemical, aloe-emodin, has been shown to enhance resistance to *Staphylococcus aureus* invasion by enhancing tight junction protein expression and reducing endometrial epithelial cell apoptosis [187]. This suggests that modulation of host–microbe interaction at the epithelial level is another plausible mechanism by which microbiome-based interventions exert protective effects on the uterus. These findings collectively suggest that well-designed microbial interventions can beneficially modulate local inflammatory responses in the bovine uterus, potentially improving reproductive outcomes. Compared to systemic drug delivery, microbial and phytochemical interventions offer more practical routes in cattle, such as intravaginal or intrauterine administration.

These strategies show advantages for field-level adoption, especially in dairy systems facing postpartum uterine disorders.

5. Conclusions

Extensive research in humans, primates, and rodents has provided insights into uterine fibrosis and biological aging. However, the pathophysiological mechanisms underlying uterine biological aging and fibrosis in cows, particularly in high-producing dairy cows, remain insufficiently characterized, and species-specific mitigation strategies have yet to be established. Addressing uterine biological aging and fibrosis in cattle requires an approach that considers species-specific reproductive physiology and the physiological consequences of intensive production systems. Research strategies include (1) nutritional modulation tailored for ruminant digestion and metabolism to alleviate metabolic stress, (2) epigenetic interventions such as SIRT1 activation or histone deacetylase inhibition adapted to bovine uterine tissue, and (3) microbiome-targeted approaches to regulate the immune environment and reduce inflammatory signaling in the bovine uterus. These strategies should be validated in both in vitro and in vivo bovine models to assess their effects on implantation, pregnancy maintenance, and reproductive longevity.

Understanding the mechanisms of biological aging is essential for enhancing reproductive efficiency in cattle. Identifying reliable biological aging markers, such as epigenetic clock, fibrosis levels, and immune alterations, will support the development of targeted interventions. For example, therapeutic strategies that aim to reduce fibrosis or rejuvenate uterine tissue could improve implantation and pregnancy outcomes. Given the strong link between metabolic stress and biological aging, nutritional status must be carefully considered to avoid chronic energy surplus and support endocrine homeostasis. Excessive energy intake, required for high-producing cows, and metabolic dysregulation can accelerate uterine biological aging, promoting fibrosis and impaired fertility.

While the rumen microbiome helps the cattle digest feed to obtain nutrients and maintain overall bovine health, its influence on the uterine environment and reproductive function remain understudied. Investigating the microbial population associated with reproductive health and identifying strategies to manipulate them, either through nutrition or probiotic interventions, may offer a novel approach to support uterine function and fecundity. Recent studies in sheep have shown that increased cervical microbial abundance and diversity are associated with reduced fertility following artificial insemination, suggesting a link between microbial homeostasis and reproductive efficiency in livestock [188]. While these findings provide valuable insights, species-specific validation is needed to determine whether similar microbial mechanisms influence bovine uterine function. A deeper understanding of these interconnected mechanisms could help optimize breeding strategies that mitigate the negative effects of intensive production systems, although significant challenges remain. These include a lack of bovine-specific longitudinal data, variability in microbial responses to interventions, and practical implementation hurdles in commercial farm settings. Ultimately, improving uterine health through integrative approaches, such as metabolic regulation, anti-fibrotic therapies, and microbiome modulation, holds promise, but will require further research and careful field-level validation to support fertility while maintaining production sustainability.

Advancing knowledge of biological aging and fibrosis is essential to address the reproductive limitations imposed by high productivity demands in modern livestock production, such as fertility management, metabolic stress, and sustainability. Such insights can inform the development of sustainable breeding strategies that balance reproductive longevity with production efficiency. Ultimately, improving uterine health through integrative approaches, including metabolic regulation, anti-fibrotic therapies, and microbial modulation, will support fertility while maintaining production sustainability in modern livestock systems (Figure 4).

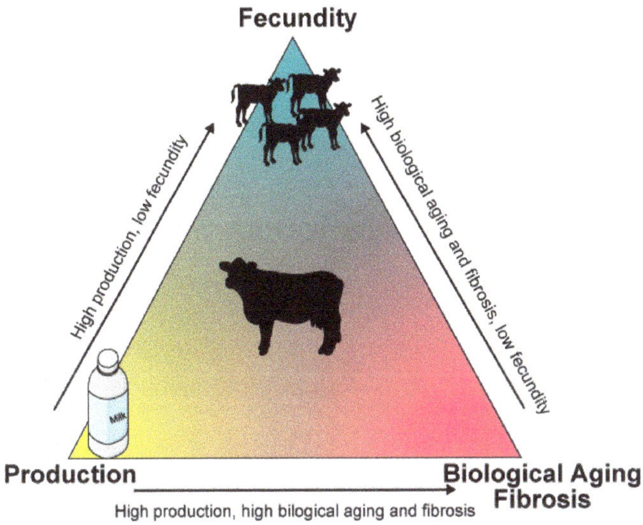

Figure 4. Production–Fecundity Trade-off in Cattle. Higher productivity reduces fecundity due to increased metabolic demands. Higher productivity increases the risk of biological aging and fibrosis. Biological aging and fibrosis contribute to decreased fecundity. It is essential to balance productivity with reproductive health while minimizing biological aging and fibrosis risks.

Author Contributions: Y.M.: writing—original draft, editing and preparing figures; K.I.: conceptualization, writing, editing, and review. All authors have read and agreed to the published version of the manuscript.

Funding: Our research has been in part supported by Livestock Promotional Funds of Japan Racing Association (JRA) to K.I. This work was in part supported by JSPS KAKNHI Grant Number 21K14965.

Institutional Review Board Statement: Not applicable.

Informed Consent Statement: Not applicable.

Data Availability Statement: No new data was generated.

Acknowledgments: The authors are grateful to the members of the laboratory of molecular reproduction (Tokai University) for their helpful suggestions during the preparation of this manuscript. The illustrations are, in part, from NIAID NIH BIOART (bioart.niaid.nih.gov/bioart/) (accessed on 1 April 2025).

Conflicts of Interest: The authors declare no conflicts of interest.

Abbreviations

The following abbreviations are used in this manuscript:

ECM	Extra-cellular matrix
FSH	Follicle-stimulating hormone
SASP	Senescence-associated secretory phenotype
LMNB1	Lamin B1
SIRT1	Sirtuin 1
NAD	Nicotinamide adenine dinucleotide

ROS	Reactive oxygen species
mTOR	Mammalian target of rapamycin
AGEs	Advanced glycation end products
ncRNA	Non-coding RNA
cfDNA	Cell-free DNA
AMH	Anti-mullerian hormone
TGF-β	Transforming growth factor beta
TGFBR3	Transforming growth factor beta receptor 3
Smad	Suppressor of mother against decapentaplegic
GPER1	G protein-coupled estrogen receptor 1
HSP47	Heat shock protein 47
WNT	Wingless-type MMTV integration site family
MST1/2	Mammalian Ste20-like kinase 1 and 2
LATS1/2	Large tumor suppressor 1 and 2
YAP	Yes-associated protein
TAZ	Transcriptional coactivator with PDZ-binding motif
TEAD	TEA domain family member
DUSP4	Dual specificity protein phosphatase 4
GSK3β	Glycogen synthase kinase-3 beta
SNAI1	Snail family transcriptional repressor 1
MAPK	Mitogen-activated protein kinase
PIK3	Phosphatidylinositol-3 kinase
ERα	Estrogen receptor alpha
Notch1	Neurogenic locus notch homolog protein 1
IL6	Interleukin 6
VEGF	Vascular endothelial growth factor
IgG	Immunoglobulin G
PAI-1	Plasminogen activator inhibitor-1
PAR	Protease-activated receptor
DHEA	Dehydroepiandrosterone
RB1	RB transcriptional corepressor 1
D+Q	Dasatinib and Quercetin combination
NKG2-CAR	Natural killer group 2 member D chimeric antigen receptor
MitoTam	Mitochondria-targeted tamoxifen
BCL-W	B-cell lymphoma-w
BCL-XL	B-cell lymphoma-extra-large
SGLT2	Sodium–glucose transport protein 2
PD-L1	Programmed cell death ligand 1
NFE2L2	Nuclear factor, erythroid derived 2, like 2
HSC	Hepatic stellate cell
NLRP3	NLR family pyrin domain containing 3
NF-κB	Nuclear factor kappa-light-chain-enhancer of activated B cells
LAB	Lactic acid bacteria
uNK	uterine natural killer

References

1. Diskin, M.G.; Waters, S.M.; Parr, M.H.; Kenny, D.A. Pregnancy Losses in Cattle: Potential for Improvement. *Reprod. Fertil. Dev.* **2016**, *28*, 83. [CrossRef] [PubMed]
2. Kidder, H.E.; Black, W.G.; Wiltbank, J.N.; Ulberg, L.C.; Casida, L.E. Fertilization Rates and Embryonic Death Rates in Cows Bred to Bulls of Different Levels of Fertility. *J. Dairy Sci.* **1954**, *37*, 691–697. [CrossRef]
3. Artificial Insemination Association of Japan. Survey of Bovine Conception Rates in Japan (2023–2024). Available online: https://aiaj.lin.gr.jp/3/tyosa.html (accessed on 31 March 2025). (In Japanese)
4. Mann, G.; Lamming, G. The Influence of Progesterone During Early Pregnancy in Cattle. *Reprod. Domest. Anim.* **1999**, *34*, 269–274. [CrossRef]

5. Reese, S.T.; Franco, G.A.; Poole, R.K.; Hood, R.; Fernadez Montero, L.; Oliveira Filho, R.V.; Cooke, R.F.; Pohler, K.G. Pregnancy Loss in Beef Cattle: A Meta-Analysis. *Anim. Reprod. Sci.* **2020**, *212*, 106251. [CrossRef]
6. Velarde, M.C.; Menon, R. Positive and Negative Effects of Cellular Senescence during Female Reproductive Aging and Pregnancy. *J. Endocrinol.* **2016**, *230*, R59–R76. [CrossRef]
7. Zavatta, A.; Parisi, F.; Mandò, C.; Scaccabarozzi, C.; Savasi, V.M.; Cetin, I. Role of Inflammaging on the Reproductive Function and Pregnancy. *Clin. Rev. Allergy Immunol.* **2022**, *64*, 145–160. [CrossRef]
8. Hurwitz, J.M.; Santoro, N. Inhibins, Activins, and Follistatin in the Aging Female and Male. *Semin. Reprod. Med.* **2004**, *22*, 209–217. [CrossRef]
9. Yaffe, H.; Ron, M.; Polishuk, W.Z. Amenorrhea, Hypomenorrhea, and Uterine Fibrosis. *Am. J. Obstet. Gynecol.* **1978**, *130*, 599–601. [CrossRef] [PubMed]
10. de Ziegler, D.; Pirtea, P.; Ayoubi, J.M. Inflammation and Uterine Fibrosis: The Possible Role of Chronic Endometritis. *Fertil. Steril.* **2019**, *111*, 890–891. [CrossRef]
11. Hu, J.; Zeng, B.; Jiang, X.; Hu, L.; Meng, Y.; Zhu, Y.; Mao, M. The Expression of Marker for Endometrial Stem Cell and Fibrosis Was Increased in Intrauterine Adhesious. *Int. J. Clin. Exp. Pathol.* **2015**, *8*, 1525.
12. Winkler, I.; Tolkachov, A.; Lammers, F.; Lacour, P.; Daugelaite, K.; Schneider, N.; Koch, M.-L.; Panten, J.; Grünschläger, F.; Poth, T.; et al. The Cycling and Aging Mouse Female Reproductive Tract at Single-Cell Resolution. *Cell* **2024**, *187*, 981–998.e25. [CrossRef]
13. Epel, E.S. Psychological and Metabolic Stress: A Recipe for Accelerated Cellular Aging? *Hormones* **2009**, *8*, 7–22. [CrossRef] [PubMed]
14. Tzanetakou, I.P.; Katsilambros, N.L.; Benetos, A.; Mikhailidis, D.P.; Perrea, D.N. Is Obesity Linked to Aging? *Ageing Res. Rev.* **2012**, *11*, 220–229. [CrossRef]
15. Ratan, P.; Rubbi, L.; Thompson, M.; Naresh, K.; Waddell, J.; Jones, B.; Pellegrini, M. Epigenetic Aging in Cows Is Accelerated by Milk Production. *Epigenetics* **2023**, *18*, 2240188. [CrossRef] [PubMed]
16. Carvalho, P.D.; Santos, V.G.; Giordano, J.O.; Wiltbank, M.C.; Fricke, P.M. Development of Fertility Programs to Achieve High 21-Day Pregnancy Rates in High-Producing Dairy Cows. *Theriogenology* **2018**, *114*, 165–172. [CrossRef] [PubMed]
17. Cole, J.B.; VanRaden, P.M. Symposium Review: Possibilities in an Age of Genomics: The Future of Selection Indices. *J. Dairy Sci.* **2018**, *101*, 3686–3701. [CrossRef]
18. Lucy, M.C. Symposium Review: Selection for Fertility in the Modern Dairy Cow—Current Status and Future Direction for Genetic Selection. *J. Dairy Sci.* **2019**, *102*, 3706–3721. [CrossRef]
19. Guinan, F.L.; Wiggans, G.R.; Norman, H.D.; Dürr, J.W.; Cole, J.B.; Van Tassell, C.P.; Misztal, I.; Lourenco, D. Changes in Genetic Trends in US Dairy Cattle since the Implementation of Genomic Selection. *J. Dairy Sci.* **2023**, *106*, 1110–1129. [CrossRef]
20. Wiltbank, M.C.; Baez, G.M.; Garcia-Guerra, A.; Toledo, M.Z.; Monteiro, P.L.J.; Melo, L.F.; Ochoa, J.C.; Santos, J.E.P.; Sartori, R. Pivotal Periods for Pregnancy Loss during the First Trimester of Gestation in Lactating Dairy Cows. *Theriogenology* **2016**, *86*, 239–253. [CrossRef]
21. Khan, S.S.; Singer, B.D.; Vaughan, D.E. Molecular and Physiological Manifestations and Measurement of Aging in Humans. *Aging Cell* **2017**, *16*, 624–633. [CrossRef]
22. Wu, Y.; Li, M.; Zhang, J.; Wang, S. Unveiling Uterine Aging: Much More to Learn. *Ageing Res. Rev.* **2023**, *86*, 101879. [CrossRef]
23. Tinelli, A.; Andjić, M.; Morciano, A.; Pecorella, G.; Malvasi, A.; D'Amato, A.; Sparić, R. Uterine Aging and Reproduction: Dealing with a Puzzle Biologic Topic. *Int. J. Mol. Sci.* **2023**, *25*, 322. [CrossRef]
24. López-Otín, C.; Blasco, M.A.; Partridge, L.; Serrano, M.; Kroemer, G. Hallmarks of Aging: An Expanding Universe. *Cell* **2023**, *186*, 243–278. [CrossRef]
25. Balough, J.L.; Dipali, S.S.; Velez, K.; Kumar, T.R.; Duncan, F.E. Hallmarks of Female Reproductive Aging in Physiologic Aging Mice. *Nat. Aging* **2024**, *4*, 1711–1730. [CrossRef] [PubMed]
26. Kong, S.; Zhang, S.; Chen, Y.; Wang, W.; Wang, B.; Chen, Q.; Duan, E.; Wang, H. Determinants of Uterine Aging: Lessons from Rodent Models. *Sci. China Life Sci.* **2012**, *55*, 687–693. [CrossRef]
27. Ananda, H.M.; Wurlina, W.; Hidajati, N.; Hariadi, M.; Samik, A.; Restiadi, T.I. Hubungan antara umur dengan calving interval, days open, dan service per conseption sapifriesian holstein (FH). *Ovozoa J. Anim. Reprod.* **2020**, *8*, 94. [CrossRef]
28. Osoro, K.; Wright, I.A. The Effect of Body Condition, Live Weight, Breed, Age, Calf Performance, and Calving Date on Reproductive Performance of Spring-Calving Beef Cows. *J. Anim. Sci.* **1992**, *70*, 1661–1666. [CrossRef] [PubMed]
29. Butler, W.R. Nutritional Interactions with Reproductive Performance in Dairy Cattle. *Anim. Reprod. Sci.* **2000**, *60–61*, 449–457. [CrossRef]
30. Malhi, P.S.; Adams, G.P.; Mapletoft, R.J.; Singh, J. Oocyte Developmental Competence in a Bovine Model of Reproductive Aging. *Reproduction* **2007**, *134*, 233–239. [CrossRef]
31. Malhi, P.S.; Adams, G.P.; Singh, J. Bovine Model for the Study of Reproductive Aging in Women: Follicular, Luteal, and Endocrine Characteristics1. *Biol. Reprod.* **2005**, *73*, 45–53. [CrossRef]

32. Bernadotte, A.; Mikhelson, V.M.; Spivak, I.M. Markers of Cellular Senescence. Telomere Shortening as a Marker of Cellular Senescence. *Aging* **2016**, *8*, 3–11. [CrossRef] [PubMed]
33. Wei, W.; Ji, S. Cellular Senescence: Molecular Mechanisms and Pathogenicity. *J. Cell Physiol.* **2018**, *233*, 9121–9135. [CrossRef]
34. Hernandez-Segura, A.; Nehme, J.; Demaria, M. Hallmarks of Cellular Senescence. *Trends Cell Biol.* **2018**, *28*, 436–453. [CrossRef] [PubMed]
35. Marquez, C.M.D.; Ibana, J.A.; Velarde, M.C. The Female Reproduction and Senescence Nexus. *Am. J. Reprod. Immunol.* **2017**, *77*, e12646. [CrossRef]
36. Gong, G.-S.; Muyayalo, K.P.; Zhang, Y.-J.; Lin, X.-X.; Liao, A.-H. Flip a Coin: Cell Senescence at the Maternal–Fetal Interface. *Biol. Reprod.* **2023**, *109*, 244–255. [CrossRef] [PubMed]
37. Baar, M.P.; Perdiguero, E.; Muñoz-Cánoves, P.; de Keizer, P.L. Musculoskeletal Senescence: A Moving Target Ready to Be Eliminated. *Curr. Opin. Pharmacol.* **2018**, *40*, 147–155. [CrossRef]
38. Tomari, H.; Kawamura, T.; Asanoma, K.; Egashira, K.; Kawamura, K.; Honjo, K.; Nagata, Y.; Kato, K. Contribution of Senescence in Human Endometrial Stromal Cells during Proliferative Phase to Embryo Receptivity. *Biol. Reprod.* **2020**, *103*, 104–113. [CrossRef]
39. Garcia, D.N.; Hense, J.D.; Zanini, B.M.; Isola, J.V.; Prosczek, J.B.; Ashiqueali, S.; Oliveira, T.L.; Mason, J.B.; Schadock, I.C.; Barros, C.C.; et al. Senolytic Treatment Fails to Improve Ovarian Reserve or Fertility in Female Mice. *Geroscience* **2024**, *46*, 3445–3455. [CrossRef]
40. Malvezzi, H.; Dobo, C.; Filippi, R.Z.; Mendes do Nascimento, H.; Palmieri da Silva e Sousa, L.; Meola, J.; Piccinato, C.A.; Podgaec, S. Altered P16Ink4a, IL-1β, and Lamin B1 Protein Expression Suggest Cellular Senescence in Deep Endometriotic Lesions. *Int. J. Mol. Sci.* **2022**, *23*, 2476. [CrossRef]
41. Malvezzi, H.; Cestari, B.A.; Meola, J.; Podgaec, S. Higher Oxidative Stress in Endometriotic Lesions Upregulates Senescence-Associated P16ink4a and β-Galactosidase in Stromal Cells. *Int. J. Mol. Sci.* **2023**, *24*, 914. [CrossRef]
42. Lin, X.; Dai, Y.; Tong, X.; Xu, W.; Huang, Q.; Jin, X.; Li, C.; Zhou, F.; Zhou, H.; Lin, X.; et al. Excessive Oxidative Stress in Cumulus Granulosa Cells Induced Cell Senescence Contributes to Endometriosis-Associated Infertility. *Redox Biol.* **2020**, *30*, 101431. [CrossRef] [PubMed]
43. Malvezzi, H.; Hernandes, C.; de Azevedo, B.C.; Piccinato, C.A.; Meola, J.; Podgaec, S. Increased cellular senescence in deep infiltrating endometriosis. *Fertil. Steril.* **2020**, *114*, e79. [CrossRef]
44. Wang, M.; Wu, Y.; He, Y.; Liu, J.; Chen, Y.; Huang, J.; Qi, G.; Li, P. SIRT1 Upregulation Promotes Epithelial-Mesenchymal Transition by Inducing Senescence Escape in Endometriosis. *Sci. Rep.* **2022**, *12*, 12302. [CrossRef]
45. Vassiliadis, S. Premature Immunosenescence Impairs Immune Surveillance Allowing the Endometriotic Stem Cell to Migrate: The Cytokine Profile as a Common Denominator. *J. Endometr.* **2010**, *2*, 7–18. [CrossRef]
46. Lucas, E.S.; Vrljicak, P.; Muter, J.; Diniz-da-Costa, M.M.; Brighton, P.J.; Kong, C.-S.; Lipecki, J.; Fishwick, K.J.; Odendaal, J.; Ewington, L.J.; et al. Recurrent Pregnancy Loss Is Associated with a Pro-Senescent Decidual Response during the Peri-Implantation Window. *Commun. Biol.* **2020**, *3*, 37. [CrossRef]
47. Deryabin, P.I.; Borodkina, A. V Stromal Cell Senescence Contributes to Impaired Endometrial Decidualization and Defective Interaction with Trophoblast Cells. *Hum. Reprod.* **2022**, *37*, 1505–1524. [CrossRef]
48. Nakamura, Y.; Iwata, H.; Kuwayama, T.; Shirasuna, K. S100A8, Which Increases with Age, Induces Cellular Senescence-like Changes in Bovine Oviduct Epithelial Cells. *Am. J. Reprod. Immunol.* **2019**, *82*, e13163. [CrossRef]
49. Nakamura, Y.; Aihara, R.; Iwata, H.; Kuwayama, T.; Shirasuna, K. IL1B Triggers Inflammatory Cytokine Production in Bovine Oviduct Epithelial Cells and Induces Neutrophil Accumulation via CCL2. *Am. J. Reprod. Immunol.* **2021**, *85*, e13365. [CrossRef]
50. Tanaka, H.; Ohtsu, A.; Shiratsuki, S.; Kawahara-Miki, R.; Iwata, H.; Kuwayama, T.; Shirasuna, K. Age-Dependent Changes in Inflammation and Extracellular Matrix in Bovine Oviduct Epithelial Cells during the Post-Ovulatory Phase. *Mol. Reprod. Dev.* **2016**, *83*, 815–826. [CrossRef]
51. Tanikawa, N.; Ohtsu, A.; Kawahara-Miki, R.; Kimura, K.; Matsuyama, S.; Iwata, H.; Kuwayama, T.; Shirasuna, K. Age-Associated MRNA Expression Changes in Bovine Endometrial Cells in Vitro. *Reprod. Biol. Endocrinol.* **2017**, *15*, 63. [CrossRef]
52. Hori, K.; Matsuyama, S.; Nakamura, S.; Iwata, H.; Kuwayama, T.; Miyamoto, A.; Shirasuna, K. Age-related Changes in the Bovine Corpus Luteum Function and Progesterone Secretion. *Reprod. Domest. Anim.* **2019**, *54*, 23–30. [CrossRef]
53. Jin, W.; Zheng, J.; Xiao, Y.; Ju, L.; Chen, F.; Fu, J.; Jiang, H.; Zhang, Y. A Universal Molecular Mechanism Driving Aging. *bioRxiv* **2024**. [CrossRef]
54. Lenart, P.; Krejci, L. DNA, the Central Molecule of Aging. *Mutat. Res./Fundam. Mol. Mech. Mutagen.* **2016**, *786*, 1–7. [CrossRef] [PubMed]
55. Blagosklonny, M. V Aging: ROS or TOR. *Cell Cycle* **2008**, *7*, 3344–3354. [CrossRef] [PubMed]
56. Blagosklonny, M.V.; Hall, M.N. Growth and Aging: A Common Molecular Mechanism. *Aging* **2009**, *1*, 357–362. [CrossRef]
57. Golubev, A.; Hanson, A.D.; Gladyshev, V.N. Non-Enzymatic Molecular Damage as a Prototypic Driver of Aging. *J. Biol. Chem.* **2017**, *292*, 6029–6038. [CrossRef]

58. Chaudhuri, J.; Bains, Y.; Guha, S.; Kahn, A.; Hall, D.; Bose, N.; Gugliucci, A.; Kapahi, P. The Role of Advanced Glycation End Products in Aging and Metabolic Diseases: Bridging Association and Causality. *Cell Metab.* **2018**, *28*, 337–352. [CrossRef]
59. Bürkle, A. Mechanisms of Ageing. *Eye* **2001**, *15*, 371–375. [CrossRef]
60. Hayflick, L. Biological Aging Is No Longer an Unsolved Problem. *Ann. N. Y. Acad. Sci.* **2007**, *1100*, 1–13. [CrossRef]
61. Cao, X.; Yang, G.; Li, X.; Fu, J.; Mohedaner, M.; Danzengzhuoga; Høj Jørgensen, T.S.; Agogo, G.O.; Wang, L.; Zhang, X.; et al. Weight Change across Adulthood and Accelerated Biological Aging in Middle-Aged and Older Adults. *Am. J. Clin. Nutr.* **2023**, *117*, 1–11. [CrossRef]
62. Zheng, Y.; Manson, J.E.; Yuan, C.; Liang, M.H.; Grodstein, F.; Stampfer, M.J.; Willett, W.C.; Hu, F.B. Associations of Weight Gain From Early to Middle Adulthood With Major Health Outcomes Later in Life. *JAMA* **2017**, *318*, 255. [CrossRef] [PubMed]
63. Kim, S.; Parks, C.G.; DeRoo, L.A.; Chen, H.; Taylor, J.A.; Cawthon, R.M.; Sandler, D.P. Obesity and Weight Gain in Adulthood and Telomere Length. *Cancer Epidemiol. Biomark. Prev.* **2009**, *18*, 816–820. [CrossRef] [PubMed]
64. Bonomini, F.; Rodella, L.F.; Rezzani, R. Metabolic Syndrome, Aging and Involvement of Oxidative Stress. *Aging Dis.* **2015**, *6*, 109. [CrossRef]
65. Shirasuna, K.; Iwata, H. Effect of Aging on the Female Reproductive Function. *Contracept. Reprod. Med.* **2017**, *2*, 23. [CrossRef] [PubMed]
66. Berry, D.P.; Wall, E.; Pryce, J.E. Genetics and Genomics of Reproductive Performance in Dairy and Beef Cattle. *Animal* **2014**, *8*, 105–121. [CrossRef]
67. Oltenacu, P.A.; Algers, B. Selection for Increased Production and the Welfare of Dairy Cows: Are New Breeding Goals Needed? *AMBIO A J. Hum. Environ.* **2005**, *34*, 311–315. [CrossRef]
68. Sordillo, L.M.; Mavangira, V. The Nexus between Nutrient Metabolism, Oxidative Stress and Inflammation in Transition Cows. *Anim. Prod. Sci.* **2014**, *54*, 1204. [CrossRef]
69. Ward, W.R.; Parker, C.S. Field Evidence of Metabolic Stress in Dairy Cows? *BSAP Occas. Publ.* **1999**, *24*, 21–26. [CrossRef]
70. Klein, M.S.; Almstetter, M.F.; Schlamberger, G.; Nürnberger, N.; Dettmer, K.; Oefner, P.J.; Meyer, H.H.D.; Wiedemann, S.; Gronwald, W. Nuclear Magnetic Resonance and Mass Spectrometry-Based Milk Metabolomics in Dairy Cows during Early and Late Lactation. *J. Dairy Sci.* **2010**, *93*, 1539–1550. [CrossRef]
71. Sheldon, I.M.; Cronin, J.G.; Pospiech, M.; Turner, M.L. Symposium Review: Mechanisms Linking Metabolic Stress with Innate Immunity in the Endometrium. *J. Dairy Sci.* **2018**, *101*, 3655–3664. [CrossRef]
72. Yonekura, S. The Role of Endoplasmic Reticulum Stress in Metabolic Diseases and Mammary Epithelial Cell Homeostasis in Dairy Cows. *Anim. Sci. J.* **2024**, *95*, e13935. [CrossRef] [PubMed]
73. Berry, D.P. Genetics—A Tool to Improve Productivity and Profitability. *Int. J. Dairy. Technol.* **2008**, *61*, 30–35. [CrossRef]
74. Gonzalo, S. Epigenetic Alterations in Aging. *J. Appl. Physiol.* **2010**, *109*, 586–597. [CrossRef]
75. Brunet, A.; Berger, S.L. Epigenetics of Aging and Aging-Related Disease. *J. Gerontol. A Biol. Sci. Med. Sci.* **2014**, *69*, S17–S20. [CrossRef]
76. Yu, M.; Hazelton, W.D.; Luebeck, G.E.; Grady, W.M. Epigenetic Aging: More Than Just a Clock When It Comes to Cancer. *Cancer Res.* **2020**, *80*, 367–374. [CrossRef] [PubMed]
77. Saul, D.; Kosinsky, R.L. Epigenetics of Aging and Aging-Associated Diseases. *Int. J. Mol. Sci.* **2021**, *22*, 401. [CrossRef]
78. Liang, X.; Ma, J.; Schatten, H.; Sun, Q. Epigenetic Changes Associated with Oocyte Aging. *Sci. China Life Sci.* **2012**, *55*, 670–676. [CrossRef]
79. Klutstein, M.; Gonen, N. Epigenetic Aging of Mammalian Gametes. *Mol. Reprod. Dev.* **2023**, *90*, 785–803. [CrossRef]
80. Kordowitzki, P.; Haghani, A.; Zoller, J.A.; Li, C.Z.; Raj, K.; Spangler, M.L.; Horvath, S. Epigenetic Clock and Methylation Study of Oocytes from a Bovine Model of Reproductive Aging. *Aging Cell* **2021**, *20*, e13349. [CrossRef]
81. Knight, A.K.; Spencer, J.B.; Smith, A.K. DNA Methylation as a Window into Female Reproductive Aging. *Epigenomics* **2024**, *16*, 175–188. [CrossRef]
82. Ryan, C.P. "Epigenetic Clocks": Theory and Applications in Human Biology. *Am. J. Hum. Biol.* **2021**, *33*, e23488. [CrossRef] [PubMed]
83. Bell, C.G.; Lowe, R.; Adams, P.D.; Baccarelli, A.A.; Beck, S.; Bell, J.T.; Christensen, B.C.; Gladyshev, V.N.; Heijmans, B.T.; Horvath, S.; et al. DNA Methylation Aging Clocks: Challenges and Recommendations. *Genome Biol.* **2019**, *20*, 249. [CrossRef]
84. Duan, R.; Fu, Q.; Sun, Y.; Li, Q. Epigenetic Clock: A Promising Biomarker and Practical Tool in Aging. *Ageing Res. Rev.* **2022**, *81*, 101743. [CrossRef] [PubMed]
85. Caulton, A.; Dodds, K.G.; McRae, K.M.; Couldrey, C.; Horvath, S.; Clarke, S.M. Development of Epigenetic Clocks for Key Ruminant Species. *Genes* **2021**, *13*, 96. [CrossRef] [PubMed]
86. Higgins-Chen, A.T.; Thrush, K.L.; Wang, Y.; Minteer, C.J.; Kuo, P.-L.; Wang, M.; Niimi, P.; Sturm, G.; Lin, J.; Moore, A.Z.; et al. A Computational Solution for Bolstering Reliability of Epigenetic Clocks: Implications for Clinical Trials and Longitudinal Tracking. *Nat. Aging* **2022**, *2*, 644–661. [CrossRef]

87. Kabacik, S.; Lowe, D.; Fransen, L.; Leonard, M.; Ang, S.-L.; Whiteman, C.; Corsi, S.; Cohen, H.; Felton, S.; Bali, R.; et al. The Relationship between Epigenetic Age and the Hallmarks of Aging in Human Cells. *Nat. Aging* **2022**, *2*, 484–493. [CrossRef]
88. Liu, Z.; Leung, D.; Thrush, K.; Zhao, W.; Ratliff, S.; Tanaka, T.; Schmitz, L.L.; Smith, J.A.; Ferrucci, L.; Levine, M.E. Underlying Features of Epigenetic Aging Clocks in Vivo and in Vitro. *Aging Cell* **2020**, *19*, e13229. [CrossRef]
89. Wagner, W. The Link Between Epigenetic Clocks for Aging and Senescence. *Front. Genet.* **2019**, *10*, 303. [CrossRef]
90. Teo, Y.V.; Capri, M.; Morsiani, C.; Pizza, G.; Faria, A.M.C.; Franceschi, C.; Neretti, N. Cell-free DNA as a Biomarker of Aging. *Aging Cell* **2019**, *18*, e12890. [CrossRef]
91. Ryan, C.P.; Hayes, M.G.; Lee, N.R.; McDade, T.W.; Jones, M.J.; Kobor, M.S.; Kuzawa, C.W.; Eisenberg, D.T.A. Reproduction Predicts Shorter Telomeres and Epigenetic Age Acceleration among Young Adult Women. *Sci. Rep.* **2018**, *8*, 11100. [CrossRef]
92. Ryan, C.P.; Lee, N.R.; Carba, D.B.; MacIsaac, J.L.; Lin, D.T.S.; Atashzay, P.; Belsky, D.W.; Kobor, M.S.; Kuzawa, C.W. Pregnancy Is Linked to Faster Epigenetic Aging in Young Women. *Proc. Natl. Acad. Sci. USA* **2024**, *121*, e2317290121. [CrossRef] [PubMed]
93. Kresovich, J.K.; Harmon, Q.E.; Xu, Z.; Nichols, H.B.; Sandler, D.P.; Taylor, J.A. Reproduction, DNA Methylation, and Biological Age. *Obstet. Gynecol. Surv.* **2020**, *75*, 106–107. [CrossRef]
94. Suvakov, S.; Ghamrawi, R.; Cubro, H.; Tu, H.; White, W.M.; Tobah, Y.S.B.; Milic, N.M.; Grande, J.P.; Cunningham, J.M.; Chebib, F.T.; et al. Epigenetic and Senescence Markers Indicate an Accelerated Ageing-like State in Women with Preeclamptic Pregnancies. *EBioMedicine* **2021**, *70*, 103536. [CrossRef] [PubMed]
95. Ross, K.M.; Carroll, J.E.; Horvath, S.; Hobel, C.J.; Coussons-Read, M.E.; Dunkel Schetter, C. Epigenetic Age and Pregnancy Outcomes: GrimAge Acceleration Is Associated with Shorter Gestational Length and Lower Birthweight. *Clin. Epigenet.* **2020**, *12*, 120. [CrossRef]
96. Salilew-Wondim, D.; Tholen, E.; Held-Hoelker, E.; Shellander, K.; Blaschka, C.; Drillich, M.; Iwersen, M.; Suess, D.; Gebremedhn, S.; Tesfaye, D.; et al. Endometrial DNA Methylation Signatures during the Time of Breeding in Relation to the Pregnancy Outcome in Postpartum Dairy Cows Fed a Control Diet or Supplemented with Rumen-Protected Methionine. *Front. Genet.* **2024**, *14*, 1267053. [CrossRef]
97. Walker, C.G.; Littlejohn, M.D.; Meier, S.; Roche, J.R.; Mitchell, M.D. DNA Methylation Is Correlated with Gene Expression during Early Pregnancy in Bos Taurus. *Physiol. Genom.* **2013**, *45*, 276–286. [CrossRef]
98. Landry, D.A.; Vaishnav, H.T.; Vanderhyden, B.C. The Significance of Ovarian Fibrosis. *Oncotarget* **2020**, *11*, 4366–4370. [CrossRef] [PubMed]
99. Umehara, T.; Winstanley, Y.E.; Andreas, E.; Morimoto, A.; Williams, E.J.; Smith, K.M.; Carroll, J.; Febbraio, M.A.; Shimada, M.; Russell, D.L.; et al. Female Reproductive Life Span Is Extended by Targeted Removal of Fibrotic Collagen from the Mouse Ovary. *Sci. Adv.* **2022**, *8*, eabn4564. [CrossRef]
100. Briley, S.M.; Jasti, S.; McCracken, J.M.; Hornick, J.E.; Fegley, B.; Pritchard, M.T.; Duncan, F.E. Reproductive Age-Associated Fibrosis in the Stroma of the Mammalian Ovary. *Reproduction* **2016**, *152*, 245–260. [CrossRef]
101. Landry, D.A.; Yakubovich, E.; Cook, D.P.; Fasih, S.; Upham, J.; Vanderhyden, B.C. Metformin Prevents Age-Associated Ovarian Fibrosis by Modulating the Immune Landscape in Female Mice. *Sci. Adv.* **2022**, *8*, eabq1475. [CrossRef]
102. Mara, J.N.; Zhou, L.T.; Larmore, M.; Johnson, B.; Ayiku, R.; Amargant, F.; Pritchard, M.T.; Duncan, F.E. Ovulation and Ovarian Wound Healing Are Impaired with Advanced Reproductive Age. *Aging* **2020**, *12*, 9686–9713. [CrossRef]
103. Machlin, J.H.; Barishansky, S.J.; Kelsh, J.; Larmore, M.J.; Johnson, B.W.; Pritchard, M.T.; Pavone, M.E.; Duncan, F.E. Fibroinflammatory Signatures Increase with Age in the Human Ovary and Follicular Fluid. *Int. J. Mol. Sci.* **2021**, *22*, 4902. [CrossRef] [PubMed]
104. Leppert, P.C.; Jayes, F.L.; Segars, J.H. The Extracellular Matrix Contributes to Mechanotransduction in Uterine Fibroids. *Obstet. Gynecol. Int.* **2014**, *2014*, 783289. [CrossRef]
105. Islam, M.S.; Ciavattini, A.; Petraglia, F.; Castellucci, M.; Ciarmela, P. Extracellular Matrix in Uterine Leiomyoma Pathogenesis: A Potential Target for Future Therapeutics. *Hum. Reprod. Update* **2018**, *24*, 59–85. [CrossRef] [PubMed]
106. Xiao, L.-F.; Fang, Z.; Chen, X.; Qiao, L.; Jiang, Y.; Wang, X.-G.; Chang, D.; Song, L.; Yao, H. Melatonin Alleviates Endometrial Fibrosis in Bovine Endometritis by Regulating Tgf-B/Smad and Mapk Signaling Pathways Via Mt2. *SSRN Electron. J.* **2025**, *169*, 104519. [CrossRef]
107. Song, L.; Wang, L.; Li, X.; Xiao, L. Ginsenoside Rg1 Alleviates Lipopolysaccharide-Induced Fibrosis of Endometrial Epithelial Cells in Dairy Cows by Inhibiting Reactive Oxygen Species-Activated NLRP3. *Animals* **2023**, *13*, 3723. [CrossRef]
108. Fang, Z.; Wang, Q.; Duan, H.; Sheng, X.; Qi, X.; Xing, K.; Liu, B.; Chang, D.; Guo, Y.; Wang, X.; et al. 17β-Estradiol Mediates TGFBR3/Smad2/3 Signaling to Attenuate the Fibrosis of TGF-β1-induced Bovine Endometrial Epithelial Cells via GPER. *J. Cell Physiol.* **2024**, *239*, 166–179. [CrossRef]
109. Ramos, I.S.; Caldeira, M.O.; Poock, S.E.; Moraes, J.G.N.; Lucy, M.C.; Patterson, A.L. Adenomyosis and Fibrosis Define the Morphological Memory of the Postpartum Uterus of Dairy Cows Previously Exposed to Metritis. *JDS Commun.* **2025**, *6*, 250–255. [CrossRef]

110. Wanjiru, D.K.; Niyonzima, Y.B.; Kadokawa, H. Lower Expression of Colony-Stimulating Factor 2, an Embryokine, in the Endometrial Epithelium of Old Cows. *Reprod. Fertil. Dev.* **2025**, *37*, RD24163. [CrossRef]
111. Ferdousy, R.N.; Kadokawa, H. Specific Locations and Amounts of Denatured Collagen and Collagen-Specific Chaperone HSP47 in the Oviducts and Uteri of Old Cows as Compared with Those of Heifers. *Reprod. Fertil. Dev.* **2022**, *34*, 619–632. [CrossRef]
112. Ferdousy, R.N.; Suong, N.T.; Kadokawa, H. Specific Locations and Amounts of Denatured Collagen and Collagen-Specific Chaperone HSP47 in the Uterine Cervices of Old Cows Compared with Those of Heifers. *Theriogenology* **2023**, *196*, 10–17. [CrossRef]
113. Ferdousy, R.N.; Kereilwe, O.; Kadokawa, H. Anti-Müllerian Hormone Receptor Type 2 (AMHR2) Expression in Bovine Oviducts and Endometria: Comparison of AMHR2 MRNA and Protein Abundance between Old Holstein and Young and Old Wagyu Females. *Reprod. Fertil. Dev.* **2020**, *32*, 738. [CrossRef]
114. Ferdousy, R.N.; Kadokawa, H. Anti-Müllerian Hormone Stimulates Expression of the Collagen-specific Chaperone 47-kDa Heat Shock Protein in Bovine Uterine Epithelial Cells. *Anim. Sci. J.* **2022**, *93*, e13787. [CrossRef] [PubMed]
115. Bogado Pascottini, O.; Hostens, M.; Dini, P.; Vandepitte, J.; Ducatelle, R.; Opsomer, G. Distribution of Inflammation and Association between Active and Chronic Alterations within the Endometrium of Dairy Cows. *Reprod. Domest. Anim.* **2016**, *51*, 751–757. [CrossRef] [PubMed]
116. Massoud Talebkhan, G.; Farhang, S.; Parviz, H. The Histopathological Survey of Uterine Tissue in Holstein Dairy Cows with or without Recorded Reproductive Disorders. *Iran. J. Vet. Sci. Technol.* **2010**, *2*, 100–108.
117. Palomares, R.A.; Gutiérrez-Añez, J.C.; Zambrano, S.; Boscan-Ocando, J.C.; Montero, M.; Camacho, J.; Perea-Ganchou, F.P.; Ferrer, M.S.; Rodríguez-Márquez, J.M.; Portillo-Martinez, G.; et al. Chronic Inflammatory and Degenerative Endometrial Lesions in Subfertile Criollo Limonero Cattle; a B. Taurus Latin-American Breed Threatened with Extinction; A Case-Control Study. *Anim. Reprod. Sci.* **2018**, *197*, 22–32. [CrossRef]
118. Doğan, İ.; Sönmez, G.; Sağırkaya, H. Histopathological Investigation of Endometrium in Repeat Breeder Cows. *Indian J. Anim. Sci.* **2002**, *72*, 223–226.
119. Bonnett, B.N.; Miller, R.B.; Martin, S.W.; Etherington, W.G.; Buckrell, B.C. Correlations between Histological Criteria. *Can. J. Vet. Res.* **1991**, *2*, 162.
120. Hanada, M.; Maeda, Y.; Oikawa, M. Histopathological Characteristics of Endometrosis in Thoroughbred Mares in Japan: Results from 50 Necropsy Cases. *J. Equine Sci.* **2014**, *25*, 45–52. [CrossRef]
121. Walter, I.; Handler, J.; Reifinger, M.; Aurich, C. Association of Endometriosis in Horses with Differentiation of Periglandular Myofibroblasts and Changes of Extracellular Matrix Proteins. *Reproduction* **2001**, *121*, 581–586. [CrossRef]
122. Rebordão, M.; Galvão, A.; Szóstek, A.; Amaral, A.; Mateus, L.; Skarzynski, D.; Ferreira-Dias, G. Physiopathologic Mechanisms Involved in Mare Endometrosis. *Reprod. Domest. Anim.* **2014**, *49*, 82–87. [CrossRef] [PubMed]
123. Bai, X.; Liu, J.; Cao, S.; Wang, L. A Mechanisms of Endometrial Fibrosis and the Potential Application of Stem Cell Therapy. *Discov. Med.* **2019**, *27*, 267–279.
124. Zhu, H.-Y.; Ge, T.-X.; Pan, Y.-B.; Zhang, S.-Y. Advanced Role of Hippo Signaling in Endometrial Fibrosis. *Chin. Med. J.* **2017**, *130*, 2732–2737. [CrossRef]
125. Fujisawa, C.; Castellot, J.J. Matrix Production and Remodeling as Therapeutic Targets for Uterine Leiomyoma. *J. Cell Commun. Signal* **2014**, *8*, 179–194. [CrossRef] [PubMed]
126. Heldin, C.-H.; Moustakas, A. Signaling Receptors for TGF-β Family Members. *Cold Spring Harb. Perspect. Biol.* **2016**, *8*, a022053. [CrossRef]
127. Fu, M.; Hu, Y.; Lan, T.; Guan, K.-L.; Luo, T.; Luo, M. The Hippo Signalling Pathway and Its Implications in Human Health and Diseases. *Signal Transduct. Target. Ther.* **2022**, *7*, 376. [CrossRef] [PubMed]
128. Yarmolinskaya, M.I.; Polenov, N.I.; Kunitsa, V.V. Uterine Fibroids: The Role of Signaling Pathways in the Pathogenesis. A Literature Review. *J. Obstet. Women's Dis.* **2020**, *69*, 113–124. [CrossRef]
129. El Sabeh, M.; Afrin, S.; Miyashita-Ishiwata, M.; Borahay, M.A. TGF-B signaling regulates wnt/β-catenin pathway in uterine leiomyoma cells. *Fertil. Steril.* **2021**, *116*, e414. [CrossRef]
130. Zhao, G.; Li, R.; Cao, Y.; Song, M.; Jiang, P.; Wu, Q.; Zhou, Z.; Zhu, H.; Wang, H.; Dai, C.; et al. ΔNp63α-Induced DUSP4/GSK3β/SNAI1 Pathway in Epithelial Cells Drives Endometrial Fibrosis. *Cell Death Dis.* **2020**, *11*, 449. [CrossRef]
131. Swanepoel, A.C.; Lindeque, B.G.; Swart, P.J.; Abdool, Z.; Pretorius, E. Estrogen Causes Ultrastructural Changes of Fibrin Networks during the Menstrual Cycle: A Qualitative Investigation. *Microsc. Res. Tech.* **2014**, *77*, 594–601. [CrossRef]
132. Coleman, J.R.; Moore, E.E.; Schmitt, L.; Hansen, K.; Dow, N.; Freeman, K.; Cohen, M.J.; Silliman, C.C. Estradiol Provokes Hypercoagulability and Affects Fibrin Biology: A Mechanistic Exploration of Sex Dimorphisms in Coagulation. *J. Trauma. Acute Care Surg.* **2023**, *94*, 179–186. [CrossRef] [PubMed]
133. Borahay, M.A.; Asoglu, M.R.; Mas, A.; Adam, S.; Kilic, G.S.; Al-Hendy, A. Estrogen Receptors and Signaling in Fibroids: Role in Pathobiology and Therapeutic Implications. *Reprod. Sci.* **2017**, *24*, 1235–1244. [CrossRef]

134. Ito, I.; Hanyu, A.; Wayama, M.; Goto, N.; Katsuno, Y.; Kawasaki, S.; Nakajima, Y.; Kajiro, M.; Komatsu, Y.; Fujimura, A.; et al. Estrogen Inhibits Transforming Growth Factor β Signaling by Promoting Smad2/3 Degradation. *J. Biol. Chem.* **2010**, *285*, 14747–14755. [CrossRef]
135. Xu, Q.; Zhang, W.; Liu, X.; Yan, W.; Lu, L.; Song, S.; Wei, S.; Liu, Y.; Kang, J.; Su, R. Notch1 Signaling Enhances Collagen Expression and Fibrosis in Mouse Uterus. *BioFactors* **2021**, *47*, 852–864. [CrossRef]
136. Protic, O.; Islam, M.; Greco, S.; Giannubilo, S.; Lamanna, P.; Petraglia, F.; Ciavattini, A.; Castellucci, M.; Hinz, B.; Ciarmela, P. Activin A in Inflammation, Tissue Repair, and Fibrosis: Possible Role as Inflammatory and Fibrotic Mediator of Uterine Fibroid Development and Growth. *Semin. Reprod. Med.* **2017**, *35*, 499–509. [CrossRef]
137. Kendall, R.T.; Feghali-Bostwick, C.A. Fibroblasts in Fibrosis: Novel Roles and Mediators. *Front. Pharmacol.* **2014**, *5*, 123. [CrossRef] [PubMed]
138. van Nieuwenhoven, F.A.; Turner, N.A. The Role of Cardiac Fibroblasts in the Transition from Inflammation to Fibrosis Following Myocardial Infarction. *Vasc. Pharmacol.* **2013**, *58*, 182–188. [CrossRef] [PubMed]
139. Turner, J.D.; Naylor, A.J.; Buckley, C.; Filer, A.; Tak, P.-P. Fibroblasts and Osteoblasts in Inflammation and Bone Damage. In *Stromal Immunology. Advances in Experimental Medicine and Biology*; Springer: Cham, Switzerland, 2018; Volume 1060, pp. 37–54.
140. Zhu, S.; Wang, A.; Xu, W.; Hu, L.; Sun, J.; Wang, X. The Heterogeneity of Fibrosis and Angiogenesis in Endometriosis Revealed by Single-Cell RNA-Sequencing. *Biochim. et Biophys. Acta (BBA)-Mol. Basis Dis.* **2023**, *1869*, 166602. [CrossRef] [PubMed]
141. Feng, L.; Jayes, F.; Johnson, L.; Schomberg, D.; Leppert, P. Biochemical Pathways and Myometrial Cell Differentiation Leading to Nodule Formation Containing Collagen and Fibronectin. *Curr. Protein Pept. Sci.* **2016**, *18*, 155–166. [CrossRef]
142. Barragan, F.; Irwin, J.C.; Balayan, S.; Erikson, D.W.; Chen, J.C.; Houshdaran, S.; Piltonen, T.T.; Spitzer, T.L.B.; George, A.; Rabban, J.T.; et al. Human Endometrial Fibroblasts Derived from Mesenchymal Progenitors Inherit Progesterone Resistance and Acquire an Inflammatory Phenotype in the Endometrial Niche in Endometriosis1. *Biol. Reprod.* **2016**, *94*, 118. [CrossRef]
143. Ma, S.; Ji, Z.; Zhang, B.; Geng, L.; Cai, Y.; Nie, C.; Li, J.; Zuo, Y.; Sun, Y.; Xu, G.; et al. Spatial Transcriptomic Landscape Unveils Immunoglobin-Associated Senescence as a Hallmark of Aging. *Cell* **2024**, *187*, 7025–7044.e34. [CrossRef] [PubMed]
144. Yu, L.; Wan, Q.; Liu, Q.; Fan, Y.; Zhou, Q.; Skowronski, A.A.; Wang, S.; Shao, Z.; Liao, C.-Y.; Ding, L.; et al. IgG Is an Aging Factor That Drives Adipose Tissue Fibrosis and Metabolic Decline. *Cell Metab.* **2024**, *36*, 793–807.e5. [CrossRef]
145. Levi, M.; Dik, K. Fibrophilia: A New Disease Entity? *J. Thromb. Haemost.* **2008**, *6*, 1334–1335. [CrossRef]
146. Hattori, N.; Degen, J.L.; Sisson, T.H.; Liu, H.; Moore, B.B.; Pandrangi, R.G.; Simon, R.H.; Drew, A.F. Bleomycin-Induced Pulmonary Fibrosis in Fibrinogen-Null Mice. *J. Clin. Investig.* **2000**, *106*, 1341–1350. [CrossRef] [PubMed]
147. Wilberding, J.A.; Ploplis, V.A.; McLennan, L.; Liang, Z.; Cornelissen, I.; Feldman, M.; Deford, M.E.; Rosen, E.D.; Castellino, F.J. Development of Pulmonary Fibrosis in Fibrinogen-Deficient Mice. *Ann. N. Y. Acad. Sci.* **2001**, *936*, 542–548. [CrossRef] [PubMed]
148. Shea, B.S.; Probst, C.K.; Brazee, P.L.; Rotile, N.J.; Blasi, F.; Weinreb, P.H.; Black, K.E.; Sosnovik, D.E.; Van Cott, E.M.; Violette, S.M.; et al. Uncoupling of the Profibrotic and Hemostatic Effects of Thrombin in Lung Fibrosis. *JCI Insight* **2017**, *2*, e86608. [CrossRef]
149. Loskutoff, D.J.; Quigley, J.P. PAI-1, Fibrosis, and the Elusive Provisional Fibrin Matrix. *J. Clin. Investig.* **2000**, *106*, 1441–1443. [CrossRef]
150. Ruppert, C.; Markart, P.; Wygrecka, M.; Preissner, K.T.; Günther, A. Role of Coagulation and Fibrinolysis in Lung and Renal Fibrosis. *Hamostaseologie* **2008**, *28*, 30–36. [CrossRef]
151. Schuliga, M.; Grainge, C.; Westall, G.; Knight, D. The Fibrogenic Actions of the Coagulant and Plasminogen Activation Systems in Pulmonary Fibrosis. *Int. J. Biochem. Cell Biol.* **2018**, *97*, 108–117. [CrossRef]
152. Zhang, X.; Zhang, C.; Shen, S.; Xia, Y.J.; Yi, L.; Gao, Q.; Wang, Y. Dehydroepiandrosterone Induces Ovarian and Uterine Hyperfibrosis in Female Rats. *Hum. Reprod.* **2013**, *28*, 3074–3085. [CrossRef]
153. Matsuno, Y.; Amin, Y.A.; Kusama, K.; Imakawa, K. Formation of Fibrin at Sites of Conceptus Adhesion in the Ewe. *Reproduction* **2021**, *161*, 709–720. [CrossRef]
154. Yamada, O.; Todoroki, J.; Takahashi, T.; Hashizume, K. The Dynamic Expression of Extracellular Matrix in the Bovine Endometrium at Implantation. *J. Vet. Med. Sci.* **2002**, *64*, 207–214. [CrossRef] [PubMed]
155. Song, G.; Spencer, T.E.; Bazer, F.W. Cathepsins in the Ovine Uterus: Regulation by Pregnancy, Progesterone, and Interferon Tau. *Endocrinology* **2005**, *146*, 4825–4833. [CrossRef]
156. Salamonsen, L.; Nagase, H.; Woolley, D. Matrix Metalloproteinases and Their Tissue Inhibitors at the Ovine Trophoblast-Uterine Interface. *Biosci. Proc.* **2019**, *3*, 29–37. [CrossRef]
157. Egashira, M.; Hirota, Y.; Shimizu-Hirota, R.; Saito-Fujita, T.; Haraguchi, H.; Matsumoto, L.; Matsuo, M.; Hiraoka, T.; Tanaka, T.; Akaeda, S.; et al. F4/80+ Macrophages Contribute to Clearance of Senescent Cells in the Mouse Postpartum Uterus. *Endocrinology* **2017**, *158*, 2344–2353. [CrossRef]
158. Secomandi, L.; Borghesan, M.; Velarde, M.; Demaria, M. The Role of Cellular Senescence in Female Reproductive Aging and the Potential for Senotherapeutic Interventions. *Hum. Reprod. Update* **2022**, *28*, 172–189. [CrossRef] [PubMed]
159. Yang, D.; Sun, B.; Li, S.; Wei, W.; Liu, X.; Cui, X.; Zhang, X.; Liu, N.; Yan, L.; Deng, Y.; et al. NKG2D-CAR T Cells Eliminate Senescent Cells in Aged Mice and Nonhuman Primates. *Sci. Transl. Med.* **2023**, *15*, eadd1951. [CrossRef] [PubMed]

160. Hubackova, S.; Davidova, E.; Rohlenova, K.; Stursa, J.; Werner, L.; Andera, L.; Dong, L.; Terp, M.G.; Hodny, Z.; Ditzel, H.J.; et al. Selective Elimination of Senescent Cells by Mitochondrial Targeting Is Regulated by ANT2. *Cell Death Differ.* **2019**, *26*, 276–290. [CrossRef]
161. Yosef, R.; Pilpel, N.; Tokarsky-Amiel, R.; Biran, A.; Ovadya, Y.; Cohen, S.; Vadai, E.; Dassa, L.; Shahar, E.; Condiotti, R.; et al. Directed Elimination of Senescent Cells by Inhibition of BCL-W and BCL-XL. *Nat. Commun.* **2016**, *7*, 11190. [CrossRef]
162. Katsuumi, G.; Shimizu, I.; Suda, M.; Yoshida, Y.; Furihata, T.; Joki, Y.; Hsiao, C.-L.; Jiaqi, L.; Fujiki, S.; Abe, M.; et al. SGLT2 Inhibition Eliminates Senescent Cells and Alleviates Pathological Aging. *Nat. Aging* **2024**, *4*, 926–938. [CrossRef]
163. Katsuumi, G.; Suda, M.; Shimizu, I.; Yoshida, Y.; Furihata, T.; Joki, Y.; Hsiao, C.L.L.; Minamino, T. Abstract 14285: Sodium Glucose Co-Transporter 2 Inhibition Eliminates Senescent Cells Through Downregulation of PD-L1 and Improves Pathological Aging. *Circulation* **2023**, *148*, A14285. [CrossRef]
164. Guo, G.; Amor, C. SGLT2 Regulates Immune-Mediated Senolysis. *Nat. Aging* **2024**, *4*, 909–910. [CrossRef] [PubMed]
165. Kim, M.N.; Moon, J.H.; Cho, Y.M. Sodium-glucose Cotransporter-2 Inhibition Reduces Cellular Senescence in the Diabetic Kidney by Promoting Ketone Body-induced NRF2 Activation. *Diabetes Obes. Metab.* **2021**, *23*, 2561–2571. [CrossRef] [PubMed]
166. Wang, G.; Zhu, Y.; Feng, D.; Yao, J.; Cao, Y.; Deng, L. Hepatic Gluconeogenesis and Regulatory Mechanisms in Lactating Ruminants: A Literature Review. *Anim. Res. One Health* **2024**, 1–10. [CrossRef]
167. Kirkland, J.L.; Tchkonia, T.; Zhu, Y.; Niedernhofer, L.J.; Robbins, P.D. The Clinical Potential of Senolytic Drugs. *J. Am. Geriatr. Soc.* **2017**, *65*, 2297–2301. [CrossRef]
168. Cavalcante, M.B.; Saccon, T.D.; Nunes, A.D.C.; Kirkland, J.L.; Tchkonia, T.; Schneider, A.; Masternak, M.M. Dasatinib plus Quercetin Prevents Uterine Age-Related Dysfunction and Fibrosis in Mice. *Aging* **2020**, *12*, 2711–2722. [CrossRef]
169. Kusama, K.; Yamauchi, N.; Yoshida, K.; Azumi, M.; Yoshie, M.; Tamura, K. Senolytic Treatment Modulates Decidualization in Human Endometrial Stromal Cells. *Biochem. Biophys. Res. Commun.* **2021**, *571*, 174–180. [CrossRef]
170. Luo, M.; Cai, X.; Yan, D.; Liu, X.; Guo, S.-W. Sodium Tanshinone IIA Sulfonate Restrains Fibrogenesis through Induction of Senescence in Mice with Induced Deep Endometriosis. *Reprod. Biomed. Online* **2020**, *41*, 373–384. [CrossRef]
171. Gasek, N.S.; Yan, P.; Zhu, J.; Purushothaman, K.-R.; Kim, T.; Wang, L.; Wang, B.; Flynn, W.F.; Sun, M.; Guo, C.; et al. Clearance of P21 Highly Expressing Senescent Cells Accelerates Cutaneous Wound Healing. *Nat. Aging* **2024**, *5*, 21–27. [CrossRef]
172. Cohn, R.L.; Gasek, N.S.; Kuchel, G.A.; Xu, M. The Heterogeneity of Cellular Senescence: Insights at the Single-Cell Level. *Trends Cell Biol.* **2023**, *33*, 9–17. [CrossRef]
173. Glasser, S.W.; Hagood, J.S.; Wong, S.; Taype, C.A.; Madala, S.K.; Hardie, W.D. Mechanisms of Lung Fibrosis Resolution. *Am. J. Pathol.* **2016**, *186*, 1066–1077. [CrossRef] [PubMed]
174. Caligiuri, A.; Gentilini, A.; Pastore, M.; Gitto, S.; Marra, F. Cellular and Molecular Mechanisms Underlying Liver Fibrosis Regression. *Cells* **2021**, *10*, 2759. [CrossRef]
175. Iredale, J.P.; Benyon, R.C.; Pickering, J.; McCullen, M.; Northrop, M.; Pawley, S.; Hovell, C.; Arthur, M.J. Mechanisms of Spontaneous Resolution of Rat Liver Fibrosis. Hepatic Stellate Cell Apoptosis and Reduced Hepatic Expression of Metalloproteinase Inhibitors. *J. Clin. Investig.* **1998**, *102*, 538–549. [CrossRef] [PubMed]
176. Ramachandran, P.; Iredale, J.; Fallowfield, J. Resolution of Liver Fibrosis: Basic Mechanisms and Clinical Relevance. *Semin. Liver Dis.* **2015**, *35*, 119–131. [CrossRef] [PubMed]
177. Pei, Q.; Yi, Q.; Tang, L. Liver Fibrosis Resolution: From Molecular Mechanisms to Therapeutic Opportunities. *Int. J. Mol. Sci.* **2023**, *24*, 9671. [CrossRef]
178. Ward-Caviness, C.K.; Huffman, J.E.; Everett, K.; Germain, M.; van Dongen, J.; Hill, W.D.; Jhun, M.A.; Brody, J.A.; Ghanbari, M.; Du, L.; et al. DNA Methylation Age Is Associated with an Altered Hemostatic Profile in a Multiethnic Meta-Analysis. *Blood* **2018**, *132*, 1842–1850. [CrossRef]
179. Chamani, I.J.; Keefe, D.L. Epigenetics and Female Reproductive Aging. *Front. Endocrinol.* **2019**, *10*, 473. [CrossRef]
180. Li Piani, L.; Vigano', P.; Somigliana, E. Epigenetic Clocks and Female Fertility Timeline: A New Approach to an Old Issue? *Front. Cell Dev. Biol.* **2023**, *11*, 1121231. [CrossRef]
181. Dai, Z.; Wu, Z.; Hang, S.; Zhu, W.; Wu, G. Amino Acid Metabolism in Intestinal Bacteria and Its Potential Implications for Mammalian Reproduction. *Mol. Hum. Reprod.* **2015**, *21*, 389–409. [CrossRef]
182. Van Tran, L.; Malla, B.A.; Kumar, S.; Tyagi, A.K. Polyunsaturated Fatty Acids in Male Ruminant Reproduction—A Review. *Asian-Australas. J. Anim. Sci.* **2016**, *30*, 622–637. [CrossRef]
183. Genís, S.; Bach, À.; Fàbregas, F.; Arís, A. Potential of Lactic Acid Bacteria at Regulating Escherichia Coli Infection and Inflammation of Bovine Endometrium. *Theriogenology* **2016**, *85*, 625–637. [CrossRef] [PubMed]
184. Adnane, M.; Whiston, R.; Tasara, T.; Bleul, U.; Chapwanya, A. Harnessing Vaginal Probiotics for Enhanced Management of Uterine Disease and Reproductive Performance in Dairy Cows: A Conceptual Review. *Animals* **2024**, *14*, 1073. [CrossRef] [PubMed]

185. Peter, S.; Gärtner, M.A.; Michel, G.; Ibrahim, M.; Klopfleisch, R.; Lübke-Becker, A.; Jung, M.; Einspanier, R.; Gabler, C. Influence of Intrauterine Administration of Lactobacillus Buchneri on Reproductive Performance and Pro-Inflammatory Endometrial MRNA Expression of Cows with Subclinical Endometritis. *Sci. Rep.* **2018**, *8*, 5473. [CrossRef] [PubMed]
186. Deng, Q.; Odhiambo, J.F.; Farooq, U.; Lam, T.; Dunn, S.M.; Ametaj, B.N. Intravaginal Lactic Acid Bacteria Modulated Local and Systemic Immune Responses and Lowered the Incidence of Uterine Infections in Periparturient Dairy Cows. *PLoS ONE* **2015**, *10*, e0124167. [CrossRef]
187. Li, Z.; Teng, Y.; Feng, S.; Hu, Z.; Zhao, J.; Ding, H.; Fang, Y.; Liu, H.; Ma, X.; Guo, J.; et al. Microbial Responses and Changes in Metabolic Products in Bovine Uteri Infected with Staphylococcus Aureus. *Int. J. Biol. Macromol.* **2024**, *262*, 130039. [CrossRef]
188. Nørstebø, S.F.; Rodriguez-Campos, S.; Umu, Ö.C.O.; Abril-Parreño, L.; Dalland, M.; Gilfillan, G.D.; Fair, S.; Krogenaes, A. The Cervical Microbiome of Ewe Breeds with Known Divergent Fertility Following Artificial Insemination with Frozen-Thawed Semen. *Sci. Rep.* **2025**, *15*, 14614. [CrossRef]

cells MDPI

Article

Somatic *MED12* Mutations in Myometrial Cells

Yinuo Li [1], Huma Asif [2], Yue Feng [1], Julie J. Kim [2,*] and Jian-Jun Wei [1,2,*]

[1] Department of Pathology, Northwestern University, Chicago, IL 60611, USA; sd09lyn@163.com (Y.L.); yue-feng@northwestern.edu (Y.F.)
[2] Department of Obstetrics and Gynecology, Northwestern University, Chicago, IL 60611, USA; huma.asif@northwestern.edu
* Correspondence: j-kim4@northwestern.edu (J.J.K.); jianjun-wei@northwestern.edu (J.-J.W.); Tel.: +1-312-503-5377 (J.J.K.); +1-312-926-1815 (J.-J.W.)

Abstract: Over 70% of leiomyoma (LM) harbor *MED12* mutations, primarily in exon 2 at c.130-131 (GG). Myometrial cells are the cell origin of leiomyoma, but the MED12 mutation status in non-neoplastic myometrial cells is unknown. In this study, we investigated the mutation burden of *MED12* in myometrium. As traditional Sanger or even NGS sequencing may not be able to detect *MED12* mutations that are lower than 0.1% in the testing sample, we used duplex deep sequencing analysis (DDS) to overcome this limitation. Tumor-free myometria (confirmed by pathology evaluation) were dissected, and genomic DNA from *MED12* exon 2 (test) and TP53 exon 5 (control) were captured by customer-designed probe sets, followed by DDS. Notably, DDS demonstrated that myometrial cells harbored a high frequency of mutations in *MED12* exon 2 and predominantly in code c.130-131. In contrast, the baseline mutations in other coding sequences of *MED12* exon 2 as well as in the TP53 mutation hotspot, c.477-488 were comparably low in myometrial cells. This is the first report demonstrating a non-random accumulation of *MED12* mutations at c.130-131 sites in non-neoplastic myometrial cells which provide molecular evidence of early somatic mutation events in myometrial cells. This early mutation may contribute to the cell origin for uterine LM development in women of reproductive age.

Keywords: myometrium; duplex deep sequencing; MED12 mutation; leiomyoma

1. Introduction

Uterine leiomyomas (LM) occur in up to 77% of women of reproductive age [1,2]. It was until recently that the major driving genes for LM were discovered by whole genomic sequencing analysis. *MED12* (mediator complex subunit 12) is among the most common mutations found in >70% of uterine LM [3,4]. MED12 encodes a large gene with a total of 45 exons and *MED12* mutations in LM are exclusively confined in exon 2 and the adjacent intron 1–exon 2 boundary. The dominant hotspots are at codon 44 c.130G-131G [3]. *MED12* mutations are correlated with the number of LM in the uterus [5,6]. *MED12* mutations were found only in a few mesenchymal tumors, including LM and fibroadenoma of the breast [7], suggesting the selective mechanisms of MED12 in tumorigenesis of certain cell types.

To date, the molecular cause of such prevalent *MED12* mutations remains unknown. Previously, we have demonstrated that *MED12* mutations (predominantly at c.130-131 of exon 2) could be induced in myometrial cells with chronic exposure to reactive oxygen species (ROS) in vitro, suggesting that a mutation in this region could be an early event for the initiation of LM development [8]. The uterus is exposed to high levels of reactive oxidative species (ROS), mediated by cycle steroid hormone changes, hypoxic environment, and ROS-related tissue functions [9–12]. Moreover, ROS is a potent carcinogenic factor that induces DNA damage and mutations [13,14], with the accumulation of oxidative nucleotides in genomic DNA [15,16], dominantly guanine 8-OHdG (8-hydroxy-2′–deoxyguanosine). A high ROS burden promotes 8-OHdG formation in nuclei [15,16]. 8-OHdG causes misrepair and results in G transversions. This misrepair will allow 8-OHdG repairing with dATP and produces a stable G to T transversion, which is the hallmark of oxidative mutagenesis [13,17]. Detection of 8-OHdG misrepair induced by ROS has been tested at specific genes as well as globally [18] such as those found for *K-Ras* and *p53* [13,14]. Thus the association of ROS, DNA damage, and *MED12* mutations needs further investigation.

It is unclear when mutations occur in LM or whether non-neoplastic myometrial cells in reproductive-aged women gain somatic gene mutations, particularly in the *MED12* gene. Unfortunately, in order to detect mutations in a small subset of myometrial cells, conventional next-generation sequencing is insufficient in sensitivity as it is unable to detect the mutations at a frequency of <1% and almost impossible at <0.1% [19]. Deep sequencing analysis is a useful technique that can be used to detect extremely low-frequency mutations of individual cells in a large non-mutant cell population [20]. However, deep sequencing may yield a false positive error rate due to multiple steps of DNA preparation, PCR amplification, and sequencing [21]. The development of duplex deep sequencing (DDS) has enabled superior sensitivity and accuracy as it simultaneously sequences both strands of DNA and avoid errors inherent of the procedure [22,23].

In this study, we investigate *MED12* mutations in normal myometrial cells from hysterectomy specimens as an early genetic mutational event using DDS. For comparison purposes, *TP53* exon 5 was sequenced to understand the overall misrepair/mutation burden in the myometrium.

2. Materials and Methods

2.1. Study Design and Case Selection

Uterine tissues were collected from pre- and peri-menopausal women with hysterectomy at Northwestern University Prentice Women's Hospital. Fresh tissues of LM and matched tumor-free myometrium from a total of 32 women were collected after informed consent. Histologic evaluation was performed by the pathologist to ensure the myometrium was absent of microscopic LM tumors. The dominant LM was also collected for *MED12* mutation analysis. Patient age at surgery, race, uterine weight, and number of tumors were documented in Table 1. This study obtained approval from the Northwestern Institutional Review Board (IRB) (STU00212471).

Table 1. General information about the cases used in this study.

	Black	White	Total
No. of Cases	16	13	29
Age	43.12 ± 1.79	45.54 ± 0.97	44.31 ± 1.08
>5	56.25% (9/16)	30.77% (4/13)	44.83% (13/29)
<5	43.75% (7/16)	69.23% (9/13)	55.17% (16/29)
MED12 Mut in LM	87.50% (14/16)	69.23% (9/13)	79.31% (23/29)
MED12 Mut in MM	50.00% (8/16)	76.92% (10/13)	62.07% (18/29)

2.2. Histologic Evaluation Criteria

Fresh tissue samples from an area of tumor-free myometrium were bisected. One-half of the tissue sample was snap-frozen for molecular testing, and the other half was processed for histologic evaluation. Routine hematoxylin and eosin-stained slides were prepared from formalin-fixed and paraffin-embedded (FFPE) tissue samples to ensure no LM contamination. All selected myometrial tissue samples for this study were LM-free at the microscopic level. The dominant LM from each case were also collected for histology evaluations, and they were all usual type LM.

2.3. DNA Extraction and Sanger Sequencing

DNA extraction was performed in fresh/frozen and FFPE tissues with aid of ZYMO RESEARCH Quick-DNA™ MiniPrep Plus Kit and ZYMO RESEARCH Quick-DNA™ FFPE Kit (ZYMO RESEARCH, D4068, D3067, lrvine, CA, USA), respectively, following the manufacturer's instruction. In brief, 10 μm FFPE tissue sections were collected into 1.5 mL Eppendorf tubes and deparaffinized. For fresh/frozen tissues, grinded tissue were digested with a proteinase K at 56 °C for 1–4 h followed by incubation at 90 °C for 20 min. DNA was purified by column purification and was quantitated using NanoDrop Spectrophotometer (Thermo fisher, Waltham, MA, USA) and DNA quality was further assessed by A260/A280 ratio and its integrity by 1% agarose gel electrophoresis. For the Sanger sequencing, *MED12* exon 2 was amplified by PCR with 50 ng of genomic DNA with primers 5′-ACAACTAAACGCCGCTTTCCT-3′ (forward) and 5′-GGGCCTTTGCTCCTTCTTAG-3′ (reverse). The purified PCR products by ExoSAP-IT (Affymetrix, Inc.) were sent to NU sequence core for Sanger DNA sequencing with Applied Biosystem's 3730xl DNA Analyzer (Torrance, CA, USA). Utations/variations were analyzed by software FinchTV 1.5 and Indigo (https://www.gear-genomics.com/indigo/).

DNA Library Preparation and Duplex Deep Sequencing

Genomic DNA from myometrium was prepared by the ZYMO RESEARCH Quick-DNA™ MiniPrep Plus Kit. In preparation for duplex sequencing [22,23], DNA probe sets for MED12 and p53 were purchased from IDT, which covered the entire target sequences for the regions of interest and captured *MED12* exon 2 and *P53* exon 5. The genomic DNA was sonicated, end-repaired, A-tailed, and ligated with UMI (IDT) and index adapters (IDT) using the KAPA HyperPrep library kit (Roche Sequencing, Basel, Switzerland). The DNA quality is shown in Figure S1. After genomic DNA were amplified by 5 PCR cycles, the biotinylated oligonucleotide probes (IDT) were used to capture the DNA fragments from *MED12* exon2, *p53* exon 5 with xGen® Lockdown® Reagents (IDT) following the manufacturer's instruction. After two rounds of target sequence captures, the sufficient target enrichment will be obtained. Then, the libraries of capture DNA fragments were prepared and proceeded to sequenced in 2 × 150 paired-end reads with eight-base indexing read on an Illumina NovaSeq 6000 (Illumina, San Diego, CA, USA).

For duplex sequencing, data generated from deep sequencing were processed based on the published method [23]. Briefly, FASTQ files were converted to unaligned BAM using Picard FastqToSam. Reads were grouped based on the unique tagged sequences to generate read families needed to make single-strand consensus sequences (SSCS) and duplex consensus sequences (DCS) using Unified Consensus Maker (UCM). Consensus sequences were aligned to the reference hg38 genome using Burrows-Wheeler aligner (BWA-mem) and variants calling was performed using VarDict (Java). The sequence was assembled into an error-corrected Duplex Consensus Sequence by Samtools-Mpileup and GATK4 (detail methods described [24]). Double-stranded molecular barcodes, which used to identify the specific sequence and reads derived from both strands of DNA molecule. Mutations were recorded when they were detected in both DNA strands. This would eliminate any errors that occurred in single strand. *MED12* and *p53* of each nucleotide mutation frequencies were calculated according to the identified mutations divided by the total number of Duplex nucleotides sequenced. The mutant allele frequency (MAF) was calculated based on the mutated Duplex bases divided by the total DS reads at a given nucleotide position [25].

2.4. Statistical Analysis

Statistical analysis was performed using GraphPad Prism software v9 (GraphPad Software). An unpaired *t*-test was applied to compare two groups. A paired t-test was applied for matched data from two groups. When necessary, the ordinary one-way ANOVA, Brown–Forsythe, and Welch ANOVA, or Kruskal–Wallis tests might be used for multiple group comparisons depending upon the distribution and variances of the data set. The mean ± SEM from three or more independent experiments were calculated. Data or comparison with p-value < 0.05 were considered statistically significant.

3. Results

3.1. Case Selection and Sanger Sequencing Analysis of MED12 in LM

A total of 32 cases for LM, including 16 black and 16 white age-matched women, from hysterectomies, were selected for this study. Patients' ages ranged from 27 to 55 years old. Overall, 56% of cases contained five or more LM and 44% were less than five LM. The dissected LM and matched myometrium were subjected to histologic evaluation to differentiate between LM and tumor-free myometrial tissues. Dissected tissue samples were snap-frozen for sequencing analysis. The general information of the selected cases for this study is summarized in Table 1. The dominant LM from each case was subjected to Sanger sequencing analysis for *MED12* mutation. Overall, 79% (23/29) of LM harbored *MED12* mutations (Table S1). This is consistent with the *MED12* mutation pattern found in most LM cohort studies.

3.2. Duplex Deep Sequencing of Myometrial Tissues

This study used Duplex deep sequencing technique (DDS) as it is more sensitive than traditional deep sequencing [22,23] by eliminating tens of thousands of erroneous mutations and capturing true point mutations on both DNA strands of the target genes (Figure S1B) [23]. Overall, the estimated error rate of DDS is <1 in 10 million [22], and this allows us to examine the low rate of mutations with extreme sensitivity and accuracy in myometrial tissue samples. When size-fractioned genomic DNA passed quality evaluation (Figure S1A), DNA libraries were prepared with adaptors specifically designed for DDS (Figure S1B). The target DNA components from coding regions exon 2 of *MED12* and exon 5 of *TP53* were then captured and subjected to DDS analysis. *TP53* exon 5 was used as a comparative internal DDS control.

After eliminating 1 sample that did not reach the passing scale of required sequence reads, 31 cases were subjected to the first round of DDS analysis (Figure 1A). Initial DDS mutation analysis revealed that two cases from myometrium had significantly higher mutation reads in comparison to the rest of the specimens (Figure 1B). The higher mutants from these two samples were from c.130-131 sites (Figure 1B). Further analysis revealed these two cases were obtained from myometrium with ≥5 LMs (Figure 1C). Since it is difficult to determine with absolute certainty whether these two were from potential LM contamination, to ensure the integrity of the DDS analysis, these two samples were eliminated from further analysis.

3.3. MED12 Mutation Analysis in Tumor-Free Myometrium

DDS covered DNA sequences in *MED12* exon 2 of nucleotides c.100 to c.204. The detected mutations had generally low frequency, with a mean of 0.81/10,000 reads/base in this region. Notably, there was a significantly different mutation distribution in exon 2, with the highest in c.130-131 (mean 1.24/10,000) and the lowest in other nucleotides (mean 0.34/10,000) (Figure 1F). As high as 62% of myometrial tissue (18/29) had detectable mutations at c.130-131 sites in the rate range of 0.18–1.76/1000 (Figure 1D). The mutation distributions in 29 cases across exon 2 were illustrated in Figure 1E. Apparently, c.130-131 enriched a high rate of mutations in tumor-free myometrium. Such a non-random distribution of specific nucleotide alterations was similar to our previous study of primary myometrial cells treated with chronic oxidative stress [8]. Findings strongly suggest that there is an unproportional high rate of *MED12* mutations in exon 2 code c.130-131 in myometrial cells.

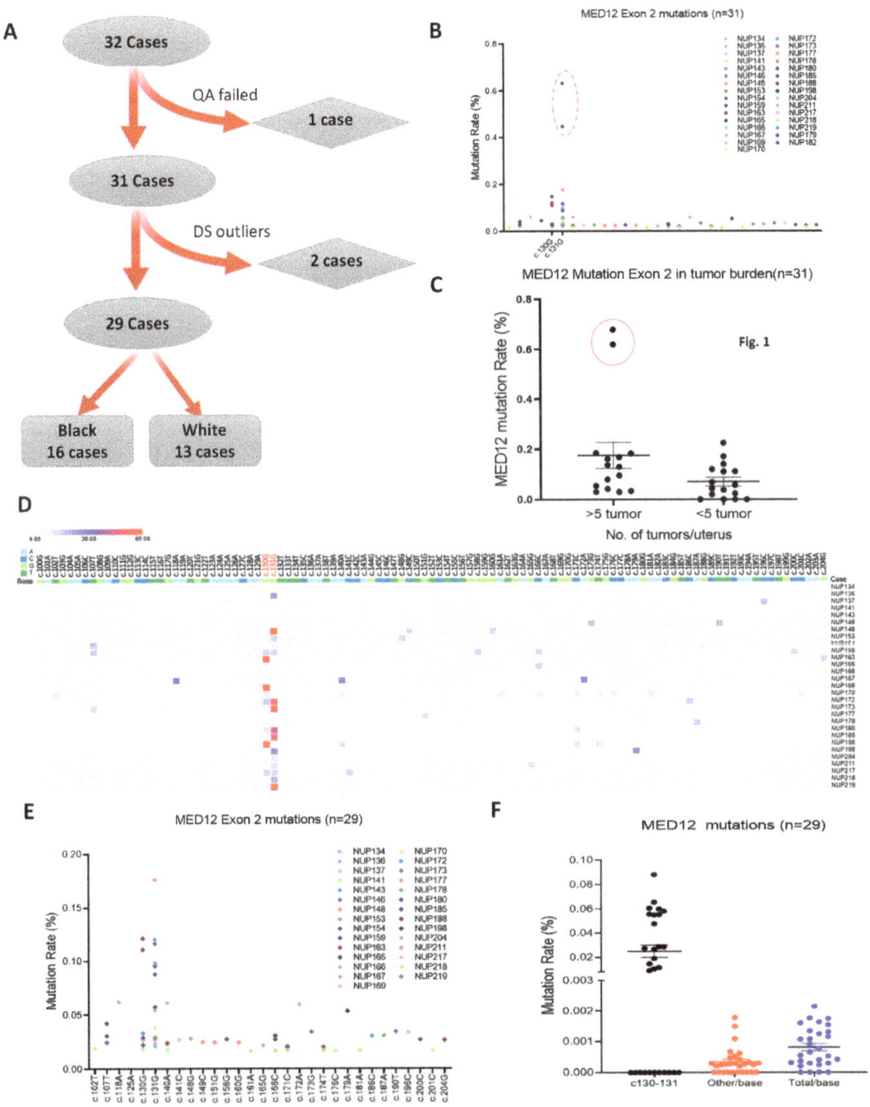

Figure 1. Point mutations of MED12 exon 2 detected by DDS in 29 cases. (**A**) A diagram illustrating the myometrial samples used for DDS. (**B**) Mutation plot showing the distribution of point mutations in exon 2 for 31 cases. Red circle highlights the 2 outliers. (**C**) Dot plot highlighting 2 outliers in the myometrium of 2 cases with >5 LM. (**D**) Mutation plot shows exon 2 mutation distribution in 29 cases by excluding 2 outliers. (**E**) Two-dimensional map illustrating the distribution of MED12 exon 2 mutations in each of 29 cases. Mutation frequency is marked from blue (low) to high (red). (**F**) Dot plot illustrating the MED12 mutations in c.130-131 and other sites.

3.4. MED12 Mutations in Association with Pathological Parameters

To investigate the *MED12* mutation in association with the risk factors for LM development, we compared the mutation rate with several clinical and pathologic factors. *MED12* mutations of exon 2 and specifically c.130-131 in the myometrium of White and Black patients showed no significant difference (Figure 2A,B). Although there was a trend of higher *MED12* mutations in myometrium with ≥5 LM than with <5 LM, no statistical significance was reached (Figure 2C). The myometrium tissues were collected from patients with hysterectomies with ages ranging from 27 to 55 (Tables 1 and S1). Mutations can be detected in all tissues of all ages (Figure 2D). When comparing the total mutations with age and c.130-131 with age, there was a slight increase in mutation rate in older age but did not reach statistical significance ($p > 0.05$, Figure 2E). The findings suggest that the mutation burdens in myometrial cells are similar in different races, ages, and numbers of tumors in this cohort of tissue samples.

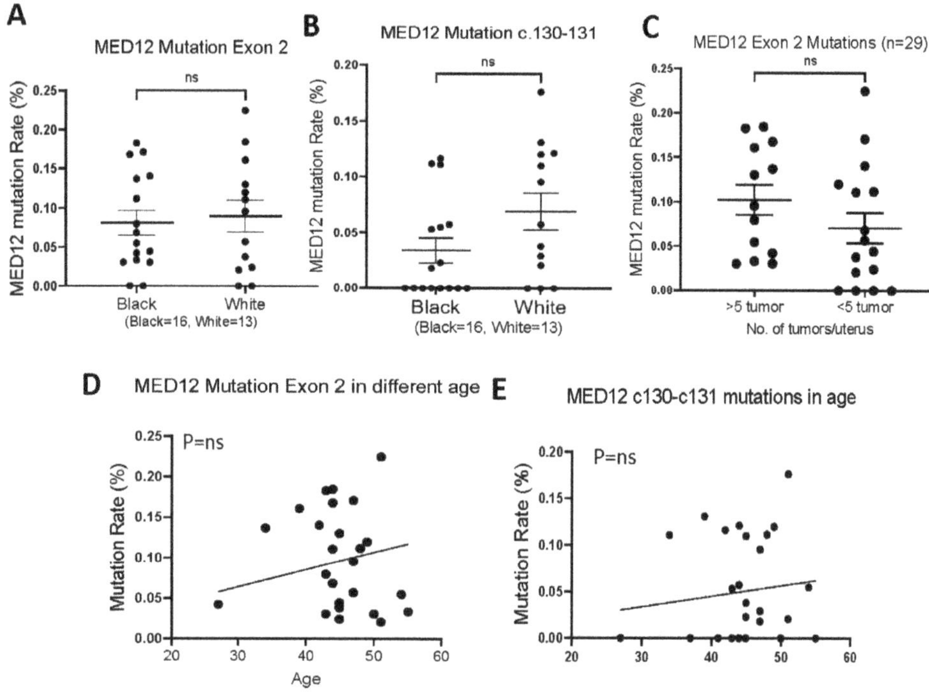

Figure 2. Mutation analysis of *MED12* exon 2 in association with race, number of tumors and age. (**A**,**B**) Mutation rate of *MED12* exon 2 (**A**) and c.130-131 (**B**) in myometrium of black (n = 16) and white (n = 13) women. (**C**) Mutation rate of MED12 exon 2 in myometrium with ≥5 and <5 LM. (**D**,**E**). Mutation rate of *MED12* exon 2 (**D**) and c.130-c.131 (**E**) in association with patient's age. ns: no statistical significance.

3.5. MED12 Mutational Signature Analysis

The spectrum of base changes identified in myometrial cells of 29 cases was further analyzed for mutational signatures from the matrix of mutation counts across each tissue sample. Among 8 mutational signatures, 3 signatures of single base pair change were identified (Figure 3A). They were represented by multiple G > A, G > C, and G > T, with higher peaks at c.131 site, followed by c.130 and rarely at other sites (Figure 3A). Overall, mutations were highly enriched in G, accounting for 66%, followed by A (14%), C (12%), and T (8%) (Figure 3B,C). The mutational signature showed its unique set of peaks, which were similar but remained different from those of LM signatures [8]. The specific G transversion and conversion ratio are summarized in Figure 3D. The high mutation signatures of G presented in myometrial cells suggest that mutation selection was non-random among different nucleotides, and such a signature pattern may be highly relevant to oxidative stress-induced misrepair, previously identified in our in vitro study [8]. Furthermore, high mutations of G enriched in c.130-131 indicate the mutation selections in myometrial cells.

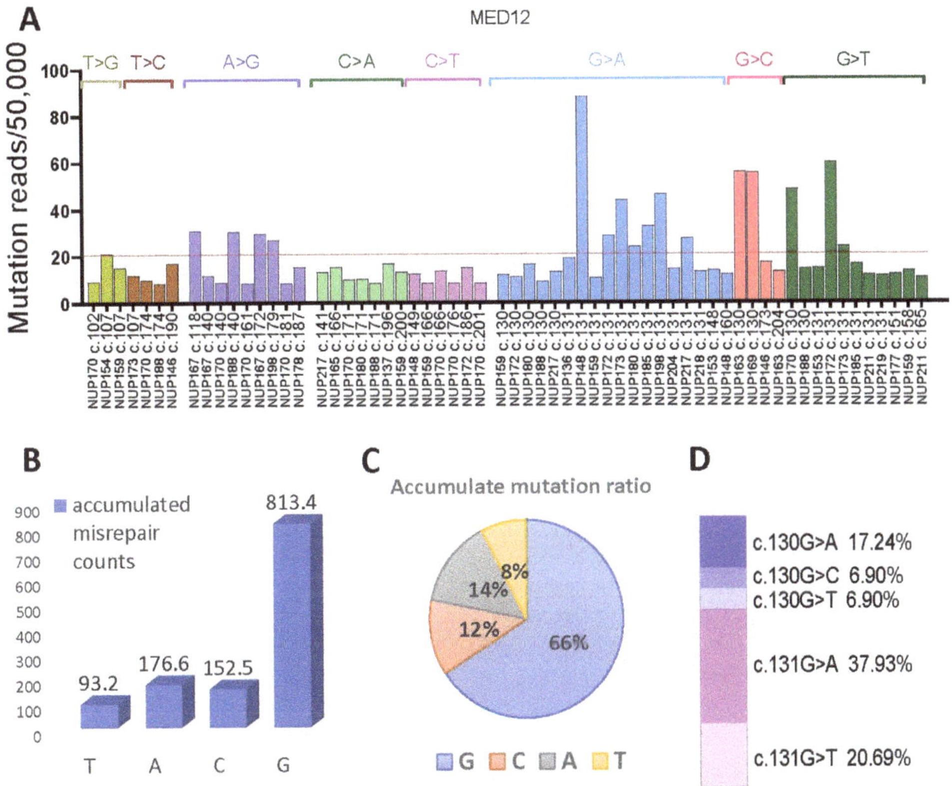

Figure 3. *MED12* exon 2 mutation signature analysis. (**A**) Distribution of mutation frequency in each of the 8 mutation signatures in 29 cases. (**B**,**C**) Accumulated point mutations and percentage in G, C, A, and T nucleotides. (**D**) Mutation rate of G > A, C, T in c.130-c.131.

3.6. P53 Mutations in Exon 5 Detected by DDS

The high rate of *MED12* exon 2 mutations found in myometrial cells was notable. To determine the general burden of single base misrepair or mutations in myometrial cells, we included *P53* exon 5 for mutation analysis by DDS. *P53* exon 5 harbors a high rate of mutations in Mullerian carcinoma in comparison to other exons [14,26] and it is thus a good target sequence for comparison purposes. As illustrated in Figure 4A, the rate of single nucleotide mutations in *TP53* was observed as a mean mutation rate of 0.57/100,000/base which was low. Notably, there were frequent somatic mutations enriched in c.477-c.478 and found in 11/29 cases of myometrium. There were 7 mutation signatures identified (Figure 4B). The dominant signatures were C > A and A > C enriched in c.477-c.478, accounting for over 65% of mutations (Figure 4C,D). Such dominant point mutations of A in *TP53* exon 5 and of G in *MED12* strongly suggest that despite the same genotoxic conditions, the types of mutations can vary for different genes.

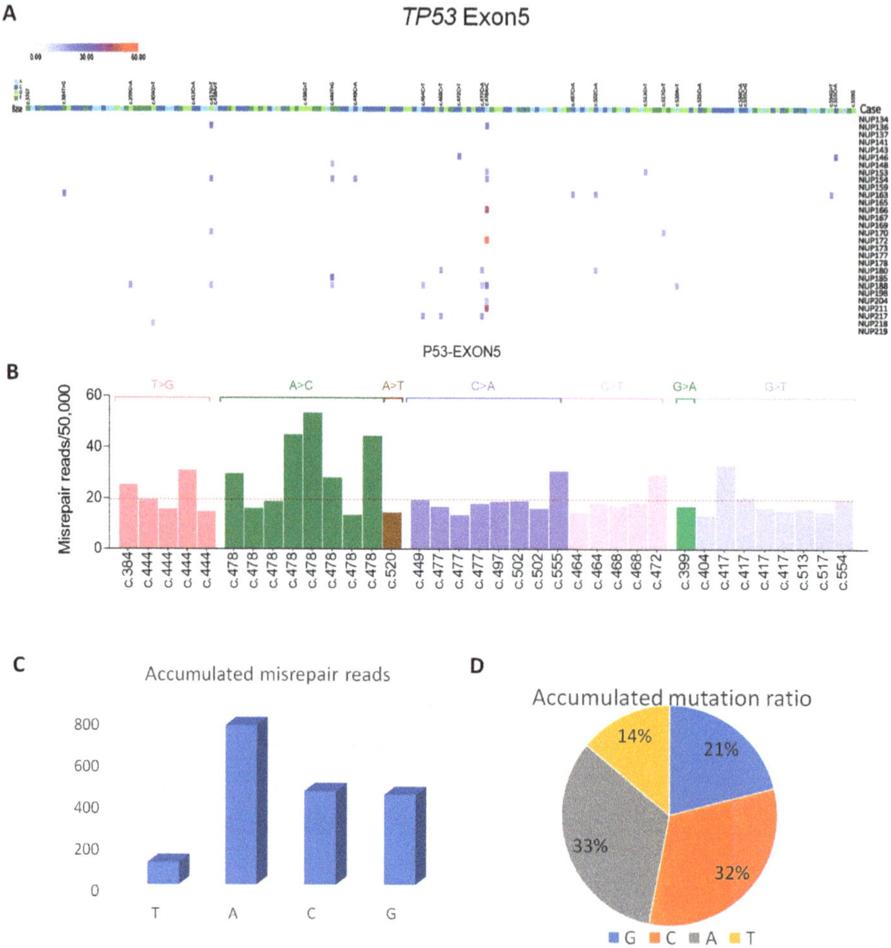

Figure 4. Mutation analysis of *TP53* exon 5 in 29 cases. (**A**) Two-dimensional distribution of base mutations in each of 29 cases. Mutation rate was indicated from low (blue) to high (red) bar. (**B**) 7 mutation signatures detected in *TP53* exon 5. (**C**,**D**) Accumulated point mutations and percentage in G, C, A, and T nucleotides in *TP53* exon 5.

4. Discussion

In this study, we demonstrated that *MED12* mutations occur in myometrial cells from hysterectomy specimens, giving insight into its role as a genetic driver of LM development. Our study showed there were detectable non-random *MED12* mutations in exon 2 enriched at c.130-131 nucleotides, in myometrial cells. This is the first study to uncover the somatic mutations in non-neoplastic myometrial cells of freshly collected tissue samples providing further insight into the mutational pressure in myometrial tissue.

The high rate leiomyomata in reproductive aged women harbor a high rate of *MED12* mutation. We hypothesize that there may be presence of *MED12* mutations in some non-neoplasmic myometrial cells. However, detection of *MED12* mutations in normal myometrial tissue is technically challenging. This is largely due to the fact that (1) standard next-generation sequencing NGS has low sensitivity in detections of low rate of mutations [19] and (2) the estimated error rate in standard NGS is 10^{-2} to 10^{-3}/per base [22]. These errors can be induced by PCR amplification misincorporation, tissue-processing stress, DNA damage during extraction, and probe hybridization and capture errors. The low rate of sequencing errors from general NGS is acceptable for some research purposes or molecular analyses, but it has a major limitation in conditions that require greater accuracy to detect extremely low-frequency mutation, such as searching for true mutations in normal myometrial cells in this study [27].

Duplex deep sequencing (DDS) is a new platform for next-generation sequencing (NGS). A DNA library is prepared by adding barcoded tags in both stranded DNA which will facilitate in mutation detection with higher accuracy and lower error rates. The degenerate barcoded tags in each strand will aid in identification of reads originating from each DNA strand. Since the two strands are complementary, true mutations that are found at the same location on both strands were recorded. Therefore, any errors or mismatch in one strand can be eliminated. Studies showed that DDS can detect mutations at frequencies as low as 5×10^{-8}, which is more than 10,000-fold specificity in comparison to traditional next-generation sequencing methods [22]. Therefore, we used the DDS technique in this study and were able to reliably detect and calculate the *MED12* mutations in normal myometrial cells.

LMs are the monoclonal smooth muscle tumor which are derived from single myometrial cell. The causes of neoplastic transformation from myometrial cells are largely unknown. It warrants further investigation into the biology and molecular nature of myometrial cells and this will aid our understanding of the high burden of LM in reproductive-aged women. The human myometrium is organized by an intricate network with terminally differentiated mullerian smooth muscle linages and with rich extracellular matrix including connective tissues of vessels, fibroblasts, and niche of inflammatory cell network [28]. In response to tissue regeneration, myometrium can adapt to dynamic changes, such as expansion/hypertrophy during puberty, cell proliferation or apoptosis during cyclic hormonal change, and finally in involution at postpartum [29].

The causes for mutations in non-neoplastic myometrial cells remain unknown. The patterns of *MED12* mutations found in myometrial cells were similar to those found in primary myometrial cells treated with oxidative stress as reported in our previous study [8], suggesting that ROS may be one of the critical local microenvironmental factors promoting *MED12* mutations. Specifically, the high single G and C transversion/conversion we observed for *MED12* mutational signature in myometrial cells are commonly seen in oxidative-related carcinogenesis [13,14]. Oxidative stress is common in functional uterus and is tightly regulated by the local microenvironment [9–12]. The functional myometrium in reproductive-aged women encounters the cyclic contractions which promote local hypoxia [30], and increased inflammation during the menstrual cycle and menses [31]. Oxidative burden can be further exacerbated by lifestyle factors such as diet, physical activity, and psychosocial variables. This higher oxidative burden in the myometrium can result in oxidative DNA damage and misrepair [32]. This may partially explain the increased MED12 and p53 mutations in tumor-free myometrial cells. However, the rate of mutations remains to be quite low, as 40% of myometrial tissue samples did not detect *MED12* mutations with our cutoff of 50,000 cell detection levels.

Our findings of MED12 mutations in normal myometrial cells suggest that ROS burden may be a major risk factor for *MED12* mutation [8,32]. This may be supported by a finding of age-matched myometrial tissues from women with LM have higher immunoreactivity for 8-OHdG compared to the myometrial tissues without LM [8]. The expression of proteins associated with oxidative DNA damage and repair, were higher in LM and matched myometrium [32]. Along with increased oxidative stress in the myometrium, there are increases in DNA damage through 8-OHdG formation and misrepair. Interestingly, the *P53* mutational signatures were different from *MED12* in this current study. *P53* harbors more mutations in A and C in contrast to G found in *MED12*, indicating mutation selection to certain genes. The exact molecular mechanism remains unknown.

LMs are monoclonal tumors from myometrium. Study showed that early-life with adverse environmental exposures can impose a risk for LM development. The uterine smooth muscle cells are the target for developmental programming during early-life hormone exposure [33,34]. Many environmental factors, different ethnic backgrounds and lifestyle can also impose a high oxidative stress in the myometrium [35–38]. Due to the small sample size of this study, we did not observe a significant difference of *MED12* exon 2 mutations in myometrium of different races, and also no association with tumor burden or patients' age.

To date, data on the molecular pathogenesis of LM remain insufficient to elucidate the reasons for the high incidence of LM burden. Detection of the single-cell gene mutation burden by DDS allows us to explore the early stages of *MED12* mutations in non-neoplastic myometrium and may significantly improve our understanding of the high incidence of LM in the uterus. The discovery of the G/C rich mutational signatures further supports the potential carcinogenic role of ROS in promoting *MED12* mutations. Notably, our observation of non-random mutational signatures and target hot spot mutations in *P53* and *MED12* are indicative of different mutation selections for different genes, for cell survival and tumor development. Limitations of this study include the small sample size, lack of mutation analysis in other exons of *MED12*, and technical challenges in examining different anatomic sites of myometrial tissues. For the first time, we have demonstrated similar mutations in exon 2 of the *MED12* gene found in LM, in myometrial cells, strongly implicating the *MED12* mutation, specifically at c.130-c.131 as an early driver of LM.

5. Conclusions

This study is the first to report MED12 mutations in non-neoplasm myometrial cells. Our findings will help to understand the molecular basis for a high incidence of uterine leiomyoma in women of reproductive age.

Supplementary Materials: The following supporting information can be downloaded at: https://www.mdpi.com/article/10.3390/cells13171432/s1, Figure S1: A sketch diagram of duplex deep sequencing analysis; Table S1: Case details for clinical and molecular findings.

Author Contributions: Conceptualization: J.-J.W. and J.J.K.; methodology: Y.L. and J.-J.W.; experiments: Y.L., Y.F. and J.-J.W.; materials: J.-J.W. and J.J.K.; data analysis: Y.L. and H.A.; writing—original: J.-J.W. and Y.L.; writing—review and editing: J.-J.W., J.J.K. and Y.L.; supervision: J.-J.W. and J.J.K.; funding acquisition: J.J.K. and J.-J.W. All authors have read and agreed to the published version of the manuscript.

Funding: This study was supported by NCI (R01CA254367).

Institutional Review Board Statement: This study was approved by the Northwestern Institutional Review Board (IRB) (STU00212471).

Informed Consent Statement: Informed consent was obtained from all subjects involved in the study.

Data Availability Statement: All data are available upon request.

Acknowledgments: We thank Brendan F. Kohrn, and Scott R. Kennedy for technical support of duplex sequencing. We thank the NUSeq Core for their technical support for deep sequencing.

Conflicts of Interest: The authors declare no conflicts of interest.

References

1. Bulun, S.E. Uterine fibroids. *N. Engl. J. Med.* **2013**, *369*, 1344–1355. [CrossRef]
2. Catherino, W.H.; Parrott, E.; Segars, J. Proceedings from theNational Institute of Child Health and Human Development conference on the Uterine Fibroid Research Update Workshop. *Fertil. Steril.* **2011**, *95*, 9–12. [CrossRef] [PubMed]
3. Makinen, N.; Mehine, M.; Tolvanen, J.; Kaasinen, E.; Li, Y.; Lehtonen, H.J.; Gentile, M.; Yan, J.; Enge, M.; Taipale, M.; et al. MED12, the mediator complex subunit 12 gene, is mutated at high frequency in uterine leiomyomas. *Science* **2011**, *334*, 252–255. [CrossRef] [PubMed]
4. Bertsch, E.; Qiang, W.; Zhang, Q.; Espona-Fiedler, M.; Druschitz, S.; Liu, Y.; Mittal, K.; Kong, B.; Kurita, T.; Wei, J.J. MED12 and HMGA2 mutations: Two independent genetic events in uterine leiomyoma and leiomyosarcoma. *Mod. Pathol.* **2014**, *27*, 1144–1153. [CrossRef]

5. Heinonen, H.R.; Pasanen, A.; Heikinheimo, O.; Tanskanen, T.; Palin, K.; Tolvanen, J.; Vahteristo, P.; Sjoberg, J.; Pitkanen, E.; Butzow, R.; et al. Multiple clinical characteristics separate MED12-mutation-positive and -negative uterine leiomyomas. *Sci. Rep.* **2017**, *7*, 1015. [CrossRef] [PubMed]
6. Bullerdiek, J.; Rommel, B. Factors targeting MED12 to drive tumorigenesis? *F1000Research* **2018**, *7*, 359. [CrossRef]
7. Lim, W.K.; Ong, C.K.; Tan, J.; Thike, A.A.; Ng, C.C.; Rajasegaran, V.; Myint, S.S.; Nagarajan, S.; Nasir, N.D.; McPherson, J.R.; et al. Exome sequencing identifies highly recurrent MED12 somatic mutations in breast fibroadenoma. *Nat. Genet.* **2014**, *46*, 877–880. [CrossRef]
8. Li, Y.; Xu, X.; Asif, H.; Feng, Y.; Kohrn, B.F.; Kennedy, S.R.; Kim, J.J.; Wei, J.J. Myometrial oxidative stress drives MED12 mutations in leiomyoma. *Cell Biosci.* **2022**, *12*, 111. [CrossRef]
9. Fletcher, N.M.; Abusamaan, M.S.; Memaj, I.; Saed, M.G.; Al-Hendy, A.; Diamond, M.P.; Saed, G.M. Oxidative stress: A key regulator of leiomyoma cell survival. *Fertil. Steril.* **2017**, *107*, 1387–1394.e1. [CrossRef]
10. Foksinski, M.; Kotzbach, R.; Szymanski, W.; Olinski, R. The level of typical biomarker of oxidative stress 8-hydroxy-2′-deoxyguanosine is higher in uterine myomas than in control tissues and correlates with the size of the tumor. *Free Radic. Biol. Med.* **2000**, *29*, 597–601. [CrossRef]
11. Torres, M.J.; Kew, K.A.; Ryan, T.E.; Pennington, E.R.; Lin, C.T.; Buddo, K.A.; Fix, A.M.; Smith, C.A.; Gilliam, L.A.; Karvinen, S.; et al. 17β-Estradiol Directly Lowers Mitochondrial Membrane Microviscosity and Improves Bioenergetic Function in Skeletal Muscle. *Cell Metab.* **2018**, *27*, 167–179 e167. [CrossRef] [PubMed]
12. Vidimar, V.; Gius, D.; Chakravarti, D.; Bulun, S.E.; Wei, J.J.; Kim, J.J. Dysfunctional MnSOD leads to redox dysregulation and activation of prosurvival AKT signaling in uterine leiomyomas. *Sci. Adv.* **2016**, *2*, e1601132. [CrossRef] [PubMed]
13. Ziech, D.; Franco, R.; Pappa, A.; Panayiotidis, M.I. Reactive oxygen species (ROS)--induced genetic and epigenetic alterations in human carcinogenesis. *Mutat. Res.* **2011**, *711*, 167–173. [CrossRef]
14. Olivier, M.; Hollstein, M.; Hainaut, P. TP53 mutations in human cancers: Origins, consequences, and clinical use. *Cold Spring Harb. Perspect. Biol.* **2010**, *2*, a001008. [CrossRef]
15. Poetsch, A.R. The genomics of oxidative DNA damage, repair, and resulting mutagenesis. *Comput. Struct. Biotechnol. J.* **2020**, *18*, 207–219. [CrossRef] [PubMed]
16. Sekiguchi, M.; Tsuzuki, T. Oxidative nucleotide damage: Consequences and prevention. *Oncogene* **2002**, *21*, 8895–8904. [CrossRef] [PubMed]
17. Cheng, K.C.; Cahill, D.S.; Kasai, H.; Nishimura, S.; Loeb, L.A. 8-Hydroxyguanine, an abundant form of oxidative DNA damage, causes G-T and A-C substitutions. *J. Biol. Chem.* **1992**, *267*, 166–172. [CrossRef]
18. Kucab, J.E.; Zou, X.; Morganella, S.; Joel, M.; Nanda, A.S.; Nagy, E.; Gomez, C.; Degasperi, A.; Harris, R.; Jackson, S.P.; et al. A Compendium of Mutational Signatures of Environmental Agents. *Cell* **2019**, *177*, 821–836 e816. [CrossRef]
19. Chin, E.L.; da Silva, C.; Hegde, M. Assessment of clinical analytical sensitivity and specificity of next-generation sequencing for detection of simple and complex mutations. *BMC Genet.* **2013**, *14*, 6. [CrossRef]
20. Goldman, D.; Domschke, K. Making sense of deep sequencing. *Int. J. Neuropsychopharmacol.* **2014**, *17*, 1717–1725. [CrossRef]
21. Peng, Q.; Xu, C.; Kim, D.; Lewis, M.; DiCarlo, J.; Wang, Y. Targeted Single Primer Enrichment Sequencing with Single End Duplex-UMI. *Sci. Rep.* **2019**, *9*, 4810. [CrossRef] [PubMed]
22. Schmitt, M.W.; Kennedy, S.R.; Salk, J.J.; Fox, E.J.; Hiatt, J.B.; Loeb, L.A. Detection of ultra-rare mutations by next-generation sequencing. *Proc. Natl. Acad. Sci. USA* **2012**, *109*, 14508–14513. [CrossRef] [PubMed]
23. Kennedy, S.R.; Schmitt, M.W.; Fox, E.J.; Kohrn, B.F.; Salk, J.J.; Ahn, E.H.; Prindle, M.J.; Kuong, K.J.; Shen, J.C.; Risques, R.A.; et al. Detecting ultralow-frequency mutations by Duplex Sequencing. *Nat. Protoc.* **2014**, *9*, 2586–2606. [CrossRef]
24. Dilliott, A.A.; Farhan, S.M.K.; Ghani, M.; Sato, C.; Liang, E.; Zhang, M.; McIntyre, A.D.; Cao, H.; Racacho, L.; Robinson, J.F.; et al. Targeted Next-generation Sequencing and Bioinformatics Pipeline to Evaluate Genetic Determinants of Constitutional Disease. *J. Vis. Exp.* **2018**, *134*, e57266. [CrossRef]
25. Salk, J.J.; Loubet-Senear, K.; Maritschnegg, E.; Valentine, C.C.; Williams, L.N.; Higgins, J.E.; Horvat, R.; Vanderstichele, A.; Nachmanson, D.; Baker, K.T.; et al. Ultra-Sensitive TP53 Sequencing for Cancer Detection Reveals Progressive Clonal Selection in Normal Tissue over a Century of Human Lifespan. *Cell Rep.* **2019**, *28*, 132–144 e133. [CrossRef]
26. Cole, A.J.; Dwight, T.; Gill, A.J.; Dickson, K.A.; Zhu, Y.; Clarkson, A.; Gard, G.B.; Maidens, J.; Valmadre, S.; Clifton-Bligh, R.; et al. Assessing mutant p53 in primary high-grade serous ovarian cancer using immunohistochemistry and massively parallel sequencing. *Sci. Rep.* **2016**, *6*, 26191. [CrossRef]
27. McGranahan, N.; Swanton, C. Biological and therapeutic impact of intratumor heterogeneity in cancer evolution. *Cancer Cell* **2015**, *27*, 15–26. [CrossRef]
28. Goad, J.; Rudolph, J.; Zandigohar, M.; Tae, M.; Dai, Y.; Wei, J.J.; Bulun, S.E.; Chakravarti, D.; Rajkovic, A. Single-cell sequencing reveals novel cellular heterogeneity in uterine leiomyomas. *Hum. Reprod.* **2022**, *37*, 2334–2349. [CrossRef]
29. Shynlova, O.; Oldenhof, A.; Dorogin, A.; Xu, Q.; Mu, J.; Nashman, N.; Lye, S.J. Myometrial apoptosis: Activation of the caspase cascade in the pregnant rat myometrium at midgestation. *Biol. Reprod.* **2006**, *74*, 839–849. [CrossRef] [PubMed]
30. Mayer, A.; Hockel, M.; Wree, A.; Leo, C.; Horn, L.C.; Vaupel, P. Lack of hypoxic response in uterine leiomyomas despite severe tissue hypoxia. *Cancer Res.* **2008**, *68*, 4719–4726. [CrossRef]

31. Wu, B.; Chen, X.; He, B.; Liu, S.; Li, Y.; Wang, Q.; Gao, H.; Wang, S.; Liu, J.; Zhang, S.; et al. ROS are critical for endometrial breakdown via NF-kappaB-COX-2 signaling in a female mouse menstrual-like model. *Endocrinology* **2014**, *155*, 3638–3648. [CrossRef]
32. Li, Y.; McNally, R.P.; Feng, Y.; Kim, J.J.; Wei, J.-J. Racial differences in transcriptomics and reactive oxygen species burden in myometrium and leiomyoma. *Hum. Reprod.* **2023**, *38*, 609–620. [CrossRef]
33. Silbergeld, E.K.; Patrick, T.E. Environmental exposures, toxicologic mechanisms, and adverse pregnancy outcomes. *Am. J. Obstet. Gynecol.* **2005**, *192*, S11–S21. [CrossRef] [PubMed]
34. Mas, A.; Prusinski, L.; Yang, Q.; Diaz-Gimeno, P.; Stone, L.; Diamond, M.P.; Simon, C.; Al-Hendy, A. Role of Stro1+/CD44+ stem cells in myometrial physiology and uterine remodeling during pregnancy. *Biol. Reprod.* **2017**, *96*, 70–80. [CrossRef] [PubMed]
35. Williams, D.R.; Mohammed, S.A. Discrimination and racial disparities in health: Evidence and needed research. *J. Behav. Med.* **2009**, *32*, 20–47. [CrossRef]
36. Wise, L.A.; Palmer, J.R.; Cozier, Y.C.; Hunt, M.O.; Stewart, E.A.; Rosenberg, L. Perceived racial discrimination and risk of uterine leiomyomata. *Epidemiology* **2007**, *18*, 747–757. [CrossRef]
37. Geronimus, A.T.; Hicken, M.; Keene, D.; Bound, J. "Weathering" and age patterns of allostatic load scores among blacks and whites in the United States. *Am. J. Public Health* **2006**, *96*, 826–833. [CrossRef]
38. Van Dyke, M.E.; Baumhofer, N.K.; Slopen, N.; Mujahid, M.S.; Clark, C.R.; Williams, D.R.; Lewis, T.T. Pervasive Discrimination and Allostatic Load in African American and White Adults. *Psychosom. Med.* **2020**, *82*, 316–323. [CrossRef] [PubMed]

cells

MDPI

Review

Molecular Mechanisms of Cellular Senescence in Age-Related Endometrial Dysfunction

Hiroshi Kobayashi [1,2,*], Mai Umetani [1], Miki Nishio [1], Hiroshi Shigetomi [2,3], Shogo Imanaka [1,2] and Hiratsugu Hashimoto [1]

1 Department of Gynecology and Reproductive Medicine, Ms.Clinic MayOne, 871-1 Shijo-cho, Kashihara 634-0813, Japan; mai_umetani@yahoo.co.jp (M.U.); mikin.yaruki10000.boshi@gmail.com (M.N.); shogo_0723@naramed-u.ac.jp (S.I.); hiratsugu_hashimoto@yahoo.co.jp (H.H.)
2 Department of Obstetrics and Gynecology, Nara Medical University, 840 Shijo-cho, Kashihara 634-8522, Japan; hshige35@gmail.com
3 Department of Gynecology and Reproductive Medicine, Aska Ladies Clinic, 3-3-17 Kitatomigaoka-cho, Nara 634-0001, Japan
* Correspondence: hirokoba@naramed-u.ac.jp

Abstract: The endometrium is essential for reproductive function, supporting implantation and pregnancy through mechanisms such as hormonal responsiveness, immune regulation, and tissue regeneration. Aging disrupts these processes, with cellular senescence—marked by irreversible cell cycle arrest due to DNA damage and oxidative stress—being a key contributor. While senescence aids in tumor suppression and tissue repair, its dysregulation impairs endometrial function. Central to this regulation are p53, AMPK, and mTOR, which coordinate stress responses, metabolic regulation, and proliferation control. p53 activates AMPK and inhibits mTOR, promoting energy conservation and limiting senescence. AMPK also suppresses mTOR, reducing age-related dysfunction. This p53–AMPK–mTOR axis, along with autophagy, governs cell fate in response to stress and nutrient status. Although moderate senescence supports endometrial function, excessive accumulation can hinder fertility. Understanding these molecular interactions may advance fertility treatments and strategies to counteract reproductive aging.

Keywords: AMPK; autophagy; endometrium; mTOR; p53; senescence

1. Introduction

The endometrium is in direct contact with the external environment and is continuously exposed to various stimuli, such as infections and physical damage. To maintain homeostasis, it employs a multi-layered defense system, which includes protection against microbial invasion [1], remodeling in response to hormonal fluctuations, repair mechanisms against mechanical stress [2], and immune regulation [3]. These coordinated actions enable the endometrium to adapt to diverse external stimuli and fulfill its essential role in reproduction. Throughout a woman's life, the endometrium undergoes significant physiological changes during key milestones such as menstruation, pregnancy, childbirth, and menopause. These changes are governed by a delicate balance between cellular regeneration and senescence, influenced primarily by hormonal activity. In particular, the regenerative capacity of the endometrium reflects the functional interplay between hormonal signals and endometrial cells and is critical for maintaining reproductive function [4]. Conversely, when cellular senescence associated with aging becomes dysregulated, it can impair endometrial receptivity and fertility and may contribute to conditions such as infertility, endometriosis, and endometrial cancer [4].

 https://doi.org/10.3390/cells14120858

Cellular senescence is a tightly regulated physiological process triggered by factors such as DNA damage, oxidative stress, telomere shortening, oncogene activation, and nutritional stress [5,6]. Senescent cells exhibit several characteristic features, including irreversible cell cycle arrest; morphological alterations such as hypertrophy and cellular flattening [7]; increased activity of senescence-associated β-galactosidase (SA-β-gal); elevated expression of cell cycle inhibitors ($p16^{INK4a}$ [*CDKN2A*] and $p21^{CIP1}$ [*CDKN1A*]); and the accumulation of DNA damage markers such as γ-H2AX [7]. In addition, senescent cells acquire a senescence-associated secretory phenotype (SASP), characterized by the secretion of pro-inflammatory cytokines, chemokines, growth factors, and proteases [8], which can influence neighboring cells and the tissue microenvironment. The SASP contributes to wound healing, tissue repair, and tumor suppression [8–10], as well as tissue development during embryogenesis [8,11]. However, when senescent cells are not effectively cleared, their accumulation can result in chronic inflammation and promote the progression of age-related diseases such as atherosclerosis, osteoarthritis, and cancer [12].

The regulation of cellular senescence involves numerous genes and signaling networks associated with autophagy, the cell cycle, and metabolism [8]. Autophagy, a process that degrades and recycles unnecessary intracellular components, plays a pivotal role in maintaining cellular homeostasis and tissue remodeling. In the endometrium, cellular senescence is modulated by stress responses, inflammatory signaling, and intercellular communication, and it exerts a profound influence on endometrial function, embryo receptivity, and fertility [13,14]. Senescent cells also undergo significant metabolic alterations, including changes in mitochondrial function and energy metabolism [15]. Although signaling molecules such as the tumor suppressor p53, mechanistic target of rapamycin (mTOR), and AMP-activated protein kinase (AMPK) are known to be involved in cellular senescence, the mechanisms underlying the initiation of senescence and metabolic changes in the endometrium remain incompletely understood.

In this review, we outline the fundamental characteristics of autophagy and energy metabolism associated with cellular senescence, examine the roles and interactions of p53, mTOR, and AMPK, present the latest findings on endometrial cellular senescence, and discuss potential therapeutic strategies for addressing age-related declines in reproductive function.

2. Methods

Search Strategy and Selection Criteria

In this narrative review, we performed an extensive literature survey centered on autophagy and energy homeostasis in the context of endometrial cellular senescence, with particular emphasis on the roles of p53, AMPK, and mTOR. Relevant studies published up to 31 March 2025 were identified through searches in two electronic databases: PubMed (https://pubmed.ncbi.nlm.nih.gov/ accessed on 19 May 2025) and Google Scholar (https://scholar.google.com/ accessed on 19 May 2025), utilizing a combination of the following keywords: "endometrium", "reproduction", "cellular senescence", "autophagy", "p53", "AMPK", and "mTOR". This review does not constitute a systematic review employing a predefined research framewo rk but rather seeks to provide a comprehensive and integrative overview of the current state of knowledge on this subject.

3. Factors Involved in Endometrial Cellular Senescence

Endometrial senescence is regulated by the interplay of multiple molecular mechanisms, including inflammatory cytokines, intracellular signaling pathway regulators, and cell cycle modulators. Among these, the SASP plays a central role in both the progression of cellular senescence and its effects on surrounding tissues. Inflammatory cytokines such as interleukin-17 receptor B (IL17RB) and interleukin-1β (IL-1β) activate the c-Jun N-terminal kinase (JNK) pathway, thereby promoting senescence in endometrial stromal cells and stimulating the secretion of SASP factors [16,17]. Plasminogen activator inhibitor-1 (PAI-1), secreted by senescent stromal cells, acts as a master regulator of paracrine senescence signals, amplifying SASP expression in neighboring cells [18]. Moreover, insulin-like growth factor-binding protein 3 (IGFBP3), which is upregulated in aged mesenchymal stem cells, has been shown to be regulated by the PI3K (phosphoinositide 3-kinase)/Akt (protein kinase B) pathway and to contribute to SASP-independent paracrine senescence [19]. These inflammatory mediators play critical roles in the progression of the chronic inflammatory state characteristic of aging tissues, commonly referred to as "inflammaging" [19]. Inflammaging, characterized by chronic, low-grade inflammation linked to aging, plays a pivotal role in promoting endometrial senescence, thereby contributing to endometrial dysfunction [20]. This process involves the secretion of pro-inflammatory mediators collectively known as the SASP, and the resulting sustained inflammatory milieu disrupts tissue homeostasis and diminishes endometrial receptivity, ultimately impairing fertility and pregnancy outcomes [20]. To date, several comprehensive review articles have explored the relationship between cellular senescence and inflammaging. Saito et al. discuss the dynamic interplay between SASP-mediated cellular senescence and inflammaging [21], while Lee et al. use the skin as a model to study these phenomena [22]. Other studies have examined the involvement of cellular senescence and inflammaging in traumatic brain injury [23] and their impact on bone health [24]. Li et al. provide a broad overview of cellular senescence, its role in aging and disease, and potential therapeutic strategies [25]. Collectively, these studies offer important insights into how cellular senescence and inflammaging contribute to aging and age-related pathologies.

Additionally, Cbl proto-oncogene-like 1 (CBLL1) and cell division cycle 42 (CDC42) have emerged as significant regulators of signaling pathways involved in senescence. Elevated CBLL1 expression in senescent endometrial stromal cells influences cellular senescence and proliferation by downregulating the tumor suppressor phosphatase and tensin homolog (PTEN) and increasing intracellular reactive oxygen species (ROS) levels [26]. Furthermore, the upregulation of CDC42—a member of the Rho GTPase family—induces premature senescence in stromal cells and is associated with reduced endometrial receptivity and implantation capacity [27]. Umbayev et al. have discussed the involvement of CDC42 in cellular senescence and its potential as a therapeutic target for age-related diseases [28]. Moreover, Ito et al. demonstrated that CDC42 mediates chronic inflammation associated with endothelial senescence and contributes to atherosclerosis development [29].

The tumor suppressor p53 and the signaling molecules AMPK and mTOR are also key regulators of cellular senescence. To better understand their functional roles in controlling senescence, current research has focused on several key areas. The identification of autophagy machinery components and metabolism-related determinants that influence whether p53 activation leads to cellular senescence or reversible quiescence, as well as the elucidation of the underlying mechanisms. Additionally, a detailed investigation of the context-dependent role of AMPK in senescence, particularly its crosstalk with the p53 and mTOR signaling pathways under various cellular stress conditions. Furthermore, a deeper understanding of the molecular mechanisms through which mTOR activity induces and sustains senescence-associated phenotypes remains a pressing challenge.

In the following section, we individually examine the molecular mechanisms by which major senescence-associated genes—including p53, mTOR, and AMPK—regulate endometrial cell senescence.

3.1. p53

The p53 protein, encoded by the TP53 gene located on chromosome 17p13.1, responds to a wide range of endogenous and exogenous stress stimuli—such as DNA damage, oxidative stress, and oncogene activation—and plays a critical role in maintaining genomic stability by arresting the cell cycle and inducing the transcription of DNA repair-related genes [30]. p53 is widely recognized as a tumor suppressor [31] and functions as a central transcription factor that induces cellular senescence, an irreversible form of cell cycle arrest [32–34]. In cases where DNA damage is irreparable, p53 activates the apoptotic pathway to eliminate damaged cells.

3.1.1. Functional Role of p53 in Cellular Senescence and Autophagy Regulation

The activity of p53 is tightly regulated by negative modulators, such as the E3 ubiquitin ligase mouse double minute 2 homolog (MDM2), which suppresses p53 protein accumulation and transcriptional activity to prevent overactivation [35]. Activation of p53 has been reported to regulate SASP expression via mTOR inhibition, thereby suppressing the secretion of inflammatory cytokines [36–38] (Figure 1 ①). In addition, p53 halts the cell cycle at the G1/S checkpoint through cyclin-dependent kinase inhibitors, allowing DNA damage to be repaired prior to cell division [39,40]. It also supports genome integrity by promoting the transcription of DNA repair factors such as damage-specific DNA-binding protein 2 (DDB2) and xeroderma pigmentosum group C (XPC) [41]. When DNA damage is reversible, p53-mediated senescence induction is circumvented (Figure 1 ②). Conversely, in the presence of irreparable DNA damage, p53 activates the transcription of pro-apoptotic genes such as BAX (BCL2-associated X, apoptosis regulator), PUMA (p53 upregulated modulator of apoptosis), and PMAIP1 (phorbol-12-myristate-13-acetate-induced protein 1), triggering cell death and preventing the accumulation of abnormal cells [36,42] (Figure 1 ③). Thus, as a master regulator of the cellular stress response, p53 orchestrates diverse cell fate decisions—ranging from DNA repair and cell cycle control to apoptosis and senescence—thereby preserving cellular and tissue homeostasis and suppressing tumorigenesis [30]. Furthermore, p53 plays a crucial role in metabolic regulation. It suppresses glycolysis [43], promotes oxidative phosphorylation (OXPHOS) [44], and induces the expression of antioxidant enzymes such as superoxide dismutase (SOD) and catalase. It also regulates glutathione metabolism and maintains mitochondrial function, thereby contributing to the suppression of cellular senescence [45] (Figure 1 ④). Consequently, p53 has been suggested to possess a context-dependent dual role—either promoting or suppressing senescence depending on the cellular environment [32,46]. Excessive senescence may lead to tissue dysfunction and age-related diseases, whereas inadequate senescence responses may permit abnormal cell proliferation and tumorigenesis.

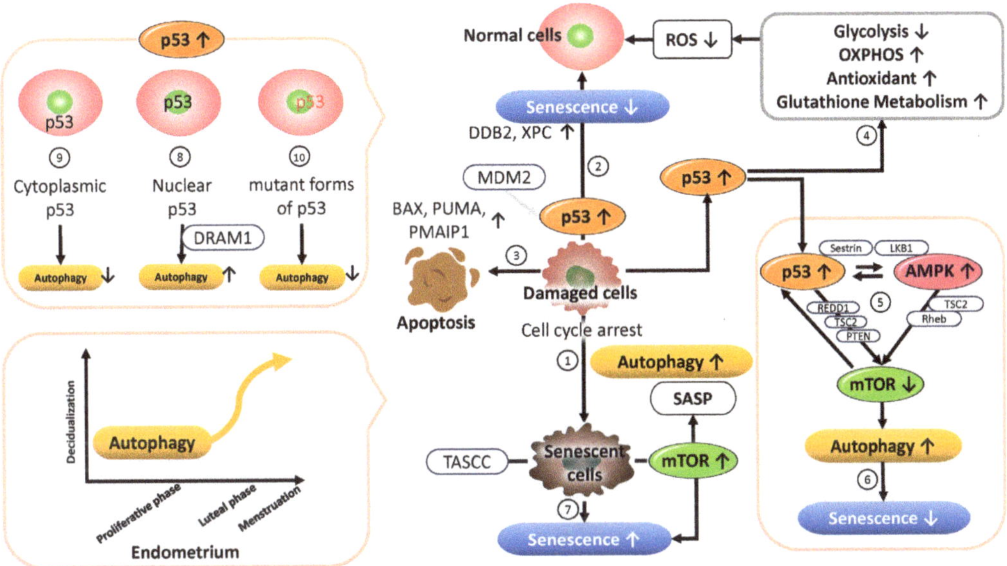

Figure 1. Functional role of p53 in cellular senescence and autophagy regulation. Stress activates p53, causing cell cycle arrest (①); reversible damage avoids senescence (②), while severe damage leads to apoptosis (③). p53 also boosts OXPHOS and antioxidant defenses (④), activates autophagy via AMPK and mTOR (⑤,⑥), but excess autophagy promotes SASP and senescence (⑦). Localization affects p53's role in autophagy (⑧–⑩). **Top-left bubble**: The subcellular localization of p53 critically modulates its regulatory function in autophagy (⑧–⑩). **Bottom-left bubble**: Autophagic activity in the endometrium exhibits dynamic fluctuations, increasing from the proliferative to the secretory phase, and is intricately linked to the process of decidualization. While some studies report a concomitant increase in both p53 expression and autophagy activity during the menstrual cycle, others present contrasting findings. Black arrows indicate facilitative or activating effects, whereas gray arrows denote inhibitory actions. AMPK, AMP-activated protein kinase; BAX, BCL2-associated X, apoptosis regulator; DDB2, damage-specific DNA-binding protein 2; DRAM1, DNA damage-regulated autophagy modulator 1; LKB1, liver kinase B1; mTOR, mechanistic target of rapamycin; OXPHOS, oxidative phosphorylation; PMAIP1, phorbol-12-myristate-13-acetate-induced protein 1; PTEN, phosphatase and tensin homolog; PUMA, p53 upregulated modulator of apoptosis; REDD1, regulated in development and DNA damage responses 1; Rheb, Ras homolog, mTORC1 binding; ROS, reactive oxygen species; SASP, senescence-associated secretory phenotype; TASCC, TOR-Autophagy Spatial Coupling Compartments; TSC2, TSC complex subunit 2; XPC, xeroderma pigmentosum group C.

Additionally, autophagy plays an important role in regulating cellular senescence. This intracellular homeostatic mechanism is activated by various stressors, including nutrient deprivation, hypoxia, DNA damage, and cytotoxicity. Autophagy maintains mitochondrial and overall cellular function by degrading and recycling organelles and abnormal proteins via a lysosome-dependent pathway [47,48], thus mitigating the accumulation of intracellular damage and slowing the progression of cellular senescence. The key regulators of autophagy include p53, AMPK, and mTOR. p53 is known to inhibit mTOR signaling through both transcription-dependent and -independent mechanisms [32,49] (Figure 1 ⑤). As a transcription factor, p53 induces the expression of several genes—such as REDD1 (regulated in development and DNA damage responses 1), TSC2 (TSC complex subunit 2), PTEN, LKB1 (liver kinase B1), the AMPKβ subunit, and Sestrin1/2—which act directly or indirectly to suppress mTORC1 (mechanistic Target of Rapamycin Complex 1) activity [50–52]. Moreover, p53 has been shown to directly bind the promoters of autophagy-related genes (ATGs), thereby positively regulating autophagy [32]. In terms of transcription-independent mechanisms, p53 suppresses the translation of mTOR pathway components (e.g., mTOR, ribosomal S6 kinase (S6K1), and Rictor) by modulating the expression of microRNAs such as miR-34, miR-145, and miR-155 [32]. AMPK, activated by intracellular adenosine 5′ triphosphate (ATP) depletion or increased AMP levels, phosphorylates TSC2 to enhance its GTPase-activating protein (GAP) activity, thereby inactivating Rheb (Ras homolog, mTORC1 binding) and inhibiting mTORC1. Through this molecular network, autophagy contributes to the regulation of energy metabolism and the suppression of cellular senescence [53] (Figure 1 ⑥). In addition, the p53 family members p63 and p73 also play important roles in maintaining cellular metabolic homeostasis. Berkers et al. demonstrated that the p53 family mediates the link between energy stress responses and fate-determining processes such as cellular senescence by activating AMPK and suppressing mTORC1 [54]. The p53 family is involved in a variety of metabolic processes, including regulation of the AMPK/mTOR pathway, glucose and lipid metabolism, mitochondrial function, redox balance, and autophagy. Cellular senescence suppresses cell growth and proliferation through p53 activation [52] (Figure 1 ⑦). Meanwhile, amino acids and metabolites generated by autophagy can activate mTORC1 within intracellular structures called TASCCs (TOR-Autophagy Spatial Coupling Compartments), thereby promoting the mass production of SASP factors [55] (Figure 1 ⑦). The formation of TASCCs spatially integrates the catabolic activity of autophagy with the anabolic signaling of mTORC1, enhancing SASP production in senescent cells. These findings suggest that autophagy may also contribute to the promotion of cellular senescence in a context-dependent manner [36,56].

Furthermore, p53 is known to both promote and suppress autophagy depending on its subcellular localization. Nuclear p53 acts as an autophagy activator by enhancing the transcription of autophagy-inducing genes such as DNA damage-regulated autophagy modulator 1 (DRAM1), and by regulating histone-modifying enzymes to promote the transcription of ATG genes [57] (Figure 1 ⑧). Conversely, cytoplasmic p53 has been reported to suppress autophagy [58], highlighting the dual and complex role of p53 in autophagy regulation (Figure 1 ⑨). Additionally, certain mutant forms of p53 can suppress autophagy via activation of the mTOR pathway [59] (Figure 1 ⑩). Targeting these mutant forms may alleviate autophagy suppression and restore its functional activation. Collectively, these findings indicate that autophagy is a condition-dependent process that may either promote or suppress cellular senescence. Its biological effects vary significantly depending on factors such as the subcellular localization of p53, mutation status, cellular environment, and stress conditions [60].

3.1.2. Functional Role of p53 in Metabolism

Cellular senescence differs from the reversible quiescent state (G_0 phase), in which cells temporarily cease division in response to external stresses such as nutrient deprivation or growth factor withdrawal, and from terminal differentiation, in which nerve and muscle cells permanently exit the cell cycle and acquire specialized functions [8]. Senescent cells permanently lose their proliferative capacity due to various stress responses—such as DNA damage, oxidative stress, and telomere shortening—but remain metabolically active and undergo distinctive metabolic reprogramming [8]. Senescent cells often exhibit increased glycolytic flux and enhanced lactate production [8]. In addition, significant changes occur in mitochondrial function and lipid metabolism, including impaired mitochondrial function despite an increase in mitochondrial mass [61,62]. Reports on fatty acid β-oxidation are inconsistent, with some studies showing an increase [63] and others a decrease [62], suggesting that metabolic imbalance may lead to intracellular lipid droplet accumulation.

These metabolic adaptations are thought to compensate for the energy demands associated with sustaining the SASP [8].

Below, we outline the metabolic regulatory mechanisms of p53, a central transcription factor in the progression of cellular senescence (Figure 2, left). p53 modulates energy metabolism primarily by suppressing glycolysis and enhancing mitochondrial OXPHOS [43,64–66]. Specifically, p53 represses the expression of glucose transporters GLUT1 (glucose transporter type 1), GLUT3, and GLUT4, thereby limiting glucose uptake and reducing substrate availability for glycolysis [65,67]. Moreover, p53 induces the expression of TP53-induced glycolysis and apoptosis regulator (TIGAR), which lowers the activity of phosphofructokinase-1 (PFK1), a key glycolytic enzyme [45]. As a result, fructose-6-phosphate (F6P) accumulates in the cell. Under oxidative stress, the increased consumption of nicotinamide adenine dinucleotide phosphate (NADPH) raises the $NADP^+$/NADPH ratio, enhancing the activity of glucose-6-phosphate dehydrogenase (G6PDH) and diverting glucose metabolism from glycolysis to the pentose phosphate pathway (PPP). The PPP provides ribose-5-phosphate for nucleotide synthesis, supporting the biosynthetic demands of senescent cells. Additionally, increased NADPH production through the PPP enhances antioxidant capacity and protects against oxidative damage. However, excessive PPP activity can disturb metabolic homeostasis, and p53 counteracts this by directly inhibiting G6PDH, the rate-limiting enzyme of the PPP [68].

p53 also suppresses the activity of phosphoglycerate mutase (PGM), thereby inhibiting the conversion of glucose to pyruvate [69]. Pyruvate metabolism is regulated by pyruvate dehydrogenase (PDH) and its inhibitor, pyruvate dehydrogenase kinase (PDK). PDH converts pyruvate to acetyl-CoA, linking glycolysis to the tricarboxylic acid (TCA) cycle and OXPHOS [70]. In contrast, PDK phosphorylates and inactivates PDH, redirecting pyruvate toward lactate production via lactate dehydrogenase (LDH) [70]. The activity of the PDK/PDH axis varies among senescent cells depending on the type and cause of senescence. For example, in senescent human prostate stromal cells, upregulation of PDK4 inhibits PDH and promotes aerobic glycolysis (the Warburg effect), increasing glucose uptake and lactate production [71]. In contrast, in cancer cells undergoing senescence induced by the $BRAF^{V600E}$ mutation, PDK1 expression is suppressed, leading to enhanced oxidative metabolism via the TCA cycle [70,72]. These findings highlight the metabolic flexibility of senescent cells, which remodel their pathways in response to environmental and intracellular cues to balance glycolysis and OXPHOS according to changing energy and biosynthetic needs [73].

Furthermore, p53 promotes energy-efficient metabolic pathways to maintain energy homeostasis under stress. It induces the expression of Lipin1, a regulator of fatty acid β-oxidation [74], and GAMT (guanidinoacetate methyltransferase), which participates in the creatine-phosphocreatine shuttle, an alternative system for ATP production [75]. These adaptive metabolic changes support cell survival, particularly under nutrient-limited conditions.

In addition, p53 promotes OXPHOS by activating the transcription of synthesis of cytochrome c oxidase 2 (SCO2), a gene essential for the assembly of complex IV, thereby ensuring efficient mitochondrial ATP production [76]. When SCO2 expression is reduced due to p53 deficiency, mitochondrial respiration becomes impaired, leading to a compensatory shift toward glycolysis [76]. Similarly, p53 induces the expression of the mitochondrial enzyme glutaminase 2 (GLS2), which catalyzes the conversion of glutamine to glutamate and subsequently to α-ketoglutarate (α-KG), a key intermediate in the TCA cycle. This pathway not only supports oxidative metabolism but also contributes to the antioxidant defense mechanism [77,78]. Apoptosis-inducing factor (AIF), while known for its role in promoting cell death, is also essential for maintaining mitochondrial function. p53 may enhance AIF expression, thereby stabilizing mitochondrial complex I and improving the efficiency of OXPHOS, which in turn suppresses excessive ROS production and prevents oxidative damage [79]. Additionally, the p53-inducible ribonucleotide reductase subunit p53R2 (RRM2B) plays a crucial role in deoxyribonucleotide synthesis and mitochondrial DNA (mtDNA) replication, supporting mitochondrial function and sustaining energy production under stress conditions [80]. p53 also activates the transcription of Parkin (PARK2), an E3 ubiquitin ligase, thereby promoting mitochondrial respiration and inhibiting glycolysis [81]. Notably, nuclear-localized p53 enhances Parkin transcription, whereas cytoplasmic p53 inhibits Parkin activity in mitophagy [82]. Through Parkin, p53 contributes to the regulation of mitochondrial quality control and cellular metabolism.

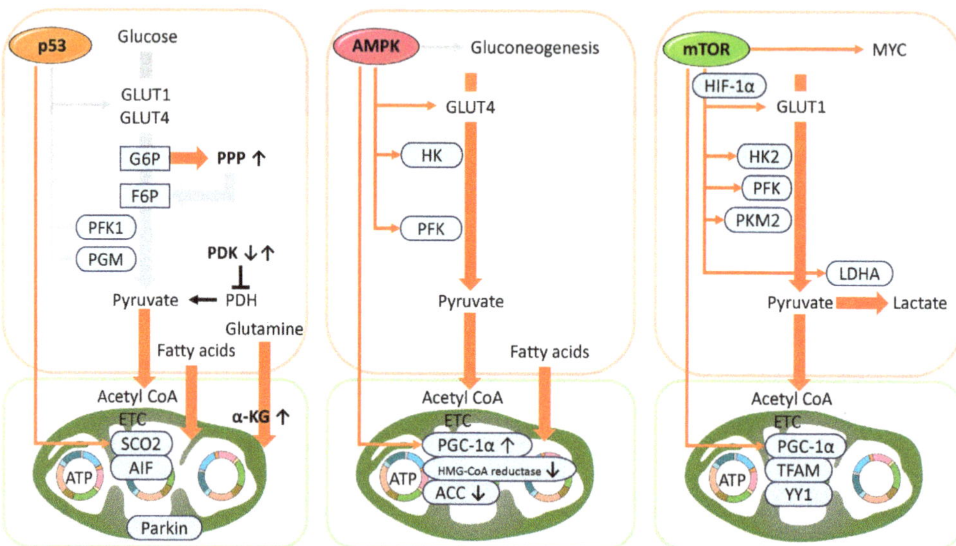

Figure 2. Functional role of p53 (**left**), AMPK (**middle**), and mTOR (**right**) in cellular metabolism. **Left panel**, p53 acts as a metabolic checkpoint by suppressing glycolysis through downregulation of glucose transporters and glycolytic enzymes. It promotes OXPHOS by inducing genes like SCO2 and enhances β-oxidation via activation of genes such as GAMT and Lipin1, thereby supporting efficient ATP production and reducing oxidative stress. **Middle panel**, AMPK stimulates glycolysis and β-oxidation to generate ATP and inhibits anabolic processes by suppressing mTOR activity, thus conserving energy and promoting catabolic pathways. **Right panel**, mTOR serves as a central regulator of cell growth and metabolism, promoting glycolysis and lipid synthesis under nutrient-rich conditions. However, its activation can inhibit autophagy and β-oxidation, potentially leading to energy imbalance and contributing to cellular senescence if not properly regulated. Together, these pathways coordinate to maintain energy balance, adapting cellular metabolism in response to varying energy demands and environmental conditions. Orange arrows indicate facilitative or activating effects, whereas gray arrows denote inhibitory actions. Orange boxes primarily denote glycolytic processes, whereas green boxes correspond to oxidative phosphorylation pathways. The colored rings depicted within the mitochondria represent mitochondrial DNA. ACC, acetyl-CoA carboxylase; AIF, apoptosis-inducing factor; AMPK, AMP-activated protein kinase; ATP, adenosine 5′ triphosphate; α-KG, α-ketoglutarate; ETC, electron transport chain; F6P, fructose-6-phosphate; G6P, glucose-6-phosphate; GLUT1, glucose transporter type 1; HIF-1α, hypoxia-inducible factor 1-alpha; HMG-CoA reductase, 3-hydroxy-3-methylglutaryl-CoA reductase; LDHA, lactate dehydrogenase A; MYC, MYC proto-oncogene, bHLH transcription factor; PDH, pyruvate dehydrogenase; PDK, pyruvate dehydrogenase kinase; PGC-1α, peroxisome proliferator-activated receptor gamma coactivator 1-alpha; PGM, phosphoglycerate mutase; PKM2, pyruvate kinase M2; PPP, pentose phosphate pathway; TFAM, mitochondrial transcription factor A; YY1, Yin Yang 1.

Therefore, p53 activation in response to DNA damage under metabolic stress often induces cellular senescence, reflecting its role as an energy sensor. In general, p53 functions to suppress glycolysis and enhance mitochondrial metabolism [64]. However, exceptions exist—for instance, in pancreatic β-cells and hepatocytes, p53 has been reported to promote glycolysis and inhibit OXPHOS [83]. These observations underscore the context-dependent nature of p53-mediated metabolic regulation, which varies with cell type and physiological or pathological conditions. Moreover, p53 regulates the oxidative stress response and influences cell lifespan by modulating intracellular ROS levels. Through its integrated roles in metabolic homeostasis and cellular senescence, p53 acts as a key regulator of cellular function and a critical suppressor of tumorigenesis [80]. In addition, it has been suggested that p53 may contribute to the maintenance of reproductive function by regulating autophagy, cellular senescence, energy metabolism, and redox balance.

3.1.3. Functional Role of p53 in Maintaining Reproductive Function

In the endometrium, the role of p53 in regulating autophagy and cellular senescence has been increasingly clarified based on recent findings. In normal endometrial tissue, autophagy is a fundamental physiological process required for maintaining cellular homeostasis, tissue remodeling, and adaptation to fluctuations in the hormonal environment. Estrogen, in particular, regulates the transcription of autophagy-related genes through its receptors (ERα, ERβ, and GPER (G protein-coupled estrogen receptor 1)). Moreover, estrogen modulates autophagy activity through multiple layers of epigenetic and post-transcriptional mechanisms, including the action of transcription factors, microRNAs, and histone modifications [84]. Endometrial cells dynamically modulate autophagy activity in response to cyclic changes in estrogen and progesterone concentrations during the menstrual cycle. This regulation enables the functional restructuring of the endometrium in preparation for embryo implantation and pregnancy [47]. In particular, the decline in ovarian steroid hormone levels during the late menstrual phase is believed to activate autophagy, contributing to endometrial breakdown, clearance of cellular debris, and immune system activation (Figure 1, bottom left). Autophagy activity markedly increases during the transition from the secretory phase to the menstrual phase [84], underscoring its importance in the cyclic regeneration of the endometrium. Conversely, impaired or dysregulated autophagy can result in defective decidualization and severely compromise endometrial function [85]. Abnormalities in autophagy have also been associated with pathological conditions such as infertility [86], endometriosis [87], and endometrial cancer [88]. Furthermore, estrogen deficiency has been shown to enhance autophagy and induce apoptosis in endometrial epithelial cells in both pre- and postmenopausal women [89] and in ovariectomized rat models [90]. These findings suggest that the coordinated regulation of autophagy and apoptosis plays a role in endometrial atrophy. Age-related declines in estrogen levels are thought to alter autophagy activity, contributing to reduced endometrial function. Thus, ovarian steroid hormones are critically involved in maintaining the homeostasis of reproductive tissues through the regulation of autophagy [89,90].

In parallel with these shifts in autophagy, p53 expression in the endometrium is also known to vary in response to hormonal cues. Estrogen has been reported to enhance p53 promoter activity via transcription factors such as Nuclear Factor kappa B (NF-κB) and Sp1, leading to increased transcription of the p53 gene [91]. Notably, p53 expression rises significantly during the proliferative phase of the menstrual cycle, suggesting a role for p53 in cell cycle regulation and DNA damage repair during periods of active cell proliferation and tissue remodeling [92]. These findings imply that hormone-dependent regulation of p53 is critical for maintaining endometrial functional integrity. Moreover, studies using a uterus-specific p53 knockout mouse model have demonstrated reduced decidual development and increased cellular senescence, supporting the notion that p53 is essential for endometrial functional maturation and the regulation of senescence [93]. However, the reported dynamics of p53 expression across the menstrual cycle are inconsistent. For example, one study reported higher p53 expression during the secretory phase compared to the proliferative phase [94]. Therefore, further research is needed to clarify the relationship between hormonal fluctuations and p53 expression in the endometrium.

p53 also plays a critical role in maintaining reproductive function. In p53-deficient female mice, the expression of leukemia inhibitory factor (LIF)—a cytokine essential for embryo implantation—was reduced, resulting in significantly lower implantation and pregnancy rates. These findings suggest that p53 contributes to successful implantation by promoting the transcription of LIF [95]. Feng et al. demonstrated that the p53 family (p53, p63, and p73) plays a central role in female reproductive function, particularly in embryo implantation and pregnancy maintenance [96]. In p63-deficient mice, diminished ovarian function and reduced fertility were observed, indicating that p63 is involved in the formation of mature oocytes and the regulation of apoptosis within the ovary. In p73-deficient mice, increased spindle abnormalities and chromosomal aneuploidy were found in oocytes, suggesting that p73 contributes to the spindle assembly checkpoint that ensures accurate chromosome segregation during meiosis. Furthermore, in humans, a specific single nucleotide polymorphism (SNP) in the p73 gene was found to be significantly more prevalent among in vitro fertilization (IVF) patients aged 35 years or older, suggesting a link to poor oocyte quality and implantation failure [96]. Collectively, these findings indicate that the p53 family acts as a key regulator of reproductive processes, including embryo implantation and the preservation of oocyte quality. In addition, a study by Delenko et al. reported that quercetin—a flavonoid with antioxidant and senolytic properties—promotes the decidualization of human endometrial stromal cells via the AKT–ERK–p53 signaling pathway [97]. Their findings suggest that quercetin may enhance decidualization in patients with endometriosis and thereby contribute to improved endometrial function. Experimental studies further support the essential role of p53 in decidualization, showing that p53 activation promotes this process while its inhibition suppresses it. Taken together, these studies strongly suggest that p53 is intricately involved in the regulation of cellular senescence, autophagy, energy metabolism, and decidualization within the endometrium and that it plays a pivotal role in maintaining female reproductive function.

3.2. AMPK

AMPK is a crucial metabolic sensor that maintains cellular energy homeostasis by responding to fluctuations in intracellular energy status. It is activated under energy stress conditions such as glucose deprivation, reduced ATP levels, and increased concentrations of AMP and ADP [98]. To restore cellular energy balance, AMPK promotes glucose uptake and fatty acid β-oxidation and facilitates ATP resynthesis while simultaneously inhibiting ATP-consuming anabolic pathways, including lipid and protein synthesis. Additionally, AMPK contributes to metabolic reprogramming by inducing mitochondrial biogenesis and autophagy, thereby supporting cellular adaptation and survival under stress conditions. AMPK activation is mediated by several upstream kinases, primarily LKB1 [99], CaMKKβ (calcium/calmodulin-dependent protein kinase kinase β) [99], and ATM (ataxia telangiectasia mutated) [100]. LKB1 phosphorylates and activates AMPK in response to energy depletion, such as decreased ATP or elevated AMP levels. CaMKKβ is involved in AMPK activation in response to increased intracellular Ca^{2+} concentrations ($[Ca^{2+}]_i$), while ATM, a serine/threonine kinase primarily involved in the DNA damage response, activates AMPK in the context of oxidative stress and DNA damage-induced cellular stress. These kinases dynamically regulate AMPK activity in response to diverse stress signals, playing a key role in preserving metabolic homeostasis and promoting cell survival. Moreover, AMPK modulates the nutrient-sensitive sensor mTORC1, either directly or indirectly, thereby inhibiting anabolic processes and promoting catabolic pathways to reestablish metabolic balance [101,102] (see Figure 1 ⑤). AMPK also induces autophagy while preventing its excessive activation, contributing to the proper regulation and maintenance of the autophagic machinery and overall cellular homeostasis [102]. Thus, AMPK functions not only as an energy sensor but also as a central regulator of autophagy, making it indispensable for stress adaptation and cellular equilibrium. Additionally, AMPK is thought to regulate autophagy through crosstalk with the p53 and mTOR pathways and to play a significant role in reproductive cell functions, including cellular senescence and the decidualization of endometrial stromal cells.

3.2.1. Functional Roles of AMPK in Cellular Senescence and Autophagy Regulation

For detailed information regarding the signaling pathways involving upstream regulators and downstream targets of AMPK, please refer to previous studies [98,102]. This subsection focuses on the dual role of AMPK in the regulation of cellular senescence. When intracellular energy levels decline, AMPK becomes activated and regulates the activity of mTORC1 via downstream targets such as TSC2 and Rheb [103] (Figure 3 ①). mTORC1 is a key regulator of cell growth and protein synthesis, and its suppression contributes to energy conservation and metabolic reprogramming. Under energy-deprived conditions such as glucose starvation and ATP depletion, AMPK activation leads to the phosphorylation of ULK1 (Unc-51-Like Autophagy-Activating Kinase 1), inducing autophagy. This facilitates the selective removal of damaged organelles and misfolded proteins, thereby contributing to the maintenance of cellular homeostasis and the suppression of aging [104,105] (Figure 3 ②). Conversely, recent studies have suggested that AMPK may also negatively regulate ULK1 activity, thereby suppressing autophagy [106,107]. It has also been reported that AMPK can differentially regulate both the early (autophagosome formation) and late (lysosomal activation) phases of autophagy through distinct mechanisms [107]. These findings suggest that AMPK may either promote or inhibit autophagy via ULK1, depending on the cell's energy status, nutrient availability, and type of stress, indicating a dynamic and context-dependent interaction. In this way, the AMPK–ULK1 interaction appears to play a "dual" role, maintaining basal autophagic function while preventing its excessive activation [102]. Moreover, AMPK enhances the activity of nicotinamide phosphoribosyltransferase (NAMPT), thereby promoting the biosynthesis of NAD^+ (oxidized nicotinamide adenine dinucleotide), which serves a protective role against oxidative stress-induced cellular senescence [108,109] (Figure 3 ③). Elevated NAD^+ levels enhance the activity of the NAD^+-dependent deacetylase SIRT1 (sirtuin 1), which in turn induces the deacetylation and transcriptional activation of PGC-1α (peroxisome proliferator-activated receptor gamma coactivator 1-alpha) and the FOXO (Forkhead box O) transcription factors [110]. This pathway promotes mitochondrial biogenesis and function, contributing significantly to the extension of cellular lifespan and enhancement of stress resistance. Under acute stress or energy-deprived conditions, AMPK also suppresses inflammation by inhibiting the NF-κB (Nuclear Factor kappa B)–SASP–STAT3 (Signal Transducer and Activator of Transcription 3) signaling axis, thereby mitigating the progression of aging [111,112] (Figure 3 ④). In many cell types, AMPK activation has demonstrated anti-aging effects through the maintenance of metabolic homeostasis and enhancement of oxidative stress resistance.

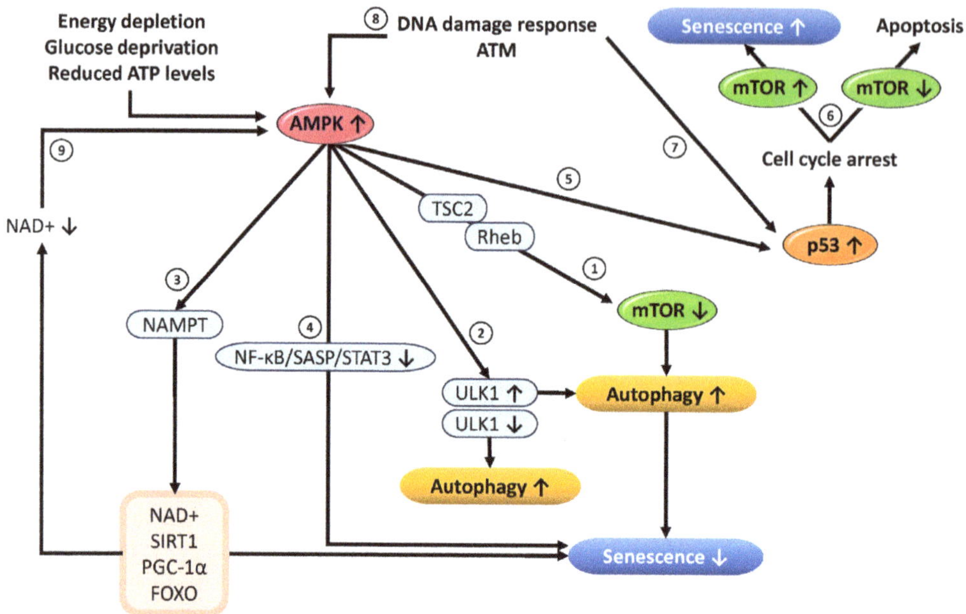

Figure 3. Functional role of AMPK in cellular senescence and autophagy regulation. AMPK regulates senescence via mTORC1 inhibition (①) and autophagy through ULK1 (②), though it may also suppress autophagy. It boosts NAD$^+$ via NAMPT (③), reducing stress and inflammation through NF-κB–SASP–STAT3 inhibition (④). AMPK activates p53 under stress (⑤), inducing p21 (⑥). DNA damage activates ATM→p53 (⑦), ATM→AMPK (⑧), and NAD$^+$/NADH shifts trigger AMPK→p53→p21 (⑨). AMPK, AMP-activated protein kinase; FOXO, Forkhead box O; mTOR, mechanistic target of rapamycin; NAD+, nicotinamide adenine dinucleotide; NAMPT, nicotinamide phosphoribosyltransferase; NF-κB, Nuclear Factor kappa B; PGC-1α, peroxisome proliferator-activated receptor gamma coactivator 1-alpha; Rheb, Ras homolog, mTORC1 binding; SASP, senescence-associated secretory phenotype; SIRT1, sirtuin 1; STAT3, Signal Transducer and Activator of Transcription 3; TSC2, TSC complex subunit 2; ULK1, Unc-51-Like Autophagy-Activating Kinase 1.

On the other hand, under metabolic stress conditions such as glucose deprivation, AMPK activates p53 (tumor protein p53), leading to cell cycle arrest [113,114] (Figure 3 ⑤). Activated p53 induces the transcription of the cyclin-dependent kinase inhibitor p21, triggering permanent cell cycle arrest (senescence) or apoptosis [113,115] (Figure 3 ⑥). Mathematical modeling studies by Zhou and Liu have proposed a mechanism in which the AMPK–p53 axis governs cell cycle arrest: if glucose levels recover early, cells can resume proliferation, but if energy depletion persists, moderate p53 activation sustains senescence, and prolonged stress may lead to high p53 expression and inhibition of mTOR and Akt, resulting in apoptosis [113]. The biphasic activation of p53 acts as a critical switch in cell fate determination. The mTOR pathway may further regulate p53 activity either positively or negatively, depending on cell type and stress context [32]. Thus, under excessive metabolic stress, AMPK may also promote aging. In the DNA damage response, ATM kinase phosphorylates p53 and induces downstream expression of p21 [116] (Figure 3 ⑦). AMPK can also be activated by ATM [117], potentially contributing to the induction of cellular senescence (Figure 3 ⑧). Furthermore, a reduced intracellular NAD$^+$/NADH ratio can activate AMPK via ATP depletion, which in turn promotes p53 phosphorylation and accelerates senescence [118] (Figure 3 ⑨→⑤→⑥). Interestingly, in human dental follicle cells, both excessive activation and suppression of AMPK have been reported to induce cellular senescence [119]. Additionally, long-term administration of AMPK activators such as metformin or AICAR (5-aminoimidazole-4-carboxamide ribonucleoside) may promote aging through autophagy dysfunction and mitochondrial abnormalities [120]. Meanwhile, p53 also has anti-inflammatory effects, such as suppressing the expression of inflammatory cytokines like IL-1 [121]. Therefore, while the AMPK–p53 axis contributes to the establishment of senescence via growth arrest, it also suppresses inflammatory components of the SASP, serving a dual function. This may lead to a less inflammatory phenotype—referred to as "silent senescence"—which could help prevent the progression of diseases driven by chronic inflammation [121].

In summary, under conditions of energy stress or oxidative damage, AMPK functions to suppress cellular senescence by regulating autophagy, maintaining NAD^+ homeostasis, and inhibiting inflammatory responses. However, depending on the duration of its activation, cell type, and nature of the stimuli, AMPK can also promote senescence by inducing mitochondrial dysfunction or the expression of senescence-associated genes. Thus, AMPK is emerging as a dual-faced and context-dependent modulator of cellular senescence.

3.2.2. Functional Role of AMPK in Metabolism

AMPK functions as a central sensor for maintaining intracellular energy homeostasis, and its activation is closely linked to cellular metabolic adaptation. Upon activation, AMPK orchestrates multiple metabolic pathways, particularly by promoting ATP-generating processes such as glycolysis, thereby eliciting a metabolic response to energy deficiency [122,123]. Specifically, AMPK suppresses gluconeogenesis by downregulating the expression of gluconeogenic enzymes such as phosphoenolpyruvate carboxykinase (PEPCK) and glucose-6-phosphatase (G6Pase) [124,125] while simultaneously enhancing the translocation of the glucose transporter GLUT4 to the plasma membrane, thereby increasing glucose uptake into the cell [126,127]. In addition, AMPK activation enhances the activity of key glycolytic enzymes such as hexokinase, PFK1, and PFK2, which rapidly restore intracellular ATP levels [127,128]. Concurrently, AMPK inhibits anabolic pathways to conserve energy. For instance, it phosphorylates and inactivates 3-hydroxy-3-methylglutaryl-CoA reductase (HMG-CoA reductase), a rate-limiting enzyme in the cholesterol biosynthesis pathway, thereby suppressing the mevalonate pathway and downstream cholesterol synthesis, ultimately reducing ATP consumption [129]. Similarly, AMPK inhibits acetyl-CoA carboxylase (ACC), a key enzyme in fatty acid synthesis, through phosphorylation, contributing further to the conservation of cellular energy [126,127]. Conversely, under conditions of sufficient energy availability or in scenarios requiring long-term metabolic restructuring, AMPK suppresses glycolysis and promotes a metabolic shift toward oxidative pathways. This shift involves enhanced mitochondrial β-oxidation and the activation of transcriptional coactivators such as PGC-1α, which facilitate mitochondrial biogenesis and enhance oxidative phosphorylation [126,130]. Through this mechanism, AMPK enables cells to transition from glycolysis to the more efficient ATP production pathway of oxidative phosphorylation. Moreover, AMPK activation strongly supports oxidative phosphorylation by modulating transcription factors, phosphorylating substrate proteins, regulating mitochondrial dynamics, and inducing autophagy, particularly selective mitochondrial degradation (mitophagy) [130,131]. Through these actions, AMPK serves as a master regulator of metabolism, promoting ATP-generating pathways while suppressing energy-consuming anabolic processes to maintain energy homeostasis. Notably, AMPK exhibits bidirectional regulation of glycolysis, promoting or suppressing it depending on the metabolic context. This dual capability underpins its role as a "metabolic master switch," allowing cells to swiftly adapt to acute energy deprivation while also facilitating long-term metabolic optimization and homeostatic maintenance. Through this flexible regulatory capacity, AMPK plays a pivotal role in ensuring cellular survival and functional integrity under various environmental changes and stress conditions.

3.2.3. Functional Role of AMPK in the Maintenance of Reproductive Function

AMPK plays a central role in suppressing the progression of cellular senescence through the regulation of autophagy and intracellular metabolic homeostasis. In skeletal muscle, cardiac muscle, and the central nervous system, the responsiveness to AMPK activation has been reported to decline markedly with age [98]. One proposed mechanism for this phenomenon is that age-related chronic inflammation—referred to as "inflammaging"—may inhibit AMPK activation [98]. Inducers of cellular senescence include oxidative stress, chronic inflammation, aberrant accumulation of $[Ca^{2+}]_i$, and the activation of DNA damage response pathways. These factors are associated with the aging of endometrial stem cells and defective decidualization [132]. In particular, $[Ca^{2+}]_i$ is known to be involved in cell fate determination processes such as autophagy, apoptosis, and necrosis under oxidative stress conditions, and its precise regulation is critically important for the suppression of cellular senescence [133]. The removal of intracellular calcium at high concentrations has been shown to induce AMPK-dependent autophagy and may suppress premature senescence in human endometrial stromal cells [133].

Several studies have demonstrated that AMPK plays an important role in female reproductive function, particularly in maintaining metabolic homeostasis, suppressing cellular senescence, and regulating decidualization. For example, Kawano et al. reported that inflammatory cytokine IL-1β enhances the production of IL-8, MCP-1 (monocyte chemoattractant protein-1), PGE_2 (prostaglandin E_2), and $PGF_2\alpha$(prostaglandin $F_2\alpha$) in endometrial stromal cells. However, treatment with the AMPK activator AICAR significantly reduced the production of these inflammatory mediators [134]. This finding suggests that AMPK activation may mitigate excessive inflammatory responses in the endometrium. Furthermore, experiments using genetically modified mice lacking the catalytic AMPK subunits Prkaa1 and Prkaa2, specifically in the uterus (Prkaa1/2 d/d mice), revealed impaired endometrial regeneration postpartum, accompanied by fibrosis and structural disorganization [135]. Notably, AMPK inhibited hormonally induced artificial decidualization without affecting ovarian function or hormone production, indicating that AMPK is indispensable for decidualization and uterine tissue repair after parturition [135]. Another study demonstrated that AMPK deficiency impairs the proliferation of endometrial epithelial cells and disrupts the decidualization process, suggesting a crucial role for AMPK in hormonal responsiveness, endometrial receptivity, and embryo implantation during early pregnancy [136]. A study by Zhang et al. further reported that folate deficiency suppresses autophagy via the AMPK/mTOR signaling pathway, thereby impairing the decidualization process [137]. This finding supports the notion that AMPK contributes to the functional differentiation of the endometrium via autophagy, playing a critical role in the establishment of pregnancy.

Collectively, these findings clearly demonstrate that AMPK functions as a key regulatory factor in hormonal response modulation, endometrial regeneration, decidualization, and overall female reproductive function. Therefore, AMPK dysfunction is likely to contribute to the pathogenesis of infertility and pregnancy-related complications. Elucidating the molecular mechanisms underlying AMPK activity may provide important insights into the development of novel therapeutic targets in reproductive medicine.

3.3. *mTOR*

mTOR is classified as a serine/threonine-specific protein kinase. It functions as a central intracellular signaling molecule that integratively regulates diverse cellular processes—such as protein synthesis, lipid metabolism, and autophagy—in response to extracellular and intracellular nutritional status, energy levels, and the presence of growth factors [138]. As a "master regulator" of cellular growth, metabolism, proliferation, and survival, mTOR forms two functionally distinct multi-protein complexes: mTORC1 and mTORC2, each of which governs specific biological processes in response to different stimuli [32,113,139]. mTORC1 is activated by growth factors such as insulin and insulin-like growth factors (IGFs), essential amino acids (especially leucine and arginine), and high-energy states (e.g., elevated intracellular ATP levels). Once activated, mTORC1 phosphorylates ribosomal S6K1 and eIF4E-binding protein 1 (4E-BP1), thereby promoting mRNA translation and inducing protein synthesis [139,140]. Additionally, mTORC1 promotes lipid biosynthesis by regulating the expression and activity of enzymes involved in lipid synthesis, and it activates de novo purine and pyrimidine nucleotide synthesis pathways, supporting the supply of biosynthetic substrates essential for cell proliferation. In this way, mTORC1 contributes to cell growth and homeostasis by promoting anabolic pathways and simultaneously suppressing catabolic processes such as autophagy. On the other hand, mTORC2 is primarily activated by growth factor stimulation and T-cell receptor signaling. It induces the activation of Akt through phosphorylation at Ser473, thereby regulating processes including cell survival, energy metabolism, and actin cytoskeleton remodeling [139]. Moreover, mTORC2 plays a critical role in insulin signaling and is also involved in metabolic regulation, such as lipogenesis [141].

3.3.1. Functional Role of mTOR in Cellular Senescence and Autophagy Regulation

This subsection focuses on the functional significance of mTOR in the regulation of autophagy and cellular senescence. The PI3K/AKT/mTOR pathway is a pivotal intracellular signaling cascade that orchestrates a wide spectrum of cellular functions, including growth, proliferation, metabolism, and survival. Upon activation by growth factors and other extracellular cues, the PI3K/AKT axis phosphorylates and inhibits the tuberous sclerosis complex (TSC1/2), thereby relieving its suppressive effect on mTORC1. Subsequently, activated mTORC1 phosphorylates downstream effectors such as S6K1 [142] and eukaryotic translation initiation factor 4E-BP1 [143], thereby facilitating protein synthesis and promoting cell proliferation. This process contributes to cell hypertrophy, a characteristic morphological feature of senescent cells [144] (Figure 4 ①). The excessive accumulation of biomass under non-proliferative conditions is also involved in maintaining permanent cell cycle arrest, a key feature in the development of the senescence phenotype. mTOR further induces the expression of SASP factors, including pro-inflammatory cytokines and growth factors secreted by senescent cells. In particular, the translation of IL-1$\alpha$is mTOR-dependent, and activation of mTOR signaling enhances IL-1α expression and promotes SASP induction (Figure 4 ②). Moreover, mTORC1 negatively regulates autophagy through phosphorylation of the ULK1 complex (Unc-51-Like Autophagy-Activating Kinase 1, ATG13, FIP200 (FAK family kinase-interacting protein of 200 kDa), and ATG101), which is involved in the initiation of autophagy [105]. This phosphorylation-dependent inhibition directly suppresses autophagy initiation, supporting the role of mTORC1 as a negative regulator of autophagy (Figure 4 ③). Therefore, excessive activation of mTORC1 impairs the proper clearance of misfolded proteins and damaged organelles, leading to the accumulation of intracellular stress, which in turn contributes to the progression of cellular senescence and an increased risk of age-related diseases. In contrast, under nutrient-deprived or starvation conditions, inhibition of mTORC1 allows activation of the ULK1 complex, thereby inducing autophagy. In senescent cells, however, mTORC1 activity remains persistently high, and sensitivity to nutrient and growth factor signals is diminished. This sustained mTORC1 activation is closely associated with the emergence of senescence-related phenotypes. Interestingly, even under conditions of extracellular nutrient depletion, intracellular amino acids produced by autophagy have been reported to maintain mTORC1 activity [145] (Figure 4 ④), suggesting a potential positive feedback loop between autophagy and mTORC1. This dynamic interaction plays a central role in metabolic adaptation, maintenance of homeostasis, and regulation of senescence. In addition, mTOR functions as a key regulatory factor that determines cell fate—proliferation, senescence, or apoptosis—in response to external stimuli, such as cellular stress and nutrient availability, through interactions with p53 and AMPK. For instance, in response to stress stimuli such as DNA damage or energy depletion, p53 is known to suppress mTORC1 via activation of AMPK and induction of REDD1 expression [140] (Figure 4 ⑤). This inhibition of mTORC1 leads to attenuation of cell growth and protein synthesis and promotes autophagy. Conversely, treatment with the mTORC1 inhibitor rapamycin reduces the binding of p53 mRNA to polysomes and decreases its translational efficiency, suggesting that mTORC1 directly regulates p53 protein synthesis [146] (Figure 4 ⑥). Moreover, genotoxic stress, such as DNA damage, activates the mTOR-S6K1 signaling axis. The activated S6K1 exhibits increased affinity for MDM2, thereby attenuating MDM2-mediated ubiquitination of p53 and promoting its stabilization and accumulation [147] (Figure 4 ⑦). Under specific conditions, such as in PTEN-deficient cells, both mTORC1 and mTORC2 have been reported to compete with MDM2 for binding to p53, promoting phosphorylation at Ser15 and increasing its transcriptional activity [148]. In this context, mTOR may positively regulate p53 stability and activation depending on the cellular conditions. Conversely, activation of the PI3K/AKT pathway induces phosphorylation of specific serine residues within MDM2, enhancing its nuclear translocation and protein stability. This post-translational modification augments MDM2's E3 ubiquitin ligase activity toward p53, facilitating its proteasomal degradation [149] (Figure 4 ⑧). Furthermore, mTORC1 activation has been shown to upregulate MDM2 translation, thereby reinforcing its role as a key negative regulator of p53 [150]. This regulatory mechanism is governed via the PI3K/AKT/mTOR axis. For instance, growth factors such as insulin-like growth factor 1 (IGF-1) and hepatocyte growth factor (HGF) activate the PI3K/AKT pathway, which in turn stimulates mTORC1. Activated mTORC1 enhances the translational efficiency of MDM2, resulting in elevated protein levels. The increased abundance of MDM2 promotes accelerated p53 degradation, thereby modulating cell proliferation and survival. This intricate regulatory interplay between the mTOR and p53 pathways under stress conditions has been elucidated in detail in prior investigations [32]. In summary, activation of the PI3K/AKT/mTOR pathway facilitates p53 degradation by promoting MDM2 translation (Figure 4 ⑧).

Collectively, these findings demonstrate that the mTOR pathway exerts multifaceted effects on p53 expression and activity and plays a crucial role in determining cell fate. Conversely, p53 negatively regulates mTORC1 through AMPK activation and REDD1 induction, thereby maintaining cellular homeostasis by suppressing cell growth and survival [32]. Thus, mTOR acts as a bidirectional regulatory factor that modulates p53 activity positively or negatively depending on the cellular context through mechanisms involving translational control, regulation of protein stability, and post-translational modifications. The interaction between p53 and the mTOR pathway serves as a central control mechanism in the decision between cellular adaptation and proliferation or survival under stress stimuli. As shown in Figure 3 ①, mTOR is also functionally connected to AMPK [103]. Together, p53, AMPK, and mTOR independently regulate autophagy, cell growth, senescence, and apoptosis while also forming an integrated regulatory network through close interactions. This network serves as a fundamental control system that enables cells to appropriately respond to environmental stress and make precise decisions regarding survival, repair, or cell death.

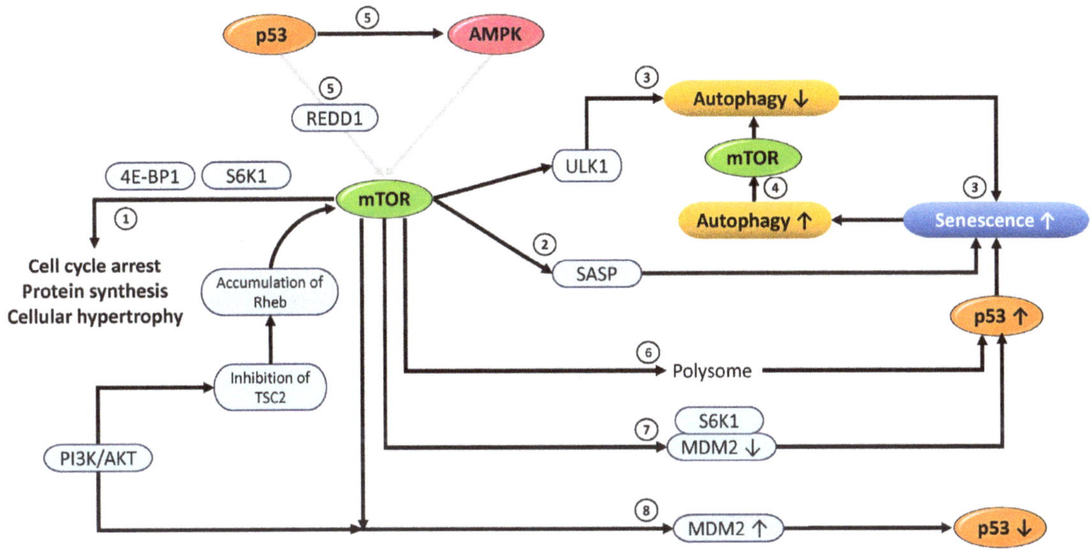

Figure 4. Functional role of mTOR in cellular senescence and autophagy regulation. mTORC1 promotes hypertrophy via S6K1/4E-BP1 (①) and induces SASP through IL-1α translation (②). It suppresses autophagy by ULK1 complex phosphorylation (③). Even during starvation, autophagy-derived amino acids sustain mTORC1 (④). p53 AMPK inhibits mTORC1 (⑤), while mTORC1 regulates p53 via translation (⑥), stabilization (⑦), or degradation (⑧). Black arrows indicate facilitative or activating effects, whereas gray arrows denote inhibitory actions. 4E-BP1, eIF4E-binding protein 1; AMPK, AMP-activated protein kinase; Akt, protein kinase B; MDM2, mouse double minute 2 homolog; mTOR, mechanistic target of rapamycin; PI3K, phosphoinositide 3-kinase; REDD1, regulated in development and DNA damage responses 1; S6K1, ribosomal S6 kinase; SASP, senescence-associated secretory phenotype; ULK1, Unc-51-Like Autophagy-Activating Kinase 1.

Autophagy is a tightly regulated intracellular catabolic mechanism that progresses through several distinct phases: initiation, nucleation, elongation and maturation, fusion, and ultimately degradation and recycling [151]. This subsection delineates the specific stages at which p53, AMPK, and mTORC1 exert regulatory influence. The process is triggered by the activation of the ULK1 complex, initiating the formation of phagophores. During the nucleation phase, the activated ULK1 complex stimulates the class III PI3K complex, promoting the generation of phosphatidylinositol 3-phosphate (PI3P) and thereby facilitating phagophore initiation. Subsequent expansion of the phagophore leads to the sequestration of cytoplasmic components. This stage involves ubiquitin-like conjugation systems that drive membrane elongation and closure, culminating in the formation of autophagosomes. Mature autophagosomes fuse with lysosomes to form autophagolysosomes, wherein the enclosed materials are enzymatically degraded. The resulting macromolecules are then recycled into the cytosol for reuse in cellular metabolism (Figure 5). Nuclear p53 functions as a transcriptional activator of several autophagy-related genes, including DRAM1, Sestrin1/2, TSC2, PTEN, and members of the ATG family, particularly under stress conditions such as genotoxic damage or nutrient scarcity [50–52,57]. Conversely, gain-of-function mutations in p53 have been shown to repress the transcription of key autophagy genes such as BECN1, DRAM1, and ATG12 while concurrently activating the mTOR pathway, thereby inhibiting autophagy and facilitating tumor cell survival and proliferation [152]. AMPK directly phosphorylates ULK1 at serine residues Ser317 and Ser777, activating the ULK1 complex and promoting autophagosome formation [105]. However, under nutrient-rich conditions, mTORC1 phosphorylates ULK1 at Ser757, thereby disrupting its interaction with AMPK and suppressing autophagy initiation [105].

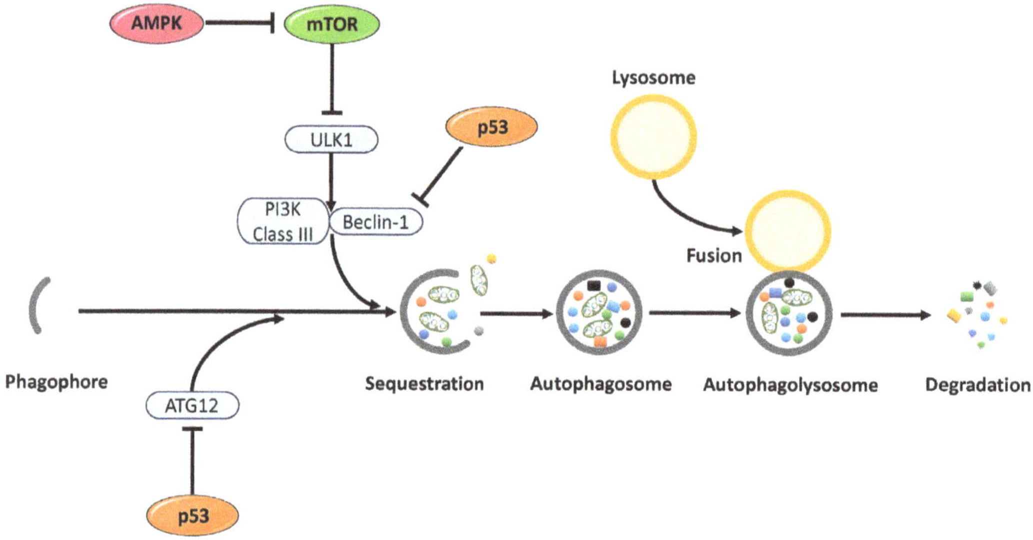

Figure 5. Regulatory involvement of p53, AMPK, and mTORC1 in the autophagy pathway. p53, AMPK, and mTORC1 primarily function at the initiation stage of autophagy. AMPK, AMP-activated protein kinase; ATG12, autophagy-related 12; mTOR, mechanistic target of rapamycin; PI3K, phosphoinositide 3-kinase; ULK1, Unc-51-Like Autophagy-Activating Kinase 1.

3.3.2. Functional Role of mTOR in Metabolism

The mTOR, similar to p53 and AMPK, plays a critical role in cellular energy homeostasis and metabolic regulation, with its activation closely associated with the metabolic status of the cell. In particular, mTORC1 serves as a major metabolic regulator that promotes cell growth and biosynthesis through the activation of glycolysis [153] (right panel of Figure 2). mTORC1 strongly enhances glycolytic flux via multiple mechanisms, including upregulation of the glucose transporter GLUT1, increased expression and activity of glycolytic enzymes, stimulation of hypoxia-inducible factor 1-alpha (HIF-1α) translation, and activation of the transcription factor MYC [153]. Specifically, mTORC1 promotes the localization of GLUT1 to the plasma membrane through the PI3K-Akt signaling pathway, thereby enhancing cellular glucose uptake. It also induces the expression of key glycolytic enzymes, including hexokinase 2 (HK2), PFKs, pyruvate kinase M2 (PKM2), and LDH, leading to increased glycolytic flux [154]. Additionally, mTORC1 facilitates the translation of HIF-1α, which in turn activates the transcription of glycolysis-related genes such as GLUT1, HK2, PKM2, and LDHA (lactate dehydrogenase A) [138,155]. Through the enhancement of MYC (MYC proto-oncogene, bHLH transcription factor) transcriptional activity, mTORC1 further augments the expression of glucose transporters and glycolytic enzymes, thereby promoting glucose metabolism [156]. Moreover, mTORC1 stimulates protein translation via phosphorylation of the 4E-BP, activating the translation of nuclear-encoded mitochondrial genes such as PGC-1α, TFAM (mitochondrial transcription factor A), YY1 (Yin Yang 1), mitochondrial ribosomal components, and subunits of the electron transport chain complexes I, II, IV, and V [140,157–159]. These actions enhance mitochondrial biogenesis and function, leading to the promotion of OXPHOS and increased ATP production capacity. Additionally, mTORC1 has been reported to regulate the transcription of glycolysis-related genes via NEAT1 (nuclear paraspeckle assembly transcript 1), a long non-coding RNA (lncRNA) [160]. This metabolic activation contributes to a reprogramming of the metabolic profile in senescent cells, suggesting the presence of a feedback loop between metabolism and senescence that can further modulate mTOR activity. Such a mechanism may be involved in the progression of cellular senescence and deterioration of metabolic function [118]. However, the glycolysis-promoting function of mTOR is known to be context-dependent. Under environmental conditions such as nutrient deprivation or oxidative stress, adaptive suppression of mTOR activity occurs to balance energy demand and survival strategies [161]. Thus, mTOR can be considered a central regulatory factor in metabolic sensing, plastically modulated in response to intra- and extracellular metabolic cues.

3.3.3. Functional Role of mTOR in the Maintenance of Reproductive Function

Reduced autophagic activity has been reported to be closely associated with the progression of cellular senescence in various cell types, including hepatocytes [162], cardiomyocytes [163], and microglia [164]. For instance, in human cardiac progenitor cells, administration of the mTOR inhibitor rapamycin has been shown to alleviate replicative senescence and improve cellular function [165]. The mTOR signaling pathway functions as an aging-promoting factor in multiple model organisms, including Saccharomyces cerevisiae (budding yeast), Caenorhabditis elegans (nematodes), Drosophila melanogaster (fruit flies), and mice [46,166–169]. Moreover, genetic and pharmacological interventions such as overexpression of ATG genes, inhibition of mTOR signaling, or administration of autophagy-inducing compounds have been demonstrated to enhance autophagy, contributing to lifespan extension and delayed aging. In particular, inhibition strategies targeting both mTORC1 and mTORC2 complexes have garnered attention as potential therapeutic approaches for age regulation and age-related diseases, as they may enable reversible control of senescence-associated phenotypes in aging cells [169,170]. These findings underscore the protective role of autophagy in mitigating aging through the maintenance of cellular homeostasis and structural integrity. Dysregulation of mTOR signaling has been widely implicated in the onset and progression of various pathological conditions, including cancer, obesity, type 2 diabetes, metabolic syndrome, and neurodegenerative diseases such as Alzheimer's disease [138,171,172]. Consequently, suppression of aging via mTOR inhibition remains an area of intense ongoing research.

In addition, mTOR has been shown to play a crucial regulatory role in maintaining reproductive function, including processes such as cellular senescence and decidualization. Baek et al. reported that, in human endometrial stromal cells, cAMP-induced decidualization is associated with increased activity of mTORC1 and decreased activity of mTORC2 [173]. In this study, mTORC1 activation was found to promote the expression of decidualization markers—prolactin (PRL) and insulin-like growth factor-binding protein 1 (IGFBP1)—through the transcription factor FOXO1, identifying mTORC1 as a key regulator of decidualization. Similarly, Hirota et al. demonstrated using a mouse model that hyperactivation of uterine mTORC1 signaling induces premature decidual senescence, resulting in preterm birth and fetal loss [13]. This pathological condition was shown to be ameliorated by low-dose administration of rapamycin, suggesting that proper regulation of mTORC1 signaling is essential for maintaining a normal pregnancy duration. Furthermore, a review by Guo et al. outlined the central role of mTOR signaling in female reproductive functions, including ovarian activity, cyclic endometrial remodeling, embryo implantation, and placental formation [14]. Notably, dysregulation of the mTOR pathway has been linked to gynecological disorders such as polycystic ovary syndrome (PCOS), premature ovarian failure (POF), and endometriosis. Collectively, these findings clearly demonstrate that mTOR signaling is indispensable for female reproductive function, particularly in the regulation of cellular senescence and decidualization.

3.4. Mechanisms of Cellular Stress Response and Senescence Regulation via Autophagy

This subsection outlines how nutritional status and cellular stress influence cellular senescence through the regulation of autophagy, mediated by the p53, AMPK, and mTOR signaling pathways. In cellular and microenvironmental conditions characterized by an abundance of glucose, amino acids, and growth factors such as insulin, the mTOR signaling pathway becomes activated. This activation promotes protein synthesis, glycolysis, and lipid synthesis and, consequently, enhances cell proliferation (Figure 6, left panel). Concurrently, AMPK is suppressed, and in the absence of cellular stress, p53 activity is also maintained at a low level. Under these conditions, autophagy is generally inhibited, which supports short-term cell growth by promoting anabolic metabolism and suppressing the catabolic process of autophagy. However, sustained inhibition of autophagy can lead to the accumulation of damaged organelles and misfolded proteins, which may impair cellular function over time [174]. Since autophagy plays a crucial role in the removal of these intracellular damaged components and in the prevention of senescence, prolonged autophagy inhibition in nutrient-rich conditions may contribute to the breakdown of cellular homeostasis and accelerate aging processes [53]. Therefore, while suppression of autophagy under nutrient-excess conditions may support cell proliferation in the short term, it could also promote cellular senescence in the long term through the accumulation of cellular damage.

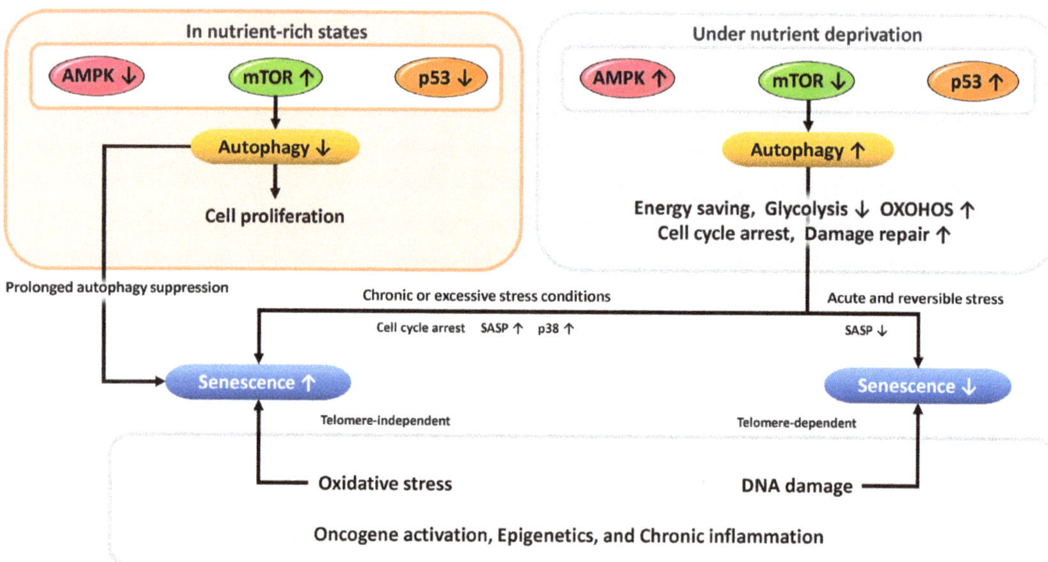

Figure 6. Mechanisms of cellular stress response and senescence regulation via autophagy. Nutrient-abundant conditions activate mTOR signaling while suppressing AMPK and p53 pathways, thereby inhibiting autophagy and promoting cellular growth. However, prolonged autophagy suppression may result in the accumulation of cellular damage, ultimately leading to the induction of senescence. Under nutrient deprivation, mTOR is suppressed, AMPK and p53 are activated, and autophagy is induced, aiding survival. Chronic stress or p38 MAPK activation promotes SASP and senescence. Autophagy has dual roles. AMPK, AMP-activated protein kinase; mTOR, mechanistic target of rapamycin; OXPHOS, oxidative phosphorylation; SASP, senescence-associated secretory phenotype.

Conversely, under nutrient-deprived conditions, cells halt proliferation and activate a metabolic program that prioritizes damage repair and energy conservation (Figure 6, right panel). In such states, mTOR activity is suppressed, leading to inhibition of protein synthesis and cell growth. Meanwhile, AMPK is activated, promoting fatty acid β-oxidation and the induction of autophagy via mTOR inhibition. Additionally, p53 is activated, inducing the expression of TIGAR, which suppresses glycolysis, and SCO2, which promotes mitochondrial oxidative phosphorylation. These changes, including cell cycle arrest and autophagy activation, facilitate the recycling of intracellular resources and the supply of essential metabolites, thereby contributing to cell survival and maintenance of function. Under specific conditions, autophagy induced through such metabolic adaptation has been shown to suppress the activation of senescence pathways. In other words, during acute and reversible stress, autophagy may function as a mechanism that counteracts senescence [175]. However, under chronic or excessive stress conditions, autophagy activation may instead promote cellular senescence [176]. In such aging processes, autophagy can provide the substrates necessary for the production of SASP factors, which act on the tissue microenvironment to reinforce and sustain the senescent state [177]. Furthermore, in response to stress signals such as oxidative stress, DNA damage, or oncogenic activation (e.g., Ras), p38 mitogen-activated protein kinase (p38 MAPK) is activated and has been shown to induce persistent senescence-associated phenotypes [178–180]. Activated p38 MAPK contributes to cellular senescence through the induction of cell cycle arrest and promotion of SASP. The involvement of p38 MAPK in senescence has been demonstrated in human endometrial-derived mesenchymal stem cells [181], human somatic stem cells [15], and fetal-derived cells [182].

There are at least two major types of cellular senescence. The first is telomere-dependent senescence, which is triggered by DNA damage responses caused by telomere shortening during cell division and involves activation of the p53/p21 pathway. The second is telomere-independent senescence, which is induced by growth factor stimuli such as insulin or oxidative stress, including damage from ROS, without telomere shortening [183]. In addition to these, factors such as oncogene activation, epigenetic modifications, and chronic inflammation also influence the regulation of cellular senescence. The effects of autophagy on senescence are context-dependent, varying with the nature of the stress, cellular differentiation stage or status, and the specific signaling pathways involved. Autophagy thus plays a dual role, capable of both suppressing and promoting senescence.

3.5. The Relationship Between Endometrial Cellular Senescence and Reproductive Capacity

Aging is a phenomenon characterized by the accumulation of widespread physiological changes over time, encompassing irreversible functional decline at the organismal level, including systemic deterioration of bodily functions, disruption of metabolic homeostasis, and reduced immune competence. In contrast, cellular senescence refers to a state in which cells undergo an irreversible arrest of proliferation, accompanied by distinctive morphological and functional changes. A hallmark of this state is the secretion of the SASP, which includes pro-inflammatory cytokines, chemokines, and growth factors. Cellular senescence is considered a contributing factor to aging, as the accumulation of senescent cells leads to impaired tissue function and chronic inflammation, thereby promoting the development of age-related diseases [184]. As illustrated in Figure 1, cellular senescence involves not only cell cycle arrest but also complex biological changes, including metabolic reprogramming. At the organismal level, age-related changes in metabolic profiles are also observed. For instance, in skeletal muscle, mitochondrial dysfunction leads to impaired OXPHOS, contributing to a decline in muscle mass and strength. Similarly, in endometrial cells, aging causes significant shifts in the balance between glycolysis and OXPHOS, leading to metabolic reprogramming. Specifically, aged endometrial cells exhibit hyperactivation of glycolysis [185], reduced OXPHOS, and mitochondrial dysfunction [186]. These metabolic abnormalities may impair decidualization and foster a pro-inflammatory microenvironment, ultimately diminishing endometrial receptivity and contributing to implantation failure, thereby negatively affecting reproductive function [187].

Several studies suggest that moderate cellular senescence may play a physiologically beneficial role in reproductive function, while excessive senescence could be associated with reduced implantation rates [13]. In fact, regulated cellular senescence plays an essential role in reproductive processes. For example, during the decidualization of endometrial stromal cells, the production of SASP factors is thought to enhance embryonic receptivity [188]. This controlled senescent state appears to be indispensable for establishing a receptive endometrial environment. However, excessive accumulation of senescent cells in the endometrium may exert detrimental effects on reproduction. According to a study by Tomari et al., human endometrial stromal cells from women with recurrent implantation failure exhibited elevated expression of the senescence markers p16 and p21, along with increased levels of SASP factors [189]. These changes were associated with decreased expression of stemness markers such as ABCG2 (ATP binding cassette subfamily G member 2) and ALDH1A1 (aldehyde dehydrogenase 1 family member A1), suggesting that excessive cellular senescence impairs endometrial regenerative capacity and reduces embryonic receptivity. Moreover, the disruption of interactions between senescent and immune cells is closely linked to implantation failure. Specifically, in women experiencing implantation failure, a functional disconnection has been observed between senescent endometrial cells and $CD4^+$ T helper cells, indicating that impaired immune modulation may hinder successful implantation [190]. In summary, while moderate cellular senescence is essential for normal decidualization and the establishment of endometrial receptivity, excessive senescence may contribute to implantation failure through immune dysregulation and diminished regenerative potential.

Cellular senescence and aging are closely intertwined. In particular, key signaling pathways involved in the regulation of cellular senescence—such as p53, AMPK, and mTOR—serve as important modulators of age-related decline in reproductive function. Therapeutic strategies targeting these pathways to regulate cellular senescence may offer promising approaches for the treatment of age-related infertility.

4. Discussion

With advancing age, the functional capacity of the endometrium gradually declines, leading to a concomitant reduction in reproductive potential. This functional alteration is closely associated with cellular senescence in the endometrium. Endometrial cells are chronically exposed to cellular stressors such as DNA damage and oxidative stress, which force them into a binary fate—either entering an irreversible state of growth arrest known as senescence or maintaining survival. This process is primarily regulated by the DNA damage response pathway centered on p53 and is further modulated through crosstalk with AMPK and mTOR signaling pathways, which are involved in cellular energy metabolism and nutrient sensing [191]. In recent years, the complex interplay between cellular senescence and signaling pathways such as p53, AMPK, and mTOR has been actively investigated by many researchers. These molecules function as key regulators that coordinately control energy homeostasis, autophagy, and the progression of senescence, each via distinct mechanisms. Notably, their actions are context-dependent and often dualistic in nature.

For instance, under nutrient-deprived conditions, p53 and AMPK are activated, leading to the suppression of the mTOR pathway and a reduction in energy consumption.

Conversely, in the presence of severe DNA damage or accumulated metabolic stress, persistent cell cycle arrest, induction of the SASP, and activation of the p38 MAPK pathway promote cellular senescence or apoptosis. On the other hand, if the cellular stress is mild and the damage is reversible, autophagy is induced, enabling the cell to escape senescence through transient cell cycle arrest [192]. Thus, p53, AMPK, and mTOR serve as central regulators that integrate cellular energy metabolism with the balance of growth, survival, senescence, and cell death. In addition to nutrient status, environmental factors such as chronic inflammation, oxidative stress involving epigenetic modifications, and DNA damage also play crucial roles in determining the cellular fate of endometrial cells—whether toward survival, senescence, or apoptosis.

Understanding the dualistic role of cellular senescence in the endometrium is essential for establishing therapeutic interventions targeting age-related reproductive decline. While appropriately regulated senescence responses contribute to optimizing endometrial receptivity, dysregulated senescence can impair endometrial function and reduce reproductive capacity. Recent attention has focused on the therapeutic potential of senolytics—agents such as quercetin and dasatinib—that selectively eliminate senescent cells. These compounds are expected to improve endometrial function and enhance implantation rates by removing dysfunctional senescent cells [193]. Findings from the present analysis suggest that p53 and AMPK activation, mTOR inhibition, the expression levels of local SASP factors, and the activation status of p38 MAPK in endometrial cells may serve as valuable biomarkers for assessing reproductive potential. For example, oxidative stress, exemplified by hydrogen peroxide exposure, activates the p38MAPK signaling pathway in human endometrial mesenchymal stem cells, thereby initiating an irreversible cell cycle arrest characteristic of cellular senescence [15]. Experimental evidence indicates that pharmacological inhibition of p38MAPK can partially reverse the senescent phenotype and reinstate proliferative capacity [15]. Age-related reproductive aging is believed to progress in a stepwise manner from a reversible to an irreversible pathological state; however, currently, no reliable biomarkers are available to distinguish these stages. Therefore, future research aimed at identifying and functionally characterizing such markers is anticipated to contribute to the prevention and treatment of reproductive aging.

The fundamental role of regulators such as p53, AMPK, and mTOR is thought to lie in their ability to suppress the proliferation of abnormal cells by inducing senescence, thereby functioning as a protective mechanism for tumor suppression [52]. At the same time, processes such as cell proliferation, differentiation, migration, tissue remodeling, and immune evasion, which are critical during cancer progression, also occur during embryogenesis—albeit under strict regulation. In embryogenesis, these processes are precisely controlled to ensure proper development and differentiation, whereas in cancer, the regulatory mechanisms are disrupted, leading to uncontrolled proliferation and tissue destruction. From this perspective, cancer can be regarded as a disease resulting from the aberrant reactivation of developmental regulatory pathways. Therefore, the fundamental difference between development and cancer may lie in the presence or absence of appropriate senescence control mechanisms. A deeper understanding of the similarities between embryogenesis and tumorigenesis may shed light on the molecular mechanisms underlying reproductive aging and aid in the development of novel therapeutic strategies. In particular, targeting signaling pathways specifically activated during early development may lead to innovative treatments capable of mitigating reproductive aging and preserving endometrial function. Continued elucidation of the mechanisms governing cellular senescence is expected to accelerate the development of effective interventions for age-related reproductive decline.

5. Conclusions

Signaling pathways such as p53, AMPK, and mTOR play pivotal roles in the regulation of endometrial cellular senescence and exert profound influences on reproductive function and the pathogenesis of associated disorders. A comprehensive understanding of the molecular crosstalk among these pathways holds promise for the development of novel therapeutic strategies aimed at preserving endometrial homeostasis and preventing or treating related pathologies. p53 acts as a key responder to various cellular stresses, including DNA damage and oxidative insults, and is instrumental in orchestrating cell cycle arrest, apoptosis, and the induction of cellular senescence. Within the endometrial context, p53 facilitates the suppression of mTORC1 signaling by activating AMPK, thereby modulating the progression of cellular senescence. AMPK, functioning as an intracellular energy sensor, is activated under conditions of ATP depletion and mitigates senescence by inhibiting cell proliferation and protein synthesis through the downregulation of mTORC1 activity. Conversely, mTOR represents a central regulatory axis that governs cellular growth and metabolic activity. Its hyperactivation has been implicated in the acceleration of cellular aging and oncogenic transformation. Activation of mTORC1 promotes anabolic processes, including cell growth and protein biosynthesis, yet its activity is stringently modulated by upstream regulators such as AMPK and p53. These interconnected pathways form a tightly regulated network that is essential for the maintenance of endometrial cellular integrity and homeostasis. Nevertheless, the functional roles of these molecular circuits are highly context-dependent, exhibiting pleiotropic effects that vary with environmental cues such as nutrient availability and oxidative stress. As a result, their regulatory functions are intricate and multifaceted, impacting a broad spectrum of cellular processes, including fate determination, metabolic control, and stress responses. Age-related progression of endometrial cellular senescence contributes significantly to the decline in reproductive capacity and the onset of infertility. Moreover, the accumulation of senescent cells is associated with an increased risk for pathological conditions such as endometriosis and endometrial carcinoma. Notably, dysregulation of p53 activity and aberrant mTOR signaling are critically implicated in abnormal cellular proliferation, tumorigenesis, and compromised reproductive outcomes, thereby making them attractive targets for therapeutic intervention. Looking forward, the integration of these molecular insights into clinical practice is anticipated to advance the realization of personalized medicine. By tailoring treatment strategies to the individual molecular profiles of patients, it will become feasible to implement precision therapies that optimize clinical efficacy while minimizing adverse effects.

Author Contributions: Conceptualization, H.K.; methodology, H.K.; software, H.K., M.N., M.U., H.S. and S.I.; validation, S.I.; formal analysis, H.K. and S.I.; investigation, M.N., M.U. and S.I.; resources, H.H.; data curation, S.I. and H.K.; writing—original draft preparation, H.K.; writing—review and editing, H.K., M.N., M.U., H.S., S.I. and H.H.; visualization, H.K., M.N., M.U., H.S., S.I. and H.H.; supervision, H.H.; project administration, H.K.; funding acquisition, none. All authors have read and agreed to the published version of the manuscript.

Funding: This research received no external funding.

Institutional Review Board Statement: Not applicable.

Informed Consent Statement: Not applicable.

Data Availability Statement: No new data were created.

Acknowledgments: Figures were created by Toyomi Kobayashi (Ms.Clinic MayOne, Nara, Japan); https://www.mscl-mayone.com/ (accessed on 30 April 2025). All individuals have consented to the acknowledgement.

Conflicts of Interest: The authors declare no conflicts of interest.

References

1. Sheldon, I.M.; Bromfield, J.J. Innate immunity in the human endometrium and ovary. *Am. J. Reprod. Immunol.* **2011**, *66*, 63–71. [CrossRef] [PubMed]
2. Sternberg, A.K.; Buck, V.U.; Classen-Linke, I.; Leube, R.E. How Mechanical Forces Change the Human Endometrium during the Menstrual Cycle in Preparation for Embryo Implantation. *Cells* **2021**, *10*, 2008. [CrossRef]
3. Agostinis, C.; Mangogna, A.; Bossi, F.; Ricci, G.; Kishore, U.; Bulla, R. Uterine Immunity and Microbiota: A Shifting Paradigm. *Front. Immunol.* **2019**, *10*, 2387. [CrossRef] [PubMed]
4. Maruyama, T.; Yoshimura, Y. Molecular and cellular mechanisms for differentiation and regeneration of the uterine endometrium. *Endocr. J.* **2008**, *55*, 795–810. [CrossRef]
5. Varela-Eirín, M.; Demaria, M. Cellular senescence. *Curr. Biol.* **2022**, *32*, R448–R452. [CrossRef] [PubMed]
6. Di Micco, R.; Krizhanovsky, V.; Baker, D.; d'Adda di Fagagna, F. Cellular senescence in ageing: From mechanisms to therapeutic opportunities. *Nat. Rev. Mol. Cell Biol.* **2021**, *22*, 75–95. [CrossRef]
7. Dodig, S.; Čepelak, I.; Pavić, I. Hallmarks of senescence and aging. *Biochem. Med.* **2019**, *29*, 030501. [CrossRef]
8. Wiley, C.D.; Campisi, J. From Ancient Pathways to Aging Cells-Connecting Metabolism and Cellular Senescence. *Cell Metab.* **2016**, *23*, 1013–1021. [CrossRef]
9. Regulski, M.J. Cellular Senescence: What, Why, and How. *Wounds* **2017**, *29*, 168–174.
10. Campisi, J.; d'Adda di Fagagna, F. Cellular senescence: When bad things happen to good cells. *Nat. Rev. Mol. Cell Biol.* **2007**, *8*, 729–740. [CrossRef]
11. Muñoz-Espín, D.; Serrano, M. Cellular senescence: From physiology to pathology. *Nat. Rev. Mol. Cell Biol.* **2014**, *15*, 482–496. [CrossRef] [PubMed]
12. Zhu, Y.; Armstrong, J.L.; Tchkonia, T.; Kirkland, J.L. Cellular senescence and the senescent secretory phenotype in age-related chronic diseases. *Curr. Opin. Clin. Nutr. Metab. Care* **2014**, *17*, 324–328. [CrossRef] [PubMed]
13. Hirota, Y.; Cha, J.; Yoshie, M.; Daikoku, T.; Dey, S.K. Heightened uterine mammalian target of rapamycin complex 1 (mTORC1) signaling provokes preterm birth in mice. *Proc. Natl. Acad. Sci. USA* **2011**, *108*, 18073–18078. [CrossRef] [PubMed]
14. Guo, Z.; Yu, Q. Role of mTOR Signaling in Female Reproduction. *Front. Endocrinol.* **2019**, *10*, 692. [CrossRef]
15. Borodkina, A.; Shatrova, A.; Abushik, P.; Nikolsky, N.; Burova, E. Interaction between ROS dependent DNA damage, mitochondria and p38 MAPK underlies senescence of human adult stem cells. *Aging* **2014**, *6*, 481–495. [CrossRef]
16. Kawamura, K.; Matsumura, Y.; Kawamura, T.; Araki, H.; Hamada, N.; Kuramoto, K.; Yagi, H.; Onoyama, I.; Asanoma, K.; Kato, K. Endometrial senescence is mediated by interleukin 17 receptor B signaling. *Cell Commun. Signal.* **2024**, *22*, 363. [CrossRef]
17. Taylor, R.N.; Berga, S.L.; Zou, E.; Washington, J.; Song, S.; Marzullo, B.J.; Bagchi, I.C.; Bagchi, M.K.; Yu, J. Interleukin-1beta induces and accelerates human endometrial stromal cell senescence and impairs decidualization via the c-Jun N-terminal kinase pathway. *Cell Death Discov.* **2024**, *10*, 288. [CrossRef]
18. Griukova, A.; Deryabin, P.; Shatrova, A.; Burova, E.; Severino, V.; Farina, A.; Nikolsky, N.; Borodkina, A. Molecular basis of senescence transmitting in the population of human endometrial stromal cells. *Aging* **2019**, *11*, 9912–9931. [CrossRef]
19. Vassilieva, I.; Kosheverova, V.; Vitte, M.; Kamentseva, R.; Shatrova, A.; Tsupkina, N.; Skvortsova, E.; Borodkina, A.; Tolkunova, E.; Nikolsky, N.; et al. Paracrine senescence of human endometrial mesenchymal stem cells: A role for the insulin-like growth factor binding protein 3. *Aging* **2020**, *12*, 1987–2004. [CrossRef]
20. Pathare, A.D.S.; Loid, M.; Saare, M.; Gidlöf, S.B.; Zamani Esteki, M.; Acharya, G.; Peters, M.; Salumets, A. Endometrial receptivity in women of advanced age: An underrated factor in infertility. *Hum. Reprod. Update* **2023**, *29*, 773–793. [CrossRef]
21. Saito, Y.; Yamamoto, S.; Chikenji, T.S. Role of cellular senescence in inflammation and regeneration. *Inflamm. Regen.* **2024**, *44*, 28. [CrossRef] [PubMed]
22. Lee, Y.I.; Choi, S.; Roh, W.S.; Lee, J.H.; Kim, T.G. Cellular Senescence and Inflammaging in the Skin Microenvironment. *Int. J. Mol. Sci.* **2021**, *22*, 3849. [CrossRef]
23. Lu, Y.; Jarrahi, A.; Moore, N.; Bartoli, M.; Brann, D.W.; Baban, B.; Dhandapani, K.M. Inflammaging, cellular senescence, and cognitive aging after traumatic brain injury. *Neurobiol. Dis.* **2023**, *180*, 106090. [CrossRef] [PubMed]
24. Lawrence, M.; Goyal, A.; Pathak, S.; Ganguly, P. Cellular Senescence and Inflammaging in the Bone: Pathways, Genetics, Anti-Aging Strategies and Interventions. *Int. J. Mol. Sci.* **2024**, *25*, 7411. [CrossRef]
25. Li, C.; Yuan, Y.; Jia, Y.; Zhou, Q.; Wang, Q.; Jiang, X. Cellular senescence: From homeostasis to pathological implications and therapeutic strategies. *Front. Immunol.* **2025**, *16*, 1534263. [CrossRef] [PubMed]
26. Liu, X.; Wei, X.; Wu, J.; Xu, Y.; Hu, J.; Qin, C.; Chen, C.; Lin, Y. CBLL1 promotes endometrial stromal cell senescence via inhibiting PTEN in recurrent spontaneous abortion. *FASEB J.* **2024**, *38*, e23833. [CrossRef]
27. Tang, X.; Zhu, Y.; Cao, Z.; Wang, X.; Cai, X.; Tang, Y.; Zhou, J.; Wu, M.; Zhen, X.; Ding, L.; et al. CDC42 deficiency leads to endometrial stromal cell senescence in recurrent implantation failure. *Hum. Reprod.* **2024**, *39*, 2768–2784. [CrossRef]
28. Umbayev, B.; Safarova Yantsen, Y.; Yermekova, A.; Nessipbekova, A.; Syzdykova, A.; Askarova, S. Role of a small GTPase Cdc42 in aging and age-related diseases. *Biogerontology* **2023**, *24*, 27–46. [CrossRef]

29. Ito, T.K.; Yokoyama, M.; Yoshida, Y.; Nojima, A.; Kassai, H.; Oishi, K.; Okada, S.; Kinoshita, D.; Kobayashi, Y.; Fruttiger, M.; et al. A crucial role for CDC42 in senescence-associated inflammation and atherosclerosis. *PLoS ONE* **2014**, *9*, e102186. [CrossRef]
30. Johmura, Y.; Shimada, M.; Misaki, T.; Naiki-Ito, A.; Miyoshi, H.; Motoyama, N.; Ohtani, N.; Hara, E.; Nakamura, M.; Morita, A.; et al. Necessary and sufficient role for a mitosis skip in senescence induction. *Mol. Cell* **2014**, *55*, 73–84. [CrossRef]
31. Soussi, T. The p53 tumor suppressor gene: From molecular biology to clinical investigation. *Ann. N. Y. Acad. Sci.* **2000**, *910*, 121–137. [CrossRef] [PubMed]
32. Cui, D.; Qu, R.; Liu, D.; Xiong, X.; Liang, T.; Zhao, Y. The Cross Talk Between p53 and mTOR Pathways in Response to Physiological and Genotoxic Stresses. *Front. Cell Dev. Biol.* **2021**, *9*, 775507. [CrossRef] [PubMed]
33. Vousden, K.H.; Lu, X. Live or let die: The cell's response to p53. *Nat. Rev. Cancer* **2002**, *2*, 594–604. [CrossRef]
34. Qian, Y.; Chen, X. Senescence regulation by the p53 protein family. *Methods Mol. Biol.* **2013**, *965*, 37–61. [CrossRef]
35. Moll, U.M.; Petrenko, O. The MDM2-p53 interaction. *Mol. Cancer Res.* **2003**, *1*, 1001–1008. [PubMed]
36. Mijit, M.; Caracciolo, V.; Melillo, A.; Amicarelli, F.; Giordano, A. Role of p53 in the Regulation of Cellular Senescence. *Biomolecules* **2020**, *10*, 420. [CrossRef]
37. Coppé, J.P.; Desprez, P.Y.; Krtolica, A.; Campisi, J. The senescence-associated secretory phenotype: The dark side of tumor suppression. *Annu. Rev. Pathol.* **2010**, *5*, 99–118. [CrossRef]
38. Christy, B.; Demaria, M.; Campisi, J.; Huang, J.; Jones, D.; Dodds, S.G.; Williams, C.; Hubbard, G.; Livi, C.B.; Gao, X.; et al. p53 and rapamycin are additive. *Oncotarget* **2015**, *6*, 15802–15813. [CrossRef]
39. Wang, H.; Guo, M.; Wei, H.; Chen, Y. Targeting p53 pathways: Mechanisms, structures, and advances in therapy. *Signal Transduct. Target. Ther.* **2023**, *8*, 92. [CrossRef]
40. Giono, L.E.; Manfredi, J.J. The p53 tumor suppressor participates in multiple cell cycle checkpoints. *J. Cell. Physiol.* **2006**, *209*, 13–20. [CrossRef]
41. Adimoolam, S.; Ford, J.M. p53 and DNA damage-inducible expression of the xeroderma pigmentosum group C gene. *Proc. Natl. Acad. Sci. USA* **2002**, *99*, 12985–12990. [CrossRef]
42. Ozaki, T.; Nakagawara, A. Role of p53 in Cell Death and Human Cancers. *Cancers* **2011**, *3*, 994–1013. [CrossRef] [PubMed]
43. Puzio-Kuter, A.M. The Role of p53 in Metabolic Regulation. *Genes Cancer* **2011**, *2*, 385–391. [CrossRef]
44. Simabuco, F.M.; Morale, M.G.; Pavan, I.C.B.; Morelli, A.P.; Silva, F.R.; Tamura, R.E. p53 and metabolism: From mechanism to therapeutics. *Oncotarget* **2018**, *9*, 23780–23823. [CrossRef] [PubMed]
45. Bensaad, K.; Tsuruta, A.; Selak, M.A.; Vidal, M.N.; Nakano, K.; Bartrons, R.; Gottlieb, E.; Vousden, K.H. TIGAR, a p53-inducible regulator of glycolysis and apoptosis. *Cell* **2006**, *126*, 107–120. [CrossRef]
46. Demidenko, Z.N.; Korotchkina, L.G.; Gudkov, A.V.; Blagosklonny, M.V. Paradoxical suppression of cellular senescence by p53. *Proc. Natl. Acad. Sci. USA* **2010**, *107*, 9660–9664. [CrossRef]
47. Devis-Jauregui, L.; Eritja, N.; Davis, M.L.; Matias-Guiu, X.; Llobet-Navàs, D. Autophagy in the physiological endometrium and cancer. *Autophagy* **2021**, *17*, 1077–1095. [CrossRef] [PubMed]
48. Kim, J.; Lee, S.; Kim, H.; Lee, H.; Seong, K.M.; Youn, H.; Youn, B. Autophagic Organelles in DNA Damage Response. *Front. Cell Dev. Biol.* **2021**, *9*, 668735. [CrossRef]
49. Maiuri, M.C.; Galluzzi, L.; Morselli, E.; Kepp, O.; Malik, S.A.; Kroemer, G. Autophagy regulation by p53. *Curr. Opin. Cell Biol.* **2010**, *22*, 181–185. [CrossRef]
50. Deng, W.; Cha, J.; Yuan, J.; Haraguchi, H.; Bartos, A.; Leishman, E.; Viollet, B.; Bradshaw, H.B.; Hirota, Y.; Dey, S.K. p53 coordinates decidual sestrin 2/AMPK/mTORC1 signaling to govern parturition timing. *J. Clin. Investig.* **2016**, *126*, 2941–2954. [CrossRef]
51. Budanov, A.V.; Karin, M. p53 target genes sestrin1 and sestrin2 connect genotoxic stress and mTOR signaling. *Cell* **2008**, *134*, 451–460. [CrossRef] [PubMed]
52. Chan, A.S.; Mowla, S.N.; Arora, P.; Jat, P.S. Tumour suppressors and cellular senescence. *IUBMB Life* **2014**, *66*, 812–822. [CrossRef] [PubMed]
53. Kwon, Y.; Kim, J.W.; Jeoung, J.A.; Kim, M.S.; Kang, C. Autophagy Is Pro-Senescence When Seen in Close-Up, but Anti-Senescence in Long-Shot. *Mol. Cells* **2017**, *40*, 607–612. [CrossRef]
54. Berkers, C.R.; Maddocks, O.D.; Cheung, E.C.; Mor, I.; Vousden, K.H. Metabolic regulation by p53 family members. *Cell Metab.* **2013**, *18*, 617–633. [CrossRef]
55. Narita, M.; Young, A.R.; Arakawa, S.; Samarajiwa, S.A.; Nakashima, T.; Yoshida, S.; Hong, S.; Berry, L.S.; Reichelt, S.; Ferreira, M.; et al. Spatial coupling of mTOR and autophagy ments secretory phenotypes. *Science* **2011**, *332*, 966–970. [CrossRef]
56. Lee, Y.; Kim, J.; Kim, M.S.; Kwon, Y.; Shin, S.; Yi, H.; Kim, H.; Chang, M.J.; Chang, C.B.; Kang, S.B.; et al. Coordinate regulation of the senescent state by selective autophagy. *Dev. Cell* **2021**, *56*, 1512–1525.e7. [CrossRef]
57. Shim, D.; Duan, L.; Maki, C.G. P53-regulated autophagy and its impact on drug resistance and cell fate. *Cancer Drug Resist.* **2021**, *4*, 85–95. [CrossRef] [PubMed]
58. Tasdemir, E.; Maiuri, M.C.; Galluzzi, L.; Vitale, I.; Djavaheri-Mergny, M.; D'Amelio, M.; Criollo, A.; Morselli, E.; Zhu, C.; Harper, F.; et al. Regulation of autophagy by cytoplasmic p53. *Nat. Cell Biol.* **2008**, *10*, 676–687. [CrossRef]

59. Cordani, M.; Butera, G.; Pacchiana, R.; Donadelli, M. Molecular interplay between mutant p53 proteins and autophagy in cancer cells. *Biochim. Biophys. Acta Rev. Cancer* **2017**, *1867*, 19–28. [CrossRef]
60. Hajimohammadebrahim-Ketabforoush, M.; Zali, A.; Shahmohammadi, M.; Hamidieh, A.A. Metformin and its potential influence on cell fate decision between apoptosis and senescence in cancer, with a special emphasis on glioblastoma. *Front. Oncol.* **2024**, *14*, 1455492. [CrossRef]
61. Stabenow, L.K.; Zibrova, D.; Ender, C.; Helbing, D.L.; Spengler, K.; Marx, C.; Wang, Z.Q.; Heller, R. Oxidative Glucose Metabolism Promotes Senescence in Vascular Endothelial Cells. *Cells* **2022**, *11*, 2213. [CrossRef] [PubMed]
62. Sabbatinelli, J.; Prattichizzo, F.; Olivieri, F.; Procopio, A.D.; Rippo, M.R.; Giuliani, A. Where Metabolism Meets Senescence: Focus on Endothelial Cells. *Front. Physiol.* **2019**, *10*, 1523. [CrossRef]
63. Liu, B.; Meng, Q.; Gao, X.; Sun, H.; Xu, Z.; Wang, Y.; Zhou, H. Lipid and glucose metabolism in senescence. *Front. Nutr.* **2023**, *10*, 1157352. [CrossRef] [PubMed]
64. Abukwaik, R.; Vera-Siguenza, E.; Tennant, D.; Spill, F. p53 Orchestrates Cancer Metabolism: Unveiling Strategies to Reverse the Warburg Effect. *Bull. Math. Biol.* **2024**, *86*, 124. [CrossRef] [PubMed]
65. Yu, L.; Wu, M.; Zhu, G.; Xu, Y. Emerging Roles of the Tumor Suppressor p53 in Metabolism. *Front. Cell Dev. Biol.* **2022**, *9*, 762742. [CrossRef]
66. Kruiswijk, F.; Labuschagne, C.F.; Vousden, K.H. p53 in survival, death and metabolic health: A lifeguard with a licence to kill. *Nat. Rev. Mol. Cell Biol.* **2015**, *16*, 393–405. [CrossRef] [PubMed]
67. Schwartzenberg-Bar-Yoseph, F.; Armoni, M.; Karnieli, E. The tumor suppressor p53 down-regulates glucose transporters GLUT1 and GLUT4 gene expression. *Cancer Res.* **2004**, *64*, 2627–2633. [CrossRef]
68. Jiang, P.; Du, W.; Wang, X.; Mancuso, A.; Gao, X.; Wu, M.; Yang, X. p53 regulates biosynthesis through direct inactivation of glucose-6-phosphate dehydrogenase. *Nat. Cell Biol.* **2011**, *13*, 310–316. [CrossRef]
69. Kondoh, H.; Lleonart, M.E.; Gil, J.; Wang, J.; Degan, P.; Peters, G.; Martinez, D.; Carnero, A.; Beach, D. Glycolytic enzymes can modulate cellular life span. *Cancer Res.* **2005**, *65*, 177–185. [CrossRef]
70. Liang, Y.; Liu, J.; Feng, Z. The regulation of cellular metabolism by tumor suppressor p53. *Cell Biosci.* **2013**, *3*, 9. [CrossRef]
71. Dou, X.; Fu, Q.; Long, Q.; Liu, S.; Zou, Y.; Fu, D.; Xu, Q.; Jiang, Z.; Ren, X.; Zhang, G.; et al. PDK4-dependent hypercatabolism and lactate production of senescent cells promotes cancer malignancy. *Nat. Metab.* **2023**, *5*, 1887–1910. [CrossRef] [PubMed]
72. Olenchock, B.A.; Vander Heiden, M.G. Pyruvate as a pivot point for oncogene-induced senescence. *Cell* **2013**, *153*, 1429–1430. [CrossRef]
73. Gomes, A.S.; Ramos, H.; Soares, J.; Saraiva, L. p53 and glucose metabolism: An orchestra to be directed in cancer therapy. *Pharmacol. Res.* **2018**, *131*, 75–86. [CrossRef] [PubMed]
74. Assaily, W.; Rubinger, D.A.; Wheaton, K.; Lin, Y.; Ma, W.; Xuan, W.; Brown-Endres, L.; Tsuchihara, K.; Mak, T.W.; Benchimol, S. ROS-mediated p53 induction of Lpin1 regulates fatty acid oxidation in response to nutritional stress. *Mol. Cell* **2011**, *44*, 491–501. [CrossRef]
75. Ide, T.; Chu, K.; Aaronson, S.A.; Lee, S.W. GAMT joins the p53 network: Branching into metabolism. *Cell Cycle* **2010**, *9*, 1706–1710. [CrossRef] [PubMed]
76. Matoba, S.; Kang, J.G.; Patino, W.D.; Wragg, A.; Boehm, M.; Gavrilova, O.; Hurley, P.J.; Bunz, F.; Hwang, P.M. p53 regulates mitochondrial respiration. *Science* **2006**, *312*, 1650–1653. [CrossRef]
77. Kang, R.; Kroemer, G.; Tang, D. The tumor suppressor protein p53 and the ferroptosis network. *Free Radic. Biol. Med.* **2019**, *133*, 162–168. [CrossRef]
78. Vousden, K.H. Alternative fuel--another role for p53 in the regulation of metabolism. *Proc. Natl. Acad. Sci. USA* **2010**, *107*, 7117–7118. [CrossRef]
79. Stambolsky, P.; Weisz, L.; Shats, I.; Klein, Y.; Goldfinger, N.; Oren, M.; Rotter, V. Regulation of AIF expression by p53. *Cell Death Differ.* **2006**, *13*, 2140–2149. [CrossRef]
80. Kuo, M.L.; Sy, A.J.; Xue, L.; Chi, M.; Lee, M.T.; Yen, T.; Chiang, M.I.; Chang, L.; Chu, P.; Yen, Y. RRM2B suppresses activation of the oxidative stress pathway and is up-regulated by p53 during senescence. *Sci. Rep.* **2012**, *2*, 822. [CrossRef]
81. Zhang, C.; Lin, M.; Wu, R.; Wang, X.; Yang, B.; Levine, A.J.; Hu, W.; Feng, Z. Parkin, a p53 target gene, mediates the role of p53 in glucose metabolism and the Warburg effect. *Proc. Natl. Acad. Sci. USA* **2011**, *108*, 16259–16264. [CrossRef]
82. Huang, W.; Xie, W.; Zhong, H.; Cai, S.; Huang, Q.; Liu, Y.; Zeng, Z.; Liu, Y. Cytosolic p53 Inhibits Parkin-Mediated Mitophagy and Promotes Acute Liver Injury Induced by Heat Stroke. *Front. Immunol.* **2022**, *13*, 859231. [CrossRef] [PubMed]
83. Lacroix, M.; Riscal, R.; Arena, G.; Linares, L.K.; Le Cam, L. Metabolic functions of the tumor suppressor p53: Implications in normal physiology, metabolic disorders, and cancer. *Mol. Metab.* **2020**, *33*, 2–22. [CrossRef]
84. Shen, H.H.; Zhang, T.; Yang, H.L.; Lai, Z.Z.; Zhou, W.J.; Mei, J.; Shi, J.W.; Zhu, R.; Xu, F.Y.; Li, D.J.; et al. Ovarian hormones-autophagy-immunity axis in menstruation and endometriosis. *Theranostics* **2021**, *11*, 3512–3526. [CrossRef]
85. Mestre Citrinovitz, A.C.; Strowitzki, T.; Germeyer, A. Decreased Autophagy Impairs Decidualization of Human Endometrial Stromal Cells: A Role for ATG Proteins in Endometrial Physiology. *Int. J. Mol. Sci.* **2019**, *20*, 3066. [CrossRef]

86. Park, H.; Cho, M.; Do, Y.; Park, J.K.; Bae, S.J.; Joo, J.; Ha, K.T. Autophagy as a Therapeutic Target of Natural Products Enhancing Embryo Implantation. *Pharmaceuticals* **2021**, *15*, 53. [CrossRef] [PubMed]
87. Huang, J.; Chen, X.; Lv, Y. HMGB1 Mediated Inflammation and Autophagy Contribute to Endometriosis. *Front. Endocrinol.* **2021**, *12*, 616696. [CrossRef] [PubMed]
88. Fukuda, T.; Wada-Hiraike, O. The Two-Faced Role of Autophagy in Endometrial Cancer. *Front. Cell Dev. Biol.* **2022**, *10*, 839416. [CrossRef]
89. Yang, S.; Wang, H.; Li, D.; Li, M. Role of Endometrial Autophagy in Physiological and Pathophysiological Processes. *J. Cancer* **2019**, *10*, 3459–3471. [CrossRef]
90. Zhou, S.; Zhao, L.; Yi, T.; Wei, Y.; Zhao, X. Menopause-induced uterine epithelium atrophy results from arachidonic acid/prostaglandin E2 axis inhibition-mediated autophagic cell death. *Sci. Rep.* **2016**, *6*, 31408. [CrossRef]
91. Berger, C.; Qian, Y.; Chen, X. The p53-estrogen receptor loop in cancer. *Curr. Mol. Med.* **2013**, *13*, 1229–1240. [CrossRef] [PubMed]
92. Maia, H., Jr.; Maltez, A.; Studart, E.; Athayde, C.; Coutinho, E.M. Ki-67, Bcl-2 and p53 expression in endometrial polyps and in the normal endometrium during the menstrual cycle. *BJOG Int. J. Obstet. Gynaecol.* **2004**, *111*, 1242–1247. [CrossRef]
93. Hirota, Y.; Daikoku, T.; Tranguch, S.; Xie, H.; Bradshaw, H.B.; Dey, S.K. Uterine-specific p53 deficiency confers premature uterine senescence and promotes preterm birth in mice. *J. Clin. Investig.* **2010**, *120*, 803–815. [CrossRef] [PubMed]
94. Sang, L.; Fang, Q.J.; Zhao, X.B. A research on the protein expression of p53, p16, and MDM2 in endometriosis. *Medicine* **2019**, *98*, e14776. [CrossRef]
95. Hu, W.; Feng, Z.; Teresky, A.K.; Levine, A.J. p53 regulates maternal reproduction through LIF. *Nature* **2007**, *450*, 721–724. [CrossRef]
96. Feng, Z.; Zhang, C.; Kang, H.J.; Sun, Y.; Wang, H.; Naqvi, A.; Frank, A.K.; Rosenwaks, Z.; Murphy, M.E.; Levine, A.J. Regulation of female reproduction by p53 and its family members. *FASEB J.* **2011**, *25*, 2245–2255. [CrossRef]
97. Delenko, J.; Xue, X.; Chatterjee, P.K.; Hyman, N.; Shih, A.J.; Adelson, R.P.; Safaric Tepes, P.; Gregersen, P.K.; Metz, C.N. Quercetin enhances decidualization through AKT-ERK-p53 signaling and supports a role for senescence in endometriosis. *Reprod. Biol. Endocrinol.* **2024**, *22*, 100. [CrossRef]
98. Salminen, A.; Kaarniranta, K. AMP-activated protein kinase (AMPK) controls the aging process via an integrated signaling network. *Ageing Res. Rev.* **2012**, *11*, 230–241. [CrossRef] [PubMed]
99. Garcia, D.; Shaw, R.J. AMPK: Mechanisms of Cellular Energy Sensing and Restoration of Metabolic Balance. *Mol. Cell* **2017**, *66*, 789–800. [CrossRef]
100. Ditch, S.; Paull, T.T. The ATM protein kinase and cellular redox signaling: Beyond the DNA damage response. *Trends Biochem. Sci.* **2012**, *37*, 15–22. [CrossRef]
101. Alao, J.P.; Legon, L.; Dabrowska, A.; Tricolici, A.M.; Kumar, J.; Rallis, C. Interplays of AMPK and TOR in Autophagy Regulation in Yeast. *Cells* **2023**, *12*, 519. [CrossRef] [PubMed]
102. Kim, D.H. Contrasting views on the role of AMPK in autophagy. *Bioessays* **2024**, *46*, e2300211. [CrossRef]
103. Ranek, M.J.; Kokkonen-Simon, K.M.; Chen, A.; Dunkerly-Eyring, B.L.; Vera, M.P.; Oeing, C.U.; Patel, C.H.; Nakamura, T.; Zhu, G.; Bedja, D.; et al. PKG1-modified TSC2 regulates mTORC1 activity to counter adverse cardiac stress. *Nature* **2019**, *566*, 264–269. [CrossRef] [PubMed]
104. Egan, D.F.; Shackelford, D.B.; Mihaylova, M.M.; Gelino, S.; Kohnz, R.A.; Mair, W.; Vasquez, D.S.; Joshi, A.; Gwinn, D.M.; Taylor, R.; et al. Phosphorylation of ULK1 (hATG1) by AMP-activated protein kinase connects energy sensing to mitophagy. *Science* **2011**, *331*, 456–461. [CrossRef]
105. Kim, J.; Kundu, M.; Viollet, B.; Guan, K.L. AMPK and mTOR regulate autophagy through direct phosphorylation of Ulk1. *Nat. Cell Biol.* **2011**, *13*, 132–141. [CrossRef] [PubMed]
106. Park, J.M.; Lee, D.H.; Kim, D.H. Redefining the role of AMPK in autophagy and the energy stress response. *Nat. Commun.* **2023**, *14*, 2994. [CrossRef]
107. Nwadike, C.; Williamson, L.E.; Gallagher, L.E.; Guan, J.L.; Chan, E.Y.W. AMPK Inhibits ULK1-Dependent Autophagosome Formation and Lysosomal Acidification via Distinct Mechanisms. *Mol. Cell. Biol.* **2018**, *38*, e00023-18. [CrossRef]
108. Gong, H.; Chen, H.; Xiao, P.; Huang, N.; Han, X.; Zhang, J.; Yang, Y.; Li, T.; Zhao, T.; Tai, H.; et al. miR-146a impedes the anti-aging effect of AMPK via NAMPT suppression and NAD(+)/SIRT inactivation. *Signal Transduct. Target. Ther.* **2022**, *7*, 66. [CrossRef]
109. Han, X.; Tai, H.; Wang, X.; Wang, Z.; Zhou, J.; Wei, X.; Ding, Y.; Gong, H.; Mo, C.; Zhang, J.; et al. AMPK activation protects cells from oxidative stress-induced senescence via autophagic flux restoration and intracellular NAD(+) elevation. *Aging Cell* **2016**, *15*, 416–427. [CrossRef]
110. Han, Y.; Liu, Y.; Zhang, Y.; Wang, W.; Lv, T.; Huang, J.; Peng, X. The Role and Application of the AMPK-Sirtuins Network in Cellular Senescence. *Front. Biosci.-Landmark* **2023**, *28*, 250. [CrossRef]
111. Han, X.; Zhang, T.; Zhang, X.; Zhang, R.; Lao, K.; Mi, Y.; Gou, X. AMPK alleviates oxidative stress-induced premature senescence via inhibition of NF-kappaB/STAT3 axis-mediated positive feedback loop. *Mech. Ageing Dev.* **2020**, *191*, 111347. [CrossRef] [PubMed]

112. Cheng, X.Y.; Li, Y.Y.; Huang, C.; Li, J.; Yao, H.W. AMP-activated protein kinase reduces inflammatory responses and cellular senescence in pulmonary emphysema. *Oncotarget* **2017**, *8*, 22513–22523. [CrossRef] [PubMed]
113. Zhou, Y.; Liu, F. Coordination of the AMPK, Akt, mTOR, and p53 Pathways under Glucose Starvation. *Int. J. Mol. Sci.* **2022**, *23*, 14945. [CrossRef]
114. Jones, R.G.; Plas, D.R.; Kubek, S.; Buzzai, M.; Mu, J.; Xu, Y.; Birnbaum, M.J.; Thompson, C.B. AMP-activated protein kinase induces a p53-dependent metabolic checkpoint. *Mol. Cell* **2005**, *18*, 283–293. [CrossRef] [PubMed]
115. Weinberg, R.L.; Veprintsev, D.B.; Bycroft, M.; Fersht, A.R. Comparative binding of p53 to its promoter and DNA recognition elements. *J. Mol. Biol.* **2005**, *348*, 589–596. [CrossRef]
116. Liebl, M.C.; Hofmann, T.G. Cell Fate Regulation upon DNA Damage: p53 Serine 46 Kinases Pave the Cell Death Road. *Bioessays* **2019**, *41*, e1900127. [CrossRef]
117. Sun, Y.; Connors, K.E.; Yang, D.Q. AICAR induces phosphorylation of AMPK in an ATM-dependent, LKB1-independent manner. *Mol. Cell. Biochem.* **2007**, *306*, 239–245. [CrossRef]
118. Wiley, C.D.; Campisi, J. The metabolic roots of senescence: Mechanisms and opportunities for intervention. *Nat. Metab.* **2021**, *3*, 1290–1301. [CrossRef]
119. Pieles, O.; Hartmann, M.; Morsczeck, C. AMP-activated protein kinase and the down-stream activated process of autophagy regulate the osteogenic differentiation of human dental follicle cells. *Arch. Oral Biol.* **2021**, *122*, 104951. [CrossRef]
120. Morsczeck, C.; Reck, A.; Reichert, T.E. Changes in AMPK activity induces cellular senescence in human dental follicle cells. *Exp. Gerontol.* **2023**, *172*, 112071. [CrossRef]
121. Wiley, C.D.; Velarde, M.C.; Lecot, P.; Liu, S.; Sarnoski, E.A.; Freund, A.; Shirakawa, K.; Lim, H.W.; Davis, S.S.; Ramanathan, A.; et al. Mitochondrial Dysfunction Induces Senescence with a Distinct Secretory Phenotype. *Cell Metab.* **2016**, *23*, 303–314. [CrossRef] [PubMed]
122. Hardie, D.G.; Ross, F.A.; Hawley, S.A. AMPK: A nutrient and energy sensor that maintains energy homeostasis. *Nat. Rev. Mol. Cell Biol.* **2012**, *13*, 251–262. [CrossRef]
123. Guo, S.; Zhang, C.; Zeng, H.; Xia, Y.; Weng, C.; Deng, Y.; Wang, L.; Wang, H. Glycolysis maintains AMPK activation in sorafenib induced Warburg effect. *Mol. Metab.* **2023**, *77*, 101796. [CrossRef] [PubMed]
124. Lee, J.M.; Seo, W.Y.; Song, K.H.; Chanda, D.; Kim, Y.D.; Kim, D.K.; Lee, M.W.; Ryu, D.; Kim, Y.H.; Noh, J.R.; et al. AMPK-dependent repression of hepatic gluconeogenesis via disruption of CREB.CRTC2 complex by orphan nuclear receptor small heterodimer partner. *J. Biol. Chem.* **2010**, *285*, 32182–32191. [CrossRef]
125. Kim, Y.D.; Park, K.G.; Lee, Y.S.; Park, Y.Y.; Kim, D.K.; Nedumaran, B.; Jang, W.G.; Cho, W.J.; Ha, J.; Lee, I.K.; et al. Metformin inhibits hepatic gluconeogenesis through AMP-activated protein kinase-dependent regulation of the orphan nuclear receptor SHP. *Diabetes* **2008**, *57*, 306–314. [CrossRef] [PubMed]
126. Steinberg, G.R.; Carling, D. AMP-activated protein kinase: The current landscape for drug development. *Nat. Rev. Drug Discov.* **2019**, *18*, 527–551. [CrossRef]
127. Jeon, S.M. Regulation and function of AMPK in physiology and diseases. *Exp. Mol. Med.* **2016**, *48*, e245. [CrossRef]
128. Zhang, P.; Fu, H.J.; Lv, L.X.; Liu, C.F.; Han, C.; Zhao, X.F.; Wang, J.X. WSSV exploits AMPK to activate mTORC2 signaling for proliferation by enhancing aerobic glycolysis. *Commun. Biol.* **2023**, *6*, 361. [CrossRef]
129. Clarke, P.R.; Hardie, D.G. Regulation of HMG-CoA reductase: Identification of the site phosphorylated by the AMP-activated protein kinase in vitro and in intact rat liver. *EMBO J.* **1990**, *9*, 2439–2446. [CrossRef]
130. Chaube, B.; Malvi, P.; Singh, S.V.; Mohammad, N.; Viollet, B.; Bhat, M.K. AMPK maintains energy homeostasis and survival in cancer cells via regulating p38/PGC-1alpha-mediated mitochondrial biogenesis. *Cell Death Discov.* **2015**, *1*, 15063. [CrossRef]
131. Herzig, S.; Shaw, R.J. AMPK: Guardian of metabolism and mitochondrial homeostasis. *Nat. Rev. Mol. Cell Biol.* **2018**, *19*, 121–135. [CrossRef] [PubMed]
132. Wu, Y.; Li, M.; Zhang, J.; Wang, S. Unveiling uterine aging: Much more to learn. *Ageing Res. Rev.* **2023**, *86*, 101879. [CrossRef]
133. Borodkina, A.V.; Shatrova, A.N.; Deryabin, P.I.; Griukova, A.A.; Abushik, P.A.; Antonov, S.M.; Nikolsky, N.N.; Burova, E.B. Calcium alterations signal either to senescence or to autophagy induction in stem cells upon oxidative stress. *Aging* **2016**, *8*, 3400–3418. [CrossRef]
134. Kawano, Y.; Sato, H.; Goto, K.; Nishida, M.; Nasu, K. The inhibitory effect of AMP-activated protein kinase (AMPK) on chemokine and prostaglandin production in human endometrial stromal cells. *Reprod. Biol. Endocrinol.* **2021**, *19*, 188. [CrossRef]
135. McCallum, M.L.; Pru, C.A.; Smith, A.R.; Kelp, N.C.; Foretz, M.; Viollet, B.; Du, M.; Pru, J.K. A functional role for AMPK in female fertility and endometrial regeneration. *Reproduction* **2018**, *156*, 501–513. [CrossRef] [PubMed]
136. Griffiths, R.M.; Pru, C.A.; Behura, S.K.; Cronrath, A.R.; McCallum, M.L.; Kelp, N.C.; Winuthayanon, W.; Spencer, T.E.; Pru, J.K. AMPK is required for uterine receptivity and normal responses to steroid hormones. *Reproduction* **2020**, *159*, 707–717. [CrossRef] [PubMed]

137. Zhang, Y.; Gao, R.; Zhang, L.; Geng, Y.; Chen, Q.; Chen, X.; Liu, X.; Mu, X.; Ding, Y.; Wang, Y.; et al. AMPK/mTOR downregulated autophagy enhances aberrant endometrial decidualization in folate-deficient pregnant mice. *J. Cell. Physiol.* **2021**, *236*, 7376–7389. [CrossRef]
138. Saxton, R.A.; Sabatini, D.M. mTOR Signaling in Growth, Metabolism, and Disease. *Cell* **2017**, *168*, 960–976. [CrossRef]
139. Szwed, A.; Kim, E.; Jacinto, E. Regulation and metabolic functions of mTORC1 and mTORC2. *Physiol. Rev.* **2021**, *101*, 1371–1426. [CrossRef]
140. Hasty, P.; Sharp, Z.D.; Curiel, T.J.; Campisi, J. mTORC1 and p53: Clash of the gods? *Cell Cycle* **2013**, *12*, 20–25. [CrossRef]
141. Guri, Y.; Colombi, M.; Dazert, E.; Hindupur, S.K.; Roszik, J.; Moes, S.; Jenoe, P.; Heim, M.H.; Riezman, I.; Riezman, H.; et al. mTORC2 Promotes Tumorigenesis via Lipid Synthesis. *Cancer Cell* **2017**, *32*, 807–823.e12. [CrossRef] [PubMed]
142. Wu, X.; Xie, W.; Xie, W.; Wei, W.; Guo, J. Beyond controlling cell size: Functional analyses of S6K in tumorigenesis. *Cell Death Dis.* **2022**, *13*, 646. [CrossRef] [PubMed]
143. Tsai, S.; Sitzmann, J.M.; Dastidar, S.G.; Rodriguez, A.A.; Vu, S.L.; McDonald, C.E.; Academia, E.C.; O'Leary, M.N.; Ashe, T.D.; La Spada, A.R.; et al. Muscle-specific 4E-BP1 signaling activation improves metabolic parameters during aging and obesity. *J. Clin. Investig.* **2015**, *125*, 2952–2964. [CrossRef]
144. Bodine, S.C. The role of mTORC1 in the regulation of skeletal muscle mass. *Fac. Rev.* **2022**, *11*, 32. [CrossRef] [PubMed]
145. Carroll, B.; Nelson, G.; Rabanal-Ruiz, Y.; Kucheryavenko, O.; Dunhill-Turner, N.A.; Chesterman, C.C.; Zahari, Q.; Zhang, T.; Conduit, S.E.; Mitchell, C.A.; et al. Persistent mTORC1 signaling in cell senescence results from defects in amino acid and growth factor sensing. *J. Cell Biol.* **2017**, *216*, 1949–1957. [CrossRef]
146. Lee, C.H.; Inoki, K.; Karbowniczek, M.; Petroulakis, E.; Sonenberg, N.; Henske, E.P.; Guan, K.L. Constitutive mTOR activation in TSC mutants sensitizes cells to energy starvation and genomic damage via p53. *EMBO J.* **2007**, *26*, 4812–4823. [CrossRef]
147. Lai, K.P.; Leong, W.F.; Chau, J.F.; Jia, D.; Zeng, L.; Liu, H.; He, L.; Hao, A.; Zhang, H.; Meek, D.; et al. S6K1 is a multifaceted regulator of Mdm2 that connects nutrient status and DNA damage response. *EMBO J.* **2010**, *29*, 2994–3006. [CrossRef]
148. Jung, S.H.; Hwang, H.J.; Kang, D.; Park, H.A.; Lee, H.C.; Jeong, D.; Lee, K.; Park, H.J.; Ko, Y.G.; Lee, J.S. mTOR kinase leads to PTEN-loss-induced cellular senescence by phosphorylating p53. *Oncogene* **2019**, *38*, 1639–1650. [CrossRef]
149. Chibaya, L.; Karim, B.; Zhang, H.; Jones, S.N. Mdm2 phosphorylation by Akt regulates the p53 response to oxidative stress to promote cell proliferation and tumorigenesis. *Proc. Natl. Acad. Sci. USA* **2021**, *118*, e2003193118. [CrossRef]
150. Pavlova, J.A.; Guseva, E.A.; Dontsova, O.A.; Sergiev, P.V. Natural Activators of Autophagy. *Biochemistry* **2024**, *89*, 1–26. [CrossRef]
151. Yu, L.; Chen, Y.; Tooze, S.A. Autophagy pathway: Cellular and molecular mechanisms. *Autophagy* **2018**, *14*, 207–215. [CrossRef] [PubMed]
152. Cordani, M.; Oppici, E.; Dando, I.; Butturini, E.; Dalla Pozza, E.; Nadal-Serrano, M.; Oliver, J.; Roca, P.; Mariotto, S.; Cellini, B.; et al. Mutant p53 proteins counteract autophagic mechanism sensitizing cancer cells to mTOR inhibition. *Mol. Oncol.* **2016**, *10*, 1008–1029. [CrossRef] [PubMed]
153. Lin, J.; Fan, L.; Han, Y.; Guo, J.; Hao, Z.; Cao, L.; Kang, J.; Wang, X.; He, J.; Li, J. The mTORC1/eIF4E/HIF-1alpha Pathway Mediates Glycolysis to Support Brain Hypoxia Resistance in the Gansu Zokor, Eospalax cansus. *Front. Physiol.* **2021**, *12*, 626240. [CrossRef] [PubMed]
154. Fan, H.; Wu, Y.; Yu, S.; Li, X.; Wang, A.; Wang, S.; Chen, W.; Lu, Y. Critical role of mTOR in regulating aerobic glycolysis in carcinogenesis (Review). *Int. J. Oncol.* **2021**, *58*, 9–19. [CrossRef]
155. Sun, Q.; Chen, X.; Ma, J.; Peng, H.; Wang, F.; Zha, X.; Wang, Y.; Jing, Y.; Yang, H.; Chen, R.; et al. Mammalian target of rapamycin up-regulation of pyruvate kinase isoenzyme type M2 is critical for aerobic glycolysis and tumor growth. *Proc. Natl. Acad. Sci. USA* **2011**, *108*, 4129–4134. [CrossRef]
156. Magaway, C.; Kim, E.; Jacinto, E. Targeting mTOR and Metabolism in Cancer: Lessons and Innovations. *Cells* **2019**, *8*, 1584. [CrossRef]
157. Morita, M.; Gravel, S.P.; Hulea, L.; Larsson, O.; Pollak, M.; St-Pierre, J.; Topisirovic, I. mTOR coordinates protein synthesis, mitochondrial activity and proliferation. *Cell Cycle* **2015**, *14*, 473–480. [CrossRef]
158. de la Cruz López, K.G.; Toledo Guzmán, M.E.; Sánchez, E.O.; García Carrancá, A. mTORC1 as a Regulator of Mitochondrial Functions and a Therapeutic Target in Cancer. *Front. Oncol.* **2019**, *9*, 1373. [CrossRef]
159. Chrienova, Z.; Nepovimova, E.; Kuca, K. The role of mTOR in age-related diseases. *J. Enzym. Inhib. Med. Chem.* **2021**, *36*, 1679–1693. [CrossRef]
160. Zhang, H.; Su, X.; Burley, S.K.; Zheng, X.F.S. mTOR regulates aerobic glycolysis through NEAT1 and nuclear paraspeckle-mediated mechanism in hepatocellular carcinoma. *Theranostics* **2022**, *12*, 3518–3533. [CrossRef]
161. Levy, T.; Voeltzke, K.; Hruby, L.; Alasad, K.; Bas, Z.; Snaebjörnsson, M.; Marciano, R.; Scharov, K.; Planque, M.; Vriens, K.; et al. mTORC1 regulates cell survival under glucose starvation through 4EBP1/2-mediated translational reprogramming of fatty acid metabolism. *Nat. Commun.* **2024**, *15*, 4083. [CrossRef]
162. Li, Q.; Lin, Y.; Liang, G.; Xiao, N.; Zhang, H.; Yang, X.; Yang, J.; Liu, A. Autophagy and Senescence: The Molecular Mechanisms and Implications in Liver Diseases. *Int. J. Mol. Sci.* **2023**, *24*, 16880. [CrossRef]

163. Zhai, P.; Sung, E.A.; Shiheido-Watanabe, Y.; Takayama, K.; Tian, Y.; Sadoshima, J. Suppression of autophagy induces senescence in the heart. *J. Mol. Cell. Cardiol.* **2024**, *195*, 83–96. [CrossRef] [PubMed]
164. Amin, S.; Liu, B.; Gan, L. Autophagy prevents microglial senescence. *Nat. Cell Biol.* **2023**, *25*, 923–925. [CrossRef] [PubMed]
165. Park, J.H.; Lee, N.K.; Lim, H.J.; Ji, S.T.; Kim, Y.J.; Jang, W.B.; Kim, D.Y.; Kang, S.; Yun, J.; Ha, J.S.; et al. Pharmacological inhibition of mTOR attenuates replicative cell senescence and improves cellular function via regulating the STAT3-PIM1 axis in human cardiac progenitor cells. *Exp. Mol. Med.* **2020**, *52*, 615–628. [CrossRef] [PubMed]
166. Gelino, S.; Hansen, M. Autophagy—An Emerging Anti-Aging Mechanism. *J. Clin. Exp. Pathol.* **2012**, *S4*, 006. [CrossRef]
167. Weichhart, T. mTOR as Regulator of Lifespan, Aging, and Cellular Senescence: A Mini-Review. *Gerontology* **2018**, *64*, 127–134. [CrossRef]
168. Wilkinson, J.E.; Burmeister, L.; Brooks, S.V.; Chan, C.C.; Friedline, S.; Harrison, D.E.; Hejtmancik, J.F.; Nadon, N.; Strong, R.; Wood, L.K.; et al. Rapamycin slows aging in mice. *Aging Cell* **2012**, *11*, 675–682. [CrossRef]
169. Xu, S.; Cai, Y.; Wei, Y. mTOR Signaling from Cellular Senescence to Organismal Aging. *Aging Dis.* **2013**, *5*, 263–273. [CrossRef]
170. Walters, H.E.; Deneka-Hannemann, S.; Cox, L.S. Reversal of phenotypes of cellular senescence by pan-mTOR inhibition. *Aging* **2016**, *8*, 231–244. [CrossRef]
171. Zou, Z.; Tao, T.; Li, H.; Zhu, X. mTOR signaling pathway and mTOR inhibitors in cancer: Progress and challenges. *Cell Biosci.* **2020**, *10*, 31. [CrossRef] [PubMed]
172. Panwar, V.; Singh, A.; Bhatt, M.; Tonk, R.K.; Azizov, S.; Raza, A.S.; Sengupta, S.; Kumar, D.; Garg, M. Multifaceted role of mTOR (mammalian target of rapamycin) signaling pathway in human health and disease. *Signal Transduct. Target. Ther.* **2023**, *8*, 375. [CrossRef] [PubMed]
173. Baek, M.O.; Song, H.I.; Han, J.S.; Yoon, M.S. Differential regulation of mTORC1 and mTORC2 is critical for 8-Br-cAMP-induced decidualization. *Exp. Mol. Med.* **2018**, *50*, 1–11. [CrossRef] [PubMed]
174. Wang, R.C.; Levine, B. Autophagy in cellular growth control. *FEBS Lett.* **2010**, *584*, 1417–1426. [CrossRef]
175. Luo, Y.; Zou, P.; Zou, J.; Wang, J.; Zhou, D.; Liu, L. Autophagy regulates ROS-induced cellular senescence via p21 in a p38 MAPKalpha dependent manner. *Exp. Gerontol.* **2011**, *46*, 860–867. [CrossRef]
176. Kang, C.; Elledge, S.J. How autophagy both activates and inhibits cellular senescence. *Autophagy* **2016**, *12*, 898–899. [CrossRef]
177. Pantelis, P.; Theocharous, G.; Lagopati, N.; Veroutis, D.; Thanos, D.F.; Lampoglou, G.P.; Pippa, N.; Gatou, M.A.; Tremi, I.; Papaspyropoulos, A.; et al. The Dual Role of Oxidative-Stress-Induced Autophagy in Cellular Senescence: Comprehension and Therapeutic Approaches. *Antioxidants* **2023**, *12*, 169. [CrossRef]
178. Xu, Y.; Li, N.; Xiang, R.; Sun, P. Emerging roles of the p38 MAPK and PI3K/AKT/mTOR pathways in oncogene-induced senescence. *Trends Biochem. Sci.* **2014**, *39*, 268–276. [CrossRef]
179. Slobodnyuk, K.; Radic, N.; Ivanova, S.; Llado, A.; Trempolec, N.; Zorzano, A.; Nebreda, A.R. Autophagy-induced senescence is regulated by p38alpha signaling. *Cell Death Dis.* **2019**, *10*, 376. [CrossRef]
180. Kwong, J.; Hong, L.; Liao, R.; Deng, Q.; Han, J.; Sun, P. p38α and p38γ mediate oncogenic ras-induced senescence through differential mechanisms. *J. Biol. Chem.* **2009**, *284*, 11237–11246. [CrossRef]
181. Borodkina, A.V.; Shatrova, A.N.; Nikolsky, N.N.; Burova, E.B. Role of P38 Map-Kinase in the Stress-Induced Senescence Progression of Human Endometrium-Derived Mesenchymal Stem Cells. *Tsitologiia* **2016**, *58*, 429–435.
182. Jin, J.; Richardson, L.; Sheller-Miller, S.; Zhong, N.; Menon, R. Oxidative stress induces p38MAPK-dependent senescence in the feto-maternal interface cells. *Placenta* **2018**, *67*, 15–23. [CrossRef] [PubMed]
183. Ohtani, N.; Mann, D.J.; Hara, E. Cellular senescence: Its role in tumor suppression and aging. *Cancer Sci.* **2009**, *100*, 792–797. [CrossRef] [PubMed]
184. Childs, B.G.; Durik, M.; Baker, D.J.; van Deursen, J.M. Cellular senescence in aging and age-related disease: From mechanisms to therapy. *Nat. Med.* **2015**, *21*, 1424–1435. [CrossRef] [PubMed]
185. Toniyan, K.A.; Malkov, A.A.; Biryukov, N.S.; Gorbacheva, E.Y.; Boyarintsev, V.V.; Ogneva, I.V. The Cellular Respiration of Endometrial Biopsies from Patients with Various Forms of Endometriosis. *Int. J. Mol. Sci.* **2024**, *25*, 3680. [CrossRef]
186. Miwa, S.; Kashyap, S.; Chini, E.; von Zglinicki, T. Mitochondrial dysfunction in cell senescence and aging. *J. Clin. Investig.* **2022**, *132*, e158447. [CrossRef]
187. Huang, Y.; Zhu, Q.; Sun, Y. Glucose metabolism and endometrium decidualization. *Front. Endocrinol.* **2025**, *16*, 1546335. [CrossRef]
188. Velarde, M.C.; Menon, R. Positive and negative effects of cellular senescence during female reproductive aging and pregnancy. *J. Endocrinol.* **2016**, *230*, R59–R76. [CrossRef]
189. Tomari, H.; Kawamura, T.; Asanoma, K.; Egashira, K.; Kawamura, K.; Honjo, K.; Nagata, Y.; Kato, K. Contribution of senescence in human endometrial stromal cells during proliferative phase to embryo receptivity. *Biol. Reprod.* **2020**, *103*, 104–113. [CrossRef]
190. Parvanov, D.; Ganeva, R.; Arsov, K.; Decheva, I.; Handzhiyska, M.; Ruseva, M.; Vidolova, N.; Scarpellini, F.; Metodiev, D.; Stamenov, G. Association between endometrial senescent cells and immune cells in women with repeated implantation failure. *J. Assist. Reprod. Genet.* **2023**, *40*, 1631–1638. [CrossRef]

191. Takasugi, M.; Yoshida, Y.; Hara, E.; Ohtani, N. The role of cellular senescence and SASP in tumour microenvironment. *FEBS J.* **2023**, *290*, 1348–1361. [CrossRef] [PubMed]
192. Xiao, S.; Qin, D.; Hou, X.; Tian, L.; Yu, Y.; Zhang, R.; Lyu, H.; Guo, D.; Chen, X.Z.; Zhou, C.; et al. Cellular senescence: A double-edged sword in cancer therapy. *Front. Oncol.* **2023**, *13*, 1189015. [CrossRef] [PubMed]
193. Kusama, K.; Yamauchi, N.; Yoshida, K.; Azumi, M.; Yoshie, M.; Tamura, K. Senolytic treatment modulates decidualization in human endometrial stromal cells. *Biochem. Biophys. Res. Commun.* **2021**, *571*, 174–180. [CrossRef] [PubMed]

cells

MDPI

Review

The Estrogen–Immune Interface in Endometriosis

Emily Greygoose [1,2], Pat Metharom [1,2], Hakan Kula [3], Timur K. Seckin [4], Tamer A. Seckin [5,*], Ayse Ayhan [6,7,*] and Yu Yu [1,2,8,*]

1 Curtin Health Innovation Research Institute (CHIRI), Faculty of Health Sciences, Curtin University, Perth, WA 6102, Australia
2 Curtin Medical School, Faculty of Health Sciences, Curtin University, Perth, WA 6102, Australia
3 Department of Obstetrics and Gynecology, Faculty of Medicine, Dokuz Eylul University, Izmir 35340, Turkey
4 Burnett School of Medicine, Texas Christian University, Fort Worth, TX 76104, USA; timurseckin@me.com
5 Department of Gynecology, Lenox Hill Hospital, and Hofstra University, New York, NY 10075, USA
6 Department of Tumor Pathology, Hamamatsu University School of Medicine, Hamamatsu 431-3192, Japan
7 Department of Gynecology and Obstetrics, Johns Hopkins University School of Medicine, Baltimore, MD 21205, USA
8 Discipline of Obstetrics and Gynaecology, Medical School, University of Western Australia, Crawley, WA 6009, Australia
* Correspondence: tamer.seckin@gmail.com (T.A.S.); ayseayhanjp@gmail.com (A.A.); yu.yu@curtin.edu.au (Y.Y.)

Abstract: Endometriosis is a gynecologic condition characterized by the growth of endometrium-like stroma and glandular elements outside of the uterine cavity. The involvement of hormonal dysregulation, specifically estrogen, is well established in the initiation, progression, and maintenance of the condition. Evidence also highlights the association between endometriosis and altered immune states. The human endometrium is a highly dynamic tissue that undergoes frequent remodeling in response to hormonal regulation during the menstrual cycle. Similarly, endometriosis shares this propensity, compounded by unclear pathogenic mechanisms, presenting unique challenges in defining its etiology and pathology. Here, we provide a lens to understand the interplay between estrogen and innate and adaptive immune systems throughout the menstrual cycle in the pathogenesis of endometriosis. Estrogen is closely linked to many altered inflammatory and immunomodulatory states, affecting both tissue-resident and circulatory immune cells. This review summarizes estrogenic interactions with specific myeloid and lymphoid cells, highlighting their implications in the progression of endometriosis.

Keywords: endometriosis; estrogen; immune cells

1. Introduction

The endometrium is a highly complex organ characterized by significant heterogeneity amongst individuals and across various physiological functions. Understanding its dynamic changes throughout the menstrual cycle and across various physiological functions in healthy conditions is essential, particularly in the context of intricate and complex illnesses like endometriosis.

Despite advancements in single-cell and spatial multiomics technologies, which have enhanced our insights into endometrial cell interactions, challenges persist. Issues such as cell isolation, low sensitivity to rare ribonucleic acid (RNA) populations embedded within extracellular matrices or tightly associated with neighboring cells, and non-uniform annotation continue to hinder our ability to assess transcript distribution and RNA processing roles. Integrating multiomics data, particularly combining transcriptomics and proteomics, may offer significant promise for identifying novel biomarkers and elucidating the pathological mechanisms of endometriosis. By analyzing mRNA expression (from transcriptomics) and protein abundance (from proteomics) in endometriosis tissues, differentially expressed genes and proteins can be identified, advancing diagnostic markers and insights into disease progression. This approach also enables pathway analysis, revealing disrupted molecular pathways and protein–protein interaction networks, which are crucial for understanding the disease pathophysiology and identifying potential therapeutic targets. Moreover, multiomics can help address disease heterogeneity by uncovering distinct RNA and protein expression patterns across endometriosis subtypes. However, several challenges must be addressed to fully realize the potential of multiomics in endometriosis research. Data complexity, tissue availability, and the inherent heterogeneity of the disease complicate the interpretation of results. In particular, the integration of diverse omics data requires robust technical standardization. Developing standardized protocols for sample collection, processing, and data analysis is essential to ensure reproducibility and enable cross-study comparisons. These efforts are critical for advancing both the diagnostic and therapeutic potential of multiomics in endometriosis. Still, without a comprehensive grasp of normal physiological states, elucidating disease mechanisms, especially in endometriosis, becomes exceedingly difficult.

This gap in knowledge is likely a significant factor contributing to our limited understanding of endometriosis pathophysiology, which is crucial for developing effective treatments and identifying specific pathological alterations. Therefore, there is an urgent need for multi-institutional, multidisciplinary studies to explore spatial and temporal expression patterns that govern differentiation, homeostasis, and pathological states, ultimately advancing our understanding and treatment of this prevalent condition.

Recognition of the molecular mechanisms underlying endometriosis is vital, particularly the factors that trigger lesion formation. Investigating the genes and pathways involved in the attachment, growth, and persistence of endometrial tissue at ectopic sites, alongside those in the eutopic endometrium of affected individuals, is a critical focus area, especially regarding immune system dysfunction. While documented abnormalities in immune cells and cytokine activity are known to contribute to the disease, it remains unclear whether these changes are causative or resultant.

This literature review aims to enhance our understanding of hormonal–immune interactions in endometriosis by critically analyzing and summarizing the existing research. We have used the following search terms in PubMed database: endometriosis AND immune cells; estrogen OR oestrogen AND Uterus; endometriosis NOT (estrogen OR oestrogen); (endometriosis OR estrogen OR oestrogen) AND immune cells; (estrogen OR oestrogen OR estrogen receptor OR oestrogen receptor) AND (endometriosis OR immune cells); endometriosis lesions AND immune cells; endometriosis AND immune cells NOT (estrogen OR oestrogen OR estrogen receptor OR oestrogen receptor); (estrog* OR oestro*) AND immune cells; ("immune cells" NEAR/estrogen) OR ("immune cells" NEAR/oestrogen). The searches also included additional terms (in the above same formatting): 'autoimmunity', 'monocytes', 'leukocytes', 'uterus', 'menstrual cycle', 'ovarian cycle', 'endometrial', and 'female immunity'. Additional specific searches include for endometrium OR endometriosis AND specific immune cells (Neutrophil, Basophil, T cells, etc.). This review presents relevant theories, concepts, and findings to identify research gaps and to provide a foundation for future studies.

We will first explore the structure, function, and physiological changes in the normal endometrium throughout the ovulatory menstrual cycle, followed by current knowledge gaps related to immune–hormonal interactions in both healthy and endometriosis-affected tissues, including the eutopic endometrium. Endometriotic lesions comprise endometrial-like tissue fragments and immune cells that likely evade destruction during retrograde menstruation. The survival and proliferation of ectopic endometrial cells is sustained by impaired immune surveillance and abnormal cytokine expression; this facilitates invasion into the mesothelial lining whilst promoting growth and neovascularization. Macrophages play a pivotal role, with their normal functions altered during both the initiation and persistence of the disease.

After providing an overview of estrogens and their isoforms, we will conclude by discussing hormone interactions with specific myeloid and lymphoid cells, highlighting the recent literature on hormone–immune interactions in endometriosis.

2. Human Endometrium, Physiology, and Anatomy

The human endometrium features the following two distinct yet interdependent layers: the luminal "functionalis" and the foundational "basalis". The functionalis, a dynamic layer lining the uterine cavity, evolves throughout the menstrual cycle, comprising epithelial and stromal components, intricate glands, resilient stromal cells, spiral arteries, and various immune cells that create a nurturing environment. During pregnancy, the functionalis transforms into a supportive bed for the embryo, establishing a vital fetal–maternal connection. Below this active surface lies the basalis layer, home to progenitor stem cells that ensures self-renewal and which enables monthly regeneration. Together, these layers showcase the endometrium's remarkable adaptability and regenerative capacity, inviting the deeper exploration of reproductive health. Since the 1950s, it has been understood that endometrial morphology corresponds directly to circulating changes in estrogen (E2) and progesterone (P4) across three distinct phases in a 28-day cycle—a phenomenon that recurs approximately 451 times during a woman's reproductive years (assuming a typical reproductive duration of about 34.7 years) [1,2]. Each phase—menstruation, proliferation, and secretory—unfolds with rhythmic precision, driven primarily by E2 and 4, along with regulators like the hypothalamic–pituitary–ovarian (HPO) axis. In the proliferative phase, rising E2 levels stimulate rapid growth by increased mitotic activity in epithelial and stromal components and angiogenesis, creating delicate-walled blood vessels. Following ovulation, the secretory phase is marked by increased P4, thickening the endometrium through stromal decidualization, enhancing the epithelial secretory function, resulting in a convoluted structure, and blood vessel remodeling into prominent thick-walled spiral arteries that is vital for nourishing the enriched endometrium, priming it for potential implantation [3,4]. The symphony of postovulatory hormonal changes is so unique that a postovulatory day can be accurately, precisely, and reliably assessed histopathologically, based on stromal and epithelial alterations.

The adaptive response is crucial for supplying blood to the thickened endometrium, facilitating fertilized egg implantation. Conversely, if fertilization does not occur, declining P4 levels trigger tissue ischemia and necrosis, leading to menstruation and restarting the cycle. This dynamic hormonal interplay not only prepares the body for potential new life but also highlights the complex biological processes governing fertility, weaving a compelling narrative of resilience and renewal.

3. Immune Cells of Human Endometrium in Physiological Conditions

Endocrine and immune factors are crucial in regulating endometrial functions throughout the menstrual cycle, but the underlying dynamic cellular mechanisms remain poorly understood. At the heart of harmonious changes is a complex interplay between the endocrine and immune systems. The immune environment of the endometrium is significantly influenced by E2 and P4, reflecting the complex interplay; in fact, it is also regulated by the HPO axis and its associated immune microenvironment. Throughout the cycle, E2 promotes endometrial proliferation and selectively recruits specific immune cell populations, while P4 provides crucial immunosuppression during the preimplantation phase, creating a conducive environment for potential pregnancy, even though many menstrual cycles do not culminate in successful implantation. Moreover, E2 promotes an additional immune tolerance during pregnancy, ensuring optimal conditions for fetal development. This intricate balance between hormonal action and immune regulation underscores the remarkable adaptability of the endometrium, setting the stage for successful reproduction [5].

The composition of immune cells in the endometrium, particularly the tissue-resident ones, along with the balance and interactions of pro- and anti-inflammatory cytokines essential for proper function, is subject to fluctuations throughout the menstrual cycle (Figure 1). During the proliferative phase, counts of natural killer (NK) cells and certain T cells gradually increase, in addition to enhanced angiogenesis and tissue regeneration. However, lymphocyte numbers appear to be fewer, leading to the interpretation of a less active immune environment compared to the secretory phase. Immune cell counts rise significantly during the late secretory phase and menstruation, contributing to tissue remodeling, menstrual breakdown, and repair [6].

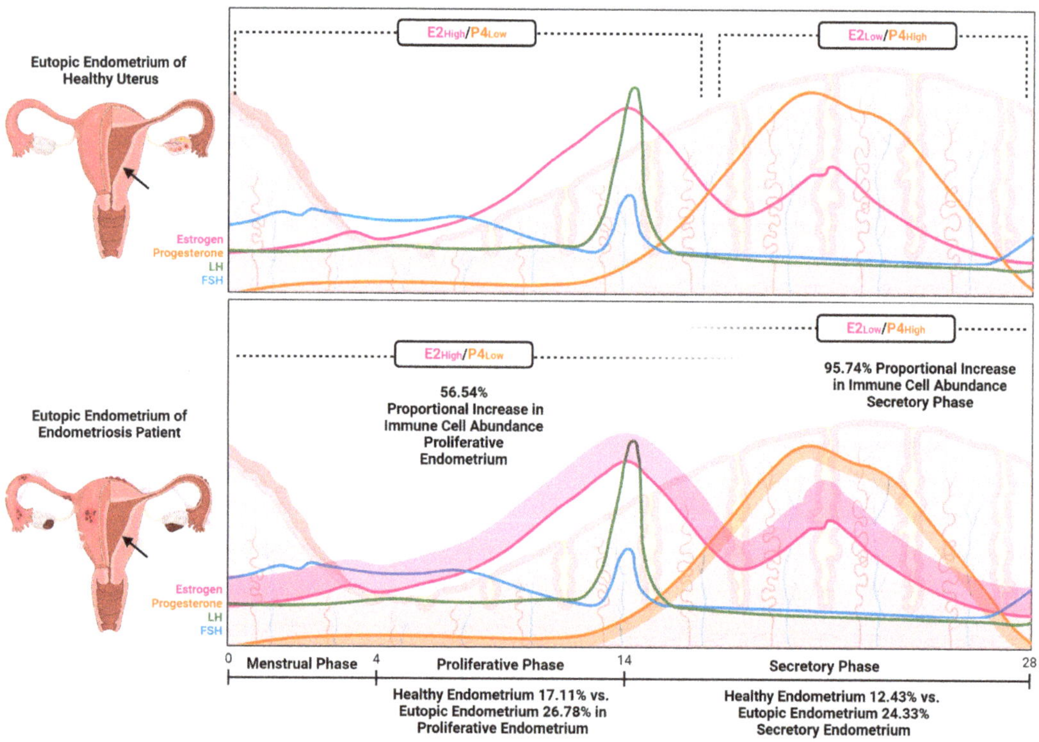

Figure 1. A summary of the hormonal changes in endometrial tissue, with reference to the proportional abundance of immune cells between the healthy endometrium versus eutopic endometrium in endometriosis. The menstrual cycle (0 to 28 days) is annotated to scale given the expected endometrial thickness, with corresponding estrogen (E2, pink), progesterone (P4, orange), luteinizing hormone (LH, green), and follicle-stimulating hormone (FSH, blue) levels overlaid. Wider lines indicate hormonal variance, while dashed lines to fade show uncertainty in hormonal crossovers in reference to the 28-day cycle. Proportional Increases were calculated from statistics derived from works reported in Huang, et al. [7].

Endometrial immune cells, marked by the CD45 surface protein found on most hematopoietic cells, make up 10–20% of endometrial cells, peaking during the secretory phase. T cells, identified by CD3 positivity, dominate in the proliferative phase, but by the mid-to-late secretory phase, uterine NK (uNK) cells identified by CD56 positivity become predominant, comprising 70–80% of endometrial leukocytes [8]. T cells are essential for maintaining the mucosal barrier and local immune defense [9]. Key endometrial leukocytes, other than T and uNK cells, include macrophages (Mφs, CD68 positivity), along with dendritic cells (DCs, ITGAX positivity), B cells (CD20 positivity), mast cells (MS4A2 positivity), and neutrophils, which are absent in the proliferative and early secretory phases but surge before menstruation, influenced by steroid hormones [10,11].

Research underscores the critical role of immune responses during menstruation, where Mφs, NK cells, and cytotoxic T cells orchestrate tissue destruction and regeneration. These cells are carefully regulated to target menstrual remnants while preserving the integrity of regenerating endometrial tissue, thereby maintaining homeostasis. Uterine macrophages (uMφs) and uNK cells engage extensively with stromal, endothelial, and perivascular cells to support endometrial regeneration. uMφs play a pivotal role in clearing debris, sustaining tissue homeostasis, and exhibiting phenotypic plasticity to adapt to the local microenvironment. The traditional M1 pro-inflammatory and M2 anti-inflammatory classification of Mφ simplifies their diverse roles. In pregnancy, the balance of Mφ polarization at the maternal–fetal interface is crucial for maintaining a healthy pregnancy. uNK cells, briefly divided into uNK1-3, provide localized immune protection and, distinct from their peripheral counterparts, are classified as endometrial or decidual based on their location and pregnancy status, and contribute to uterine remodeling and immune tolerance so that the abnormalities in their function or numbers could lead to pregnancy complications. Endometrial stromal cells express matrix metalloproteinases (MMPs) during the menstrual phase, along with integrins and key cytokines like CCL5 (RANTES, Regulated on Activation, Normal T cell Expressed and Secreted, a pro-inflammatory chemokine that recruits leukocytes to inflammation, also chemotactic for eosinophils, basophils, monocytes, T cells, NK cells, DCs, and mastocytes) and Retinoic Acid Receptor Responder Protein 2 (RARRES2, a cell surface receptor that stimulates Mφ and DC chemotaxis while inhibiting growth) to facilitate the recruitment of uMφs to the tissue. Both uMφ1 and uMφ2 upregulate Platelet-Derived Growth Factor Subunit B (PDGFB), which is essential for wound healing, as it binds to PDGFRB on stromal cells post-menstruation [12,13].

uMφ1 upregulates tumor necrosis factor-α (TNF-α), which activates ER signaling in endometrial epithelial cells to regulate the secretion of inflammatory substances—a process inhibited by P4. It also increases Epiregulin (EREG), a member of the EGF family involved in inflammation, wound healing, oocyte maturation, and cell proliferation. In contrast, uMφ2 elevates insulin-like growth factor 1 (IGF1). Together, these factors promote the proliferation and survival of endometrial stromal cells. Additionally, immunoregulatory genes such as IL10, Galectin-9 (LGALS9, which modulate T and B cell responses and support Treg differentiation through TGF-b), and triggering receptor expressed on myeloid cells 2 (TREM2, another protein that functions in immune response) are upregulated by uMφs. This fosters an anti-inflammatory environment that is crucial for the scarless regeneration of endometrial tissue during the proliferative phase [14]. These immune cells and their modulators undergo dynamic shifts to maintain endometrial homeostasis. Disruptions in their populations or functions can create an inhospitable environment for embryo implantation and potentially lead to endometrial pathologies like endometriosis. Exploring these intricate interactions sparks fascinating questions about the immune system's influence on reproductive health [15,16].

4. Immune Cells in Endometriosis with a Comparison to Typical Human Endometrium

The information above underscores the critical role of immune cells in the human endometrium for maintaining tissue homeostasis and regeneration throughout the menstrual cycle. In endometriosis, often described as 'endometrium-like tissue outside the endometrium', these immune cells exhibit significant alterations in distribution and signaling pathways, contributing to both the initiation and progression of the condition. Notably, women with early-stage endometriosis exhibit elevated pro-inflammatory cytokines in their endometrial immune cells, raising intriguing questions about their role in disease development [7].

In addition to endometriotic tissue, the "eutopic endometrium" in patients with endometriosis shows significant hormonal and immune alterations, including elevated levels of ERα. By enhancing estrogen activity, this leads to abnormal proliferation and disrupts normal function, particularly when tissue is displaced during retrograde menstruation. Elevated ERα levels not only aid the adhesion and survival of endometriotic lesions but also promote inflammation and angiogenesis [17].

Moreover, progesterone resistance in the eutopic endometrium interferes with normal decidualization, potentially reducing receptivity to embryos. While significant changes in PRs (PR-B and PR-A) are not typically found in endometriotic lesions, mechanisms like promoter hypermethylation and microRNA dysregulation—particularly miR-29c, which affects FKBP4 expression—may explain the loss of PR-B and weakened P4 signaling, further contributing to this resistance [18,19].

The immune environment (Figure 2) in the eutopic endometrium of women with endometriosis differs notably from that of healthy women, particularly in the immune cell composition and activation of inflammatory pathways [20]. However, the precise proportions and distribution of immune cell subtypes in their eutopic endometria remain debated. These variations contribute to secondary epigenetic changes that can alter critical biological pathways. In the eutopic endometria of women with endometriosis, a defect in the suppressive function of non-T regulatory (non-Treg) cells, combined with the polarization of T helper 17 (Th17) cells driven by activated DCs, may promote a shift toward pro-inflammatory M1φs. This shift is further enhanced by elevated interleukin-17 (IL-17) levels, which are known to stimulate cytotoxic CD8+ T cells and aberrantly activate NKs. The IL-17 signaling pathway is significantly enriched in the eutopic endometrium of endometriosis patients, characterized by increased CD8+ T cells, NK cells, and follicular T helper cells compared to healthy controls. Additionally, the IL-17-producing non-Treg subset may contribute to the observed B cell enrichment in endometriosis, highlighting the complex interactions among different immune cell types [21–23]. Studies show that women with endometriosis have increased populations of CD8+ T cells and CD56+ NK cells, alongside a decrease in CD163+ Mφs in their eutopic endometrium. This shift reinforces a pro-inflammatory immune environment, with elevated CD8+ T cells potentially increasing the risk of infertility associated with the condition [22].

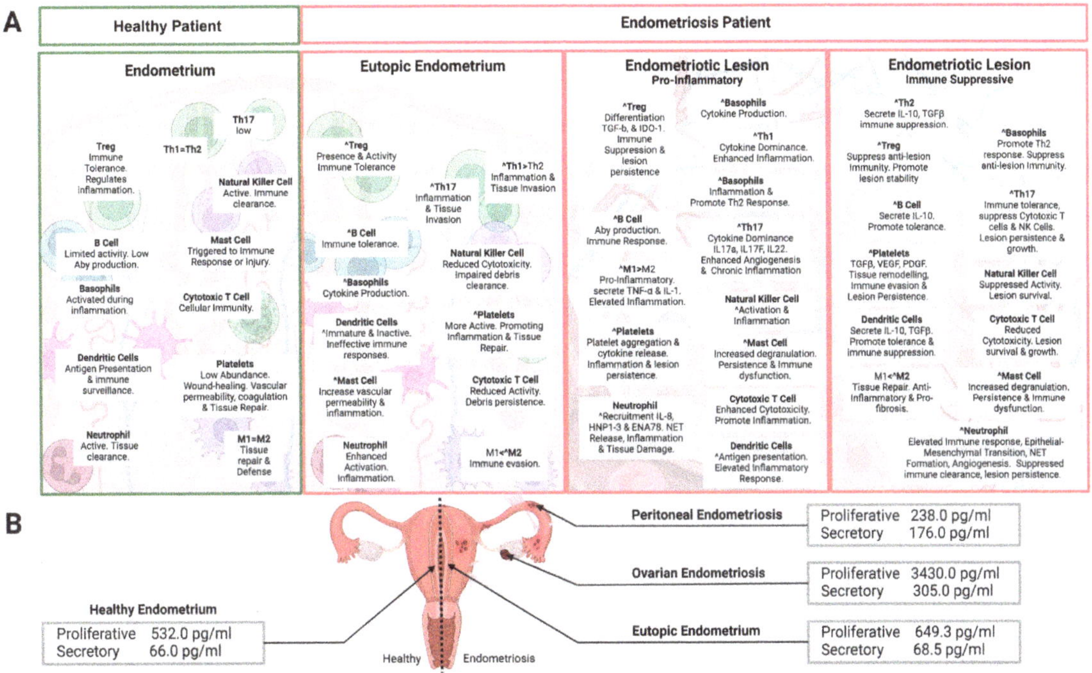

Figure 2. Immune cell dynamics and local estrogen concentrations in endometriosis. (**A**) Comparative immune cell profiles in the healthy endometrium, eutopic endometrium, and peritoneal lesions: pro-inflammatory vs. immune-suppressive microenvironments. The corresponding endometriosis patient panels, although not depicted directly, are under the pretext of a dysregulated endocrine milieu of estrogen dominance and progesterone resistance. The exact proportion and distribution of immune cell subtypes is still debated; however, there remain notable differences in activation states and function. '^' indicates an elevation of relative cell abundance or activity. Key acronyms: macrophage phenotypes 1 or 2 (M1 and M2); T helper (Th); T regulatory (Treg). (**B**) Local E2 concentration in women with and without endometriosis. Local (intratissue) E2 concentrations derived from Huhtinen et al., 2012 [24]. Tissue-specific E2 levels illustrate the differences in focal concentrations between endometriotic lesions and control tissues. Notable differences in E2 concentrations across endometriotic subtypes are also evident. These differences in E2 concentrations highlight the importance of investigating the constitutive or induced expression profiles of specific estrogen receptors in focal immune cells.

A recent study on cell clusters from menstrual effluent rich in endometrial tissue revealed striking differences between endometriosis patients and healthy controls during the menstrual phase. Endometriosis patients showed a significant depletion of uNK cells and an unexpected increase in B cells. Additionally, endometrial stromal cells from women with endometriosis exhibited a deficiency in P4-sensitive markers linked to decidualization. These key markers include insulin-like growth factor-binding protein 1 (IGFBP1, also known as placental protein 12, which binds insulin-like growth factors in plasma to prolong their half-life and alter interactions with surface receptors to produce different transcriptional splice variants), left–right determination factor 2 (LEFTY2, a member of the TGF-β family that plays a role in the developmental asymmetry and which is also implicated in endometrial bleeding), lumican (LUM, an extracellular matrix protein for collagen fibril organization, epithelial cell migration, and tissue repair), and decorin (DCN, a proteoglycan that modulates TGF-β and autophagy, and inhibits angiogenesis), all of which play important roles in tissue function and repair [25].

Mφs, vital for pathogen defense, also show significant differences in activity within the eutopic endometrium of women with endometriosis compared to healthy individuals, and recent single-cell studies indicate a pronounced pro-inflammatory phenotype among Mφs in endometriosis patients [7,14,26,27]. The predominant Mφ population is Mφ1 in these patients, which contributes to a detrimental inflammatory milieu affecting embryo implantation and endometrial functionalis layer regeneration. Interestingly, Mφs and DCs do not increase during the secretory phase, allowing cellular debris to accumulate and potentially migrate to the peritoneal cavity, where they tend to implant, allowing for the growth of ectopic lesions [8]. Recent data also suggest that the populations of Mφ1 and Mφ2 are key players in creating the abnormal inflammatory environment in endometriosis. At the mRNA level, sustained WNT signaling and dysregulated insulin signaling have been observed in early/mid-stromal cell populations in endometriosis, aligning with previous reports of IGF2 downregulation and impaired WNT inhibition during the secretory phase. This indicates altered WNT signaling in stromal cells, which is essential for the differentiation of the glandular epithelium in response to P4 [14]. These immune dysfunctions in the eutopic endometrium not only hinder anti-inflammatory responses but also enable endometrial cells to survive and migrate to ectopic sites, paving the way for lesion development.

Novel perspectives have attributed the imbalance of immune cell populations in endometriosis to the improper expression of immune checkpoints (ICPs) and their ligands. ICPs function in maintaining self-tolerance and the modulation (initiation, duration, and magnitude) of immune responses of effector cells, with co-stimulatory and co-inhibitory signaling between antigen-presenting cells and effector cells, a possible source of immune dysregulation in endometriosis [28]. These complex interactions highlight the urgent need for the further investigation into how immune and signaling pathways intertwine in endometriosis, potentially opening new avenues for targeted therapies that could restore normal immune function and improve reproductive outcomes.

Although our primary focus is endometriosis, it is important to also address adenomyosis—endometriosis located within the uterine myometrial muscles. Previous studies have underscored the crucial role of the immune profile in the development and progression of adenomyosis. The eutopic endometrium in adenomyosis exhibits immune-related alterations that distinguish it from the endometrial tissue of individuals without the disease. These changes contribute to a pro-inflammatory microenvironment characterized by impaired immune surveillance and dysregulated immune responses, similar to those of endometriosis. In particular, immune cells, such as Mφs and NK cells, display altered functionality, resulting in a reduced ability to clear abnormal cells. Additionally, the cytokine profiles in the eutopic endometrium reveal elevated levels of pro-inflammatory mediators, which further contribute to chronic inflammation and tissue remodeling. This immune dysfunction likely facilitates the migration of basal endometrial cells into the myometrium, promoting the progression of adenomyosis.

Dysregulated immune responses in the peritoneal environment may fail to eliminate misplaced endometrial cells, allowing them to implant [29,30]. Endometriotic cells that implant superficially on the peritoneum can proliferate and thrive in the peritoneal fluid, a specialized environment rich in estrogens, progestins, cytokines, growth factors, and other bioactive molecules. This setting supports the survival, growth, and invasion of ectopic endometrial cells, leading to chronic inflammation, neovascularization, and the progression of peritoneal endometriosis lesions [31].

Additionally, hormonal fluctuations during the menstrual cycle significantly alter steroid hormone levels in the peritoneal fluid, influencing the local microenvironment and potentially impacting the progression and development of endometriosis. Notably, during the follicular phase, E2 and P4 levels in the peritoneal cavity are significantly elevated compared to plasma, with P4 levels even reaching those observed during the luteal phase. At ovulation, the ruptured follicle releases steroid hormones into the peritoneal cavity, resulting in hormone concentrations that can be 100 to 1000 times higher than in plasma [29,32,33]. Ovulation also triggers a cascade of inflammation and resultant cell damage. Superficial peritoneal implants on the ovarian cortex might affect the ovulation site by altering the surrounding inflammatory microenvironment [34]. Key players in this process are cytokines such as interleukin-1, interleukin-6, and tumor necrosis factor, which are vital in both the inflammatory response associated with endometriosis and the mechanisms governing ovulation. This interplay raises intriguing questions about how these factors shape reproductive health [35].

In women with endometriosis, NK cells of the innate immune system appear to play a crucial role in the impaired clearance of endometriotic tissue within the peritoneal cavity. While there is some disagreement among studies regarding the quantitative alterations of peritoneal NK cells, a consensus has emerged that NK cells from both peripheral and peritoneal sources in endometriosis patients exhibit significantly reduced cytotoxic function. This diminished activity is closely linked to an increased expression of killer inhibitory receptors, which interact with HLA class I molecules on target cells, ultimately suppressing NK cell activity and function [36,37]. Compounding this issue, the cytotoxic function of CD8+ T lymphocytes in peritoneal fluid is also impaired, allowing endometriotic cells to evade immune surveillance, which is achieved by inducing T lymphocyte apoptosis through the Fas–FasL pathway [38]. Previous research has shown that endometriotic cells not only develop resistance to Fas-mediated NK apoptosis but also exploit this mechanism to launch a "Fas counterattack" against the host immune response. By upregulating FasL expression, these endometriotic cells induce apoptosis in immune cells, effectively shielding themselves from Fas-mediated cell death and enhancing their ability to evade immune detection, providing another nuanced interaction between immune evasion and tissue persistence, highlighting the complex nature of the underlying mechanisms at play in endometriosis [39].

In a prior study, researchers delved into what they referred to as endometriosis-associated immune cell infiltrates (EMaICIs) across various forms of endometriosis (including peritoneal–ovarian) and adenomyosis, and found that EMaICIs were most abundant in peritoneal and ovarian endometriosis, while adenomyosis exhibited a significantly lower presence of these infiltrates. The analysis identified key immune cell types, predominantly CD3+ T lymphocytes, which included both helper (CD4+) and cytotoxic (CD8+) T cells, as well as antigen-experienced memory T cells (CD45RO+). Additionally, Mφs and B lymphocytes were part of the infiltrate. Interestingly, regulatory T cells (CD25+ high and FoxP3+) and NKs (CD56+) were significantly absent [40].

Conversely, some studies suggest that the immune response action may be a protective mechanism that limits disease progression. Based on an investigation of immune activation in lymph nodes associated with the bowel lesions of deep endometriosis (DIE), lymph nodes showed increased T cell proliferation, particularly rich in CD4+, along with a diverse array of paracortical immune cells, including helper and regulatory T cells, B cells, DCs, Mφs, and plasma cells. The lymph nodes in question exhibited a notable decrease in CD10+ endometrial–stromal-like cells compared to other pelvic nodes. This observation suggests that endometriotic lesions or the movement of endometrial–stromal-like cells may trigger and activate the immune response, which drain them, potentially playing a role in controlling the disease's spread and progression [41]. However, it is crucial to acknowledge the significant variability in tissue localization and the stages of endometriosis among the studies mentioned. This variability complicates the efforts to draw definitive conclusions, as many hormonal and environmental factors, along with mediators of pathogenesis, remain poorly understood and could influence these complex interactions.

5. Estrogen, Isoforms, and Their Transcriptional Regulation

Estrogens are a group of structurally similar steroid hormones derived from cholesterol, including estrone (E1), 17β-estradiol (E2), estriol (E3), estetrol (E4), and 27-hydroxycholesterol (27HC). E2 is the most potent one, with the highest affinity for estrogen receptors (ERs), and plays a crucial role in various biological processes [42,43]. E1 predominates in postmenopausal women, produced by gonads and adipose tissue, while estriol and E4 are primarily present during pregnancy. E2 exerts its effects through ER binding, regulating the cellular metabolism, proliferation, ion transport, and contractility [44].

Estrogen receptors, ERα and ERβ, encoded by the *ESR1* and *ESR2* genes, can form homo- or heterodimers with distinct functional roles. Their primary difference lies in the N-terminal domains, specifically the 'activation-function 1 (AF-1)' and the 'ligand-binding domain (AF-2)', which exhibit low sequence similarity (Figure 3). This affects their recruitment of coregulators and gene transcription regulation, with ERα demonstrating greater transcriptional activity due to its more potent AF-1 [45,46]. In the absence of E2, ERs remain inactive, bound to heat-shock proteins (HSPs) in the cytoplasm or nucleus. E2 binding induces a conformational change, releasing HSPs, dimerizing the receptors, and facilitating their translocation to the nucleus to interact with estrogen-response elements (EREs) in target gene promoters [47,48].

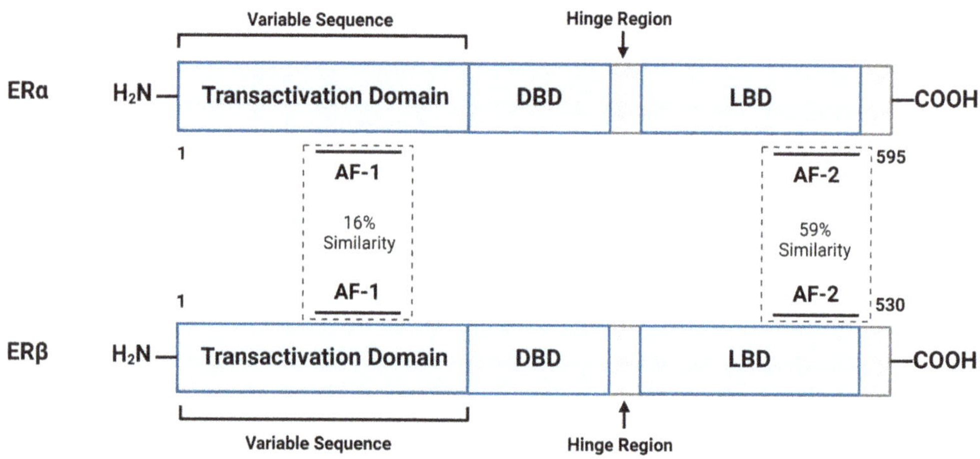

Figure 3. Structural summary of ER isoforms and their role in transcriptional regulation: differences between ERα and ERβ in AF1 and AF2 activity. The general structure of the ER includes the transactivation domain containing activation-function 1 (AF1), the DNA-binding domain (DBD), hinge region, ligand-binding domain (LBD), and activation-function 2 (AF2). ERs exist in two isoforms, ERα and ERβ, which form functionally distinct homo- and heterodimers (αα, αβ, and ββ) that play non-redundant roles in transcriptional regulation. The primary differences between ERα and ERβ are in the N-terminal domain, which contains the ligand-independent AF1, with a 16% similarity between the isoforms. The AF2 at the COOH-terminal ligand-binding domain (LBD) exhibits a 59% amino acid sequence similarity between the isoforms. Upon E2 binding, conformational change ensues to the receptor, enabling the recruitment of coregulators to the activation-function domains. The AF1 region, differing between ERα and ERβ, results in distinct transcriptional activities. In contrast, AF2 activity is similar in both isoforms.

E2 also activates the non-genomic signaling pathway by binding to membrane-bound ERα and ERβ. This rapidly triggers nuclear transcription factors through ion channel regulation and the activation of enzymes like Ca^{2+} mobilization, PI3K, and MAPK. This process occurs within seconds to minutes, independent of gene regulation, and is known as the rapid "non-genomic effect" [49].

ERα and ERβ, the primary nuclear receptors for estrogens, are encoded by separate genes and perform various roles that can overlap, contrast, or act independently, depending on the cell type they are found in. ERα is more prevalent and is predominantly found in immune cells. In contrast, a definitive understanding of ERβ expression remains elusive, as many previous studies have been called into question due to issues with the accuracy of the antibodies used for detecting ERβ in cells [46,50].

These receptors are vital in regulating the bone metabolism, cardiovascular health, and central nervous system functions. Accordingly, the dysregulation of estrogen signaling may be associated with various diseases, including cancer and metabolic disorders. ERα expression typically increases in early breast cancer stages, facilitating tumor growth, and antiestrogens are often employed in therapies. Conversely, ERβ expression generally declines during carcinogenesis, suggesting a suppressive role and the potential as a therapeutic target. However, conflicting evidence regarding ERβ's role in cancer arises from patient variability, tissue heterogeneity, and weak correlations between ERβ mRNA and protein levels, highlighting the need for further research to better elucidate its function in cancer as well [51].

Additionally, 'G protein-coupled estrogen receptor 1 (GPER1)', previously known as 'G protein coupled receptor 30 (GPR30)', has emerged as another ER gene. Studies have shown that E2 can activate rapid signaling through the GPER1 protein. GPER1 has a unique pattern of expression in endometriosis. A study consisting of 74 ovarian, peritoneal, and deep infiltrating endometriosis showed high cytoplasmic GPER1 expression levels in the epithelial component of endometriosis and none in the normal endometrium. There was no significant difference in nuclear epithelial GPER1. The endometrioma has the highest frequency of cytoplasmic epithelial GPER1 as compared to peritoneal or deep infiltrating endometriotic lesions. GPER overexpression in endometriosis suggests roles in hormonal regulation. The inhibition of GPER in ectopic endometrial stromal cells using miRNAs reduced the proliferation, migration, and invasion [24,52,53]. These data suggest the importance of GPER in endometriosis progression.

6. Estrogen–Immunity Relationship and Alterations in Endometriosis

The immune system exhibits notable sexual dimorphism, with females generally displaying stronger immune responses and a higher incidence of autoimmune diseases than males. This disparity is linked to differences in estrogen levels and their effects [54]. Research has established a connection between fluctuations in circulating estrogen and the onset and severity of various autoimmune diseases, highlighting estrogens' role as key immune modulators [55,56].

Specifically, E2 plays a crucial role in shaping the development and function of immune cells, as summarized in Table 1. Specifically, E2 is essential for the self-renewal of hematopoietic stem cells (HSCs) and further impacts the quantity and type of immune cells involved in both innate and adaptive immune responses, with studies demonstrating that ERα and ERβ mRNAs and proteins are expressed in hematopoietic progenitors and mature immune cells [44,57].

ESR1 expression is the highest in B cells, with moderate levels in helper T cells, cytotoxic T cells, NK cells, and plasmacytoid DCs. Interestingly, while monocytes exhibit the lowest *ESR1* RNA levels, this expression increases in monocyte-derived DCs during differentiation. *ESR2* is primarily expressed in B cells and plasmacytoid DCs, while *GPER1* is also found in various immune cells, including T cells, B cells, monocytes, Mφs, and neutrophils [57–60]. Moreover, estrogen protects primary B cells from apoptosis by repressing PD-1 and promotes the differentiation, proliferation, and survival of early B-lineage precursors by upregulating CD22, a transmembrane protein that regulates B cell function; SHP1, a protein tyrosine phosphatase expressed in all hematopoietic cells and a key regulator of immune cell function; and Bcl-2, the protein which controls cell fate by blocking apoptosis [61,62]. These findings suggest that estrogens modulate immune responses based on ER expression, and can significantly impact the differentiation and function of neutrophils, Mφs, and NK cells [57].

In endometriosis, estrogen significantly impacts both immune responses and disease progression. ERβ plays a crucial role in mediating E2 effects, supporting the survival of endometriotic lesions, promoting pelvic peritoneum remodeling, and inducing inflammatory mediator release. A unique known challenge in endometriosis, P4 resistance, stems from the deficiency of P4 in the stromal cells of endometriotic tissue, complicating the hormonal and immune interplay characteristic of the condition [63,64].

The expression levels of ERs vary markedly between the healthy endometrium and ectopic lesions or the eutopic endometrium. In endometriotic stromal cells, ERβ is significantly overexpressed, while ERα levels are notably lower compared to endometrial stromal cells of the healthy endometrium. This receptor imbalance is linked to endometriosis pathogenesis and is influenced by aberrant epigenetic modifications, such as DNA methylation in endometriotic cells. Despite the considerable data on ERs in endometriosis, the specific contributions of ERα and ERβ to disease development remain inadequately understood [65]. In particular, the ERα distribution in superficial peritoneal lesions shows significant variability, even within single biopsy samples [66]. This heterogeneity also obscures the influence of menstrual cycle fluctuations on receptor expression, complicating the accurate classification of the ER status in the disease.

The abnormal activity of ERs, particularly their elevated expression in localized endometriotic tissues, likely drives their growth and persistence. Additionally, local de novo estrogen production within endometriotic tissues further disrupts the microenvironment, altering steroid hormone dynamics and triggering secondary immune responses. These shifts in the tissue microenvironment and immune activity are critical to the development and persistence of endometriosis [67]. It is also important to recognize that these hormone receptors may be upregulated in response to hormonal fluctuations or tissue-specific signals. While their roles under physiological conditions are not fully elucidated, understanding how they affect immune cells is essential, prompting further investigation.

Table 1. The role of estrogen on immune cell function in the human endometrium and endometriosis.

Immune Cell	Immune Cell Phenotype Marker	Estrogenic Effect on Immune Cell	General Role of Immune Cell in Healthy Endometrium	Role of Immune Cell in Endometriotic Lesions	Refs.
Macrophages (Mφ1s)	CD14+, CD68+, HLA-DR+, CD11c−, and CD86+, CD80+, or iNOS+ (pro-inflammatory)	17β-estradiol significantly impairs the gene expression of inflammatory markers and the production of IL-1β.	- Proliferative: Mφ1 activity is balanced with Mφ2, maintaining immune tolerance and preparing for potential implantation. - Secretory: Mφ1 cells help in controlling inflammatory response. - Menses: Mφ1 cells dominate to Mφ2, and contribute to tissue shedding by increasing the activity through estrogen drop.	Persistent Mφ1s exacerbate inflammation in lesions and may interfere with effective lesion clearance.	[68–72]

Table 1. *Cont.*

Immune Cell	Immune Cell Phenotype Marker	Estrogenic Effect on Immune Cell	General Role of Immune Cell in Healthy Endometrium	Role of Immune Cell in Endometriotic Lesions	Refs.
Macrophages (Mφ2s)	CD14+, CD68+, HLA-DR+, CD11c−, and CD163+ or CD206+ (anti-inflammatory)	17β-estradiol significantly upregulates the expression of anti-inflammatory markers and enhances migration.	- Proliferative: High estrogen shifts the phenotype towards Mφ2 polarization, promoting tissue repair and immune tolerance. Increased activity supports endometrial regeneration. - Secretory: The Mφ 1 phenotype dominates to Mφ1. Gland remodeling and angiogenesis through VEGF secretion, and function in tissue preparation for implantation. - Menses: Mφ2 activity increases, reducing inflammatory responses.	Reduction in the wound-healing phenotype CD163+. Persistent Mφ2 activity supports immune tolerance in lesions.	[71,72]
Neutrophils	CD16+ and CD66b+ or CD15/SSEA1+	17β-estradiol; reduced adhesion (CD11b, CD18, and CD62L); selective inhibition of phagocytosis; E2 concentration-dependent suppression of ROS. High E2 levels increase NET formation, enhancing inflammation.	- Proliferative: Elevated estrogen increases neutrophil influx, with a moderate inflammatory role. - Secretory: Estrogen suppresses neutrophil activity, aiding in tissue maintenance and avoiding excessive inflammation. - Menses: Neutrophils contribute to tissue shedding and the clearance of cell debris through neutrophil extracellular trap (NET) formation.	Increased sensitivity to estrogen, leading to impaired immune clearance and exacerbated tissue damage, and inflammation around the lesions. Increase in NET formation may exacerbate local inflammation, contributing to lesion persistence.	[73–76]
Eosinophils	CD11b+, CD16+, CD193+, and Siglec-8+, and CD16−	Primarily enacted through Erα; GPER increases CCL1 chemotaxis, modulates apoptosis (blocks caspase3), and increases IL-5-stimulated apoptosis.	- Proliferative/Secretory: Eosinophils are not prominent. - Menses: Eosinophil numbers increase, facilitating tissue breakdown during menstruation. GPER aids in tissue remodeling through blocking caspase-3 and enhancing chemotaxis.	Eosinophils are often decreased, and they exhibit altered function, which may contribute to the failure in immune surveillance and inflammation.	[77,78]
NK Cells	CD56low, CD16+; CD56+, NCR+, and CD69+	Suppresses NK cell cytotoxicity.	- Proliferative: Estrogen supports NK cell function, promoting activation and immune surveillance, and aiding in tissue homeostasis. - Secretory: Estrogen reduces NK cytotoxicity, and promotes secretion of factors to maintain uterine vascular stability and tolerance. - Menses: NK cell cytotoxic activity peaks to assist with tissue breakdown.	NK cytotoxicity declines as the disease progresses, supporting lesion persistence and inflammation.	[79–81]
T Cells (CD8+ Cytotoxic)	CD3+, CD8+, Granzyme+, or Perforin+	Enhances effector functions (cytokine secretion and lytic activity) predominantly through Erα, with Erβ having a more inhibitory effect.	- Proliferative: Cytotoxic T cell activity is downregulated by estrogen, promoting tissue growth. - Secretory: Cytotoxic T cell function further suppressed to avoid damaging the endometrium. - Menses: Cytotoxicity increases to aid in shedding and tissue breakdown.	Cytotoxic T cell activity is low in lesions, promoting immune tolerance and the persistence of ectopic tissue.	[82–84]
T Cells (CD4+ Helper)	CD4+, T-Bet+, and IFNγ+; GATA-3+ and IL-4+; RORγt+ and IL-17+; FoxP3+ and CD25+	Supports the balance between Th1 and Th2, where elevated E2 favors Th2 in human unstimulated PBMCs.	- Proliferative: Estrogen supports a balance between Th1 and Th2 cells for tissue homeostasis. - Secretory: Higher estrogen promotes Th2 differentiation, which suppresses inflammatory responses, which supports embryo implantation. - Menses: Low estrogen promotes Th1 differentiation, allowing for inflammation and immune activation processes. Th17 cell activity increases, potentially aiding in uterine shedding.	Th1/Th17 cells drive inflammation in lesions, contributing to disease pathogenesis. Their persistence in lesions may exacerbate tissue damage and chronic inflammation.	[83,85]

Table 1. *Cont.*

Immune Cell	Immune Cell Phenotype Marker	Estrogenic Effect on Immune Cell	General Role of Immune Cell in Healthy Endometrium	Role of Immune Cell in Endometriotic Lesions	Refs.
Regulatory T Cells (Tregs)	CD4+ and FoxP3+	Elevation influences Treg expansion, promoting immune tolerance and tissue homeostasis.	- Proliferative: Tregs help suppress excessive immune responses by regulating Th1/Th2. - Secretory: Tregs promote immune tolerance by secreting immunosuppressive cytokines (IL-10 and TGF-β), preventing inflammation to support implantation. - Menses: Reduced Treg function increases inflammation but prevents excessive immune activation.	Tregs are often dysregulated in lesions, contributing to the persistence of inflammation and lesion survival. Estrogen influences Treg expansion, supporting lesion survival.	[21,83,86]
B Cells	CD19+	Impacts B cell differentiation, survival, and function primarily by upregulating CD22, SHP-1, and BCL-2.	- Proliferative: Lower estrogen stimulates B cell differentiation, allowing for the secretion of antibodies and cytokines for the regulation of tolerance and defense in tissue. - Secretory: Estrogen upregulates CD22, SHP-1, and BCL-2 factors, enhancing B cell survival, and conferring tolerance during embryo implantation. - Menses: Estrogen levels decrease, altering B cell populations; reduction in cell survival factors, causing an increase in immature B cell populations and the apoptosis of Breg cells.	Estrogen suppresses B cell activity in lesions. Estrogen may promote autoreactive B cell survival, contributing to inflammation.	[62,87]
Dendritic Cells (DCs)	CD11c+, HLA-DR+, and CD80/86+	17β-estradiol inhibits DC maturation, reducing antigen presentation.	- Proliferative: DCs assist in the immune response to potential pathogens. - Secretory: Reduced DC maturation, reducing the antigen presentation capacity and supporting immune tolerance. - Menses: DCs modulate local immune activity and prevent excessive inflammation, supporting tissue turnover.	Reduced DC activity aids in lesion survival. Estrogen modulates DC function, promoting the immune tolerance of lesions. DCs might increase inflammation, supporting lesion breakdown.	[88]
Conventional DC1s	CD11c+, HLA-DR+, and XCR1 or CLEC9α+	Promotes the differentiation of cDC1 but reduces cDC1 maturation, impairing antigen presentation.	- Proliferative: Differentiation is promoted to aid in antigen presentation, immune surveillance, and priming T cell responses. - Secretory: Suppression of inflammation and inducement of immune tolerance during implantation. - Menses: DCs help to promote shedding.	DC1s are less active in lesions, with impaired antigen presentation and immune responses. This promotes immune suppression in lesions, aiding in tissue survival and persistence.	[88]
Myeloid-Derived Suppressor Cells (MDSCs)	CD11b+ and Gr1+	Positive correlation between E2 and MDSC in a dose–response manner, resulting in expansion (*S100A8* and *S100A9* upregulation), functioning in suppressing T cells, and elevated ROS production.	- Proliferative: Elevation of E2 results in the expansion and activation of MDSCs to regulate immune responses to support endometrial tissue. - Secretory: MDSCs suppress inflammation through secreting anti-inflammatory cytokines (IL-10) and suppress T cell activity for a tolerogenic state to support embryo implantation. - Menses: Estrogen declines, allowing for a tolerogenic shift, allowing for controlled inflammation and the shedding of endometrial lining.	MDSCs promote immune suppression in lesions, aiding in tissue survival and persistence.	[89–91]

Table 1. *Cont.*

Immune Cell	Immune Cell Phenotype Marker	Estrogenic Effect on Immune Cell	General Role of Immune Cell in Healthy Endometrium	Role of Immune Cell in Endometriotic Lesions	Refs.
Mast Cells	CD117+ and FcεRI+	17β-estradiol increases the infiltration and differentiation of mast cells by upregulating CPA3, VCAM1, CCL2, CMA1, CCR1, and KITLG.	- Proliferative: Mast cells are involved in immune modulation through the release of cytokines, growth factors, and histamine, facilitating an early immune response for tissue remodeling. - Secretory: Mast cell factors are repressed reducing the inflammation of the endometrium and confer tolerance for embryo implantation. - Menses: The upregulation of inflammatory mediators to facilitate menstruation and immune response.	Increased, and believed to contribute to chronic inflammation, enhancing the growth and survival of lesions.	[5,78]

7. Immune–Endocrine Interactions in Endometriosis: Cellular Mechanisms and Pathogenic Pathways Through the Immune Cell Lineage

The development of endometriosis cannot be attributed to a single pathway, as it results from the interplay of multiple physiological systems, with the immune and endocrine systems playing particularly influential roles. Understanding the mechanisms of immune–endocrine crosstalk in endometriosis is crucial for improving patient outcomes and developing more effective therapeutic strategies. ERα directly impacts HSCs, lymphoid progenitors, and myeloid progenitors, promoting various developmental pathways. A key consequence of ER action in these progenitors is the potential induction of epigenetic changes in precursor cells, influencing both developmental pathways and the functional responses of mature cells. Given this complexity, our focus in this part of our review will be on immune cells, seeking to re-evaluate their individual roles within the endometriotic milieu to better understand the disease, as summarized in Figure 4.

7.1. Myeloid Lineage

The deletion of a conditional ERα allele specifically in myeloid cells has revealed that E2 enhances Toll-like receptor 4 (TLR4) signaling in Mφs in vivo, a critical component in the innate immune response, resulting in increased levels of IL-6, IL-1β, and inducible nitric oxide synthase (iNOS), which contribute to a pro-inflammatory response; E2 amplifies this pathway by suppressing PI3K signaling and reducing Akt phosphorylation in Mφs [92]. These findings, along with related studies, highlight the heightened immune response often seen in females, which may also play a role in the increased prevalence of inflammatory diseases [60].

GPER-E2 signaling also activates the PI3K pathway, enhancing cell survival through the Akt signaling cascade. This response helps cells to counteract the stress caused by endoplasmic reticulum dysfunction by enhancing protein folding and preventing apoptosis, thus facilitating proliferation and migration. The activation of the tyrosine kinase Src pathway relocates chaperone proteins to the cell surface, further promoting cell survival. These mechanisms culminate in the activation of heparin-bound EGF, which transactivates its receptor, EGFR. The subsequent EGFR activation stimulates both the PI3K/Akt and ERK/MAPK pathways, again contributing to the regulation of cell proliferation and migration [74,93,94].

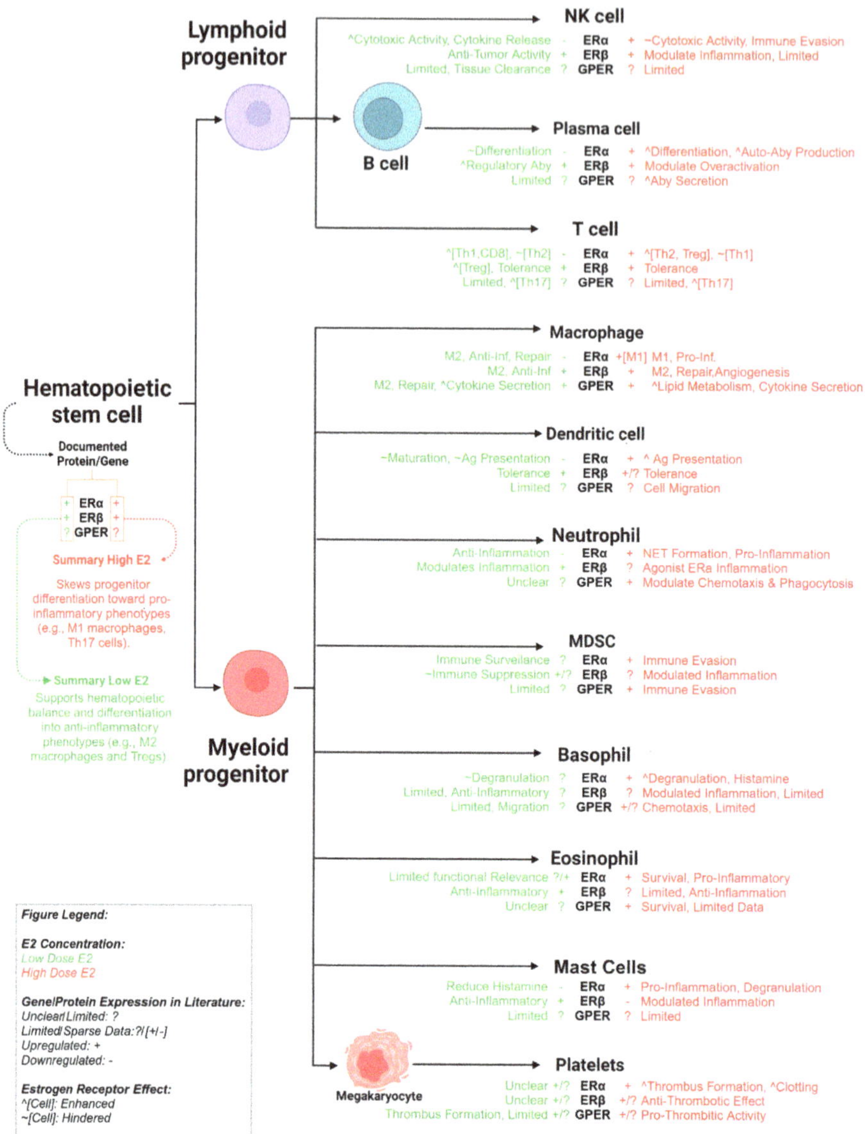

Figure 4. Effects of high and low estradiol (E2) on estrogen receptor expression and their respective roles in immune cells: A generalized snapshot of the current literature derived from endometrium and endometriosis tissue experiments. This figure provides a generalized overview of the effects of E2 on immune cells derived from endometrium and endometriosis tissues. The figure highlights how high and low E2 concentrations influence the expression of estrogen receptor alpha (Erα), beta (Erβ), and G protein-coupled estrogen receptor (GPER), as well as summarizes the resulting functions. Figure legend: E2 concentrations: high (red) and low (green); Gene/protein expression (respective to Erα/Erβ/GPER): unclear/limited information ('?'); limited/sparse data with minimal findings ('?' alongside '+' OR '−'); upregulated (+); downregulated (−). Note: up/downregulation predominantly refers to compared tissue regions inclusive of the internal (matched) endometrium and the healthy control endometrium. Estrogen receptor effect/function: enhanced (^'Cell Name'); hindered (~'Cell Name').

7.2. Myeloid-Derived Suppressor Cells (MDSCs)

MDSCs are a diverse group of immature and mature myeloid cells, including precursors for DCs, Mφs, and granulocyte-like neutrophils, eosinophils, basophils, and mast cells (MCs). MDSCs interact with various lymphoid cell types, such as NK cells, and T and B lymphocytes, and are classified into the following three main subsets: monocytic (M-MDSCs), polymorphonuclear (PMN-MDSCs), and early stage (eMDSCs) [90,95].

The immunosuppressive functions of MDSCs are mediated through multiple mechanisms. They suppress T cell activity by inducing ARGinase-1 (ARG-1), which depletes arginine and impairs T cell activation. MDSCs also produce nitric oxide (NO) and reactive oxygen species (ROS) via iNOS, further inhibiting T cell function. Additionally, they secrete TGF-β, which not only inhibits T cell activation but also reduces the cytotoxicity of cytotoxic T cells and suppresses the differentiation of Th17 cells [90,96,97].

MDSCs play a pivotal role in creating an immunosuppressive microenvironment and are associated with poor prognosis in cancers. In cancer immunology, E2 has been shown to influence the differentiation of E2-insensitive tumors by modulating MDSCs and exerting a significant inhibitory effect through the inhibition of ERα [98]. Moreover, MDSCs were suggested to serve as biomarkers in endometriosis, where elevated levels correlate with peritoneal lesions in patients with advanced disease, reflecting a sustained immunosuppressed niche [99].

7.3. Macrophages

Mφs recruited to the uterus primarily originate from the bone marrow, responding to tissue-specific signals related to damage and inflammation. In healthy females, tissue-resident Mφs are found in the perimetrium, myometrium, and endometrium, with densities varying throughout the menstrual cycle [100]. During the proliferative phase, under the influence of an E2 surge, Mφs (expressing markers CD54+, CD69+, and CD71+) constitute 1–2% of endometrial cells, playing a key role in supporting tissue proliferation and regeneration [68]. In the secretory phase, Mφ numbers rise to 3–5%, contributing to gland remodeling and angiogenesis through VEGF secretion. During menstruation, Mφs constitute 6–15% of the endometrial cell population, releasing matrix metalloproteinases, including MMP-12, MMP-9, and MMP-14, which are essential for the breakdown of the functionalis layer [68,101]. Throughout these phases, Mφs facilitate the resolution of immune responses and restore tissue homeostasis after inflammation and remodeling.

Estrogen significantly influences Mφ function, although the expression of estrogen ERα and ERβ in these cells is debated [69,70,102,103]. Some studies indicate that E2 regulates Mφ activation and cholesterol homeostasis, induces alternative activation in vitro, and affects the expression of the AF1-deficient form of ERα (the so-called ERα46) in human Mφs [70]. Additionally, ERα deficiency has been linked to increased TNFα secretion in murine Mφ in response to microbial stimuli [104]. Chronic E2 administration in ovariectomized mice triggers a pro-inflammatory response in resident peritoneal Mφ, while E2 surges promote cutaneous repair responses by enhancing Bcl-2 expression through ERK phosphorylation. In contrast, prolonged E2 exposure leads to a pro-inflammatory Mφ response via PI3K phosphorylation [92,105,106].

When it comes to the eutopic endometrium of endometriosis patients, as mentioned above, there are fewer "wound-healing" marker CD163+ Mφs and increased CCL2, indicating greater monocyte influx. Pro-inflammatory M1 Mφs accumulate due to defective non-T regulatory cells and Th17 polarization [14,61–63].

This altered Mφ activity creates a harmful inflammatory environment that hinders embryo implantation and endometrial regeneration. Additionally, Mφs do not increase during the secretory phase, leading to debris accumulation that can promote ectopic lesions, emphasizing the important roles of Mφ1 and Mφ2 [101].

In endometriosis, ectopic lesions increase while blood monocyte counts remain unchanged in patients with endometriosis; there is a fractional increase in ectopic lesions compared to the eutopic endometrium. However, the number of Mφs in peritoneal fluid rises significantly, with populations classified into the following two states: Mφ1 (pro-inflammatory) and Mφ2 (anti-inflammatory). The shift toward Mφ2, often seen during infection, relies on an ERα-mediated pathway that promotes interleukin-4 (IL-4) secretion, enhancing Mφ2 differentiation while inhibiting NFκB signaling and nitric oxide production, thus suppressing Mφ1 activity. Mφ1 is characterized by CD40+/CD80+/CD86+/HLADR+ markers, while Mφ2 expresses CD163+/CD206+/CD204+ markers [107].

Mφ polarization is influenced by various stimuli: Mφ1 develops from interferon-gamma (IFN-γ), TNF-α, or lipopolysaccharides, while Mφ2 is driven by IL-4, IL-10, IL-13, or TGF-β [107]. As both subsets increase regardless of the disease stage, the predominant Mφ phenotype in endometriosis remains debated, with some studies suggesting an Mφ1 dominance that may hinder embryo implantation [107,108]. However, the relationship between endometriosis and endometrial implantation may be the subject of another separate review [109].

The current understanding depicts Mφ polarization in endometriosis as dynamic, shifting from a pro-inflammatory Mφ1 state in early disease to an immune-suppressive-Mφ2 state in advanced stages [110]. Early ectopic endometrial cells secrete monocyte chemotactic protein 1 (MCP-1), recruiting additional Mφs. These local Mφs release TNF-α, IL-1β, IL-6, and MCP-1, promoting Mφ1 differentiation. As the lesions progress, additional factors drive the transition to the Mφ2 phenotype, facilitating angiogenesis, neurogenesis, and invasive lesion characteristics [8,111].

Based on research by Burns et al., early endometriosis development is primarily driven by immune responses, with minimal involvement of E2 and ERα. Significant infiltration of neutrophils and Mφs was observed, along with a reduced IL-6 response in lesions lacking ERα, suggesting two phases of the disease, an immune-dependent phase followed by a hormone-dependent phase, indicating that targeting the immune system may help prevent lesion attachment linked to Mφ activity [112].

7.4. Neutrophils

Neutrophils play a crucial role in the early progression of ectopic endometriotic lesions, with their recruitment driven by factors such as IL-8, human neutrophil peptides 1-3 (HNP1-3), and the Epithelial Neutrophil-Activating peptide, ENA-78 [30,113]. E2 not only increases neutrophil counts but also enhances their inflammatory activity by promoting the secretion of pro-inflammatory cytokines, including CXCL-10, IL-8, and VEGF [75]. E2 and its receptors regulate neutrophil function. Both ERα and ERβ are upregulated in pre-menopausal women during ovulation, coinciding with elevated serum E2 levels [73].

As mentioned above, neutrophils express GPER1, which interacts with key signaling pathways, including cAMP/Protein Kinase A (PKA), involved in regulating cell growth, migration, invasion, and metabolism, p38 Mitogen-Activated Protein Kinase (p38 MAPK), which coordinates cellular responses to stress and inflammation, and Extracellular Signal-Regulated Kinase (ERK), which has role in proliferation, differentiation, apoptosis, and survival. Activating GPER1 in neutrophils with the nonsteroidal agonist G1 promotes pro-inflammatory differentiation, prolongs lifespan, enhances respiratory burst activity, and modifies gene expression. Rodenas et al. propose that targeting GPER1 experimentally helps to reduce this pro-inflammatory neutrophil state [74].

In endometriosis, MyeloPerOxidase (MPO) levels are significantly elevated during the proliferative phase. Ectopic endometriotic lesions show the significantly increased mRNA expression of various chemokines such as CCL2, CXCL2, CXCL3, and CCL5; cytokines such as TNF, IL-16, and IL-7; Damage-Associated Molecular Patterns (DAMPs) such as S100A9; cell adhesion molecules like ITGAM, ITGB2, and SELP; and inflammasome markers such as NLRP3 compared to the eutopic endometrium and healthy controls. Furthermore, stage III/IV endometriosis lesions exhibit significantly higher levels of IL-8, a potent neutrophil chemoattractant, compared to controls [75].

7.5. Eosinophils

Eosinophils play a crucial role in the development and progression of endometriosis-associated fibrosis. Eosinophil peroxidase serves as an alternative marker for early-stage endometriosis-related fibrosis. Blumenthal et al. demonstrated that eosinophils co-localize with fibrotic connective tissues in endometriotic lesions, where they infiltrate and degranulate, potentially contributing to tissue remodeling and wound repair [114]. Furthermore, eosinophils enhance pro-tumorigenic processes by increasing the levels of VEGF, FGF, and PDGF, thereby promoting angiogenesis, tissue healing, and the secretion of extracellular matrix-degrading enzymes, while also inducing the Mφ2 phenotype through IL-13 secretion [115]. Most insights into E2-modulated eosinophil functions are derived from mouse models, with limited human studies. Evidence suggests that ERα is the primary mediator of E2 effects on eosinophils, facilitating their accumulation in the uterus; however, it remains uncertain whether this effect originates directly from eosinophils or eosinophil-responsive cells [116]. Eosinophils, initiators of inflammatory responses, are reported in an enriched manner in all phases of the cycle from the eutopic endometrium of endometriosis patients compared to the control endometrium, where they appear mainly pre- and during menses, to contribute to the pro-inflammatory phenotype [78,116].

7.6. Basophils

While IL-8, a chemoattractant for granulocytes, including basophils, is significantly elevated in endometriosis, literature confirming a substantial increase in basophil levels is limited [117]. However, a 2022 study by Bunis et al. employing whole tissue and single-cell RNA sequencing revealed a notable elevation of basophils in the eutopic endometrium of endometriotic patients throughout the uterine cycle, with these cells largely absent in healthy endometrial tissue.

7.7. Mast Cells

MCs, like Mϕs, are innate immune cells derived from common MDSCs with a unique ability to neutralize and degrade toxic proteins. Additionally, similar to Mϕs, MCs can adopt two distinct polarization profiles, which may serve opposing functions [101].

They are an integral part of the endometrial cycle, playing multifaceted roles in fibrosis, angiogenesis, inflammation, wound healing, and tissue remodeling. While the overall quantity of MCs relative to stromal cells remains relatively constant throughout the menstrual cycle, their morphology, granule composition, and activation states exhibit distinct variations at different stages. Just before menstruation, resident MCs become activated, facilitating tissue degradation through the ischemic vasospasm of spiral arteries via the release of heparin and histamine, and by enzymatically degrading the matrix in stromal fibroblasts. MCs are additionally involved in neuroinflammatory processes and pain perception, interacting with the neuropeptide substance P and the Mas-related G protein-coupled receptor (MRGPR). This interaction stimulates the release of histamine, IL-6, and TNFα [78,101,118]. The MRGPR family of receptors is specifically characterized by unique expression profiles and a range of ligands to br, which serve as crucial players in inflammatory responses and present potential therapeutic targets for various conditions. Immune cell-specific MRGPRs have been implicated in modulating adverse drug reactions, inflammatory diseases, bacterial defense mechanisms, and reactions to environmental irritants [119].

In the setting of endometriosis, MCs are associated with pain and are contributors to lesion fibrosis, angiogenesis, and cellular proliferation. Their activation correlates with the formation of fibrous adhesions and the upregulation of genes related to MC activity. Endometriotic lesions are also marked by heightened levels of stem cell factor (SCF). Studies involving the inhibition of the MC/stem cell growth factor receptor KIT (cKIT) with pexidartinib—an agent that disrupts pathways linked to inflammation and cell survival—have shown a reduction in MC activation [120,121]. While cKIT expression remains consistent across the menstrual phases of the healthy endometrium, it is significantly elevated in peritoneal and ovarian endometriotic lesions. This suggests that elevated E2 levels in endometriosis promote MC infiltration and activation within the lesions, with SCF contributing to a positive feedback mechanism that enhances their recruitment and maturation [120,121].

7.8. Dendritic Cells

DCs arise from CD34+ hematopoietic stem cells in the bone marrow, maturing into the following two primary subsets: HLA-DR+/CD11c+ conventional DCs (cDCs) and HLA-DR+/CD123+ plasmacytoid DCs (pDCs). These versatile cells permeate various tissues and traverse the blood and lymphatic systems [122]. Paharkova-Vatchkova et al. demonstrated that E2 fosters the differentiation of functional DCs from murine bone marrow precursors, notably enhancing the intermediate CD11c(+) CD11b populations, which exhibit elevated MHC class II and CD86 expression [123]. This process is predominantly mediated by ERα, which is essential for maintaining normal DC function [99,117]. Additionally, IL-27 has been shown to upregulate *ESR1* gene expression in cDCs [88,124], while ERβ mRNA is also present, indicating that estrogen plays a significant role in their development and function [88,124].

Emerging research underscores the essential function of DCs throughout the menstrual cycle, across all phases, although their distribution remains contentious. Laginha et al. reported an increase in immature DCs (imDCs) during the proliferative and secretory phases of the healthy endometrium, with a subsequent decline during menstruation [122]. This initial influx is believed to facilitate embryonic implantation and support post-menstruation tissue repair by generating matrix metalloproteinases (MMPs), cytokines, and chemokines (IL-6, IL-10, IL-12, TNFα, and RANTES) [73]. In endometriosis, Schulke et al. reported pronounced imDCs during the proliferative and menstrual phases compared to controls, while mature DCs (mDCs) remained sparse [125]. Conversely, Lundberg et al. found no significant disparity in peritoneal fluid DCs between endometriosis and control samples; however, they reported an elevation in mannose receptor-positive (MR+) type 1 DCs, crucial for antigen recognition and Th2 differentiation, in endometriosis [126].

Furthermore, Izumi et al. indicated that a decrease in DC populations correlated with increased lesion burden and diminished T lymphocyte activation, suggesting that DCs in endometriosis may adopt a more phagocytic, tissue-clearing role. This highlights their potential in modulating immune responses and shaping the inflammatory landscape of endometriosis lesions, ultimately influencing disease progression [127]. Furthermore, E2 has role in altering DC behavior and cytokine production, potentially through neutrophil recruitment and activation, expanding the endometriosis–immunological dynamic complexity [35,128].

7.9. Platelets

Investigations have unveiled ERs on activated platelets and their precursor megakaryocytes, explaining how hormonal shifts might modulate platelet dynamics. Activated platelets significantly amplify E2 synthesis from endometriotic stromal cells through NF-κB and TGF-β1 pathways, facilitating the upregulation of pivotal hormone synthesis proteins like StAR, HSD3B2, aromatase, and HSD17B1, indicating a feedback loop that exacerbates endometriosis [129,130]. Additionally, platelets interact with endometriotic lesions, particularly in highly vascularized regions, where they become activated and cluster around VEGF expression [131].

Based on in vitro co-culture experiments, platelets not only boost the proliferation of stromal cells in endometriosis, but also elevate pro-angiogenic factors, including VEGF, COX-2, and MMP-9, and the anti-apoptotic protein Bcl-2 [88]. The critical involvement of platelets was further highlighted by in vivo studies showing that platelet depletion markedly reduced the lesion size and alleviated hyperalgesia in a mouse endometriosis model [132]. Additionally, Ding et al. reported increased markers of coagulation and platelet activity, as reflected by elevated surface P-selectin, enhanced platelet aggregation, higher plasma D-dimer, fibrinogen, fibrin degradation products (FDPs), soluble P-selectin, and a shorter thrombin time, signifying faster hemostasis in patients with ovarian endometriomas [133].

7.10. Lymphoid Cells

Lymphocytes are essential to the adaptive immune response, consisting of T cells, B cells, and NKs. T cells include helper T cells (CD4+), which secrete cytokines, and cytotoxic T cells (CD8+), which eliminate infected cells. B cells, derived from the bone marrow, produce antibodies. Both T and B cells generate memory cells that "remember" previous antigens, enabling a rapid response to future infections, a key aspect of acquired immunity.

Lymphocytes are pivotal, not only in immunity, but also in autoimmune disorders, facilitating the proliferation of autoreactive T and B cells, which can result in organ damage. Their sex hormone receptor expression induces nuanced epigenetic and transcriptional alterations that delicately modulate immune responses by activating immune-related genes, either directly or indirectly [134]. In the context of autoimmunity, E2 downregulates the AutoImmuneREgulator (AIRE, a transcription factor expressed in the medulla of the thymus) via DNA methylation, diminishing the expression of tissue-specific antigens and hindering Treg differentiation. This allows self-reactive T cells to evade negative selection, increasing autoimmune vulnerability in females [135].

Estrogen's influence on T cell populations is dose-dependent: low levels promote CD4+ T cell expansion, while elevated concentrations, such as those during pregnancy, lead to a reduction in both CD4 and CD8 T cell counts [136,137]. Thymocytes also express ERs, and high levels of exogenous E2 decrease thymic cellularity, resulting in thymic atrophy [138].

7.11. Cytotoxic T Cells (CD8+)

Cytotoxic (CD8+) T cells are essential for adaptive immunity and immune surveillance, with ERβ mRNA expression. In the human endometrium, E2 inhibits the secretion of perforin and granzymes, diminishing the cytotoxic function of cytotoxic T cells [139]. This regulation is vital for reproductive success, as their suppression cells foster a tolerogenic environment that supports implantation and conception

In endometriosis, cytotoxic T cells are more prevalent within endometriotic lesions, yet no significant differences are found in peripheral blood. However, the presence of these cells in peritoneal fluid and eutopic endometrium remains inconsistent [140]. To enhance our understanding of eutopic immune cells, two studies explored immune cell variability in menstrual effluent, revealing no significant differences in CD8+ or CD4+ T cell populations, or their ratio. Interestingly, they reported a decrease in perforin high-expressing cytotoxic T cells [141]. Additionally, Shih et al. reported the identification of four unique cytotoxic T cell clusters, without further characterization [25].

7.12. CD4(+) T Helper (Th1 and 2) Cells

In the human eutopic endometrium, CD4+ T helper (Th)1 and 2 coexist, with their balance influenced by E2 levels. Low E2 levels favor the Th1 phenotype, associated with cell-mediated immunity and inflammation, while higher physiological doses promote the Th2 phenotype, linked to antibody production and anti-inflammatory responses [56]. These pathways inhibit each other: Th1 is characterized by IFNγ, which stimulates pro-inflammatory cytokines like IL-1β and TNF-α, whereas Th2 is marked by IL-4, inducing anti-inflammatory cytokines such as IL-10 and TGF-β [85]. In endometriosis, the immune environment is shaped by both Th1- and Th2-related inflammatory cytokines, although the precise effects of Th1/Th2 differentiation on the condition is not yet clear. Research shows that various signaling pathways regulate cytokines, fostering immune tolerance within ectopic lesions. In advanced stages, the Th2 response predominates, elevating the levels of IL-4, IL-6, and IL-13, while Th1 cytokines persist throughout the disease [142].

Gonadotropin-releasing hormone agonists (GnRH-as) competitively bind to pituitary GnRH receptors, leading to increased TNF-α and decreased IL-10 levels in vitro, thus raising the Th1/Th2 ratio. Beyond alleviating pelvic pain, studies suggest that GnRH-a may inhibit cell proliferation and suppress inflammation in the peritoneal environment [143]. The timing and mechanisms underlying Th2 dominance warrant further exploration. Future therapies may aim to target these cellular and molecular pathways, restoring normal immune regulation and overcoming immune tolerance, potentially revolutionizing endometriosis treatment.

7.13. Th17

These cells are a subset of T cells that produce cytokines such as IL-17, GM-CSF, and IFNγ, playing a crucial role in immune responses and tissue inflammation, particularly against extracellular bacterial and fungal pathogens. Studies indicate a strong correlation between elevated Th17 cells, along with increased IL-17 in peritoneal fluid and worsened inflammatory profiles, as well as increased disease severity in women with endometriosis. These cells release a range of cytokines, including IL-17A, IL-22, and IL-21, which are integral to host defense, autoimmunity, inflammation, and tumor development. In the context of endometriosis, Th17 cells not only promote cell survival by conferring a resistance to NK cell cytotoxicity through of ERK1/2 signal activation, but IL-17 and IL-10 also simultaneously drive angiogenesis and the perpetuation of inflammatory processes [21,144–146].

7.14. Treg

Regulatory T cells (Tregs), once known as suppressor T cells, play a crucial role in modulating the immune system and maintaining self-tolerance, which helps to prevent autoimmune diseases. Functionally, Tregs act as immunosuppressors by suppressing effector T cells, and are characterized by the expression of CD4, FOXP3, and CD25, sharing a lineage with naïve CD4 cells [147].

Within the eutopic endometrium of endometriosis patients, Tregs help to suppress the immune system's recognition of endometriotic cells, enabling these cells to evade immune clearance. The elevation of Tregs during the secretory phase is thought to contribute to the ability of ectopic cells to evade immune surveillance. While the healthy secretory endometrium typically shows an increased infiltration of effector immune cells, in endometriosis, the continued elevation of Tregs from the proliferative to the secretory phase may limit the ability of immune cells to target endometrial antigens during menstruation, thereby facilitating immune evasion and the persistence of ectopic lesions [148].

Research indicates that E2 plays a significant role in Treg dynamics, and encourages Treg differentiation by increasing IDO1 expression in ectopic tissues [149]. When combined with CD3+/CD28+ stimulation, E2 further boosts Treg expansion without compromising its suppressive function, as with what occurs in pregnancy [150]. Patients with sustained high Treg levels often face fertility challenges, aligning with the E2 dominance characteristic of endometriosis [108].

7.15. NK Cells

In vitro and in vivo analyses of uNK cell differentiation reveal that this process continuously occurs in response to endometrial regeneration, driven by IL-15. Differentiated uNK cells exhibit reduced proliferation and immunomodulatory functions, including enhanced angiogenesis. Additionally, the uNK cell niche can be replenished from the circulation, which is under genetic control, indicating that the differentiation of human NK cells in the uterus is linked to significant functional changes associated with local tissue regeneration [146].

In healthy women, peripheral NK (pNK) cell populations fluctuate significantly during the menstrual cycle and play a crucial role in menstruation [137]. However, studies in endometriosis present conflicting results regarding pNK cell numbers; some show no change or a slight decrease, while others report increases, particularly in advanced disease stages. Despite this variability, it is widely acknowledged that NK cell cytotoxicity diminishes in women with endometriosis, with the impairment becoming more pronounced in advanced stages. While pNK cells are more abundant in early-stage endometriosis, their cytotoxic function declines as the disease progresses. Research indicates a decrease in activating receptors like NKG2D and NKP46, alongside an increase in inhibitory receptors such as KIRs and LILRB1, which facilitate immune evasion by ectopic endometrial cells [81,113].

The interaction between NK cells and estrogen is critical in endometriosis pathogenesis. E2 suppresses NK activity, impairing immune surveillance and promoting the survival of ectopic lesions. Moreover, NK cells are involved in regulating endometrial receptivity, follicle development, and immune evasion, all of which may contribute to implantation failure and infertility in affected patients [151].

NK cell dysfunction significantly influences the adhesion, proliferation, and growth of endometriosis, a central aspect of its pathophysiology. While NK dysfunction is recognized as a major contributing factor, the specific relationships among NK cell activity, symptomatology, and various endometriosis subtypes remain poorly understood [152].

Although the percentage of NK cells in the peritoneal fluid, endometrium, and peripheral tissue compartments is similar in women with and without endometriosis, the cytotoxic activity of NK cells is notably reduced in affected individuals. This reduction is linked to altered receptor expression and cytokine modulation, contributing to immune evasion and the persistence of ectopic lesions, underscoring the importance of NK cell cytotoxicity in both the endometriosis pathophysiology and potential therapeutic approaches [152,153].

7.16. B Cells

In the healthy endometrium, B cells constitute less than 5% of the leukocyte population and less than 2% of total cells, primarily located in the basal and stromal compartments. In ectopic endometrial lesions, B cells account for 17% of leukocytes, compared to 12% in the eutopic endometrium and 10% in controls [154,155]. Despite being classified as mucosal tissue, B cells in the endometrium are understudied compared to their role in peritoneal fluid in endometriosis.

E2 plays a dual role in B cell dynamics by suppressing lymphopoiesis while enhancing B cell function through interactions with the estrogen receptors ERα and ERβ. Grimaldi et al. found that E2 upregulates the critical molecules in B cells, including CD22, SHP-1, VCAM, and BCL-2. The increased expression of CD22 and SHP-1 diminishes the receptor-mediated signaling, while higher levels of BCL-2, an anti-apoptotic factor, promote the survival of pro/pre- and immature B cells. Additionally, E2-induced SHP-2 and VCAM are linked to the survival of autoreactive B cells, potentially increasing the risk of autoimmunity in females [62]. This estrogenic influence on B cell development correlates with the immune tolerance observed in endometriosis.

8. Conclusions

Immune dysregulations play pivotal roles in the development and progression of endometriosis, although the precise molecular mechanisms underlying their dysregulations remain poorly understood. The hormonal imbalances in endometriosis, particularly the overexpression of aromatase and aberrant steroid biosynthesis, lead to an estrogen-dominant and progesterone-resistant microenvironment. This endocrine shift perturbs immune cell behaviors and function, yet the full extent of these interactions is still unclear.

Estrogen is a significant immunomodulator and affects many immune cell types. While studies have shown the presence of ERs in various immune cells, there is still uncertainty regarding their expression levels and functional significance in the context of endometriosis. The expression of ERs and PRs in tissue-resident immune cells within the endometriotic lesions remains poorly characterized. Additionally, immune cell subsets, their polarization, and activation states, for example, for Mφs, DCs, NKs, T helpers, Treg cells, and platelets, should be more closely studied. Epigenetic modifications affect ER and PR expression in endometriosis, adding another layer of complexity.

The exact pathways of how immune cells sense the estrogenic environment is not particularly clear. It is intriguing that immune cells may utilize alternative pathways to respond to hormonal changes in an estrogen-dominated environment, potentially involving non-genomic signaling pathways and GPER signaling. Understanding how these pathways differ to classical ER signaling, and their contributions to the differentiation and survival of endometrial-resident immune cell populations, require further investigation.

Despite these insights, there are several limitations that need to be considered. First, extrapolating the findings from in vitro studies to clinical practice remains a challenge, as laboratory conditions may not fully replicate the complexity of in vivo environments. Additionally, while hormonal imbalances are well-documented in endometriosis, the precise relationship between hormonal levels and their clinical effects—such as symptom severity and disease progression—remains unclear. Not only the heterogeneity of endometriosis itself, but also the heterogeneity of immune responses in endometriosis, further complicates the interpretation of results, as different subtypes of the disease may exhibit distinct immune signatures even in different phases of the disease. Moreover, the availability and accessibility of patient tissue samples limit the scope of some studies.

Finally, the applicability of current findings in clinical practice is limited by the lack of standardized protocols for sample collection, processing, and analysis, which hampers the reproducibility and comparison across studies. Moving forward, future research should aim to clarify the immunological heterogeneity of the disease, focusing on both the functional roles of immune cell subsets and the influence of hormones on their behavior. Additionally, studies should seek to integrate clinical data with molecular findings to better understand the relationship between immune dysregulation and disease symptoms. These efforts will be essential for developing more effective and personalized therapeutic strategies for endometriosis.

This review underscores the need for further well-planned collaborative multicenter research to elucidate the role of hormone receptors in immune cells and their interactions within the endometriotic microenvironment. These insights may pave the way for targeted therapeutic approaches aimed at modulating immune responses to alleviate the symptoms and progression of endometriosis.

Author Contributions: Conceptualization, Y.Y., T.A.S. and A.A.; investigation, E.G., P.M. and H.K.; resources, T.A.S.; writing—original draft preparation, E.G., A.A., T.A.S. and Y.Y.; writing—review and editing, P.M., H.K. and T.K.S.; visualization, E.G.; supervision, Y.Y. and A.A.; funding acquisition, Y.Y., T.A.S. and A.A. All authors have read and agreed to the published version of the manuscript.

Funding: This research was funded by Endometriosis Foundation of America.

Institutional Review Board Statement: Not applicable.

Informed Consent Statement: Not applicable.

Data Availability Statement: No new data were created or analyzed in this study.

Conflicts of Interest: The authors declare no conflicts of interest.

References

1. Noyes, R.W.; Hertig, A.T.; Rock, J. Dating the endometrial biopsy. *Am. J. Obstet. Gynecol.* **1975**, *122*, 262–263. [CrossRef] [PubMed]
2. Chavez-MacGregor, M.; van Gils, C.H.; van der Schouw, Y.T.; Monninkhof, E.; van Noord, P.A.; Peeters, P.H. Lifetime cumulative number of menstrual cycles and serum sex hormone levels in postmenopausal women. *Breast Cancer Res. Treat.* **2008**, *108*, 101–112. [CrossRef] [PubMed]
3. Talbi, S.; Hamilton, A.E.; Vo, K.C.; Tulac, S.; Overgaard, M.T.; Dosiou, C.; Le Shay, N.; Nezhat, C.N.; Kempson, R.; Lessey, B.A.; et al. Molecular phenotyping of human endometrium distinguishes menstrual cycle phases and underlying biological processes in normo-ovulatory women. *Endocrinology* **2006**, *147*, 1097–1121. [CrossRef]
4. Wira, C.R.; Rodriguez-Garcia, M.; Patel, M.V. The role of sex hormones in immune protection of the female reproductive tract. *Nat. Rev. Immunol.* **2015**, *15*, 217–230. [CrossRef]
5. Jensen, F.; Woudwyk, M.; Teles, A.; Woidacki, K.; Taran, F.; Costa, S.; Malfertheiner, S.F.; Zenclussen, A.C. Estradiol and progesterone regulate the migration of mast cells from the periphery to the uterus and induce their maturation and degranulation. *PLoS ONE* **2010**, *5*, e14409. [CrossRef]
6. Izumi, G.; Koga, K.; Takamura, M.; Makabe, T.; Satake, E.; Takeuchi, A.; Taguchi, A.; Urata, Y.; Fujii, T.; Osuga, Y. Involvement of immune cells in the pathogenesis of endometriosis. *J. Obstet. Gynaecol. Res.* **2018**, *44*, 191–198. [CrossRef]
7. Huang, X.; Wu, L.; Pei, T.; Liu, D.; Liu, C.; Luo, B.; Xiao, L.; Li, Y.; Wang, R.; Ouyang, Y.; et al. Single-cell transcriptome analysis reveals endometrial immune microenvironment in minimal/mild endometriosis. *Clin. Exp. Immunol.* **2023**, *212*, 285–295. [CrossRef] [PubMed]
8. Vallve-Juanico, J.; Houshdaran, S.; Giudice, L.C. The endometrial immune environment of women with endometriosis. *Hum. Reprod. Update* **2019**, *25*, 564–591. [CrossRef]
9. Lund, J.M.; Hladik, F.; Prlic, M. Advances and challenges in studying the tissue-resident T cell compartment in the human female reproductive tract. *Immunol. Rev.* **2023**, *316*, 52–62. [CrossRef]
10. Male, V.; Moffett, A. Natural Killer Cells in the Human Uterine Mucosa. *Annu. Rev. Immunol.* **2023**, *41*, 127–151. [CrossRef]
11. Radovic Janosevic, D.; Trandafilovic, M.; Krtinic, D.; Colovic, H.; Milosevic Stevanovic, J.; Pop-Trajkovic Dinic, S. Endometrial immunocompetent cells in proliferative and secretory phase of normal menstrual cycle. *Folia Morphol.* **2020**, *79*, 296–302. [CrossRef] [PubMed]
12. Werner, S.; Grose, R. Regulation of wound healing by growth factors and cytokines. *Physiol. Rev.* **2003**, *83*, 835–870. [CrossRef] [PubMed]
13. Buechler, M.B.; Fu, W.; Turley, S.J. Fibroblast-macrophage reciprocal interactions in health, fibrosis, and cancer. *Immunity* **2021**, *54*, 903–915. [CrossRef] [PubMed]
14. Mareckova, M.; Garcia-Alonso, L.; Moullet, M.; Lorenzi, V.; Petryszak, R.; Sancho-Serra, C.; Oszlanczi, A.; Icoresi Mazzeo, C.; Wong, F.C.K.; Kelava, I.; et al. An integrated single-cell reference atlas of the human endometrium. *Nat. Genet.* **2024**, *56*, 1925–1937. [CrossRef] [PubMed]

15. Kitazawa, J.; Kimura, F.; Nakamura, A.; Morimune, A.; Takahashi, A.; Takashima, A.; Amano, T.; Tsuji, S.; Kaku, S.; Kasahara, K.; et al. Endometrial Immunity for Embryo Implantation and Pregnancy Establishment. *Tohoku J. Exp. Med.* **2020**, *250*, 49–60. [CrossRef]
16. Liu, Z.; Sun, Z.; Liu, H.; Niu, W.; Wang, X.; Liang, N.; Wang, X.; Wang, Y.; Shi, Y.; Xu, L.; et al. Single-cell transcriptomic analysis of eutopic endometrium and ectopic lesions of adenomyosis. *Cell Biosci.* **2021**, *11*, 51. [CrossRef]
17. Vannuccini, S.; Clemenza, S.; Rossi, M.; Petraglia, F. Hormonal treatments for endometriosis: The endocrine background. *Rev. Endocr. Metab. Disord.* **2022**, *23*, 333–355. [CrossRef] [PubMed]
18. Mear, L.; Herr, M.; Fauconnier, A.; Pineau, C.; Vialard, F. Polymorphisms and endometriosis: A systematic review and meta-analyses. *Hum. Reprod. Update* **2020**, *26*, 73–102. [CrossRef]
19. Joshi, N.R.; Miyadahira, E.H.; Afshar, Y.; Jeong, J.W.; Young, S.L.; Lessey, B.A.; Serafini, P.C.; Fazleabas, A.T. Progesterone Resistance in Endometriosis Is Modulated by the Altered Expression of MicroRNA-29c and FKBP4. *J. Clin. Endocrinol. Metab.* **2017**, *102*, 141–149. [CrossRef]
20. Khan, K.N.; Guo, S.W.; Ogawa, K.; Fujishita, A.; Mori, T. The role of innate and adaptive immunity in endometriosis. *J. Reprod. Immunol.* **2024**, *163*, 104242. [CrossRef] [PubMed]
21. Knez, J.; Kovacic, B.; Goropevsek, A. The role of regulatory T-cells in the development of endometriosis. *Hum. Reprod.* **2024**, *39*, 1367–1380. [CrossRef] [PubMed]
22. Wu, X.G.; Chen, J.J.; Zhou, H.L.; Wu, Y.; Lin, F.; Shi, J.; Wu, H.Z.; Xiao, H.Q.; Wang, W. Identification and Validation of the Signatures of Infiltrating Immune Cells in the Eutopic Endometrium Endometria of Women with Endometriosis. *Front. Immunol.* **2021**, *12*, 671201. [CrossRef]
23. Alanazi, H.; Zhang, Y.; Fatunbi, J.; Luu, T.; Kwak-Kim, J. The impact of reproductive hormones on T cell immunity; normal and assisted reproductive cycles. *J. Reprod. Immunol.* **2024**, *165*, 104295. [CrossRef] [PubMed]
24. Huhtinen, K.; Desai, R.; Stahle, M.; Salminen, A.; Handelsman, D.J.; Perheentupa, A.; Poutanen, M. Endometrial and endometriotic concentrations of estrone and estradiol are determined by local metabolism rather than circulating levels. *J. Clin. Endocrinol. Metab.* **2012**, *97*, 4228–4235. [CrossRef] [PubMed]
25. Shih, A.J.; Adelson, R.P.; Vashistha, H.; Khalili, H.; Nayyar, A.; Puran, R.; Herrera, R.; Chatterjee, P.K.; Lee, A.T.; Truskinovsky, A.M.; et al. Single-cell analysis of menstrual endometrial tissues defines phenotypes associated with endometriosis. *BMC Med.* **2022**, *20*, 315. [CrossRef]
26. Fonseca, M.A.S.; Haro, M.; Wright, K.N.; Lin, X.; Abbasi, F.; Sun, J.; Hernandez, L.; Orr, N.L.; Hong, J.; Choi-Kuaea, Y.; et al. Single-cell transcriptomic analysis of endometriosis. *Nat. Genet.* **2023**, *55*, 255–267. [CrossRef] [PubMed]
27. Lin, J.; Liu, L.; Zheng, F.; Chen, S.; Yang, W.; Li, J.; Mo, S.; Zeng, D.Y. Exploration the global single-cell ecological landscape of adenomyosis-related cell clusters by single-cell RNA sequencing. *Front. Genet.* **2022**, *13*, 1020757. [CrossRef] [PubMed]
28. Suszczyk, D.; Skiba, W.; Pawlowska-Lachut, A.; Dymanowska-Dyjak, I.; Wlodarczyk, K.; Paduch, R.; Wertel, I. Immune Checkpoints in Endometriosis-A New Insight in the Pathogenesis. *Int. J. Mol. Sci.* **2024**, *25*, 6266. [CrossRef] [PubMed]
29. Koninckx, P.R.; Ussia, A.; Adamyan, L.; Wattiez, A.; Gomel, V.; Martin, D.C. Pathogenesis of endometriosis: The genetic/epigenetic theory. *Fertil. Steril.* **2019**, *111*, 327–340. [CrossRef]
30. Abramiuk, M.; Grywalska, E.; Małkowska, P.; Sierawska, O.; Hrynkiewicz, R.; Niedźwiedzka-Rystwej, P. The Role of the Immune System in the Development of Endometriosis. *Cells* **2022**, *11*, 2028. [CrossRef] [PubMed]
31. Liang, Y.; Wu, J.; Wang, W.; Xie, H.; Yao, S. Pro-endometriotic niche in endometriosis. *Reprod. Biomed. Online* **2019**, *38*, 549–559. [CrossRef] [PubMed]
32. Koninckx, P.R.; Heyns, W.; Verhoeven, G.; Van Baelen, H.; Lissens, W.D.; De Moor, P.; Brosens, I.A. Biochemical characterization of peritoneal fluid in women during the menstrual cycle. *J. Clin. Endocrinol. Metab.* **1980**, *51*, 1239–1244. [CrossRef] [PubMed]
33. Koninckx, P.R.; Kennedy, S.H.; Barlow, D.H. Endometriotic disease: The role of peritoneal fluid. *Hum. Reprod. Update* **1998**, *4*, 741–751. [CrossRef]
34. Vercellini, P.; Merli, C.E.M.; Viganò, P. 'There will be blood'(†) A proof of concept for the role of haemorrhagic corpora lutea in the pathogenesis of endometriosis. *Hum. Reprod. Open* **2024**, *2024*, hoae035. [CrossRef] [PubMed]
35. Herington, J.L.; Bruner-Tran, K.L.; Lucas, J.A.; Osteen, K.G. Immune interactions in endometriosis. *Expert. Rev. Clin. Immunol.* **2011**, *7*, 611–626. [CrossRef] [PubMed]
36. Oosterlynck, D.J.; Meuleman, C.; Waer, M.; Vandeputte, M.; Koninckx, P.R. The natural killer activity of peritoneal fluid lymphocytes is decreased in women with endometriosis. *Fertil. Steril.* **1992**, *58*, 290–295. [CrossRef] [PubMed]
37. Kamiński, K.; Kotarski, J.; Gogacz, M. Characterization of cellular components in the peritoneal fluid and in the endometrial tissue of women with endometriosis. *Arch. Immunol. Ther. Exp.* **1995**, *43*, 247–252.
38. Semino, C.; Semino, A.; Pietra, G.; Mingari, M.C.; Barocci, S.; Venturini, P.L.; Ragni, N.; Melioli, G. Role of major histocompatibility complex class I expression and natural killer-like T cells in the genetic control of endometriosis. *Fertil. Steril.* **1995**, *64*, 909–916. [CrossRef] [PubMed]

39. Harada, T.; Kaponis, A.; Iwabe, T.; Taniguchi, F.; Makrydimas, G.; Sofikitis, N.; Paschopoulos, M.; Paraskevaidis, E.; Terakawa, N. Apoptosis in human endometrium and endometriosis. *Hum. Reprod. Update* **2004**, *10*, 29–38. [CrossRef]
40. Scheerer, C.; Bauer, P.; Chiantera, V.; Sehouli, J.; Kaufmann, A.; Mechsner, S. Characterization of endometriosis-associated immune cell infiltrates (EMaICI). *Arch. Gynecol. Obstet.* **2016**, *294*, 657–664. [CrossRef] [PubMed]
41. Jerman, L.F.; Anderson, L.; Markham, R.; Hey-Cunningham, A.J. The Lymphatic System in Endometriosis: A Pilot Study of Endometrial-like Cells and Immune Cell Populations in Lymph Nodes Associated with Deep Infiltrating Bowel Lesions. *Reprod. Sci.* **2020**, *27*, 977–987. [CrossRef]
42. Qureshi, R.; Picon-Ruiz, M.; Aurrekoetxea-Rodriguez, I.; Nunes de Paiva, V.; D'Amico, M.; Yoon, H.; Radhakrishnan, R.; Morata-Tarifa, C.; Ince, T.; Lippman, M.E.; et al. The Major Pre- and Postmenopausal Estrogens Play Opposing Roles in Obesity-Driven Mammary Inflammation and Breast Cancer Development. *Cell Metab.* **2020**, *31*, 1154–1172.e1159. [CrossRef]
43. Holinka, C.F.; Diczfalusy, E.; Coelingh Bennink, H.J. Estetrol: A unique steroid in human pregnancy. *J. Steroid Biochem. Mol. Biol.* **2008**, *110*, 138–143. [CrossRef] [PubMed]
44. Chakraborty, B.; Byemerwa, J.; Krebs, T.; Lim, F.; Chang, C.Y.; McDonnell, D.P. Estrogen Receptor Signaling in the Immune System. *Endocr. Rev.* **2023**, *44*, 117–141. [CrossRef]
45. Cowley, S.M.; Parker, M.G. A comparison of transcriptional activation by ER alpha and ER beta. *J. Steroid Biochem. Mol. Biol.* **1999**, *69*, 165–175. [CrossRef] [PubMed]
46. Paterni, I.; Granchi, C.; Katzenellenbogen, J.A.; Minutolo, F. Estrogen receptors alpha (ERalpha) and beta (ERbeta): Subtype-selective ligands and clinical potential. *Steroids* **2014**, *90*, 13–29. [CrossRef]
47. Brundin, P.M.A.; Landgren, B.M.; Fjallstrom, P.; Shamekh, M.M.; Gustafsson, J.A.; Johansson, A.F.; Nalvarte, I. Expression of Sex Hormone Receptor and Immune Response Genes in Peripheral Blood Mononuclear Cells During the Menstrual Cycle. *Front. Endocrinol.* **2021**, *12*, 721813. [CrossRef] [PubMed]
48. Vrtacnik, P.; Ostanek, B.; Mencej-Bedrac, S.; Marc, J. The many faces of estrogen signaling. *Biochem. Med.* **2014**, *24*, 329–342. [CrossRef] [PubMed]
49. Chen, P.; Li, B.; Ou-Yang, L. Role of estrogen receptors in health and disease. *Front. Endocrinol.* **2022**, *13*, 839005. [CrossRef]
50. Heldring, N.; Pike, A.; Andersson, S.; Matthews, J.; Cheng, G.; Hartman, J.; Tujague, M.; Ström, A.; Treuter, E.; Warner, M.; et al. Estrogen receptors: How do they signal and what are their targets. *Physiol. Rev.* **2007**, *87*, 905–931. [CrossRef] [PubMed]
51. Jia, M.; Dahlman-Wright, K.; Gustafsson, J. Estrogen receptor alpha and beta in health and disease. *Best. Pract. Res. Clin. Endocrinol. Metab.* **2015**, *29*, 557–568. [CrossRef] [PubMed]
52. Shi, H.; Xu, K.; Huang, M.; Mao, M.; Ou, J. Regulatory mechanism of GPER in the invasion and migration of ectopic endometrial stromal cells in endometriosis. *Women Health* **2024**, *64*, 109–120. [CrossRef]
53. Samartzis, N.; Samartzis, E.P.; Noske, A.; Fedier, A.; Dedes, K.J.; Caduff, R.; Fink, D.; Imesch, P. Expression of the G protein-coupled estrogen receptor (GPER) in endometriosis: A tissue microarray study. *Reprod. Biol. Endocrinol.* **2012**, *10*, 30. [CrossRef]
54. Fairweather, D.; Rose, N.R. Women and autoimmune diseases. *Emerg. Infect. Dis.* **2004**, *10*, 2005–2011. [CrossRef] [PubMed]
55. Khan, D.; Ansar Ahmed, S. The Immune System Is a Natural Target for Estrogen Action: Opposing Effects of Estrogen in Two Prototypical Autoimmune Diseases. *Front. Immunol.* **2016**, *6*, 635. [CrossRef]
56. Straub, R.H. The complex role of estrogens in inflammation. *Endocr. Rev.* **2007**, *28*, 521–574. [CrossRef] [PubMed]
57. Kovats, S. Estrogen receptors regulate innate immune cells and signaling pathways. *Cell Immunol.* **2015**, *294*, 63–69. [CrossRef] [PubMed]
58. Pelekanou, V.; Kampa, M.; Kiagiadaki, F.; Deli, A.; Theodoropoulos, P.; Agrogiannis, G.; Patsouris, E.; Tsapis, A.; Castanas, E.; Notas, G. Estrogen anti-inflammatory activity on human monocytes is mediated through cross-talk between estrogen receptor ERalpha36 and GPR30/GPER1. *J. Leukoc. Biol.* **2016**, *99*, 333–347. [CrossRef]
59. Rothenberger, N.J.; Somasundaram, A.; Stabile, L.P. The Role of the Estrogen Pathway in the Tumor Microenvironment. *Int. J. Mol. Sci.* **2018**, *19*, 611. [CrossRef] [PubMed]
60. Klein, S.L.; Flanagan, K.L. Sex differences in immune responses. *Nat. Rev. Immunol.* **2016**, *16*, 626–638. [CrossRef] [PubMed]
61. Medina, K.L.; Strasser, A.; Kincade, P.W. Estrogen influences the differentiation, proliferation, and survival of early B-lineage precursors. *Blood* **2000**, *95*, 2059–2067. [CrossRef] [PubMed]
62. Grimaldi, C.M.; Cleary, J.; Dagtas, A.S.; Moussai, D.; Diamond, B. Estrogen alters thresholds for B cell apoptosis and activation. *J. Clin. Investig.* **2002**, *109*, 1625–1633. [CrossRef]
63. Attar, E.; Tokunaga, H.; Imir, G.; Yilmaz, M.B.; Redwine, D.; Putman, M.; Gurates, B.; Attar, R.; Yaegashi, N.; Hales, D.B.; et al. Prostaglandin E2 via steroidogenic factor-1 coordinately regulates transcription of steroidogenic genes necessary for estrogen synthesis in endometriosis. *J. Clin. Endocrinol. Metab.* **2009**, *94*, 623–631. [CrossRef] [PubMed]
64. Bulun, S.E.; Yilmaz, B.D.; Sison, C.; Miyazaki, K.; Bernardi, L.; Liu, S.; Kohlmeier, A.; Yin, P.; Milad, M.; Wei, J. Endometriosis. *Endocr. Rev.* **2019**, *40*, 1048–1079. [CrossRef]

65. Garcia-Gomez, E.; Vazquez-Martinez, E.R.; Reyes-Mayoral, C.; Cruz-Orozco, O.P.; Camacho-Arroyo, I.; Cerbon, M. Regulation of Inflammation Pathways and Inflammasome by Sex Steroid Hormones in Endometriosis. *Front. Endocrinol.* **2019**, *10*, 935. [CrossRef]
66. Colgrave, E.M.; Keast, J.R.; Healey, M.; Rogers, P.A.; Girling, J.E.; Holdsworth-Carson, S.J. Extensive heterogeneity in the expression of steroid receptors in superficial peritoneal endometriotic lesions. *Reprod. Biomed. Online* **2024**, *48*, 103409. [CrossRef] [PubMed]
67. Mori, T.; Ito, F.; Koshiba, A.; Kataoka, H.; Takaoka, O.; Okimura, H.; Khan, K.N.; Kitawaki, J. Local estrogen formation and its regulation in endometriosis. *Reprod. Med. Biol.* **2019**, *18*, 305–311. [CrossRef]
68. Hogg, C.; Horne, A.W.; Greaves, E. Endometriosis-Associated Macrophages: Origin, Phenotype, and Function. *Front. Endocrinol.* **2020**, *11*, 7. [CrossRef]
69. Barish, G.D.; Downes, M.; Alaynick, W.A.; Yu, R.T.; Ocampo, C.B.; Bookout, A.L.; Mangelsdorf, D.J.; Evans, R.M. A Nuclear Receptor Atlas: Macrophage activation. *Mol. Endocrinol.* **2005**, *19*, 2466–2477. [CrossRef] [PubMed]
70. Murphy, A.J.; Guyre, P.M.; Wira, C.R.; Pioli, P.A. Estradiol regulates expression of estrogen receptor ERalpha46 in human macrophages. *PLoS ONE* **2009**, *4*, e5539. [CrossRef]
71. Enright, S.; Werstuck, G.H. Investigating the Effects of Sex Hormones on Macrophage Polarization. *Int. J. Mol. Sci.* **2024**, *25*, 951. [CrossRef] [PubMed]
72. Dama, A.; Baggio, C.; Boscaro, C.; Albiero, M.; Cignarella, A. Estrogen Receptor Functions and Pathways at the Vascular Immune Interface. *Int. J. Mol. Sci.* **2021**, *22*, 4254. [CrossRef] [PubMed]
73. Molero, L.; Garcia-Duran, M.; Diaz-Recasens, J.; Rico, L.; Casado, S.; Lopez-Farre, A. Expression of estrogen receptor subtypes and neuronal nitric oxide synthase in neutrophils from women and men: Regulation by estrogen. *Cardiovasc. Res.* **2002**, *56*, 43–51. [CrossRef] [PubMed]
74. Rodenas, M.C.; Tamassia, N.; Cabas, I.; Calzetti, F.; Meseguer, J.; Cassatella, M.A.; Garcia-Ayala, A.; Mulero, V. G Protein-Coupled Estrogen Receptor 1 Regulates Human Neutrophil Functions. *Biomed. Hub.* **2017**, *2*, 1–13. [CrossRef] [PubMed]
75. Symons, L.K.; Miller, J.E.; Tyryshkin, K.; Monsanto, S.P.; Marks, R.M.; Lingegowda, H.; Vanderbeck, K.; Childs, T.; Young, S.L.; Lessey, B.A.; et al. Neutrophil recruitment and function in endometriosis patients and a syngeneic murine model. *FASEB J.* **2020**, *34*, 1558–1575. [CrossRef]
76. Yasuda, H.; Sonoda, A.; Yamamoto, M.; Kawashima, Y.; Takishita, Y.; Morita, A.; Tsutsumi, T.; Tsuchiya, M.; Sato, E.F. 17-beta-estradiol enhances neutrophil extracellular trap formation by interaction with estrogen membrane receptor. *Arch. Biochem. Biophys.* **2019**, *663*, 64–70. [CrossRef]
77. Douin-Echinard, V.; Calippe, B.; Billon-Gales, A.; Fontaine, C.; Lenfant, F.; Tremollieres, F.; Bayard, F.; Guery, J.C.; Arnal, J.F.; Gourdy, P. Estradiol administration controls eosinophilia through estrogen receptor-alpha activation during acute peritoneal inflammation. *J. Leukoc. Biol.* **2011**, *90*, 145–154. [CrossRef] [PubMed]
78. Jeziorska, M.; Salamonsen, L.A.; Woolley, D.E. Mast cell and eosinophil distribution and activation in human endometrium throughout the menstrual cycle. *Biol. Reprod.* **1995**, *53*, 312–320. [CrossRef] [PubMed]
79. Drury, J.A.; Parkin, K.L.; Coyne, L.; Giuliani, E.; Fazleabas, A.T.; Hapangama, D.K. The dynamic changes in the number of uterine natural killer cells are specific to the eutopic but not to the ectopic endometrium in women and in a baboon model of endometriosis. *Reprod. Biol. Endocrinol.* **2018**, *16*, 67. [CrossRef]
80. Godoy-Pacheco, A.; Garcia-Chagollan, M.; Ramirez-De-Arellano, A.; Hernandez-Silva, C.D.; Villegas-Pineda, J.C.; Ramirez-Lopez, I.G.; Zepeda-Nuno, J.S.; Aguilar-Lemarroy, A.; Pereira-Suarez, A.L. Differential modulation of natural killer cell cytotoxicity by 17beta-estradiol and prolactin through the NKG2D/NKG2DL axis in cervical cancer cells. *Oncol. Lett.* **2022**, *24*, 288. [CrossRef]
81. Yang, S.; Wang, H.; Li, D.; Li, M. An Estrogen-NK Cells Regulatory Axis in Endometriosis, Related Infertility, and Miscarriage. *Int. J. Mol. Sci.* **2024**, *25*, 3362. [CrossRef] [PubMed]
82. Yuan, B.; Clark, C.A.; Wu, B.; Yang, J.; Drerup, J.M.; Li, T.; Jin, V.X.; Hu, Y.; Curiel, T.J.; Li, R. Estrogen receptor beta signaling in CD8(+) T cells boosts T cell receptor activation and antitumor immunity through a phosphotyrosine switch. *J. Immunother. Cancer* **2021**, *9*, e001932. [CrossRef]
83. Navarro, F.C.; Watkins, S.K. Estrogen Stimulation Differentially Impacts Human Male and Female Antigen-Specific T Cell Anti-Tumor Function and Polyfunctionality. *Gend. Genome* **2017**, *1*, 1–13. [CrossRef]
84. Zhou, W.J.; Zhang, J.; Xie, F.; Wu, J.N.; Ye, J.F.; Wang, J.; Wu, K.; Li, M.Q. CD45RO(-)CD8(+) T cell-derived exosomes restrict estrogen-driven endometrial cancer development via the ERβ/miR-765/PLP2/Notch axis. *Theranostics* **2021**, *11*, 5330–5345. [CrossRef] [PubMed]
85. Takamura, M.; Koga, K.; Izumi, G.; Hirata, T.; Harada, M.; Hirota, Y.; Hiraike, O.; Fujii, T.; Osuga, Y. Simultaneous Detection and Evaluation of Four Subsets of CD4+ T Lymphocyte in Lesions and Peripheral Blood in Endometriosis. *Am. J. Reprod. Immunol.* **2015**, *74*, 480–486. [CrossRef] [PubMed]

86. Fujii, M.; Tanaka, Y.; Okimura, H.; Maeda, E.; Hamaguchi, M.; Fukui, M.; Kitawaki, J.; Mori, T. Decrease in activated regulatory T cell populations in the endometrium during ovulation in endometriosis. *J. Reprod. Immunol.* **2023**, *156*, 103825. [CrossRef] [PubMed]
87. Sciarra, F.; Campolo, F.; Franceschini, E.; Carlomagno, F.; Venneri, M.A. Gender-Specific Impact of Sex Hormones on the Immune System. *Int. J. Mol. Sci.* **2023**, *24*, 6302. [CrossRef]
88. Laffont, S.; Seillet, C.; Guery, J.C. Estrogen Receptor-Dependent Regulation of Dendritic Cell Development and Function. *Front. Immunol.* **2017**, *8*, 108. [CrossRef] [PubMed]
89. De Leo, B.; Esnal-Zufiaurre, A.; Collins, F.; Critchley, H.O.D.; Saunders, P.T.K. Immunoprofiling of human uterine mast cells identifies three phenotypes and expression of ERβ and glucocorticoid receptor. *F1000Research* **2017**, *6*, 667. [CrossRef] [PubMed]
90. Suszczyk, D.; Skiba, W.; Jakubowicz-Gil, J.; Kotarski, J.; Wertel, I. The Role of Myeloid-Derived Suppressor Cells (MDSCs) in the Development and/or Progression of Endometriosis-State of the Art. *Cells* **2021**, *10*, 677. [CrossRef]
91. Pan, T.; Liu, Y.; Zhong, L.M.; Shi, M.H.; Duan, X.B.; Wu, K.; Yang, Q.; Liu, C.; Wei, J.Y.; Ma, X.R.; et al. Myeloid-derived suppressor cells are essential for maintaining feto-maternal immunotolerance via STAT3 signaling in mice. *J. Leukoc. Biol.* **2016**, *100*, 499–511. [CrossRef] [PubMed]
92. Calippe, B.; Douin-Echinard, V.; Laffargue, M.; Laurell, H.; Rana-Poussine, V.; Pipy, B.; Guery, J.C.; Bayard, F.; Arnal, J.F.; Gourdy, P. Chronic estradiol administration in vivo promotes the proinflammatory response of macrophages to TLR4 activation: Involvement of the phosphatidylinositol 3-kinase pathway. *J. Immunol.* **2008**, *180*, 7980–7988. [CrossRef]
93. Thomas, P.; Pang, Y.; Filardo, E.J.; Dong, J. Identity of an estrogen membrane receptor coupled to a G protein in human breast cancer cells. *Endocrinology* **2005**, *146*, 624–632. [CrossRef] [PubMed]
94. Revankar, C.M.; Cimino, D.F.; Sklar, L.A.; Arterburn, J.B.; Prossnitz, E.R. A transmembrane intracellular estrogen receptor mediates rapid cell signaling. *Science* **2005**, *307*, 1625–1630. [CrossRef]
95. Wu, L.; Deng, Z.; Peng, Y.; Han, L.; Liu, J.; Wang, L.; Li, B.; Zhao, J.; Jiao, S.; Wei, H. Ascites-derived IL-6 and IL-10 synergistically expand CD14(+)HLA-DR(-/low) myeloid-derived suppressor cells in ovarian cancer patients. *Oncotarget* **2017**, *8*, 76843–76856. [CrossRef] [PubMed]
96. Gabrilovich, D.I.; Nagaraj, S. Myeloid-derived suppressor cells as regulators of the immune system. *Nat. Rev. Immunol.* **2009**, *9*, 162–174. [CrossRef] [PubMed]
97. Marvel, D.; Gabrilovich, D.I. Myeloid-derived suppressor cells in the tumor microenvironment: Expect the unexpected. *J. Clin. Investig.* **2015**, *125*, 3356–3364. [CrossRef]
98. Ren, J.; Hou, Y.; Wang, T. Roles of estrogens on myeloid-derived suppressor cells in cancer and autoimmune diseases. *Cell Mol. Immunol.* **2018**, *15*, 724–726. [CrossRef] [PubMed]
99. Chen, H.; Qin, S.; Lei, A.; Li, X.; Gao, Q.; Dong, J.; Xiao, Q.; Zhou, J. Expansion of monocytic myeloid-derived suppressor cells in endometriosis patients: A pilot study. *Int. Immunopharmacol.* **2017**, *47*, 150–158. [CrossRef]
100. Ogawa, K.; Tanida, T. Mixed-Culture Propagation of Uterine-Tissue-Resident Macrophages and Their Expression Properties of Steroidogenic Molecules. *Biomedicines* **2023**, *11*, 985. [CrossRef] [PubMed]
101. Lampiasi, N. Interactions between Macrophages and Mast Cells in the Female Reproductive System. *Int. J. Mol. Sci.* **2022**, *23*, 5414. [CrossRef] [PubMed]
102. Kramer, P.R.; Wray, S. 17-Beta-estradiol regulates expression of genes that function in macrophage activation and cholesterol homeostasis. *J. Steroid Biochem. Mol. Biol.* **2002**, *81*, 203–216. [CrossRef]
103. Khan, K.N.; Masuzaki, H.; Fujishita, A.; Kitajima, M.; Sekine, I.; Matsuyama, T.; Ishimaru, T. Estrogen and progesterone receptor expression in macrophages and regulation of hepatocyte growth factor by ovarian steroids in women with endometriosis. *Hum. Reprod.* **2005**, *20*, 2004–2013. [CrossRef]
104. Lambert, K.C.; Curran, E.M.; Judy, B.M.; Lubahn, D.B.; Estes, D.M. Estrogen receptor-alpha deficiency promotes increased TNF-alpha secretion and bacterial killing by murine macrophages in response to microbial stimuli in vitro. *J. Leukoc. Biol.* **2004**, *75*, 1166–1172. [CrossRef] [PubMed]
105. Subramanian, M.; Shaha, C. Up-regulation of Bcl-2 through ERK phosphorylation is associated with human macrophage survival in an estrogen microenvironment. *J. Immunol.* **2007**, *179*, 2330–2338. [CrossRef] [PubMed]
106. Campbell, L.; Emmerson, E.; Williams, H.; Saville, C.R.; Krust, A.; Chambon, P.; Mace, K.A.; Hardman, M.J. Estrogen receptor-alpha promotes alternative macrophage activation during cutaneous repair. *J. Invest. Dermatol.* **2014**, *134*, 2447–2457. [CrossRef] [PubMed]
107. Ramirez-Pavez, T.N.; Martinez-Esparza, M.; Ruiz-Alcaraz, A.J.; Marin-Sanchez, P.; Machado-Linde, F.; Garcia-Penarrubia, P. The Role of Peritoneal Macrophages in Endometriosis. *Int. J. Mol. Sci.* **2021**, *22*, 10792. [CrossRef] [PubMed]
108. Hudson, Q.J.; Ashjaei, K.; Perricos, A.; Kuessel, L.; Husslein, H.; Wenzl, R.; Yotova, I. Endometriosis Patients Show an Increased M2 Response in the Peritoneal CD14(+low)/CD68(+low) Macrophage Subpopulation Coupled with an Increase in the T-helper 2 and T-regulatory Cells. *Reprod. Sci.* **2020**, *27*, 1920–1931. [CrossRef]

109. Somigliana, E.; Li Piani, L.; Paffoni, A.; Salmeri, N.; Orsi, M.; Benaglia, L.; Vercellini, P.; Vigano, P. Endometriosis and IVF treatment outcomes: Unpacking the process. *Reprod. Biol. Endocrinol.* **2023**, *21*, 107. [CrossRef]
110. Kobayashi, H.; Imanaka, S. Understanding the molecular mechanisms of macrophage polarization and metabolic reprogramming in endometriosis: A narrative review. *Reprod. Med. Biol.* **2022**, *21*, e12488. [CrossRef]
111. Bouet, P.E.; Chao de la Barca, J.M.; Boucret, L.; Descamps, P.; Legendre, G.; Hachem, H.E.; Blanchard, S.; Jeannin, P.; Reynier, P.; May-Panloup, P. Elevated Levels of Monocyte Chemotactic Protein-1 in the Follicular Fluid Reveals Different Populations among Women with Severe Endometriosis. *J. Clin. Med.* **2020**, *9*, 1306. [CrossRef] [PubMed]
112. Burns, K.A.; Thomas, S.Y.; Hamilton, K.J.; Young, S.L.; Cook, D.N.; Korach, K.S. Early Endometriosis in Females Is Directed by Immune-Mediated Estrogen Receptor α and IL-6 Cross-Talk. *Endocrinology* **2018**, *159*, 103–118. [CrossRef] [PubMed]
113. Chen, S.; Liu, Y.; Zhong, Z.; Wei, C.; Liu, Y.; Zhu, X. Peritoneal immune microenvironment of endometriosis: Role and therapeutic perspectives. *Front. Immunol.* **2023**, *14*, 1134663. [CrossRef]
114. Blumenthal, R.D.; Samoszuk, M.; Taylor, A.P.; Brown, G.; Alisauskas, R.; Goldenberg, D.M. Degranulating eosinophils in human endometriosis. *Am. J. Pathol.* **2000**, *156*, 1581–1588. [CrossRef] [PubMed]
115. Ghaffari, S.; Rezaei, N. Eosinophils in the tumor microenvironment: Implications for cancer immunotherapy. *J. Transl. Med.* **2023**, *21*, 551. [CrossRef] [PubMed]
116. Artham, S.; Chang, C.Y.; McDonnell, D.P. Eosinophilia in cancer and its regulation by sex hormones. *Trends Endocrinol. Metab.* **2023**, *34*, 5–20. [CrossRef] [PubMed]
117. Poli-Neto, O.B.; Meola, J.; Rosa, E.S.J.C.; Tiezzi, D. Transcriptome meta-analysis reveals differences of immune profile between eutopic endometrium from stage I-II and III-IV endometriosis independently of hormonal milieu. *Sci. Rep.* **2020**, *10*, 313. [CrossRef]
118. Green, D.P.; Limjunyawong, N.; Gour, N.; Pundir, P.; Dong, X. A Mast-Cell-Specific Receptor Mediates Neurogenic Inflammation and Pain. *Neuron* **2019**, *101*, 412–420.E3. [CrossRef] [PubMed]
119. Lui, B.; Chang, J.; Sun, J.S.; Billings, J.; Steves, A.; Ackerhalt, R.; Molnar, M.; Kung, H.F. Radioactive iodine exchange reaction of HIPDM: Kinetics and mechanism. *J. Nucl. Med.* **1987**, *28*, 360–365.
120. Mwaura, A.N.; Marshall, N.; Anglesio, M.S.; Yong, P.J. Neuroproliferative dyspareunia in endometriosis and vestibulodynia. *Sex. Med. Rev.* **2023**, *11*, 323–332. [CrossRef] [PubMed]
121. McCallion, A.; Nasirzadeh, Y.; Lingegowda, H.; Miller, J.; Khalaj, K.; Ahn, S.H.; Monsanto, S.P.; Bidarimath, M.; Sisnett, D.; Craig, A.; et al. Estrogen mediates inflammatory role of mast cells in endometriosis pathophysiology. *Front. Immunol.* **2022**, *13*, 961599. [CrossRef]
122. Laginha, P.A.; Arcoverde, F.V.L.; Riccio, L.G.C.; Andres, M.P.; Abrao, M.S. The role of dendritic cells in endometriosis: A systematic review. *J. Reprod. Immunol.* **2022**, *149*, 103462. [CrossRef] [PubMed]
123. Paharkova-Vatchkova, V.; Maldonado, R.; Kovats, S. Estrogen preferentially promotes the differentiation of CD11c+ CD11b(intermediate) dendritic cells from bone marrow precursors. *J. Immunol.* **2004**, *172*, 1426–1436. [CrossRef]
124. Carreras, E.; Turner, S.; Paharkova-Vatchkova, V.; Mao, A.; Dascher, C.; Kovats, S. Estradiol acts directly on bone marrow myeloid progenitors to differentially regulate GM-CSF or Flt3 ligand-mediated dendritic cell differentiation. *J. Immunol.* **2008**, *180*, 727–738. [CrossRef]
125. Schulke, L.; Berbic, M.; Manconi, F.; Tokushige, N.; Markham, R.; Fraser, I.S. Dendritic cell populations in the eutopic and ectopic endometrium of women with endometriosis. *Hum. Reprod.* **2009**, *24*, 1695–1703. [CrossRef]
126. Lundberg, K.; Rydnert, F.; Broos, S.; Andersson, M.; Greiff, L.; Lindstedt, M. Allergen-Specific Immunotherapy Alters the Frequency, as well as the FcR and CLR Expression Profiles of Human Dendritic Cell Subsets. *PLoS ONE* **2016**, *11*, e0148838. [CrossRef]
127. Izumi, G.; Koga, K.; Takamura, M.; Makabe, T.; Nagai, M.; Urata, Y.; Harada, M.; Hirata, T.; Hirota, Y.; Fujii, T.; et al. Mannose receptor is highly expressed by peritoneal dendritic cells in endometriosis. *Fertil. Steril.* **2017**, *107*, 167–173.e2. [CrossRef]
128. Seillet, C.; Rouquie, N.; Foulon, E.; Douin-Echinard, V.; Krust, A.; Chambon, P.; Arnal, J.F.; Guery, J.C.; Laffont, S. Estradiol promotes functional responses in inflammatory and steady-state dendritic cells through differential requirement for activation function-1 of estrogen receptor alpha. *J. Immunol.* **2013**, *190*, 5459–5470. [CrossRef] [PubMed]
129. Qi, Q.; Liu, X.; Zhang, Q.; Guo, S.W. Platelets induce increased estrogen production through NF-κB and TGF-β1 signaling pathways in endometriotic stromal cells. *Sci. Rep.* **2020**, *10*, 1281. [CrossRef]
130. Dupuis, M.; Severin, S.; Noirrit-Esclassan, E.; Arnal, J.F.; Payrastre, B.; Valéra, M.C. Effects of Estrogens on Platelets and Megakaryocytes. *Int. J. Mol. Sci.* **2019**, *20*, 3111. [CrossRef] [PubMed]
131. Guo, S.W. The Role of Platelets in the Pathogenesis and Pathophysiology of Adenomyosis. *J. Clin. Med.* **2023**, *12*, 842. [CrossRef] [PubMed]
132. Ding, D.; Liu, X.; Duan, J.; Guo, S.W. Platelets are an unindicted culprit in the development of endometriosis: Clinical and experimental evidence. *Hum. Reprod.* **2015**, *30*, 812–832. [CrossRef] [PubMed]

133. Ding, D.; Liu, X.; Guo, S.W. Further Evidence for Hypercoagulability in Women With Ovarian Endometriomas. *Reprod. Sci.* **2018**, *25*, 1540–1548. [CrossRef] [PubMed]
134. Dodd, K.C.; Menon, M. Sex bias in lymphocytes: Implications for autoimmune diseases. *Front. Immunol.* **2022**, *13*, 945762. [CrossRef] [PubMed]
135. Moulton, V.R. Sex Hormones in Acquired Immunity and Autoimmune Disease. *Front. Immunol.* **2018**, *9*, 2279. [CrossRef]
136. Maret, A.; Coudert, J.D.; Garidou, L.; Foucras, G.; Gourdy, P.; Krust, A.; Dupont, S.; Chambon, P.; Druet, P.; Bayard, F.; et al. Estradiol enhances primary antigen-specific CD4 T cell responses and Th1 development in vivo. Essential role of estrogen receptor alpha expression in hematopoietic cells. *Eur. J. Immunol.* **2003**, *33*, 512–521. [CrossRef]
137. Lee, S.; Kim, J.; Jang, B.; Hur, S.; Jung, U.; Kil, K.; Na, B.; Lee, M.; Choi, Y.; Fukui, A.; et al. Fluctuation of peripheral blood T, B, and NK cells during a menstrual cycle of normal healthy women. *J. Immunol.* **2010**, *185*, 756–762. [CrossRef] [PubMed]
138. Rosenzweig, R.; Gupta, S.; Kumar, V.; Gumina, R.J.; Bansal, S.S. Estrogenic bias in T-Lymphocyte biology: Implications for cardiovascular disease. *Pharmacol. Res.* **2021**, *170*, 105606. [CrossRef]
139. Shen, Z.; Rodriguez-Garcia, M.; Patel, M.V.; Wira, C.R. Direct and Indirect endocrine-mediated suppression of human endometrial CD8+T cell cytotoxicity. *Sci. Rep.* **2021**, *11*, 1773. [CrossRef]
140. Kisovar, A.; Becker, C.M.; Granne, I.; Southcombe, J.H. The role of CD8+ T cells in endometriosis: A systematic review. *Front. Immunol.* **2023**, *14*, 1225639. [CrossRef] [PubMed]
141. Schmitz, T.; Hoffmann, V.; Olliges, E.; Bobinger, A.; Popovici, R.; Nößner, E.; Meissner, K. Reduced frequency of perforin-positive CD8+ T cells in menstrual effluent of endometriosis patients. *J. Reprod. Immunol.* **2021**, *148*, 103424. [CrossRef]
142. Chang, L.Y.; Shan, J.; Hou, X.X.; Li, D.J.; Wang, X.Q. Synergy between Th1 and Th2 responses during endometriosis: A review of current understanding. *J. Reprod. Immunol.* **2023**, *158*, 103975. [CrossRef]
143. Khan, K.N.; Kitajima, M.; Hiraki, K.; Fujishita, A.; Nakashima, M.; Ishimaru, T.; Masuzaki, H. Cell proliferation effect of GnRH agonist on pathological lesions of women with endometriosis, adenomyosis and uterine myoma. *Hum. Reprod.* **2010**, *25*, 2878–2890. [CrossRef]
144. Shi, J.L.; Zheng, Z.M.; Chen, M.; Shen, H.H.; Li, M.Q.; Shao, J. IL-17: An important pathogenic factor in endometriosis. *Int. J. Med. Sci.* **2022**, *19*, 769–778. [CrossRef]
145. Kang, Y.J.; Cho, H.J.; Lee, Y.; Park, A.; Kim, M.J.; Jeung, I.C.; Jung, Y.W.; Jung, H.; Choi, I.; Lee, H.G.; et al. IL-17A and Th17 Cells Contribute to Endometrial Cell Survival by Inhibiting Apoptosis and NK Cell Mediated Cytotoxicity of Endometrial Cells via ERK1/2 Pathway. *Immune Netw.* **2023**, *23*, e14. [CrossRef]
146. Strunz, B.; Bister, J.; Jonsson, H.; Filipovic, I.; Crona-Guterstam, Y.; Kvedaraite, E.; Sleiers, N.; Dumitrescu, B.; Brannstrom, M.; Lentini, A.; et al. Continuous human uterine NK cell differentiation in response to endometrial regeneration and pregnancy. *Sci. Immunol.* **2021**, *6*, eabb7800. [CrossRef]
147. Ashby, K.M.; Hogquist, K.A. A guide to thymic selection of T cells. *Nat. Rev. Immunol.* **2024**, *24*, 103–117. [CrossRef]
148. Podgaec, S.; Barbeiro, D.F.; Gueuvoghlanian-Silva, B.Y.; Bellelis, P.; Abrão, M.S.; Baracat, E.C. Foxp3 expression in deep rectosigmoid endometriosis lesions and its association with chronic pelvic pain. *J. Reprod. Immunol.* **2014**, *104-105*, 96–99. [CrossRef]
149. Wei, C.; Mei, J.; Tang, L.; Liu, Y.; Li, D.; Li, M.; Zhu, X. 1-Methyl-tryptophan attenuates regulatory T cells differentiation due to the inhibition of estrogen-IDO1-MRC2 axis in endometriosis. *Cell Death Dis.* **2016**, *7*, e2489. [CrossRef]
150. Prieto, G.A.; Rosenstein, Y. Oestradiol potentiates the suppressive function of human CD4 CD25 regulatory T cells by promoting their proliferation. *Immunology* **2006**, *118*, 58–65. [CrossRef]
151. Azeze, G.G.; Wu, L.; Alemu, B.K.; Wang, C.C.; Zhang, T. Changes in the number and activity of natural killer cells and its clinical association with endometriosis: Systematic review and meta-analysis. *FS Rev.* **2024**, *5*, 100072. [CrossRef]
152. Reis, J.L.; Rosa, N.N.; Ângelo-Dias, M.; Martins, C.; Borrego, L.M.; Lima, J. Natural Killer Cell Receptors and Endometriosis: A Systematic Review. *Int. J. Mol. Sci.* **2022**, *24*, 331. [CrossRef]
153. Wang, L.; Li, L.; Li, Y.; Huang, C.; Lian, R.; Wu, T.; Ma, J.; Zhang, Y.; Cheng, Y.; Diao, L.; et al. A History of Endometriosis Is Associated With Decreased Peripheral NK Cytotoxicity and Increased Infiltration of Uterine CD68(+) Macrophages. *Front. Immunol.* **2021**, *12*, 711231. [CrossRef]
154. Antsiferova, Y.S.; Sotnikova, N.Y.; Posiseeva, L.V.; Shor, A.L. Changes in the T-helper cytokine profile and in lymphocyte activation at the systemic and local levels in women with endometriosis. *Fertil. Steril.* **2005**, *84*, 1705–1711. [CrossRef]
155. Shen, M.; O'Donnell, E.; Leon, G.; Kisovar, A.; Melo, P.; Zondervan, K.; Granne, I.; Southcombe, J. The role of endometrial B cells in normal endometrium and benign female reproductive pathologies: A systematic review. *Hum. Reprod. Open* **2022**, *2022*, hoab043. [CrossRef]

cells

MDPI

Article

Endometriosis Cell Spheroids Undergo Mesothelial Clearance in a Similar Manner to Ovarian Cancer Cell Spheroids

Allison A. Kloeckner and Sarah R. Walker *

Department of Molecular, Cellular and Biomedical Sciences, University of New Hampshire, Durham, NH 03824, USA
* Correspondence: sarah.walker@unh.edu

Abstract: Endometriosis is a gynecological disease characterized by the presence of endometrium-like cells located outside the uterus. The most widely accepted theory for endometriosis development, retrograde menstruation, does not account for extra-pelvic lesions or ones found on other organs in the peritoneal cavity. Similar to ovarian cancer, endometriosis cells can interact with the mesothelial cells of the peritoneal cavity. In ovarian cancer metastasis, ovarian cancer cell spheroids attach and push away the mesothelial cells lining the peritoneal cavity, clearing the mesothelial layer. Since endometriosis cells are known to interact with the mesothelium, we hypothesized that endometriosis cells would be able to form spheroids capable of undergoing mesothelial clearance. To test this, we designed an in vitro mesothelial clearance assay using endometriosis spheroids and a mesothelial cell monolayer. Our results demonstrate that normal and endometriotic epithelial cell spheroids can perform mesothelial clearance similar to ovarian cancer spheroids, though normal endometrial cells do not clear as well as endometriosis cells. Additionally, we demonstrated that our mesothelial clearance assay can test potential pharmacological therapies for endometriosis prior to clinical trials. These results give insight into the development of endometriosis lesions, but further research is needed to determine the mechanisms behind mesothelial clearance in endometriosis.

Keywords: endometriosis; mesothelial clearance; epithelial cells; in vitro model

1. Introduction

Endometriosis is a gynecological disease that affects around 190 million women and people assigned female at birth during their reproductive years [1]. This disease is defined by the presence of endometrial-like stroma and epithelial cells in any extra-uterine location [1,2]. The endometriotic lesions that form result in a chronic, inflammatory condition with many patients experiencing debilitating pain [1–3]. Common symptoms include pelvic pain, infertility, and many others that negatively affect physical, mental, and social well-being [3,4]. While most lesions are found within the abdomen, lesions can be found in parts of the gastrointestinal system and outside the abdominal area as well [1]. The varying locations of endometriotic lesions present an interesting perspective through which to research endometriosis. Determining how extra-abdominal lesions form could shed light on the pathogenesis of the disease.

As the endometrium, the lining of the uterus, is a very complex tissue, understanding its function in the body is necessary to fully comprehend how it is involved in the development of endometriosis [5–7]. The main role of the endometrium is to prepare itself for embryo implantation and, if implantation is successful, to maintain the pregnancy [6]. Since endometrium plays an important part in sexual reproduction, it is heavily influenced

https://doi.org/10.3390/cells14100742

by ovarian steroid hormones [6]. Estrogen and progesterone prompt modifications of the endometrium, causing it to thicken or thin depending on the menstrual phase [5,6]. When pregnancy does not occur, the change in steroid hormones causes the endometrium to shed through menstruation [5,6]. Sampson's theory on how endometriosis occurs suggests retrograde menstruation is responsible [8,9]. Instead of flowing out through the vagina, during retrograde menstruation, shed endometrium cells travel back up the uterus and into the fallopian tubes, implanting in the peritoneal cavity, thus leading to the formation of endometriotic lesions [7,8,10]. However, retrograde menstruation is estimated to occur in almost 90% of women, and therefore, it cannot fully explain all instances of endometriosis and the varying locations where endometriotic lesions have been found [7,10,11]. This led the way for alternative theories on the pathogenesis of endometriosis, yet none fully encapsulates all instances of endometriosis and the interaction between the lesions and the peritoneum [12]. Specifically, the attachment of endometriosis lesions to the mesothelial layer of the peritoneal cavity left many questions unanswered. How the lesions were able to attach to the mesothelial layer to initiate implantation was not well understood.

In order to investigate the interaction between the monolayer of mesothelial cells that line the peritoneal cavity, an adequate in vitro model of endometriotic lesions that can represent the complexities of the disease is essential [13]. Although benign, endometriotic lesions share similarities with tumors, such as increased cell proliferation, neoangiogenesis, and reduced apoptosis [10,14]. Notably, endometriosis has been associated with ovarian cancer, the deadliest gynecological malignancy for women [15,16]. These similarities between endometriotic lesions and ovarian tumors warrant the application of ovarian cancer research approaches for the study of endometriosis.

Previous studies have highlighted the benefits of utilizing three-dimensional (3D) models of cancer to better represent what is occurring in vivo [17,18]. While two-dimensional (2D) models have been a vital part of furthering biological research, they often have differences in gene expression and drug response due to their inability to represent the complex microenvironment that surrounds tumors in vivo [17]. Recently, 3D cell culture methods have become more popular due to the need for a complex in vitro system that accurately represents cell–cell interactions [13,18]. Three-dimensional spheroids are clusters of cells that can comprise either exclusively tumor cells or a combination of tumor, stromal, and immune cells [13,18]. These spheroids self-assemble into circular aggregates that do not attach to any plate surfaces [13]. A review of publications in which 3D spheroid models of ovarian cancer were used showed that the most common spheroid-formation method was to culture ovarian cancer cells in pre-coated ultra-low-attachment (ULA) plates [17,19]. These ULA-plate-generated spheroids showed similarities to the spheroids that are typically found in the ascites of ovarian cancer patients [18].

The concept for 3D spheroid models does not only apply to ovarian cancer. In 2014, researchers established in vitro 3D spheroid models of endometriosis using a well-characterized endometriosis epithelial cell line, 12Z, and an ovarian endometriosis epithelial cell line, EEC16, isolated from a patient with severe endometriosis [20]. Spheroids from this study were made in non-adherent conditions using polyHEMA-coated cell culture plastics [20]. The spheroids were compared to 2D monolayer cell cultures and were found to have differential gene expression of endometriosis-related genes [20]. Furthermore, the molecular and histological features of these spheroids were more similar to human endometriosis lesions than the 2D monolayer cultures [20]. In addition, Song et al. demonstrated the formation of endometriosis spheroids comprising 12Z epithelial cells and either uterine stromal cells or endometriotic stromal cells [13]. In both combination spheroids, the stromal cells aggregated in the center of the spheroid, with the 12Z cells surrounding them on the outside [13].

The establishment of 3D spheroid models has been integral in enhancing endometriosis research, but a need still exists for 3D models that enable researchers to study the interactions between the microenvironment and these 3D spheroids [13]. Specifically, models that can show the interactions between endometriosis cells and the peritoneal mesothelial cells are needed to better understand the generation of peritoneal endometriotic lesions. The mesothelium is a monolayer of cells that acts as a barrier around all of the organs in the peritoneal cavity [21]. It was initially believed that endometriotic lesion attachment to the peritoneal mesothelium was only able to occur when the mesothelial layer was damaged [9]. However, Witz et al. demonstrated endometrial stromal cells could attach to undamaged mesothelium [9]. They mechanically dispersed endometrial stromal and epithelial cells onto an intact mesothelium of peritoneal explants and saw adhesion to the mesothelium occurred only with stromal cells [9]. To build on this, Witz et al. have also shown endometrial stromal cell adherence to the peritoneal mesothelium and provided evidence to support that they are capable of invading past the mesothelium [22]. While these studies did not use endometriosis cells, they provided data to support the possibility that the formation of endometriotic lesions on the peritoneal mesothelial layer could occur.

In the study by Song et al. mentioned previously, the invasion capability of the combination endometriosis spheroids was also assessed [13]. In their 3D invasion model, the spheroids invaded a Matrigel matrix and penetrated a layer of mesothelial cells [13]. Interestingly, the endometriotic stromal cells were consistently located at the leading edge of invasion and were the first cells to breach the mesothelial layer, dragging the 12Z cells behind them [13]. These data concurred with the data from non-endometriotic mesothelial invasion tests in which the stromal cells were the driving force of attachment and invasion.

Our current research aims to study the ability of endometriotic epithelial cells themselves to disrupt mesothelial cell monolayers. Our methods do not investigate the invasive properties once the mesothelial layer is breached, which has previously been shown. We believe that epithelial endometriotic cells can disrupt or clear the mesothelial cell layer, similar to how ovarian cancer metastasis occurs. Mesothelial cells act as a barrier for endometriotic lesion formation; thus, the endometriotic cells must interact with this monolayer for lesion implantation. The ability of ovarian cancer spheroids to clear the mesothelial layer of cells in the peritoneal cavity gives them access to the underlying extracellular matrix, which they must breach if the cancer is to spread to locations outside the peritoneal cavity. Showing that endometriosis spheroids can clear a mesothelial monolayer offers a potential explanation for the mechanisms that lead to extra-pelvic endometriotic lesions.

To study this, we have developed a straightforward in vitro mesothelial clearance assay that eliminates the need for scaffolds and stable fluorescently labeled cells, making it a low-cost and less time-consuming method. An immortalized endometriosis cell line, 12Z, along with an immortalized endometrium cell line, hEM3, were used to generate 3D spheroids. Spheroids were also made using the ovarian cancer cell line OVCAR8 as a positive control for mesothelial clearance. Our clearance method permits both the study of genes that promote mesothelial clearance and the effectiveness of drugs that inhibit mesothelial clearance.

2. Materials and Methods

2.1. Cell Lines

Two endometrial epithelial cell lines, 12Z (Applied Biological Materials, Ferndale, WA, USA) and hEM3 (kindly provided by Dr. Ie-Ming Shih [23]), were used. 12Z cells are an immortalized endometriosis cell line, and hEM3 cells are an immortalized normal endometrium cell line. 12Z cells were grown in DMEM High-Glucose media and supplemented with 10% FBS and 1x Antibiotic–Antimycotic. hEM3 cells were grown in RPMI 1640 media and supplemented with 15% FBS, 1x Antibiotic–Antimycotic, and 1x MEM Non-Essential Amino Acids Solution. LP9 Mesothelial cells (a kind gift from Martin Sattler, Dana-Farber Cancer Institute, Boston, MA USA) were used to represent the monolayer of mesothelial cells that line the peritoneal cavity in our in vitro assay. LP9 cells were grown in media containing a 1:1 ratio of Human Fibroblast Expansion Media (Gibco, Grand Island, NY, USA) and Media 199, supplemented with 15% Iron-Enriched Calf Serum, 1x Antibiotic–Antimycotic, 4 μg/mL Hydrocortisone, and 10 ng/mL Human EGF (Peprotech, Cranbury, NJ, USA). The ovarian cancer cell line OVCAR8 (kindly provided by Ronnie Drapkin, University of Pennsylvania) was used as a comparison for the activity of the endometriosis cell lines. OVCAR8 cells were cultured in RPMI 1640 media that was supplemented with 10% FBS and 1x Antibiotic–Antimycotic. Ishikawa cells (Millipore Sigma, Burlington, MA, USA) were grown in MEM media and supplemented with 5% FBS, 1x Antibiotic–Antimycotic, 1x MEM Non-Essential Amino Acid Solution, and 2 mM L-Glutamine. ZR-75 (kindly provided by Kornelia Polyak, Dana-Farber Cancer Institute) and mouse embryonic fibroblasts (*fl/fl STAT3*) (kindly provided by David Frank, Emory University) were grown in DMEM with 10% FBS and 1x Antibiotic–Antimycotic. All cells were grown in a 37 °C environment supplied with 5% CO_2 and tested for mycoplasma contamination.

2.2. Drug Treatment

Atorvastatin (Selleckchem, Houston, TX, USA), simvastatin (MedChemExpress, Monmouth Junction, NJ, USA), and lovastatin (MedChemExpress, Monmouth Junction, NJ, USA) were dissolved in dimethyl sulfoxide (DMSO) to stock concentrations of 10 mM. All drugs were stored at −20 °C. For the mesothelial clearance assay, spheroids were treated with either 10 μM Atorvastatin, 10 μM Simvastatin, or 10 μM Lovastatin and incubated at 37 °C for 24 h before spheroids were transferred on top of LP9 cells.

2.3. 3D Spheroid Culture

12Z, hEM3, and OVCAR8 spheroids were generated by plating 1×10^4 cells in 100 μL media in U-bottom ULA plates (Greiner Bio-One, Pittsburgh, PA, USA) for 72 h unless otherwise indicated.

2.4. Cell Viability Staining

NucBlue and propidium iodide (PI) cell stains (ReadyProbes^{TM} Cell Viability Imaging Kit, ThermoFisher Scientific, Waltham, MA USA) were used to visualize the viability of cells that made up the spheroids. NucBlue stains the nucleus of all cells, dead or alive, causing them to fluoresce blue. To visualize these stained cells, images were taken on an EVOS FL Auto 2 using the DAPI channel. PI is a red cell stain that was used to differentiate the dead cells from the live cells, as it stains only the nuclei of dead cells that have damaged membranes. The RFP channel on the EVOS was used to capture images of the cells stained with PI.

2.5. Mesothelial Clearance Assay and Analysis

A schematic of this assay can be seen in Figure 1. A total of 4×10^6 LP9 cells were resuspended in 4 mL of media before being stained red with the Invitrogen Vybrant^{TM} DiI cell-labeling stain (Waltham, MA, USA). A total of 3×10^6 OVCAR8 or endometrial cells were resuspended in 3 mL of their respective media and stained green with the Invitrogen Vybrant^{TM} DiO cell-labeling stain (Waltham, MA, USA). All cells were then incubated at 37° for 20 min. Cells were centrifuged and washed with PBS twice. After washing, the PBS was removed, and LP9s were resuspended in 15 mL media while the spheroid cells were resuspended in 3 mL of their respective media. LP9 cells were plated at a volume of 100 μL per well in a clear, flat-bottom, tissue-culture-treated 96-well plate for 72 h. For spheroid plating, 1 mL of the stained cell solution was combined with 9 mL of media and plated at a volume of 100 μL or 50 μL if adding drugs, as described above. The cells were incubated for 48–72 h until spheroids formed. After incubation, the media was removed from the LP9 mesothelial cells, and the spheroids, along with their media, were transferred on top of the mesothelial monolayer using an Integra VIAFLO electronic pipette with the Integra Wide-Bore pipette tips (Hudson, NH, USA). The cells were incubated for 30 min to allow for spheroid adhesion to the monolayer. The Invitrogen EVOS FL Auto 2 microscope (Waltham, MA, USA) was used to image the wells using the GFP and RFP channels and the 4× objective at 0 and 24 h by creating a scan protocol. To measure the size of the spheroids and the area cleared by them, the 3DSlicer image computing platform was used (www.slicer.org, accessed on 22 February 2023, version 5.2.2). The Segment Editor was used to trace the outlines of the spheroids and cleared area, while the Segment Statistics were used to quantify the measured areas. The initial spheroid size at 0 h and the clearance area at 24 h were measured in mm^2. The clearance ratio was calculated as the area cleared at 24 h divided by the area of the spheroid at 0 h. Each experiment represents 5–8 spheroids per condition per experiment replicate (per N), as indicated in the figure legends. Prism by GraphPad (version 10.4.2) was used to run one-way ANOVA analyses on the clearance ratios ($p < 0.05$), followed by Dunnett's multiple comparison tests. Once analyzed, ImageJ (version 1.54g) was used to enhance clearance images for visualization purposes only. The brightness and contrast of the images were increased, and the color balance values were adjusted to augment the different cell stains. Unenhanced images can be found at FigShare (https://figshare.com/articles/figure/Complete_Image_Sets/28796033).

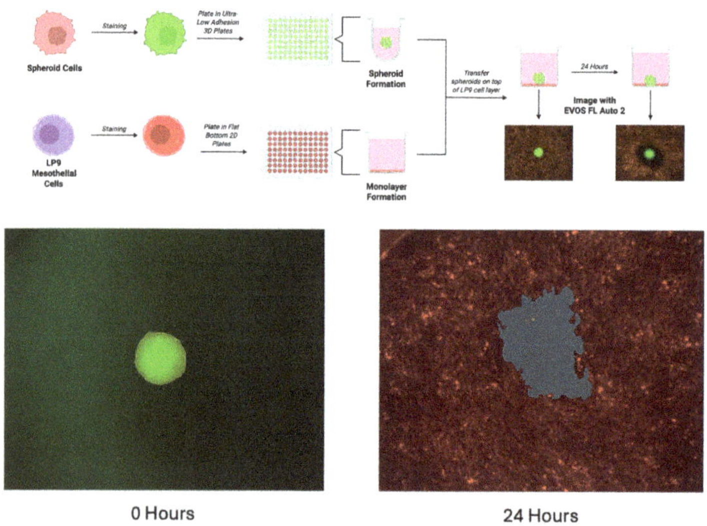

Figure 1. Mesothelial Clearance Assay and Analysis Methods. Top: Schematic of mesothelial clearance. Endometriosis cells are stained green and plated in U-bottom ULA plates to form spheroids. LP9 mesothelial cells are stained red and plated in flat-bottom tissue-culture-treated plates to form a monolayer. After 72 h, the media is removed from the LP9 monolayers, and spheroids are transferred on top. After a 30 min incubation to promote spheroid attachment, images are taken using the EVOS microscope at the initial timepoint and again 24 h later. Bottom: 3D Slicer software (version 5.2.2) is used to measure the area of the spheroid at the initial image timepoint (left, yellow line) and the clearance area at the 24 h timepoint (right, blue line) to find the clearance ratio of cleared area/spheroid area.

3. Results

3.1. 12Z Endometriosis and hEM3 Endometrial Cells Form Spheroids in 3D Cell Culture

To study endometriosis, we wanted to establish a 3D in vitro model. It is well-accepted that 3D cell cultures better represent the interactions between cells as well as interactions between cells and their environment. OVCAR8 ovarian cancer cells form spheroids when grown in ULA plates and were used as a control to compare against the 12Z and hEM3 cells. To determine if the endometrial 12Z and hEM3 cells would be able to form spheroids in 3D cell culture, cells were plated in 2D and 3D and imaged. Both cell lines were able to form spheroids (Supplemental Figure S1). Next, 12Z and hEM3 cells were plated at varying cell densities in U-bottom ULA plates. The spheroids were then stained with NucBlue and PI to assess their morphology and composition of live versus dead cells (Figure 2). At all cell densities, hEM3 cells form smaller spheroids compared to 12Z and OVCAR8 cells. All spheroids were round with smooth outer borders. These results are consistent with ovarian cancer cell spheroids. PI staining of 12Z and OVCAR8 spheroids demonstrated that the spheroid centers were composed of dead cells (red), while the NucBlue staining in combination with the PI showed live cells (blue) located mainly on the outer portions of the spheroid. However, hEM3 spheroids had more dispersed dead cells (red) that were sometimes found along the outer portions of the spheroid in addition to the center. Increasing cell numbers appeared to increase the proportion of dead cells in comparison to live cells in all cell types.

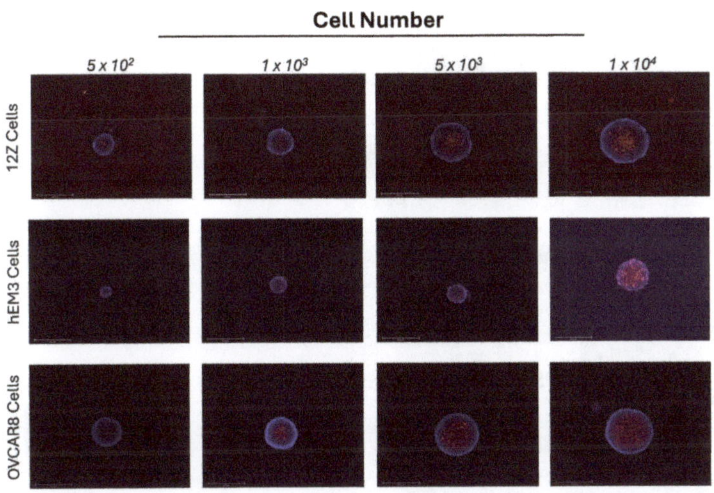

Figure 2. Increasing cell number increases the proportion of dead cells to live cells in spheroids. 12Z, hEM3, and OVCAR8 cells were plated in 3D at varying cell densities per well. On day five, NucBlue (all cells) and PI (dead cells) stains were added, and spheroids were imaged on the EVOS using the 4× objective.

3.2. 12Z and hEM3 Spheroids Can Execute Mesothelial Clearance

Most endometriosis lesions are attached to the pelvic peritoneum. We aimed to expand on past research that showed that only endometrial stromal cells were the driving force of attachment and invasion to mesothelial cell layers. Therefore, we developed an in vitro mesothelial clearance assay to determine if spheroids made up of either 12Z or hEM3 epithelial cells could clear a monolayer of mesothelial cells similar to ovarian cancer spheroids. OVCAR8 ovarian cancer cells were used as a positive control as they are capable of undergoing mesothelial clearance [24].

We found that both 12Z and hEM3 endometrial epithelial cell spheroids cleared the monolayer of LP9 cells after 24 h (Figure 3). The dark space surrounding the spheroid is the area that the spheroid has cleared by pushing away the mesothelial cells to create a space. We found that after 24 h, the spheroids changed morphology and switched from a tight circular-shaped spheroid to a more oval shape with cells that were not as tightly packed. In some 24 h images, the spheroid cells can be seen dispersed throughout the cleared area. Time-lapse videos of 12Z, hEM3, and OVCAR8 cells suggest that endometriosis cells clear mesothelial cells in a manner similar to ovarian cancer spheroids (Supplemental Figures S2–S4). Interestingly, we also analyzed the ability of Ishikawa endometrial cancer cells to undergo mesothelial clearance, as these cells are often used in endometriosis studies. Unlike the other cells, Ishikawa spheroids are unable to clear the mesothelial cell monolayer beyond the boundary of the sphere (Supplemental Figure S5).

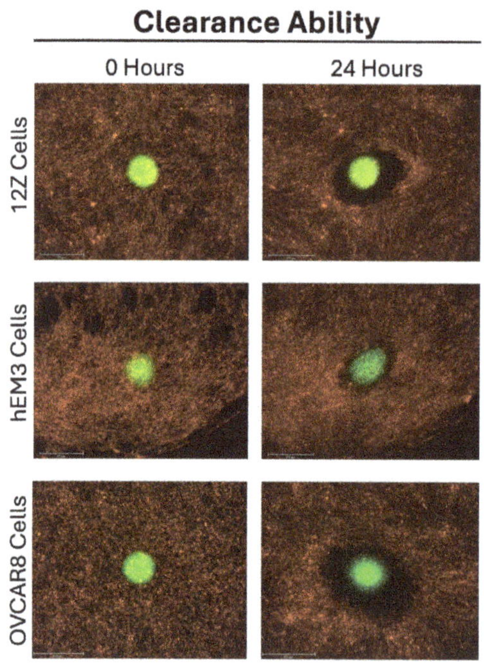

Figure 3. 12Z, hEM3, and OVCAR8 spheroids undergo mesothelial clearance. LP9 mesothelial cells were stained red and grown as a monolayer. 12Z, hEM3, and OVCAR8 cells were stained green and plated in 3D to induce spheroid formation. After 72 h, spheroids were transferred on top of the LP9 monolayer, and images were taken at 0 and 24 h using the 4× objective.

3.3. Clearance Ratio Has an Inverse Relationship with Initial Spheroid Size

After determining that endometrial spheroids could perform mesothelial clearance, we wanted to investigate what impact spheroid size had on the clearance ratio. The results of the NucBlue and PI staining (Figure 2) suggest that the proportion of dead cells to live cells increased as the spheroid size increased. Therefore, we plated different spheroid cell densities per well, and the mesothelial clearance assay was then executed.

The results of the mesothelial clearance assay using differing cell numbers were consistent across all cell lines used. We found that increasing the number of cells in a spheroid increased the cleared area in 12Z cells (Figure 4), hEM3 cells (Figure 5), and OVCAR8 cells (Figure 6). Most notably, the results demonstrated that larger spheroids plated with a higher cell count had the smallest clearance ratio, whereas spheroids with lower cell counts had the highest clearance ratio. These results suggest that the extent to which a spheroid can perform mesothelial clearance is dependent on its initial size and the number of cells it is made up of. Based on the NucBlue and PI staining of all spheroids (Figure 2), the larger spheroids appear to have more dead cells than smaller ones, which could account for the inverse relationship between initial spheroid size and clearance ratio.

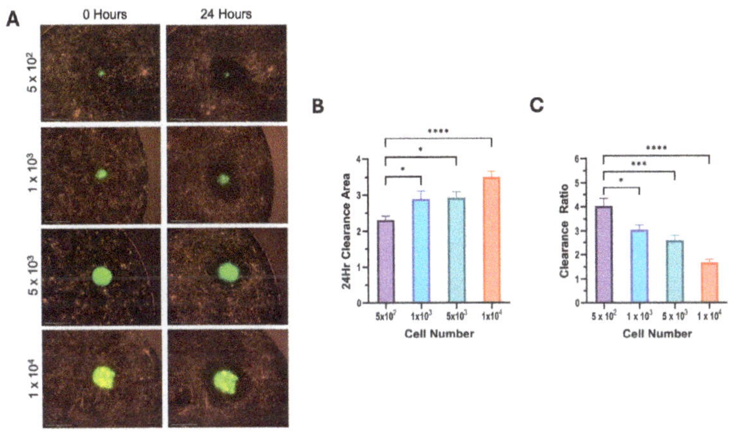

Figure 4. Increasing 12Z cell number increases cleared area but decreases clearance ratio. A mesothelial clearance assay was performed, plating 12Z cells at the indicated spheroid cell numbers. (**A**) Representative images of 0 and 24 h of mesothelial clearance using the 4× objective. (**B**) Clearance area and (**C**) clearance ratio were determined. Data are shown as mean +/− SEM. Data represent a minimum of 5 spheroids per N, N = 3; * $p \leq 0.05$, *** $p \leq 0.001$, **** $p \leq 0.0001$.

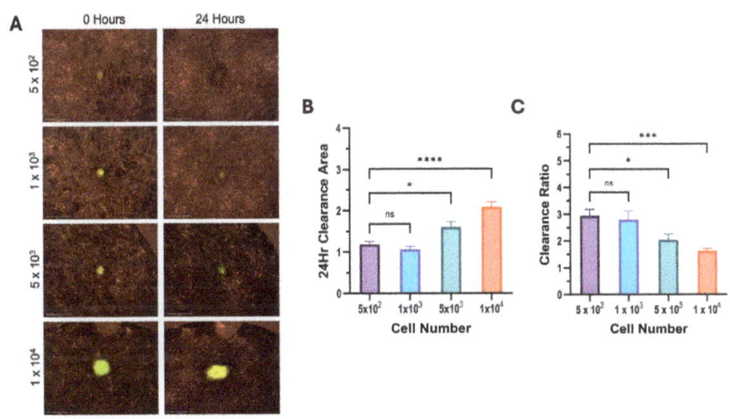

Figure 5. Increasing hEM3 cell number increases cleared area but decreases the clearance ratio. A mesothelial clearance assay was performed, plating hEM3 cells at the indicated spheroid cell numbers. (**A**) Representative images of 0 and 24 h of mesothelial clearance using the 4× objective. (**B**) Clearance area and (**C**) clearance ratio were determined. Data represent a minimum of 5 spheroids per assay, N = 3; ns > 0.05, * $p \leq 0.05$, *** $p \leq 0.001$, **** $p \leq 0.0001$.

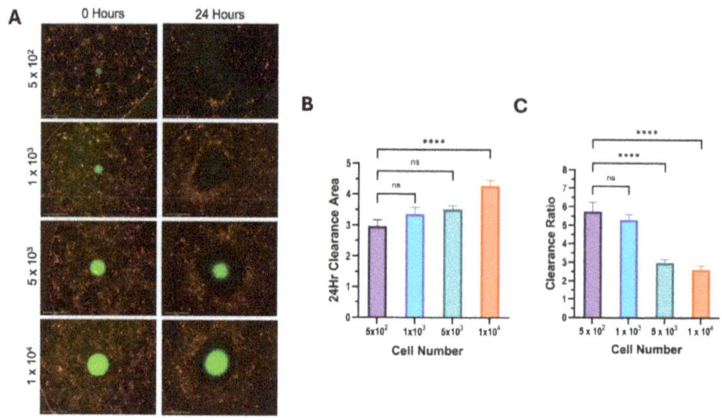

Figure 6. Increasing OVCAR8 cell number increases cleared area but decreases clearance ratio. A mesothelial clearance assay was performed, plating OVCAR8 cells at the indicated spheroid cell numbers. (**A**) Representative images of 0 and 24 h of mesothelial clearance using the 4× objective. (**B**) Clearance area and (**C**) clearance ratio were determined. Data represent a minimum of 5 spheroids per assay, N = 3; ns > 0.05, **** $p \leq 0.0001$.

3.4. Analysis of Potential Pharmacological Inhibitors of Mesothelial Clearance

Many 2D cell culture models fail to accurately represent the complexity of the physical environment surrounding tissues in the body. This lack of sophistication is believed to be a main contributing factor to the high failure rate in drug discovery. We wanted to ensure our in vitro mesothelial clearance assay would be able to assess the ability of potential clearance inhibitors for endometriosis lesions. Statins have been suggested as a possible treatment for endometriosis [25]; therefore, we wanted to determine the effects of statin treatment on spheroids in our mesothelial clearance assay. To accomplish this, we treated 12Z and hEM3 spheroids with 10 mM of either atorvastatin, lovastatin, or simvastatin for 24 h prior to adding them on top of the LP9 mesothelial cell monolayer. The results for the 12Z spheroids demonstrated a significant reduction in clearance ratio when treated with lovastatin but no significant reduction in clearance ratio when treated with atorvastatin or simvastatin (Figure 7). In contrast, hEM3 spheroids showed a significant decrease in clearance ratio when treated with all three statins (Figure 8).

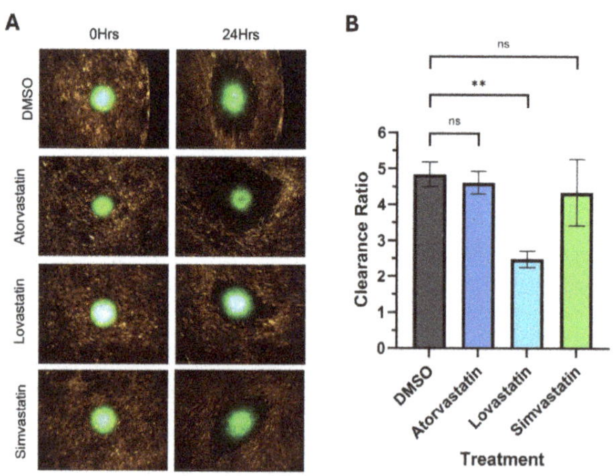

Figure 7. Treating 12Z spheroids with lovastatin reduces the amount of mesothelial clearance. 12Z cells were treated with 10 μM of either atorvastatin, lovastatin, or simvastatin for 24 h prior to starting

the clearance assay. DMSO was used as a vehicle control. 12Z cells were allowed to clear for 24 h. Mesothelial clearance was imaged using the 4× objective (**A**), and clearance ratio was determined (**B**). Data represent a minimum of 8 spheroids per N, N = 3; ns > 0.05, ** $p \leq 0.01$.

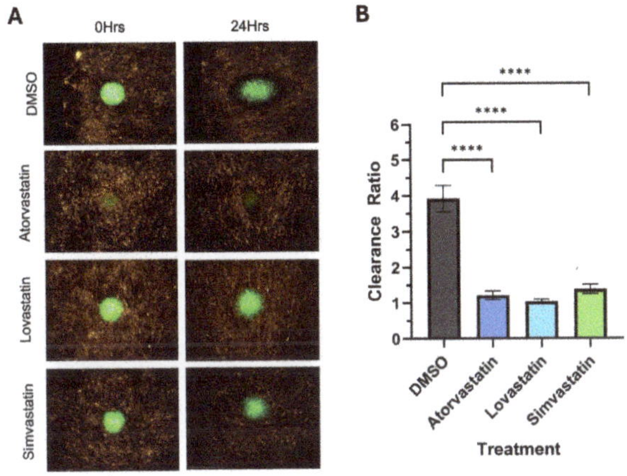

Figure 8. Treating hEM3 spheroids with statins significantly reduces the ability of the spheroids to undergo mesothelial clearance. hEM3 cells were treated with 10 µM of either atorvastatin, lovastatin, or simvastatin for 24 h prior to starting the clearance assay. DMSO was used as a control. hEM3 cells were allowed to clear for 24 h. Mesothelial clearance was imaged using the 4× objective (**A**), and clearance ratio was determined (**B**). Data represent a minimum of 8 spheroids per assay, N = 3; **** $p \leq 0.0001$.

4. Discussion

Endometriosis is a painful gynecological disease whose pathology is poorly understood. Endometriotic lesions bypass the peritoneal mesothelial cell layer in order to facilitate their attachment to organs within the female reproductive system [26]. While most endometriotic lesions are found within the pelvic area, there have been some documented cases of extra-abdominal lesions [1]. The poorly understood process of endometriosis development presents an important area in the endometriosis field that has yet to be fully investigated. Previous studies have generated 3D spheroids made up of various combinations of endometriotic and normal endometrial stromal cells [13]. In these studies, researchers have demonstrated the ability of spheroids, either made up of normal endometrial stromal cells alone or in combination with endometriotic cells, to attach and invade past a monolayer of mesothelial cells [13]. What these studies did not investigate was the direct interaction of endometriotic cells with the mesothelial cell layer. In addition, the mechanisms responsible for the attachment of endometriotic lesions to the peritoneal mesothelial cell layer have yet to be uncovered.

We have developed an adaptable mesothelial clearance assay that can be used to model the interaction of spheroids with a monolayer of mesothelial cells. First, we cultured 12Z endometriotic epithelial cells and hEM3 epithelial cells in 3D cell culture using low-adhesion U-bottom plates. Both cell lines formed round spheroids with smooth outer borders, similar to OVCAR8 ovarian cancer cells, which we used as a reference since we previously determined they form spheroids. After we determined that our 12Z and hEM3 cells could form spheroids, we tested their ability to clear a monolayer of LP9 mesothelial cells and found that they were indeed able to disrupt the LP9 cell monolayer and push apart the mesothelial cells. The spheroids created a gap in the monolayer, which the cells that made up the spheroids started to disperse into. The results from this study using our mesothelial clearance assay have provided evidence to support our theory that endometriosis cell spheroids can undergo mesothelial clearance.

To ensure the clearance that occurred was not due to a co-culture effect, we tested the ability of Ishikawa endometrial cancer cells and ZR-75 breast cancer cells to clear the LP9 monolayer. Ishikawa spheroids were unable to disperse into the monolayer and only cleared the area beneath the spheroid. Notably, the Ishikawa cells formed irregular spheroids, which may have influenced their clearance capability. We also examined the breast cancer cell line ZR-75, which makes uniform spheroids, and these were able to create a gap in the monolayer and appeared to disperse slightly beyond the spheroid boundary (Supplemental Figure S5). This may not be surprising as a small number of breast cancer patients have been found to have metastases in the peritoneal cavity [27]. To further assess the potential co-culture effects, we created a monolayer of mouse embryonic fibroblasts and assessed the ability of the 12Z and hEM3 cells to clear these cells (Supplemental Figure S6). In all cases, the spheroids were unable to clear beyond the spheroid boundary. They remained tightly packed and did not disperse outward like the spheroid cells did with mesothelial cells. This suggests that the clearance by the endometrial spheroids is not due solely to a co-culture effect. Moreover, a study conducted by Davidowitz et al. found that not all ovarian cancer cell spheroids could clear a mesothelial cell monolayer [28]. The ability of ovarian cancer spheroids to clear the mesothelial layer was correlated with the expression of genes relating to the epithelial-to-mesenchymal transition (EMT) [28]. It would be interesting to further investigate the expression of EMT-related genes in the 12Z and hEM3 spheroids during clearance to unveil any similarities they have with the ovarian cancer spheroids and differences between the 12Z and hEM3 cell lines.

Interestingly, our data determined that the ratio of clearance area compared to the initial spheroid area had an inverse relationship to the cell seeding density and, therefore, the size of the original spheroid. Lastly, we tested the clearance ability of 12Z and hEM3 spheroids after treatment with three different statins. This demonstrated that our clearance assay could be used as a model for different drug treatments on endometriosis lesions. Our assay provides a straightforward approach that can be modified to investigate the role of different molecular pathways involved in the progression of endometriosis.

Previously, Witz et al. compared the ability of antegrade to shed menstrual endometrium from women without endometriosis and demonstrated that both endometrial stromal and epithelial cells were able to attach to human peritoneum explants [29]. Kavoussi et al. used the endometrial epithelioid cell line EM42 and reported that attachment to a monolayer of LP9 mesothelial cells occurred [30]. These studies demonstrated that epithelial cells could attach to mesothelial cells, but again, they did not use endometriosis cells. Song et al. published a study where they used endometriotic epithelial cells (12Zs) and endometriotic stromal cells to form spheroids [13]. These cells self-organized into spheroids that had a stromal cell core that was surrounded by multiple layers of epithelial cells after one week of growing [13]. Interestingly, using a Matrigel/LP9 cell layer to mimic invasion, the endometriotic stromal cells were the driving force in the disruption of the mesothelial cell layer even though the 12Z epithelial cells were located on the outermost layers of the spheroid [13]. However, they did not measure the ability of the endometriotic epithelial cells to interact with the LP9 cell layer on their own. We determined that when transferred on top of a mesothelial cell monolayer, both 12Z and hEM3 cell spheroids can push away the mesothelial cells, creating a space that they can then colonize. These results reveal an important element in the development of endometriosis that augments what other studies have previously published.

Endometriosis is a complex disease whose presence is due to an unknown combination of multiple factors [7]. Endometriosis lesion formation and heterogeneity can differ greatly from case to case [1,7,11]. It is therefore possible that epithelial cells from normal and endometriotic endometrium may have an intrinsic ability to clear the mesothelial cell layer that lines the peritoneal cavity. In different conditions, different cell types could potentially drive mesothelial clearance, with some cell types having a stronger ability than others. In our clearance assays, 12Z cells made consistently larger spheroids compared to hEM3 cells. In addition, the area cleared by the 12Z spheroids was also larger than the area cleared by hEM3 spheroids. This resulted in an overall larger clearance ratio value for 12Z spheroids than hEM3 spheroids. Overall, our results support our hypothesis that 12Z and hEM3 epithelial cells can form spheroids in 3D culture that are capable of performing mesothelial clearance. The differences in clearance ratio values between these two cell lines could potentially be due to one cell line being endometriotic cells and the other being normal cells. Regardless, we have demonstrated that endometrial epithelial cells can perform mesothelial clearance, which can help us better understand how endometriosis lesions occur not only within the female reproductive system but around the peritoneal cavity and may suggest that mesothelial clearance is needed prior to moving to extra-pelvic locations.

Solid tumors have unique structures and arrangements of cells that result in multiple layers of cells [31]. The external layer of cells has more access to oxygen and nutrients within the tumor microenvironment, which causes them to proliferate more rapidly [31]. In contrast, the middle layer of cells has less access to oxygen and nutrients and is in a senescent state [31]. The innermost layer of solid tumors often has no access to oxygen or nutrients, causing them to become necrotic [31]. To assess the composition of the 12Z and hEM3 cell spheroids, we stained them with NucBlue and propidium iodine. As expected, the OVCAR8 cell spheroids had an outer layer of living cells, while the cells making up the inner layer of the spheroid were dead. This pattern was also observed in the 12Z cell spheroids. Interestingly, the spheroids formed from hEM3 cells showed no definitive separation of dead and living cells within the inner and outer layers of the spheroid. This could potentially explain why hEM3 cells formed smaller spheroids and had an overall smaller clearance ratio when compared to 12Z cell spheroids. The difference between these cell lines is that 12Z cells are endometriotic cells, whereas hEM3 cells are not, which could explain the differences we see in the 3D spheroid formation. To explore why these differences in spheroid composition occur, future research efforts should focus on 3D spheroid gene expression.

Three-dimensional spheroids have become popular in vitro models for drug testing in cancer fields due to the similarities they show with not only the morphology of solid tumor spheroids but gene expression and drug response as well [31]. After determining that endometriotic spheroids could undergo mesothelial clearance, we wanted to determine if our assay could be used in combination with drug treatments to demonstrate the effectiveness of the drugs on inhibiting mesothelial clearance. Statins, a class of medication typically prescribed to help lower cholesterol, are also being tested as a potential treatment for cancer in clinical trials [32]. The effectiveness of statins for cancer treatment seems to vary by statin type and cancer type [32]. Due to their antioxidant characteristics, statins have also been investigated as a potential treatment for endometriosis, as it is a highly inflammatory disease with increased levels of oxidative stress [1,33]. Using a rat model of endometriosis, Oktem et al. reported that a high dose of atorvastatin lowered the mean area of lesion size compared to the control group [33]. Another statin, lovastatin, was found to inhibit the proliferation and angiogenesis of endometrial stromal cells in a dose-dependent manner [25]. In another study using nude mice, simvastatin also decreased the number and size of endometriotic lesions in a dose-dependent manner [34]. Acknowledging that these statins were administered after endometriosis was induced in the mouse and rat models and after endometriotic tissues from patients were collected and cultured in vitro, these results may speak more to the treatment of pre-established endometriosis lesions rather than as a method of endometriosis prevention.

In a literature review by Gibran et al., several studies highlighted simvastatin treatment on endometriotic stromal cells, resulting in a decrease in proliferation and increases in apoptosis, cell shrinkage, and the breakdown of the cytoskeleton [35–37]. These data suggest that statins could be a useful treatment for endometriosis. Therefore, we treated 12Z and hEM3 spheroids with atorvastatin, lovastatin, and simvastatin to ascertain if they had any impact on the clearance ability of these cell lines. All statins significantly decreased the clearance ratio of hEM3 spheroids, but only lovastatin significantly reduced the clearance ratio of 12Z spheroids. This outcome may reflect the differences between the cell lines, 12Z cells being endometriosis cells and hEM3 cells being normal endometrium cells, or it may reflect differences between the statins themselves. While some data suggest that statins may be beneficial for endometriosis, there is some evidence that statin use is associated with endometriosis. Jiao et al. investigated ovary- and uterus-related adverse events associated with statin use based on available data from the FDA Adverse Event Reporting System (FAERS) [38]. This analysis took into consideration not only statins as a class but also seven individual statins and their reported adverse events. For example, statins as a class were reported to reduce the proliferation, viability, and migration of endometrial cells [38]. Individually, atorvastatin and simvastatin were reported to be associated with an increased risk of endometriosis, while lovastatin use was not [38]. This could explain why only lovastatin significantly reduced the mesothelial clearance done by our 12Z spheroids when all three statins significantly reduced the amount of mesothelial clearance the hEM3 spheroids did. It is also interesting to note that compared to atorvastatin and simvastatin, Jiao et al. found that lovastatin did not result in as many adverse events overall [38]. Therefore, more research is needed to understand the benefits and risks of statins and endometriosis. Regardless, we have shown that the mesothelial clearance model we developed has a promising use in testing possible pharmacological therapies for endometriosis prior to clinical trials.

There are currently no definitive biomarkers for endometriosis [1]. Endometriosis is an inflammatory disease, which has led researchers to study altered levels of pro-inflammatory cytokines in endometriosis [1,39]. Cytokines can influence cell proliferation and differentiation and are important for the progression of endometriosis [40]. While multiple studies have found altered cytokine levels in patients with endometriosis, these results are not consistent across multiple studies and the menstrual phases of patients [2]. Instead of focusing solely on the altered levels of cytokines, investigating related pathways and key proteins involved with the cytokines of interest could shed light on endometriotic lesion formation and attachment. Our mesothelial clearance model enables researchers to accomplish this. For example, increased levels of the cytokine interleukin-6 (IL-6) have been reported in the serum of patients with endometriosis, leading to persistent activation of the signal transducer and activator of transcription 3 (STAT3) protein [40,41]. Our assay could be used to assess IL-6-activated STAT3 importance in mesothelial clearance in two ways. First, by stimulating the spheroid cells with IL-6 before transferring them onto the LP9 mesothelial layer and comparing the amount of clearance between stimulated and unstimulated spheroids, and second, through knocking down *STAT3* in the spheroids for the mesothelial clearance assay. This will allow researchers to determine if STAT3 activated via IL-6 is required for mesothelial clearance or not. This will help to gain a deeper understanding of how endometriosis lesions are able to attach to and clear the mesothelial cell layer in the peritoneal cavity, resulting in lesions on organs within the peritoneal cavity and in extra-pelvic regions as well. Lastly, as mesothelial clearance and invasion are two separate steps, our method can be used to separate the genes specific to mesothelial clearance versus invasion.

5. Conclusions

Overall, the data presented here demonstrate that endometriotic and normal endometrial epithelial cells are capable of undergoing mesothelial clearance. Importantly, spheroids made up of fewer numbers of cells had a smaller clearance area but a larger clearance ratio, whereas spheroids made up of a greater number of cells had a larger clearance area but a smaller clearance ratio. This highlights the importance of comparing the clearance ratio instead of just analyzing the cleared area. Here, we have shown that endometriotic epithelial cells can also clear a mesothelial cell monolayer. This is important for endometriotic lesions as they are found attached to organs within the peritoneal cavity. Some cases of extra-pelvic endometriotic lesions have been documented; therefore, further studies should be conducted to determine if endometriotic lesions can break through the extracellular matrix below, as ovarian cancer cells can. This will shed light on the elusive pathology of endometriosis and how lesions spread and better inform research aimed at treating this disease.

Supplementary Materials: The following supporting information can be downloaded at https://www.mdpi.com/article/10.3390/cells14100742/s1. Figure S1: Phase-contrast images of cells in 2D and 3D. Figures S2–S4: Time-lapse images of 12Z, hEM3, and OVCAR8 cells, respectively, undergoing mesothelial clearance. Figure S5: Mesothelial clearance by Ishikawa and ZR-75 cells. Figure S6: Clearance of MEFs by 12Z, hEM3, and OVCAR8 spheroids.

Author Contributions: Conceptualization, A.A.K. and S.R.W.; methodology, validation, and investigation, A.A.K.; writing—original draft preparation, A.A.K.; writing—review and editing, A.A.K. and S.R.W.; supervision, project administration, and funding acquisition, S.R.W. All authors have read and agreed to the published version of the manuscript.

Funding: The research reported in this publication was supported by the University of New Hampshire's Center for Integrated Biomedical and Bioengineering Research (CIBBR) through a grant from the National Institute of General Medical Sciences of the National Institutes of Health under Award Number P20GM113131.

Institutional Review Board Statement: Not applicable.

Informed Consent Statement: Not applicable.

Data Availability Statement: Data are available at https://figshare.com/articles/figure/Complete_Image_Sets/28796033.

Acknowledgments: The graphical abstract and part of Figure 1 were created in BioRender. Walker, S. (2025). https://BioRender.com/ezehomz.

Conflicts of Interest: The authors declare no conflicts of interest. The funders had no role in the design of this study; in the collection, analyses, or interpretation of the data; in the writing of the manuscript; or in the decision to publish the results.

Abbreviations

The following abbreviations are used in this manuscript:

3D	Three-dimensional
2D	Two-dimensional
ULA	Ultra-low attachment
PI	Propidium iodide
DMSO	Dimethyl sulfoxide

References

1. Horne, A.W.; Missmer, S.A. Pathophysiology, Diagnosis, and Management of Endometriosis. *BMJ* **2022**, *379*, e070750. [CrossRef] [PubMed]
2. May, K.E.; Villar, J.; Kirtley, S.; Kennedy, S.H.; Becker, C.M. Endometrial Alterations in Endometriosis: A Systematic Review of Putative Biomarkers. *Hum. Reprod. Updat.* **2011**, *17*, 637–653. [CrossRef] [PubMed]
3. Kennedy, S.; Bergqvist, A.; Chapron, C.; D'Hooghe, T.; Dunselman, G.; Greb, R.; Hummelshoj, L.; Prentice, A.; Saridogan, E.; Koninckx, P.; et al. ESHRE Guideline for the Diagnosis and Treatment of Endometriosis. *Hum. Reprod.* **2005**, *20*, 2698–2704. [CrossRef]
4. Bulun, S.E. Endometriosis. *N. Engl. J. Med.* **2009**, *360*, 268–279. [CrossRef]
5. Ramírez-González, J.A.; Vaamonde-Lemos, R.; Cunha-Filho, J.S.; Varghese, A.C.; Swanson, R.J. Overview of the Female Reproductive System. In *Exercise and Human Reproduction: Induced Fertility Disorders and Possible Therapies*; Springer: Berlin/Heidelberg, Germany, 2016; pp. 19–46. [CrossRef]
6. Critchley, H.O.D.; Maybin, J.A.; Armstrong, G.M.; Williams, A.R.W. Physiology of the Endometrium and Regulation of Menstruation. *Physiol. Rev.* **2020**, *100*, 1149–1179. [CrossRef]
7. Laganà, A.S.; Garzon, S.; Götte, M.; Viganò, P.; Franchi, M.; Ghezzi, F.; Martin, D.C. The Pathogenesis of Endometriosis: Molecular and Cell Biology Insights. *Int. J. Mol. Sci.* **2019**, *20*, 5615. [CrossRef]
8. Sampson, J.A. Peritoneal Endometriosis Due to the Menstrual Dissemination of Endometrial Tissue into the Peritoneal Cavity. *Am. J. Obstet. Gynecol.* **1927**, *14*, 422–469. [CrossRef]
9. Witz, C.A.; Monotoya-Rodriguez, I.A.; Schenken, R.S. Whole Explants of Peritoneum and Endometrium: A Novel Model of the Early Endometriosis Lesion. *Fertil. Steril.* **1999**, *71*, 56–60. [CrossRef]
10. Young, V.J.; Brown, J.K.; Saunders, P.T.K.; Horne, A.W. The Role of the Peritoneum in the Pathogenesis of Endometriosis. *Hum. Reprod. Update* **2013**, *19*, 558–569. [CrossRef]
11. Koninckx, P.R.; Fernandes, R.; Ussia, A.; Schindler, L.; Wattiez, A.; Al-Suwaidi, S.; Amro, B.; Al-Maamari, B.; Hakim, Z.; Tahlak, M. Pathogenesis Based Diagnosis and Treatment of Endometriosis. *Front. Endocrinol.* **2021**, *12*, 745548. [CrossRef]
12. Nair, A.S.; Nair, H.B.; Lucidi, R.S.; Kirchner, A.J.; Schenken, R.S.; Tekmal, R.R.; Witz, C.A. Modeling the Early Endometriotic Lesion: Mesothelium-Endometrial Cell Co-Culture Increases Endometrial Invasion and Alters Mesothelial and Endometrial Gene Transcription. *Fertil. Steril.* **2008**, *90*, 1487–1495. [CrossRef] [PubMed]
13. Song, Y.; Burns, G.W.; Joshi, N.R.; Arora, R.; Kim, J.J.; Fazleabas, A.T. Spheroids as a Model for Endometriotic Lesions. *JCI Insight* **2023**, *8*, e160815. [CrossRef] [PubMed]
14. Nezhat, F.R.; Pejovic, T.; Reis, F.M.; Guo, S.W. The Link Between Endometriosis and Ovarian Cancer: Clinical Implications. *Int. J. Gynecol. Cancer* **2014**, *24*, 623–628. [CrossRef]
15. Guo, S.W. The Association of Endometriosis with Ovarian Cancer: A Critical Review of Epidemiological Data. In *Endometriosis: Pathogenesis and Treatment*; Springer: Berlin/Heidelberg, Germany, 2014; pp. 363–384. [CrossRef]
16. Pearce, C.L.; Templeman, C.; Rossing, M.A.; Lee, A.; Near, A.M.; Webb, P.M.; Nagle, C.M.; Doherty, J.A.; Cushing-Haugen, K.L.; Wicklund, K.G.; et al. Association between Endometriosis and Risk of Histological Subtypes of Ovarian Cancer: A Pooled Analysis of Case-Control Studies. *Lancet Oncol.* **2012**, *13*, 385–394. [CrossRef]
17. Yee, C.; Dickson, K.A.; Muntasir, M.N.; Ma, Y.; Marsh, D.J. Three-Dimensional Modelling of Ovarian Cancer: From Cell Lines to Organoids for Discovery and Personalized Medicine. *Front. Bioeng. Biotechnol.* **2022**, *10*, 836984. [CrossRef]
18. Kapałczyńska, M.; Kolenda, T.; Przybyła, W.; Zajączkowska, M.; Teresiak, A.; Filas, V.; Ibbs, M.; Bliźniak, R.; Łuczewski, Ł.; Lamperska, K. 2D and 3D Cell Cultures—A Comparison of Different Types of Cancer Cell Cultures. *Arch. Med. Sci.* **2018**, *14*, 910. [CrossRef]
19. Kerslake, R.; Belay, B.; Panfilov, S.; Hall, M.; Kyrou, I.; Randeva, H.S.; Hyttinen, J.; Karteris, E.; Sisu, C. Transcriptional Landscape of 3D vs. 2D Ovarian Cancer Cell Models. *Cancers* **2023**, *15*, 3350. [CrossRef]
20. Brueggmann, D.; Templeman, C.; Starzinski-Powitz, A.; Rao, N.P.; Gayther, S.A.; Lawrenson, K. Novel Three-Dimensional In Vitro Models of Ovarian Endometriosis. *J. Ovarian Res.* **2014**, *7*, 17. [CrossRef]
21. Davidowitz, R.A.; Iwanicki, M.P.; Brugge, J.S. In Vitro Mesothelial Clearance Assay That Models the Early Steps of Ovarian Cancer Metastasis. *J. Vis. Exp.* **2012**, *60*, e3888. [CrossRef]
22. Witz, C.A.; Cho, S.; Centonze, V.E.; Montoya-Rodriguez, I.A.; Schenken, R.S. Time Series Analysis of Transmesothelial Invasion by Endometrial Stromal and Epithelial Cells Using Three-Dimensional Confocal Microscopy. *Fertil. Steril.* **2003**, *79* (Suppl. S1), 770–778. [CrossRef]
23. Park, Y.; Jung, J.G.; Yu, Z.C.; Asaka, R.; Shen, W.; Wang, Y.; Jung, W.H.; Tomaszewski, A.; Shimberg, G.; Chen, Y.; et al. A Novel Human Endometrial Epithelial Cell Line for Modeling Gynecological Diseases and for Drug Screening. *Lab. Investig.* **2021**, *101*, 1505–1512. [CrossRef] [PubMed]
24. Zhang, J.; Yang, N.; Kreeger, P.K.; Notbohm, J. Topological Defects in the Mesothelium Suppress Ovarian Cancer Cell Clearance. *APL Bioeng.* **2021**, *5*, 36103. [CrossRef] [PubMed]

25. Esfandiari, N.; Khazaei, M.; Ai, J.; Bielecki, R.; Gotlieb, L.; Ryan, E.; Casper, R.F. Effect of a Statin on an in Vitro Model of Endometriosis. *Fertil. Steril.* **2007**, *87*, 257–262. [CrossRef]
26. Young, V.J.; Brown, J.K.; Saunders, P.T.K.; Duncan, W.C.; Horne, A.W. The Peritoneum Is Both a Source and Target of TGF-β in Women with Endometriosis. *PLoS ONE* **2014**, *9*, e106773. [CrossRef] [PubMed]
27. Beniey, M. Peritoneal Metastases from Breast Cancer: A Scoping Review. *Cureus* **2019**, *11*, e5367. [CrossRef]
28. Davidowitz, R.A.; Selfors, L.M.; Iwanicki, M.P.; Elias, K.M.; Karst, A.; Piao, H.; Ince, T.A.; Drage, M.G.; Dering, J.; Konecny, G.E.; et al. Mesenchymal Gene Program–Expressing Ovarian Cancer Spheroids Exhibit Enhanced Mesothelial Clearance. *J. Clin. Investig.* **2014**, *124*, 2611–2625. [CrossRef]
29. Witz, C.A.; Allsup, K.T.; Montoya-Rodriguez, I.A.; Vaughn, S.L.; Centonze, V.E.; Schenken, R.S. Culture of Menstrual Endometrium with Peritoneal Explants and Mesothelial Monolayers Confirms Attachment to Intact Mesothelial Cells. *Hum. Reprod.* **2002**, *17*, 2832–2838. [CrossRef]
30. Kavoussi, S.K.; Witz, C.A.; Binkley, P.A.; Nair, A.S.; Lebovic, D.I. Peroxisome-Proliferator Activator Receptor-Gamma Activation Decreases Attachment of Endometrial Cells to Peritoneal Mesothelial Cells in an in Vitro Model of the Early Endometriotic Lesion. *Mol. Hum. Reprod.* **2009**, *15*, 687–692. [CrossRef]
31. Costa, E.C.; Moreira, A.F.; de Melo-Diogo, D.; Gaspar, V.M.; Carvalho, M.P.; Correia, I.J. 3D Tumor Spheroids: An Overview on the Tools and Techniques Used for Their Analysis. *Biotechnol. Adv.* **2016**, *34*, 1427–1441. [CrossRef]
32. Jiang, W.; Hu, J.W.; He, X.R.; Jin, W.L.; He, X.Y. Statins: A Repurposed Drug to Fight Cancer. *J. Exp. Clin. Cancer Res.* **2021**, *40*, 241. [CrossRef]
33. Oktem, M.; Esinler, I.; Eroglu, D.; Haberal, N.; Bayraktar, N.; Zeyneloglu, H.B. High-Dose Atorvastatin Causes Regression of Endometriotic Implants: A Rat Model. *Hum. Reprod.* **2007**, *22*, 1474–1480. [CrossRef] [PubMed]
34. Bruner-Tran, K.L.; Osteen, K.G.; Duleba, A.J. Simvastatin Protects against the Development of Endometriosis in a Nude Mouse Model. *J. Clin. Endocrinol. Metab.* **2009**, *94*, 2489–2494. [CrossRef] [PubMed]
35. Nasu, K.; Yuge, A.; Tsuno, A.; Narahara, H. Simvastatin Inhibits the Proliferation and the Contractility of Human Endometriotic Stromal Cells: A Promising Agent for the Treatment of Endometriosis. *Fertil. Steril.* **2009**, *92*, 2097–2099. [CrossRef]
36. Sokalska, A.; Wong, D.H.; Cress, A.; Piotrowski, P.C.; Rzepczynska, I.; Villanueva, J.; Duleba, A.J. Simvastatin Induces Apoptosis and Alters Cytoskeleton in Endometrial Stromal Cells. *J. Clin. Endocrinol. Metab.* **2010**, *95*, 3453–3459. [CrossRef]
37. Gibran, L.; Maranhão, R.C.; Abrão, M.S.; Baracat, E.C.; Podgaec, S. Could Statins Constitute a Novel Treatment for Endometriosis? Systematic Review of the Literature. *Eur. J. Obstet. Gynecol. Reprod. Biol.* **2014**, *179*, 153–158. [CrossRef]
38. Jiao, X.F.; Li, H.L.; Jiao, X.Y.; Guo, Y.C.; Zhang, C.; Yang, C.S.; Zeng, L.N.; Bo, Z.Y.; Chen, Z.; Song, H.B.; et al. Ovary and Uterus Related Adverse Events Associated with Statin Use: An Analysis of the FDA Adverse Event Reporting System. *Sci. Rep.* **2020**, *10*, 11955. [CrossRef]
39. Cicinelli, E.; Vitagliano, A.; Loizzi, V.; De Ziegler, D.; Fanelli, M.; Bettocchi, S.; Nardelli, C.; Trojano, G.; Cicinelli, R.; Minervini, C.F.; et al. Altered Gene Expression Encoding Cytochines, Grow Factors and Cell Cycle Regulators in the Endometrium of Women with Chronic Endometritis. *Diagnostics* **2021**, *11*, 471. [CrossRef]
40. Oală, I.E.; Mitranovici, M.-I.; Chiorean, D.M.; Irimia, T.; Crișan, A.I.; Melinte, I.M.; Cotruș, T.; Tudorache, V.; Moraru, L.; Moraru, R.; et al. Endometriosis and the Role of Pro-Inflammatory and Anti-Inflammatory Cytokines in Pathophysiology: A Narrative Review of the Literature. *Diagnostics* **2024**, *14*, 312. [CrossRef]
41. Matsuzaki, S.; Pouly, J.-L.; Canis, M. Persistent Activation of Signal Transducer and Activator of Transcription 3 via Interleukin-6 Trans-Signaling Is Involved in Fibrosis of Endometriosis. *Hum. Reprod.* **2022**, *37*, 1489–1504. [CrossRef]

cells

MDPI

Review

Molecular Basis of Impaired Decidualization in the Eutopic Endometrium of Endometriosis Patients

Alejandra Monserrat Retis-Resendiz [1,†], Sandra Karen Gómez-Suárez [1,†], Elizabeth García-Gómez [2] and Edgar Ricardo Vázquez-Martínez [1,*]

1 Unidad de Investigación en Reproducción Humana, Instituto Nacional de Perinatología (INPer)-Facultad de Química, Universidad Nacional Autónoma de México (UNAM), Mexico City 11000, Mexico; monser9610@gmail.com (A.M.R.-R.); sk-gs@hotmail.com (S.K.G.-S.)

2 Secretaría de Ciencia, Humanidades, Tecnologías e Innovación (SECIHTI)-Unidad de Investigación en Reproducción Humana, Instituto Nacional de Perinatología (INPer)-Facultad de Química, Universidad Nacional Autónoma de México (UNAM), Mexico City 11000, Mexico; egarciag1982@gmail.com

* Correspondence: vamer@comunidad.unam.mx; Tel.: +52-55-55209900 (ext. 346)

† These authors contributed equally to this work.

Abstract: Endometriosis is a chronic gynecological disorder characterized by the presence of endometrial tissue outside the uterine cavity. A common feature of this pathology is the impaired decidualization of endometrial stromal cells, a critical process that prepares the uterus for embryo implantation. This decidualization defect has been mechanistically linked to progesterone resistance in endometriotic lesions. However, the presence and underlying mechanisms of decidualization defects in the eutopic endometrium of women with endometriosis remain controversial. The aim of the present study is to integrate and discuss molecular evidence from both in vivo and in vitro studies examining decidualization alterations in the eutopic endometrium of patients with endometriosis. Multiple studies have demonstrated impaired decidualization in the eutopic endometrium of women with endometriosis. These alterations have been reported on multiple genes, signaling pathways, and epigenetic processes. However, additional functional studies are warranted to elucidate whether these decidualization defects directly contribute to endometriosis-associated infertility. A better understanding of the decidualization process and its dysregulation in endometriosis will not only advance the development of targeted fertility treatments but also facilitate the design of more effective therapeutic strategies for managing this chronic condition.

Keywords: endometriosis; decidualization; eutopic endometrium; endometrial stromal cells; gene expression; signaling pathway; epigenetics

1. Introduction

The endometrium, the innermost layer of the uterus, is essential for reproductive function, particularly during pregnancy [1]. This tissue comprises multiple cell types: luminal and glandular epithelial cells (EECs), stromal cells (ESCs), immune cells such as lymphocytes and macrophages, endothelial cells, smooth muscle cells, and ciliated epithelial cells [2]. In response to cyclic ovarian steroid hormones, the endometrium undergoes dynamic remodeling throughout the menstrual cycle [2–4].

The endometrial cycle consists of three distinct phases: proliferative, secretory, and menstrual [5]. During the proliferative phase, the endometrium undergoes estrogen-mediated growth and thickening, accompanied by spiral artery elongation [5,6]. The subsequent secretory phase is characterized by a slight decrease in plasma estradiol (E2)

https://doi.org/10.3390/cells14050326

levels and markedly elevated progesterone (P4) levels [5,7,8]. This phase involves enhanced vascular supply, increased mucus secretion, and cessation of proliferation. In addition, the endometrium also undergoes a critical transformation for implantation, a process known as decidualization [7,8]. In non-conception cycles, declining E2 and P4 levels trigger spiral artery constriction, resulting in functional layer ischemia and endometrial shedding during the menstrual phase [5].

Decidualization, critical for pregnancy establishment and maintenance, encompasses functional and morphological modifications that occur within the endometrium to generate the decidua, the specialized tissue supporting pregnancy [7,8]. During this process, ESCs differentiate into specialized secretory cells known as decidual stromal cells (DSCs) in response to post-ovulatory P4 elevation and local cyclic adenosine monophosphate (cAMP) production, involving complex signaling cascades and extensive genetic reprogramming [7–10].

During decidualization, fibroblast-like ESCs undergo marked morphological changes, becoming enlarged and rounded with prominent nuclei and expanded cytoplasm [8]. Functionally, DSCs acquire a specialized secretory phenotype, producing key implantation factors, including prolactin (PRL) and insulin-like growth factor binding protein-1 (IGFBP1), which serve as decidualization biomarkers [11]. Mounting evidence indicates that cellular or molecular decidualization defects may contribute to pregnancy complications and endometrial disorders, such as endometriosis [12–14]. Impaired decidualization has been mainly associated with P4 resistance in endometriotic lesions of endometriosis patients [15,16]. However, decidualization defects in the eutopic endometrium (the normal endometrial tissue located in the proper place in the uterus) of endometriosis patients remain a subject of considerable debate. Evidence suggests that decidualization and implantation-related processes are dysregulated in endometriosis patients [17–19]; however, conflicting studies report no significant alterations [17–19]. While endometriosis-associated infertility is well documented, the molecular basis of eutopic endometrial dysfunction remains incompletely understood. This review synthesizes current evidence from both in vivo and in vitro studies examining decidualization defects in eutopic endometrium of patients with endometriosis. By integrating recent molecular and cellular findings, we aim to provide comprehensive insights into this underexplored aspect of reproductive biology and identify potential therapeutic targets for improving fertility outcomes in endometriosis patients.

2. Symptoms and Pathogenesis of Endometriosis

Endometriosis is an estrogen-dependent inflammatory disorder affecting approximately 10% of reproductive-age women, representing an estimated 176 million women worldwide. This condition is characterized by the presence of endometrial-like tissue outside the uterine cavity (ectopic endometrium), predominantly in reproductive organs, including the ovaries, fallopian tubes, and uterosacral ligaments [20]. Less frequently, lesions may develop in the rectum, bladder, intestines, and vagina [20,21].

Endometriosis manifests with a spectrum of symptoms, including chronic pelvic pain, dyspareunia, and urinary issues; however, symptom severity does not consistently correlate with the extent of lesions [20,21]. This condition is strongly associated with infertility, affecting up to 50% of patients, and notably, endometriosis is implicated in approximately 80% of idiopathic infertility cases [22]. Despite this well-established association between endometriosis and infertility, the underlying biological mechanisms, particularly in mild and moderate disease, remain incompletely understood [22]. In fact, extensive research efforts have yet to fully elucidate the pathogenic mechanisms of endometriosis. Recent reviews describe three main theories regarding its origin: implantation through retrograde menstruation, metaplasia of multipotent mesenchymal stem cells, and lymphatic metastasis of endometrial cells [23].

Similar to the eutopic endometrium, ectopic lesions express receptors for estrogens, P4, and androgens, rendering them responsive to both endogenous and exogenous hormonal stimulation [18,24]. As a result, the ectopic endometrium undergoes cyclic proliferation and breakdown, mirroring the changes observed in the eutopic endometrium; however, its differentiation capacity is reduced [24]. Interestingly, the eutopic endometrium of women with endometriosis exhibits significant alterations in proliferation, adhesion, angiogenesis, and decidualization compared to the endometrium of women without the disease [20,21]. Particularly, impaired decidualization and other endometrial alterations have been associated with P4 resistance and dysregulated P4 receptor (PGR) expression in both eutopic and ectopic endometrial tissues [15,16,25–27].

In this review, we clearly differentiate between findings from eutopic and ectopic endometrium in endometriosis. While our primary focus is on eutopic endometrium, we include relevant findings from ectopic lesions to provide a comparative context and highlight the current knowledge gaps regarding eutopic endometrial dysfunction in endometriosis.

3. Sex-Steroid Hormonal Signaling, Dysregulation, and Impaired Decidualization in Endometriosis

Since decidualization is a P4-dependent process [8] and the eutopic endometrium of endometriosis patients exhibits P4 resistance [15,28], impaired endometrial function may contribute to the reproductive failure in these patients. This decidualization defect could partially explain the high prevalence of infertility observed in women with endometriosis [20].

The decidualization process is regulated by steroid hormones (SH), primarily E2 and P4, along with other factors that increase intracellular cAMP levels [7–10]. E2 and P4 exert distinct genomic and non-genomic effects through their respective receptors. Genomic effects are mediated through their respective nuclear steroid receptors (nSR): estrogen receptors (ESR1 and ESR2) [29,30], and PGR [31], which primarily function as transcription factors. Additionally, both hormones can activate membrane receptors, initiating signaling cascades that modulate both nSR-dependent and -independent pathways [31,32]. Particularly, estrogens signal through the membrane G protein-coupled estrogen receptor (GPER, also known as GPR30) [33], while P4 acts via the P4 receptor membrane components (PGRMC) [34] and the class II members of the P4 and adipoQ receptor (PAQR) family [35,36].

PGR has two main isoforms: PGR-A and PGR-B, each regulated by distinct promoter regions and translation start sites [37]. Both isoforms are essential for normal endometrium function, with their expression varying through the menstrual cycle [38]. PGR-A and PGR-B expression is elevated proliferative and early secretory phases but decreases markedly in functional glands during the mid and late secretory phases. In contrast, stromal expression of both isoforms in the functionalis and basalis layers remains constant through the cycle in human endometrial biopsies [38].

Studies using chromatin immunoprecipitation sequencing (ChIP-seq) and RNA sequencing (RNA-seq) in E2- and progestin-treated ESCs have revealed that PGR-B regulates a broader spectrum of endometrial receptivity genes compared to PGR-A, although both isoforms contribute to in vitro decidualization [39].

Initial studies demonstrated that eutopic ESCs isolated from endometriosis patients show impaired in vitro differentiation compared to cells from women without the disease [40]. Additionally, Nisolle and Donnez documented differential expression patterns of ESRs and PGR isoforms between eutopic and ectopic endometrium [40].

Recent evidence demonstrates altered *PGR* expression in the endometrium of endometriosis patients. Attia et al. reported undetectable PGR-B and markedly reduced PGR-A levels in ectopic tissue compared to eutopic endometrium [25]. Igarashi et al. reported a decreased PGR-B: PGR-A ratio in eutopic endometrium from infertile women with endometriosis [41]. Similarly, Shen et al. observed reduced PGR-B protein levels in infertile endometriosis patients compared to infertile controls [42]. Additionally, Wölfler et al. reported decreased *PGR-B* expression during the mid to late secretory phase in the eutopic endometrium of women with endometriosis, suggesting that impaired PGR signaling may contribute to decidualization defects in these patients [43].

PGR isoforms expression is regulated by promoter-specific DNA methylation. Rocha et al. demonstrated that while *PGR-A* promoter DNA methylation remains unchanged, the *PGR-B* promoter shows increased DNA methylation levels in the eutopic endometrium from endometriosis infertile patients compared to infertile women without the disease during the secretory phase [44]. This increase in DNA methylation and the decreased *PGR-B* expression reported in previous studies may contribute to impaired endometrial receptivity and function in endometriosis (Figure 1A).

MicroRNAs (miRNAs) are a class of small non-coding regulatory RNAs involved in post-transcriptional gene regulation [45]. miR-194-3p shows elevated expression in the eutopic endometrium from endometriosis patients during the mid-secretory phase. In vitro studies demonstrate that miR-194-3p directly regulates *PGR* expression through interaction with its 3′-untranslated region, with miR-194-3p mimics suppressing and inhibitors enhancing *PGR* levels. Moreover, miR-194-3p overexpression impairs ESC decidualization [46], suggesting that this miRNA contributes to P4 resistance and decidualization defects in the eutopic endometrium of women with endometriosis.

Vázquez-Martínez et al. reported reduced *PAQR7* and *PAQR5* in both eutopic and ectopic endometrium from endometriosis patients compared to control endometrium. In contrast, *PAQR8* and *PAQR6* expression was decreased exclusively in the eutopic endometrium. However, reduced PAQR7 and PAQR8 protein levels were only observed in the ectopic endometrium [36]. This downregulation of PAQRs in endometriosis may contribute to the P4 resistance observed in these patients (Figure 1B). Further studies are needed to determine whether PAQR expression is dysregulated in the eutopic endometrium of infertile endometriosis patients.

On the other hand, endometriosis is an estrogen-dependent disorder, with E2 being essential for ectopic tissue [20]. In endometriosis women, E2 reaches both eutopic and ectopic endometrium through circulation but is predominantly synthetized locally due to elevated expression of key steroidogenic enzymes: aromatase (CYP19A1, the rate-limiting enzyme in estrogen biosynthesis) and steroidogenic acute regulatory protein (StAR, mediates cholesterol transport to the inner mitochondrial membrane). This local steroidogenic capacity contrasts with normal endometrium, which shows almost undetectable levels of these biosynthetic enzymes [47,48]. In vitro studies demonstrated that 8-Br-cAMP treatment of ESCs from the eutopic endometrium of endometriosis patients increases CYP19A1 expression while decreasing HSD3B1 (which encodes the pregnenolone to P4 converting enzyme). Additionally, 8-Br-cAMP upregulated HSD17B1 (encoding the estrone to E2 converting enzyme) in endometriotic ESCs [49]. These alterations in steroidogenic enzyme expression promote an estrogen-dominant environment within the endometrium, potentially contributing to impaired decidualization in endometriosis.

Local E2 accumulation contributes to lesion development and progression through the activation of ESRs [50]. In healthy endometrium, ESR1 expression shows cyclic variation, with higher levels during the proliferative phase compared to the secretory phase (in both functional and basalis stroma), while ESR2 expression remains relatively constant throughout the menstrual cycle [38]. This pattern, combined with significantly higher ESR1 expression compared to ESR2, suggests a predominant role for ESR1 in mediating endometrial estrogen responses [51].

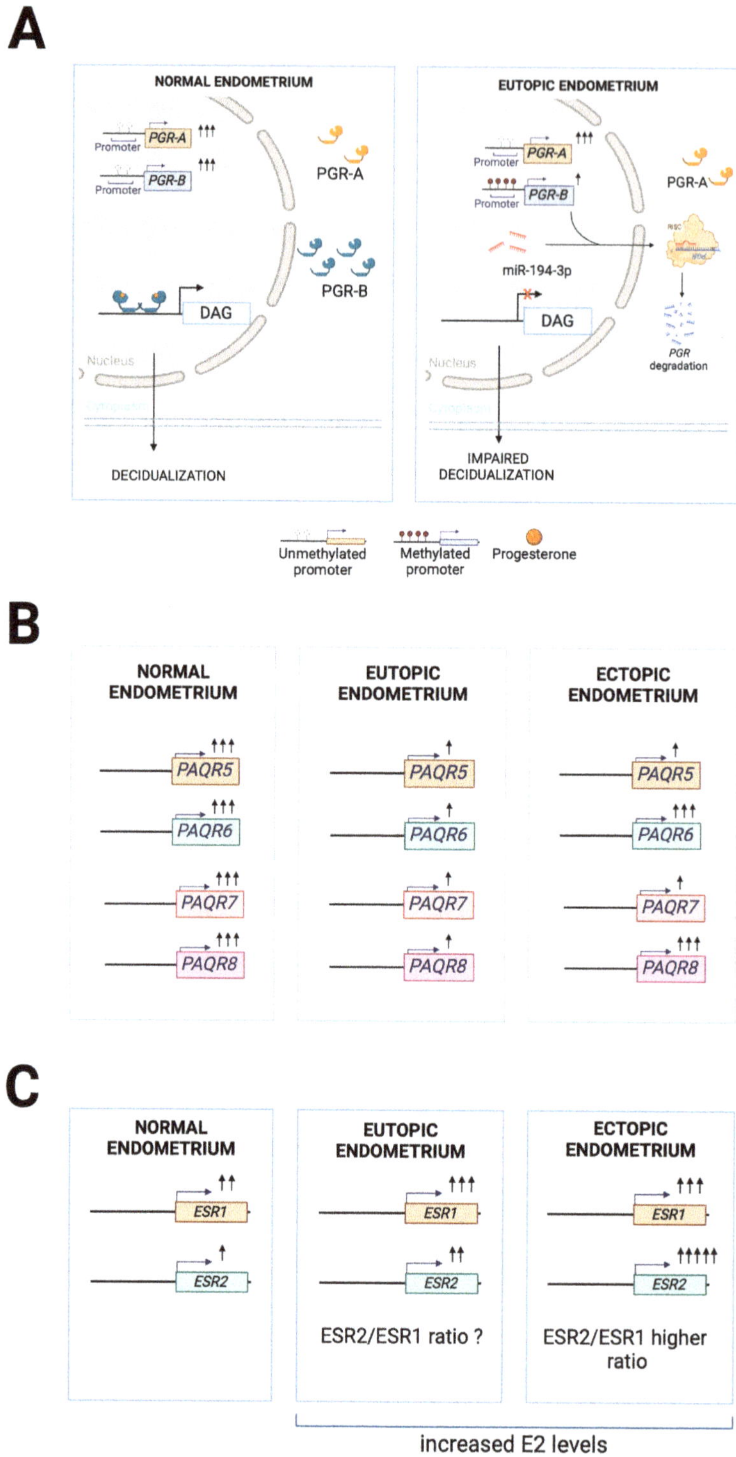

Figure 1. Molecular Alterations in Sex-Steroid Hormone Receptors and Signaling in Endometriosis. (**A**) Progesterone receptor (PGR) signaling in the normal and eutopic endometrium of endometriosis

patients. In a healthy endometrium, PGR-B and PGR-A isoforms are expressed, and ligand-bound PGR-B regulates the expression of decidualization-associated genes (DAGs). In endometriosis, *PGR-B* promoter hypermethylation and miR-194-3 mediated degradation reduce its protein levels, impairing decidualization. (**B**) Membrane progesterone receptor (PAQR) expression patterns across endometrial tissues. Differential expression of membrane progesterone receptors in normal endometrium compared to eutopic and ectopic endometriotic tissues. Expression levels are represented by the number of arrows. (**C**) Estrogen receptor dysregulation in endometriosis. *ESR1* and *ESR2* show increased expression in both eutopic and ectopic endometrial tissues in endometriosis patients compared to control endometrium, with ectopic lesions exhibiting the highest ESR2/ESR1 ratio. However, whether the ESR2/ESR1 ratio is altered in eutopic endometrium remains to be determined.

ESR1 and *ESR2* expression is upregulated in the eutopic and ectopic endometrium from endometriosis patients compared to control endometrium [52], with ectopic lesions showing higher expression of both genes compared to eutopic tissue [51]. Dysregulation of the ESR2/ESR1 ratio in ectopic endometrial stromal cells has been implicated in endometriosis pathogenesis and disease severity [53] (Figure 1C). Further investigation of the ESR2/ESR2 ratio in eutopic endometrium from endometriosis patients may reveal its potential role in disease pathogenesis and progression.

An enhanced understanding of endometriosis-associated molecular mechanisms provides crucial insights for developing hormone signaling-targeted therapies. Specifically, elucidating progesterone resistance mechanisms will facilitate the development of selective progesterone receptor modulators that account for PGR isoform-specific alterations and epigenetic regulation of receptor expression. Novel therapeutic strategies targeting membrane progesterone receptors could include PAQR family-specific modulators and pathway-selective interventions. Additionally, interventions aimed at normalizing estrogen signaling through ESR2/ESR1 ratio modulation, regulation of local estrogen production, and integration with progesterone signaling pathways represent promising therapeutic approaches. These mechanistic insights will enable the development of more targeted and effective treatments for endometriosis-associated infertility.

4. Dysregulation of *IGFBP1* and *PRL* Expression During Decidualization in Endometriosis

Multiple molecular and genetic studies have identified aberrant regulation of decidualization pathways in endometriosis. In this section, we focus on *IGFBP1* and *PRL* expression dysregulation during the decidualization of ESCs from patients with endometriosis (Table 1).

Minici et al. demonstrated reduced decidualization capacity in eutopic ESCs from endometriosis patients, evidenced by significantly reduced IGFBP1 and PRL secretion compared to control ESCs following E2 and 6α-methyl-17α-hydroxyprogesterone acetate (MPA) treatment for 7 and 13 days [17]. Although extended treatment (16 days) resulted in increased PRL secretion compared to controls, suggesting potential recovery of decidualization capacity in vitro [17], these findings must be interpreted cautiously given the absence of in vivo regulatory factors. Notably, treatment of control ESCs with peritoneal fluid from endometriosis patients (EPF), characterized by elevated cytokines, growth factors, and adhesion molecules, reproduced the decidualization defects [17]. Interestingly, when control ESCs were treated with EPF, they showed similar defects in decidualization markers. This suggests that both intrinsic molecular alterations and the inflammatory microenvironment contribute to impaired endometrial function in endometriosis.

Klemmt et al. further demonstrated reduced decidualization capacity in both ectopic and eutopic ESCs from endometriosis patients, with decreased PRL and IGFBP1 secretion following 8-Bromo-cAMP (8-Br-cAMP) treatment [24]. Similarly, Barragán et al. reported undetectable IGFBP1 secretion in endometrial mesenchymal stem cells (eMSCs) and endometrial stromal fibroblasts progenitors (eSFs) from endometriosis patients following a 14-day E2 and P4 treatment, contrasting with significant IGFBP1 secretion in cells derived from women without endometriosis [19]. Studies using menstrual effluent-derived stromal fibroblast cells (dME-SFCs) derived from endometriosis patients and controls have demonstrated reduced IGFBP1 secretion in endometriosis group at 6, 24, and 48 h post-8-Br-cAMP stimulation [54], with similar findings reported by Nayyar et al. at 24 h [55]. Single-cell RNA sequencing analysis by Shih et al. revealed significant *IGFBP1* expression in unstimulated control dME-SFCs compared to endometriotic cells [56].

Table 1. Effects of different in vitro decidualization treatments on IGFBP1 and PRL secretion/expression in eutopic and control * endometrial stromal cells.

Cells Type	Treatment	Days of Treatment	Results	References
ESCs	1×10^{-8} M E2 2×10^{-7} M MPA	6 and 13 days	Decreased secretion of IGFBP1 and PRL in eutopic endometrium compared to control.	Minici et al. [17]
		16 days	Increased secretion of PRL in eutopic endometrium compared to control.	
ESCs	5×10^{-4} M 8-Br-cAMP	3 to 20 days	Decreased secretion of IGFBP1 and PRL in eutopic endometrium compared to control.	Klemmt et al. [24]
eMSC and eSF	1×10^{-8} M E2 1×10^{-6} M P4	14 days	Undetectable expression of IGFBP1 in eutopic endometrium.	Barragan et al. [19]
ESCs	1×10^{-6} M P4	14 days	Increased expression of *IGFBP1* and *PRL* in both eutopic and control endometrium.	Aghajanova et al. [49]
	5×10^{-4} M 8-Br-cAMP	14 days	Decreased expression of *IGFBP1* and *PRL* in eutopic endometrium compared to control.	
ESCs	3.6×10^{-8} M E2 1×10^{-6} M P4 1×10^{-4} M db-cAMP	6 and 8 days	Decreased expression of *IGFBP1* and *PRL* in eutopic endometrium compared to control.	Su et al. [57]
ESCs	5×10^{-4} M 8-Br-cAMP 1×10^{-6} M MPA/dienogest	12 days	No significant changes in *IGFBP1* and *PRL* expression between eutopic and control endometrium.	Tsuno et al. [58]
dME-SFCs	5×10^{-4} M 8-Br-cAMP	6, 24 and 48 h	Decreased secretion of IGFBP1 in eutopic compared to control.	Warren et al. [54]
dME-SFCs	5×10^{-4} M 8-Br-cAMP	24 h	Decreased secretion of IGFBP1 in eutopic compared to control.	Nayyar et al. [55]

* Eutopic endometrium refers to the endometrium properly located in the uterus in patients with endometriosis, and the endometrium from women without the disease is referred to control endometrium.

In ESCs isolated from both endometriosis patients and controls, P4 treatment (14 days) upregulates *IGFBP1* and *PRL* expression. However, when treated with 8-Br-cAMP (96 h), control ESCs show significantly higher expression of both markers compared to endometriotic ESCs [49], suggesting PKA pathway involvement in this differential response. Similarly, Su et al. demonstrated that combined E2, P4, and dibutryl-cAMP (db-cAMP) treatment for 8 days resulted in significantly higher *IGFBP1* and *PRL* expression in control compared to eutopic (endometriosis) ESCs [57]. Tsuno et al. reported that while *PRL* and

IGFBP1 expression remained unchanged in eutopic ESCs from women with endometriosis treated with db-cAMP and medroxyprogesterone (MPA) or dienogest (12 days), ectopic ESCs showed significantly lower expression compared to both eutopic (endometriosis) and control cells [58]. These variable results highlight the need for standardized in vitro decidualization protocols that better reflect physiological conditions.

Collectively, these studies demonstrate impaired decidualization capacity in eutopic ESCs from endometriosis patients, characterized by reduced expression and secretion of the decidualization markers PRL and IGFBP1. While more studies on eutopic (endometriosis) ESCs and standardized protocols are needed for more reproducible results, current evidence indicates that both intrinsic molecular alterations and environmental factors contribute to defective decidualization in the endometrium of women with endometriosis.

5. Molecular and Cellular Disruptions of Decidualization in Endometriosis: Altered Gene Expression, Pathway Dysregulation, and Inflammatory Influences

Klemmt et al. demonstrated delayed morphological transition during in vitro decidualization in eutopic ESCs from endometriosis patients, requiring 6 days of 8-Br-cAMP treatment compared to 3 days in ESCs from women without the disease [24,59], supporting the presence of molecular defects affecting decidualization capacity. This altered temporal response may have significant implications, given that successful implantation requires precise temporal regulation of endometrial receptivity.

Recent molecular analyses revealed that E2, P4, and cAMP-stimulated eutopic ESCs from endometriosis patients exhibit distinct transcriptional signatures compared to control cells [59]. Major dysregulation was observed in the TGFβ signaling pathway and its key regulators, including SMAD4 and bone morphogenetic proteins (BMPs) [59]. Typically, BMPs signal through SMAD1/5/4 and TGFβ through SMAD2/3/4 [60]. In this context, eutopic ESCs show altered SMAD4 and H3K27ac binding patterns during decidualization (Figure 2). These findings suggest that impaired TGFβ-mediated transcriptional responses are SMAD4-dependent. Notably, exogenous BMP2 treatment restored the expression of the downregulated decidualization markers IGFBP1 and FOXO1 in both eutopic (endometriosis) ESCs and epithelial assembloids [59].

FOXO1, a transcription factor that regulates cellular metabolism, cell-cycle progression, oxidative stress response, and apoptosis [61], plays a crucial role in decidualization, such as the regulation of *IGFBP1* and *PRL* expression [62]. The eutopic endometrium of endometriosis patients exhibit reduced *FOXO1* expression compared to controls, suggesting that its dysregulation contributes to impaired decidualization [57]

In endometriosis, multiple pathways converge to disrupt FOXO1 function. Su et al. demonstrated downregulation of Notch signaling pathway genes (*NOTCH1*, *NOTCH4*, *JAGGED2*, and *DLL4*) in both the eutopic endometrium and in vitro decidualized eutopic ESCs of endometriosis patients. Moreover, *NOTCH1* silencing reduced *FOXO1* expression more markedly in ESCs from endometriosis patients compared to controls [57], indicating that impaired Notch signaling contributes to decidualization defects through FOXO1 downregulation (Figure 2). Kang et al. revealed elevated levels of phosphorylated AKT1 (pAKT1), Calpain-7 (CAPN7), and phosphorylated FOXO1 (Ser319) in the endometrium of endometriosis patients compared to controls. Their findings suggest that CAPN7 promotes AKT1 phosphorylation, leading to FOXO1 phosphorylation and nuclear exclusion, thereby inhibiting decidualization (Figure 2B) [63]. Additionally, NEK2 is upregulated in both eutopic and ectopic endometrium from endometriosis patients compared to controls, correlating with decreased FOXO1 levels [64]. NEK2 phosphorylates FOXO1 at Ser184, reducing protein stability and promoting cell proliferation while impairing decidualization (Figure 2B) [64].

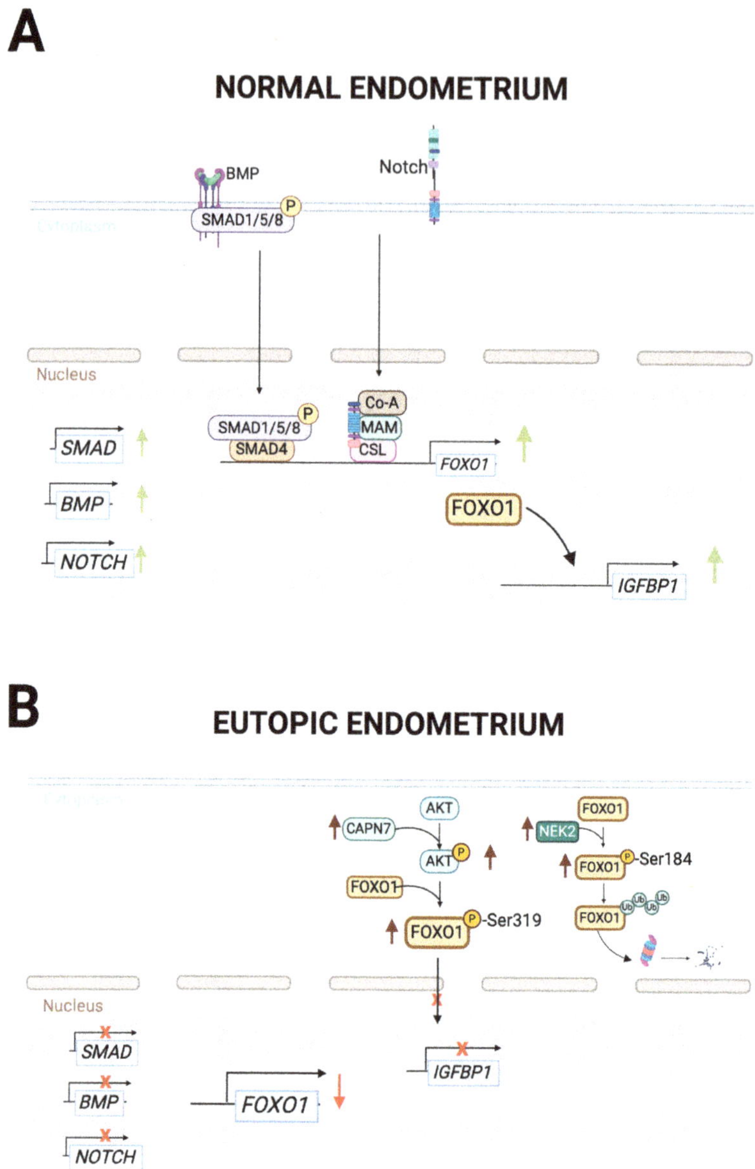

Figure 2. FOXO1 Regulation in the Normal and Eutopic Endometrium of Endometriosis Patients. Signaling pathways involved in the regulation of FOXO1 during in vitro decidualization in ESCs from control endometrium (**A**) and from eutopic endometrium of endometriosis patients (**B**). (**A**) During in vitro decidualization, *BMP*, *SMAD*, and *NOTCH* genes are upregulated. BMP-SMAD and Notch signaling independently promote *FOXO1* transcription through SMAD and NICD (Notch Intracellular Domain) activation, respectively. FOXO1 then regulates decidualization

markers, including *IGFBP1*, enabling normal endometrial function. (**B**) Multiple pathway dysregulation reduces FOXO1 levels during decidualization in the eutopic endometrium of endometriosis patients. Downregulation of SMAD-BMP and Notch signaling decreases *FOXO1* expression. Enhanced CAPN7 and pAKT1 activity promotes FOXO1 phosphorylation (Ser319), triggering nuclear exclusion. Overexpression of NEK2 induces FOXO1 phosphorylation (Ser184) that, in turn, promotes ubiquitination and proteasomal degradation, ultimately impairing decidualization.

Collectively, these studies demonstrate that reduced FOXO1 expression and the associated defective decidualization in eutopic endometrium of endometriosis patients results from multiple dysregulated pathways, including impaired Notch signaling and enhanced CAPN7 and NEK activity [64].

Connexin 43 (Cx43), a gap junction protein essential for cellular communication, plays a critical role in endometrial receptivity and decidualization through the regulation of angiogenesis [65,66]. Yu et al. demonstrated reduced Cx43 expression in the eutopic endometrium in endometriosis patients. Furthermore, E2, P4, and cAMP-induced decidualization of eutopic ESCs from endometriosis patients for 7 days showed impaired epithelioid transformation and significantly reduced Cx43 protein upregulation compared to controls [67]. This gap junction dysfunction may disrupt intercellular transport of essential small metabolites (<1 kDa), potentially affecting DSC morphological transformation and endometrial receptivity through altered angiogenic processes [65,66]. Interestingly, under the previously mentioned decidualization protocol, IL-1β suppresses Cx43, PRL, and VEGF expression through ERK1/2 and p38 MAP kinase pathways [68]. Notably, the eutopic endometrium of women with endometriosis exhibits elevated levels of inflammatory cytokines, including IL-1β, TNF-α, and IL-8 compared to the control [69], suggesting that this pro-inflammatory environment contributes to impaired decidualization in endometriosis.

IL-11 expression is reduced in the eutopic endometrium of endometriosis patients compared to controls [70]. It has been demonstrated that IL-11 plays essential roles in decidualization in both mouse and human systems, although its regulation differs between species [71,72]. In mice, IL-11 expression increases post-implantation, while in humans, P4 drives its upregulation during the secretory phase of the menstrual cycle [73,74]. Despite these regulatory differences, IL-11 dysregulation is associated with impaired decidualization in both species [74]. Karpovich et al. demonstrated reduced IL-11 levels in eutopic ESCs during in vitro decidualization with 8-Br-cAMP compared to controls, correlating with decreased PRL and IGFBP1 levels [75]. This reduction in IL-11 expression in the eutopic endometrium, likely resulting from progesterone resistance and altered PKA signaling, may contribute to endometriosis-associated infertility through multiple mechanisms: impaired decidualization, defective implantation, compromised placentation, altered Natural Killer cell differentiation, and decreased decidual cell survival [75–81]. However, the precise mechanisms linking IL-11 signaling to decidualization in the eutopic endometrium of endometriosis patients require further investigation [17].

As previously established, environmental factors may also contribute to impaired decidualization in endometriosis. Minici et al. demonstrated that EPF suppresses PRL secretion in decidualizing ESCs from controls treated with E2+MPA for 16 days [17]. Given the elevated TNF-α levels in EPF [81], the addition of soluble TNF-α receptor 1 (sTNFR-1) partially rescued PRL secretion in eutopic (endometriosis) ESCs, suggesting that TNF-α mediates the inhibitory effects of EPF [17].

Barragan et al. also reported that fibroblasts of ESCs from the eutopic endometrium of endometriosis patients exhibit a distinct pro-inflammatory transcriptional profile compared to controls [19]. Collectively, these studies demonstrate that both impaired Cx43 expression and elevated pro-inflammatory mediators contribute to decidualization defects in endometriosis. These molecular alterations represent potential therapeutic targets, though additional research is needed to fully characterize the pathways involved and develop targeted interventions.

Tiberi et al. demonstrated reduced expression of prokineticin (PROK1), a critical mediator of implantation and decidualization, in the endometrium of women with endometriosis compared to controls [82]. They also reported that during in vitro decidualization with E2+MPA, gene expression of both *PROK1* and *PGR* was undetectable in eutopic (endometriosis) ESCs at 72 and 96 h, while control ESCs showed significant upregulation. However, after extended treatment (16 days), gene expression patterns were similar in both groups [82]. These findings support the idea that eutopic ESCs from endometriosis patients display an initial resistance to decidualization signals but can eventually respond to prolonged hormonal stimulation.

6. Epigenetic Mechanisms and Their Role in Endometrial Dysfunction in Endometriosis

Epigenetics encompasses heritable modifications in gene expression mediated through alterations in chromatin structure without changes in DNA sequence [83]. These regulatory mechanisms operate through DNA base modifications, chromosomal remodeling, and modifications at multiple molecular levels, influenced by both genetic and environmental factors. These modifications can be transmitted across generations and are implicated in various pathological conditions [83,84].

Recent studies have focused on characterizing epigenetic dysregulation in endometriosis. Here, we examine its contribution to impaired decidualization (Figure 3 and Table 2). The participation of DNA methylation in the regulation of *PGR* isoforms expression was described in the "Hormonal Signaling, Dysregulation and Impaired Decidualization in Endometriosis" section [39,41,43,44,46,85–87].

Histone modifications regulate gene expression through alterations in chromatin structure and nucleosome dynamics [88]. Eutopic and ectopic endometrium exhibit distinct histone methylation patterns at H3K4, H3K9, and H3K27 compared to control endometrium [89–91]. These modifications show menstrual cycle-dependent changes that may contribute to impaired decidualization [92] and endometriosis-associated infertility.

Zang et al. demonstrated elevated H3K27me3 and H3K9me3 levels in both ectopic and eutopic endometrium of endometriosis patients compared to control endometrium [89]. These epigenetic alterations show lesion-specific patterns, with deep infiltrating endometriosis (DIE) exhibiting higher levels of EZH2, H3K27me3, and H3K9me3 compared to ovarian endometriomas (OMA), suggesting progressive epigenetic reprogramming [93]. However, the role of these histone modifications in the decidualization of eutopic ESCs of women with endometriosis remains to be elucidated.

In a healthy endometrium, EZH2 levels progressively decrease throughout the menstrual cycle. This reduction is paralleled in ESCs during in vitro decidualization, accompanied by decreased H3K27me3 enrichment at *PRL* and *IGFBP1* transcription start sites in response to 8-Br-cAMP and MPA [94]. Simultaneously, H3K27ac enrichment increases at a distal enhancer region upstream of *IGFBP1*, modulating its expression [95]. *IGFBP1* upregulation involves cAMP-mediated recruitment of C/EBPβ, FOXO1, and p300 to its enhancer region. Notably, C/EBPβ inhibition disrupts multiple regulatory mechanisms, including H3K27ac modification, chromatin remodeling, and p300 recruitment at the *IGFBP1* enhancer [95,96].

Endometriotic lesions exhibit elevated EZH2 and H3K27me3 levels [93], a characteristic influenced by the hypoxic microenvironment. Hypoxia induces *EZH2* expression in eutopic (endometriosis) ESCs, contributing to impaired decidualization [97]. Lin et al.

demonstrated that in normal ESCs, the N6-Methyladenosine (m^6A) reader YTHDF2 targets *EZH2* mRNA for degradation, reducing H3K27me3 levels at the *IGFBP1* promoter and enabling *IGFBP1* expression during decidualization. However, hypoxia upregulates the m^6A demethylase ALKBH5, which stabilizes *EZH2* mRNA by removing m^6A modifications, consequently suppressing *IGFBP1* expression. Supporting these findings, mouse studies demonstrate that *Ezh2* depletion enhances endometrial receptivity and fertility through *IGFBP1* upregulation [98].

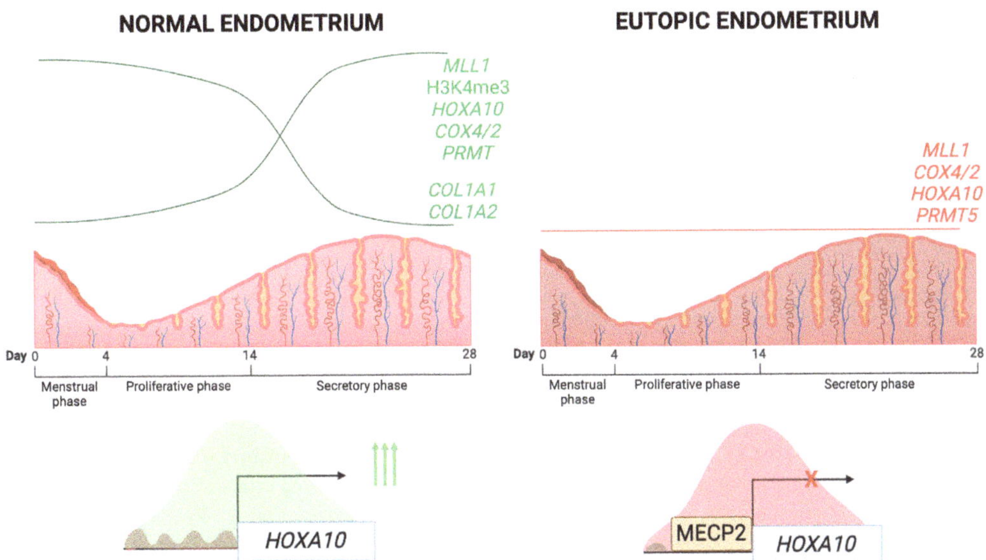

Figure 3. Epigenetic regulation of gene expression in normal and eutopic endometrium tissues of endometriosis patients. The menstrual cycle is divided into three phases: menstrual, proliferative, and secretory. Normal endometrium (**left panel**): *MLL1*, *HOXA10*, *COX4/2*, *PRMT5*, and H3K4me3 levels show progressive upregulation during the menstrual cycle, while *COL1A1* and *COL1A2* maintain low expression levels. During the secretory phase, *HOXA10* expression is enhanced by a transcriptional active chromatin state at its promoter. Eutopic endometrium from endometriosis patients (**right panel**).

Prostaglandin E2 (PGE2) signaling through EP2 and EP4 receptors cooperates with steroid hormones via PKA signaling to induce decidualization in both human ESCs and rodent models [10,99–104]. In endometriosis patients, PGE2 levels are elevated in peritoneal fluid [105], with increased PGE2 production and cyclooxygenase-2 (COX-2) expression in both ectopic and eutopic endometrium compared to healthy endometrium [106]. Remarkably, selective COX-2 inhibitors reduce endometriosis-associated pain [107–109]. EP2 and EP4 receptors are expressed in various endometriosis lesions (ovarian, adenomyotic, and peritoneal), and their specific antagonists also alleviate endometriosis-associated pain [110,111]. The eutopic endometrium of endometriosis patients exhibits increased expression of COX-2, EP2, and EP4 compared to controls [112]. Notably, selective EP2/EP4 antagonists suppress cAMP signaling, proliferation, proinflammatory cytokine production, and aromatase expression in endometriotic ESCs [110].

Arosh et al. demonstrated that pharmacological inhibition of EP2/EP4 in endometriotic cells decreases H3K9 and H3K27 methylation while increasing their acetylation [113]. Additionally, EP2/EP4 inhibition selectively increases H3K4 methylation in stromal but not epithelial endometriotic cells [113]. These findings suggest that PGE2 signaling modulates chromatin state in endometriosis, leading to the establishment of aberrant euchromatin states. The complex interaction between PGE2, proinflammatory cytokines, estrogen synthesis, and lesion innervation identify EP receptors as promising therapeutic targets. While these pathways likely impact fertility in endometriosis patients, their precise role in eutopic endometrial decidualization requires further investigation.

Key regulatory genes (*MLL1, COX4/2, HOXA10, PRMT5*) show reduced expression compared to the normal endometrium. *HOXA10* regulation is particularly altered during the secretory phase, characterized by increased MeCP2 occupancy and a transcriptional repressive chromatin state at its promoter, resulting in transcriptional repression.

Epigenetic regulation requires precise coordination of histone-modifying enzymes, including histone lysine methyltransferases (KMTs) [114], histone acetyltransferases (HATs) [115], or histone deacetylases (HDACs). The dysregulation of these enzymes has been implicated in various pathologies, including endometrial cancer and endometriosis.

Mixed lineage leukemia 1 (MLL1), which catalyzes H3K4 trimethylation at gene promoters [116], shows endometrial cycle-dependent expression in the normal endometrium but is downregulated in the eutopic endometrium of endometriosis patients, correlating with reduced H3K4me3 levels [90]. During in vitro decidualization, *MLL1* expression and H3K4me3 levels increase in control ESCs, an effect blocked by the P4 antagonist mifepristone (PGR antagonist). *MLL1* silencing prevents *IGFBP1* and *PRL* induction and inhibits characteristic decidual morphological changes, demonstrating its essential role in decidualization [117]. Additionally, PGR-P4 complexes bind directly to the *MLL1* promoter, demonstrating transcriptional regulation of *MLL1* by P4 signaling. In the eutopic endometrium of endometriosis patients, decreased *PGR* expression correlates with reduced MLL1 levels, resulting in diminished H3K4me3 enrichment at promoters of decidualization-related genes [117].

Cytochrome c oxidase COX4/2, normally upregulated during the secretory phase, remains suppressed throughout the menstrual cycle in the eutopic endometrium of patients with endometriosis [117]. In vitro decidualization induces MLL1 recruitment to the *COX4/2* promoter, increasing H3K4me3 enrichment in immortalized ESCs. *MLL1* silencing prevents both MLL1 recruitment and H3K4me3 enrichment, correlating with COX4/2 downregulation in endometriosis [117]. This dysregulation may promote a hypoxic endometrial microenvironment through reduced COX4/2-mediated oxygen affinity.

Cai et al. demonstrated reduced expression of protein arginine methyltransferase 5 (PRMT5) in the eutopic endometrium (only in the stroma compartment) of endometriosis patients compared to fertile controls during the mid-secretory phase of the menstrual cycle. While *PRMT5* expression increases during normal in vitro decidualization, pharmacological inhibition with GSK591 prevents *IGFBP1* induction and PRL secretion, establishing PRMT5's essential role in ESC decidualization. Transcriptomic analysis revealed that decreased PRMT5 activity enhances NF-κB signaling through nuclear p65 translocation, a feature observed in endometriotic tissue. Notably, *PRMT5* overexpression restores *IGFBP1* and *PRL* expression in eutopic ESCs of endometriosis patients [118], suggesting that PRMT5 dysregulation contributes to impaired decidualization in endometriosis. These findings identify histone-modifying enzymes as potential therapeutic targets for improving decidualization in endometriosis-associated infertility.

HDAC3 shows reduced protein levels in the eutopic endometrium of infertile women with endometriosis compared to controls. *HDAC3* silencing in ESCs from women without endometriosis impairs in vitro decidualization, evidenced by decreased *IGFBP1* and *PRL* expression. Additionally, while *COL1A1* and *COL1A2* are normally downregulated during decidualization, HDAC3 silencing induces their expression [119], suggesting that HDAC3 deficiency contributes to decidualization defects in endometriosis.

Table 2. Epigenetic alterations in the eutopic endometrial cells of endometriosis patients.

Epigenetic Mechanism	Alteration in Endometriosis	Effect on Gene Expression and Decidualization	Reference
DNA methylation of *PGR-B* Promoter	Increased methylation in eutopic endometrium during the secretory phase in endometriosis	Associated with decreased *PGR-B* expression, potentially affecting decidualization and P4 responsiveness	Rocha-Junior et al. [44]
H3K27me3 and H3K9me3 levels	Significantly higher levels in eutopic endometrium from endometriosis patients	Not reported	Zhang et al. [89]
EZH2, H3K27me3, and m^6A levels	Hypoxia induces ALKBH5 that promotes *EZH2* mRNA stability by removing m^6A and increases H3K27me3 levels.	Increased EZH2 and H3K27me3 levels that downregulate *IGFBP1* expression, impairing decidualization and contributing to eutopic endometrial dysfunction in endometriosis	Lin et al. [98]
MLL1 and H3K4me3 levels	Decreased *MLL1* and H3K4me3 levels in the eutopic endometrium of endometriosis patients	Possibly associated with lower enrichment of H3K4me3 in promoter regions of decidualization-associated genes such as *COX4/2*	Wen et al. [117]
HDAC3 levels	Lower HDAC3 levels in the endometrium of infertile women with endometriosis	Reduced HDAC3 levels may contribute to aberrant gene expression of decidualization-associated genes such as *IGFBP1* and *PRL*	Kim et al. [119]
SIRT1 levels	Significantly higher levels of SIRT1 in both eutopic ESCs and EECs of endometriosis patients.	Overexpression of SIRT1 may lead to suppressed *IGFBP1* and *PRL* expression during decidualization	Kim et al. [120] Yu et al. [121]
H3K9me3 and H3K9ac levels on *HOXA10* promoter	Decreased expression of *HOXA10* is associated with increased H3K9me3 and reduced H3K9ac levels in eutopic endometrium from endometriosis patients	Reduced HOXA10 expression may impair endometrial receptivity and decidualization, contributing to infertility in endometriosis patients	Samadieh et al. [122] Gui et al. [123]
PCAF levels	Abnormal expression of *PCAF* in eutopic endometrium of endometriosis patients	Reduced PCAF activity leads to decreased acetylation of HOXA10-regulated genes, impairing gene expression needed for endometrial receptivity and function.	Zhu et al. [124]

SIRT1, another histone modifier, is significantly upregulated in both the eutopic ESCs and EECs of endometriosis patients compared to controls [120,121]. While normal decidualization reduces SIRT1 expression, its overexpression in T-hESC suppresses *PRL* and *IGFBP1* expression [121]. These findings indicate that elevated SIRT1 levels in the eutopic endometrium of endometriosis patients may contribute to impaired decidualization and disease pathogenesis.

HOXA10, a key regulator of uterine development and endometrial differentiation, shows altered expression in endometriosis [125–129]. In healthy endometrium, *HOXA10* expression is regulated by sexual steroid hormones [130] and exhibits cycle-dependent patterns [131], with significant upregulation during the secretory phase correlating with elevated E2 and P4 levels. However, this physiological increase is absent in endometriosis [123,129]. This reduced *HOXA10* expression in the eutopic endometrium during the secretory phase correlates with specific epigenetic alterations: decreased H3K9 acetylation, increased H3K9 methylation, and enhanced MeCP2 enrichment at the *HOXA10* promoter [122]. Furthermore, dysregulation of p300/CREB-binding protein-associated factor (PCAF) impairs HOXA10-mediated gene regulation through altered acetylation patterns in the eutopic endometrium of women with endometriosis [124].

Together, these studies suggest that several epigenetic modifications contribute to impaired endometrial receptivity and decidualization in endometriosis. Further functional studies are needed to validate these findings and investigate additional histone modifications and epigenetic mechanisms in eutopic endometrium, as current research has primarily focused on ectopic lesions. Understanding these molecular mechanisms will suggest several promising therapeutic strategies, including targeted epigenetic interventions to modulate DNA methylation, the regulation of histone-modifying enzymes, and the development of selective EP2/EP4 receptor antagonists to normalize histone modifications. Future therapeutic approaches could incorporate tissue-specific delivery systems to maximize efficacy while minimizing off-target effects.

7. Therapeutic Approaches Targeting Dysregulated Pathways in the Eutopic Endometrium of Endometriosis Patients

As epigenetic changes are reversible, developing therapies targeting enzymes or molecules involved in altering methylome and histone patterns seems a viable alternative for treating diseases such as endometriosis [132], particularly as our understanding of epigenetic regulation in this disease continues to expand.

While clinical trials targeting the molecular pathways described in this review are currently lacking, some drugs have potential as epigenetic therapeutics for endometriosis-associated infertility.

Nucleoside and non-nucleoside DNMT inhibitors, which provide sustained epigenetic modulation, represent promising therapeutic candidates [133]. Specifically, selective DNMT3B inhibitors have been proposed as potential therapeutic targets for endometriosis treatment [133]. While further investigation in the context of endometriosis is needed, structure-based drug design approaches targeting DNMT3B have already identified promising compounds for cancer therapy, suggesting potential applications for other epigenetic disorders [134]. However, potential adverse effects must be considered, as demethylating agents such as 5-aza-20-deoxycytidine (5-Aza-dC) increase aromatase expression in patient-derived endometrial stromal cells, potentially promoting endometriotic-like characteristics [135].

HDAC inhibitors represent another promising therapeutic approach for endometriosis. These compounds modulate gene expression in endometriotic cyst stromal cells, inhibiting cell proliferation and inducing cell cycle arrest and apoptosis [136]. Specific HDAC1 inhibitors, particularly benzamides like Chinamid, warrant further investigation as potential therapeutic agents [137].

While various compounds have been evaluated in preclinical and clinical studies, most current therapeutic approaches for endometriosis focus on suppressing ectopic lesions or managing pain symptoms [138]. However, GnRH antagonists (Elagolix) and aromatase inhibitors (Letrozole) have emerged as promising treatments specifically addressing endometriosis-associated infertility [139]. Nevertheless, further research is required to develop targeted therapies that address the molecular mechanisms underlying decidualization defects in the eutopic endometrium.

8. Conclusions and Perspectives

Current evidence demonstrates that endometriosis is characterized by profound alterations in the decidualization of eutopic endometrium, involving dysregulation of steroid hormone signaling, transcriptional networks, and epigenetic mechanisms. The eutopic endometrium of endometriosis patients exhibits reduced *PGR* expression, particularly *PGR-B*. This progesterone resistance is further compounded by alterations in membrane progesterone receptors and disrupted estrogen signaling.

Decidualization defects are evidenced by impaired expression of key biomarkers such as IGFBP1 and PRL, altered cellular morphology, and dysregulated inflammatory responses. These abnormalities involve multiple pathways, including disrupted TGFβ signaling, altered gap junction communication through Cx43, and compromised epigenetic regulation through histone-modifying enzymes such as MLL1, PRMT5, and HDAC3.

While significant progress has been made in understanding decidualization defects in endometriosis, several key areas require further investigation, such as the comprehensive molecular characterization using advanced omics approaches, functional studies linking decidualization impairment to infertility, standardization of in vitro decidualization protocols, and investigation of potential therapeutic targets.

An enhanced understanding of these mechanisms will be crucial for developing targeted therapies to improve both fertility outcomes and disease management in endometriosis patients.

Author Contributions: A.M.R.-R. and S.K.G.-S.: investigation, writing, and editing; E.G.-G.: review, editing, and supervision; and E.R.V.-M.: conceptualization, review, editing, supervision, and funding acquisition. All authors have read and agreed to the published version of the manuscript.

Funding: A.M.R.-R. receives a Ph.D. scholarship from SECIHTI with CVU number 956173. We acknowledge SECIHTI (Grant Number A1-S-26749), INPer (Grant Number 571, 3000-20109-01-571-17), and UNAM-PAPIIT (Grant Number IA209520).

Institutional Review Board Statement: Not applicable.

Informed Consent Statement: Not applicable.

Data Availability Statement: No new data were created or analyzed in this study. Data sharing is not applicable to this article.

Conflicts of Interest: The authors declare no conflicts of interest.

Abbreviations

The following abbreviations are used in this manuscript:

8-Br-cAMP	8-Bromo-cAMP
BMP	Bone morphogenetic proteins
CAPN7	Calpain-7
ChIP-seq	Chromatin immunoprecipitation sequencing
Cx43	Connexin 43
DIE	Deep infiltrating endometriosis
DSCs	Decidual stromal cells
E2	Estradiol
ECSs	Endometrial stromal cells
EECs	Endometrial epithelial cells

eMSCs	Endometrial mesenchymal stem cells
eSFs	Endometrial stromal fibroblasts progenitors
EPF	Peritoneal fluid from endometriosis patients
ESR	Estrogen receptor
GPER	Membrane G protein-coupled estrogen receptor
IGFBP1	Insulin-like growth factor binding protein-1
KMTs	Histone lysine methyltransferases
HATs	Histone acetyltransferases
HDACs	Histone deacetylases
m^6A	RNA methyladenosine
MPA	6α-methyl-17α-hydroxyprogesterone acetate
MLL1	Mixed lineage leukemia 1
nSR	Nuclear steroid receptors
OMA	Ovarian endometriomas
PAQR	Progestin and adipoQ receptor family members
P4	Progesterone
PGE2	Prostaglandin E2
pAKT1	Phosphorylated AKT1
PRL	Prolactin
PRMT5	Protein arginine methyltransferase 5
PROK1	Prokineticin
PGR	Progesterone receptor
PGRMC	Membrane-associated P4 receptor components
RNA-seq	RNA sequencing
STFs	Stromal cells

References

1. Rosner, J.; Samardzic, T.; Sarao, M.S. *Physiology, Female Reproduction*; StatPearls Publishing: Treasure Island, FL, USA, 2021.
2. Wang, W.; Vilella, F.; Alama, P.; Moreno, I.; Mignardi, M.; Isakova, A.; Pan, W.; Simon, C.; Quake, S.R. Single-Cell Transcriptomic Atlas of the Human Endometrium during the Menstrual Cycle. *Nat. Med.* **2020**, *26*, 1644–1653. [CrossRef] [PubMed]
3. Lessey, B.A.; Young, S.L. What Exactly Is Endometrial Receptivity? *Fertil. Steril.* **2019**, *111*, 611–617. [CrossRef] [PubMed]
4. Singh, M.; Chaudhry, P.; Asselin, E. Bridging Endometrial Receptivity and Implantation: Network of Hormones, Cytokines, and Growth Factors. *J. Endocrinol.* **2011**, *210*, 5–14. [CrossRef] [PubMed]
5. Thiyagarajan, D.K.; Basit, H.; Jeanmonod, R. *Physiology, Menstrual Cycle*; StatPearls Publishing: Treasure Island, FL, USA, 2021.
6. Yilmaz, B.D.; Sison, C.A.M.; Yildiz, S.; Miyazaki, K.; Coon, V.J.; Yin, P.; Bulun, S.E. Genome-Wide Estrogen Receptor-α Binding and Action in Human Endometrial Stromal Cells. *F S Sci.* **2020**, *1*, 59–66. [CrossRef]
7. Gellersen, B.; Brosens, J.J. Cyclic Decidualization of the Human Endometrium in Reproductive Health and Failure. *Endocr. Rev.* **2014**, *35*, 851–905. [CrossRef]
8. Okada, H.; Tsuzuki, T.; Murata, H. Decidualization of the Human Endometrium. *Reprod. Med. Biol.* **2018**, *17*, 220. [CrossRef]
9. Gellersen, B.; Brosens, J. Cyclic AMP and Progesterone Receptor Cross-Talk in Human Endometrium: A Decidualizing Affair. *J. Endocrinol.* **2003**, *178*, 357–372. [CrossRef]
10. Yoshie, M.; Kusama, K.; Tamura, K. Molecular Mechanisms of Human Endometrial Decidualization Activated by Cyclic Adenosine Monophosphate Signaling Pathways. *J. Mamm. Ova Res.* **2015**, *32*, 95–102. [CrossRef]
11. Pittari, D.; Dalla Torre, M.; Borini, E.; Hummel, B.; Sawarkar, R.; Semino, C.; van Anken, E.; Panina-Bordignon, P.; Sitia, R.; Anelli, T. CREB3L1 and CREB3L2 Control Golgi Remodelling during Decidualization of Endometrial Stromal Cells. *Front. Cell Dev. Biol.* **2022**, *10*, 986997. [CrossRef]
12. Zhang, P.; Wang, G. Progesterone Resistance in Endometriosis: Current Evidence and Putative Mechanisms. *Int. J. Mol. Sci.* **2023**, *24*, 6992. [CrossRef]
13. Garrido-Gomez, T.; Castillo-Marco, N.; Clemente-Ciscar, M.; Cordero, T.; Muñoz-Blat, I.; Amadoz, A.; Jimenez-Almazan, J.; Monfort-Ortiz, R.; Climent, R.; Perales-Marin, A.; et al. Disrupted PGR-B and ESR1 Signaling Underlies Defective Decidualization Linked to Severe Preeclampsia. *eLife* **2021**, *10*, e70753. [CrossRef] [PubMed]
14. Bahia, W.; Finan, R.R.; Al-Mutawa, M.; Haddad, A.; Soua, A.; Janhani, F.; Mahjoub, T.; Almawi, W.Y. Genetic Variation in the Progesterone Receptor Gene and Susceptibility to Recurrent Pregnancy Loss: A Case-Control Study. *BJOG* **2018**, *125*, 729–735. [CrossRef] [PubMed]

15. Aghajanova, L.; Tatsumi, K.; Horcajadas, J.A.; Zamah, A.M.; Esteban, F.J.; Herndon, C.N.; Conti, M.; Giudice, L.C. Unique Transcriptome, Pathways, and Networks in the Human Endometrial Fibroblast Response to Progesterone in Endometriosis. *Biol. Reprod.* **2011**, *84*, 801–815. [CrossRef] [PubMed]
16. Burney, R.O.; Talbi, S.; Hamilton, A.E.; Kim, C.V.; Nyegaard, M.; Nezhat, C.R.; Lessey, B.A.; Giudice, L.C. Gene Expression Analysis of Endometrium Reveals Progesterone Resistance and Candidate Susceptibility Genes in Women with Endometriosis. *Endocrinology* **2007**, *148*, 3814–3826. [CrossRef] [PubMed]
17. Minici, F.; Tiberi, F.; Tropea, A.; Orlando, M.; Gangale, M.F.; Romani, F.; Campo, S.; Bompiani, A.; Lanzone, A.; Apa, R. Endometriosis and Human Infertility: A New Investigation into the Role of Eutopic Endometrium. *Hum. Reprod.* **2008**, *23*, 530–537. [CrossRef]
18. Klemmt, P.A.; Starzinski-Powitz, A. Molecular and Cellular Pathogenesis of Endometriosis. *Curr. Womens Health Rev.* **2018**, *14*, 106–116. [CrossRef]
19. Barragan, F.; Irwin, J.C.; Balayan, S.; Erikson, D.W.; Chen, J.C.; Houshdaran, S.; Piltonen, T.T.; Spitzer, T.L.B.; George, A.; Rabban, J.T.; et al. Human Endometrial Fibroblasts Derived from Mesenchymal Progenitors Inherit Progesterone Resistance and Acquire an Inflammatory Phenotype in the Endometrial Niche in Endometriosis. *Biol. Reprod.* **2016**, *94*, 118. [CrossRef]
20. Bulun, S.E.; Yilmaz, B.D.; Sison, C.; Miyazaki, K.; Bernardi, L.; Liu, S.; Kohlmeier, A.; Yin, P.; Milad, M.; Wei, J.J. Endometriosis. *Endocr. Rev.* **2019**, *40*, 1048–1079. [CrossRef]
21. Zondervan, K.T.; Becker, C.M.; Koga, K.; Missmer, S.A.; Taylor, R.N.; Viganò, P. Endometriosis. *Nat. Rev. Dis. Primers* **2018**, *4*, 9. [CrossRef]
22. Lessey, B.A.; Lebovic, D.I.; Taylor, R.N. Eutopic Endometrium in Women with Endometriosis: Ground Zero for the Study of Implantation Defects. *Semin. Reprod. Med.* **2013**, *31*, 109–124. [CrossRef]
23. Lamceva, J.; Uljanovs, R.; Strumfa, I. The Main Theories on the Pathogenesis of Endometriosis. *Int. J. Mol. Sci.* **2023**, *24*, 4254. [CrossRef] [PubMed]
24. Klemmt, P.A.B.; Carver, J.G.; Kennedy, S.H.; Koninckx, P.R.; Mardon, H.J. Stromal Cells from Endometriotic Lesions and Endometrium from Women with Endometriosis Have Reduced Decidualization Capacity. *Fertil. Steril.* **2006**, *85*, 564–572. [CrossRef] [PubMed]
25. Attia, G.R.; Zeitoun, K.; Edwards, D.; Johns, A.; Carr, B.R.; Bulun, S.E. Progesterone Receptor Isoform A but Not B Is Expressed in Endometriosis. *J. Clin. Endocrinol. Metab.* **2000**, *85*, 2897–2902. [CrossRef] [PubMed]
26. Zeitoun, K.; Takayama, K.; Sasano, H.; Suzuki, T.; Moghrabi, N.; Andersson, S.; Johns, A.; Meng, L.; Putman, M.; Carr, B.; et al. Deficient 17beta-Hydroxysteroid Dehydrogenase Type 2 Expression in Endometriosis: Failure to Metabolize 17beta-Estradiol. *J. Clin. Endocrinol. Metab.* **1998**, *83*, 4474–4480. [CrossRef]
27. Kim, M.; Park, H.J.; Seol, J.W.; Jang, J.Y.; Cho, Y.S.; Kim, K.R.; Choi, Y.; Lydon, J.P.; Demayo, F.J.; Shibuya, M.; et al. VEGF-A Regulated by Progesterone Governs Uterine Angiogenesis and Vascular Remodelling during Pregnancy. *EMBO Mol. Med.* **2013**, *5*, 1415–1430. [CrossRef]
28. Patel, B.; Peters, G.A.; Skomorovska-Prokvolit, Y.; Yi, L.; Tan, H.; Yousef, A.; Wang, J.; Mesiano, S. Control of Progesterone Receptor-A Transrepressive Activity in Myometrial Cells: Implications for the Control of Human Parturition. *Reprod. Sci.* **2018**, *25*, 214–221. [CrossRef]
29. Métivier, R.; Penot, G.; Hübner, M.R.; Reid, G.; Brand, H.; Koš, M.; Gannon, F. Estrogen Receptor-Alpha Directs Ordered, Cyclical, and Combinatorial Recruitment of Cofactors on a Natural Target Promoter. *Cell* **2003**, *115*, 751–763. [CrossRef]
30. Kastner, P.; Krust, A.; Turcotte, B.; Stropp, U.; Tora, L.; Gronemeyer, H.; Chambon, P. Two Distinct Estrogen-Regulated Promoters Generate Transcripts Encoding the Two Functionally Different Human Progesterone Receptor Forms A and B. *EMBO J.* **1990**, *9*, 1603–1614. [CrossRef]
31. Beato, M.; Wright, R.H.G.; Le Dily, F. 90 YEARS OF PROGESTERONE: Molecular Mechanisms of Progesterone Receptor Action on the Breast Cancer Genome. *J. Mol. Endocrinol.* **2020**, *65*, T65–T79. [CrossRef]
32. Pedroza, D.A.; Subramani, R.; Tiula, K.; Do, A.; Rashiraj, N.; Galvez, A.; Chatterjee, A.; Bencomo, A.; Rivera, S.; Lakshmanaswamy, R. Crosstalk between Progesterone Receptor Membrane Component 1 and Estrogen Receptor α Promotes Breast Cancer Cell Proliferation. *Lab. Investig.* **2021**, *101*, 733–744. [CrossRef]
33. Clusan, L.; Ferrière, F.; Flouriot, G.; Pakdel, F. A Basic Review on Estrogen Receptor Signaling Pathways in Breast Cancer. *Int. J. Mol. Sci.* **2023**, *24*, 6834. [CrossRef] [PubMed]
34. Velázquez Hernández, D.M.; Vázquez-Martínez, E.R.; Camacho-Arroyo, I. The Role of Progesterone Receptor Membrane Component (PGRMC) in the Endometrium. *Steroids* **2022**, *184*, 109040. [CrossRef]
35. Valadez-Cosmes, P.; Vázquez-Martínez, E.R.; Cerbón, M.; Camacho-Arroyo, I. Membrane Progesterone Receptors in Reproduction and Cancer. *Mol. Cell. Endocrinol.* **2016**, *434*, 166–175. [CrossRef] [PubMed]

36. Vázquez-Martínez, E.R.; Bello-Alvarez, C.; Hermenegildo-Molina, A.L.; Solís-Paredes, M.; Parra-Hernández, S.; Cruz-Orozco, O.; Silvestri-Tomassoni, J.R.; Escobar-Ponce, L.F.; Hernández-López, L.A.; Reyes-Mayoral, C.; et al. Expression of Membrane Progesterone Receptors in Eutopic and Ectopic Endometrium of Women with Endometriosis. *Biomed. Res. Int.* **2020**, *2020*, 2196024. [CrossRef] [PubMed]
37. Graham, J.D.; Clarke, C.L. Physiological Action of Progesterone in Target Tissues. *Endocr. Rev.* **1997**, *18*, 502–519. [CrossRef] [PubMed]
38. Mehasseb, M.K.; Panchal, R.; Taylor, A.H.; Brown, L.; Bell, S.C.; Habiba, M. Estrogen and Progesterone Receptor Isoform Distribution through the Menstrual Cycle in Uteri with and without Adenomyosis. *Fertil. Steril.* **2011**, *95*, 2228–2235.e1. [CrossRef]
39. Kaya, H.S.; Hantak, A.M.; Stubbs, L.J.; Taylor, R.N.; Bagchi, I.C.; Bagchi, M.K. Roles of Progesterone Receptor A and B Isoforms during Human Endometrial Decidualization. *Mol. Endocrinol.* **2015**, *29*, 882–895. [CrossRef]
40. Nisolle, M.; Donnez, J. Peritoneal Endometriosis, Ovarian Endometriosis, and Adenomyotic Nodules of the Rectovaginal Septum Are Three Different Entities. *Fertil. Steril.* **1997**, *68*, 585–596. [CrossRef]
41. Igarashi, T.M.; Bruner-Tran, K.L.; Yeaman, G.R.; Lessey, B.A.; Edwards, D.P.; Eisenberg, E.; Osteen, K.G. Reduced Expression of Progesterone Receptor-B in the Endometrium of Women with Endometriosis and in Cocultures of Endometrial Cells Exposed to 2,3,7,8-Tetrachlorodibenzo-p-Dioxin. *Fertil. Steril.* **2005**, *84*, 67–74. [CrossRef]
42. Shen, F.; Yan, C.; Liu, M.; Feng, Y.; Chen, Y. Decreased Expression of Mucin-1 in Endometriosis Endometrium Correlated with Progesterone Receptor B Involved in Infertility. *Arch. Gynecol. Obstet.* **2015**, *291*, 439–445. [CrossRef]
43. Wölfler, M.M.; Küppers, M.; Rath, W.; Buck, V.U.; Meinhold-Heerlein, I.; Classen-Linke, I. Altered Expression of Progesterone Receptor Isoforms A and B in Human Eutopic Endometrium in Endometriosis Patients. *Ann. Anat.* **2016**, *206*, 1–6. [CrossRef] [PubMed]
44. Rocha-Junior, C.V.; Da Broi, M.G.; Miranda-Furtado, C.L.; Navarro, P.A.; Ferriani, R.A.; Meola, J. Progesterone Receptor B (PGR-B) Is Partially Methylated in Eutopic Endometrium From Infertile Women With Endometriosis. *Reprod. Sci.* **2019**, *26*, 1568–1574. [CrossRef]
45. Chuang, J.C.; Jones, P.A. Epigenetics and MicroRNAs. *Pediatr. Res.* **2007**, *61*, 24–29. [CrossRef] [PubMed]
46. Pei, T.; Liu, C.; Liu, T.; Xiao, L.; Luo, B.; Tan, J.; Li, X.; Zhou, G.; Duan, C.; Huang, W. MiR-194-3p Represses the Progesterone Receptor and Decidualization in Eutopic Endometrium From Women With Endometriosis. *Endocrinology* **2018**, *159*, 2554–2562. [CrossRef] [PubMed]
47. Tsai, S.J.; Wu, M.H.; Lin, C.C.; Sun, H.S.; Chen, H.M. Regulation of Steroidogenic Acute Regulatory Protein Expression and Progesterone Production in Endometriotic Stromal Cells. *J. Clin. Endocrinol. Metab.* **2001**, *86*, 5765–5773. [CrossRef]
48. Kitawaki, J.; Noguchi, T.; Amatsu, T.; Maeda, K.; Tsukamoto, K.; Yamamoto, T.; Fushiki, S.; Osawa, Y.; Honjo, H. Expression of Aromatase Cytochrome P450 Protein and Messenger Ribonucleic Acid in Human Endometriotic and Adenomyotic Tissues but Not in Normal Endometrium. *Biol. Reprod.* **1997**, *57*, 514–519. [CrossRef]
49. Aghajanova, L.; Hamilton, A.; Kwintkiewicz, J.; Vo, K.C.; Giudice, L.C.; Jaffe, R.B. Steroidogenic Enzyme and Key Decidualization Marker Dysregulation in Endometrial Stromal Cells from Women with Versus Without Endometriosis. *Biol. Reprod.* **2009**, *80*, 105–114. [CrossRef]
50. Chantalat, E.; Valera, M.C.; Vaysse, C.; Noirrit, E.; Rusidze, M.; Weyl, A.; Vergriete, K.; Buscail, E.; Lluel, P.; Fontaine, C.; et al. Estrogen Receptors and Endometriosis. *Int. J. Mol. Sci.* **2020**, *21*, 2815. [CrossRef]
51. Enmark, E.; Pelto-Huikko, M.; Grandien, K.; Lagercrantz, S.; Lagercrantz, J.; Fried, G.; Nordenskjöld, M.; Gustafsson, J.Å. Human Estrogen Receptor Beta-Gene Structure, Chromosomal Localization, and Expression Pattern. *J. Clin. Endocrinol. Metab.* **1997**, *82*, 4258–4265. [CrossRef]
52. Pellegrini, C.; Gori, I.; Achtari, C.; Hornung, D.; Chardonnens, E.; Wunder, D.; Fiche, M.; Canny, G.O. The Expression of Estrogen Receptors as Well as GREB1, c-MYC, and Cyclin D1, Estrogen-Regulated Genes Implicated in Proliferation, Is Increased in Peritoneal Endometriosis. *Fertil. Steril.* **2012**, *98*, 1200–1208. [CrossRef]
53. Hu, L.; Zhang, J.; Lu, Y.; Fu, B.; Hu, W. Estrogen Receptor Beta Promotes Endometriosis Progression by Upregulating CD47 Expression in Ectopic Endometrial Stromal Cells. *J. Reprod. Immunol.* **2022**, *151*, 103513. [CrossRef] [PubMed]
54. Warren, L.A.; Shih, A.; Renteira, S.M.; Seckin, T.; Blau, B.; Simpfendorfer, K.; Lee, A.; Metz, C.N.; Gregersen, P.K. Analysis of Menstrual Effluent: Diagnostic Potential for Endometriosis. *Mol. Med.* **2018**, *24*, 1. [CrossRef] [PubMed]
55. Nayyar, A.; Saleem, M.I.; Yilmaz, M.; DeFranco, M.; Klein, G.; Elmaliki, K.M.; Kowalsky, E.; Chatterjee, P.K.; Xue, X.; Viswanathan, R.; et al. Menstrual Effluent Provides a Novel Diagnostic Window on the Pathogenesis of Endometriosis. *Front. Reprod. Health* **2020**, *2*, 3. [CrossRef] [PubMed]
56. Shih, A.J.; Adelson, R.P.; Vashistha, H.; Khalili, H.; Nayyar, A.; Puran, R.; Herrera, R.; Chatterjee, P.K.; Lee, A.T.; Truskinovsky, A.M.; et al. Single-Cell Analysis of Menstrual Endometrial Tissues Defines Phenotypes Associated with Endometriosis. *BMC Med.* **2022**, *20*, 315. [CrossRef]

57. Su, R.W.; Strug, M.R.; Joshi, N.R.; Jeong, J.W.; Miele, L.; Lessey, B.A.; Young, S.L.; Fazleabas, A.T. Decreased Notch Pathway Signaling in the Endometrium of Women with Endometriosis Impairs Decidualization. *J. Clin. Endocrinol. Metab.* **2015**, *100*, E433–E442. [CrossRef]
58. Tsuno, A.; Nasu, K.; Yuge, A.; Matsumoto, H.; Nishida, M.; Narahara, H. Decidualization Attenuates the Contractility of Eutopic and Ectopic Endometrial Stromal Cells: Implications for Hormone Therapy of Endometriosis. *J. Clin. Endocrinol. Metab.* **2009**, *94*, 2516–2523. [CrossRef]
59. Liao, Z.; Tang, S.; Jiang, P.; Geng, T.; Cope, D.I.; Dunn, T.N.; Guner, J.; Radilla, L.A.; Guan, X.; Monsivais, D. Impaired Bone Morphogenetic Protein (BMP) Signaling Pathways Disrupt Decidualization in Endometriosis. *Commun. Biol.* **2024**, *7*, 227. [CrossRef]
60. Katagiri, T.; Watabe, T. Bone Morphogenetic Proteins. *Cold Spring Harb. Perspect. Biol.* **2016**, *8*, a021899. [CrossRef]
61. Lu, H.; Huang, H. FOXO1: A Potential Target for Human Diseases. *Curr. Drug Targets* **2011**, *12*, 1235. [CrossRef]
62. Christian, M.; Zhang, X.; Schneider-Merck, T.; Unterman, T.G.; Gellersen, B.; White, J.O.; Brosens, J.J. Cyclic AMP-Induced Forkhead Transcription Factor, FKHR, Cooperates with CCAAT/Enhancer-Binding Protein Beta in Differentiating Human Endometrial Stromal Cells. *J. Biol. Chem.* **2002**, *277*, 20825–20832. [CrossRef]
63. Kang, N.; Shan, H.; Wang, J.; Mei, J.; Jiang, Y.; Zhou, J.; Huang, C.; Zhang, H.; Zhang, M.; Zhen, X.; et al. Calpain7 Negatively Regulates Human Endometrial Stromal Cell Decidualization in EMs by Promoting FoxO1 Nuclear Exclusion via Hydrolyzing AKT1. *Biol. Reprod.* **2022**, *106*, 1112–1125. [CrossRef] [PubMed]
64. Wang, M.; Sun, F.; Zhang, S.; Zhang, X.; Sun, Y.; Yu, T.; Li, Y.; Jiang, A.; Qiao, P.; Ren, C.; et al. NEK2 Promotes the Development of Ovarian Endometriosis and Impairs Decidualization by Phosphorylating FOXO1. *Cell. Mol. Life Sci.* **2024**, *81*, 237. [CrossRef] [PubMed]
65. Yu, J.; Wu, J.; Bagchi, I.C.; Bagchi, M.K.; Sidell, N.; Taylor, R.N. Disruption of Gap Junctions Reduces Biomarkers of Decidualization and Angiogenesis and Increases Inflammatory Mediators in Human Endometrial Stromal Cell Cultures. *Mol. Cell Endocrinol.* **2011**, *344*, 25–34. [CrossRef] [PubMed]
66. Laws, M.J.; Taylor, R.N.; Sidell, N.; DeMayo, F.J.; Lydon, J.P.; Gutstein, D.E.; Bagchi, M.K.; Bagchi, I.C. Gap Junction Communication between Uterine Stromal Cells Plays a Critical Role in Pregnancy-Associated Neovascularization and Embryo Survival. *Development* **2008**, *135*, 2659–2668. [CrossRef]
67. Yu, J.; Boicea, A.; Barrett, K.L.; James, C.O.; Bagchi, I.C.; Bagchi, M.K.; Nezhat, C.; Sidell, N.; Taylor, R.N. Reduced Connexin 43 in Eutopic Endometrium and Cultured Endometrial Stromal Cells from Subjects with Endometriosis. *Mol. Hum. Reprod.* **2014**, *20*, 260–270. [CrossRef]
68. Yu, J.; Berga, S.L.; Zou, W.; Grace Yook, D.; Pan, J.C.; Andrade, A.A.; Zhao, L.; Sidell, N.; Bagchi, I.C.; Bagchi, M.K.; et al. IL-1β Inhibits Connexin 43 and Disrupts Decidualization of Human Endometrial Stromal Cells Through ERK1/2 and P38 MAP Kinase. *Endocrinology* **2017**, *158*, 4270–4285. [CrossRef]
69. Matteo, M.; Cicinelli, E.; Neri, M.; Carrubba, R.; Carpagnano, F.A.; Romeo, F.; Scutiero, G.; Greco, P.; Garlanda, C.; Vendemiale, G.; et al. Pro-Inflammatory M1/Th1 Type Immune Network and Increased Expression of TSG-6 in the Eutopic Endometrium from Women with Endometriosis. *Eur. J. Obstet. Gynecol. Reprod. Biol.* **2017**, *218*, 99–105. [CrossRef]
70. Dimitriadis, E.; Stoikos, C.; Stafford-Bell, M.; Clark, I.; Paiva, P.; Kovacs, G.; Salamonsen, L.A. Interleukin-11, IL-11 Receptoralpha and Leukemia Inhibitory Factor Are Dysregulated in Endometrium of Infertile Women with Endometriosis during the Implantation Window. *J. Reprod. Immunol.* **2006**, *69*, 53–64. [CrossRef]
71. Robb, L.; Li, R.; Hartley, L.; Nandurkar, M.H.; Koentgen, F.; Glenn Begley, C. Infertility in Female Mice Lacking the Receptor for Interleukin 11 Is Due to a Defective Uterine Response to Implantation. *Nat. Med.* **1998**, *4*, 303–308. [CrossRef]
72. White, C.A.; Dimitriadis, E.; Sharkey, A.M.; Salamonsen, L.A. Interleukin-11 Inhibits Expression of Insulin-like Growth Factor Binding Protein-5 MRNA in Decidualizing Human Endometrial Stromal Cells. *Mol. Hum. Reprod.* **2005**, *11*, 649–658. [CrossRef]
73. Dimitriadis, E.; Nie, G.; Hannan, N.J.; Paiva, P.; Salamonsen, L.A. Local Regulation of Implantation at the Human Fetal-Maternal Interface. *Int. J. Dev. Biol.* **2010**, *54*, 313–322. [CrossRef] [PubMed]
74. Ramathal, C.Y.; Bagchi, I.C.; Taylor, R.N.; Bagchi, M.K. Endometrial Decidualization: Of Mice and Men. *Semin. Reprod. Med.* **2010**, *28*, 17–26. [CrossRef] [PubMed]
75. Karpovich, N.; Klemmt, P.; Jung, H.H.; McVeigh, J.E.; Heath, J.K.; Barlow, D.H.; Mardon, H.J. The Production of Interleukin-11 and Decidualization Are Compromised in Endometrial Stromal Cells Derived from Patients with Infertility. *J. Clin. Endocrinol. Metab.* **2005**, *90*, 1607–1612. [CrossRef] [PubMed]
76. Dimitriadis, E.; Robb, L.; Liu, Y.X.; Enders, A.C.; Martin, H.; Stoikos, C.; Wallace, E.; Salamonse, L.A. IL-11 and IL-11Rα Immunolocalisation at Primate Implantation Sites Suports a Role for IL-11 in Placentation and Fetal Development. *Reprod. Biol. Endocrinol.* **2003**, *1*, 1–10. [CrossRef]
77. Dimitriadis, E.; Robb, L.; Salamonsen, L.A. Interleukin 11 Advances Progesterone-Induced Decidualization of Human Endometrial Stromal Cells. *Mol. Hum. Reprod.* **2002**, *8*, 636–643. [CrossRef]
78. Tanaka, T.; Sakamoto, T.; Miyama, M.; Ogita, S.; Umesaki, N. Interleukin-11 Enhances Cell Survival of Decidualized Normal Human Endometrial Stromal Cells. *Gynecol. Endocrinol.* **2001**, *15*, 272–278. [CrossRef]

79. Ain, R.; Trinh, M.L.; Soares, M.J. Interleukin-11 Signaling Is Required for the Differentiation of Natural Killer Cells at the Maternal–Fetal Interface. *Dev. Dyn.* **2004**, *231*, 700–708. [CrossRef]
80. Alzaidi, Z.; Yildiz, Ş.M.; Saatçi, Ç.; Akalin, H.Ü.; Muderris, I.I.; Aynekin, B.; Şahin, I.O.; Dündar, M. The Effect of Cytokine Leukemia-Inhibitory Factor (LIF) and Interleukin-11 (IL-11) Gene Expression on the Primary Infertility Related to Polycystic Ovary Syndrome, Tubal Factor, and Unexplained Infertility in Turkish Women. *Egypt. J. Med. Human. Genet.* **2021**, *22*, 85. [CrossRef]
81. Rana, N.; Braun, D.P.; House, R.; Gebel, H.; Rotman, C.; Dmowski, W.P. Basal and Stimulated Secretion of Cytokines by Peritoneal Macrophages in Women with Endometriosis. *Fertil. Steril.* **1996**, *65*, 925–930. [CrossRef]
82. Tiberi, F.; Tropea, A.; Apa, R.; Romani, F.; Lanzone, A.; Marana, R. Prokineticin 1 MRNA Expression in the Endometrium of Healthy Women and in the Eutopic Endometrium of Women with Endometriosis. *Fertil. Steril.* **2010**, *93*, 2145–2149. [CrossRef]
83. Berger, S.L.; Kouzarides, T.; Shiekhattar, R.; Shilatifard, A. An Operational Definition of Epigenetics. *Genes. Dev.* **2009**, *23*, 781–783. [CrossRef] [PubMed]
84. Moosavi, A.; Ardekani, A.M. Role of Epigenetics in Biology and Human Diseases. *Iran. Biomed. J.* **2016**, *20*, 246. [CrossRef]
85. Mazur, E.C.; Vasquez, Y.M.; Li, X.; Kommagani, R.; Jiang, L.; Chen, R.; Lanz, R.B.; Kovanci, E.; Gibbons, W.E.; DeMayo, F.J. Progesterone Receptor Transcriptome and Cistrome in Decidualized Human Endometrial Stromal Cells. *Endocrinology* **2015**, *156*, 2239–2253. [CrossRef] [PubMed]
86. Pathiraja, T.N.; Shetty, P.B.; Jelinek, J.; He, R.; Hartmaier, R.; Margossian, A.L.; Hilsenbeck, S.G.; Issa, J.P.J.; Oesterreich, S. Progesterone Receptor Isoform-Specific Promoter Methylation—Association of PRA Promoter Methylation with Worse Outcome in Breast Cancer Patients. *Clin. Cancer Res.* **2011**, *17*, 4177. [CrossRef]
87. Bjorkman, S.; Taylor, H.S. MicroRNAs in Endometriosis: Biological Function and Emerging Biomarker Candidates. *Biol. Reprod.* **2019**, *100*, 1135. [CrossRef]
88. Bannister, A.J.; Kouzarides, T. Regulation of Chromatin by Histone Modifications. *Cell Res.* **2011**, *21*, 381–395. [CrossRef]
89. Zhang, Q.; Dong, P.; Liu, X.; Sakuragi, N.; Guo, S.W. Enhancer of Zeste Homolog 2 (EZH2) Induces Epithelial-Mesenchymal Transition in Endometriosis. *Sci. Rep.* **2017**, *7*, 6804. [CrossRef]
90. Monteiro, J.B.; Colón-Díaz, M.; García, M.; Gutierrez, S.; Colón, M.; Seto, E.; Laboy, J.; Flores, I. Endometriosis Is Characterized by a Distinct Pattern of Histone 3 and Histone 4 Lysine Modifications. *Reprod. Sci.* **2014**, *21*, 305. [CrossRef]
91. Xiaomeng, X.; Ming, Z.; Jiezhi, M.; Xiaoling, F. Aberrant Histone Acetylation and Methylation Levels in Woman with Endometriosis. *Arch. Gynecol. Obstet.* **2013**, *287*, 487–494. [CrossRef]
92. Retis-Resendiz, A.M.; González-García, I.N.; León-Juárez, M.; Camacho-Arroyo, I.; Cerbón, M.; Vázquez-Martínez, E.R. The Role of Epigenetic Mechanisms in the Regulation of Gene Expression in the Cyclical Endometrium. *Clin. Epigenetics* **2021**, *13*, 116. [CrossRef]
93. Liu, X.; Zhang, Q.; Guo, S.W. Histological and Immunohistochemical Characterization of the Similarity and Difference Between Ovarian Endometriomas and Deep Infiltrating Endometriosis. *Reprod. Sci.* **2018**, *25*, 329–340. [CrossRef] [PubMed]
94. Grimaldi, G.; Christian, M.; Steel, J.H.; Henriet, P.; Poutanen, M.; Brosens, J.J. Down-Regulation of the Histone Methyltransferase EZH2 Contributes to the Epigenetic Programming of Decidualizing Human Endometrial Stromal Cells. *Mol. Endocrinol.* **2011**, *25*, 1892–1903. [CrossRef] [PubMed]
95. Tamura, I.; Jozaki, K.; Sato, S.; Shirafuta, Y.; Shinagawa, M.; Maekawa, R.; Taketani, T.; Asada, H.; Tamura, H.; Sugino, N. The Distal Upstream Region of Insulin-like Growth Factor-Binding Protein-1 Enhances Its Expression in Endometrial Stromal Cells during Decidualization. *J. Biol. Chem.* **2018**, *293*, 5270–5280. [CrossRef] [PubMed]
96. Tamura, I.; Asada, H.; Maekawa, R.; Tanabe, M.; Lee, L.; Taketani, T.; Yamagata, Y.; Tamura, H.; Sugino, N. Induction of IGFBP-1 Expression by CAMP Is Associated with Histone Acetylation Status of the Promoter Region in Human Endometrial Stromal Cells. *Endocrinology* **2012**, *153*, 5612–5621. [CrossRef] [PubMed]
97. Li, W.N.; Wu, M.H.; Tsai, S.J. Hypoxia and reproductive health: The Role of Hypoxia in the Development and Progression of Endometriosis. *Reproduction* **2021**, *161*, F19–F31. [CrossRef]
98. Lin, X.; Dai, Y.; Gu, W.; Zhang, Y.; Zhuo, F.; Zhao, F.; Jin, X.; Li, C.; Huang, D.; Tong, X.; et al. The Involvement of RNA N6-methyladenosine and Histone Methylation Modification in Decidualization and Endometriosis-associated Infertility. *Clin. Transl. Med.* **2024**, *14*, e1564. [CrossRef]
99. Stadtmauer, D.J.; Wagner, G.P. Single-Cell Analysis of Prostaglandin E2-Induced Human Decidual Cell in Vitro Differentiation: A Minimal Ancestral Deciduogenic Signal. *Biol. Reprod.* **2022**, *106*, 155–172. [CrossRef]
100. Kennedy, T.G.; Ross, H.E. Effect of Prostaglandin E2 on Rate of Decidualization in Rats. *Prostaglandins* **1993**, *46*, 243–250. [CrossRef]
101. Yee, G.M.; Kennedy, T.G. Prostaglandin E2, CAMP and CAMP-Dependent Protein Kinase Isozymes during Decidualization of Rat Endometrial Stromal Cells in Vitro. *Prostaglandins* **1993**, *46*, 117–138. [CrossRef]
102. Frank, G.R.; Brar, A.K.; Cedars, M.I.; Handwerger, S. Prostaglandin E2 Enhances Human Endometrial Stromal Cell Differentiation. *Endocrinology* **1994**, *134*, 258–263. [CrossRef]

103. Brighton, P.J.; Walker, A.R.; Mann, O.; Kong, C.-S.; Lucas, E.S.; Vrljicak, P.; Brosens, J.J.; Hanyaloglu, A.C. Spatial-Temporal Regulation of the Prostanoid Receptor EP2 Co-Ordinates PGE2-Mediated CAMP Signaling in Decidualizing Human Endometrium. *iScience* **2024**, *27*, 111170. [CrossRef] [PubMed]
104. Dimitriadis, E.; Stoikos, C.; Baca, M.; Fairlie, W.D.; McCoubrie, J.E.; Salamonsen, L.A. Relaxin and Prostaglandin E2 Regulate Interleukin 11 during Human Endometrial Stromal Cell Decidualization. *J. Clin. Endocrinol. Metab.* **2005**, *90*, 3458–3465. [CrossRef] [PubMed]
105. De Leon, F.D.; Vijayakumar, R.; Brown, M.; Rao, C.V.; Yussman, M.A.; Schultz, G. Peritoneal Fluid Volume, Estrogen, Progesterone, Prostaglandin, and Epidermal Growth Factor Concentrations in Patients with and without Endometriosis. *Obstet. Gynecol.* **1986**, *68*, 189–194. [CrossRef] [PubMed]
106. Rakhila, H.; Carli, C.; Daris, M.; Lemyre, M.; Leboeuf, M.; Akoum, A. Identification of Multiple and Distinct Defects in Prostaglandin Biosynthetic Pathways in Eutopic and Ectopic Endometrium of Women with Endometriosis. *Fertil. Steril.* **2013**, *100*, 1650–1659.e2. [CrossRef]
107. Cobellis, L.; Razzi, S.; De Simone, S.; Sartini, A.; Fava, A.; Danero, S.; Gioffrè, W.; Mazzini, M.; Petraglia, F. The Treatment with a COX-2 Specific Inhibitor Is Effective in the Management of Pain Related to Endometriosis. *Eur. J. Obstet. Gynecol. Reprod. Biol.* **2004**, *116*, 100–102. [CrossRef]
108. Ota, H.; Igarashi, S.; Sasaki, M.; Tanaka, T. Distribution of Cyclooxygenase-2 in Eutopic and Ectopic Endometrium in Endometriosis and Adenomyosis. *Hum. Reprod.* **2001**, *16*, 561–566. [CrossRef]
109. Chishima, F.; Hayakawa, S.; Sugita, K.; Kinukawa, N.; Aleemuzzaman, S.; Nemoto, N.; Yamamoto, T.; Honda, M. Increased Expression of Cyclooxygenase-2 in Local Lesions of Endometriosis Patients. *Am. J. Reprod. Immunol.* **2002**, *48*, 50–56. [CrossRef]
110. Makabe, T.; Koga, K.; Nagabukuro, H.; Asada, M.; Satake, E.; Taguchi, A.; Takeuchi, A.; Miyashita, M.; Harada, M.; Hirata, T.; et al. Use of Selective PGE2 Receptor Antagonists on Human Endometriotic Stromal Cells and Peritoneal Macrophages. *Mol. Hum. Reprod.* **2021**, *27*, gaaa077. [CrossRef]
111. Arosh, J.A.; Lee, J.H.; Starzinski-Powitz, A.; Banu, S.K. Selective Inhibition of Prostaglandin E2 Receptors EP2 and EP4 Modulates DNA Methylation and Histone Modification Machinery Proteins in Human Endometriotic Cells. *Mol. Cell. Endocrinol.* **2015**, *409*, 51. [CrossRef]
112. Banu, S.K.; Lee, J.H.; Starzinski-Powitz, A.; Arosh, J.A. Gene Expression Profiles and Functional Characterization of Human Immortalized Endometriotic Epithelial and Stromal Cells. *Fertil. Steril.* **2008**, *90*, 972–987. [CrossRef]
113. Arosh, J.A.; Lee, J.H.; Balasubbramanian, D.; Stanley, J.A.; Long, C.R.; Meagher, M.W.; Osteen, K.G.; Bruner-Tran, K.L.; Burghardt, R.C.; Starzinski-Powitz, A.; et al. Molecular and Preclinical Basis to Inhibit PGE2 Receptors EP2 and EP4 as a Novel Nonsteroidal Therapy for Endometriosis. *Proc. Natl. Acad. Sci. USA* **2015**, *112*, 9716–9721. [CrossRef] [PubMed]
114. Bhat, K.P.; Ümit Kaniskan, H.; Jin, J.; Gozani, O. Epigenetics and beyond: Targeting Writers of Protein Lysine Methylation to Treat Disease. *Nat. Rev. Drug Discov.* **2021**, *20*, 265. [CrossRef] [PubMed]
115. Wapenaar, H.; Dekker, F.J. Histone Acetyltransferases: Challenges in Targeting Bi-Substrate Enzymes. *Clin. Epigenetics* **2016**, *8*, 1–11. [CrossRef] [PubMed]
116. Cao, F.; Townsend, E.C.; Karatas, H.; Xu, J.; Li, L.; Lee, S.; Liu, L.; Chen, Y.; Ouillette, P.; Zhu, J.; et al. Targeting MLL1 H3K4 Methyltransferase Activity in Mixed-Lineage Leukemia. *Mol. Cell* **2014**, *53*, 247–261. [CrossRef]
117. Wen, X.; Xiong, Y.; Liu, H.; Geng, T.; Jin, L.; Zhang, M.; Ma, L.; Zhang, Y. Decreased Mixed Lineage Leukemia 1 Is Involved in Endometriosis-Related Infertility. *J. Mol. Endocrinol.* **2021**, *66*, 45–57. [CrossRef]
118. Cai, X.; Xu, M.; Zhang, H.; Zhang, M.; Wang, J.; Mei, J.; Zhang, Y.; Zhou, J.; Zhen, X.; Kang, N.; et al. Endometrial Stromal PRMT5 Plays a Crucial Role in Decidualization by Regulating NF-KB Signaling in Endometriosis. *Cell Death Discov.* **2022**, *8*, 408. [CrossRef]
119. Kim, T.H.; Yoo, J.Y.; Choi, K.C.; Shin, J.H.; Leach, R.E.; Fazleabas, A.T.; Young, S.L.; Lessey, B.A.; Yoon, H.G.; Jeong, J.W. Loss of HDAC3 Results in Nonreceptive Endometrium and Female Infertility. *Sci. Transl. Med.* **2019**, *11*, 7533. [CrossRef]
120. Kim, T.H.; Young, S.L.; Sasaki, T.; Deaton, J.L.; Schammel, D.P.; Palomino, W.A.; Jeong, J.W.; Lessey, B.A. Role of SIRT1 and Progesterone Resistance in Normal and Abnormal Endometrium. *J. Clin. Endocrinol. Metab.* **2022**, *107*, 788–800. [CrossRef]
121. Yu, S.L.; Lee, S.I.; Park, H.W.; Lee, S.K.; Kim, T.H.; Kang, J.; Park, S.R. SIRT1 Suppresses in Vitro Decidualization of Human Endometrial Stromal Cells through the Downregulation of Forkhead Box O1 Expression. *Reprod. Biol.* **2022**, *22*, 100672. [CrossRef]
122. Samadieh, Y.; Favaedi, R.; Ramezanali, F.; Afsharian, P.; Aflatoonian, R.; Shahhoseini, M. Epigenetic Dynamics of HOXA10 Gene in Infertile Women With Endometriosis. *Reprod. Sci.* **2019**, *26*, 88–96. [CrossRef]
123. Gui, Y.; Zhang, J.; Yuan, L.; Lessey, B.A. Regulation of HOXA-10 and Its Expression in Normal and Abnormal Endometrium. *Mol. Hum. Reprod.* **1999**, *5*, 866–873. [CrossRef] [PubMed]
124. Zhu, L.H.; Sun, L.H.; Hu, Y.L.; Jiang, Y.; Liu, H.Y.; Shen, X.Y.; Jin, X.Y.; Zhen, X.; Sun, H.X.; Yan, G.J. PCAF Impairs Endometrial Receptivity and Embryo Implantation by Down-Regulating B3-Integrin Expression via HOXA10 Acetylation. *J. Clin. Endocrinol. Metab.* **2013**, *98*, 4417–4428. [CrossRef] [PubMed]
125. Browne, H.; Taylor, H. HOXA10 Expression in Ectopic Endometrial Tissue. *Fertil. Steril.* **2006**, *85*, 1386–1390. [CrossRef]

126. Taylor, H.S.; Vanden Heuvel, G.B.; Igarashi, P. A Conserved Hox Axis in the Mouse and Human Female Reproductive System: Late Establishment and Persistent Adult Expression of the Hoxa Cluster Genes. *Biol. Reprod.* **1997**, *57*, 1338–1345. [CrossRef] [PubMed]
127. Jiang, Y.-P.; Peng, Y.-Q.; Wang, L.; Qin, J.; Zhang, Y.; Zhao, Y.-Z.; Tan, A.-L.; Wang, S.-J.; Pi, J. RNA-Sequencing Identifies Differentially Expressed Genes in T Helper 17 Cells in Peritoneal Fluid of Patients with Endometriosis. *J. Reprod. Immunol.* **2022**, *149*, 103453. [CrossRef]
128. Akter, S.; Xu, D.; Nagel, S.C.; Bromfield, J.J.; Pelch, K.; Wilshire, G.B.; Joshi, T. Machine Learning Classifiers for Endometriosis Using Transcriptomics and Methylomics Data. *Front. Genet.* **2019**, *10*, 466838. [CrossRef]
129. Zanatta, A.; Rocha, A.M.; Carvalho, F.M.; Pereira, R.M.A.; Taylor, H.S.; Motta, E.L.A.; Baracat, E.C.; Serafini, P.C. The Role of the Hoxa10/HOXA10 Gene in the Etiology of Endometriosis and Its Related Infertility: A Review. *J. Assist. Reprod. Genet.* **2010**, *27*, 701–710. [CrossRef]
130. Taylor, H.S.; Arici, A.; Olive, D.; Igarashi, P. HOXA10 Is Expressed in Response to Sex Steroids at the Time of Implantation in the Human Endometrium. *J. Clin. Investig.* **1998**, *101*, 1379–1384. [CrossRef]
131. Modi, D.; Godbole, G. HOXA10 Signals on the Highway through Pregnancy. *J. Reprod. Immunol.* **2009**, *83*, 72–78. [CrossRef]
132. Adamczyk, M.; Wender-Ozegowska, E.; Kedzia, M. Epigenetic Factors in Eutopic Endometrium in Women with Endometriosis and Infertility. *Int. J. Mol. Sci.* **2022**, *23*, 3804. [CrossRef]
133. Ding, L.; Yang, L.; Ren, C.; Zhang, H.; Lu, J.; Wang, S.; Wu, Z.; Yang, Y. A Review of Aberrant DNA Methylation and Epigenetic Agents Targeting DNA Methyltransferases in Endometriosis. *Curr. Drug Targets* **2020**, *21*, 1047–1055. [CrossRef] [PubMed]
134. Newton, A.S.; Faver, J.C.; Micevic, G.; Muthusamy, V.; Kudalkar, S.N.; Bertoletti, N.; Anderson, K.S.; Bosenberg, M.W.; Jorgensen, W.L. Structure-Guided Identification of DNMT3B Inhibitors. *ACS Med. Chem. Lett.* **2020**, *11*, 971–976. [CrossRef] [PubMed]
135. Izawa, M.; Taniguchi, F.; Uegaki, T.; Takai, E.; Iwabe, T.; Terakawa, N.; Harada, T. Demethylation of a Nonpromoter Cytosine-Phosphate-Guanine Island in the Aromatase Gene May Cause the Aberrant up-Regulation in Endometriotic Tissues. *Fertil. Steril.* **2011**, *95*, 33–39. [CrossRef] [PubMed]
136. Kawano, Y.; Nasu, K.; Li, H.; Tsuno, A.; Abe, W.; Takai, N.; Narahara, H. Application of the Histone Deacetylase Inhibitors for the Treatment of Endometriosis: Histone Modifications as Pathogenesis and Novel Therapeutic Target. *Hum. Reprod.* **2011**, *26*, 2486–2498. [CrossRef]
137. Lu, X.; Ning, Z.; Li, Z.; Cao, H.; Wang, X. Development of Chidamide for Peripheral T-Cell Lymphoma, the First Orphan Drug Approved in China. *Intractable Rare Dis. Res.* **2016**, *5*, 185–191. [CrossRef]
138. Saunders, P.T.K.; Horne, A.W. Endometriosis: Etiology, Pathobiology, and Therapeutic Prospects. *Cell* **2021**, *184*, 2807–2824. [CrossRef]
139. Shi, J.; Tan, X.; Feng, G.; Zhuo, Y.; Jiang, Z.; Banda, S.; Wang, L.; Zheng, W.; Chen, L.; Yu, D.; et al. Research Advances in Drug Therapy of Endometriosis. *Front. Pharmacol.* **2023**, *14*, 1199010. [CrossRef]

cells MDPI

Article

Expression of Toll-like Receptors on Lymphocyte Subpopulations and Their Soluble Forms in Serum and Urine of Women with Endometriosis

Anna Sobstyl [1], Paulina Mertowska [1,*], Sebastian Mertowski [1,*], Rafał Tarkowski [2], Dominik Dudziński [1], Michał Kotowski [3], Krzysztof Bojarski [4], Bogusława Stelmach [5], Błażej Chermuła [5], Maciej Brązert [5] and Ewelina Grywalska [1]

1 Department of Experimental Immunology, Medical University of Lublin, Chodźki 4a St., 20-093 Lublin, Poland; sobstylanna.1@gmail.com (A.S.); d.dudzinski1996@gmail.com (D.D.); ewelina.grywalska@umlub.pl (E.G.)
2 I Chair and Department of Oncological Gynaecology and Gynaecology, Medical University of Lublin, 20-059 Lublin, Poland; rafal.tarkowski@umlub.pl
3 Chair and Department of Paediatric Otolaryngology, Phoniatry and Audiology, Medical University of Lublin, Antoniego Gębali 6 Street, 20-093 Lublin, Poland; michal.kotowski@umlub.pl
4 General Surgery Department, SP ZOZ in Leczna, 52 Krasnystawska Street, 21-010 Leczna, Poland; k.bojarski@szpital.leczna.pl
5 Department of Diagnostics and Treatment of Infertility, Poznan University of Medical Sciences, 61-701 Poznan, Poland; b_stelmach@wp.pl (B.S.); blazej.chermula@wp.pl (B.C.); maciejbrazert@ump.edu.pl (M.B.)
* Correspondence: paulina.mertowska@umlub.pl (P.M.); sebastian.mertowski@umlub.pl (S.M.)

Abstract: Introduction: Endometriosis is a chronic inflammatory disease affecting women of reproductive age, often accompanied by chronic pelvic pain and infertility. Despite numerous studies, its pathogenesis remains incompletely understood. Increasing evidence indicates the important role of immunological disorders, especially in the mechanisms of innate immunity and Toll-like receptors (TLRs). Study objective: This study aimed to assess the expression of selected TLRs (TLR2, TLR3, TLR4, TLR7, TLR8, and TLR9) on peripheral blood lymphocyte subpopulations (CD4+, CD8+, and CD19+ cells) in patients diagnosed with endometriosis and to quantify the levels of their soluble forms in serum and urine. This study was conducted on patients who were not undergoing hormonal bridging therapy and were not using any form of hormonal contraception to eliminate potential confounding effects on immune parameters. Methods: Flow cytometric analysis of TLR expression on peripheral blood lymphocytes was performed, and the levels of their soluble forms in serum and urine samples were determined. Additionally, ROC curve analysis was used to evaluate the diagnostic potential of the studied parameters. Results: We found increased expression of TLRs in lymphocyte populations in patients with endometriosis compared to the control group, suggesting their involvement in both local and systemic immune responses. In addition, ROC analysis showed the diagnostic potential of TLR expression in differentiating patients with endometriosis from healthy women, and it may also identify disease subtypes. Conclusions: The findings support the role of TLRs in the immunopathogenesis of endometriosis and highlight their promise as diagnostic biomarkers and therapeutic targets. Further studies on larger patient cohorts and functional signaling analyses are warranted.

Cells **2025**, *14*, 1273 https://doi.org/10.3390/cells14161273

Keywords: endometriosis; Toll-like receptors (TLRs); innate immunity; immune dysregulation; lymphocytes; TLR expression; biomarkers; chronic inflammation; immunopathogenesis

1. Introduction

Endometriosis is a chronic, estrogen-dependent gynecological disease with complex etiopathogenesis, affecting approximately 10% of women of reproductive age, of whom up to 70% experience chronic pelvic pain and fertility disorders [1–4]. This disease is characterized by the presence of active endometrial foci outside the uterine cavity, which leads to a persistent inflammatory reaction, angiogenesis, fibrosis, and damage to the surrounding tissues [5,6]. Despite numerous studies, the molecular mechanisms underlying the development and progression of endometriosis remain incompletely understood. Current data indicate that dysregulation of the immune system plays a key role in the pathogenesis of the disease, both in the innate and acquired responses [7–12]. In particular, abnormalities in the functioning of T lymphocytes (especially the Th17 and Treg subpopulations), B lymphocytes, impaired phagocytosis by macrophages, and reduced cytotoxic activity of NK cells have been described, which result in the ineffective elimination of endometrial cells from ectopic locations.

Among the molecular mediators of the nonspecific response, an important role is attributed to Toll-like receptors (TLRs), which serve as key sensors of pathogens and signals of endogenous threats. The expression of TLR2, TLR3, TLR4, and TLR9 has been demonstrated in ectopic endometrium foci, in stromal and epithelial cells, and in infiltrating macrophages [13–17]. Their activation initiates signaling cascades dependent on the MyD88 and TRIF adaptors, leading to the activation of the transcription factor NF-κB and increased production of pro-inflammatory cytokines (IL-6, IL-8, TNF-α). Chronic, sterile TLR stimulation—induced, for example, by endogenous DAMPs or microbiome components—can sustain inflammation, promote angiogenesis, and support the survival of endometrial cells [14,15,18,19]. Additionally, disturbances in the negative regulation of these pathways have been observed, contributing to persistent low-intensity inflammation. Increasing importance is also attributed to interactions between TLRs and the intestinal and endometrial microbiota, which can modulate the local immune response and affect immunotolerance [20–28].

Immune regulation in women is closely related to the action of sex hormones, mainly estrogens and progesterone. Estrogens enhance the activation of immune cells, increasing the production of pro-inflammatory cytokines and the expression of TLRs, especially TLR2 and TLR4. At the same time, progesterone has an inhibitory effect, promoting tolerance and limiting the activation of the NF-κB pathway [29–34]. In patients with endometriosis, a disturbance of this balance is observed—estrogen dominance and progesterone resistance, which promote the perpetuation of inflammation and the pathological stimulation of the immune system, including excessive activation of TLRs. These complex hormonal–immunological relationships, further reinforced by oxidative and microbiological factors, promote the survival, migration, and invasiveness of endometrial cells, thereby facilitating the clinical progression of the disease.

A review of the scientific literature available in the PubMed database indicates that out of 37,427 publications containing the term "endometriosis" (1927–2025), only about 4% (approx. 1482 works) directly concern the involvement of the immune system, and the issue related to TLRs is addressed by only 0.12% of available publications. This state of affairs highlights a significant research gap in the immunobiology of endometriosis, with particular emphasis on the innate response and the role of TLRs in the molecular mechanisms of this disease.

In light of the above, this study aimed to assess the expression of selected TLRs (TLR2, TLR3, TLR4, TLR7, TLR8, and TLR9) on subpopulations of T lymphocytes (CD4+ and CD8+) and B lymphocytes (CD19+) in the peripheral blood of patients with newly diagnosed, untreated endometriosis. Additionally, an analysis of their soluble form concentrations in serum and urine was performed and compared to the control group. The key methodological assumption of this study was to exclude patients using any hormonal therapy, including hormonal contraception or bridging treatment. Thanks to this, the influence of exogenous hormones on immunological parameters was eliminated, allowing for a more objective picture of the natural immune mechanisms associated with endometriosis and a more accurate assessment of the potential role of TLRs in the disease's pathogenesis.

2. Materials and Methods

2.1. Patient Characteristics

This study included a total of 74 women, comprising 50 patients diagnosed with endometriosis and 24 individuals without endometriosis, who served as the control group. The diagnosis of endometriosis was established based on the criteria of the American Society for Reproductive Medicine (ASRM) [35] and the clinical guidelines of the Polish Society of Gynecologists and Obstetricians [36] for the management of women with endometriosis. The control group included women undergoing surgery for benign ovarian cysts, diagnostic laparoscopy, or tubal ligation, as well as healthy volunteers (HV).

Inclusion criteria for the study group were as follows:

- Confirmed diagnosis of endometriosis by histopathological and clinical evaluation;
- Age $\geq$ 18 years;
- No immunosuppressive treatment within the past 12 months;
- No use of hormonal contraception (including intrauterine systems);
- Provision of written informed consent to participate in the study.

Exclusion criteria were as follows:

- Presence of uterine fibroids or adenomyosis;
- Active bacterial, viral, or fungal infections;
- Diabetes mellitus or other metabolic/endocrine disorders;
- Chronic wounds, severe allergic diseases, or significant systemic illnesses affecting the kidneys, gastrointestinal tract, heart, or lungs, as well as adherence to strict elimination diets;
- History of hematopoietic stem cell or organ transplantation;
- Active malignancy or autoimmune disorders;
- Pregnancy or breastfeeding;
- Use of immunosuppressive medications or participation in other clinical trials;
- Presence of CNS metastatic disease or diagnosed psychiatric disorders.

An experienced gynecologist specializing in the diagnosis and treatment of endometriosis established the diagnosis. Initial diagnostic steps included transvaginal and transabdominal ultrasound examinations of the pelvis and other anatomical sites suspected of endometrial lesions. In more complex cases, additional imaging with computed tomography (CT) or magnetic resonance imaging (MRI) was performed to enhance diagnostic accuracy. The final diagnosis was confirmed through histopathological evaluation of tissue samples obtained during laparoscopy.

In the control group, no evidence of ectopic endometrial tissue was identified in pelvic or abdominal ultrasound imaging. From each participant, peripheral blood (10 mL) was collected via the basilic vein into EDTA-coated tubes (Sarstedt, Nümbrecht, Germany) for immunophenotyping. An additional 5 mL of blood was obtained using tubes (Sarstedt, Nümbrecht, Germany) with a clot activator for serum isolation and analysis of soluble TLRs. Furthermore, 10 mL of midstream urine was collected for analysis of urinary biomarkers. All biological samples were obtained during the follicular phase of the menstrual cycle (days 5–11) to minimize hormonal variability. Based on the location and type of endometrial lesions—composed of ectopic endometrial glands and stroma—patients diagnosed with endometriosis were classified into the following subgroups:

- Peritoneal endometriosis (PE);
- Ovarian endometriosis (OE);
- Deeply infiltrating endometriosis (DIE);
- Cesarean section scar endometriosis (CC).

This stratification was used in subsequent analyses to evaluate potential immunological and molecular differences across disease subtypes.

Basic blood parameters were determined from fasting venous blood collected into two types of tubes: one with EDTA (for complete blood count) (Sarstedt, Nümbrecht, Germany) and one with a clot activator (Sarstedt, Nümbrecht, Germany) (for CRP and immunoglobulins). Complete blood counts, including RBC [M/µL], WBC [K/µL], LYM [K/µL], MON [K/µL], NEU [K/µL], EOS [K/µL], BAS [K/µL], HGB [g/dL], and PLT [K/µL], were determined by the optical impedance method using an automated hematology analyzer (Sysmex XN-1000, Sysmex Corporation, Kobe, Japan) employing standard laboratory procedures. CRP concentration [mg/L] was determined in serum by the immunoturbidimetric method on a biochemistry analyzer (Cobas, Abbott, Chicago, IL, USA). IgG, IgA, and IgM immunoglobulin levels [g/L] were determined by the nephelometric method using an immunochemical analyzer (BN ProSpec System, Siemens Healthineers, Marburg, Germany). All tests were performed at the time of patient recruitment in the hospital laboratory.

The study was conducted in accordance with the Declaration of Helsinki and approved by the Bioethics Committee of the Medical University of Lublin (KE-0254/10/01/2023). All participants signed informed consent and were informed of the nature and purpose of the study.

2.2. Immunophenotyping

Phenotypic characterization of peripheral blood lymphocytes was conducted using multiparameter flow cytometry. For surface and intracellular staining, a panel of monoclonal antibodies specific to selected lineage markers and TLRs was employed. Detailed information regarding the antibodies used is provided in Table S1 of the Supplementary Materials.

To minimize spectral overlap when using fluorochromes excited by violet lasers, BD Horizon™ Brilliant Stain Buffer Plus (BD Biosciences Franklin Lakes, NJ, USA, Cat. No. 566385) was incorporated into the antibody cocktail, as recommended by the manufacturer. After adding the antibodies, samples were incubated for 20 min at room temperature in the dark. Subsequently, cells were washed using BD Pharmingen™ Stain Buffer (FBS) (BD Biosciences, Franklin Lakes, NJ, USA Cat. No. 554656) to remove unbound antibodies.

For the intracellular staining of TLRs, the BD Cytofix/Cytoperm™ Fixation/Permeabilization Kit (BD Biosciences, Franklin Lakes, NJ, USA Cat. No. 554714) was employed according to the manufacturer's protocol, ensuring adequate cell fixation and membrane permeabilization. Flow cytometric acquisition was performed using the CytoFLEX LX platform (Beckman Coulter, Brea, CA, USA), and data were analyzed with Kaluza software v 2.1 (Beckman Coulter, Brea, CA, USA). Daily instrument performance was validated with

CytoFLEX Daily QC Fluorospheres (Beckman Coulter, Brea, CA, USA, Cat. No. C65719). Compensation of spectral overlap across fluorochrome channels was conducted using the VersaComp Antibody Capture Kit (Beckman Coulter, Brea, CA, USA Cat. No. B22804).

To ensure accurate gating and minimize false-positive signals, fluorescence-minus-one (FMO) controls were prepared individually for each anti-TLR antibody. Each FMO sample was derived from the same biological specimen as the fully stained sample and underwent identical staining, fixation, and analysis procedures. Gates for marker positivity were set based on these FMO controls, accounting for background fluorescence and autofluorescence specific to each detector channel.

A detailed gating strategy and representative plots are provided in the Supplementary Materials (Figures S1–S10).

2.3. Determination of the Concentration of Soluble Forms of TLRs in the Tested Material

The concentration of soluble forms of Toll-like receptors (sTLRs) was analyzed in serum and urine samples collected from patients with endometriosis and the control group. Serum was obtained by centrifugation of whole blood collected in a clot tube (10 min, 2000× *g*, 4 °C) and then stored at −80 °C until analysis. Urine samples were centrifuged (10 min, 3000× *g*, 4 °C) to remove cell sediment, and the supernatant was frozen at −80 °C. Commercially available immunoenzymatic ELISA kits (Thermo Fisher Scientific, Waltham, MA, USA and Wuhan Fine Biotech Co., Ltd., Wuhan, China) were used for the determinations according to the protocols provided by the manufacturers. Soluble TLR2 and TLR4 concentrations were measured using Invitrogen assays (Thermo Fisher Scientific, Waltham, MA, USA), with a sensitivity of 0.32 ng/mL (TLR2, Cat. No. EH459RB) and 0.4 ng/mL (TLR4, Cat. No. EH460RB), respectively. TLR3 (Cat. No. EH0301; with a sensitivity of 0.094 ng/mL), TLR7 (Cat. No. EH2015; with a sensitivity of 0.094 ng/mL), TLR8 (Cat. No. EH4931; with a sensitivity of 0.094 ng/mL), and TLR9 (Cat. No. EH1019; with a sensitivity of 0.094 ng/mL) were measured using FineTest kits (Wuhan Fine Biotech Co., Ltd., Wuhan, China). Absorbance was measured at the appropriate wavelengths using a VICTOR3™ microplate reader (PerkinElmer, Waltham, MA, USA) and dedicated software.

2.4. Statistical Analyses

The data obtained in the study were statistically analyzed using GraphPad Prism (GraphPad Software, v. 9.4.1, San Diego, CA, USA) and Statistica (TIBCO Software, v. 13.3 Palo Alto, CA, USA). Before the analysis, the normality of the variable distributions was assessed using the Shapiro–Wilk test. Due to the lack of compliance with the normal distribution for most of the analyzed parameters, nonparametric statistical tests were used. The Mann–Whitney U test compared two groups (patients with endometriosis vs. the control group), whereas the analysis of differences among more than two groups was performed using the Kruskal–Wallis test, followed by a post hoc test.

A receiver operating characteristic (ROC) curve analysis was performed to evaluate the diagnostic potential of the studied parameters. The analysis was performed using GraphPad Prism (GraphPad Software, v. 9.4.1, San Diego, CA, USA). ROC curves were constructed by plotting the relationship between sensitivity and 1-specificity for different threshold values of the biomarkers studied. The results were visualized as ROC plots indicating the AUC (Area Under the Curve) values.

3. Results

3.1. Comparison of Immunological and Biochemical Parameters Between Patients with Endometriosis and Healthy Women

The state of chronic inflammation associated with endometriosis affects several hematological, biochemical, and immunological parameters, modifying both local and systemic inflammatory responses and lymphocyte activity [9,36]. Therefore, the first stage of this study involved analyzing differences in basic peripheral blood parameters, inflammatory markers, and subpopulations of immune cells between women with endometriosis (EM) and healthy volunteers (HV).

A total of 74 women were included in the study: 50 patients with newly diagnosed, clinically and histopathologically confirmed endometriosis (according to current diagnostic recommendations; details in Section 2), with a median age of 35 years (range, 24–47 years), and 24 healthy women in the control group, with a median age of 33 years (range, 18–57 years). All participants—both patients and the control group—did not use any form of hormonal therapy, including hormonal contraception, which allowed avoiding the influence of exogenous hormonal modulation on the assessed immunological parameters.

Particular attention was paid to the concentrations of C-reactive protein (CRP), selected immunoglobulins (IgG, IgA, and IgM), and the subpopulation composition of leukocytes, with particular emphasis on T lymphocytes (CD3+, CD4+, and CD8+), B lymphocytes (CD19+), and NK cells. These parameters are crucial in assessing the ongoing inflammation, immune system activity, and potential mechanisms of impaired immunoregulation that underlie the pathogenesis of endometriosis. The data are summarized in Table 1.

Table 1. Comparison of basic hematological parameters, inflammatory markers, and the percentage distribution of immune cell subpopulations between patients with endometriosis and healthy controls.

	Parameters	EM (n = 50)		HV (n = 24)		p-Value
		Median	Q1–Q3	Median	Q1–Q3	
Hematological, inflammatory, and immunoglobulin parameters	RBC [M/µL]	4.52	4.10–4.71	4.53	4.18–4.64	0.850
	WBC [K/µL]	6.10	5.04–7.73	6.01	5.71–7.15	0.718
	LYM [K/µL]	1.79	1.43–2.29	1.68	1.39–1.83	0.180
	MON [K/µL]	0.52	0.42–0.60	0.52	0.49–0.63	0.626
	BAS [K/µL]	0.03	0.02–0.06	0.04	0.03–0.05	0.367
	EOS [K/µL]	0.02	0.01–0.07	0.16	0.11–0.23	<0.0001 *
	NEU [K/µL]	4.43	3.68–5.43	3.26	2.38–4.16	<0.0001 *
	HGB [g/dL]	13.15	12.20–13.85	12.90	12.50–13.80	0.895
	PLT [K/µL]	256.55	206.75–335.00	270.5	249.00–302.50	0.562
	CRP [mg/L]	5.61	3.51–6.60	2.76	1.54–3.32	<0.0001 *
	IgG [g/L]	7.84	7.20–8.75	11.81	8.53–13.55	<0,0001 *
	IgM [g/L]	4.28	3.71–5.10	1.59	1.38–2.33	<0,0001 *
	IgA [g/L]	4.70	3.76–5.94	1.95	1.68–2.36	<0,0001 *
Immunophenotypic characterization of lymphocyte subpopulations	CD45+ [%]	87.86	81.86–91.80	96.28	92.83–98.05	<0.0001 *
	CD3+ [%]	66.02	59.95–74.20	72.34	70.70–76.54	0.0012 *
	CD4+ [%]	41.84	27.20–51.37	47.49	44.77–49.84	0.0144 *
	CD8+ [%]	20.51	17.31–28.39	27.23	25.88–28.52	0.0021 *
	CD19+ [%]	11.27	10.17–14.35	12.70	11.82–13.69	0.032 *
	NK cells [%]	9.24	4.26–17.72	7.90	7.67–10.89	0.042 *

The following abbreviations are used: BAS—basophils; CD—cluster of differentiation (including CD3—T lymphocytes; CD4—T helper lymphocytes; CD8—cytotoxic T lymphocytes; CD19—B lymphocytes; CD45—total leukocytes); CRP—C-reactive protein; EOS—eosinophils; HGB—hemoglobin; IgA—immunoglobulin A; IgG—immunoglobulin G; IgM—immunoglobulin M; LYM—lymphocytes; MON—monocytes; NK cells—natural killer cells; NEU—neutrophils; PLT—platelets; RBC—red blood cells; and WBC—white blood cells. [%] Refers to the percentage of cells positive for the tested marker in the total population of analyzed cells, * statistically significant difference ($p < 0.05$).

Statistical analysis revealed significant differences between patients with endometriosis and the control group in terms of the number of inflammatory, hematological, and immunological parameters. The concentration of C-reactive protein (CRP), an indicator of ongoing inflammation, was significantly higher in the study group ($p < 0.0001$), suggesting the presence of a chronic, systemic inflammatory process. Blood morphology revealed higher neutrophil counts (NEU) and significantly lower eosinophil counts (EOS) in patients with endometriosis ($p < 0.0001$), which may indicate changes in leukocyte composition typical of nonspecific inflammation. In terms of humoral immunity, a significantly reduced IgG concentration was observed, with simultaneous increases in IgM and IgA levels ($p < 0.0001$). This may indicate a different profile of the immune response associated with the presence of the disease.

Phenotypic analysis of mononuclear cells revealed a significantly lower percentage of CD45+, CD3+, CD4+, CD8+, and CD19+ cells in the patient population with endometriosis compared to the healthy women in the control group. At the same time, the percentage of NK cells was moderately but significantly higher ($p = 0.042$) in the study group.

The next element of the analysis was the assessment of the expression of Toll-like receptors (TLR2, TLR3, TLR4, TLR7, TLR8, and TLR9) on the surface of T helper lymphocytes (CD4+), cytotoxic T lymphocytes (CD8+), and B lymphocytes (CD19+). In all analyzed subpopulations, a significantly higher percentage of cells expressing the studied TLRs was found in patients with endometriosis compared to the control group. Increased expression of these receptors may reflect dysregulation of immune regulatory mechanisms and the chronic inflammatory state accompanying the disease. The results are summarized in Figure 1.

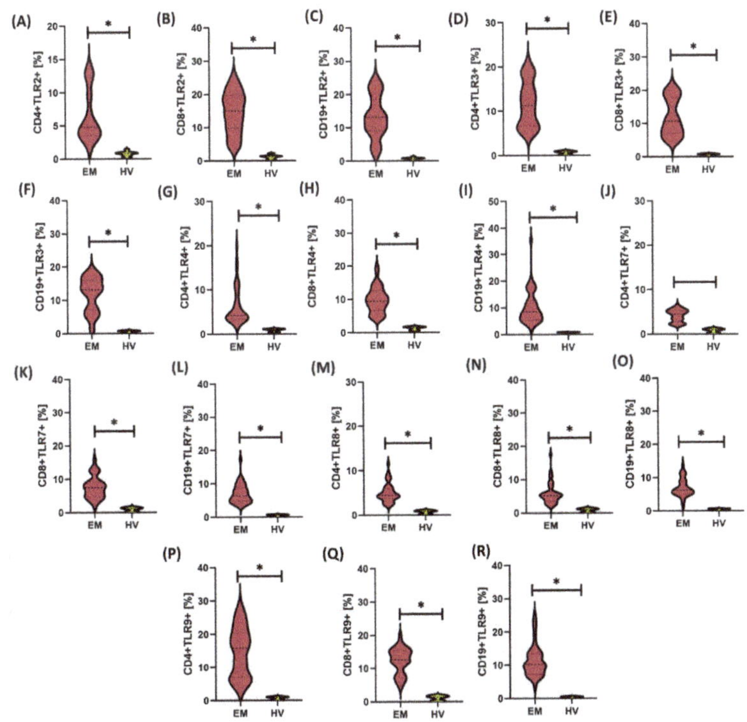

Figure 1. **(A)** Percentage of CD4+ T helper lymphocytes expressing TLR2 in patients with endometriosis (EM) and healthy volunteers (HV); **(B)** Percentage of CD8+ T helper lymphocytes expressing TLR2 in

patients with endometriosis (EM) and healthy volunteers (HV); (**C**) Percentage of CD19+ T helper lymphocytes expressing TLR2 in patients with endometriosis (EM) and healthy volunteers (HV); (**D**) Percentage of CD4+ T helper lymphocytes expressing TLR3 in patients with endometriosis (EM) and healthy volunteers (HV); (**E**) Percentage of CD8+ T helper lymphocytes expressing TLR3 in patients with endometriosis (EM) and healthy volunteers (HV); (**F**) Percentage of CD19+ T helper lymphocytes expressing TLR3 in patients with endometriosis (EM) and healthy volunteers (HV); (**G**) Percentage of CD4+ T helper lymphocytes expressing TLR4 in patients with endometriosis (EM) and healthy volunteers (HV); (**H**) Percentage of CD8+ T helper lymphocytes expressing TLR4 in patients with endometriosis (EM) and healthy volunteers (HV); (**I**) Percentage of CD19+ T helper lymphocytes expressing TLR4 in patients with endometriosis (EM) and healthy volunteers (HV); (**J**) Percentage of CD4+ T helper lymphocytes expressing TLR7 in patients with endometriosis (EM) and healthy volunteers (HV); (**K**) Percentage of CD8+ T helper lymphocytes expressing TLR7 in patients with endometriosis (EM) and healthy volunteers (HV); (**L**) Percentage of CD19+ T helper lymphocytes expressing TLR7 in patients with endometriosis (EM) and healthy volunteers (HV); (**M**) Percentage of CD4+ T helper lymphocytes expressing TLR8 in patients with endometriosis (EM) and healthy volunteers (HV); (**N**) Percentage of CD8+ T helper lymphocytes expressing TLR8 in patients with endometriosis (EM) and healthy volunteers (HV); (**O**) Percentage of CD19+ T helper lymphocytes expressing TLR8 in patients with endometriosis (EM) and healthy volunteers (HV); (**P**) Percentage of CD4+ T helper lymphocytes expressing TLR9 in patients with endometriosis (EM) and healthy volunteers (HV); (**Q**) Percentage of CD8+ T helper lymphocytes expressing TLR9 in patients with endometriosis (EM) and healthy volunteers (HV); (**R**) Percentage of CD19+ T helper lymphocytes expressing TLR9 in patients with endometriosis (EM) and healthy volunteers (HV). Data are presented as violin plots, showing the median and distribution of values. In all analyzed subpopulations, a significantly higher percentage of cells positive for a given TLR was found in the group of patients with endometriosis compared to the control group ($p < 0.05$). The results indicate an increased expression profile of innate immune response receptors in patients with endometriosis. Statistically significant differences are indicated by an asterisk (*).

Analysis of the concentrations of soluble forms of Toll-like receptors (sTLR2, sTLR3, sTLR4, sTLR7, sTLR8, and sTLR9) in the serum and urine of patients with endometriosis and healthy women revealed significant differences both between the studied groups and between the analyzed body fluids. In both groups—patients with endometriosis (EM) and healthy volunteers (HV)—sTLR concentrations in serum were significantly higher than in urine (Figures 2 and 3). The highest serum values in the EM group were noted for sTLR2, sTLR3, and sTLR4. Despite lower concentrations in urine, the levels of these same receptors were also significantly increased compared to the control group, suggesting their potential importance at the local level as well. sTLR7, sTLR8, and sTLR9 showed a similar pattern—significantly higher concentrations in serum than in urine—and these values were increased in patients with endometriosis compared to healthy individuals. An apparent difference was observed for sTLR8 and sTLR9, whose significantly increased serum levels may indicate their dominant participation in the systemic immune response and chronic inflammatory process associated with endometriosis.

3.2. Comparison of Immunological and Biochemical Parameters in Patients with Different Forms of Endometriosis: Peritoneal, Ovarian, Deeply Infiltrating, and Cesarean Section Scar

Endometriosis exhibits significant morphological and anatomical heterogeneity, which is reflected in the clinical classification of the disease into several main types, differing in location, characteristics of the lesions, and their impact on bodily functions. In the next stage of the study, a comparison of selected immunological and biochemical parameters was conducted between patients with the above-mentioned forms of the disease and the HV group (Table 2).

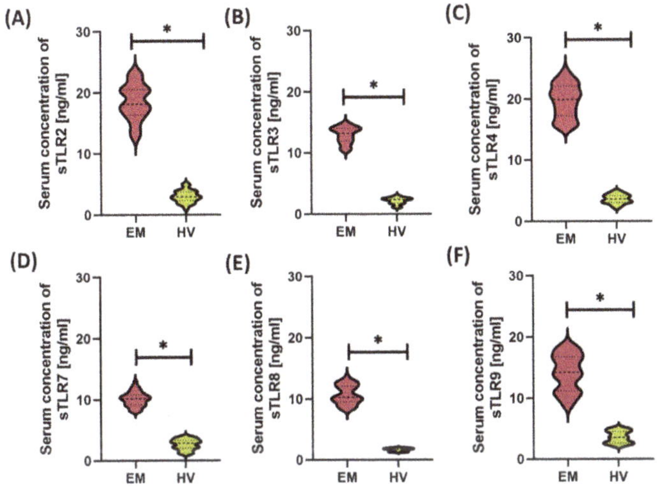

Figure 2. Concentrations of soluble forms of Toll-like receptors (sTLR2, sTLR3, sTLR4, sTLR7, sTLR8, sTLR9) in the serum of patients with endometriosis (EM) and healthy volunteers (HV). (**A**) Serum concentration of sTLR2 in EM and HV groups; (**B**) Serum concentration of sTLR3 in EM and HV groups; (**C**) Serum concentration of sTLR4 in EM and HV groups; (**D**) Serum concentration of sTLR7 in EM and HV groups; (**E**) Serum concentration of sTLR8 in EM and HV groups; (**F**) Serum concentration of sTLR9 in EM and HV groups. In each of the analyzed receptor groups, significantly higher sTLR concentrations were observed in the serum of women with endometriosis compared to the control group ($p < 0.05$). Data are presented as violin plots, showing the median and distribution of values. Statistically significant differences are indicated by an asterisk (*).

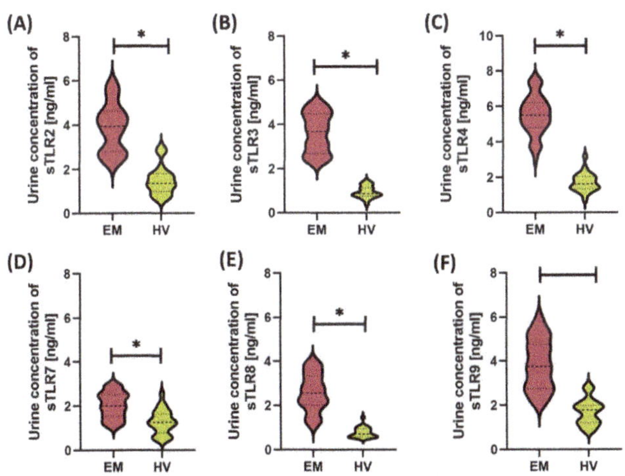

Figure 3. Concentrations of soluble forms of Toll-like receptors (sTLR2, sTLR3, sTLR4, sTLR7, sTLR8, sTLR9) in the urine of patients with endometriosis (EM) and healthy volunteers (HV). (**A**) Urine concentration of sTLR2 in EM and HV groups; (**B**) Urine concentration of sTLR3 in EM and HV groups; (**C**) Urine concentration of sTLR4 in EM and HV groups; (**D**) Urine concentration of sTLR7 in EM and HV groups; (**E**) Urine concentration of sTLR8 in EM and HV groups; (**F**) Urine concentration of sTLR9 in EM and HV groups. In all cases, significantly higher concentrations of sTLRs were observed in the urine of patients with endometriosis compared to the control group ($p < 0.05$), indicating the potential involvement of local inflammatory mechanisms in disease progression. Statistically significant differences are indicated by an asterisk (*).

Table 2. Demographic characteristics of patients with different subtypes of endometriosis and the control group.

Group	*n*	Median Age (Years)	Age Range (Years)
Peritoneal endometriosis (PE)	16	33	26–46
Ovarian endometriosis (OE)	10	35.5	26–47
Deeply infiltrating endometriosis (DIE)	10	36.5	27–47
Cesarean section scar endometriosis (CC)	14	34	24–42
HV	24	33	18–57

The analysis aimed to identify potential differences in systemic inflammatory markers, immunoglobulin profiles, and immune cell phenotypes, which could reflect varying immune activity and the severity of pathological processes depending on the form of the disease.

Detailed results of the comparative analyses are presented in Table S2 and Figure S4 of the Supplementary Materials.

In the assessment of individual hematological parameters between different forms of endometriosis—peritoneal (PE), ovarian (OE), deeply infiltrating (DIE), and cesarean section scar (CC)—no statistically significant differences were found in the number of erythrocytes (RBC), total leukocytes (WBC), lymphocytes (LYM), basophils (BAS), or hemoglobin concentration (HGB), both in comparisons between the endometriosis subtypes and in relation to the control group (HV) (Figure 4B–D,F,I).

The variability was concerned primarily with inflammatory markers and selected immunological parameters. Significantly higher concentrations of CRP were observed in patients with OE, DIE, and CC compared to healthy women, which confirms the intense nature of the inflammatory reaction in these forms of the disease (Figure 4A). Additionally, monocytes (MON) showed significantly higher values in the PE group compared to the DIE group (Figure 4E).

Granulocyte analysis showed that the eosinophil level (EOS) was significantly lower in all endometriosis subtypes compared to the HV group. In turn, the number of neutrophils (NEU) was increased in the PE and CC groups compared to the HV group, which may reflect a nonspecific inflammatory component (Figure 4G,H).

Regarding platelet count (PLT), significantly higher values were noted in patients with endometriosis in the CC group compared to the PE group, which may suggest the presence of local regenerative and inflammatory processes (Figure 4J).

In terms of humoral response, significantly reduced immunoglobulin G (IgG) levels were found in patients with PE, DIE, and CC compared to the HV group. However, the levels of immunoglobulins M and A (IgM and IgA) were significantly increased in all analyzed endometriosis subtypes compared to healthy women, which may indicate the activation of the nonspecific immune response and the mucosal component in the pathogenesis of the disease (Figure 4K–M).

In all analyzed lymphocyte subpopulations, patients with endometriosis demonstrated a significantly higher percentage of TLR-expressing cells compared to healthy volunteers (HV), indicating enhanced immune activation. Among CD4+ T cells, a markedly increased expression of all studied TLRs was observed in the endometriosis groups ($p < 0.0001$ for most comparisons), with the highest levels noted for TLR3 and TLR9. Notably, in the case of TLR4 expression, significant differences were also found between specific disease subtypes—particularly between cesarean section scar endometriosis (CC) and deeply infiltrating endometriosis (DIE)—suggesting distinct patterns of immunopathogenesis

(Supplementary Materials, Figure S11C and Table S3). A similar pattern was observed in cytotoxic CD8+ T lymphocytes, where all subtypes of endometriosis were associated with a significantly elevated percentage of $CD8^{+}$+LR^{+} cells compared to the HV group. The only statistically significant difference between disease subtypes within this population was noted for the CD8+TLR4+subset, which was more prominent in patients with CC compared to DIE (Supplementary Materials, Figure S12C and Table S3). In the case of B cells (CD19+), significantly higher percentages of cells expressing TLR2–TLR4 and TLR7–TLR9 were observed in all endometriosis subtypes compared to the HV group (Supplementary Materials, Figure S13). However, despite observable differences in medians between disease subtypes, no statistically significant intergroup differences were found within the CD19+TLR+ population (Supplementary Materials, Table S3).

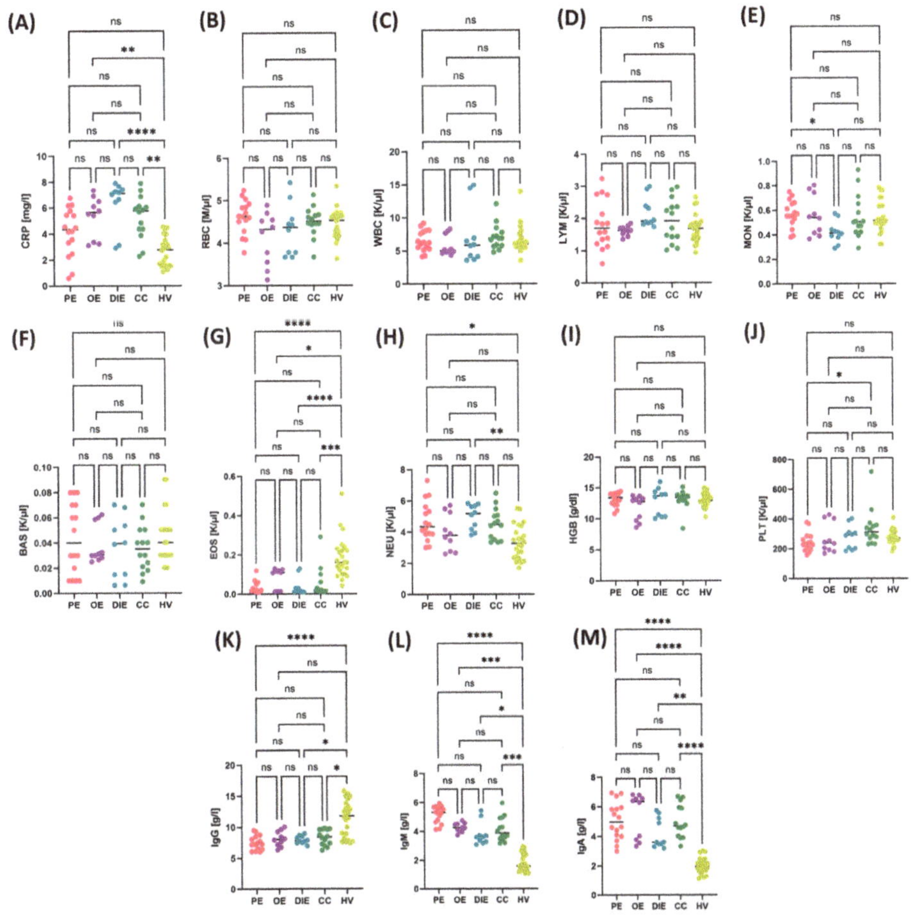

Figure 4. Comparison of selected hematological and immunological parameters between patients with different forms of endometriosis—peritoneal (PE), ovarian (OE), deeply infiltrating (DIE), cesarean section scar (CC)—and healthy volunteers (HV). Statistical significance values are marked as follows: $p \leq 0.05$ (*), $p \leq 0.01$ (**), $p \leq 0.001$ (***), $p \leq 0.0001$ (****), ns—no statistical significance. (**A**) Comparison concentration of CRP between patients with PE, OE, DIE, CC, HV; (**B**) Comparison concentration of RBC between patients with PE, OE, DIE, CC, HV; (**C**) Comparison concentration of WBC between patients with PE, OE, DIE, CC, HV; (**D**) Comparison concentration of LYM between

patients with PE, OE, DIE, CC, HV; (**E**) Comparison concentration of MON between patients with PE, OE, DIE, CC, HV; (**F**) Comparison concentration of BAS between patients with PE, OE, DIE, CC, HV; (**G**) Comparison concentration of EOS between patients with PE, OE, DIE, CC, HV; (**H**) Comparison concentration of NEU between patients with PE, OE, DIE, CC, HV; (**I**) Comparison concentration of HGB between patients with PE, OE, DIE, CC, HV; (**J**) Comparison concentration of PLT between patients with PE, OE, DIE, CC, HV; (**K**) Comparison concentration of IgG between patients with PE, OE, DIE, CC, HV; (**L**) Comparison concentration of IgM between patients with PE, OE, DIE, CC, HV; (**M**) Comparison concentration of IgA between patients with PE, OE, DIE, CC, HV.The next part of the analysis focused on assessing the concentrations of soluble forms of TLRs (sTLR2, sTLR3, sTLR4, sTLR7, sTLR8, and sTLR9) in two types of biological material—serum and urine—from patients with endometriosis. The aim was to determine whether there were differences in the concentration patterns of the analyzed receptors between the individual subtypes and in comparison to the HV group. The numerical data are included in Table S4 of the Supplementary Materials. Analysis of the results concerning serum concentrations showed that all assessed sTLRs were significantly increased in patients with endometriosis, regardless of its form, compared to the control group. At the same time, no significant differences were noted in serum sTLR concentrations between the studied endometriosis subtypes, which may suggest that the increase in these markers is systemic and independent of the location of ectopic endometrial changes (Figure 5).

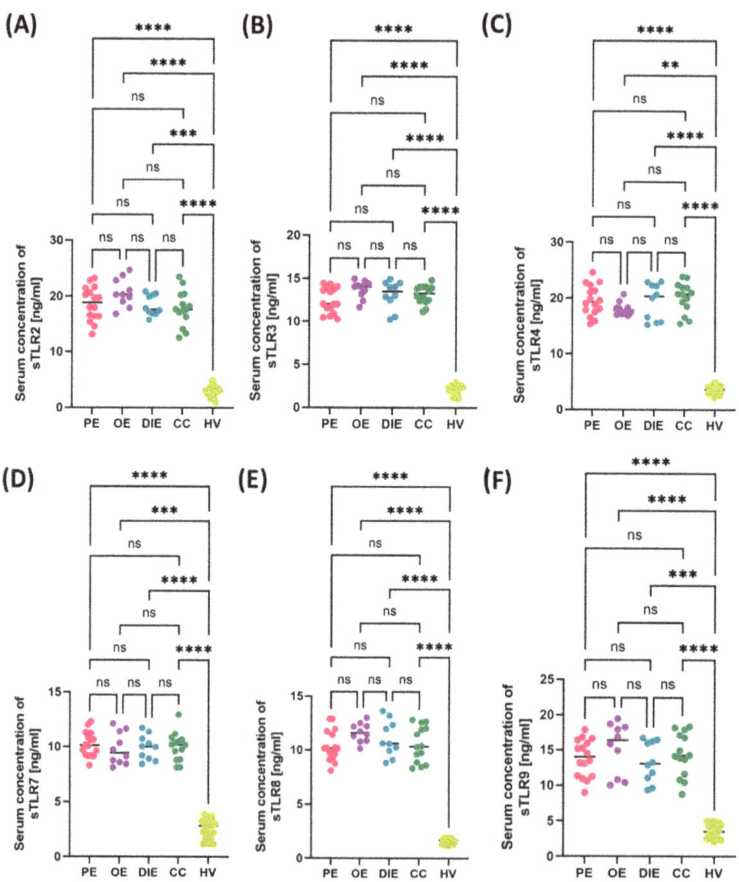

Figure 5. Concentrations of soluble forms of Toll-like receptors (sTLR2–sTLR8) in the serum of patients with different subtypes of endometriosis—peritoneal (PE), ovarian (OE), deeply infiltrating

(DIE), and cesarean section scar (CC)—compared with healthy volunteers (HV). Concentrations of (**A**) sTLR2, (**B**) sTLR3, (**C**) sTLR4, (**D**) sTLR7, (**E**) sTLR8, and (**F**) sTLR9 are presented. Statistical significance values are marked as follows: $p \leq 0.01$ (**), $p \leq 0.001$ (***), $p \leq 0.0001$ (****), ns—no statistical significance.

In the analysis of the concentrations of soluble forms of Toll-like receptors (sTLRs) in urine samples from patients with endometriosis, a general trend similar to that observed in serum was observed. In the case of sTLR2, sTLR3, sTLR4, sTLR8, and sTLR9, the concentration values were significantly increased in all subtypes of endometriosis compared to healthy volunteers, with no significant differences between the subtypes of the disease (Figure 6A–C,E,F). The obtained results indicate that the increased excretion of these soluble receptors into urine may be a general phenomenon accompanying endometriosis, regardless of the location of the disease foci.

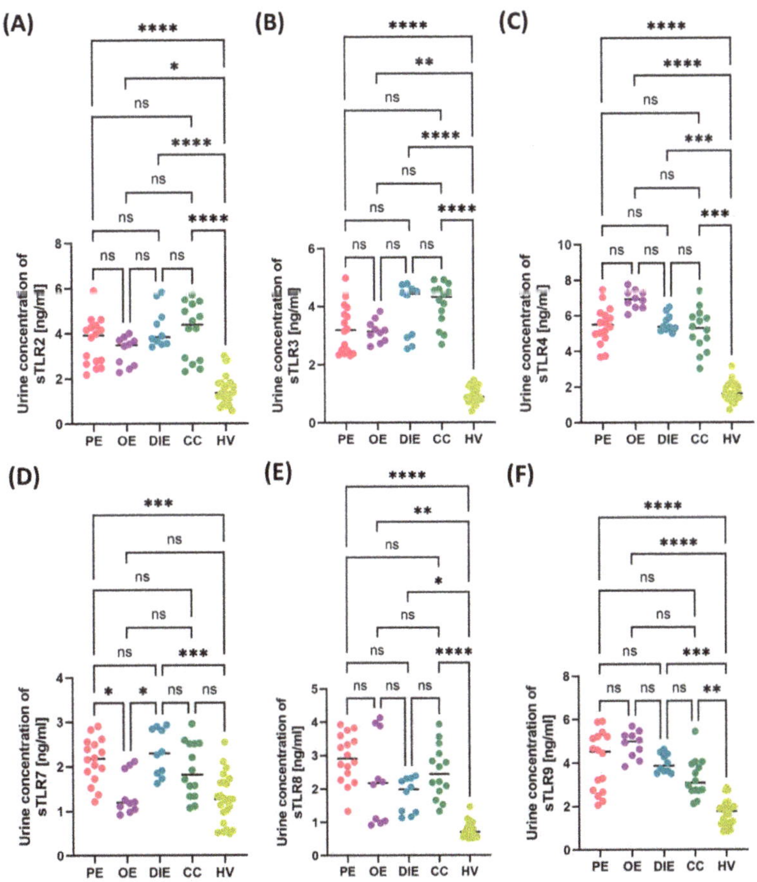

Figure 6. Concentrations of soluble forms of Toll-like receptors (sTLR2–sTLR9) in the urine of patients with different forms of endometriosis—peritoneal (PE), ovarian (OE), deeply infiltrating (DIE), and cesarean section scar (CC)—compared with healthy volunteers (HV). Concentrations of (**A**) sTLR2, (**B**) sTLR3, (**C**) sTLR4, (**D**) sTLR7, (**E**) sTLR8, and (**F**) sTLR9 are presented. Statistical significance values are marked as follows: $p \leq 0.05$ (*), p s≤ 0.01 (**), $p \leq 0.001$ (***), $p \leq 0.0001$ (****), ns—no statistical significance.

However, a different pattern was observed for sTLR7. In this case, statistical analysis revealed significant differences not only between the patient group and the HV group but also between selected subtypes of endometriosis. Urinary sTLR7 levels were significantly higher in PE and DIE patients compared to the HV group. Additionally, significant differences were observed between PE and OE, as well as between DIE and OE, which may suggest a different mechanism of TLR7 involvement in individual clinical forms of the disease (Figure 6D).

3.3. Diagnostic Value of TLR Expression and Their Soluble Forms in Endometriosis—ROC Curve Analysis

In the next stage of our research, we decided to use ROC curves to check the ability of the tested TLRs to differentiate the study group (e.g., patients with endometriosis) from the HV group based on parameters such as sensitivity and specificity. The obtained results are presented in Figure 7. All analyzed TLRs (TLR2, TLR3, TLR4, TLR7, TLR8, and TLR9) exhibited high sensitivity and specificity in the diagnosis of endometriosis, suggesting that their expression on CD4+, CD8+, and CD19+ lymphocytes may serve as a potential marker for differentiating patients with endometriosis from healthy women.

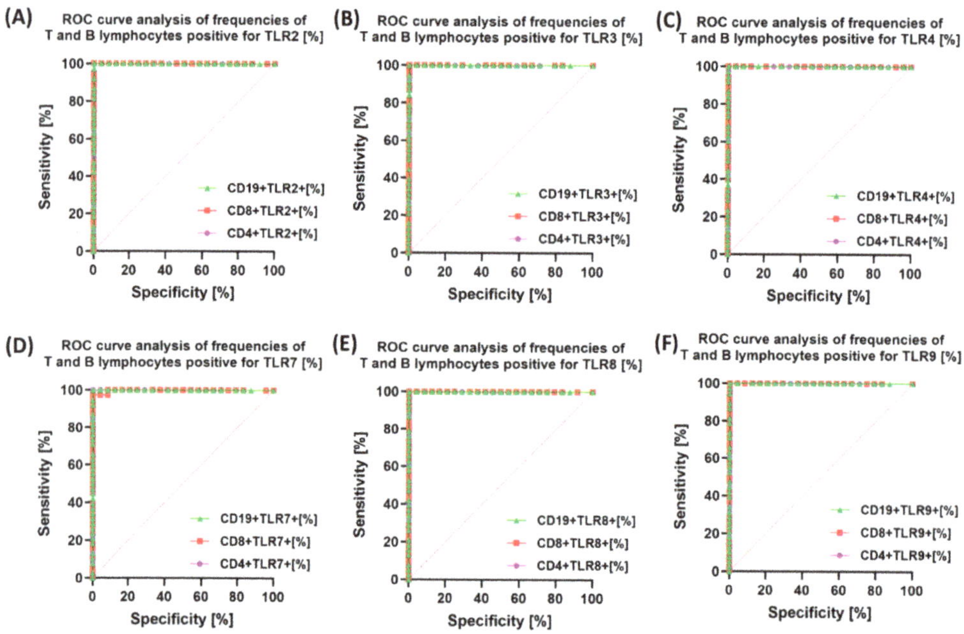

Figure 7. ROC curves illustrating the discriminatory ability of the percentage expression of the tested Toll-like receptors (TLRs) on selected subpopulations of immune system cells (**A–F**) in patients with endometriosis compared with healthy volunteers.

We observed similar trends when analyzing the serum concentrations of the studied sTLRs (Figure 8A). In urine, sTLR3 was the most sensitive, while sTLR7 was the least sensitive and specific (Figure 8B).

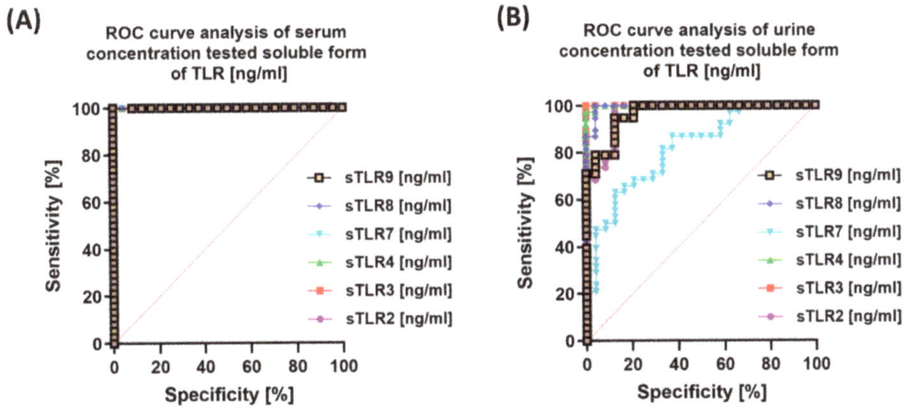

Figure 8. ROC curves showing the diagnostic value of soluble forms of Toll-like receptors (sTLRs) in serum (**A**) and urine (**B**) from patients with endometriosis in comparison to healthy volunteers.

The same analysis was performed for individual subtypes of endometriosis, and the results are presented in Table S5 of the Supplementary Materials. The highest diagnostic value (AUC = 1.0) was demonstrated by CD4 + TLR2, CD4 + TLR3, CD4 + TLR4, CD4 + TLR9, CD8 + TLR2, CD8 + TLR3, CD8 + TLR8, CD19 + TLR2, CD19 + TLR3, and CD19 + TLR9 in the comparisons between PE and HV, as well as OE and HV. High AUC values for these biomarkers confirm their efficacy in distinguishing between the groups and suggest their potential use in diagnosing endometriosis. Similarly, the expression of the same markers in the comparisons of DIE vs. HV and CC vs. HV indicates their usefulness in identifying patients with a more advanced form of the disease. In the case of differentiation between endometriosis phenotypes, AUC values $\geq$ 0.85 were obtained for CD8 + TLR3, CD4 + TLR4, and CD19 + TLR3 in the comparisons of PE vs. CC and DIE vs. CC, suggesting their importance in distinguishing these disease subtypes. Additionally, the expression of CD19 + TLR9 in the comparisons of PE vs. CC (AUC = 0.86) and OE vs. CC (AUC = 0.92) may indicate that TLR9 on B lymphocytes is mainly associated with more advanced forms of endometriosis. Several molecules showed moderate diagnostic value (AUC ranging from 0.7 to 0.85), especially in differentiating between endometriosis phenotypes. Among them, CD4 + TLR2 and CD8 + TLR2 in the comparison of PE vs. DIE (AUC = 0.80) suggest that TLR2 expression increases significantly in DIE compared to PE. Furthermore, CD4 + TLR4 in the comparison of PE vs. CC (AUC = 0.86) confirms the role of this receptor in more advanced forms of endometriosis. High AUC values were also noted for CD8+-TLR8 in the comparisons of PE vs. CC (AUC = 0.88) and OE vs. CC (AUC = 0.88), which may indicate activation of innate immunity in CC. CD19 + TLR3 also demonstrated significant diagnostic value in the comparisons of PE versus CC (AUC = 0.93) and OE versus CC (AUC = 0.95).

Analysis of the diagnostic capacity of sTLR in the serum of patients with endometriosis showed that the highest efficacy in differentiating these patients from healthy women (AUC = 1.0) was achieved by sTLR2, sTLR3, sTLR4, sTLR7, sTLR8, and sTLR9 concentrations in the comparisons of PE vs. HV, OE vs. HV, DIE vs. HV, and CC vs. HV (Supplementary Materials, Table S2). In the context of differentiating endometriosis phenotypes, moderate diagnostic value was demonstrated by sTLR2 (AUC = 0.833) in the comparison of OE vs. CC; sTLR3 (AUC = 0.813) in the comparison of PE vs. OE; and sTLR8 (AUC = 0.813) in the comparison of PE vs. OE. In the assessment of the usefulness for differentiating endometriosis phenotypes, moderate diagnostic value was obtained for sTLR2 in urine (AUC = 0.80) in the comparison of OE vs. DIE; sTLR3 in urine (AUC = 0.857) in the comparison of OE vs. CC; sTLR4 in urine (AUC = 0.922) in the comparisons of PE vs. OE and OE vs. DIE (AUC = 0.933) or OE vs. CC (AUC = 0.929); sTLR7 in urine (AUC = 0.896) in the comparisons of PE vs. OE and OE vs. DIE (AUC = 0.867); sTLR8 in urine (AUC = 0.875) in the comparison of PE vs. DIE; and sTLR9 in urine (AUC = 0.933 and 0.929) in the comparisons of OE vs. DIE and OE vs. CC, respectively.

4. Discussion

Endometriosis is a multifactorial chronic disease characterized by the ectopic presence of endometrial glands and stroma outside the uterine cavity. Its development and progression are closely linked to a dysregulated immune response [37]. Clinically, endometriosis exhibits significant morphological and anatomical heterogeneity, which underlies its classification into several main types, differing in location, characteristics of lesions, and their effect on organ function. The most commonly distinguished subtypes include the following:

- Peritoneal endometriosis (PE): Involves superficial lesions on the serosa of the abdominal and pelvic cavities. These lesions are typically multifocal, small, and vary in color (red, black, or white), often accompanied by local fibrosis and inflammatory infiltration. PE is most frequently diagnosed during laparoscopy and commonly manifests as chronic pelvic pain.
- Ovarian endometriosis (OE): Characterized by the presence of endometriotic ovarian cysts (so-called "chocolate cysts") filled with thick, hemorrhagic content. These lesions contribute to structural damage to ovarian tissue, reduced ovarian reserve, and decreased fertility.
- Deeply infiltrating endometriosis (DIE): The most severe form, defined by lesions penetrating more than 5 mm into extraperitoneal tissues. DIE may affect the sacrotuberous ligaments, rectovaginal septum, bladder wall, intestines, and, rarely, the diaphragm or lungs. This form is typically associated with severe pelvic pain, dyspareunia, gastrointestinal symptoms, and infertility.
- Cesarean section scar endometriosis (CC): A less frequent form, occurring in the subcutaneous tissue or muscle layers of the anterior abdominal wall, usually in the region of surgical scars following cesarean delivery. It typically presents as cyclic pain localized to the scar area that intensifies during menstruation.

Considering these subtypes is of high diagnostic and therapeutic relevance, as each form may differ in clinical presentation, immune response profile, rate of progression, and treatment efficacy. These variations are also believed to reflect distinct underlying pathophysiological mechanisms and immune dysregulations specific to the location and type of ectopic endometrial lesions.

Among the key players in this immune dysfunction are TLRs, which serve critical roles in innate immunity by recognizing PAMPs and DAMPs and by orchestrating downstream inflammatory signaling pathways [38,39].

In particular, TLR4 has garnered considerable attention for its significant involvement in the pathophysiology of endometriosis. Aberrant expression and polymorphisms in the *TLR4* gene may contribute to the initiation of ectopic lesions and the severity of clinical symptoms. Activation of TLR4 by lipopolysaccharides (LPS) of bacterial origin triggers the release of pro-inflammatory cytokines such as interleukin-6 (IL-6) and tumor necrosis factor-alpha (TNF-α), thus creating a chronic inflammatory microenvironment that promotes the persistence and proliferation of endometrial lesions [40–42]. The detection of bacteria such as *Escherichia coli* in menstrual blood further supports the concept that both exogenous and endogenous ligands can activate TLR4 pathways, thereby exacerbating the inflammatory responses associated with endometriosis [43,44].

intracellular TLR signaling pathways and intercellular immune interactions will be crucial to better understand the functional implications of TLR dysregulation in endometriosis.

Longitudinal studies are particularly warranted to investigate the temporal dynamics of TLR expression in disease progression, symptom severity, and treatment response. Furthermore, the inclusion of disease-specific control groups—particularly those with other inflammatory or gynecological conditions—will be essential for establishing the specificity and clinical relevance of TLR-based biomarkers.

In the long term, these insights may inform the development of novel therapeutic strategies aimed at modulating TLR activity, thereby attenuating chronic inflammation and halting the progression of endometriotic lesions, ultimately improving clinical outcomes for affected patients.

Supplementary Materials: The following supporting information can be downloaded at https://www.mdpi.com/article/10.3390/cells14161273/s1, Figure S1: Gating strategy for lymphocyte subpopulations. (A) Forward scatter (FSC-A) versus side scatter (SSC-A) plot for initial lymphocyte gating. (B) CD45 versus SSC-A plot to define leukocytes. (C) CD3 vs CD4 plot for CD4+ T cells, (violet) (D) CD3 vs CD8 plot for CD8+ T cells, (blue) (E) CD3 vs CD19 plot for B cells (orange), (F) CD3 vs CD56+CD16 plot to define natural killer (NK) cells (green), characterized by $CD3^{-}$/CD56+CD16+ phenotype. Figure S2: The figure shows the analysis of TLR2 and TLR4 expression on CD4+ T cells using flow cytometry, including a complete gating strategy and quality control of the assays. Panel (A) illustrates the CD4+ T cell population selected based on the simultaneous expression of the CD3 and CD4 surface markers. Fluorescence minus one (FMO) controls (shown in panels (B) and (D) were used to correctly define the fluorescence threshold and define the boundaries of positive expression; FMO controls for the PE dye (for TLR2) and the APC dye (for TLR4) distinguish the specific signal from autofluorescence and background. Panel (C) shows the level of TLR4 expression (APC dye) within the selected CD4+ cell population, while panel (E) shows the expression of TLR2 (PE dye) in the same population. The use of two independent FMO controls for different fluorescence channels allowed for a reliable assessment of the presence of TLRs on CD4+ T lymphocytes, minimizing the risk of misinterpretation of results resulting from overlapping fluorescence spectra or inhomogeneous background. Figure S3: The figure shows the analysis of TLR7 and TLR8 expression on CD4+ T lymphocytes using flow cytometry and fluorescence signal quality control. Panel (A) shows the gating of the CD3+CD4+ lymphocyte population, which is the starting point for further analyses. Panels (B) and (D) show FMO (fluorescence minus one) controls, respectively for the PE (TLR7) and APC (TLR8) dyes, which allow for precise determination of the positive fluorescence limit and elimination of errors resulting from autofluorescence and background. Panel (C) shows the level of TLR7 expression, while panel (E) shows TLR8 expression on the surface of CD4+ lymphocytes. The use of two independent FMO controls adapted to a specific fluorescence channel allowed for precise and reliable determination of the percentage of cells expressing the analyzed Toll-like receptors, increasing the accuracy and reproducibility of the obtained results. Figure S4: The figure shows the analysis of TLR3 and TLR9 expression on CD4+ T cells by flow cytometry using FMO control. Panel (A) shows the gating of the CD3+CD4+ subpopulation, which served as a starting point for further analysis of Toll-like receptor expression. Panels (B) and (D) show FMO (fluorescence minus one) controls for TLR3 (PE) and TLR9 (APC) detection channels, respectively, which allow precise determination of fluorescence signal positivity thresholds by eliminating the influence of autofluorescence and nonspecific background. Panel (C) shows the detected expression of TLR3, while panel (E) shows the expression of TLR9 on the surface of CD4+ T cells. The use of separate FMO controls for each fluorescence channel allowed for reliable and reproducible determination of the percentage of cells positive for a given receptor, increasing the accuracy of the analysis and the correct interpretation of results. Figure S5: The figure shows the analysis of TLR2 and TLR4 receptor expression on CD8+ T cells using flow cytometry, including FMO control. Panel (A) shows gating of CD3+CD8+ cells, which were the starting population for further analysis of Toll-like receptor expression. Panels (B) and (D), respectively, show FMO controls for TLR4 (APC) and TLR2 (PE) fluorescence channels, used to determine signal positivity thresholds by excluding nonspecific background and autofluorescence. Panel (C) shows the detected TLR4 expression, while panel (E) shows TLR2 expression on CD8+ cells. The use of FMO control allows for precise and reproducible determination of fluorescence gate boundaries for each receptor, which is crucial for reliable assessment of their presence on the analyzed effector cells of the immune system. Figure S6: The figure shows the analysis of TLR7 and TLR8 expression on CD8+ T cells using flow cytometry and the appropriate FMO controls. Panel (A) shows the gating of the CD3+CD8+ cell population, which was the basis for further analysis. Panels (B) and (D) show FMO controls for the appropriate fluorescence channels, allowing precise determination of TLR7 and TLR8 positivity thresholds by eliminating background signal and autofluorescence. Panel (C) shows the detected TLR7 expression, while panel (E) shows TLR8 expression on CD8+ T cells. The use of FMO controls allows for a reliable determination of which cells actually express the tested receptors, thus ensuring the accuracy and reproducibility of the flow cytometric analysis. Figure S7: The figure shows the analysis of TLR3 and TLR9 expression on CD8+ T cells using flow cytometry. Panel (A) shows the CD3+CD8+ cell population that was isolated for further analysis. Panels (B) and (D) show the FMO (fluorescence minus one) controls for TLR3 and TLR9 channels, respectively, which allow for determining the correct threshold of positivity by eliminating background signal and autofluorescence. Panel (C) shows TLR3 receptor expression, and panel (E) shows TLR9 receptor expression in CD8+ cells. The use of two independent FMO controls for each detection channel allows for precise determination of the range of positivity and ensures the reliability and reproducibility of the analysis of Toll-like receptor expression on cytotoxic lymphocytes. Figure S8: The figure shows the analysis of TLR2 and TLR4 expression on B cells (CD19+) using flow cytometry. Panel (A) shows gating of $CD3^{-}$CD19+ lymphocytes, which were isolated for further analysis. In panels (B) and (D) are included appropriate FMO (fluorescence minus one) controls, allowing for determining the borderline of signal positivity for TLR2 and TLR4, taking into account autofluorescence and background. Panel (C) shows TLR4 expression, and panel (E) shows TLR2 expression within selected B lymphocytes. The use of two independent FMO controls allows for reliable and precise analysis of the expression level of the tested Toll-like receptors, eliminating the risk of false positive results. Figure S9: Na przedstawionej rycinie zaprezentowano analizę ekspresji receptorów TLR7 i TLR8 na limfocytach B (CD19+) metodą cytometrii przepływowej. W panelu (A) pokazano bramkowanie komórek $CD3^{-}$CD19+, które zostały wyizolowane do dalszej analizy ekspresji TLR. Panele (B) i (D) zawierają odpowiednie kontrole FMO (fluorescence minus one) umożliwiające precyzyjne ustalenie progów pozytywności fluorescencji dla TLR7 i TLR8, z uwzględnieniem autofluorescencji i sygnału tła. Panel (C) przedstawia poziom ekspresji TLR7, natomiast panel (E) dotyczy ekspresji TLR8 na limfocytach B. Zastosowanie indywidualnych kontroli FMO dla każdego fluorochromu zapewnia wiarygodność uzyskanych wyników, pozwalając na rzetelną ocenę obecności receptorów TLR7 i TLR8 na komórkach CD19+. Figure S10: The figure shows the analysis of TLR3 and TLR9 expression on the surface of B lymphocytes (CD19+) using flow cytometry. The middle panel (A) shows the gating of the $CD3^{-}$CD19+ cell population, which was selected for further analysis. Panels (B) and (D) contain FMO (fluorescence minus one) controls, which allow for precise determination of fluorescence thresholds for the appropriate markers, eliminating the influence of autofluorescence and enabling correct setting of gates for detection of the specific signal. Panel (C) shows the level of TLR3 receptor expression, while panel (E) shows the level of TLR9 receptor expression in the selected CD19+ cell population. The use of appropriate controls and uniform analysis conditions allows for a reliable assessment of the presence of the analyzed receptors in the B lymphocyte subpopulation. Figure S11: Percentage of T helper lymphocytes (CD4+) expressing selected Toll-like receptors: TLR2 (A), TLR3 (B), TLR4 (C), TLR7 (D), TLR8 (E) and TLR9 (F) in individual subtypes of endometriosis—peritoneal (PE), ovarian (OE), deeply infiltrating (DIE), in the cesarean section scar (CC)—and in healthy women (HV). Statistical significance values were marked as follows: $p \leq 0.05$ (*), $p \leq 0.01$ (**), $p \leq 0.001$ (****), $p \leq 0.0001$ (****), ns—no statistical significance. Figure S12: Percentage of cytotoxic T lymphocytes (CD8+) expressing selected Toll-like receptors: TLR2 (A), TLR3 (B), TLR4 (C), TLR7 (D), TLR8 (E) and TLR9 (F) in patients with various forms of endometriosis—peritoneal (PE), ovarian (OE), deeply infiltrating (DIE), endometriosis in the cesarean section scar (CC)—compared to the group of healthy women (HV). Statistical significance values were marked as follows: $p \leq 0.05$ (*), $p \leq 0.01$ (**), $p \leq 0.001$ (****), $p \leq 0.0001$ (****), ns—no statistical significance. Figure S13: Percentage of B lymphocytes (CD19+) expressing selected Toll-like receptors: TLR2 (A), TLR3 (B), TLR4 (C), TLR7 (D), TLR8 (E) and TLR9 (F) in patients with various forms of endometriosis—peritoneal (PE), ovarian (OE), deeply infiltrating (DIE), endometriosis in the cesarean section scar (CC)—compared to the group of healthy women (HV). Statistical significance values were marked as follows: $p \leq 0.05$ (*), $p \leq 0.01$ (**), $p \leq 0.001$ (****), $p \leq 0.0001$ (****), ns—no statistical significance. Table S1: List of conjugated monoclonal antibodies used for flow cytometric analysis, including fluorochrome conjugates, clones, manufacturers, and catalog numbers. Table S2: Comparison of selected hematological (CRP, RBC, WBC, LYM, MON, BAS, EOS, NEU, HGB, PLT), immunological (IgG, IgM, IgA) parameters, and immune cell subpopulations (CD45+, CD3+, CD4+, CD8+, CD19+) between patients with different forms of endometriosis: peritoneal (PE; group 1), ovarian (OE; group 2), deeply infiltrating (DIE; group 3), endometriosis in the cesarean section scar (CC; group 4), and the group of healthy volunteers (HV; group 5). Data are presented as median and interquartile range (Q1–Q3). Table S3: Comparison of the percentage of immune cell subpopulations expressing individual Toll-like receptors (TLR2, TLR3, TLR4, TLR7, TLR8, TLR9) among CD4+ T cells, CD8+ T cells, and CD19+ B cells in patients with different forms of endometriosis: peritoneal (PE; group 1), ovarian (OE; group 2), deeply infiltrating (DIE; group 3), endometriosis in the cesarean section scar (CC; group 4), and healthy volunteers (HV; group 5). Data are presented as median and interquartile range (Q1–Q3). Table S4: Comparison of the concentrations of soluble Toll-like receptors (sTLR2, sTLR3, sTLR4, sTLR7, sTLR8, sTLR9) in serum and urine samples from patients with different forms of endometriosis: peritoneal (PE; group 1), ovarian (OE; group 2), deeply infiltrating (DIE; group 3), endometriosis in the cesarean section scar (CC; group 4), and healthy volunteers (HV; group 5). Data are presented as median and interquartile range (Q1–Q3). Table S5: ROC curve values of the studied parameters of TLR occurrence on individual immune cell subpopulations between individual endometriosis subtypes and healthy volunteers. Table S6: ROC curve values of the studied parameters of sTLR concentration in serum and urine between individual endometriosis subtypes and healthy volunteers.

Author Contributions: Conceptualization, P.M., A.S. and E.G.; methodology, P.M. and S.M.; software, P.M.; validation, A.S., P.M. and S.M.; formal analysis, P.M. and R.T.; investigation, A.S., S.M., P.M., D.D., R.T. and K.B.; resources, E.G., R.T. and M.B.; data curation, S.M.; writing—original draft preparation, A.S., P.M., D.D., M.K., K.B., B.S., B.C., M.B. and S.M.; writing—review and editing, E.G., R.T. and M.B.; visualization, P.M. and S.M.; supervision, R.T. and E.G.; project administration, P.M.; funding acquisition, E.G., A.S., B.S., B.C. and M.B. All authors have read and agreed to the published version of the manuscript.

Funding: This research was funded by the Medical University of Lublin, PBsd 168 and Poznan University of Medical Sciences.

Institutional Review Board Statement: This study was conducted in accordance with the Declaration of Helsinki and approved by the Research Ethics Committee of the Medical University of Lublin (KE-0254/10/01/2023).

Informed Consent Statement: All patients participating in the study signed informed consent and were appropriately informed about the nature and purpose of the study.

Data Availability Statement: The data presented in this publication are available from the first author upon written request.

Conflicts of Interest: The authors declare no conflicts of interest.

Abbreviations

The following abbreviations are used in this manuscript:

BAS	basophils
CC	cesarean section scar endometriosis

CD	cluster of differentiation
CD3	T lymphocytes
CD4	T helper lymphocytes
CD8	cytotoxic T lymphocytes
CD19	B lymphocytes
CD45	total leukocytes
CRP	C-reactive protein
DAMPs	danger-associated molecular patterns
DIE	deeply infiltrating endometriosis
EM	endometriosis
EOS	eosinophils
HGB	hemoglobin
HV	healthy volunteers
IgA	immunoglobulin A
IgG	immunoglobulin G
IgM	immunoglobulin M
IL-6	interleukin-6
IL-8	interleukin-8
LPS	lipopolysaccharides
LYM	lymphocytes
MON	monocytes
MRI	magnetic resonance imaging
MyD88	myeloid differentiation primary response 88
NEU	neutrophils
NF-κB	nuclear factor kappa-light-chain-enhancer of activated B cells
NK	natural killer cells
OE	ovarian endometriosis
PAMPs	pathogen-associated molecular patterns
PE	peritoneal endometriosis
PLT	platelets
PRRs	pattern recognition receptors
ROC	receiver operating characteristic
RBC	red blood cells
sTLR	soluble Toll-like receptor
STAT	signal transducer and activator of transcription
TLR	Toll-like receptor
TLRs	Toll-like receptors
TNF-α	tumor necrosis factor alpha
TRIF	TIR-domain-containing adapter-inducing interferon-β
WBC	white blood cells

References

1. Endometriosis. Available online: https://www.who.int/news-room/fact-sheets/detail/endometriosis (accessed on 26 March 2025).
2. Parasar, P.; Ozcan, P.; Terry, K.L. Endometriosis: Epidemiology, Diagnosis and Clinical Management. *Curr. Obstet. Gynecol. Rep.* **2017**, *6*, 34–41. [CrossRef]
3. Chmaj-Wierzchowska, K.; Kampioni, M.; Wilczak, M.; Opala, T. Do Inflammatory Factors Play a Significant Role in Etiopathogenesis of Endometrial Cysts? Part 1. *Ann. Agric. Environ. Med.* **2013**, *20*, 854–858.
4. Malec-Milewska, M.; Horosz, B.; Sękowska, A.; Kolęda, I.; Kosson, D.; Jakiel, G. Pharmacological Treatment and Regional Anesthesia Techniques for Pain Management after Completion of Both Conservative and Surgical Treatment of Endometriosis and Pelvic Adhesions in Women with Chronic Pelvic Pain as a Mandated Treatment Strategy. *Ann. Agric. Environ. Med.* **2015**, *22*, 353–356. [CrossRef]
5. Słabuszewska-Jóźwiak, A.; Ciebiera, M.; Baran, A.; Jakiel, G. Effectiveness of Laparoscopic Surgeries in Treating Infertility Related to Endometriosis. *Ann. Agric. Environ. Med.* **2015**, *22*, 329–331. [CrossRef] [PubMed]

6. Santulli, P.; Tran, C.; Gayet, V.; Bourdon, M.; Maignien, C.; Marcellin, L.; Pocate-Cheriet, K.; Chapron, C.; de Ziegler, D. Oligo-Anovulation Is Not a Rarer Feature in Women with Documented Endometriosis. *Fertil. Steril.* **2018**, *110*, 941–948. [CrossRef] [PubMed]
7. Zhang, H.; Sheng, S.; Pan, Z.; Zhao, L.; Yang, C.; Li, C.; Wang, F. Immune and Endocrine Regulation in Endometriosis: What We Know. *J. Endometr. Uterine Disord.* **2023**, *4*, 100049. [CrossRef]
8. Chopyak, V.V.; Koval, H.D.; Havrylyuk, A.M.; Lishchuk-Yakymovych, K.A.; Potomkina, H.A.; Kurpisz, M.K. Immunopathogenesis of Endometriosis—A Novel Look at an Old Problem. *Cent. Eur. J. Immunol.* **2022**, *47*, 109–116. [CrossRef]
9. Abramiuk, M.; Grywalska, E.; Małkowska, P.; Sierawska, O.; Hrynkiewicz, R.; Niedźwiedzka-Rystwej, P. The Role of the Immune System in the Development of Endometriosis. *Cells* **2022**, *11*, 2028. [CrossRef]
10. Liu, H.; Li, J.; Guan, C.; Gao, W.; Li, Y.; Wang, J.; Yang, Y.; Du, Y. Endometriosis Is a Disease of Immune Dysfunction, Which Could Be Linked to Microbiota. *Front. Genet.* **2024**, *15*, 1386411. [CrossRef]
11. Agostinis, C.; Balduit, A.; Mangogna, A.; Zito, G.; Romano, F.; Ricci, G.; Kishore, U.; Bulla, R. Immunological Basis of the Endometriosis: The Complement System as a Potential Therapeutic Target. *Front. Immunol.* **2021**, *11*, 599117. [CrossRef]
12. Khan, K.N.; Kitajima, M.; Fujishita, A.; Nakashima, M.; Masuzaki, H. Toll-like Receptor System and Endometriosis. *J. Obstet. Gynaecol. Res.* **2013**, *39*, 1281–1292. [CrossRef]
13. De Azevedo, B.C.; Mansur, F.; Podgaec, S. A Systematic Review of Toll-like Receptors in Endometriosis. *Arch. Gynecol. Obstet.* **2021**, *304*, 309–316. [CrossRef]
14. Guo, B.; Chen, J.H.; Zhang, J.H.; Fang, Y.; Liu, X.J.; Zhang, J.; Zhu, H.Q.; Zhan, L. Pattern-Recognition Receptors in Endometriosis: A Narrative Review. *Front. Immunol.* **2023**, *14*, 1161606. [CrossRef]
15. Almasi, M.Z.; Hosseini, E.; Jafari, R.; Aflatoonian, K.; Aghajanpour, S.; Ramazanali, F.; Moini, A.; Shahhoseini, M.; Afsharian, P.; Aflatoonian, R. Evaluation of Toll-like Receptor 3 (TLR3) Signaling Pathway Genes and Its Genetic Polymorphisms in Ectopic and Eutopic Endometrium of Women with Endometriosis. *J. Gynecol. Obstet. Hum. Reprod.* **2021**, *50*, 102153. [CrossRef]
16. Jiang, C.; Liu, C.; Guo, J.; Chen, L.; Luo, N.; Qu, X.; Yang, W.; Ren, Q.; Cheng, Z. The Expression of Toll-like Receptors in Eutopic and Ectopic Endometrium and Its Implication in the Inflammatory Pathogenesis of Adenomyosis. *Sci. Rep.* **2017**, *7*, 7365. [CrossRef] [PubMed]
17. Jafari, R.; Taghavi, S.A.; Amirchaghmaghi, E.; Yazdi, R.S.; Karimian, L.; Ashrafi, M.; Aflatoonian, R. Detailed Investigation of Downstream TLR Signaling in the Follicular Cells of Women with Endometriosis. *J. Reprod. Infertil.* **2020**, *21*, 231–239. [CrossRef] [PubMed]
18. Kusama, K.; Satoyoshi, A.; Azumi, M.; Yoshie, M.; Kojima, J.; Mizuno, Y.; Ono, M.; Nishi, H.; Kajihara, T.; Tamura, K. Toll-like Receptor Signaling Pathway Triggered by Inhibition of Serpin A1 Stimulates Production of Inflammatory Cytokines by Endometrial Stromal Cells. *Front. Endocrinol.* **2022**, *13*, 966455. [CrossRef]
19. Khan, K.N.; Fujishita, A.; Hiraki, K.; Kitajima, M.; Nakashima, M.; Fushiki, S.; Kitawaki, J. Bacterial Contamination Hypothesis: A New Concept in Endometriosis. *Reprod. Med. Biol.* **2018**, *17*, 125–133. [CrossRef]
20. Qin, R.; Tian, G.; Liu, J.; Cao, L. The Gut Microbiota and Endometriosis: From Pathogenesis to Diagnosis and Treatment. *Front. Cell Infect. Microbiol.* **2022**, *12*, 1069557. [CrossRef] [PubMed]
21. Talwar, C.; Singh, V.; Kommagani, R. The Gut Microbiota: A Double-Edged Sword in Endometriosis. *Biol. Reprod.* **2022**, *107*, 881–901. [CrossRef]
22. Guo, C.; Zhang, C. Role of the Gut Microbiota in the Pathogenesis of Endometriosis: A Review. *Front. Microbiol.* **2024**, *15*, 1363455. [CrossRef]
23. Malvezzi, H.; Cestari, B.A.; Mendes, H.; Hernandes, C.; Podgaec, S. Peritoneal Fluid Microbiota Profile of Patients with Deep Endometriosis. *Microb. Pathog.* **2025**, *199*, 107244. [CrossRef] [PubMed]
24. Zhang, B.; Mohd Sahardi, N.F.N.; Di, W.; Long, X.; Shafiee, M.N. The Gut–Endometrium Axis: Exploring the Role of Microbiome in the Pathogenesis and Treatment of Endometrial Cancer—A Narrative Review. *Cancers* **2025**, *17*, 1044. [CrossRef]
25. Weber, I.; Sienko, A.; Urban, A.; Szwed, C.; Czajkowski, K.; Basta, P.; Sienko, J. Relationship between the Gut Microbiome and Endometriosis and Its Role in Pathogenesis, Diagnosis, and Treatment: A Systematic Review. *Ginekol. Pol.* **2024**, *95*, 893–901. [CrossRef]
26. Uzuner, C.; Mak, J.; El-Assaad, F.; Condous, G. The Bidirectional Relationship between Endometriosis and Microbiome. *Front. Endocrinol.* **2023**, *14*, 1110824. [CrossRef]
27. Hearn-Yeates, F.; Horne, A.W.; O'Mahony, S.M.; Saunders, P.T.K. Microbiome: The Impact of the Microbiota–Gut–Brain Axis on Endometriosis-Associated Symptoms: Mechanisms and Opportunities for Personalised Management Strategies. *Reprod. Fertil.* **2024**, *5*, e230085. [CrossRef] [PubMed]
28. Greygoose, E.; Metharom, P.; Kula, H.; Seckin, T.K.; Seckin, T.A.; Ayhan, A.; Yu, Y. The Estrogen–Immune Interface in Endometriosis. *Cells* **2025**, *14*, 58. [CrossRef]

29. Eisa, M.A.; Mansour, A.M.; Salama, S.A.; Elsadek, B.E.M.; Ashour, A.A.; Abdelghany, T.M. Estrogen/Estrogen Receptor Activation Protects against DEN-Induced Liver Fibrosis in Female Rats via Modulating TLR-4/NF-Kβ Signaling. *Eur. J. Pharmacol.* **2023**, *960*, 176165. [CrossRef]
30. Zandieh, Z.; Amjadi, F.; Ashrafi, M.; Aflatoonian, A.; Fazeli, A.; Aflatoonian, R. The Effect of Estradiol and Progesterone on Toll Like Receptor Gene Expression in A Human Fallopian Tube Epithelial Cell Line. *Cell J.* **2016**, *17*, 678–691. [CrossRef]
31. Gou, Y.; Li, X.; Li, P.; Zhang, H.; Xu, T.; Wang, H.; Wang, B.; Ma, X.; Jiang, X.; Zhang, Z. Estrogen Receptor β Upregulates CCL2 via NF-κB Signaling in Endometriotic Stromal Cells and Recruits Macrophages to Promote the Pathogenesis of Endometriosis. *Hum. Reprod.* **2019**, *34*, 646–658. [CrossRef] [PubMed]
32. Huhtinen, K.; Desai, R.; Ståhle, M.; Salminen, A.; Handelsman, D.J.; Perheentupa, A.; Poutanen, M. Endometrial and Endometriotic Concentrations of Estrone and Estradiol Are Determined by Local Metabolism Rather than Circulating Levels. *J. Clin. Endocrinol. Metab.* **2012**, *97*, 4228–4235. [CrossRef]
33. Su, L.; Sun, Y.; Ma, F.; Lü, P.; Huang, H.; Zhou, J. Progesterone Inhibits Toll-like Receptor 4-Mediated Innate Immune Response in Macrophages by Suppressing NF-κB Activation and Enhancing SOCS1 Expression. *Immunol. Lett.* **2009**, *125*, 151–155. [CrossRef] [PubMed]
34. Endometriosis. Available online: https://www.asrm.org/practice-guidance/coding-resources/coding/coding-summaries/endometriosis/ (accessed on 14 May 2025).
35. Kedzia, M.; Basta, P.; Czajkowski, K.; Gogacz, M.; Spaczynski, R.; Mroczkowska, B.; Stojko, R.; Szaflik, T.; Szubert, M.; Szyllo, K.; et al. Guidelines of the Polish Society of Gynecologists and Obstetricians on the Management of Women with Endometriosis. *Ginekol. Pol.* **2024**, *95*, 729–758. [CrossRef] [PubMed]
36. Kedzia, M.; Osinski, M.; Mantaj, U.; Wender-Ozegowska, E. Endometriosis Is Associated with an Increased Whole-Blood Thrombogenicity Detected by a Novel Automated Microchip Flow-Chamber System (T-TAS®). *Ginekol. Pol.* **2023**, *94*, 291–297. [CrossRef]
37. Chen, Y.; Waseem, S.; Luo, L. Advances in the Diagnosis and Management of Endometriosis: A Comprehensive Review. *Pathol.—Res. Pract.* **2025**, *266*, 155813. [CrossRef]
38. Duan, T.; Du, Y.; Xing, C.; Wang, H.Y.; Wang, R.-F. Toll-Like Receptor Signaling and Its Role in Cell-Mediated Immunity. *Front. Immunol.* **2022**, *13*, 812774. [CrossRef]
39. Chen, Y.-H.; Wu, K.-H.; Wu, H.-P. Unraveling the Complexities of Toll-like Receptors: From Molecular Mechanisms to Clinical Applications. *Int. J. Mol. Sci.* **2024**, *25*, 5037. [CrossRef]
40. Marchionni, E.; Porpora, M.G.; Megiorni, F.; Piacenti, I.; Giovannetti, A.; Marchese, C.; Benedetti Panici, P.; Pizzuti, A. TLR4 T399I Polymorphism and Endometriosis in a Cohort of Italian Women. *Diagnostics* **2020**, *10*, 255. [CrossRef]
41. Khan, K.N.; Akira, F.; Kitawaki, J.; Mori, T. Educational Lecture 1—Role of Innate and Adaptive Immunity in Endometriosis. *J. Reprod. Immunol.* **2022**, *153*, 103696. [CrossRef]
42. Su, W.; Cui, H.; Wu, D.; Yu, J.; Ma, L.; Zhang, X.; Huang, Y.; Ma, C. Suppression of TLR4-MyD88 Signaling Pathway Attenuated Chronic Mechanical Pain in a Rat Model of Endometriosis. *J. Neuroinflammation* **2021**, *18*, 65. [CrossRef]
43. Khan, K.N.; Kitajima, M.; Hiraki, K.; Yamaguchi, N.; Katamine, S.; Matsuyama, T.; Nakashima, M.; Fujishita, A.; Ishimaru, T.; Masuzaki, H. Escherichia Coli Contamination of Menstrual Blood and Effect of Bacterial Endotoxin on Endometriosis. *Fertil. Steril.* **2010**, *94*, 2860–2863.e3. [CrossRef] [PubMed]
44. Campos, G.B.; Marques, L.M.; Rezende, I.S.; Barbosa, M.S.; Abrão, M.S.; Timenetsky, J. Mycoplasma Genitalium Can Modulate the Local Immune Response in Patients with Endometriosis. *Fertil. Steril.* **2018**, *109*, 549–560.e4. [CrossRef]
45. Hayashi, C.; Chishima, F.; Sugitani, M.; Ichikawa, G.; Nakazawa-Watanabe, T.; Sugita, K.; Suzuki, M.; Nemoto, N.; Yamamoto, T. Relationship Between Toll-like Receptor-4 and mPGES-1 Gene Expression in Local Lesions of Endometriosis Patients. *Am. J. Reprod. Immunol.* **2013**, *69*, 231–239. [CrossRef]
46. Latha, M.; Vaidya, S.; Movva, S.; Chava, S.; Govindan, S.; Govatati, S.; Banoori, M.; Hasan, Q.; Kodati, V.L. Molecular Pathogenesis of Endometriosis; Toll-Like Receptor-4 A896G (D299G) Polymorphism: A Novel Explanation. *Genet. Test. Mol. Biomark.* **2011**, *15*, 181–184. [CrossRef]
47. Sobstyl, M.; Niedźwiedzka-Rystwej, P.; Grywalska, E.; Korona-Głowniak, I.; Sobstyl, A.; Bednarek, W.; Roliński, J. Toll-Like Receptor 2 Expression as a New Hallmark of Advanced Endometriosis. *Cells* **2020**, *9*, 1813. [CrossRef] [PubMed]
48. Attaman, J.A.; Stanic, A.K.; Kim, M.; Lynch, M.P.; Rueda, B.R.; Styer, A.K. The Anti-Inflammatory Impact of Omega-3 Polyunsaturated Fatty Acids During the Establishment of Endometriosis-Like Lesions. *Am. J. Reprod. Immunol.* **2014**, *72*, 392–402. [CrossRef] [PubMed]
49. Kajihara, H.; Yamada, Y.; Kanayama, S.; Furukawa, N.; Noguchi, T.; Haruta, S.; Yoshida, S.; Sado, T.; Oi, H.; Kobayashi, H. New Insights into the Pathophysiology of Endometriosis: From Chronic Inflammation to Danger Signal. *Gynecol. Endocrinol.* **2011**, *27*, 73–79. [CrossRef]
50. Henry, L.; Tremblay-Lemoine, P.-L.; Timmermans, M.; Ravet, S.; Munaut, C.; Nisolle, M.; Henry, L. The Endometrial Microbiota: Challenges and Prospects. *Medicina* **2023**, *59*, 1540. [CrossRef]

51. Hernandes, C.; Gueuvoghlanian-Silva, B.Y.; Monnaka, V.U.; Ribeiro, N.M.; Pereira, W.d.O.; Podgaec, S. Regulatory T Cells Isolated from Endometriotic Peritoneal Fluid Express a Different Number of Toll-like Receptors. *Einstein* **2020**, *18*, eAO5294. [CrossRef]
52. Xiao, F.; Liu, X.; Guo, S.-W. Platelets and Regulatory T Cells May Induce a Type 2 Immunity That Is Conducive to the Progression and Fibrogenesis of Endometriosis. *Front. Immunol.* **2020**, *11*, 610963. [CrossRef]
53. Podgaec, S.; Rizzo, L.V.; Fernandes, L.F.C.; Baracat, E.C.; Abrao, M.S. CD4+CD25highFoxp3+ Cells Increased in the Peritoneal Fluid of Patients with Endometriosis. *Am. J. Reprod. Immunol.* **2012**, *68*, 301–308. [CrossRef]
54. Peluso, C.; Christofolini, D.M.; Goldman, C.S.; Mafra, F.A.; Cavalcanti, V.; Barbosa, C.P.; Bianco, B. TYK2 Rs34536443 Polymorphism Is Associated with a Decreased Susceptibility to Endometriosis-Related Infertility. *Hum. Immunol.* **2013**, *74*, 93–97. [CrossRef]
55. Klemmt, P.A.B.; Starzinski-Powitz, A. Molecular and Cellular Pathogenesis of Endometriosis. *Curr. Women's Health Rev.* **2018**, *14*, 106–116. [CrossRef]
56. Riccio, L.G.C.; Jeljeli, M.; Santulli, P.; Chouzenoux, S.; Doridot, L.; Nicco, C.; Reis, F.M.; Abrão, M.S.; Chapron, C.; Batteux, F. B Lymphocytes Inactivation by Ibrutinib Limits Endometriosis Progression in Mice. *Hum. Reprod.* **2019**, *34*, 1225–1234. [CrossRef] [PubMed]
57. Kisovar, A.; Becker, C.M.; Granne, I.; Southcombe, J.H. The Role of CD8+ T Cells in Endometriosis: A Systematic Review. *Front. Immunol.* **2023**, *14*, 1225639. [CrossRef] [PubMed]
58. Reis, J.L.; Rosa, N.N.; Martins, C.; Ângelo-Dias, M.; Borrego, L.M.; Lima, J. The Role of NK and T Cells in Endometriosis. *Int. J. Mol. Sci.* **2024**, *25*, 10141. [CrossRef] [PubMed]
59. Calmon, M.S.; Lemos, F.F.B.; Silva Luz, M.; Rocha Pinheiro, S.L.; de Oliveira Silva, L.G.; Correa Santos, G.L.; Rocha, G.R.; Freire de Melo, F. Immune Pathway through Endometriosis to Ovarian Cancer. *World J. Clin. Oncol.* **2024**, *15*, 496–522. [CrossRef]
60. Park, M.; Kim, Y.S.; Song, H. Macrophages: A Double-Edged Sword in Female Reproduction and Disorders. *Exp. Mol. Med.* **2025**, *57*, 285–297. [CrossRef]
61. Kobayashi, H.; Imanaka, S. Understanding the Molecular Mechanisms of Macrophage Polarization and Metabolic Reprogramming in Endometriosis: A Narrative Review. *Reprod. Med. Biol.* **2022**, *21*, e12488. [CrossRef]
62. Panir, K.; Hull, M.L.; Greaves, E. Chapter 2—Macrophages in Endometriosis: They Came, They Saw, They Conquered. In *Immunology of Endometriosis*; Koga, K., Ed.; Reproductive Immunology; Academic Press: Cambridge, MA, USA, 2022; pp. 13–41, ISBN 978-0-12-820661-4.
63. Ścieżyńska, A.; Komorowski, M.; Soszyńska, M.; Malejczyk, J. NK Cells as Potential Targets for Immunotherapy in Endometriosis. *J. Clin. Med.* **2019**, *8*, 1468. [CrossRef]
64. Meggyes, M.; Szereday, L.; Bohonyi, N.; Koppan, M.; Szegedi, S.; Marics-Kutas, A.; Marton, M.; Totsimon, A.; Polgar, B. Different Expression Pattern of TIM-3 and Galectin-9 Molecules by Peripheral and Peritoneal Lymphocytes in Women with and without Endometriosis. *Int. J. Mol. Sci.* **2020**, *21*, 2343. [CrossRef]
65. Sourial, S.; Tempest, N.; Hapangama, D.K. Theories on the Pathogenesis of Endometriosis. *Int. J. Reprod. Med.* **2014**, *2014*, 179515. [CrossRef]
66. Ding, S.; Lin, Q.; Zhu, T.; Li, T.; Zhu, L.; Wang, J.; Zhang, X. Is There a Correlation between Inflammatory Markers and Coagulation Parameters in Women with Advanced Ovarian Endometriosis? *BMC Women's Health* **2019**, *19*, 169. [CrossRef]
67. El-Zayat, S.R.; Sibaii, H.; Mannaa, F.A. Toll-like Receptors Activation, Signaling, and Targeting: An Overview. *Bull. Natl. Res. Cent.* **2019**, *43*, 187. [CrossRef]
68. Fakhry, N.; Gowily, A.; Okda, T.; Houssen, M. Serum Soluble Toll-like Receptor 2 and 4 as Diagnostic and Prognostic Biomarkers for Non-Hodgkin Lymphoma. *Contemp. Oncol.* **2020**, *24*, 157–162. [CrossRef]
69. AlQallaf, H.; Hamada, Y.; Blanchard, S.; Shin, D.; Gregory, R.; Srinivasan, M. Differential Profiles of Soluble and Cellular Toll like Receptor (TLR)-2 and 4 in Chronic Periodontitis. *PLoS ONE* **2018**, *13*, e0200231. [CrossRef]
70. Ten Oever, J.; Kox, M.; van de Veerdonk, F.L.; Mothapo, K.M.; Slavcovici, A.; Jansen, T.L.; Tweehuysen, L.; Giamarellos-Bourboulis, E.J.; Schneeberger, P.M.; Wever, P.C.; et al. The Discriminative Capacity of Soluble Toll-like Receptor (sTLR)2 and sTLR4 in Inflammatory Diseases. *BMC Immunol.* **2014**, *15*, 55. [CrossRef]
71. Song, J.; Abraham, S.N. TLR Mediated Immune Responses in the Urinary Tract. *Curr. Opin. Microbiol.* **2008**, *11*, 66–73. [CrossRef] [PubMed]
72. Jiang, L.; Zhang, M.; Wang, S.; Han, Y.; Fang, X. Common and Specific Gene Signatures among Three Different Endometriosis Subtypes. *PeerJ* **2020**, *8*, e8730. [CrossRef] [PubMed]
73. Imperiale, L.; Nisolle, M.; Noël, J.-C.; Fastrez, M. Three Types of Endometriosis: Pathogenesis, Diagnosis and Treatment. State of the Art. *J. Clin. Med.* **2023**, *12*, 994. [CrossRef] [PubMed]
74. Arkoudis, N.-A.; Moschovaki-Zeiger, O.; Prountzos, S.; Spiliopoulos, S.; Kelekis, N. Caesarean-Section Scar Endometriosis (CSSE): Clinical and Imaging Fundamentals of an Underestimated Entity. *Clin. Radiol.* **2023**, *78*, 644–654. [CrossRef] [PubMed]

75. Liu, Y.; Wang, J.; Zhang, X. An Update on the Multifaceted Role of NF-kappaB in Endometriosis. *Int. J. Biol. Sci.* **2022**, *18*, 4400–4413. [CrossRef]
76. Zdrojkowski, Ł.; Jasiński, T.; Ferreira-Dias, G.; Pawliński, B.; Domino, M. The Role of NF-κB in Endometrial Diseases in Humans and Animals: A Review. *Int. J. Mol. Sci.* **2023**, *24*, 2901. [CrossRef]
77. Inui, H.; Kawakita, T.; Murayama, M.; Nakagawa, T.; Sasada, H.; Shinohara, A.; Aragaki, R.; Kagawa, T.; Kadota, Y.; Kato, T.; et al. Effects of STAT Inhibitors in Mouse Models of Endometriosis. *Reprod. Sci.* **2023**, *30*, 2449–2456. [CrossRef] [PubMed]
78. Wilson, T.R.; Kasper, S.; Burns, K.A. An Emerging Role for Neutrophils in the Pathogenesis of Endometriosis. *npj Womens Health* **2025**, *3*, 9. [CrossRef]

cells

MDPI

Review

Mapping Human Uterine Disorders Through Single-Cell Transcriptomics

Sandra Boldu-Fernández [1,†], Carolina Lliberos [1,†], Carlos Simon [1,2,3] and Aymara Mas [1,*]

[1] Carlos Simón Foundation, INCLIVA Health Research Institute, 46010 Valencia, Spain; sboldu@incliva.es (S.B.-F.); carolina.lliberos@gmail.com (C.L.); csimon@fundacioncarlossimon.com (C.S.)
[2] Department of Obstetrics and Gynecology, Universidad de Valencia, 46010 Valencia, Spain
[3] Department of Pediatrics, Obstetrics and Gynecology, Beth Israel Deaconess Medical Center, Harvard University, Boston, MA 02215, USA
* Correspondence: amas@fundacioncarlossimon.com
† These authors contributed equally to this work.

Abstract: Disruptions in uterine tissue function contribute to disorders such as endometriosis, adenomyosis, endometrial cancer, and fibroids, which all significantly impact health and fertility. Advances in transcriptomics, particularly single-cell RNA sequencing, have revolutionized uterine biological research by revealing the cellular heterogeneity and molecular mechanisms underlying disease states. Single-cell RNA sequencing and spatial transcriptomics have mapped endometrial and myometrial cellular landscapes, which helped to identify critical cell types, signaling pathways, and phase-specific dynamics. Said transcriptomic technologies also identified stromal and immune cell dysfunctions, such as fibroblast-to-myofibroblast transitions and impaired macrophage activity, which drive fibrosis, chronic inflammation, and lesion persistence in endometriosis. For endometrial cancer, scRNA-seq uncovered tumor microenvironmental complexities, identifying cancer-associated fibroblast subtypes and immune cell profiles contributing to progression and therapeutic resistance. Similarly, studies on adenomyosis highlighted disrupted signaling pathways, including Wnt and VEGF, and novel progenitor cell populations linked to tissue invasion and neuroinflammation, while single-cell approaches characterized smooth muscle and fibroblast subpopulations in uterine fibroids, elucidating their roles in extracellular matrix remodeling and signaling pathways like ERK and mTOR. Despite challenges such as scalability and reproducibility, single-cell transcriptomic approaches may have potential applications in biomarker discovery, therapeutic target identification, and personalized medicine in gynecological disorders.

Keywords: single-cell RNA sequencing; uterine disorders; endometriosis; endometrial cancer; adenomyosis; uterine fibroids/leiomyoma

1. Introduction

The uterus, a dynamic muscular organ that undergoes cyclical changes throughout a woman's reproductive years, develops from the paramesonephric ducts at 9–10 weeks of gestation and matures to adapt to physiological processes such as pregnancy and menopause [1]. By 14 weeks, the uterus differentiates into two distinct layers: the primordial endometrium and the myometrium [2].

The endometrium, the dynamic tissue lining the uterine cavity, is divided into two layers [3]. The *functionalis* layer contains various cell types, including epithelial, stromal, immune, and vascular cells, and it undergoes cyclic shedding, repair, and regeneration in response to ovarian hormones. The *basalis* layer, housing epithelial stem/progenitor cells, regenerates the *functionalis* after menstruation [4]. Below the endometrium lies the myometrium, which is a smooth muscle layer essential for menstruation and parturition regulated by hormones like progesterone, estrogen, and oxytocin [5–7]. Disruptions in these tissues can cause disorders affecting health and fertility, emphasizing the need to understand uterine biology for therapeutic advancements (Figure 1A).

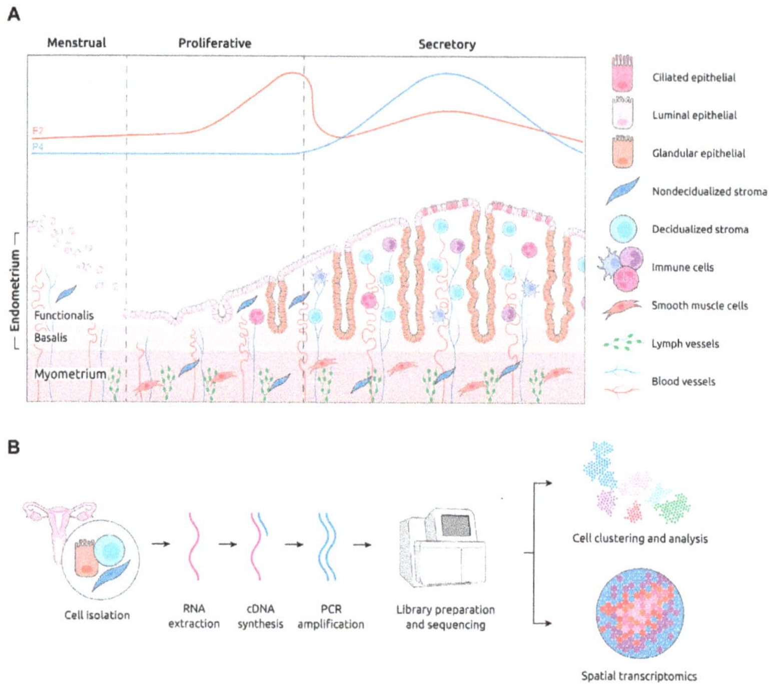

Figure 1. Illustration of (**A**) the human uterus showing the different layers and the morphological and cellular changes based on the phase of the menstrual cycle (i.e., menstrual, proliferative and secretory phase). Red and blue lines represent estradiol (E2) and progesterone (P4) fluctuations, respectively; (**B**) single-cell sequencing workflow. Briefly, the single-cell process involves the collection and dissociation of cells; extraction of total RNA, cDNA synthesis and PCR amplification; library preparation and subsequent sequencing; computational data analysis (expression profiling single-cell wise, clustering, and cell type identification, with each colored dot in the cell clusters representing a specific cell type, with each colored dot in the cell clusters representing a specific cell type).

Studies employing "bulk" transcriptomics have been pivotal in analyzing reproductive processes and identifying biomarkers for uterine disorders [8]. Microarray technology, widely used for uterine transcriptome analysis, helped to define the window of implantation to enhance assisted reproductive treatment outcomes [9,10] and aid the study of uterine disorders such as adenomyosis [11]. RNA sequencing (RNA-seq) surpasses microarray technology by offering higher resolution, greater sensitivity, reduced bias, and the ability to identify novel transcripts [12]. The application of RNA sequencing to studies of the endometrium and blastocyst revealed mechanisms affecting embryo implantation, improved assisted reproductive treatment outcomes [13,14], and advanced our understanding of gynecological disease [15–17]. While bulk RNA-seq results represent an average of gene expression values and, as such, mask cellular heterogeneity, single-cell (sc)RNA-seq addresses this limitation by profiling gene expression at the single-cell level, uncovering rare cell types and states [18]. Spatial transcriptomics further enhances these insights by integrating single-cell data with location within the tissue architecture, revealing cellular interactions and tissue organization [19]

ScRNA-seq experiments involve sample dissociation, individual cell isolation (using techniques such as microfluidics), and library construction for sequencing (Figure 1B) [19–22]. The subsequent analysis stage includes data preprocessing, quality control, dimensionality reduction, and cell type identification with optional additional analyses including cell communication or trajectory prediction [23]. Since being introduced by Tang et al. in 2009 [24], scRNA-seq has transformed reproductive medicine by elucidating the cellular compositions and gene expression profiles of healthy tissues and those affected by pathological uterine conditions [25–31].

This review highlights recent scRNA-seq-based studies of uterine disorders, which have shed light on the cellular interactions and mechanisms that significantly impact reproductive health and fertility.

2. Exploring Normal Uterine Function with Single-Cell Approaches

The comprehensive understanding of the cellular composition and molecular mechanisms at play in the human uterus remains an essential step toward unraveling the foundations of its normal functions and understanding the development of pathological conditions. This section highlights key findings that provide a basis for the advancement of knowledge regarding uterine biology.

The first scRNA-seq study of human endometrial tissue by Krjutskov et al. involved the isolation of single cells from two frozen endometrial biopsies, transcriptome sequencing, and subsequent data analysis, although it was limited to stromal and epithelial cells [32]. A more comprehensive study by Wang et al. followed, in which the authors characterized the human endometrium across the menstrual cycle [33]. Their analysis identified stromal fibroblasts, endothelial cells, macrophages, lymphocytes, unciliated epithelial cells, and a previously unknown cilium-associated epithelium and demonstrated robust correlations between cell type-specific transcriptomic profiles and the proliferative and secretory phases of the menstrual cycle. Subsequently, Garcia-Alonso et al. applied scRNA-seq and spatial transcriptomics to map cells of the endometrium, categorizing them into immune, epithelial (SOX9+, luminal, glandular, and ciliated), endothelial, stromal, and supporting (e.g., perivascular cells and smooth muscle cells [SMCs]) cells [34], which provided detailed insight into the spatial and temporal organization of these cell types within the human endometrium. Building on this work, Marečková et al. (from the same research group) further advanced the field by creating the most detailed reference atlas of the human endometrium to date [35]. They identified novel cell populations, including CDH2+ basalis cells with progenitor-like properties and unique epithelial and stromal cells present during the early secretory phase. They also described the spatiotemporal organization and functional dynamics of the endometrial epithelium throughout the menstrual cycle, significantly refining and expanding upon insights from earlier studies.

Single-cell approaches have also advanced the study of the human myometrium. Pique-Regi et al. developed the first single-cell atlas of the myometrium during term labor, revealing the cellular and molecular mechanisms involved in parturition [29]. Subsequently, Punzón-Jiménez et al. combined scRNA-seq with spatial transcriptomics to study myometrial aging in peri- and post-menopausal women, revealing the implication of novel cell types and aging-associated pathways [25]. Ji et al. provided a high-resolution comparison of non-pregnant and pregnant myometrial states, highlighting changes in SMC subpopulations and the role of M2 macrophages in pregnancy-associated anti-inflammatory responses [36]. Finally, Ulrich et al. integrated scRNA-seq datasets to create a consensus atlas of 39 uterine cell subtypes, detailing menstrual phase-specific changes and identifying a potential progenitor cell population [37].

In brief, these high-resolution "atlases" of uterine cellular and molecular landscapes provide the foundation for a deeper understanding of uterine biology and ongoing research into disorders such as endometriosis, endometrial cancer, adenomyosis, and fibroids.

3. Insights into Uterine Disorders from Single-Cell Studies

3.1. Endometriosis

Endometriosis, a condition characterized by the presence of endometrial-like tissue outside the uterine cavity, affects ~10% of women of reproductive age, impacting ~190 million women globally [38]. This chronic condition causes pelvic pain, dysmenorrhea, and infertility. It is also associated with an elevated risk of ovarian cancer [39,40] and is predominantly diagnosed through laparoscopy [41]. Endometriosis is widely attributed to retrograde menstruation with altered endometrial tissues transitioning into ectopic lesions [42]. Recent advances in single-cell transcriptomics have deepened our understanding of the cellular and molecular characteristics of endometriosis, identifying roles for stromal, epithelial, and immune cells in disease progression.

Ectopic and utopic endometrial tissues possess an increased proportion of fibroblasts with the elevated expression of genes such as DNA Topoisomerase II Alpha (*TOP2A*) and heightened anti-apoptotic signaling (e.g., through the MAPK pathway) promoting cell survival and proliferation. These fibroblasts also expressed osteoglycin and inflammation- and tumorigenesis-associated genes (i.e., *C3*, *C7*, and *S100A10*) [43,44]. However, their cellular origin remains controversial. While some studies have reported enhanced levels of fibroblast-to-myofibroblast transdifferentiation, which contributes to extracellular matrix (ECM) remodeling and fibrosis through TGF-β/Wnt signaling [42,45], others suggest that myofibroblasts in ectopic lesions may originate from endometrial mesenchymal stem cells (MSCs) under specific microenvironmental conditions [46]. Additionally, Zhang et al. demonstrated that activin A promotes the myofibroblast differentiation of endometrial MSCs through the STAT3-dependent Smad/connective tissue growth factor (CTGF) pathway [47].

Impaired decidualization, a process essential for pregnancy, further connects stromal cell dysfunction to inflammation, fibrosis, and senescence, exacerbating the condition [48–50]. Endometriotic epithelial cells exhibit disrupted homeostasis and apoptosis resistance [51] with the downregulation of *SULT1E1* leading to estradiol accumulation and enhanced MAPK signaling, which promotes survival and proliferation [44]. Emerging subpopulations of epithelial cells expressing the *SOX9* and *LGR5* stem cell markers have been implicated in endometriosis lesion growth with altered hormonal responses promoting ectopic proliferation during inappropriate menstrual phases [27]. Additionally, MUC5B+ epithelial cells display an elevated proliferative capacity, suggesting their role in establishing ectopic lesions [43].

Chronic inflammation also fosters a pro-tumorigenic microenvironment, potentially contributing to endometriosis-associated ovarian cancer (EAOC) [52]. Immune gene expression analysis showed similarities between some endometriosis patients and EAOC cases, including elevated *C3* and *C7* expression [53], in accordance with single-cell analyses [27,42,43]. Furthermore, Fonseca et al. applied multi-subject single-cell deconvolution across three independent datasets, revealing that clear cell and endometrioid ovarian cancers were enriched in ciliated endometrial-type epithelial cells, whereas high-grade serous ovarian cancers showed no epithelial cluster enrichment [27]. These findings support the hypothesis that endometrial-type epithelial cells may be precursors for clear cell and endometrioid ovarian cancers, linked to endometriosis, while high-grade serous ovarian cancers are likely unrelated [54].

Immune dysfunction represents a pivotal mechanism in endometriosis pathology with myeloid cells (particularly macrophages) driving angiogenesis, tissue remodeling, and inflammation [43,55]. Myeloid cells display a diminished capacity for antigen presentation and phagocytosis in ectopic lesions, exacerbating chronic inflammation and lesion persistence [45,56]. Other immune alterations include reduced natural killer (NK) cell activity and impaired T-cell cytotoxicity, which further facilitate the survival of ectopic lesions [42,49]. The presence of lymphocytes and tertiary lymphoid structures in lesions underscores sustained immune dysregulation [43] (Figure 2).

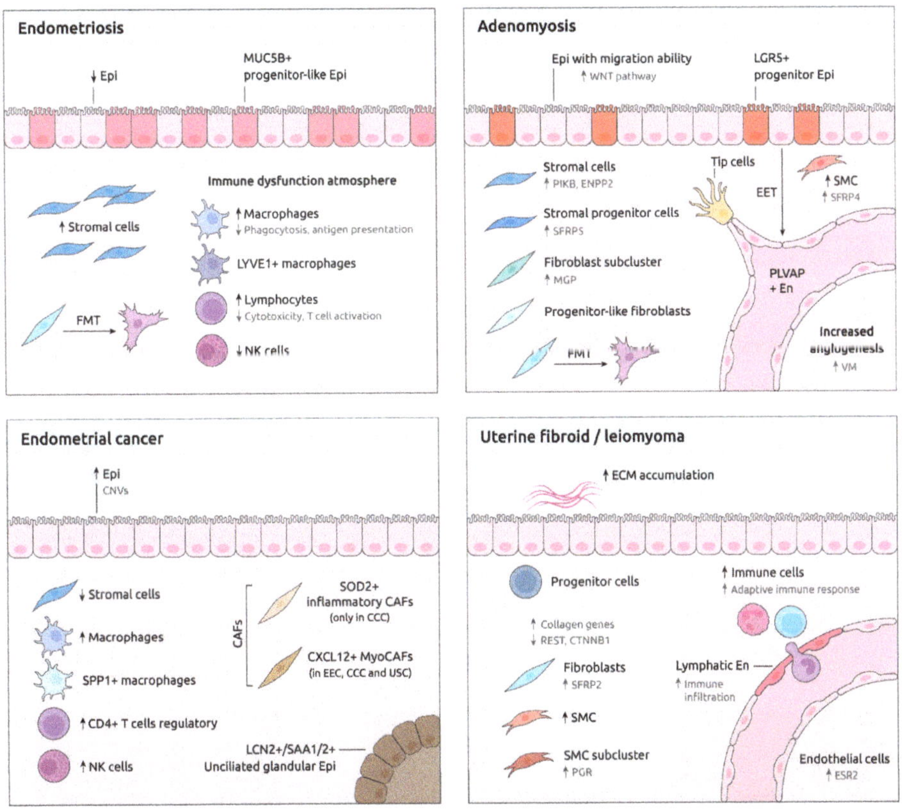

Figure 2. Representation of modified cell types, novel emerged cell types, key processes, and genes with altered expression under each condition compared to controls. Bold arrows indicate changes in cell abundance, while grey arrows denote changes in gene expression CAFs: cancer-associated fibroblasts; CCC: clear cell carcinoma; CNVs: copy number variations; Epi: epithelial cells; ECM: extracellular matrix; EEC: endometrioid endometrial cancer; EET: epithelial-to-endothelial transition; En: endothelial cells; FMT: fibroblast-to-myofibroblast transition; NK cells: natural killer cells; SMC: smooth muscle cell; USC: uterine serous carcinoma; VM: vasculogenic mimicry.

In brief, dysregulated cellular signaling pathways, immune alterations, and hormonal imbalances drive fibrosis, chronic inflammation, and lesion persistence in endometriosis. Insight into these pathways, which include fibroblast-to-myofibroblast transdifferentiation, hormonal dysregulation, and immune impairment, provide targets for the development of novel treatment strategies.

3.2. Endometrial Cancer

Endometrial cancer (EC), a malignancy originating in the uterine epithelium, represents the second most prevalent gynecological cancer worldwide (after cervical cancer) with over 400,000 new cases and nearly 100,000 deaths reported in 2022 [57,58]. Early-stage diagnosis yields a five-year survival rate of ~81%, but this figure drastically declines to ~15% in advanced stages [59]. Interestingly, endometriosis represents a potential risk factor for EC [60]. Endometrioid endometrial cancer (EEC), the most common subtype (~85% of cases), is characterized by microsatellite instability and *PTEN*, *PIK3CA*, and *CTNNB1* gene mutations [61]. Despite a generally favorable prognosis, ~20% of EEC patients experience tumor recurrence, reducing the five-year survival rate from 81% to 10% [62]. Uterine serous carcinoma (USC), the second most common subtype (~10%), is frequently associated with *TP53* mutations [61], while rarer highly aggressive subtypes such as clear cell carcinoma (CCC) and uterine carcinosarcoma account for only ~5% of cases [63].

The EC tumor microenvironment exhibits significant heterogeneity with diverse cellular populations and genetic variability influencing disease progression [64]. Single-cell technologies facilitated the characterization of said complexities, revealing critical insights into tumor biology. Epithelial cell studies identified subclusters (e.g., stem-like, secretory glandular, and ciliated cells) that display chromosomal copy number variations and upregulated cancer-related gene expression [65,66]. Marker genes such as *LCN2* and *SAA1/2* have been implicated in EEC tumorigenesis, while subtypes like USC exhibit epithelial-to-mesenchymal transition, which supports metastasis [26,67]. Stromal fibroblasts, although reduced in number in EC compared to normal tissue, play a pivotal role thanks to the activity of cancer-associated fibroblasts (CAFs). CAF subtypes (including inflammatory, vascular, and antigen presenting) act to support tumorigenesis; indeed, the presence of SOD2+ iCAFs correlates with poor survival due to their metastasis-promoting role [26,68] (Figure 2).

Immune cells within the tumor microenvironment also contribute to EC progression and aid prognosis [69]. Specific tumor-associated macrophage (TAM) subtypes facilitate tumor growth and vascularization with their presence positively correlating with therapeutic responses [70,71]. T cells, particularly CD8+ tissue-resident memory T cells (TRMs), associate with a better prognosis, although immunosuppressive FOXP3+ T regulatory cells complicate the immune landscape [66,72]. Studies have highlighted the importance of NK cells and tertiary lymphoid structures, which represent favorable prognostic markers [73]. Integrating advanced profiling techniques has enhanced our understanding of critical pathways (e.g., PI3K/AKT/mTOR) and highlighted the potential for tailored therapies targeting specific tumor characteristics. These findings underscore the complexity of EC and the need for continued research to improve diagnostic and therapeutic strategies.

3.3. Adenomyosis

Adenomyosis is a benign gynecological condition characterized by the disruption of the endometrial–myometrial junction and the presence of ectopic endometrial tissue within the myometrium, which leads to myometrial hypertrophy and hyperplasia [74,75]. While some patients remain asymptomatic, most experience abnormal menstrual bleeding, dysmenorrhea, dyspareunia, chronic pelvic pain, or infertility [76]. Furthermore, an elevated risk of ovarian cancer has been identified in patients with adenomyosis; however, this relationship has been scarcely investigated, highlighting the need for additional studies to clarify their association [77,78]. The incidence of adenomyosis—estimated at 29 cases per 10,000 women annually—varies across studies due to population differences [79]. Although the precise pathogenesis of adenomyosis remains unclear, two primary hypotheses dominate: the invagination theory, which involves endometrial–myometrial junction injury and endometrial invasion into the myometrium, and the metaplasia hypothesis, which implicates progenitor cells undergoing dysregulated differentiation/migration [80,81]. Hormonal dysregulation, inflammation, angiogenesis, and fibrosis are also implicated, although the specific mechanisms remain incompletely understood [82,83].

Advances in single-cell technologies have significantly improved our understanding of adenomyosis by enabling a high-resolution analysis of cellular and molecular features within uterine tissues (Table 1). Studies have identified distinct populations of epithelial, stromal, endothelial, fibroblast, smooth muscle, and immune cells and the upregulated expression of genes involved in angiogenesis, cell growth, migration, and fibrosis (all associated with disease progression) [84,85]. While findings can diverge due to sample variability, these studies provide a deeper understanding of adenomyosis pathogenesis and highlight potential therapeutic targets.

Epithelial and stromal cells within the ectopic endometrium exhibit increased migratory and proliferative potential, supporting the invagination hypothesis of adenomyosis. These cells display alterations in Wnt signaling with genes such as *MMP7*, *CLDN4*, and *LGR5* playing significant roles in adenomyotic lesion development [85,86]. Fibroblasts and SMCs also possess multiple distinct subclusters in ectopic tissues, which associate with fibrosis, excessive ECM deposition, and SMC proliferation. The fibroblast-to-myofibroblast transition and Wnt signaling modulation by *SFRP* family genes further underscore the role of these cells in lesion formation and tissue remodeling [85,87]. Progenitor cells in adenomyotic tissues differentiate into lesion-specific stromal cells, further implicating dysregulated signaling pathways in adenomyosis progression [88,89]. Endothelial cells in adenomyotic tissues exhibit heightened angiogenic activity with the upregulation of *VEGF*, *ANGPT*, and other angiogenesis-related genes. Single-cell analyses showed evidence of vasculogenic mimicry [90] as well as the presence of endothelial tip cells [86], which can differentiate into various endothelial subtypes [91]. Additionally, spatial transcriptomic analysis revealed the presence of PLVAP+ cells surrounding adenomyotic lesions. PLVAP expression alters endothelial cells permeability [92], suggesting modified vascular function. All these findings indicate different mechanisms contributing to abnormal vascular dynamics and symptoms such as heavy bleeding, highlighting angiogenesis as a central factor in the pathophysiology of adenomyosis [84,86,90,93]. Notably, Chen et al. [86] detected endothelial tip cells and suggested their maturation is regulated by the activation of the Notch signaling pathway. The Notch pathway's involvement in vascular alterations has been previously explored [94,95], suggesting it as a potential target for angiogenic therapy in adenomyosis. Furthermore, PLVAP has emerged as a promising target, as its specific inhibition could normalize permeability in uterine endothelial cells affected by adenomyosis [86,92].

Interestingly, Chen et al. identified a distinct subset of SFRP4+IGFBP5^{hi} Natural Killer T (NKT) cells in patients experiencing adenomyosis-related pain, which were absent in patients without such pain. SFRP4 functions as a Wnt-pathway inhibitor that suppresses cell differentiation [96], while *IGFBP5* is a gene associated with neuropathic differentiation [97], suggesting a potential role for these cells in neuroinflammatory processes and nerve fiber proliferation [98]. Single-cell analysis further indicated the potential of these cells to differentiate into neural progenitor cells. Additionally, the genetic heatmap based on pseudotime trajectory analysis revealed consistently high *IGFBP5* expression throughout differentiation along with an increased expression of genes involved in nerve growth and neuroinflammatory pathways at later stages. Among these genes, NEFM was particularly notable due to its association with neuropathic pain [99] and its positive correlation with pain duration in this study. Collectively, these findings propose that SFRP4+IGFBP5^{hi} NKT cells may drive stem cell differentiation toward neurogenic lineages through IGFBP5 expression, thereby contributing to adenomyosis-associated pain [98]. This points to IGFBP5 and NEFM as potential therapeutic targets for managing pain in adenomyosis patients (Figure 2). Furthermore, other studies have suggested that mifepristone is a safe and effective option for improving pain scores in adenomyosis patients compared to placebo [100], while non-steroidal anti-inflammatory drugs have also been reported as effective treatments for dysmenorrhea in this population [101].

Overall, single-cell analyses have revealed the complex interplay of cellular and molecular mechanisms driving adenomyosis, highlighting the potential for therapeutic interventions targeting Wnt signaling, angiogenic pathways, and pain-associated stem cell markers such as *IGFBP5* and *SFRP4*. Despite their promise, sample variability and the incomplete understanding of the biological mechanisms involved have limited the impact of single-cell studies, underscoring the need for more research [84–86].

3.4. *Uterine Fibroids/Leiomyoma*

Uterine fibroids (or leiomyomas) are benign smooth muscle tumors and the most common gynecological tumors in reproductive-age women, affecting over 80% of African Americans and ~70% of Caucasian women [102]. Although non-malignant, fibroids significantly impair quality of life by causing symptoms like abnormal uterine bleeding, pelvic pain, and pregnancy complications [103,104]. While research has linked fibroid development to hormonal and genetic factors, the precise molecular mechanisms and cellular contributors remain unclear, limiting diagnostic and therapeutic advances [104–106]. ScRNA-seq and spatial transcriptomics have provided transformative insights into fibroid biology, although applications in this area are still emerging.

Recent scRNA-seq-based studies identified critical cellular populations in fibroids, including SMCs, fibroblasts, endothelial cells, immune cells, and progenitor cells [30,107] (Table 1). These studies highlighted the role of SMCs and fibroblasts in ECM deposition, which is a hallmark feature of fibroids [108,109]. SMC subclusters contribute to collagen formation, myofibroblast function, and regulatory processes like proliferation and glycolysis [30]; similarly, fibroblasts exhibit diverse roles in ECM organization, integrin signaling, and axon guidance. The authors observed enhanced cell–cell communication between SMCs and fibroblasts in fibroids with increased signaling pathways related to collagen production, actin remodeling, and semaphorin signaling [110,111]. They also observed amplified immune-fibroblast interactions involving known enhancers (Notch and mTOR pathways) of the pro-fibrotic environment of fibroids [112–114]. Furthermore, they identified novel ligand–receptor interactions, such as IGF1-IGF1R, as contributors to disease progression [109]. Dysregulated genes, including *REST*, *CTNNB1*, and *SFRP2*, revealed potential molecular targets for future therapeutic exploration [115,116] (Figure 2).

As hormone-sensitive tumors, estrogen and progesterone play pivotal roles in fibroid development [117,118]. Spatial transcriptomics and scRNA-seq demonstrated the robust expression of estrogen receptor 1 (*ESR1*) and progesterone receptor (*PGR*) in fibroids with significant involvement of the ERK1/ERK2 pathway in mediating hormonal effects and promoting tumor progression. Pseudotime and Gene Ontology (GO) enrichment analyses further indicated that *ESR1* is predominantly expressed in smooth muscle cells (SMCs) and fibroblasts within uterine fibroids, whereas elevated *PGR* expression was detected in fibroid SMCs across all patients but was limited to fibroblasts in only one patient [107]. Conversely, Goad et al. reported an overall reduction in *ESR1* and *PGR* expression in SMCs and fibroblasts within uterine fibroids with the exception of a specific fibroid SMC subcluster that exhibited a higher proportion of *PGR* compared to the myometrium [30]. While the precise association between the differential gene expression and tumor growth remains unclear, these findings suggested paracrine mechanisms through which specific cell populations such as SMC and fibroblasts drive tumor growth, offering insights into the hormonal regulation of fibroids and potential therapeutic targets [30,119,120].

Abnormal angiogenesis, another feature of fibroids, involves abnormal endothelial cell activity also driven by estrogen-activated pathways such as ERK [121–124]. Notably, ESR2 (ERβ) was predominantly expressed in endothelial cells within fibroids [107], aligning with findings by Valladares et al., who reported high ERβ localization in these cells. Since ERβ has been shown to promote angiogenesis, the ERK1/2 pathway may mediate this process by facilitating endothelial cell migration and proliferation [125]. Immune cells, including T-cells, macrophages, and NK cells, also contribute to the inflammatory and fibrotic microenvironment of fibroids [126,127]. Enhanced interactions between immune cells and fibroblasts further highlight their role in promoting tumor development through signaling pathways such as Notch and mTOR [127].

Fibroids are traditionally considered monoclonal tumors arising from a multipotent progenitor cell that undergoes transformation and differentiation, leading to tumor proliferation [125,128]. In support of this view, Wang et al. identified a progenitor cell population within fibroids, which agreed with earlier reports [129,130]; however, Goad et al. suggested a genetically heterogeneous origin involving MED12 variant-positive progenitor cells that recruit wild-type cells [30,131]. Notably, discrepancies between these studies, particularly regarding *ESR1* and *PGR* expression and the identification of progenitor cell populations, highlight variations in tissue comparisons and sample characteristics. While Wang et al. examined fibroid tissue and its pseudocapsule [107], Goad et al. focused on fibroids and myometrium, analyzing only MED12-mutant fibroids [30]. These methodological differences highlight the need for standardized experimental designs and detailed sample descriptions to ensure reproducibility while emphasizing the complex nature of fibroids and the power of single-cell technologies in uncovering complex cellular and genetic landscapes.

Overall, advanced transcriptomic technologies have significantly enhanced our understanding of fibroid pathogenesis, shedding light on crucial pathways such as ERK1/ERK2, mTOR, and collagen signaling, as well as the roles of specific cell populations in tumor progression. Integration with genome-wide association study data has further enriched our understanding with single-cell and spatial transcriptomics offering valuable insights into fibroid biology and potential avenues for improved patient care [132].

Table 1. Summary of single-cell studies of endometriosis, endometrial cancer, adenomyosis and uterine fibroids.

Reference	Samples	Cell Number *	Key Findings
		Endometriosis	
[43]	EcE (n = 4) and EuE (n = 9) from ovarian EM; peritoneal EM (n = 8) and adjacent regions (n = 6); healthy endometrium (n = 3)	108,497	■ Identified MUC5B+ epithelial cells as possible progenitor cells of endometriotic lesions ■ Unique perivascular mural cells in peritoneal lesions promoted angiogenesis and immune cell trafficking, coinciding with increased lymphocyte proportions and upregulated macrophage–regulatory T-cell interactions
[27]	EcE (n = 8) and EuE (n = 10) from ovarian EM and unaffected ovary (n = 4); peritoneal EM (n = 23); healthy peritoneum (n = 4)	373,851	■ SOX9+/LGR5$^+$ epithelial cell subset may participate in EM ■ Epithelium, stroma, and mesothelial cells of ovarian EM exhibited dysregulated pro-inflammatory pathway activation ■ *ARID1A* somatic mutations in epithelial cells drive angiogenesis and lymphangiogenesis

Table 1. *Cont.*

Reference	Samples	Cell Number *	Key Findings
[45]	EcE and EuE from ovarian EM (n = 3); healthy endometrium (n = 3)	46,445	■ Significantly higher proportions of myofibroblasts, pericytes, endothelial cells, and macrophages in EcE compared to EuE and healthy endometrium. ■ Immunodeficient EM profile
[42]	EcE and EuE from ovarian EM (n = 3); healthy endometrium (n = 3)	55,188	■ Identified fibroblast subpopulations related to endometriosis development ■ Diminished immune function and pro-remodeling characteristics of EM
[50]	Menstrual effluent from EM (n = 11); subjects with undiagnosed chronic EM-like symptoms (n = 13); healthy controls (n = 9)	43,054	■ Reduced decidualized of an IGFBP1+ endometrial stromal cell subset in EM cases suggested impaired decidualization ■ Endometrial stromal cells in EM exhibited pro-inflammatory and senescent characteristics ■ Identification of a unique proliferating uNK cell subpopulation in controls, mostly absent in EM
[56]	Peritoneal fluid from EM (n = 1); healthy control (n = 1)	17,530	■ Identified immune cell dysfunctions in EM, including reduced phagocytosis and cytotoxicity with increased pro-inflammatory and chemotactic responses
[44]	EcE and EuE from ovarian EM (n = 23); healthy endometrium (n = 11)	7030	■ Increased proportion of ciliated cells in EuE/EcE characterized by diminished expression of estrogen sulfotransferase, likely conferring resistance to apoptosis ■ Epithelial cells in ectopic lesions stimulated CD4+ T cells, contributing to chronic inflammation
		Endometrial Cancer	
[68]	EC (n = 4); normal endometrium (n = 2)	41,358	■ CAF subsets with distinct characteristics maintained frequent communications with malignant cells, facilitating EC progression
[70]	EEC (n = 5); normal endometrium (n = 3)	46,638	■ SPP1+ TAMs associated with EC tumorigenesis and progression ■ Robust crosstalk between SPP1+ TAMs and fibroblasts, SMCs, endothelial cells, proliferating T cells, and tumor epithelial cells in the TME
[72]	EC (n = 3)	28,820	■ NK cells and CD8+ T lymphocytes represented major TIL components in EC patients ■ Transcriptionally distinct NK cell subsets identified in EC
[65]	EEC (n = 5); matched paratumor samples (n = 3)	30,780	■ Tumor epithelial cells exhibited the highest CNVs levels ■ Distinct epithelial cell clusters identified ■ Enrichment of macrophages and exhausted CD8+ T cells observed in EEC
[133]	EEC (n = 5); USC (n = 1)	150,144	■ Malignant cells from the same patients revealed variation in chromatin accessibility linked to transcriptional output, highlighting the importance of intratumoral heterogeneity

Table 1. *Cont.*

Reference	Samples	Cell Number *	Key Findings
[66]	EEC (n = 5); atypical endometrial hyperplasia tissues (n = 5); normal endometrium (n = 5)	99,215	■ Elevated proportion of epithelial cells and increased percentage of CNVs observed in EEC ■ EEC originated from unciliated glandular epithelial cells and LCN2+/SAA1/2+ cells identified as a subpopulation involved in endometrial tumorigenesis
[26]	EEC-I (n = 7), EEC-II (n = 3), USC (n = 4), and UCCC (n = 3); normal endometrium (n = 1)	146,332	■ Metabolic, immune, and proliferation pathways enriched in USC, UCCC, and ECC, respectively ■ Epithelial-to-mesenchymal transition more robustly present in USC ■ SOD2+ iCAFs exclusive to UCCC and critical for EC malignancy
		Adenomyosis	
[90]	Adenomyotic EcE and EuE (n = 1); control endometrium (n = 1)	36,781	■ Cellular cluster with epithelial and endothelial cells in EcE displayed malignant features ■ Epithelial subcluster enriched in EcE overexpressed motility-related genes ■ EET and VM found in EcE suggested as novel processes involved in AM pathogenesis
[84]	Adenomyotic EcE and EuE (n = 2); control endometrium (n = 2)	42,260	■ Endothelial cells involved in pathways related to cancer, epithelial cells in migration functions, and some SMC subclusters associated with vascular SMC contraction ■ High levels of CNVs in EcE
[98]	Adenomyotic EcE (dysmenorrhea n = 2; no dysmenorrhea n = 2)	27,924	■ SFRP4+IGFBP5^{hi} NK cells may transform multipotent stem cells into neurogenic cells, leading to AM-related pain
[87]	Adenomyotic EcE and EuE (n = 1)	21,147	■ Fibrogenesis in AM pathogenesis driven by fibroblasts and FMT alongside the contribution of SMC and immune cells
[85]	Adenomyotic EcE, EuE, and MM (n = 3)	66,000	■ Highly heterogeneous fibroblasts represented the main component of the AM endometrium ■ Wnt pathway proposed as a critical regulator of AM pathophysiology
[86]	Adenomyotic EcE, EuE, MM, and EMJ (n = 3); control EuE, MM, and EMJ (n = 1)	54,658	■ Identification of unique LGR5$^+$ epithelial cells, PKIB$^+$ invasive stromal cells, and WFDC1$^+$ progenitor cells within EcE ■ Abnormal angiogenic function and altered cell–cell communications in adenomyotic tissue
		Uterine Fibroids	
[30]	MED12$^+$ UF (n = 5); adjacent MM (n = 5)	39,209	■ Description of lymphatic endothelial cells within UF, previously reported only in MM ■ Fibroblasts drove UF microenvironmental alterations via interactions with SMCs, endothelial cells, and immune cells ■ UF genetic heterogeneity suggested a non-monoclonal origin for these tumors

Table 1. *Cont.*

Reference	Samples	Cell Number *	Key Findings
[107]	UF (n = 2), pseudocapsule (n = 2), and MM (n = 1)	26,371	■ ERK1/ERK2 pathway may mediate UF development and angiogenesis via ER and PGR signaling ■ IGF1 and IGF1R represented the most significant ligand–receptor pair identified in fibroblast interactions
[132]	MED12^{+} UF (n = 5); MM (n = 5)	48,258	■ SMCs contained the most highly expressed GWAS target genes in UF samples ■ Cell type-specific modules of enriched GWAS candidate genes identified

* Number of cells analyzed after quality control filters. Abbreviations: AM: adenomyosis; CAFs: cancer-associated fibroblasts; CNVs: copy number variations; EC: endometrial cancer; EcE: ectopic endometrium; EEC: endometrioid endometrial cancer; EEC-I: well-differentiated EEC; EEC-II: poorly differentiated ECC; EET: epithelial–endothelial transition; EM: endometriosis; EMJ: endometrial–myometrial junction; EuE: eutopic endometrium; FMT: fibroblast-to-myofibroblast transition; GWAS: genome-wide association study; MM: myometrium; NK: natural killer cells; PGR: progesterone receptor; SMCs: smooth muscle cells; TAMs: tumor-associated macrophages; TILs: tumor-infiltrating lymphocytes; TME: tumor microenvironment; UCCC: uterine clear cell carcinoma; UF: uterine fibroids; uNK: uterine natural killer cells; USC: uterine serous carcinoma; VM: vasculogenic mimicry.

4. Challenges and Constraints in Current Single-Cell Technologies

While scRNA-seq has transformed biological research by supporting the high-resolution analysis of individual cells [18], constraints remain, including the presence of biological noise (e.g., gene expression variability and cell state differences) and technical limitations (e.g., partial RNA capture and amplification bias) [134]. Said issues can lead to the generation of incomplete datasets and the underrepresentation of specific RNA species; moreover, sample handling, procurement, and processing delays can reduce RNA quality, further affecting the reliability of the results [135].

Batch effects arising from variations in sample processing conditions and sequencing protocols pose significant challenges to scRNA-seq analysis, confounding biological signals and requiring careful design and experimental replication to mitigate any impact [134]. Biological factors (e.g., cell cycle effects) further complicate analyses by obscuring real differences between cells. Technical limitations (e.g., the inability to capture certain RNA species) contribute to data sparsity with observed zeros potentially representing biological absence or technical dropouts [136]. Beyond these inherent biological and technical constraints, experimental design also plays a critical role in influencing the comparability and interpretation of findings across studies. For example, in single-cell studies of uterine fibroids, comparisons of gene expression between fibroids and either healthy myometrium [30] or surrounding pseudocapsules [107] produced conflicting results regarding *ESR1* and *PGR* expression. Similarly, studies on endometrial cancer reported inconsistent findings about CD8+ T cells with some indicating reduced proportions [66] and others suggesting an enrichment of these cells among tumor-infiltrating lymphocytes [72]. These discrepancies are partly due to the variations in the control samples used, highlighting how differences in sample selection can impact outcomes and reproducibility. Furthermore, the diversity of scRNA-seq platforms—each with unique strengths, limitations, and requirements—further complicates data integration/standardization [134,135]. Addressing these challenges through normalization, correction, and informed platform selection remains critical to generating reliable and interpretable results [137].

5. Future Perspectives: Addressing Research Gaps for Clinical Implementation

Over recent decades, advancements in omics-based technologies have significantly enhanced our understanding of the molecular mechanisms underlying various physiological and pathological processes. In particular, single-cell technologies have shown remarkable potential in studying complex disorders with applications spanning therapeutic target identification, drug screening, clinical model development, and diagnostics [138]. These advancements have shown particular promise in gynecology and reproductive medicine, where scRNA-seq and multi-omics-based approaches have offered high-resolution insight into disease pathogenesis. Notably, single-cell technologies have enabled the characterization of tumor heterogeneity, the identification of disease-specific cell and molecular subtypes, and an understanding of immune microenvironments and therapy resistance evolution in gynecological cancers such as ovarian, endometrial, and cervical cancers [138–143]. These insights have particular value when addressing challenges related to metastasis and drug resistance. Multi-omics single-cell technologies, which can integrate genomic, transcriptomic, proteomic, epigenomic and metabolomic data from the same single cell, offer a systems biology approach to deciphering the intricate mechanisms underlying gynecological disorders [144]. Additionally, scRNA-seq profiling of endometrial organoids has allowed researchers to benchmark in vitro responses against in vivo data, enhancing the relevance of these models for the study of endometrial function [34]. These advances highlight the potential of single-cell approaches to refine preclinical models and improve translational research outcomes.

Nonetheless, the analysis and interpretation of vast large-scale omics data concerning uterine diseases have been complicated by the cyclic regulation of hormones in addition to multiple other factors, such as genetic variability [145], environmental factors [146], and microbiome composition [147], among others. This challenge is further compounded by individual genetic differences, which contribute to diverse biological responses. In this context, Sengupta et al. recently proposed a tool that examines the interconnected roles of hormonal regulation, environmental factors, genetic predisposition (including DNA composition and epigenetic changes), health implications, and the resulting biological effects [144]. Moreover, wider accessibility to public repositories, such as Gene Expression Omnibus, ArrayExpress, and single-cell-focused resources like Tabula Muris, Non-Human Primate SC Atlas, Pan-Cancer Blueprint, and the Human Tumor Atlas Network, offers significant opportunities to overcome data accessibility challenges and leverage existing data for clinical applications [138].

Looking forward, the integration of single-cell technologies into clinical workflows is expected to drive significant advancements in precision medicine. These tools will enable the identification of novel therapeutic targets, including the emerging proteolysis targeting chimera (PROTAC) probe technology [148], which has been utilized to successfully degrade a variety of pathogenic proteins. Furthermore, single-cell technologies can aid in the characterization of heterogeneously manifested human traits and the refinement of cell differentiation trajectories. By elucidating the genetic drivers of metastasis, treatment resistance, and tumor progression in gynecological cancers, single-cell technologies will pave the way for personalized treatments with improved efficacy and reduced side effects. The continued development of sophisticated computational tools, standardized workflows, and interdisciplinary collaborations will be crucial for translating these innovations into routine clinical practice, heralding a new era of personalized medicine.

6. Conclusions

Gynecological disorders such as endometriosis, ovarian cancer, and uterine fibroids possess complex and multicellular etiologies, making characterization a challenging prospect. Single-cell technologies have provided a powerful approach to address this complexity by providing detailed insights into the cellular and molecular diversity within gynecological tissues. Combined with advanced multiplexed approaches, single-cell protocols enable the high-resolution analysis of various cell populations, uncovering essential cell types and signaling pathways driving disease mechanisms.

A key aspect of these disorders is the significant overlap in underlying mechanisms, including excessive collagen deposition leading to fibrosis, heightened inflammatory responses, and immune dysregulation. These shared features suggest common biomarkers and therapeutic targets that could address these conditions collectively. For instance, targeting the TGF-β signaling pathway or genes associated with extracellular matrix (ECM) remodeling, such as fibronectin, collagen, and α smooth muscle actin, holds promise for mitigating fibrosis [149]. Fibrosis represents a key pathological hallmark, impacting tissue architecture and function, ultimately influencing fertility and quality of life [150,151]. Immune dysfunction and chronic inflammation further exacerbate disease progression, impairing immune cell function and disrupting tissue homeostasis [152]. Blocking drivers of chronic inflammation, such as the NLRP3 inflammasome [153] or the NF-κB signaling pathway [154], represents a promising therapeutic strategy. However, while several NLRP3 inhibitors have been developed, their clinical use has been constrained by toxicity and safety concerns [153]. Similarly, targeting the NF-κB pathway poses challenges due to its involvement in critical biological processes, risking off-target effects [155]. Based on this premise, future efforts must focus on designing structure-based direct inhibitors with improved specificity and minimized side effects to optimize therapeutic efficacy. Additionally, profiling preclinical models at the single-cell level will help to identify those most accurately mirroring human gynecological disease biology.

This cellular-level analysis of human samples supports the development of personalized medicine by revealing novel biomarkers and aiding patient stratification based on prognosis/expected response to treatment. By dissecting disease complexity at the cellular level, single-cell technologies aid the discovery of novel biomarkers and help to clarify the fundamental biology of gynecological conditions, paving the way for targeted therapeutic interventions. Consequently, single-cell technologies will remain pivotal in identifying biological targets and carrying out their functional characterization.

Emerging evidence highlights the connection between these disorders and their impact on fertility [156]. Given the associations between these conditions, fertility preservation, and overall health, further research is required to elucidate these mechanisms and optimize therapeutic approaches.

Author Contributions: Conceptualization, A.M. and C.S.; writing—original draft preparation, S.B.-F. and C.L.; writing—review and editing, S.B.-F., C.L., A.M. and C.S.; supervision, A.M. and C.S.; funding acquisition, A.M. and C.S. All authors have read and agreed to the published version of the manuscript.

Funding: This work was supported by a Miguel Servet Spanish Program Grant CP19/00162 and Health Research Funds PI23/00536 from the Carlos III Institute, Spain (A.M.) along with support from the Valencian Scientific Foundation of the Spanish Association Against Cancer (PRDVA234168BOLD) (S.B.) and private funding from the Carlos Simon Foundation.

Data Availability Statement: No new data were created or analyzed in this study.

Acknowledgments: We acknowledge Stuart P. Atkinson for his help in manuscript construction and editing. The authors wish to thank Paula Punzón-Jiménez and Esther Lara-Cuesta for their support.

Conflicts of Interest: The authors declare no conflicts of interest.

References

1. Habiba, M.; Heyn, R.; Bianchi, P.; Brosens, I.; Benagiano, G. The Development of the Human Uterus: Morphogenesis to Menarche. *Hum. Reprod. Update* **2021**, *27*, 1–26. [CrossRef] [PubMed]
2. Robboy, S.J.; Kurita, T.; Baskin, L.; Cunha, G.R. New Insights into Human Female Reproductive Tract Development. *Differentiation* **2017**, *97*, 9–22. [CrossRef] [PubMed]
3. Yamaguchi, M.; Yoshihara, K.; Suda, K.; Nakaoka, H.; Yachida, N.; Ueda, H.; Sugino, K.; Mori, Y.; Yamawaki, K.; Tamura, R.; et al. Three-Dimensional Understanding of the Morphological Complexity of the Human Uterine Endometrium. *iScience* **2021**, *24*, 102258. [CrossRef]
4. Ruiz-Alonso, M.; Blesa, D.; Simón, C. The Genomics of the Human Endometrium. *Biochim. Biophys. Acta* **2012**, *1822*, 1931–1942. [CrossRef]
5. Renthal, N.E.; Chen, C.C.; Williams, K.C.; Gerard, R.D.; Prange-Kiel, J.; Mendelson, C.R. MiR-200 Family and Targets, ZEB1 and ZEB2, Modulate Uterine Quiescence and Contractility during Pregnancy and Labor. *Proc. Natl. Acad. Sci. USA* **2010**, *107*, 20828–20833. [CrossRef]
6. Maiti, K.; Paul, J.W.; Read, M.; Chan, E.C.; Riley, S.C.; Nahar, P.; Smith, R. G-1-Activated Membrane Estrogen Receptors Mediate Increased Contractility of the Human Myometrium. *Endocrinology* **2011**, *152*, 2448–2455. [CrossRef] [PubMed]
7. Talati, C.; Carvalho, J.C.A.; Luca, A.; Balki, M. The Effect of Intermittent Oxytocin Pretreatment on Oxytocin-Induced Contractility of Human Myometrium In Vitro. *Anesth. Analg.* **2019**, *128*, 671–678. [CrossRef]
8. Hrdlickova, R.; Toloue, M.; Tian, B. RNA-Seq Methods for Transcriptome Analysis. *Wiley Interdiscip. Rev. RNA* **2017**, *8*, e1364. [CrossRef] [PubMed]
9. Evans, G.E.; Phillipson, G.T.M.; Sykes, P.H.; McNoe, L.A.; Print, C.G.; Evans, J.J. Does the Endometrial Gene Expression of Fertile Women Vary within and between Cycles? *Hum. Reprod.* **2018**, *33*, 452–463. [CrossRef]
10. Li, F.; Gao, W.; Li, Y.; Wang, Y.; Liu, L.; Zhang, X. Potential Biomarkers and Endometrial Immune Microenvironment in Recurrent Implantation Failure. *Biomolecules* **2023**, *13*, 406. [CrossRef]
11. Herndon, C.N.; Aghajanova, L.; Balayan, S.; Erikson, D.; Barragan, F.; Goldfien, G.; Vo, K.C.; Hawkins, S.; Giudice, L.C. Global Transcriptome Abnormalities of the Eutopic Endometrium from Women with Adenomyosis. *Reprod. Sci.* **2016**, *23*, 1289–1303. [CrossRef] [PubMed]
12. Nookaew, I.; Papini, M.; Pornputtapong, N.; Scalcinati, G.; Fagerberg, L.; Uhlén, M.; Nielsen, J. A Comprehensive Comparison of RNA-Seq-Based Transcriptome Analysis from Reads to Differential Gene Expression and Cross-Comparison with Microarrays: A Case Study in Saccharomyces Cerevisiae. *Nucleic Acids Res.* **2012**, *40*, 10084–10097. [CrossRef] [PubMed]
13. Sanchez-Ribas, I.; Diaz-Gimeno, P.; Sebastián-León, P.; Mercader, A.; Quiñonero, A.; Ballesteros, A.; Pellicer, A.; Domínguez, F. Transcriptomic Behavior of Genes Associated with Chromosome 21 Aneuploidies in Early Embryo Development. *Fertil. Steril.* **2019**, *111*, 991–1001.e2. [CrossRef] [PubMed]
14. Suhorutshenko, M.; Kukushkina, V.; Velthut-Meikas, A.; Altmäe, S.; Peters, M.; Mägi, R.; Krjutškov, K.; Koel, M.; Codoñer, F.M.; Martinez-Blanch, J.F.; et al. Endometrial Receptivity Revisited: Endometrial Transcriptome Adjusted for Tissue Cellular Heterogeneity. *Hum. Reprod.* **2018**, *33*, 2074–2086. [CrossRef] [PubMed]
15. Choi, M.R.; Chang, H.J.; Heo, J.H.; Yum, S.H.; Jo, E.; Kim, M.; Lee, S.R. Expression Profiling of Coding and Noncoding RNAs in the Endometrium of Patients with Endometriosis. *Int. J. Mol. Sci.* **2024**, *25*, 10581. [CrossRef] [PubMed]
16. Xiang, Y.; Sun, Y.; Yang, B.; Yang, Y.; Zhang, Y.; Yu, T.; Huang, H.; Zhang, J.; Xu, H. Transcriptome Sequencing of Adenomyosis Eutopic Endometrium: A New Insight into Its Pathophysiology. *J. Cell. Mol. Med.* **2019**, *23*, 8381–8391. [CrossRef]
17. Machado-Lopez, A.; Alonso, R.; Lago, V.; Jimenez-Almazan, J.; Garcia, M.; Monleon, J.; Lopez, S.; Barcelo, F.; Torroba, A.; Ortiz, S.; et al. Integrative Genomic and Transcriptomic Profiling Reveals a Differential Molecular Signature in Uterine Leiomyoma versus Leiomyosarcoma. *Int. J. Mol. Sci.* **2022**, *23*, 2190. [CrossRef] [PubMed]
18. Shapiro, E.; Biezuner, T.; Linnarsson, S. Single-Cell Sequencing-Based Technologies Will Revolutionize Whole-Organism Science. *Nat. Rev. Genet.* **2013**, *14*, 618–630. [CrossRef] [PubMed]
19. Kulkarni, A.; Anderson, A.G.; Merullo, D.P.; Konopka, G. Beyond Bulk: A Review of Single Cell Transcriptomics Methodologies and Applications. *Curr. Opin. Biotechnol.* **2019**, *58*, 129–136. [CrossRef]
20. Truong, D.D.; Lamhamedi-Cherradi, S.E.; Porter, R.W.; Krishnan, S.; Swaminathan, J.; Gibson, A.; Lazar, A.J.; Livingston, J.A.; Gopalakrishnan, V.; Gordon, N.; et al. Dissociation Protocols Used for Sarcoma Tissues Bias the Transcriptome Observed in Single-Cell and Single-Nucleus RNA Sequencing. *BMC Cancer* **2023**, *23*, 488. [CrossRef] [PubMed]
21. Grindberg, R.V.; Yee-Greenbaum, J.L.; McConnell, M.J.; Novotny, M.; O'Shaughnessy, A.L.; Lambert, G.M.; Araúzo-Bravo, M.J.; Lee, J.; Fishman, M.; Robbins, G.E.; et al. RNA-Sequencing from Single Nuclei. *Proc. Natl. Acad. Sci. USA* **2013**, *110*, 19802–19807. [CrossRef]

22. Lake, B.B.; Codeluppi, S.; Yung, Y.C.; Gao, D.; Chun, J.; Kharchenko, P.V.; Linnarsson, S.; Zhang, K. A Comparative Strategy for Single-Nucleus and Single-Cell Transcriptomes Confirms Accuracy in Predicted Cell-Type Expression from Nuclear RNA. *Sci. Rep.* **2017**, *7*, 6031. [CrossRef] [PubMed]
23. Su, M.; Pan, T.; Chen, Q.Z.; Zhou, W.W.; Gong, Y.; Xu, G.; Yan, H.Y.; Li, S.; Shi, Q.Z.; Zhang, Y.; et al. Data Analysis Guidelines for Single-Cell RNA-Seq in Biomedical Studies and Clinical Applications. *Mil. Med. Res.* **2022**, *9*, 68. [CrossRef]
24. Tang, F.; Barbacioru, C.; Wang, Y.; Nordman, E.; Lee, C.; Xu, N.; Wang, X.; Bodeau, J.; Tuch, B.B.; Siddiqui, A.; et al. MRNA-Seq Whole-Transcriptome Analysis of a Single Cell. *Nat. Methods* **2009**, *6*, 377–382. [CrossRef] [PubMed]
25. Punzon-Jimenez, P.; Machado-Lopez, A.; Perez-Moraga, R.; Llera-Oyola, J.; Grases, D.; Galvez-Viedma, M.; Sibai, M.; Satorres-Perez, E.; Lopez-Agullo, S.; Badenes, R.; et al. Effect of Aging on the Human Myometrium at Single-Cell Resolution. *Nat. Commun.* **2024**, *15*, 945. [CrossRef]
26. Ren, F.; Wang, L.; Wang, Y.; Wang, J.; Wang, Y.; Song, X.; Zhang, G.; Nie, F.; Lin, S. Single-Cell Transcriptome Profiles the Heterogeneity of Tumor Cells and Microenvironments for Different Pathological Endometrial Cancer and Identifies Specific Sensitive Drugs. *Cell Death Dis.* **2024**, *15*, 571. [CrossRef] [PubMed]
27. Fonseca, M.A.S.; Haro, M.; Wright, K.N.; Lin, X.; Abbasi, F.; Sun, J.; Hernandez, L.; Orr, N.L.; Hong, J.; Choi-Kuaea, Y.; et al. Single-Cell Transcriptomic Analysis of Endometriosis. *Nat. Genet.* **2023**, *55*, 255–267. [CrossRef] [PubMed]
28. Jones, R.C.; Karkanias, J.; Krasnow, M.A.; Pisco, A.O.; Quake, S.R.; Salzman, J.; Yosef, N.; Bulthaup, B.; Brown, P.; Harper, W.; et al. The Tabula Sapiens: A Multiple-Organ, Single-Cell Transcriptomic Atlas of Humans. *Science* **2022**, *376*, eabl4896. [CrossRef] [PubMed]
29. Pique-Regi, R.; Romero, R.; Garcia-Flores, V.; Peyvandipour, A.; Tarca, A.L.; Pusod, E.; Galaz, J.; Miller, D.; Bhatti, G.; Para, R.; et al. A Single-Cell Atlas of the Myometrium in Human Parturition. *JCI Insight* **2022**, *7*, e153921. [CrossRef] [PubMed]
30. Goad, J.; Rudolph, J.; Zandigohar, M.; Tae, M.; Dai, Y.; Wei, J.J.; Bulun, S.E.; Chakravarti, D.; Rajkovic, A. Single-Cell Sequencing Reveals Novel Cellular Heterogeneity in Uterine Leiomyomas. *Hum. Reprod.* **2022**, *37*, 2334–2349. [CrossRef] [PubMed]
31. Karlsson, M.; Zhang, C.; Méar, L.; Zhong, W.; Digre, A.; Katona, B.; Sjöstedt, E.; Butler, L.; Odeberg, J.; Dusart, P.; et al. A Single-Cell Type Transcriptomics Map of Human Tissues. *Sci. Adv.* **2021**, *7*, e153921. [CrossRef] [PubMed]
32. Krjutškov, K.; Katayama, S.; Saare, M.; Vera-Rodriguez, M.; Lubenets, D.; Samuel, K.; Laisk-Podar, T.; Teder, H.; Einarsdottir, E.; Salumets, A.; et al. Single-Cell Transcriptome Analysis of Endometrial Tissue. *Hum. Reprod.* **2016**, *31*, 844–853. [CrossRef] [PubMed]
33. Wang, W.; Vilella, F.; Alama, P.; Moreno, I.; Mignardi, M.; Isakova, A.; Pan, W.; Simon, C.; Quake, S.R. Single-Cell Transcriptomic Atlas of the Human Endometrium during the Menstrual Cycle. *Nat. Med.* **2020**, *26*, 1644–1653. [CrossRef] [PubMed]
34. Garcia-Alonso, L.; Handfield, L.-F.; Roberts, K.; Nikolakopoulou, K.; Fernando, R.C.; Gardner, L.; Woodhams, B.; Arutyunyan, A.; Polanski, K.; Hoo, R.; et al. Mapping the Temporal and Spatial Dynamics of the Human Endometrium in Vivo and in Vitro. *Nat. Genet.* **2021**, *53*, 1698–1711. [CrossRef] [PubMed]
35. Marečková, M.; Garcia-Alonso, L.; Moullet, M.; Lorenzi, V.; Petryszak, R.; Sancho-Serra, C.; Oszlanczi, A.; Icoresi Mazzeo, C.; Wong, F.C.K.; Kelava, I.; et al. An Integrated Single-Cell Reference Atlas of the Human Endometrium. *Nat. Genet.* **2024**, *56*, 1925–1937. [CrossRef] [PubMed]
36. Ji, K.; Zhong, J.; Cui, L.; Wang, X.; Chen, L.N.; Wen, B.; Yang, F.; Deng, W.; Pan, X.; Wang, L.; et al. Exploring Myometrial Microenvironment Changes at the Single-Cell Level from Nonpregnant to Term Pregnant States. *Physiol. Genom.* **2024**, *56*, 32–47. [CrossRef] [PubMed]
37. Ulrich, N.D.; Vargo, A.; Ma, Q.; Shen, Y.; Hannum, D.F.; Gurczynski, S.J.; Moore, B.B.; Schon, S.; Lieberman, R.; Shikanov, A.; et al. Cellular Heterogeneity and Dynamics of the Human Uterus in Healthy Premenopausal Women. *bioRxiv* **2024**. [CrossRef] [PubMed]
38. World Health Organization. Endometriosis. Available online: https://www.who.int/news-room/fact-sheets/detail/endometriosis (accessed on 1 August 2024).
39. Barnard, M.E.; Farland, L.V.; Yan, B.; Wang, J.; Trabert, B.; Doherty, J.A.; Meeks, H.D.; Madsen, M.; Guinto, E.; Collin, L.J.; et al. Endometriosis Typology and Ovarian Cancer Risk. *JAMA* **2024**, *332*, 482–489. [CrossRef] [PubMed]
40. Linder, A.; Westbom-Fremer, S.; Mateoiu, C.; Olsson Widjaja, A.; Österlund, T.; Veerla, S.; Ståhlberg, A.; Ulfenborg, B.; Hedenfalk, I.; Sundfeldt, K. Genomic Alterations in Ovarian Endometriosis and Subsequently Diagnosed Ovarian Carcinoma. *Hum. Reprod.* **2024**, *39*, 1141–1154. [CrossRef]
41. Zondervan, K.T.; Becker, C.M.; Missmer, S.A. Endometriosis. *N. Engl. J. Med.* **2020**, *382*, 1244–1256. [CrossRef] [PubMed]
42. Ma, J.; Zhang, L.; Zhan, H.; Mo, Y.; Ren, Z.; Shao, A.; Lin, J. Single-Cell Transcriptomic Analysis of Endometriosis Provides Insights into Fibroblast Fates and Immune Cell Heterogeneity. *Cell Biosci.* **2021**, *11*, 125. [CrossRef]
43. Tan, Y.; Flynn, W.F.; Sivajothi, S.; Luo, D.; Bozal, S.B.; Davé, M.; Luciano, A.A.; Robson, P.; Luciano, D.E.; Courtois, E.T. Single-Cell Analysis of Endometriosis Reveals a Coordinated Transcriptional Programme Driving Immunotolerance and Angiogenesis across Eutopic and Ectopic Tissues. *Nat. Cell Biol.* **2022**, *24*, 1306–1318. [CrossRef] [PubMed]

44. Yan, J.; Zhou, L.; Liu, M.; Zhu, H.; Zhang, X.; Cai, E.; Xu, X.; Chen, T.; Cheng, H.; Liu, J.; et al. Single-Cell Analysis Reveals Insights into Epithelial Abnormalities in Ovarian Endometriosis. *Cell Rep.* **2024**, *43*, 113716. [CrossRef] [PubMed]
45. Zhu, S.; Wang, A.; Xu, W.; Hu, L.; Sun, J.; Wang, X. The Heterogeneity of Fibrosis and Angiogenesis in Endometriosis Revealed by Single-Cell RNA-Sequencing. *Biochim. Biophys.* **2023**, *1869*, 166602. [CrossRef] [PubMed]
46. Zhang, Z.; Suo, L.; Chen, Y.; Zhu, L.; Wan, G.; Han, X. Endometriotic Peritoneal Fluid Promotes Myofibroblast Differentiation of Endometrial Mesenchymal Stem Cells. *Stem Cells Int.* **2019**, *2019*, 6183796. [CrossRef] [PubMed]
47. Zhang, Z.; Wang, J.; Chen, Y.; Suo, L.; Chen, H.; Zhu, L.; Wan, G.; Han, X. Activin a Promotes Myofibroblast Differentiation of Endometrial Mesenchymal Stem Cells via STAT3-Dependent Smad/CTGF Pathway. *Cell Commun. Signal.* **2019**, *17*, 45. [CrossRef] [PubMed]
48. Klemmt, P.A.B.; Carver, J.G.; Kennedy, S.H.; Koninckx, P.R.; Mardon, H.J. Stromal Cells from Endometriotic Lesions and Endometrium from Women with Endometriosis Have Reduced Decidualization Capacity. *Fertil. Steril.* **2006**, *85*, 564–572. [CrossRef] [PubMed]
49. Warren, L.A.; Shih, A.; Renteira, S.M.; Seckin, T.; Blau, B.; Simpfendorfer, K.; Lee, A.; Metz, C.N.; Gregersen, P.K. Analysis of Menstrual Effluent: Diagnostic Potential for Endometriosis. *J. Mol. Med.* **2018**, *24*, 1. [CrossRef]
50. Shih, A.J.; Adelson, R.P.; Vashistha, H.; Khalili, H.; Nayyar, A.; Puran, R.; Herrera, R.; Chatterjee, P.K.; Lee, A.T.; Truskinovsky, A.M.; et al. Single-Cell Analysis of Menstrual Endometrial Tissues Defines Phenotypes Associated with Endometriosis. *BMC Med.* **2022**, *20*, 315. [CrossRef]
51. Evans, J.; Salamonsen, L.A.; Winship, A.; Menkhorst, E.; Nie, G.; Gargett, C.E.; Dimitriadis, E. Fertile Ground: Human Endometrial Programming and Lessons in Health and Disease. *Nat. Rev. Endocrinol.* **2016**, *12*, 654–667. [CrossRef]
52. Steinbuch, S.C.; Lüß, A.M.; Eltrop, S.; Götte, M.; Kiesel, L. Endometriosis-Associated Ovarian Cancer: From Molecular Pathologies to Clinical Relevance. *Int. J. Mol. Sci.* **2024**, *25*, 4306. [CrossRef]
53. Suryawanshi, S.; Huang, X.; Elishaev, E.; Budiu, R.A.; Zhang, L.; Kim, S.H.; Donnellan, N.; Mantia-Smaldone, G.; Ma, T.; Tseng, G.; et al. Complement Pathway Is Frequently Altered in Endometriosis and Endometriosis-Associated Ovarian Cancer. *Clin. Cancer Res.* **2014**, *20*, 6163–6174. [CrossRef] [PubMed]
54. Saavalainen, L.; Lassus, H.; But, A.; Tiitinen, A.; Härkki, P.; Gissler, M.; Pukkala, E.; Heikinheimo, O. Risk of Gynecologic Cancer According to the Type of Endometriosis. *Obstet. Gynecol.* **2018**, *131*, 1095–1102. [CrossRef] [PubMed]
55. Abramiuk, M.; Grywalska, E.; Małkowska, P.; Sierawska, O.; Hrynkiewicz, R.; Niedźwiedzka-Rystwej, P. The Role of the Immune System in the Development of Endometriosis. *Cells* **2022**, *11*, 2028. [CrossRef] [PubMed]
56. Zou, G.; Wang, J.; Xu, X.; Xu, P.; Zhu, L.; Yu, Q.; Peng, Y.; Guo, X.; Li, T.; Zhang, X. Cell Subtypes and Immune Dysfunction in Peritoneal Fluid of Endometriosis Revealed by Single-Cell RNA-Sequencing. *Cell Biosci.* **2021**, *11*, 98. [CrossRef] [PubMed]
57. Sung, H.; Ferlay, J.; Siegel, R.L.; Laversanne, M.; Soerjomataram, I.; Jemal, A.; Bray, F. Global Cancer Statistics 2020: GLOBOCAN Estimates of Incidence and Mortality Worldwide for 36 Cancers in 185 Countries. *CA Cancer J. Clin.* **2021**, *71*, 209–249. [CrossRef]
58. World Cancer Research Fund International. Endometrial Cancer Statistics. Available online: https://www.wcrf.org/preventing-cancer/cancer-statistics/endometrial-cancer-statistics/ (accessed on 21 December 2024).
59. Mazidimoradi, A.; Momenimovahed, Z.; Khalajinia, Z.; Allahqoli, L.; Salehiniya, H.; Alkatout, I. The Global Incidence, Mortality, and Burden of Uterine Cancer in 2019 and Correlation with SDI, Tobacco, Dietary Risks, and Metabolic Risk Factors: An Ecological Study. *Health Sci. Rep.* **2024**, *7*, e1835. [CrossRef]
60. Ye, J.; Peng, H.; Huang, X.; Qi, X. The Association between Endometriosis and Risk of Endometrial Cancer and Breast Cancer: A Meta-Analysis. *BMC Womens Health* **2022**, *22*, 455. [CrossRef] [PubMed]
61. Bianco, B.; Barbosa, C.P.; Trevisan, C.M.; Laganà, A.S.; Montagna, E. Endometrial Cancer: A Genetic Point of View. *Transl. Cancer Res.* **2020**, *9*, 7706. [CrossRef] [PubMed]
62. Cao, S.; Fan, Y.; Zhang, Y.; Ruan, J.; Mu, Y.; Li, J. Recurrence and Survival of Patients with Stage III Endometrial Cancer after Radical Surgery Followed by Adjuvant Chemo- or Chemoradiotherapy: A Systematic Review and Meta-Analysis. *BMC Cancer* **2023**, *23*, 31. [CrossRef] [PubMed]
63. Bogani, G.; Ray-Coquard, I.; Concin, N.; Ngoi, N.Y.L.; Morice, P.; Enomoto, T.; Takehara, K.; Denys, H.; Lorusso, D.; Coleman, R.; et al. Clear Cell Carcinoma of the Endometrium. *Gynecol. Oncol.* **2022**, *164*, 658–666. [CrossRef] [PubMed]
64. Jamaluddin, M.F.B.; Ko, Y.A.; Ghosh, A.; Syed, S.M.; Ius, Y.; O'Sullivan, R.; Netherton, J.K.; Baker, M.A.; Nahar, P.; Jaaback, K.; et al. Proteomic and Functional Characterization of Intra-Tumor Heterogeneity in Human Endometrial Cancer. *Cell Rep. Med.* **2022**, *3*, 100738. [CrossRef] [PubMed]
65. Guo, Y.E.; Li, Y.; Cai, B.; He, Q.; Chen, G.; Wang, M.; Wang, K.; Wan, X.; Yan, Q. Phenotyping of Immune and Endometrial Epithelial Cells in Endometrial Carcinomas Revealed by Single-Cell RNA Sequencing. *Aging* **2021**, *13*, 6565–6591. [CrossRef]
66. Ren, X.; Liang, J.; Zhang, Y.; Jiang, N.; Xu, Y.; Qiu, M.; Wang, Y.; Zhao, B.; Chen, X. Single-Cell Transcriptomic Analysis Highlights Origin and Pathological Process of Human Endometrioid Endometrial Carcinoma. *Nat. Commun.* **2022**, *13*, 6300. [CrossRef]
67. Mittal, V. Epithelial Mesenchymal Transition in Tumor Metastasis. *Annu. Rev. Pathol.* **2018**, *13*, 395–412. [CrossRef] [PubMed]

68. Yu, Z.; Zhang, J.; Zhang, Q.; Wei, S.; Shi, R.; Zhao, R.; An, L.; Grose, R.; Feng, D.; Wang, H. Single-Cell Sequencing Reveals the Heterogeneity and Intratumoral Crosstalk in Human Endometrial Cancer. *Cell Prolif.* **2022**, *55*, e13249. [CrossRef]
69. Dey, D.K.; Krause, D.; Rai, R.; Choudhary, S.; Dockery, L.E.; Chandra, V. The Role and Participation of Immune Cells in the Endometrial Tumor Microenvironment. *Pharmacol. Ther.* **2023**, *251*, 108526. [CrossRef]
70. Wu, Q.; Jiang, G.; Sun, Y.; Li, B. Reanalysis of Single-Cell Data Reveals Macrophage Subsets Associated with the Immunotherapy Response and Prognosis of Patients with Endometrial Cancer. *Exp. Cell Res.* **2023**, *430*, 113736. [CrossRef]
71. Sun, T.; Sun, B.C.; Zhao, X.L.; Zhao, N.; Dong, X.Y.; Che, N.; Yao, Z.; Ma, Y.M.; Gu, Q.; Zong, W.K.; et al. Promotion of Tumor Cell Metastasis and Vasculogenic Mimicry by Way of Transcription Coactivation by Bcl-2 and Twist1: A Study of Hepatocellular Carcinoma. *Hepatology* **2011**, *54*, 1690–1706. [CrossRef] [PubMed]
72. Jiang, F.; Jiao, Y.; Yang, K.; Mao, M.; Yu, M.; Cao, D.; Xiang, Y. Single-Cell Profiling of the Immune Atlas of Tumor-Infiltrating Lymphocytes in Endometrial Carcinoma. *Cancers* **2022**, *14*, 4311. [CrossRef]
73. Horeweg, N.; Workel, H.H.; Loiero, D.; Church, D.N.; Vermij, L.; Léon-Castillo, A.; Krog, R.T.; de Boer, S.M.; Nout, R.A.; Powell, M.E.; et al. Tertiary Lymphoid Structures Critical for Prognosis in Endometrial Cancer Patients. *Nat. Commun.* **2022**, *13*, 1373. [CrossRef] [PubMed]
74. Vercellini, P.; Viganò, P.; Somigliana, E.; Daguati, R.; Abbiati, A.; Fedele, L. Adenomyosis: Epidemiological Factors. *Best. Pract. Res. Clin. Obstet. Gynaecol.* **2006**, *20*, 465–477. [CrossRef] [PubMed]
75. Mehasseb, M.K.; Bell, S.C.; Pringle, J.H.; Habiba, M.A. Uterine Adenomyosis Is Associated with Ultrastructural Features of Altered Contractility in the Inner Myometrium. *Fertil. Steril.* **2010**, *93*, 2130–2136. [CrossRef]
76. Struble, J.; Reid, S.; Bedaiwy, M.A. Adenomyosis: A Clinical Review of a Challenging Gynecologic Condition. *J. Minim. Invasive Gynecol.* **2016**, *23*, 164–185. [CrossRef]
77. Kok, V.C.; Tsai, H.J.; Su, C.F.; Lee, C.K. The Risks for Ovarian, Endometrial, Breast, Colorectal, and Other Cancers in Women with Newly Diagnosed Endometriosis or Adenomyosis: A Population-Based Study. *Int. J. Gynecol. Cancer* **2015**, *25*, 968–976. [CrossRef] [PubMed]
78. Hermens, M.; van Altena, A.M.; Bulten, J.; van Vliet, H.A.A.M.; Siebers, A.G.; Bekkers, R.L.M. Increased Incidence of Ovarian Cancer in Both Endometriosis and Adenomyosis. *Gynecol. Oncol.* **2021**, *162*, 735–740. [CrossRef] [PubMed]
79. Yu, O.; Schulze-Rath, R.; Grafton, J.; Hansen, K.; Scholes, D.; Reed, S.D. Adenomyosis Incidence, Prevalence and Treatment: United States Population-Based Study 2006–2015. *Am. J. Obstet. Gynecol.* **2020**, *223*, 94.e1–94.e10. [CrossRef]
80. Ferenczy, A. Pathophysiology of Adenomyosis. *Hum. Reprod. Update* **1998**, *4*, 312–322. [CrossRef] [PubMed]
81. García-Solares, J.; Donnez, J.; Donnez, O.; Dolmans, M.M. Pathogenesis of Uterine Adenomyosis: Invagination or Metaplasia? *Fertil. Steril.* **2018**, *109*, 371–379. [CrossRef]
82. Maclean, A.; Barzilova, V.; Patel, S.; Bates, F.; Hapangama, D.K. Characterising the Immune Cell Phenotype of Ectopic Adenomyosis Lesions Compared with Eutopic Endometrium: A Systematic Review. *J. Reprod. Immunol.* **2023**, *157*, 103925. [CrossRef]
83. Bulun, S.E.; Yildiz, S.; Adli, M.; Wei, J.J. Adenomyosis Pathogenesis: Insights from next-Generation Sequencing. *Hum. Reprod. Update* **2021**, *27*, 1086–1097. [CrossRef] [PubMed]
84. Lin, J.; Liu, L.; Zheng, F.; Chen, S.; Yang, W.; Li, J.; Mo, S.; Zeng, D.Y. Exploration the Global Single-Cell Ecological Landscape of Adenomyosis-Related Cell Clusters by Single-Cell RNA Sequencing. *Front. Genet.* **2022**, *13*, 1020757. [CrossRef] [PubMed]
85. Yildiz, S.; Kinali, M.; Wei, J.J.; Milad, M.; Yin, P.; Adli, M.; Bulun, S.E. Adenomyosis: Single-Cell Transcriptomic Analysis Reveals a Paracrine Mesenchymal-Epithelial Interaction Involving the WNT/SFRP Pathway. *Fertil. Steril.* **2023**, *119*, 869–882. [CrossRef] [PubMed]
86. Chen, T.; Xu, Y.; Xu, X.; Wang, J.; Qiu, Z.; Yu, Y.; Jiang, X.; Shao, W.; Bai, D.; Wang, M.; et al. Comprehensive Transcriptional Atlas of Human Adenomyosis Deciphered by the Integration of Single-Cell RNA-Sequencing and Spatial Transcriptomics. *Protein Cell* **2024**, *15*, 530–546. [CrossRef]
87. Niu, W.; Zhang, Y.; Liu, H.; Liang, N.; Xu, L.; Li, Y.; Yao, W.; Shi, W.; Liu, Z. Single-Cell Profiling Uncovers the Roles of Endometrial Fibrosis and Microenvironmental Changes in Adenomyosis. *J. Inflamm. Res.* **2023**, *16*, 1949–1965. [CrossRef] [PubMed]
88. Oh, S.J.; Shin, J.H.; Kim, T.H.; Lee, H.S.; Yoo, J.Y.; Ahn, J.Y.; Broaddus, R.R.; Taketo, M.M.; Lydon, J.P.; Leach, R.E.; et al. β -Catenin Activation Contributes to the Pathogenesis of Adenomyosis through Epithelial-Mesenchymal Transition. *J. Pathol.* **2013**, *231*, 210–222. [CrossRef] [PubMed]
89. Zhang, W.J.; Chen, S.J.; Zhou, S.C.; Wu, S.Z.; Wang, H. Inflammasomes and Fibrosis. *Front. Immunol.* **2021**, *12*, 643149. [CrossRef]
90. Liu, Z.; Sun, Z.; Liu, H.; Niu, W.; Wang, X.; Liang, N.; Wang, X.; Wang, Y.; Shi, Y.; Xu, L.; et al. Single-Cell Transcriptomic Analysis of Eutopic Endometrium and Ectopic Lesions of Adenomyosis. *Cell Biosci.* **2021**, *11*, 51. [CrossRef]
91. Lee, H.W.; Xu, Y.; He, L.; Choi, W.; Gonzalez, D.; Jin, S.W.; Simons, M. Role of Venous Endothelial Cells in Developmental and Pathologic Angiogenesis. *Circulation* **2021**, *144*, 1308–1322. [CrossRef] [PubMed]
92. Denzer, L.; Muranyi, W.; Schroten, H.; Schwerk, C. The Role of PLVAP in Endothelial Cells. *Cell Tissue Res.* **2023**, *392*, 393–412. [CrossRef] [PubMed]

93. Huang, T.S.; Chen, Y.J.; Chou, T.Y.; Chen, C.Y.; Li, H.Y.; Huang, B.S.; Tsai, H.W.; Lan, H.Y.; Chang, C.H.; Twu, N.F.; et al. Oestrogen-Induced Angiogenesis Promotes Adenomyosis by Activating the Slug-VEGF Axis in Endometrial Epithelial Cells. *J. Cell. Mol. Med.* **2014**, *18*, 1358–1371. [CrossRef] [PubMed]
94. Noguera-Troise, I.; Daly, C.; Papadopoulos, N.J.; Coetzee, S.; Boland, P.; Gale, N.W.; Chieh Lin, H.; Yancopoulos, G.D.; Thurston, G. Blockade of Dll4 Inhibits Tumour Growth by Promoting Non-Productive Angiogenesis. *Nature* **2006**, *444*, 1032–1037. [CrossRef] [PubMed]
95. Li, J.L.; Sainson, R.C.A.; Shi, W.; Leek, R.; Harrington, L.S.; Preusser, M.; Biswas, S.; Turley, H.; Heikamp, E.; Hainfellner, J.A.; et al. Delta-like 4 Notch Ligand Regulates Tumor Angiogenesis, Improves Tumor Vascular Function, and Promotes Tumor Growth in Vivo. *Cancer Res.* **2007**, *67*, 11244–11253. [CrossRef] [PubMed]
96. Sundaram, S.M.; Varier, L.; Fathima, K.Z.; Dharmarajan, A.; Warrier, S. Short Peptide Domains of the Wnt Inhibitor SFRP4 Target Ovarian Cancer Stem Cells by Neutralizing the Wnt β-Catenin Pathway, Disrupting the Interaction between β-Catenin and CD24 and Suppressing Autophagy. *Life Sci.* **2023**, *316*, 121384. [CrossRef]
97. Li, J.; Diao, S.; Yang, H.; Cao, Y.; Du, J.; Yang, D. IGFBP5 Promotes Angiogenic and Neurogenic Differentiation Potential of Dental Pulp Stem Cells. *Dev. Growth Differ.* **2019**, *61*, 457–465. [CrossRef] [PubMed]
98. Chen, Y.; Zhu, J.; Chen, L.; Shen, Y.; Zhang, J.; Wang, Q. SFRP4+IGFBP5^{hi} NKT Cells Induced Neural-like Cell Differentiation to Contribute to Adenomyosis Pain. *Front. Immunol.* **2022**, *13*, 945504. [CrossRef]
99. Cui, C.Y.; Liu, X.; Peng, M.H.; Liu, Q.; Zhang, Y. Identification of Key Candidate Genes and Biological Pathways in Neuropathic Pain. *Comput. Biol. Med.* **2022**, *150*, 106135. [CrossRef] [PubMed]
100. Che, X.; Wang, J.; Sun, W.; He, J.; Wang, Q.; Zhu, D.; Zhu, W.; Zhang, J.; Dong, J.; Xu, J.; et al. Effect of Mifepristone vs Placebo for Treatment of Adenomyosis with Pain Symptoms: A Randomized Clinical Trial. *JAMA Netw. Open* **2023**, *6*, e2317860. [CrossRef] [PubMed]
101. Moldassarina, R.S. Modern View on the Diagnostics and Treatment of Adenomyosis. *Arch. Gynecol. Obstet.* **2023**, *308*, 171–181. [CrossRef]
102. Baird, D.D.; Dunson, D.B.; Hill, M.C.; Cousins, D.; Schectman, J.M. High Cumulative Incidence of Uterine Leiomyoma in Black and White Women: Ultrasound Evidence. *Am. J. Obstet. Gynecol.* **2003**, *188*, 100–107. [CrossRef] [PubMed]
103. Benson, C.B.; Chow, J.S.; Chang-Lee, W.; Iii, J.A.H.; Doubilet, P.M. Outcome of Pregnancies in Women with Uterine Leiomyomas Identified by Sonography in the First Trimester. *J. Clin. Ultrasound* **2001**, *29*, 261–264. [CrossRef] [PubMed]
104. Marsh, E.E.; Bulun, S.E. Steroid Hormones and Leiomyomas. *Obstet. Gynecol. Clin. N. Am.* **2006**, *33*, 59–67. [CrossRef] [PubMed]
105. Parikh, T.P.; Malik, M.; Britten, J.; Aly, J.M.; Pilgrim, J.; Catherino, W.H. Steroid Hormones and Hormone Antagonists Regulate the Neural Marker Neurotrimin in Uterine Leiomyoma. *Fertil. Steril.* **2020**, *113*, 176–186. [CrossRef] [PubMed]
106. Machado-Lopez, A.; Simón, C.; Mas, A. Molecular and Cellular Insights into the Development of Uterine Fibroids. *Int. J. Mol. Sci.* **2021**, *22*, 8483. [CrossRef]
107. Wang, J.; Xu, P.; Zou, G.; Che, X.; Jiang, X.; Liu, Y.; Mao, X.; Zhang, X. Integrating Spatial Transcriptomics and Single-Nucleus RNA Sequencing Reveals the Potential Therapeutic Strategies for Uterine Leiomyoma. *Int. J. Biol. Sci.* **2023**, *19*, 2515–2530. [CrossRef] [PubMed]
108. Luo, X.; Ding, L.; Xu, J.; Chegini, N. Gene Expression Profiling of Leiomyoma and Myometrial Smooth Muscle Cells in Response to Transforming Growth Factor-β. *Endocrinology* **2005**, *146*, 1097–1118. [CrossRef] [PubMed]
109. Serna, V.A.; Wu, X.; Qiang, W.; Thomas, J.; Blumenfeld, M.L.; Kurita, T. Cellular Kinetics of MED12-Mutant Uterine Leiomyoma Growth and Regression in Vivo. *Endocr. Relat. Cancer* **2018**, *25*, 747–759. [CrossRef]
110. Hung, R.J.; Terman, J.R. Extracellular Inhibitors, Repellents, and Semaphorin/Plexin/MICAL-Mediated Actin Filament Disassembly. *Cytoskeleton* **2011**, *68*, 415–433. [CrossRef]
111. Yang, Q.; Al-Hendy, A. Update on the Role and Regulatory Mechanism of Extracellular Matrix in the Pathogenesis of Uterine Fibroids. *Int. J. Mol. Sci.* **2023**, *24*, 5778. [CrossRef] [PubMed]
112. Crabtree, J.S.; Jelinsky, S.A.; Harris, H.A.; Choe, S.E.; Cotreau, M.M.; Kimberland, M.L.; Wilson, E.; Saraf, K.A.; Liu, W.; McCampbell, A.S.; et al. Comparison of Human and Rat Uterine Leiomyomata: Identification of a Dysregulated Mammalian Target of Rapamycin Pathway. *Cancer Res.* **2009**, *69*, 6171–6178. [CrossRef]
113. Bielesz, B.; Sirin, Y.; Si, H.; Niranjan, T.; Gruenwald, A.; Ahn, S.; Kato, H.; Pullman, J.; Gessler, M.; Haase, V.H.; et al. Epithelial Notch Signaling Regulates Interstitial Fibrosis Development in the Kidneys of Mice and Humans. *J. Clin. Investig.* **2010**, *120*, 4040–4054. [CrossRef]
114. Al-Hendy, A.; Diamond, M.P.; Boyer, T.G.; Halder, S.K. Vitamin D3 Inhibits Wnt/β-Catenin and MTOR Signaling Pathways in Human Uterine Fibroid Cells. *J. Clin. Endocrinol. Metab.* **2016**, *101*, 1542–1551. [CrossRef]
115. Cohen, M.L.; Brumwell, A.N.; Ho, T.C.; Garakani, K.; Montas, G.; Leong, D.; Ding, V.W.; Golden, J.A.; Trinh, B.N.; Jablons, D.M.; et al. A Fibroblast-Dependent TGFβ1/SFRP2 Noncanonical Wnt Signaling Axis Promotes Epithelial Metaplasia in Idiopathic Pulmonary Fibrosis. *J. Clin. Investig.* **2024**, *134*, e174598. [CrossRef]

116. Varghese, B.V.; Koohestani, F.; McWilliams, M.; Colvin, A.; Gunewardena, S.; Kinsey, W.H.; Nowak, R.A.; Nothnick, W.B.; Chennathukuzhi, V.M. Loss of the Repressor REST in Uterine Fibroids Promotes Aberrant G Protein-Coupled Receptor 10 Expression and Activates Mammalian Target of Rapamycin Pathway. *Proc. Natl. Acad. Sci. USA* **2013**, *110*, 2187–2192. [CrossRef] [PubMed]
117. Ishikawa, H.; Ishi, K.; Ann Serna, V.; Kakazu, R.; Bulun, S.E.; Kurita, T. Progesterone Is Essential for Maintenance and Growth of Uterine Leiomyoma. *Endocrinology* **2010**, *151*, 2433–2442. [CrossRef]
118. Li, Z.; Yin, H.; Shen, Y.; Ren, M.; Xu, X. The Influence of Phenolic Environmental Estrogen on the Transcriptome of Uterine Leiomyoma Cells: A Whole Transcriptome Profiling-Based Analysis. *Ecotoxicol. Environ. Saf.* **2021**, *211*, 111945. [CrossRef]
119. Voronin, D.; Sotnikova, N.; Rukavishnikov, K.; Malyshkina, A.; Nagornii, S.; Antsiferova, Y. Differential Regulatory Effect of Progesterone on the Proliferation and Apoptosis of Uterine Leiomyoma Tissue Explants and Primary Leiomyoma Cell Cultures. *J. Bras. Reprod. Assist.* **2021**, *25*, 540–548. [CrossRef]
120. Reschke, L.; Afrin, S.; El Sabah, M.; Charewycz, N.; Miyashita-Ishiwata, M.; Borahay, M.A. Leptin Induces Leiomyoma Cell Proliferation and Extracellular Matrix Deposition via JAK2/STAT3 and MAPK/ERK Pathways. *F&S Sci.* **2022**, *3*, 383–391. [CrossRef]
121. He, Y.Y.; Cai, B.; Yang, Y.X.; Liu, X.L.; Wan, X.P. Estrogenic G Protein-Coupled Receptor 30 Signaling Is Involved in Regulation of Endometrial Carcinoma by Promoting Proliferation, Invasion Potential, and Interleukin-6 Secretion via the MEK/ERK Mitogen-Activated Protein Kinase Pathway. *Cancer Sci.* **2009**, *100*, 1051–1061. [CrossRef]
122. Tal, R.; Segars, J.H. The Role of Angiogenic Factors in Fibroid Pathogenesis: Potential Implications for Future Therapy. *Hum. Reprod. Update* **2014**, *20*, 194–216. [CrossRef] [PubMed]
123. Afrin, S.; El Sabeh, M.; Islam, M.S.; Miyashita-Ishiwata, M.; Malik, M.; Catherino, W.H.; Akimzhanov, A.M.; Boehning, D.; Yang, Q.; Al-Hendy, A.; et al. Simvastatin Modulates Estrogen Signaling in Uterine Leiomyoma via Regulating Receptor Palmitoylation, Trafficking and Degradation. *Pharmacol. Res.* **2021**, *172*, 105856. [CrossRef] [PubMed]
124. Zhang, Y.; Cao, C.; Du, S.; Fan, L.; Zhang, D.; Wang, X.; He, M. Estrogen Regulates Endoplasmic Reticulum Stress-Mediated Apoptosis by ERK-P65 Pathway to Promote Endometrial Angiogenesis. *Reprod. Sci.* **2021**, *28*, 1216–1226. [CrossRef] [PubMed]
125. Valladares, F.; Frías, I.; Báez, D.; García, C.; López, F.J.; Fraser, J.D.; Rodríguez, Y.; Reyes, R.; Díaz-Flores, L.; Bello, A.R. Characterization of Estrogen Receptors Alpha and Beta in Uterine Leiomyoma Cells. *Fertil. Steril.* **2006**, *86*, 1736–1743. [CrossRef]
126. Li, F.; Wang, J.; Liu, W. Search for Key Genes, Key Signaling Pathways, and Immune Cell Infiltration in Uterine Fibroids by Bioinformatics Analysis. *Medicine* **2023**, *102*, E33815. [CrossRef]
127. Yang, R.; Wang, M.; Zhang, G.; Li, Y.; Wang, L.; Cui, H. POU2F2 Regulates Glycolytic Reprogramming and Glioblastoma Progression via PDPK1-Dependent Activation of PI3K/AKT/MTOR Pathway. *Cell Death Dis.* **2021**, *12*, 433. [CrossRef]
128. Zhou, S.; Yi, T.; Shen, K.; Zhang, B.; Huang, F.; Zhao, X. Hypoxia: The Driving Force of Uterine Myometrial Stem Cells Differentiation into Leiomyoma Cells. *Med. Hypotheses* **2011**, *77*, 985–986. [CrossRef] [PubMed]
129. Ono, M.; Maruyama, T.; Masuda, H.; Kajitani, T.; Nagashima, T.; Arase, T.; Ito, M.; Ohta, K.; Uchida, H.; Asada, H.; et al. Side Population in Human Uterine Myometrium Displays Phenotypic and Functional Characteristics of Myometrial Stem Cells. *Proc. Natl. Acad. Sci. USA* **2007**, *104*, 18700–18705. [CrossRef] [PubMed]
130. Mas, A.; Nair, S.; Laknaur, A.; Simón, C.; Diamond, M.P.; Al-Hendy, A. Stro-1/CD44 as Putative Human Myometrial and Fibroid Stem Cell Markers. *Fertil. Steril.* **2015**, *104*, 225–234.e3. [CrossRef]
131. Wu, X.; Serna, V.A.; Thomas, J.; Qiang, W.; Blumenfeld, M.L.; Kurita, T. Subtype-Specific Tumor-Associated Fibroblasts Contribute to the Pathogenesis of Uterine Leiomyoma. *Cancer Res.* **2017**, *77*, 6891–6901. [CrossRef] [PubMed]
132. Buyukcelebi, K.; Duval, A.J.; Abdula, F.; Elkafas, H.; Seker-Polat, F.; Adli, M. Integrating Leiomyoma Genetics, Epigenomics, and Single-Cell Transcriptomics Reveals Causal Genetic Variants, Genes, and Cell Types. *Nat. Commun.* **2024**, *15*, 1169. [CrossRef]
133. Regner, M.J.; Wisniewska, K.; Garcia-Recio, S.; Thennavan, A.; Mendez-Giraldez, R.; Malladi, V.S.; Hawkins, G.; Parker, J.S.; Perou, C.M.; Bae-Jump, V.L.; et al. A Multi-Omic Single-Cell Landscape of Human Gynecologic Malignancies. *Mol. Cell* **2021**, *81*, 4924–4941.e10. [CrossRef] [PubMed]
134. Hedlund, E.; Deng, Q. Single-Cell RNA Sequencing: Technical Advancements and Biological Applications. *Mol. Asp. Med.* **2018**, *59*, 36–46. [CrossRef] [PubMed]
135. Boakye Serebour, T.; Cribbs, A.P.; Baldwin, M.J.; Masimirembwa, C.; Chikwambi, Z.; Kerasidou, A.; Snelling, S.J.B. Overcoming Barriers to Single-Cell RNA Sequencing Adoption in Low- and Middle-Income Countries. *Eur. J. Hum. Genet.* **2024**, *32*, 1206–1213. [CrossRef] [PubMed]
136. Lähnemann, D.; Köster, J.; Szczurek, E.; McCarthy, D.J.; Hicks, S.C.; Robinson, M.D.; Vallejos, C.A.; Campbell, K.R.; Beerenwinkel, N.; Mahfouz, A.; et al. Eleven Grand Challenges in Single-Cell Data Science. *Genome Biol.* **2020**, *21*, 31. [CrossRef]
137. Heumos, L.; Schaar, A.C.; Lance, C.; Litinetskaya, A.; Drost, F.; Zappia, L.; Lücken, M.D.; Strobl, D.C.; Henao, J.; Curion, F.; et al. Best Practices for Single-Cell Analysis across Modalities. *Nat. Rev. Genet.* **2023**, *24*, 550–572. [CrossRef] [PubMed]

138. Van de Sande, B.; Lee, J.S.; Mutasa-Gottgens, E.; Naughton, B.; Bacon, W.; Manning, J.; Wang, Y.; Pollard, J.; Mendez, M.; Hill, J.; et al. Applications of Single-Cell RNA Sequencing in Drug Discovery and Development. *Nat. Rev. Drug Discov.* **2023**, *22*, 496–520. [CrossRef]
139. Elorbany, S.; Berlato, C.; Carnevalli, L.S.; Maniati, E.; Barry, S.T.; Wang, J.; Manchanda, R.; Kzhyshkowska, J.; Balkwill, F. Immunotherapy That Improves Response to Chemotherapy in High-Grade Serous Ovarian Cancer. *Nat. Commun.* **2024**, *15*, 10144. [CrossRef] [PubMed]
140. Zhou, M.; Zhang, Y.; Song, W. Single-Cell Transcriptome Analysis Identifies Subclusters and Signature with N-Glycosylation in Endometrial Cancer. *Clin. Transl. Oncol.* **2024**. [CrossRef] [PubMed]
141. González-Martínez, S.; Pérez-Mies, B.; Cortés, J.; Palacios, J. Single-Cell RNA Sequencing in Endometrial Cancer: Exploring the Epithelial Cells and the Microenvironment Landscape. *Front. Immunol.* **2024**, *15*, 1425212. [CrossRef]
142. Wang, F.; Yue, S.; Huang, Q.; Lei, T.; Li, X.; Wang, C.; Yue, J.; Liu, C. Cellular Heterogeneity and Key Subsets of Tissue-Resident Memory T Cells in Cervical Cancer. *npj Precis. Oncol.* **2024**, *8*, 145. [CrossRef] [PubMed]
143. Huang, R.; Wang, F.; Zou, W.; Li, X.; Lei, T.; Li, P.; Song, Y.; Liu, C.; Yue, J. Tumor Endothelium-Derived PODXL Correlates with Immunosuppressive Microenvironment and Poor Prognosis in Cervical Cancer Patients Receiving Radiotherapy or Chemoradiotherapy. *Biomark. Res.* **2024**, *12*, 106. [CrossRef] [PubMed]
144. Sengupta, P.; Dutta, S.; Liew, F.; Samrot, A.; Dasgupta, S.; Rajput, M.A.; Slama, P.; Kolesarova, A.; Roychoudhury, S. Reproductomics: Exploring the Applications and Advancements of Computational Tools. *Physiol. Res.* **2024**, *73*, 687–702. [CrossRef] [PubMed]
145. Altmäe, S.; Esteban, F.J.; Stavreus-Evers, A.; Simón, C.; Giudice, L.; Lessey, B.A.; Horcajadas, J.A.; Macklon, N.S.; D'Hooghe, T.; Campoy, C.; et al. Guidelines for the Design, Analysis and Interpretation of "omics" Data: Focus on Human Endometrium. *Hum. Reprod. Update* **2014**, *20*, 12–28. [CrossRef] [PubMed]
146. Gavrilescu, M.; Demnerová, K.; Aamand, J.; Agathos, S.; Fava, F. Emerging Pollutants in the Environment: Present and Future Challenges in Biomonitoring, Ecological Risks and Bioremediation. *New Biotechnol.* **2015**, *32*, 147–156. [CrossRef]
147. Montgomery, G.W.; Zondervan, K.T.; Nyholt, D.R. The Future for Genetic Studies in Reproduction. *Mol. Hum. Reprod.* **2014**, *20*, 1–14. [CrossRef] [PubMed]
148. Yan, S.; Zhang, G.; Luo, W.; Xu, M.; Peng, R.; Du, Z.; Liu, Y.; Bai, Z.; Xiao, X.; Qin, S. PROTAC Technology: From Drug Development to Probe Technology for Target Deconvolution. *Eur. J. Med. Chem.* **2024**, *276*, 116725. [CrossRef] [PubMed]
149. Györfi, A.H.; Matei, A.E.; Distler, J.H.W. Targeting TGF-β Signaling for the Treatment of Fibrosis. *Matrix Biol.* **2018**, *68–69*, 8–27. [CrossRef] [PubMed]
150. Bhat, A.S.; Singh, N.A.; Rymbai, E.; Birendra, S.; Jayaram, S.; Selvaraj, D. Importance of Fibrosis in the Pathogenesis of Uterine Leiomyoma and the Promising Anti-Fibrotic Effects of Dipeptidyl Peptidase-4 and Fibroblast Activation Protein Inhibitors in the Treatment of Uterine Leiomyoma. *Reprod. Sci.* **2023**, *30*, 1383–1398. [CrossRef] [PubMed]
151. Donnez, J.; Stratopoulou, C.A.; Marie-Madeleine, D. Endometriosis and Adenomyosis: Similarities and Differences. *Best. Pract. Res. Clin. Obstet. Gynaecol.* **2024**, *92*, 102432. [CrossRef] [PubMed]
152. Li, X.; Li, C.; Zhang, W.; Wang, Y.; Qian, P.; Huang, H. Inflammation and Aging: Signaling Pathways and Intervention Therapies. *Signal Transduct. Target. Ther.* **2023**, *8*, 239. [CrossRef]
153. Wang, Z.; Zhang, S.; Xiao, Y.; Zhang, W.; Wu, S.; Qin, T.; Yue, Y.; Qian, W.; Li, L. NLRP3 Inflammasome and Inflammatory Diseases. *Oxid. Med. Cell. Longev.* **2020**, *2020*, 4063562. [CrossRef]
154. Liu, T.; Zhang, L.; Joo, D.; Sun, S.C. NF-KB Signaling in Inflammation. *Signal Transduct. Target. Ther.* **2017**, *2*, 17023. [CrossRef] [PubMed]
155. Guo, Q.; Jin, Y.; Chen, X.; Ye, X.; Shen, X.; Lin, M.; Zeng, C.; Zhou, T.; Zhang, J. NF-KB in Biology and Targeted Therapy: New Insights and Translational Implications. *Signal Transduct. Target. Ther.* **2024**, *9*, 53. [CrossRef] [PubMed]
156. Bonavina, G.; Taylor, H.S. Endometriosis-Associated Infertility: From Pathophysiology to Tailored Treatment. *Front. Endocrinol.* **2022**, *13*, 1020827. [CrossRef] [PubMed]

cells

MDPI

Article

Unraveling the Role of Bromodomain and Extra-Terminal Proteins in Human Uterine Leiomyosarcoma

Qiwei Yang [1,*], Ali Falahati [2,3], Azad Khosh [4], Ricardo R. Lastra [5], Thomas G. Boyer [4] and Ayman Al-Hendy [1]

1 Department of Obstetrics and Gynecology, University of Chicago, Chicago, IL 60637, USA; aalhendy@bsd.uchicago.edu
2 Poundbury Cancer Institute for Personalised Medicine, Dorchester DT1 3BJ, UK; ali.falahati@nhs.net
3 DNA GTx LAB, Dubai Healthcare City, Dubai 505262, United Arab Emirates
4 Department of Molecular Medicine, Institute of Biotechnology, University of Texas Health Science Center at San Antonio, San Antonio, TX 78229, USA; khosh@livemail.uthscsa.edu (A.K.); boyer@uthscsa.edu (T.G.B.)
5 Department of Pathology, University of Chicago, Chicago, IL 60637, USA; ricardo.lastra@uchospitals.edu
* Correspondence: yangq@bsd.uchicago.edu

Abstract: Uterine leiomyosarcoma (uLMS) is the most common type of uterine sarcoma, associated with poor prognosis, high rates of recurrence, and metastasis. Currently, the molecular mechanism of the origin and development of uLMS is limited. Bromodomain and extra-terminal (BET) proteins are involved in both physiological and pathological events. However, the role of BET proteins in the pathogenesis of uLMS is unknown. Here, we show for the first time that BET protein family members, BRD2, BRD3, and BRD4, are aberrantly overexpressed in uLMS tissues compared to the myometrium, with a significant change by histochemical scoring assessment. Furthermore, inhibiting BET proteins with their small, potent inhibitors (JQ1 and I-BET 762) significantly inhibited the uLMS proliferation dose-dependently via cell cycle arrest. Notably, RNA-sequencing analysis revealed that the inhibition of BET proteins with JQ1 and I-BET 762 altered several critical pathways, including the hedgehog pathway, EMT, and transcription factor-driven pathways in uLMS. In addition, the targeted inhibition of BET proteins altered several other epigenetic regulators, including DNA methylases, histone modification, and m^6A regulators. The connections between BET proteins and crucial biological pathways provide a fundamental structure to better understand uterine diseases, particularly uLMS pathogenesis. Accordingly, targeting the vulnerable epigenome may provide an additional regulatory mechanism for uterine cancer treatment.

Keywords: uterine leiomyosarcoma; bromodomain and extra-terminal protein; transcriptome analysis; epigenome; m^6A regulators; JQ1; I-BET 762; hedgehog pathway; EMT; transcriptional factors

1. Introduction

Uterine leiomyosarcoma (uLMS) is the most common type of uterine sarcoma arising from smooth muscle cells [1,2]. The 5-year survival rate for all uLMS patients ranges between 25 and 76%, while survival for women at an advanced stage of the disease at the time of initial diagnosis approaches only 10–15% [3]. uLMS is a heterogeneous aggressive disease with a poor prognosis and a challenging clinical treatment outlook [4]. Accordingly, uLMS patients present resistance to available chemotherapies, as evidenced by high rates of both recurrence and progression [5,6]. There is an urgent need to develop effective strategies to improve the outcome for patients with uLMS.

Cumulative data strongly demonstrate that the epigenetic dysregulation of gene expression contributes to the development of various tumors [7–13]. DNA methylation profiling studies showed that uLMS displays a distinct methylation signature [14]. In addition, non-coding RNAs have been investigated and their crucial role in regulating multiple cellular signaling processes in uLMS has been demonstrated [15,16]. Histone modifications, as one of the key epigenetic regulatory features, are dynamically modulated by "writers" and "erasers". In addition, distinct effector proteins ("readers") can recognize specific histone modifications and regulate gene expression. Bromodomain-containing proteins (BRDs), as epigenetic reader proteins, are responsible for transducing regulatory signals carried by acetylated lysine residues into various biological phenotypes [17]. BRDs can act as scaffolds that enable the recruitment of large protein complexes, or they can act as transcription factors themselves. BRDs contain several catalytic domains that enable them to act as methyltransferases, ATP-dependent chromatin re-modelers, or histone acetyltransferases and helicases [18,19]. Notably, BRD9 was recently reported to contribute to the uLMS pathogenesis [5].

So far, over sixty BRDs have been identified in the human genome, and these BRDs were divided into eight subfamilies (e.g., bromodomain and extra-terminal domain [BET] subfamilies) based on protein sequence similarity [20,21]. The BET family of proteins, characterized by the presence of two tandem bromodomains and an extra-terminal domain, consists of BRD2, BRD3, BRD4, and BRDT, which bind histone acetylated lysine residues via two highly conserved amino-terminal bromodomains [19]. BRDs are specifically capable of binding acetylated lysine residues in histones, serving as chromatin-targeting modules that decipher the histone acetylation code. BET proteins regulate gene transcription through epigenetic interactions between bromodomains and acetylated histones during cellular proliferation and differentiation [22]. Studies have demonstrated that BET proteins play an oncogenic role in multiple cancer types by regulating tumor cell growth and metastasis. The connection of BET proteins with key oncogenic transcription factor pathways, chronic inflammation, and immune surveillance is implicated in cancer progression.

Gynecologic cancers include cervical cancer, ovarian cancer, vulvar cancer, vaginal cancer, fallopian tube cancer, and uterine cancer. Growing evidence demonstrates that BET BRDs play an oncogenic role in gynecological cancer, including ovarian cancer. For example, BRD4, located at chromosome 19p13, was amplified in a considerable proportion (~20%) of ovarian cancers, and the expression of BRD4 correlated with amplification status [23]. Recent reports demonstrated that BRD4 is directly involved in the metastatic process in high-grade serous ovarian carcinoma [24]. The pharmacological inhibition of BRD4 with BET inhibitors (BETis) JQ1 and I-BET151 substantially abrogated both the in vitro growth and in vivo tumorigenesis of ovarian cancer [23], and BRD4 inhibitors have been largely used in several pre-clinical studies in ovarian cancer [25–27]. Notably, combination therapy with BET inhibitors effectively overcome chemoresistance in pre-clinical settings [25,28]. Concomitant BRD4 inhibition with PARP inhibitor [29], cisplatin [30], tyrosine kinase inhibitors [31,32], and MEK inhibitors [26,33] has been shown to efficiently and synergistically suppress ovarian carcinoma growth. These studies emphasize the critical impact of abnormal BET protein function on human diseases, especially in gynecological cancer. Therefore, the targeted inhibition of BET proteins may provide a promising option for treating patients with gynecological cancers [20]. However, current knowledge regarding the role and mechanism of BET proteins in the pathogenesis of uterine cancer, such as uLMS, is limited. Accordingly, the present study aimed to investigate whether and how BET proteins contribute to aberrant uLMS cell growth, with important implications for developing novel treatment options for this highly aggressive uterine cancer.

2. Materials and Methods

2.1. Uterine Leiomyosarcoma Samples

The experimental design and flowchart of the bioinformatics analysis are shown in Figure S1. The uLMS tissues were obtained from the University of Chicago Tissue Bank with the approval and consent of the Institutional Review Board (# 20-1820) at the University of Chicago. The initial diagnosis and subsequent confirmation of uLMS cases were described previously [5].

2.2. Immunohistochemistry

Immunohistochemistry was performed as previously described [5]. Briefly, sections were deparaffinized with xylene and rehydrated passing through decreasing concentrations of ethanol in water. Then, antigen retrieval treatment (epitope retrieval solution II, AR9640, Leica Biosystems, Dear Park, IL, USA) was performed for 20 min, and the quenching of endogenous peroxidases was performed. Next, sections were incubated with primary antibodies against BRD2 (Abcam, ab139690), BRD3 (Abcam, Ab264420), and BRD4 (Abcam, Ab128874) in a humidity chamber overnight at 4 °C and developed with peroxidase labeled-dextran polymer followed by diaminobenzidine (DAKO Envision Plus System; DAKO Corporation, Carpinteria, CA, USA). Sections were counterstained with Gill's Hematoxylin (Fisher, Pittsburgh, PA, USA). To determine the percentage and intensity of BET protein-positive cells, the samples were analyzed using the positive cell detection command on QuPath software (version 0.2.3) (https://qupath.github.io, accessed on 28 November 2022). Three thresholds were set to categorize cells according to nuclei staining intensity: negative, weak, moderate, and strong intensity. The H-score captures the intensity and the proportion of the biomarker of interest from the IHC image and comprises values between 0 and 300 [34], thereby offering a dynamic range to quantify BET protein abundance between myometrium and uLMS. Human testis was used as a positive control for BRD2, BRD3, and BRD4 IHC staining.

2.3. Cells and Reagents

The uterine leiomyosarcoma (uLMS) cell line (SK-UT-1, ATCC® HTB-114™) (ATCC, Manassas, VA, USA) was cultured and maintained in ATCC-formulated Eagle's Minimum Essential Medium with 10% of fetal bovine serum. The immortalized human uterine smooth muscle (UTSM) cell line was a generous gift from Dr. Darlene Dixon. The UTSM cells were cultured and maintained in phenol red-free, 10% fetal bovine serum Dulbecco's Modified Eagle Medium: Nutrient Mixture F-12. BET protein inhibitors JQ1 and I-BET 762 were purchased from Tocris (Cat# 4499 and 6521, Minneapolis, MN, USA).

2.4. Cell Viability Assay

Cell viability was measured using trypan blue exclusion assay. The cells were seeded into 12-well tissue culture plates and treated with the BET protein inhibitors (JQ1, I-BET 762) at a dose range from 1–10 μM for 48 h. This assay was performed three times in triplicate.

2.5. Measurement of Cell Cycle Phase Distribution

Cell cycle distribution was determined by flow cytometric analysis as described previously [35]. Briefly, SK-UT-1 cells were cultured in the medium in the presence or absence of 5 μM of JQ1 and I-BET 762 for 24 h. Control cells were cultured in a medium containing an equal amount of DMSO. Cells were fixed with 70% ethanol and then hypotonically lysed in DNA staining solution (0.05 mg/mL PI (Sigma, St. Louis, MO, USA) and 0.1% Triton X-100). The flow cytometry analysis was performed to determine the difference in cell cycle distribution between treated and untreated groups.

2.6. RNA Extraction and Gene Expression

Total RNA extraction and quantitative real-time polymerase chain reaction (qRT-PCR) were performed as described previously [11]. The information about primer sequences is shown in Table S1.

2.7. RNA-Sequencing

We performed whole transcriptome analysis to determine the mechanism underlying the inhibitory effect of BET protein inhibition on the uLMS. The SK-UT-1 uLMS cells were treated with BET protein inhibitors JQ1 and I-BET 762 at a concentration of 5 μM for 48 h. Total RNA was isolated using Trizol. RNA quality and quantity were assessed using the

Agilent bio-analyzer. The RNA-Seq library preparation and assessments of library quality and quantity were described previously [5]. An Illumina NovaSEQ6000 instrument was used for RNA sequencing.

2.8. Transcriptome Profiles Analysis

2.8.1. Transcriptome Data Analysis

Transcriptome data analysis was carried out using a variety of R packages. The quality of reads was controlled using FastQC (http://www.bioinformatics.babraham.ac.uk/projects/fastqc (accessed on 21 May 2023), version 0.73, and then the reads were processed using Trimmomatic [36]. The version and parameters were chosen to trim the reads according to the FastQC results. The reads were mapped to the human reference genome, version hg38, using Hisat2 [37], version 2.2.1. Next, raw reads were mapped to the human reference transcriptome using feature counts [38], version 2.0.3, and the annotation file of Gencode [39], version V41. Gene counts were pre-processed and normalized using the DESeq2 [40] package in R, version 1.36.0. The quality of samples was examined using a comparative boxplot and PCA, and outlier samples were excluded from the analysis. Differentially expressed genes were identified using DESeq2.

2.8.2. Functional Enrichment Analysis

Gene set enrichment analysis (GSEA) preranked [41] was performed using the fgsea R package [42], version 1.22.0, with gene set collections downloaded from the Molecular Signatures Database (MSigDB v7.5.1 for H (hallmark gene sets) and C2 (curated gene sets). The significant pathways were determined based on the parameters, including $n = 1000$ permutations, where p-adjust < 0.05, and FDR < 0.05. The Enrichplot (https://yulab-smu.top/biomedical-knowledge-mining-book (accessed on 21 May 2023) R package, version 1.16.2, was utilized to visualize the results. Additionally, we conducted histone modification enrichment analysis using ENCODE Histone Modifications 2015 in EnrichR [43] through clusterProfiler [44], version 4.4.4, to uncover the epigenetic mechanisms underlying the regulation of DEGs.

2.8.3. Key Transcription Factor Identification

CeTF [45] R package, version 1.10.1, was used to identify the main TFs that control gene expression in different biological conditions, using Regulatory Impact Factors and Partial Correlation and Information Theory analysis. Co-expression networks were constructed using DEGs. To narrow down the TFs and acquire meaningful results, TFs with $|\text{FoldChange}| > 1.5$ and adjusted p-value < 0.05 in both JQ1 and I-BET 762 vs. control conditions were considered significant results.

2.8.4. Survival Analysis of BRD Genes Expression in Sarcoma Patients

We utilized Gene Expression Profiling Interactive Analysis (GEPIA) (http://gepia.cancer-pku.cn/ (accessed on 29 June 2024) to conduct survival analysis, specifically examining overall survival (OS) in sarcoma patients. Patients were categorized into two groups based on the median expression levels of BRD2, BRD3, and BRD4 in sarcoma cancer samples, resulting in high and low expression groups. Subsequently, Kaplan–Meier methods were employed to generate survival curves. Additionally, we calculated the hazard ratio (HR) and p-value, with a p-value < 0.05 denoting statistical significance.

2.9. Statistical Analysis

Statistical analysis was performed as described previously [5]. A p-value of less than 0.05 was considered statistically significant.

3. Results

3.1. The BET Proteins BRD2, BRD3, and BRD4 Are Aberrantly Overexpressed in uLMS Tissues Compared to Adjacent Myometrium from Women with uLMS

To determine the differential levels of BET proteins between uLMS and MM (myometrium), IHC staining for BRD2, BRD3, and BRD4 was performed. Figures 1 and 2 show that both staining intensity and the percentage of BRD2-, BRD3-, and BRD4-positive cells were significantly higher in uLMS than in MM. These studies reveal the dysregulation of BET protein expression in the pathogenesis and progression of uLMS.

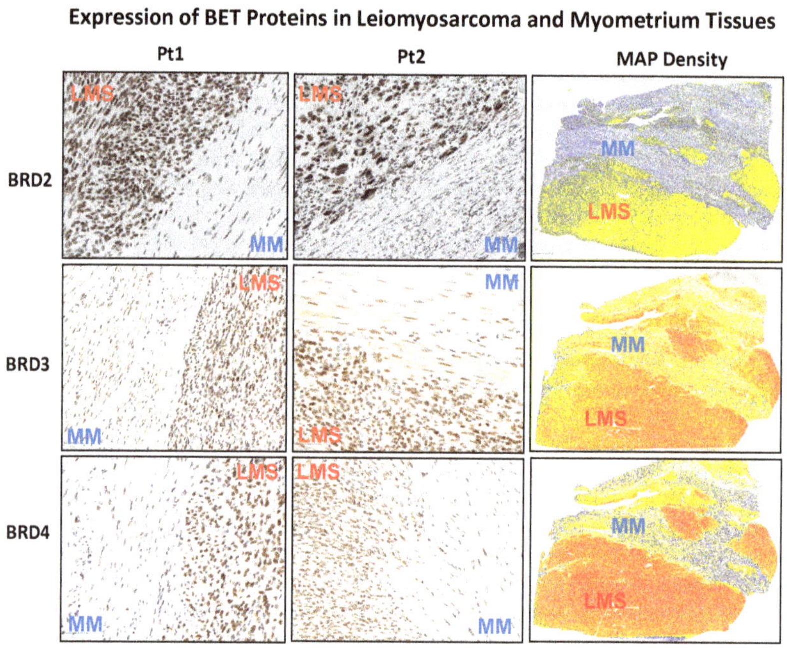

Figure 1. IHC staining of BRD2, 3, and 4 in human uLMS tissues and adjacent myometrium. IHC staining for BRD2, 3, and 4 is presented with two representative cases. The right column shows the density map of BRD2, 3, and 4 for the same representative case. Blue color: negative; yellow color: low expression; brown color: moderate expression; red color: strong expression. MM: myometrial tissues; LMS: leiomyosarcoma.

3.2. Inhibition of BET BRDs Altered uLMS Cell Viability

Two potent and selective BET protein inhibitors, JQ1 and I-BET 762, were identified in 2010 [46–48]. Both cell-permeable small inhibitors can bind to the BRD pocket in a manner competitive with acetylated peptide binding [46,47]. Furthermore, both BET inhibitors engage the bromodomain pocket competitively with acetylated peptide binding, thereby causing the displacement of all four BET proteins from chromatin in cells upon exposure to these compounds [49]. Therefore, we selected JQ1 and I-BET 762 to assess their effects on uLMS cell viability. Cell viability was evaluated using a trypan blue exclusion assay in the SK-UT-1 cell line treated for 48 h with JQ1/I-BET 762 at 1–10 μM doese. Prolonged treatment (48 h) with JQ1/I-BET 762 showed a dose-dependent inhibitory effect on the viability of SK-UT-1 cells (Figure 3A). In contrast, there was much less inhibitory effect (compared with UTSM) on the control myometrial cell line (UTSM) (Figure 3A). These results indicate that JQ1 and I-BET 762 inhibit the UF cells in a dose-dependent manner and that it preferentially targets UF cells compared with UTSM cells.

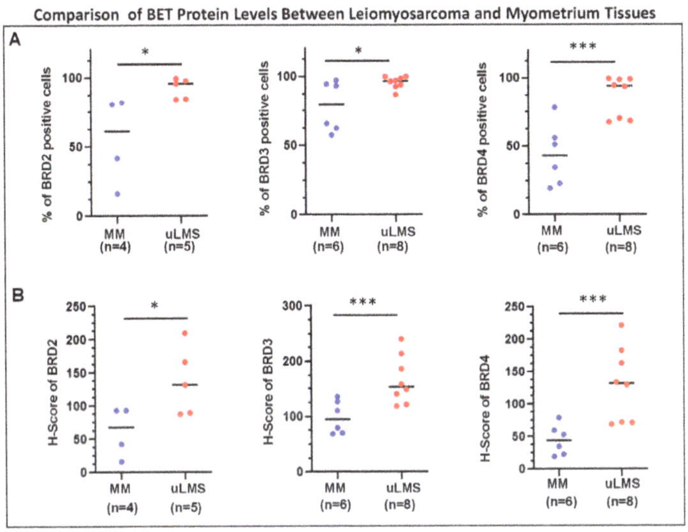

Figure 2. Percentage of BET protein-positive cells and H-score of BET protein expression in uLMS vs. myometrium. (**A**) Percentage of BRD2, BRD3, and BRD4 positive cells in uLMS and myometrium tissues. (**B**) H-score of BRD2, BRD3, and BRD4 in uLMS and myometrium tissues. * $p < 0.05$. *** $p < 0.001$.

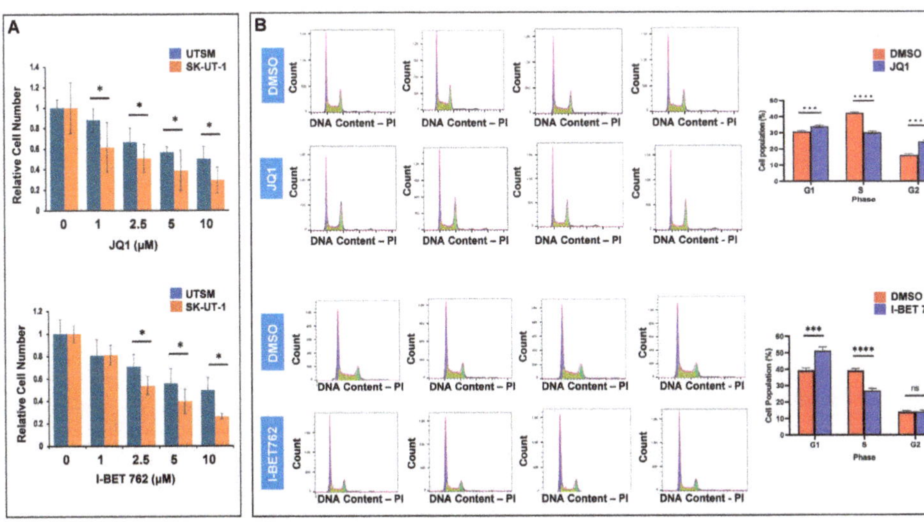

Figure 3. Treatment with JQ1 and I-BET 762 decreases uLMS cell viability and induces cell cycle arrest. (**A**) Cell viability of SK-UT-1 and UTSM cells were measured with trypan blue exclusion assay in the presence or absence of JQ1 and I-BET 762 for 48 h. (**B**) Treatment with JQ1 and I-BET 762 induces cell cycle arrest. Flow cytometric analysis was performed to measure the cell cycle phase distribution in SK-UT-1 uLMS cells in the presence or absence of JQ1 and I-BET 762. Cell cycle phases were marked as purple (G1), olive green (S), and light green (G2). Quantitative cell population analysis in response to JQ1 and I-BET 762 treatment was performed, respectively (right panel). DMSO group (n = 4), JQ1 (n = 4), I-BET 762 (n = 4). ns: no significant difference; * $p < 0.05$; *** $p < 0.001$; **** $p < 0.0001$. ns: no significant difference. n indicates the number of biological samples for each group.

3.3. Inhibition of BET BRDs Induces Cell Cycle Arrest in uLMS Cells

JQ1 treatment induced the accumulation of cells in the G1 phase and displayed a pronounced decrease of cells in the S phase, indicating the blockade of G1 progression (Figure 3B). The percentage of cells in the G1 phase increased from 30.6% to 33.9% ($p < 0.0001$) in response to 5 μM JQ1 treatment for 24 h. Accordingly, the percentage of cells in the S phase decreased from 42.3% to 30.2% ($p < 0.0001$) in response to the JQ1 treatment for 24 h. In addition, JQ1 treatment also significantly increased the cell population in the G2 phase. The percentage of cells in the G2 phase increased from 16.1% to 24.4% ($p < 0.00001$) in response to 5 μM JQ1 treatment (Figure 3B, right panel). I-BET 762 treatment also increased the accumulation of cells in the G1 phase and a corresponding decrease in the S phase, indicating the G1 progression blockade (Figure 3B). The percentage of cells in the G1 phase increased from 39.3% to 51.2% ($p < 0.0001$) in response to 5 μM I-BET 762 treatment. Accordingly, the percentage of cells in the S phase decreased from 39.2% to 26.8% ($p < 0.0001$) upon the I-BET 762 treatment (Figure 3B, right panel). These results suggest that the targeted inhibition of BET BRDs with JQ1 and I-BET 762 suppressed uLMS proliferation via cell cycle arrest (Figure 3A,B).

3.4. BET Protein Inhibition Causes Transcriptome Alterations in uLMS Cells

To examine the impact of BET protein inhibition on the uLMs transcriptome, we performed RNA-seq in SK-UT-1 cells treated without (DMSO control) or with JQ1 or I-BET 762 for 48 h. Differential gene expression analysis was done using three algorithms: limma-voom, DESeq2, and edgeR. Transcriptome analysis revealed that the treatment of SK-UT-1 cells with 5 μM JQ1 or I-BET-762 for 48 h induced 3586 DEGs (1130 up and 2456 down) and 4867 (1718 up and 3149 down), respectively. Figure 4A,B exhibit distinct expression patterns between the DMSO control group vs. JQ1 and I-BET 762 treatment groups, respectively. Figure 4C,D reveal the distribution of DEGs between treatment groups and the DMSO control group.

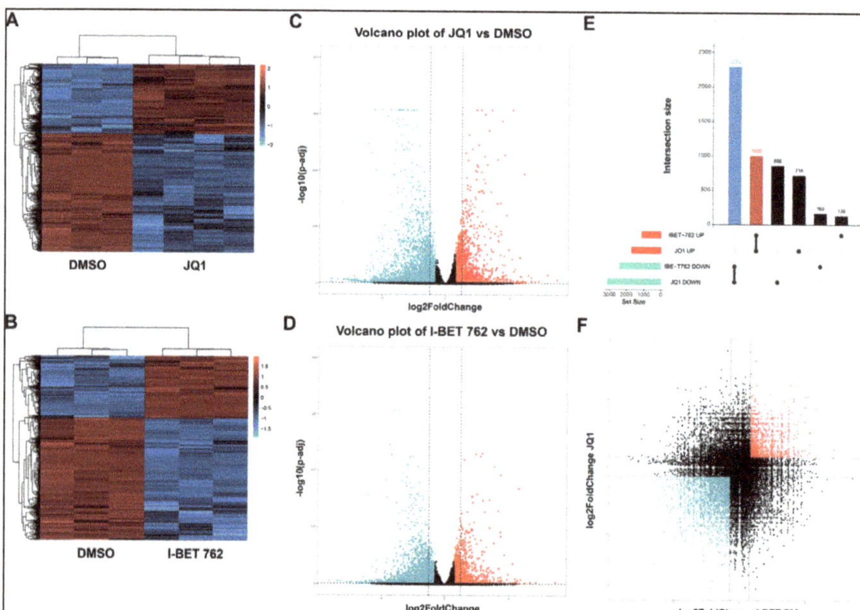

Figure 4. Treatments with JQ1 and I-BET 762 sculpt the transcriptome of uLMS cells. Heat maps are presented to cluster DEGs (JQ1 vs. control) (**A**) and (I-BET 762 vs. control) (**B**), respectively. Volcano

plots of gene expression profiles are presented for JQ1 vs. control (**C**) and I-BET 762 vs. control (**D**). The red and blue points represent upregulated and downregulated genes, respectively. The vertical dotted and the horizontal black lines represent the log (FC) cutoff and the logarithmic transformed adjusted *p*-value cutoff, respectively. (**E**) Upset diagram showing the intersection size of upregulated and downregulated genes across drug treatments. (**F**) Distribution of overlapped DEGs in response to JQ1 and I-BET 762 treatments. FC: fold-change.

To visualize the intersections of DEGs in response to treatment with either JQ1 or I-BET 762, an upset plot was utilized to present the distribution characteristics of DEGs upon either drug treatment. As shown in Figure 4E, the common up and down DEGs with JQ1 and I-BET 762 contain 1000 and 2293 genes, respectively. The latter group exhibited the most significant number of genes in all groups involved in the two types of treatments. Figure 4F shows the distribution of unique and overlapped DEGs in response to JQ1 and I-BET 762 treatments.

To determine if our RNA-seq data correlates with biological effects impacted by BETis, we initially compared the expression levels of cell cycle-related genes between control and BETi-treated uLMS cells. As shown in Figure 5, JQ1 and I-BET 762 treatments increased the expression levels of *CDKN1A* and reduced the expression levels of *CDK6*, which correlated with our finding that BETis induced cell cycle arrest. In addition, BETis induced the expression of BIK, which can stimulate apoptosis, while decreasing the expression of BCL-2, which can suppress apoptosis. Our results suggested that these cell progression regulators may be critical in BETi1-induced cell cycle arrest.

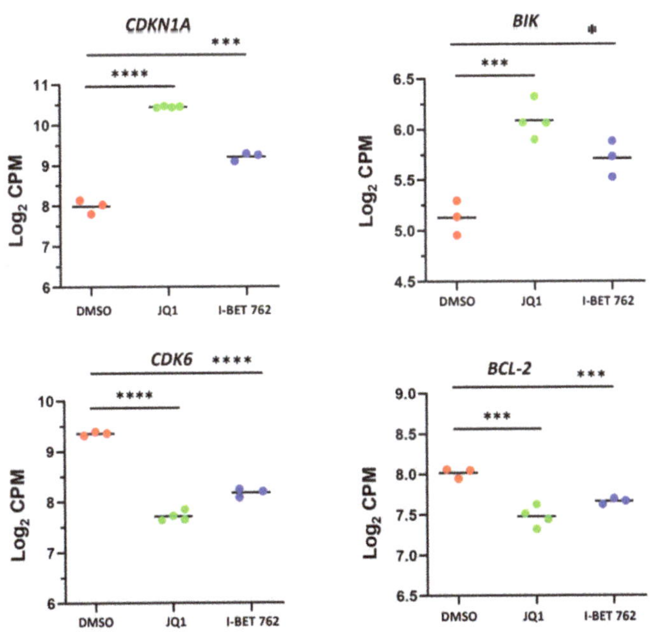

Figure 5. JQ1 and I-BET 762 altered cell cycle- and apoptosis-related gene expression in uLMS cells. RNA-seq analysis revealed the upregulation of *CDKN1A* and *BIK* and the downregulation of *CDK6* and *BCL2*, respectively, in uLMS cells upon BETis treatment. NS: no significant difference; * $p < 0.05$; *** $p < 0.001$; **** $p < 0.0001$.

3.5. Pathway Analysis of DEGs upon JQ1 and I-BET 762 Treatments

To gain insight into the biological changes by BET protein inhibition, gene set enrichment analysis (GSEA) was performed. We demonstrated that Hedgehog signaling was suppressed in response to JQ1 and I-BET 762 treatments, as shown in Figure 6A,B. Accordingly, several key components of the Hedgehog signaling pathway, including GL1, 2, 3, and SMO, were downregulated in BETi-treated uLMS cells compared to control cells treated with the vehicle (DMSO) (Figure 6C–F).

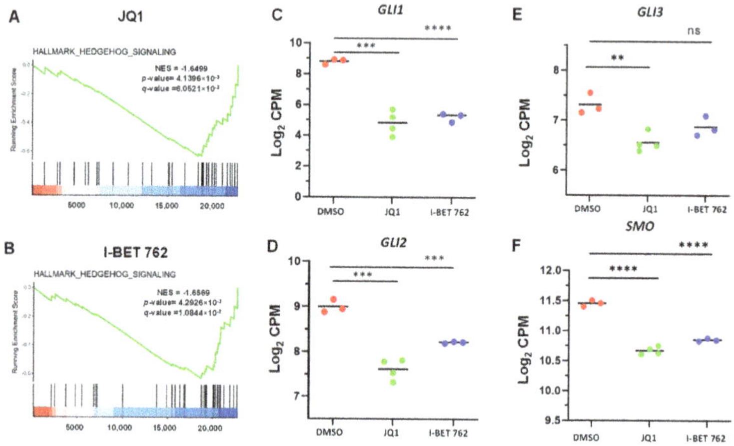

Figure 6. JQ1 and I-BET 762 altered the Hedgehog pathway in uLMS cells. Hallmark analysis demonstrated the enrichment of the Hedgehog pathway in SK-UT-1 cells in response to JQ1 (**A**) and I-BET 762 (**B**) treatments. The key components of the Hedgehog pathway, including *GLI1* (**C**), *GLI2* (**D**), *GLI3* (**E**), and *SMO* (**F**), are downregulated in response to JQ1 and I-BET 762 treatment. ** $p < 0.01$; *** $p < 0.001$; **** $p < 0.0001$. ns: no significant difference.

In addition, we demonstrated that epithelial–mesenchymal transition (EMT) was inhibited in response to BETi treatment in uLMS cells (Figure 7A,B). Accordingly, EMT inducers, including PDGFRβ, CCND2, IL32, and TNC, were downregulated in uLMS cells treated with BETis (Figure 7C–F).

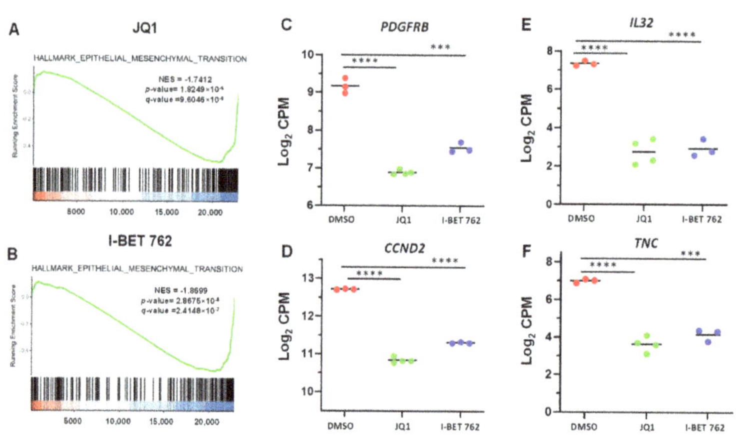

Figure 7. JQ1 and I-BET 762 altered the EMT pathway in uLMS cells. Hallmark analysis demonstrated the enrichment of the EMT pathway in SK-UT-1 cells in response to JQ1 (**A**) and I-BET 762 (**B**) treatments. The EMT inducers, including *PDGFRβ* (**C**), *CCND2* (**D**), *IL32* (**E**), and *TNC* (**F**), are downregulated in response to BETi treatment. *** $p < 0.001$; **** $p < 0.0001$.

In addition to Hedgehog signaling and EMT, JQ1 and I-BET762 altered other key pathways, including metastasis, invasive cancer, and cell migration (Figure S2).

TFs play a critical role in cancer progression. Therefore, we determined the TFs involved in the metastasis of uLMS. We identified several TFs, including GTF2I, NFIB, NFATC4, ZNF37A, KLF13, and ZNF507, involved in cancer metastasis. BETis decreased the expression of these TFs in uLMS cells. Therefore, BETis may reduce the metastasis of uLMS by inhibiting the metastasis-related TFs (Figure 8).

A

	JQ1		I-BET 762	
SYMBOL	Log2FC	padj	Log2FC	padj
NFATC4	-6.44068	2.7E-144	-4.8236	7.1E-157
IRF9	-1.20096	4.27E-24	-0.61177	1.95E-06
NFIB	-1.16186	1.22E-82	-0.66672	3.39E-24
TRIM16	-1.12299	2.59E-32	-0.74524	2.07E-13
ZNF37A	-1.10076	1.29E-78	-0.97976	7.64E-55
HMBOX1	-0.93991	1.02E-33	-0.40568	2.23E-06
KLF13	-0.89366	8.72E-05	-0.92724	0.000223
ZNF507	-0.81734	1.33E-32	-0.57007	1.99E-14
GTF2I	-0.81661	4.36E-38	-0.47002	1.3E-11
L3MBTL1	-0.73614	1.91E-10	-0.17844	0.312079
USF3	-0.73516	2.04E-21	-0.43493	3.8E-07
ZNF343	-0.64504	1.92E-12	-0.44541	1.29E-05
SMARCE1	-0.61091	9.76E-14	0.12342	0.346988
FAM200B	-0.58774	5.26E-06	-0.16933	0.436616
ZNF264	0.590992	6.94E-31	0.280007	1.55E-06
ELF1	0.598996	1.98E-19	0.283964	0.00026
ZNF75A	0.652718	5.72E-15	0.325973	0.000932
ZNF511	0.67595	1.46E-17	0.584199	1.2E-11
ZNF701	0.686487	4.7E-25	0.360572	1.94E-06
ZNF697	0.690234	4.92E-19	0.381032	1.73E-05
ZNF708	0.699107	2.6E-30	0.312036	8.92E-06
GZF1	0.761121	2.05E-39	0.523188	1.64E-16
ZNF674	0.826939	2.93E-17	0.355853	0.002682
ZNF596	0.841121	3.77E-17	0.505089	9.14E-06
BAZ1A	0.877041	2.51E-57	0.541743	2.11E-19
ZNF800	0.919093	2.44E-43	0.584401	1.37E-15
ZSCAN32	1.02539	1.43E-42	0.810646	2.09E-23
FOS	1.040717	2.03E-21	0.28493	0.047852
KLF10	1.202502	3.4E-72	0.731754	1.45E-23
ETV4	1.209967	2.21E-46	0.525811	6.69E-08
ZNF280C	1.336369	2.5E-109	0.912312	2.33E-44
ID2	1.439163	3.84E-76	0.851161	1.46E-23

B

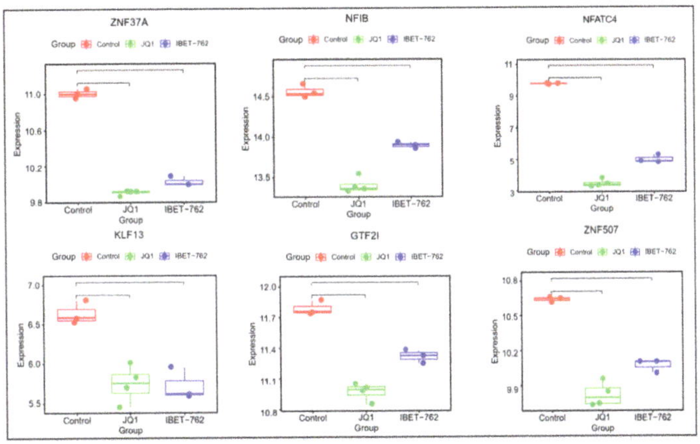

Figure 8. Targeted inhibition of BET proteins suppressed the gene expression of TFs, contributing to cancer metastasis and progression. (**A**) The list of DEGs of TFs (**B**) RNA-seq revealed the downregulation of TF genes, including *ZNF37A*, *NFIB*, *NFATC4*, *KLF13*, *GTF2I*, and *ZNF507* in uLMS cells.

3.6. JQ1 and I-BET 762 Treatment Altered the Expression of Epigenetic Regulators

Previous studies identified the functional link between DNA methylation and chromatin modifications [50–52]. In this study, we investigated whether inhibiting BET proteins altered the expression levels of DNA methylation regulators. Accordingly, we employed a targeted gene analysis approach and revealed that the expression levels of genes that regulated the dynamic status of DNA methylation were modulated in BETi-treated SK-UT-1 cells. These DNA methylation/demethylation-related DEGs included *DNMT3A*, *DNMT3B*, *DNMT1*, *TET1*, and *TET2* (Figure 9).

To determine the relationship between chromatin readers and histone modifications, we characterized the genes related to histone modifications in SK-UT-1 cells after treatment with BETis. As shown in Figure 10, the targeted inhibition of BET proteins by JQ1 and I-BET 762 significantly modulated the expression levels of *KDM1A*, *KDM1B*, *KDM4D*, *SETD4*, *SETD6*, *SETD9*, *SIRT1*, and *SIRT7*. These analyses suggest that JQ1 and I-BET 762 treatments may alter the transcriptome via histone modifications.

To better understand epigenetic-regulated transcriptional changes in response to the JQ1 and I-BET 762 treatment, we performed an enrichment analysis of epigenetic histone markers using the Enrichr web server for our further discoveries. As shown in Figure S3, we identified several histone modifications, including H4K20me1 and H3K4me3, associated with upregulated DEGs in response to JQ1 treatment (Figure S3A). In addition, we identified downregulated DEGs related to histone modifications, including H3K27me3 (Figure S3B). For the association analysis between I-BET 762-induced DEGs and histone marks, we revealed that histone modifications with up DEGs included H4K20me1 and H3K4me3, among others (Figure S3A, right panel). The histone modifications with I-BET 762-induced down DEGs included H3K27me3 (Figure S3B, right panel).

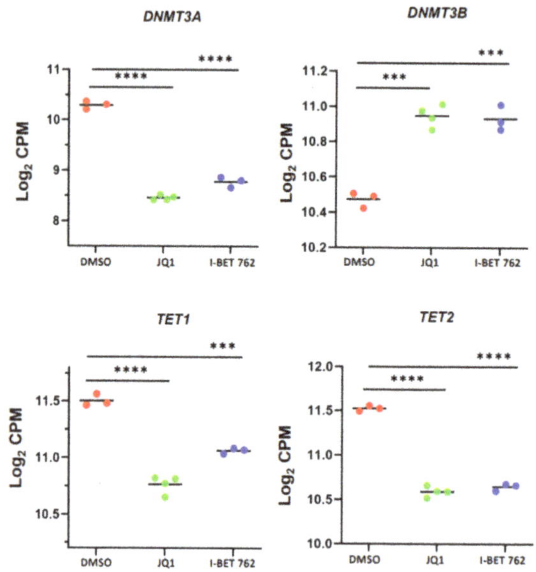

Figure 9. JQ1 and I-BET 762 treatments altered DNA methylation-related genes in uLMS cells. RNA−seq revealed the downregulation of *DNMT3A, DNMT3B, DNMT1, TET1*, and *TET2* in uLMS cells in response to JQ1 and I-BET 762 treatment. ** $p < 0.01$; *** $p < 0.001$; **** $p < 0.0001$.

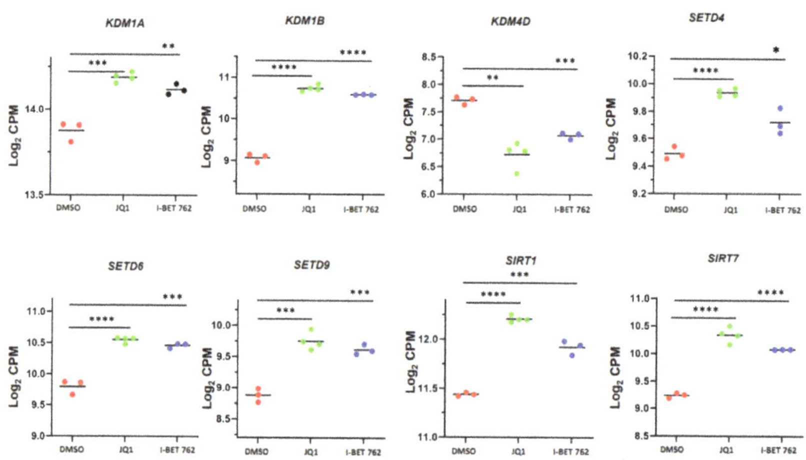

Figure 10. JQ1 and I-BET 762 altered the expression levels of histone modification-regulated genes in uLMS cells. RNA−seq revealed the altered expression of *KDM1A, KDM1B, KDM4D, SETD6, SETD9, SETDB1, SET, SETD9, SIRT1*, and *SIRT7* in uLMS cells in response to BETi treatments. * $p < 0.05$; ** $p < 0.01$; *** $p < 0.001$; **** $p < 0.0001$.

3.7. JQ1 and I-BET 762 Treatment Altered the Expression of m^{6}A Regulators

To determine the impact of BET protein inhibition on RNA epigenetics, we compared the expression of several key m^{6}A regulators between control and BETi-treated groups. As shown in Figure 11, JQ1 and I-BET 762 altered the expression levels of *FTO, YTHDC2*, and *IGF2BP1*, indicating that BET proteins may participate in reprogramming the m^{6}A epitranscriptome in uLMS.

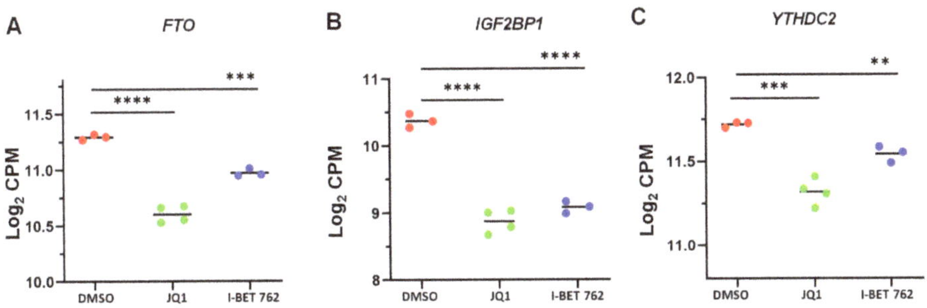

Figure 11. JQ1 and I-BET 762 altered the expression levels of m^6A regulators in uLMS cells. RNA−seq revealed the altered expression of *FTO* (**A**), *IGF2BP1* (**B**), and *YTHDC2* (**C**) in uLMS cells in response to BETi treatment. ** $p < 0.01$; *** $p < 0.001$; **** $p < 0.0001$.

To validate the DEGs related to the cell cycle, Hedgehog pathway, EMT, TF-driven pathway, and epigenetic regulators, we selected several key genes and performed RT-qPCR analysis to confirm our findings. As shown in Figure S4A–D, the expression levels of genes related to the aforementioned biological progress between control and JQ1/I-BET 762 treated cells are consistent with RNA-seq data.

3.8. The Impact of BET BRD Gene Expression on Survival Rates in Sarcoma Patients

This study demonstrated that BET proteins, BRD2, BRD3, and BRD4, are aberrantly overexpressed in uLMS tissues compared to the adjacent myometrium. Therefore, we investigated if the BRD protein expression levels are correlated with the OS of sarcoma patients. Our survival analysis revealed a significant difference in OS between the high (n = 131) and low expression (n = 131) groups of the BRD2 gene in sarcoma patients. Patients with higher expression levels of BRD2 had a significantly (p-value < 0.05) lower survival rate (1.5-fold) compared to those with lower expression levels (Figure 12A). The Kaplan–Meier survival curves demonstrated this disparity, with a calculated hazard ratio (HR) indicating a higher mortality risk in the high-expression group. The analysis for BRD3 and BRD4 gene expression exhibited a similar trend compared to BRD2 expression in terms of OS; however, the expression levels of BRD3 and BRD4 did not show a significant impact on the OS of uLMS patients (Figure 12B,C).

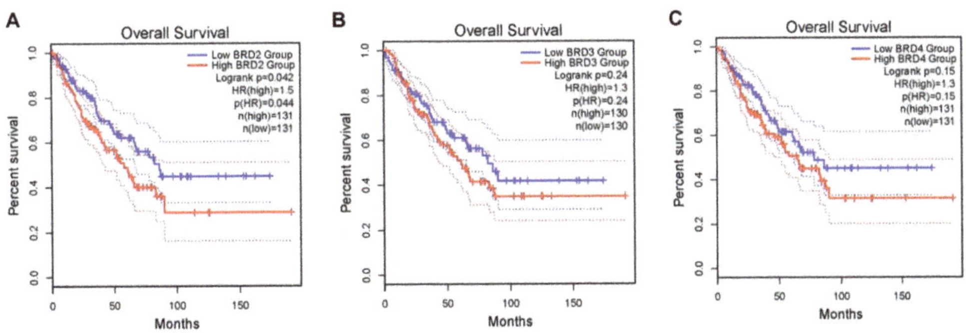

Figure 12. Kaplan–Meier curves of sarcoma patient's OS with low BRD expression (blue) versus high gene expression (red line). (**A**) BRD2; (**B**) BRD3; (**C**) BRD4. Human data from the TCGA dataset were accessed. The patients were categorized with expression levels above the median as the high group (n = 131) and those below the median as the low group (n = 131).

4. Discussion

uLMS is a highly aggressive tumor type with a high tumor recurrence rate, progression, and metastasis [4]. The origin and molecular mechanism underlying and driving its clinical and biological behavior remain unclear. Although the role of epigenetic alteration in uLMS has been investigated, the functional role and epigenetic mechanism underlying BET protein-related uLMS pathogenesis are limited. In this study, we determined the expression pattern of BET proteins in uLMS tissues and examined the biological effect of the targeted inhibition of BET proteins in uLMS cells. We demonstrated that BET proteins, including BRD2, 3, and 4, were aberrantly upregulated in uLMS and exhibited an important tumor-promoting role in uLMS. Accordingly, we utilized GEPIA to conduct survival analysis and demonstrated that BRD2 expression levels were significantly associated with the OS of sarcoma patients. Furthermore, BET proteins could alter several key pathways and reprogram the oncogenome in uLMS. In addition, specific BET protein inhibitors show promising therapeutic efficacy in treating uLMS.

The aberrant overexpression of BET proteins has been found in many cancer types and often correlates with the cancer phenotype. For instance, BRD4 expression was upregulated in pediatric primary medulloblastomas [53] and was found to be overexpressed in ovarian cancer, which is correlated with BRD4 amplification [23]. Notably, BET proteins as a therapeutic target have been reported in many cancers, including gynecological ovarian cancer [23,54–56]. We previously reported that uLMS cells grow faster than myometrial cells [57]. This study demonstrated that inhibiting BET proteins with JQ1 and I-BET 762 significantly decreased uLMS proliferation dose-dependently via cell cycle arrest. Other reports showed that JQ1 could reduce cancer cell growth in vitro and in vivo [58,59]. The underlying mechanisms include an effect of BETis on cell cycle arrest in the G1 phase and a decrease in the percentage of cells in the S phase. Accordingly, several BET protein inhibitors have been developed, and these pharmacologic inhibitors showed potent anti-tumor effects with decreased cancer phenotypes via multiple mechanisms dependent upon the cancer types and experimental conditions [60]. For example, treating medulloblastoma cell lines with JQ1 significantly decreased cell proliferation and preferentially induced apoptosis in *MYC*-overexpressed cells. Additionally, JQ1 treatment prolonged the survival of mice harboring medulloblastoma xenografts and diminished the tumor burden in these mice [53]. Integrating of genetic features with chemosensitivity data revealed a robust correlation between MYCN amplification and sensitivity to bromodomain inhibition in pediatric cancer neuroblastoma. BET inhibition conferred a significant survival advantage in neuroblastoma models via regulating MYCN, providing a compelling rationale for developing BET bromodomain inhibitors in patients with neuroblastoma [61]. In addition, BET inhibitors also provide multiple beneficial effects on anti-inflammation [47], heart failure [62], spermatogenesis [63], and chemoresistance [64,65].

To further determine the mechanisms associated with JQ1/I-BET 762-induced inhibition, we performed a genome-wide RNA-sequencing analysis comparing the profiles of DMSO-treated with BETi-treated uLMS cells. Our high-throughput sequencing analysis revealed that the targeted inhibition of BET proteins with BETis altered several critical biological pathways that may contribute to uLMS pathogenesis. Hedgehog signaling plays a fundamental role in several biological processes, including embryonic development, tissue repair, the proliferation and differentiation of various cells, hematopoiesis, as well as the pathogenesis of various types of cancer [66–70]. In this study, we demonstrated that BETi altered the Hedgehog pathway in uLMS. Furthermore, several key components in the Hedgehog pathway, including GLI1, GLI2, SMO, etc., were downregulated in response to BETi treatment. We have previously reported that the Hedgehog pathway was activated in uLMS with aberrant upregulation of the GLI family and increased nuclear translocation of GLI1 [70]. Therefore, BET inhibitors may suppress the uLMS phenotype by inhibiting the Hedgehog pathway. Interestingly, CCND2, an established GLL1 target gene [71], is also decreased in BETi-treated uLMS cells. In addition to an in vitro study on the role of BRD "reader" inhibition in LMS, a BET bromodomain inhibitor, GS-626510, has been tested in the LMS patient-derived xenograft model harboring either derangements in C-MYC and PTEN/PI3CA/AKT genes or homologous recombination deficiency signatures. The study demonstrated that GS-626510, a BET family BRD inhibitor, suppressed the LMS tumor growth in these two models [72]. These studies elucidate the critical role of histone readers in the pathogenesis of LMS (Figure S3).

EMT is a biological process crucial for tumor aggressiveness, including cancer invasion and metastasis as well as drug response [73,74]. Therefore, the cellular EMT status can be considered a reliable determinant of patient prognosis. In this study, we demonstrated that the targeted inhibition of BET proteins decreased the EMT pathway in uLMS cells. We have identified several EMT inducers in uLMS cells with a significantly decreased expression in response to BETi treatment, including *IL32*, *TNC*, and *PDGFRβ*. IL32 is a cytokine that plays a fundamental role in innate and acquired immunity through the regulation of T cells [75,76]. IL32 can trigger the onset of ECM in several types of cells [75,77]. Tenascin-C (TNC) is a large extracellular matrix glycoprotein that promotes cell adhesion and tissue remodeling and plays a critical role in the transduction of cellular signaling pathways [78]. TNC promotes EMT and relevant pathways in several types of cancer [78–80]. The inhibition of TNC by knockdown can inhibit cancer cell proliferation, migration, and invasion and suppress tumor growth in vivo [78]. PDGFRβ belongs to the type III receptor tyrosine kinase family and is known to be involved in tumor metastasis [81,82]. In addition, PDGFRβ promotes the EMT process via the activation of the PI3K/ERK pathway [83,84]. All these data suggested that BET inhibitors suppressed the uLMS phenotype by inhibiting the EMT pathway.

TFs play a central role in cancer progression by modulating the interplay between cell signaling and gene regulation. Some TFs alter multiple biological processes, including differentiation and development, DNA repair genes, cell proliferation, cellular stresses, and therapy resistance. Beyond these roles, various TFs also regulate cancer invasion, metastasis, and progression [85–87]. TFs, often in conjunction with their corresponding co-activators or co-repressors, can cause alterations in gene expression at specific genome sites [88]. In this study, we identified multiple TFs involved in cancer metastasis. For instance, ZNF37A has been found to be involved in promoting tumor metastasis via the THSD4/TGF-β axis in colorectal cancer [89]. In prostate cancer (PC), ZNF507 expression was associated with metastatic PC with a high grade. Furthermore, ZNF507 promoted the metastatic properties of PC by enhancing TGF-β signaling [90]. GTF21, as a transcriptional factor, binds to the initiator and E-box element in promoters and modulates gene expression. It is reported that GTF2I was found to have potential prognostic value for breast cancer metastasis [91]. Integrated analysis demonstrated that KLF13 may be involved in tumorigenesis and metastasis in colon cancer [92]. NFIB has been reported to promote tumor growth, metastasis, and recurrence in various cancers [93]. Another TF, NFATC4, has been reported to be aberrantly activated and is involved in initiation, proliferation, invasion, and metastasis in several types of cancer [94]. Altogether, BET inhibitors may suppress EMT via TF alterations in uLMS.

Notably, BET proteins play a central role in gene transcription in chromatin [95]. The interplay between BET proteins and chromatin regulation and organization has been reported [96]. A recent study by Zhou et al. demonstrated that BRD4 was highly expressed in gastric cancer (GC) tissues and was significantly associated with poor prognosis. JQ1 inhibits the malignant progression of GC by downregulating chromatin accessibility and inactivating RUNX2/NID1 signaling [96]. In our loss-of-function study, we demonstrated that the targeted inhibition of BET proteins altered the expression levels of histone acetylation modulators, histone methylation enzymes, and DNA-methylation-related epigenetic regulators, further demonstrating the substantial crosstalk between BET proteins and other epigenetic mechanisms. Accordingly, the targeted inhibition of BET proteins might alter the transcriptome by reprogramming the network of oncogenic epigenomes in uLMS.

N6-methyladenosine (m^6A) is the most prevalent, abundant, and conserved posttranscriptional modification in eukaryotic RNAs and plays an important role in many biological processes [97]. Recently, we reported that m^6A demethylase FTO plays an oncogenic role in the pathogenesis of uLMS [98]. Herein, we determined the link between BET proteins and RNA methylation regulators. Our study demonstrated that BETi decreased the expression of FTO m^6A RNA demethylase. In addition, BETi also altered the expression of m^6A readers, such as YTHDC2 and IGF2BP1. These data suggest that the targeted inhibition of BET proteins may alter the epitranscriptome in uLMS, emphasizing the importance of m^6A regulation in uLMS progression.

Based on our studies, we propose a mechanistic model for the targeted inhibition of BET proteins in uLMS: (1) BET proteins, including BRD2, 3, and 4 are aberrantly overexpressed in uLMS compared to the adjacent myometrium; (2) targeting BET proteins with JQ1 and I-BET 762 alters the uLMS phenotype by suppressing cell proliferation via cell cycle arrest and the modulation of cell cycle-related genes and others; (3) BETis reversed the phenotype of uLMS via different biological pathways including the Hedgehog pathway, the EMT pathway, and TF-driven signaling; (4) BET proteins constitute a potential therapeutic vulnerability in malignant uLMS, and BET protein inhibitors, such as JQ1 and I-BET 762, alter key pathways and reprogram the oncogenic profiling and epigenetic network to suppress the uLMS phenotype (Figure 13).

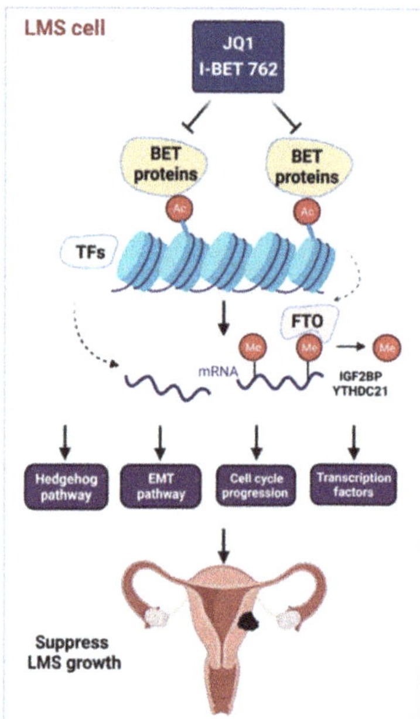

Figure 13. Experimental model. Our experimental model shows that targeting BET proteins with JQ1 and I-BET 762 induces cell cycle arrest, modulates the Hedgehog and EMT pathways, and alters the TF network as well as interactions between target genes and epigenetic regulators in uLMS cells. This figure was created using the BioRender software online app (BioRender.com).

5. Conclusions

In conclusion, our study demonstrated for the first time that uLMS tumors exhibited an aberrant upregulation of BET proteins, highlighting the critical role of histone readers in the pathogenesis of uLMS. The targeted inhibition of BET proteins may impart beneficial effects in uLMS and provide a promising and novel strategy for treating patients with this aggressive uterine cancer.

Supplementary Materials: The following supporting information can be downloaded at: https://www.mdpi.com/article/10.3390/cells13171443/s1, Figure S1: Flowchart of experimental design and bioinformatics analysis; Figure S2: Hallmark analysis demonstrated the enrichment of metastasis, invasive cancer, and cell migration in SK-UT-1 cells in response to JQ1 and I-BET 762 treatment; Figure S3: Targeted inhibition of BET proteins altered the gene expression correlating to histone modifications; Figure S4: The validation of differentially expressed genes between control (DMSO) and BETi-treated uLMS cells. Table S1: Primers used in the study.

Author Contributions: Conceptualization, Q.Y.; data curation, Q.Y., A.F. and A.K.; formal analysis, Q.Y., A.F., A.K. and R.R.L.; funding acquisition, Q.Y., T.G.B. and A.A.-H.; investigation, Q.Y., A.F., A.K. and T.G.B.; methodology, Q.Y., A.F., A.K. and R.R.L.; project administration, Q.Y. and A.A.-H.; resources, Q.Y. and A.A.-H.; software, A.F. and A.K.; supervision, Q.Y., A.F., T.G.B. and A.A.-H.; validation, Q.Y.; visualization, Q.Y.; writing—original draft, Q.Y.; writing—review and editing, Q.Y., A.F., A.K., R.R.L., T.G.B. and A.A.-H. All authors have read and agreed to the published version of the manuscript.

Funding: This study was supported in part by National Institutes of Health (NIH) grant R01 HD106285.

Institutional Review Board Statement: This study was approved by the Institutional Review Board (#20-1820) at the University of Chicago.

Informed Consent Statement: Informed consent was obtained from all subjects involved in the study.

Data Availability Statement: Raw FASTQ files were deposited in the NCBI Gene Expression Omnibus (GSE275087).

Acknowledgments: We would also like to thank The University of Chicago Genomics Facility (RRID: SCR_019196), especially Pieter W. Faber, for their assistance with Illumina RNA sequencing. We want to thank The University of Chicago Human Tissue Resource Center (RRID: SCR 019199), especially Shihong Li and Gong Can, for their assistance with the histology and immunohistochemistry study. Finally, we thank The University of Chicago Cytometry and Antibody Technology Facility, especially David Leclerc, for their help with cell cycle analysis (RRID: SCR_017760).

Conflicts of Interest: The authors declare that they have no conflicts of interest.

References

1. D'Angelo, E.; Prat, J. Uterine sarcomas: A review. *Gynecol. Oncol.* **2010**, *116*, 131–139. [CrossRef] [PubMed]
2. Yang, Q.; Madueke-Laveaux, O.S.; Cun, H.; Wlodarczyk, M.; Garcia, N.; Carvalho, K.C.; Al-Hendy, A. Comprehensive Review of Uterine Leiomyosarcoma: Pathogenesis, Diagnosis, Prognosis, and Targeted Therapy. *Cells* **2024**, *13*, 1106. [CrossRef] [PubMed]
3. Seagle, B.L.; Sobecki-Rausch, J.; Strohl, A.E.; Shilpi, A.; Grace, A.; Shahabi, S. Prognosis and treatment of uterine leiomyosarcoma: A National Cancer Database study. *Gynecol. Oncol.* **2017**, *145*, 61–70. [CrossRef]
4. Hensley, M.L.; Blessing, J.A.; Mannel, R.; Rose, P.G. Fixed-dose rate gemcitabine plus docetaxel as first-line therapy for metastatic uterine leiomyosarcoma: A Gynecologic Oncology Group phase II trial. *Gynecol. Oncol.* **2008**, *109*, 329–334. [CrossRef] [PubMed]
5. Yang, Q.; Bariani, M.V.; Falahati, A.; Khosh, A.; Lastra, R.R.; Siblini, H.; Boyer, T.G.; Al-Hendy, A. The Functional Role and Regulatory Mechanism of Bromodomain-Containing Protein 9 in Human Uterine Leiomyosarcoma. *Cells* **2022**, *11*, 2160. [CrossRef]
6. Gadducci, A.; Landoni, F.; Sartori, E.; Zola, P.; Maggino, T.; Lissoni, A.; Bazzurini, L.; Arisio, R.; Romagnolo, C.; Cristofani, R. Uterine leiomyosarcoma: Analysis of treatment failures and survival. *Gynecol. Oncol.* **1996**, *62*, 25–32. [CrossRef]
7. Zuccala, E. Epigenetics: Misdirecting methylation to drive oncogenesis. *Nat. Rev. Cancer* **2016**, *16*, 410. [CrossRef]
8. Wong, C.C.; Qian, Y.; Yu, J. Interplay between epigenetics and metabolism in oncogenesis: Mechanisms and therapeutic approaches. *Oncogene* **2017**, *36*, 3359–3374. [CrossRef]
9. Yang, Q.W.; Liu, S.; Tian, Y.; Salwen, H.R.; Chlenski, A.; Weinstein, J.; Cohn, S.L. Methylation-associated silencing of the thrombospondin-1 gene in human neuroblastoma. *Cancer Res.* **2003**, *63*, 6299–6310.
10. Yang, Q.; Zage, P.; Kagan, D.; Tian, Y.; Seshadri, R.; Salwen, H.R.; Liu, S.; Chlenski, A.; Cohn, S.L. Association of epigenetic inactivation of RASSF1A with poor outcome in human neuroblastoma. *Clin. Cancer Res.* **2004**, *10*, 8493–8500. [CrossRef]
11. Yang, Q.; Vafaei, S.; Falahati, A.; Khosh, A.; Bariani, M.V.; Omran, M.M.; Bai, T.; Siblini, H.; Ali, M.; He, C.; et al. Bromodomain-Containing Protein 9 Regulates Signaling Pathways and Reprograms the Epigenome in Immortalized Human Uterine Fibroid Cells. *Int. J. Mol. Sci.* **2024**, *25*, 905. [CrossRef] [PubMed]
12. Dennis, C. Epigenetics and disease: Altered states. *Nature* **2003**, *421*, 686–688. [CrossRef]
13. Feinberg, A.P. Phenotypic plasticity and the epigenetics of human disease. *Nature* **2007**, *447*, 433–440. [CrossRef]
14. The Cancer Genome Atlas Research Network. Comprehensive and Integrated Genomic Characterization of Adult Soft Tissue Sarcomas. *Cell* **2017**, *171*, 950–965.e28. [CrossRef]

15. Yoshida, K.; Yokoi, A.; Kitagawa, M.; Sugiyama, M.; Yamamoto, T.; Nakayama, J.; Yoshida, H.; Kato, T.; Kajiyama, H.; Yamamoto, Y. Downregulation of miR-10b-5p facilitates the proliferation of uterine leiomyosarcoma cells: A microRNA sequencing-based approach. *Oncol. Rep.* **2023**, *49*, 86. [CrossRef] [PubMed]
16. de Almeida, B.C.; dos Anjos, L.G.; Uno, M.; da Cunha, I.W.; Soares, F.A.; Baiocchi, G.; Baracat, E.C.; Carvalho, K.C. Let-7 miRNA's Expression Profile and Its Potential Prognostic Role in Uterine Leiomyosarcoma. *Cells* **2019**, *8*, 1452. [CrossRef]
17. Jain, A.K.; Barton, M.C. Bromodomain Histone Readers and Cancer. *J. Mol. Biol.* **2017**, *429*, 2003–2010. [CrossRef]
18. Fujisawa, T.; Filippakopoulos, P. Functions of bromodomain-containing proteins and their roles in homeostasis and cancer. *Nat. Rev. Mol. Cell Biol.* **2017**, *18*, 246–262. [CrossRef]
19. Cribbs, A.P.; Filippakopoulos, P.; Philpott, M.; Wells, G.; Penn, H.; Oerum, H.; Valge-Archer, V.; Feldmann, M.; Oppermann, U. Dissecting the Role of BET Bromodomain Proteins BRD2 and BRD4 in Human NK Cell Function. *Front. Immunol.* **2021**, *12*, 626255. [CrossRef]
20. Filippakopoulos, P.; Picaud, S.; Mangos, M.; Keates, T.; Lambert, J.P.; Barsyte-Lovejoy, D.; Felletar, I.; Volkmer, R.; Müller, S.; Pawson, T.; et al. Histone recognition and large-scale structural analysis of the human bromodomain family. *Cell* **2012**, *149*, 214–231. [CrossRef]
21. Wang, N.; Wu, R.; Tang, D.; Kang, R. The BET family in immunity and disease. *Signal Transduct. Target. Ther.* **2021**, *6*, 23. [CrossRef] [PubMed]
22. Taniguchi, Y. The Bromodomain and Extra-Terminal Domain (BET) Family: Functional Anatomy of BET Paralogous Proteins. *Int. J. Mol. Sci.* **2016**, *17*, 1849. [CrossRef] [PubMed]
23. Zhang, Z.; Ma, P.; Jing, Y.; Yan, Y.; Cai, M.C.; Zhang, M.; Zhang, S.; Peng, H.; Ji, Z.L.; Di, W.; et al. BET Bromodomain Inhibition as a Therapeutic Strategy in Ovarian Cancer by Downregulating FoxM1. *Theranostics* **2016**, *6*, 219–230. [CrossRef] [PubMed]
24. Ji, Y.J.; Shao, Y.; Zhang, J.; Zhang, X.; Qiang, P. Bromodomain-containing protein 4 silencing by microRNA-765 produces anti-ovarian cancer cell activity. *Aging* **2021**, *13*, 8214–8227. [CrossRef]
25. Kurimchak, A.M.; Shelton, C.; Herrera-Montávez, C.; Duncan, K.E.; Chernoff, J.; Duncan, J.S. Intrinsic Resistance to MEK Inhibition through BET Protein-Mediated Kinome Reprogramming in NF1-Deficient Ovarian Cancer. *Mol. Cancer Res.* **2019**, *17*, 1721–1734. [CrossRef] [PubMed]
26. Jing, Y.; Zhang, Z.; Ma, P.; An, S.; Shen, Y.; Zhu, L.; Zhuang, G. Concomitant BET and MAPK blockade for effective treatment of ovarian cancer. *Oncotarget* **2016**, *7*, 2545–2554. [CrossRef] [PubMed]
27. Drumond-Bock, A.L.; Bieniasz, M. The role of distinct BRD4 isoforms and their contribution to high-grade serous ovarian carcinoma pathogenesis. *Mol. Cancer* **2021**, *20*, 145. [CrossRef] [PubMed]
28. Maser, T.; Zagorski, J.; Kelly, S.; Ostrander, A.; Goodyke, A.; Nagulapally, A.; Bond, J.; Park, Y.; Saulnier Sholler, G. The MDM2 inhibitor CGM097 combined with the BET inhibitor OTX015 induces cell death and inhibits tumor growth in models of neuroblastoma. *Cancer Med.* **2020**, *9*, 8144–8158. [CrossRef]
29. Sun, C.; Yin, J.; Fang, Y.; Chen, J.; Jeong, K.J.; Chen, X.; Vellano, C.P.; Ju, Z.; Zhao, W.; Zhang, D.; et al. BRD4 Inhibition Is Synthetic Lethal with PARP Inhibitors through the Induction of Homologous Recombination Deficiency. *Cancer Cell* **2018**, *33*, 401–416.e8. [CrossRef]
30. Momeny, M.; Eyvani, H.; Barghi, F.; Ghaffari, S.H.; Javadikooshesh, S.; Hassanvand Jamadi, R.; Esmaeili, F.; Alishahi, Z.; Zaghal, A.; Bashash, D.; et al. Inhibition of bromodomain and extraterminal domain reduces growth and invasive characteristics of chemoresistant ovarian carcinoma cells. *Anticancer Drugs* **2018**, *29*, 1011–1020. [CrossRef]
31. Bauer, K.; Berger, D.; Zielinski, C.C.; Valent, P.; Grunt, T.W. Hitting two oncogenic machineries in cancer cells: Cooperative effects of the multi-kinase inhibitor ponatinib and the BET bromodomain blockers JQ1 or dBET1 on human carcinoma cells. *Oncotarget* **2018**, *9*, 26491–26506. [CrossRef] [PubMed]
32. Rhyasen, G.W.; Yao, Y.; Zhang, J.; Dulak, A.; Castriotta, L.; Jacques, K.; Zhao, W.; Gharahdaghi, F.; Hattersley, M.M.; Lyne, P.D.; et al. BRD4 amplification facilitates an oncogenic gene expression program in high-grade serous ovarian cancer and confers sensitivity to BET inhibitors. *PLoS ONE* **2018**, *13*, e0200826. [CrossRef] [PubMed]
33. Shorstova, T.; Su, J.; Zhao, T.; Dahabieh, M.; Leibovitch, M.; De Sa Tavares Russo, M.; Avizonis, D.; Rajkumar, S.; Watson, I.R.; Del Rincón, S.V.; et al. Reprogramming of Nucleotide Metabolism Mediates Synergy between Epigenetic Therapy and MAP Kinase Inhibition. *Mol. Cancer Ther.* **2021**, *20*, 64–75. [CrossRef] [PubMed]
34. Ram, S.; Vizcarra, P.; Whalen, P.; Deng, S.; Painter, C.L.; Jackson-Fisher, A.; Pirie-Shepherd, S.; Xia, X.; Powell, E.L. Pixelwise H-score: A novel digital image analysis-based metric to quantify membrane biomarker expression from immunohistochemistry images. *PLoS ONE* **2021**, *16*, e0245638. [CrossRef]
35. Yang, Q.; Dahl, M.J.; Albertine, K.H.; Ramchandran, R.; Sun, M.; Raj, J.U. Role of histone deacetylases in regulation of phenotype of ovine newborn pulmonary arterial smooth muscle cells. *Cell Prolif.* **2013**, *46*, 654–664. [CrossRef]
36. Bolger, A.M.; Lohse, M.; Usadel, B. Trimmomatic: A flexible trimmer for Illumina sequence data. *Bioinformatics* **2014**, *30*, 2114–2120. [CrossRef]
37. Kim, D.; Paggi, J.M.; Park, C.; Bennett, C.; Salzberg, S.L. Graph-based genome alignment and genotyping with HISAT2 and HISAT-genotype. *Nat. Biotechnol.* **2019**, *37*, 907–915. [CrossRef]
38. Liao, Y.; Smyth, G.K.; Shi, W. featureCounts: An efficient general purpose program for assigning sequence reads to genomic features. *Bioinformatics* **2014**, *30*, 923–930. [CrossRef]

39. Harrow, J.; Frankish, A.; Gonzalez, J.M.; Tapanari, E.; Diekhans, M.; Kokocinski, F.; Aken, B.L.; Barrell, D.; Zadissa, A.; Searle, S.; et al. GENCODE: The reference human genome annotation for The ENCODE Project. *Genome Res.* **2012**, *22*, 1760–1774. [CrossRef]
40. Love, M.I.; Huber, W.; Anders, S. Moderated estimation of fold change and dispersion for RNA-seq data with DESeq2. *Genome Biol.* **2014**, *15*, 550. [CrossRef]
41. Subramanian, A.; Tamayo, P.; Mootha, V.K.; Mukherjee, S.; Ebert, B.L.; Gillette, M.A.; Paulovich, A.; Pomeroy, S.L.; Golub, T.R.; Lander, E.S.; et al. Gene set enrichment analysis: A knowledge-based approach for interpreting genome-wide expression profiles. *Proc. Natl. Acad. Sci. USA* **2005**, *102*, 15545–15550. [CrossRef] [PubMed]
42. Korotkevich, G.; Sukhov, V.; Budin, N.; Shpak, B.; Artyomov, M.N.; Sergushichev, A. Fast gene set enrichment analysis. *BioRxiv* **2021**. [CrossRef]
43. Kuleshov, M.V.; Jones, M.R.; Rouillard, A.D.; Fernandez, N.F.; Duan, Q.; Wang, Z.; Koplev, S.; Jenkins, S.L.; Jagodnik, K.M.; Lachmann, A.; et al. Enrichr: A comprehensive gene set enrichment analysis web server 2016 update. *Nucleic Acids Res.* **2016**, *44*, W90–W97. [CrossRef] [PubMed]
44. Yu, G.; Wang, L.G.; Han, Y.; He, Q.Y. clusterProfiler: An R package for comparing biological themes among gene clusters. *OMICS* **2012**, *16*, 284–287. [CrossRef]
45. Oliveira de Biagi, C.A., Jr.; Nociti, R.P.; Brotto, D.B.; Funicheli, B.O.; Cássia Ruy, P.; Bianchi Ximenez, J.P.; Alves Figueiredo, D.L.; Araújo Silva, W., Jr. CeTF: An R/Bioconductor package for transcription factor co-expression networks using regulatory impact factors (RIF) and partial correlation and information (PCIT) analysis. *BMC Genom.* **2021**, *22*, 624. [CrossRef]
46. Filippakopoulos, P.; Qi, J.; Picaud, S.; Shen, Y.; Smith, W.B.; Fedorov, O.; Morse, E.M.; Keates, T.; Hickman, T.T.; Felletar, I.; et al. Selective inhibition of BET bromodomains. *Nature* **2010**, *468*, 1067–1073. [CrossRef]
47. Nicodeme, E.; Jeffrey, K.L.; Schaefer, U.; Beinke, S.; Dewell, S.; Chung, C.W.; Chandwani, R.; Marazzi, I.; Wilson, P.; Coste, H.; et al. Suppression of inflammation by a synthetic histone mimic. *Nature* **2010**, *468*, 1119–1123. [CrossRef] [PubMed]
48. Ferri, E.; Petosa, C.; McKenna, C.E. Bromodomains: Structure, function and pharmacology of inhibition. *Biochem. Pharmacol.* **2016**, *106*, 1–18. [CrossRef]
49. Shi, J.; Vakoc, C.R. The mechanisms behind the therapeutic activity of BET bromodomain inhibition. *Mol. Cell* **2014**, *54*, 728–736. [CrossRef]
50. Lasman, L.; Hanna, J.H.; Novershtern, N. Role of m^6A in Embryonic Stem Cell Differentiation and in Gametogenesis. *Epigenomes* **2020**, *4*, 5. [CrossRef]
51. Fiorenzano, A.; Pascale, E.; Gagliardi, M.; Terreri, S.; Papa, M.; Andolfi, G.; Galasso, M.; Tagliazucchi, G.M.; Taccioli, C.; Patriarca, E.J.; et al. An Ultraconserved Element Containing lncRNA Preserves Transcriptional Dynamics and Maintains ESC Self-Renewal. *Stem Cell Rep.* **2018**, *10*, 1102–1114. [CrossRef] [PubMed]
52. Fiorenzano, A.; Pascale, E.; Patriarca, E.J.; Minchiotti, G.; Fico, A. LncRNAs and PRC2: Coupled Partners in Embryonic Stem Cells. *Epigenomes* **2019**, *3*, 14. [CrossRef]
53. Henssen, A.; Thor, T.; Odersky, A.; Heukamp, L.; El-Hindy, N.; Beckers, A.; Speleman, F.; Althoff, K.; Schäfers, S.; Schramm, A.; et al. BET bromodomain protein inhibition is a therapeutic option for medulloblastoma. *Oncotarget* **2013**, *4*, 2080–2095. [CrossRef]
54. Gilan, O.; Rioja, I.; Knezevic, K.; Bell, M.J.; Yeung, M.M.; Harker, N.R.; Lam, E.Y.N.; Chung, C.W.; Bamborough, P.; Petretich, M.; et al. Selective targeting of BD1 and BD2 of the BET proteins in cancer and immunoinflammation. *Science* **2020**, *368*, 387–394. [CrossRef] [PubMed]
55. Kulikowski, E.; Rakai, B.D.; Wong, N.C.W. Inhibitors of bromodomain and extra-terminal proteins for treating multiple human diseases. *Med. Res. Rev.* **2021**, *41*, 223–245. [CrossRef] [PubMed]
56. Baek, M.; Yoo, E.; Choi, H.I.; An, G.Y.; Chai, J.C.; Lee, Y.S.; Jung, K.H.; Chai, Y.G. The BET inhibitor attenuates the inflammatory response and cell migration in human microglial HMC3 cell line. *Sci. Rep.* **2021**, *11*, 8828. [CrossRef]
57. Yang, Q.; Falahati, A.; Khosh, A.; Mohammed, H.; Kang, W.; Corachán, A.; Bariani, M.V.; Boyer, T.G.; Al-Hendy, A. Targeting Class I Histone Deacetylases in Human Uterine Leiomyosarcoma. *Cells* **2022**, *11*, 3801. [CrossRef]
58. Shi, X.; Liu, C.; Liu, B.; Chen, J.; Wu, X.; Gong, W. JQ1: A novel potential therapeutic target. *Pharmazie* **2018**, *73*, 491–493.
59. da Motta, L.L.; Ledaki, I.; Purshouse, K.; Haider, S.; De Bastiani, M.A.; Baban, D.; Morotti, M.; Steers, G.; Wigfield, S.; Bridges, E.; et al. The BET inhibitor JQ1 selectively impairs tumour response to hypoxia and downregulates CA9 and angiogenesis in triple negative breast cancer. *Oncogene* **2017**, *36*, 122–132. [CrossRef]
60. Fu, L.L.; Tian, M.; Li, X.; Li, J.J.; Huang, J.; Ouyang, L.; Zhang, Y.; Liu, B. Inhibition of BET bromodomains as a therapeutic strategy for cancer drug discovery. *Oncotarget* **2015**, *6*, 5501–5516. [CrossRef]
61. Puissant, A.; Frumm, S.M.; Alexe, G.; Bassil, C.F.; Qi, J.; Chanthery, Y.H.; Nekritz, E.A.; Zeid, R.; Gustafson, W.C.; Greninger, P.; et al. Targeting MYCN in neuroblastoma by BET bromodomain inhibition. *Cancer Discov.* **2013**, *3*, 308–323. [CrossRef]
62. Anand, P.; Brown, J.D.; Lin, C.Y.; Qi, J.; Zhang, R.; Artero, P.C.; Alaiti, M.A.; Bullard, J.; Alazem, K.; Margulies, K.B.; et al. BET bromodomains mediate transcriptional pause release in heart failure. *Cell* **2013**, *154*, 569–582. [CrossRef] [PubMed]
63. Matzuk, M.M.; McKeown, M.R.; Filippakopoulos, P.; Li, Q.; Ma, L.; Agno, J.E.; Lemieux, M.E.; Picaud, S.; Yu, R.N.; Qi, J.; et al. Small-molecule inhibition of BRDT for male contraception. *Cell* **2012**, *150*, 673–684. [CrossRef] [PubMed]
64. Pérez-Peña, J.; Serrano-Heras, G.; Montero, J.C.; Corrales-Sánchez, V.; Pandiella, A.; Ocaña, A. In Silico Analysis Guides Selection of BET Inhibitors for Triple-Negative Breast Cancer Treatment. *Mol. Cancer Ther.* **2016**, *15*, 1823–1833. [CrossRef]

65. Nieto-Jiménez, C.; Alcaraz-Sanabria, A.; Pérez-Peña, J.; Corrales-Sánchez, V.; Serrano-Heras, G.; Galán-Moya, E.M.; Serrano-Oviedo, L.; Montero, J.C.; Burgos, M.; Llopis, J.; et al. Targeting basal-like breast tumors with bromodomain and extraterminal domain (BET) and polo-like kinase inhibitors. *Oncotarget* **2017**, *8*, 19478–19490. [CrossRef]
66. Chen, J.; Sun, Y.; Chi, Z. Regulation of hematopoiesis by hedgehog signaling (Review). *Mol. Med. Rep.* **2023**, *27*, 100. [CrossRef] [PubMed]
67. Wu, A.; Turner, K.A.; Woolfson, A.; Jiang, X. The Hedgehog Pathway as a Therapeutic Target in Chronic Myeloid Leukemia. *Pharmaceutics* **2023**, *15*, 958. [CrossRef] [PubMed]
68. Zhang, Y.; Beachy, P.A. Cellular and molecular mechanisms of Hedgehog signalling. *Nat. Rev. Mol. Cell Biol.* **2023**, *24*, 668–687. [CrossRef]
69. Garcia, N.; Ulin, M.; Al-Hendy, A.; Yang, Q. The Role of Hedgehog Pathway in Female Cancers. *J. Cancer Sci. Clin. Ther.* **2020**, *4*, 487–498. [CrossRef]
70. Garcia, N.; Al-Hendy, A.; Baracat, E.C.; Carvalho, K.C.; Yang, Q. Targeting Hedgehog Pathway and DNA Methyltransferases in Uterine Leiomyosarcoma Cells. *Cells* **2020**, *10*, 53. [CrossRef]
71. Katoh, Y.; Katoh, M. Hedgehog target genes: Mechanisms of carcinogenesis induced by aberrant hedgehog signaling activation. *Curr. Mol. Med.* **2009**, *9*, 873–886. [CrossRef] [PubMed]
72. Choi, J.; Manzano, A.; Dong, W.; Bellone, S.; Bonazzoli, E.; Zammataro, L.; Yao, X.; Deshpande, A.; Zaidi, S.; Guglielmi, A.; et al. Integrated mutational landscape analysis of uterine leiomyosarcomas. *Proc. Natl. Acad. Sci. USA* **2021**, *118*, e2025182118. [CrossRef] [PubMed]
73. Chakraborty, P.; George, J.T.; Tripathi, S.; Levine, H.; Jolly, M.K. Comparative Study of Transcriptomics-Based Scoring Metrics for the Epithelial-Hybrid-Mesenchymal Spectrum. *Front. Bioeng. Biotechnol.* **2020**, *8*, 220. [CrossRef]
74. Cai, L.; Hu, C.; Yu, S.; Liu, L.; Zhao, J.; Zhao, Y.; Lin, F.; Du, X.; Yu, Q.; Xiao, Q. Identification of EMT-Related Gene Signatures to Predict the Prognosis of Patients with Endometrial Cancer. *Front. Genet.* **2020**, *11*, 582274. [CrossRef]
75. Gong, L.; Liu, G.; Zhu, H.; Li, C.; Li, P.; Liu, C.; Tang, H.; Wu, K.; Wu, J.; Liu, D.; et al. IL-32 induces epithelial-mesenchymal transition by triggering endoplasmic reticulum stress in A549 cells. *BMC Pulm. Med.* **2020**, *20*, 278. [CrossRef]
76. Wen, S.; Hou, Y.; Fu, L.; Xi, L.; Yang, D.; Zhao, M.; Qin, Y.; Sun, K.; Teng, Y.; Liu, M. Cancer-associated fibroblast (CAF)-derived IL32 promotes breast cancer cell invasion and metastasis via integrin beta3-p38 MAPK signalling. *Cancer Lett.* **2019**, *442*, 320–332. [CrossRef]
77. Park, J.S.; Choi, S.Y.; Lee, J.H.; Lee, M.; Nam, E.S.; Jeong, A.L.; Lee, S.; Han, S.; Lee, M.S.; Lim, J.S.; et al. Interleukin-32beta stimulates migration of MDA-MB-231 and MCF-7cells via the VEGF-STAT3 signaling pathway. *Cell Oncol.* **2013**, *36*, 493–503. [CrossRef]
78. Cheng, X.; Li, F.; Tao, Z. Tenascin-C promotes epithelial-to-mesenchymal transition and the mTOR signaling pathway in nasopharyngeal carcinoma. *Oncol. Lett.* **2021**, *22*, 570. [CrossRef]
79. Kang, X.; Xu, E.; Wang, X.; Qian, L.; Yang, Z.; Yu, H.; Wang, C.; Ren, C.; Wang, Y.; Lu, X.; et al. Tenascin-c knockdown suppresses vasculogenic mimicry of gastric cancer by inhibiting ERK- triggered EMT. *Cell Death Dis.* **2021**, *12*, 890. [CrossRef]
80. Yang, Z.; Zhang, C.; Qi, W.; Cui, C.; Cui, Y.; Xuan, Y. Tenascin-C as a prognostic determinant of colorectal cancer through induction of epithelial-to-mesenchymal transition and proliferation. *Exp. Mol. Pathol.* **2018**, *105*, 216–222. [CrossRef] [PubMed]
81. Guo, J.Q.; Wang, C.D.; Tang, H.Y.; Sang, B.T.; Liu, X.; Yi, F.P.; Wu, X.M. PDGF-BB/PDGFRbeta promotes epithelial-mesenchymal transition by affecting PI3K/AKT/mTOR-driven aerobic glycolysis in Wilms' tumor G401 cells. *Cell Biol. Int.* **2022**, *46*, 907–921. [CrossRef] [PubMed]
82. Steller, E.J.; Raats, D.A.; Koster, J.; Rutten, B.; Govaert, K.M.; Emmink, B.L.; Snoeren, N.; van Hooff, S.R.; Holstege, F.C.; Maas, C.; et al. PDGFRB promotes liver metastasis formation of mesenchymal-like colorectal tumor cells. *Neoplasia* **2013**, *15*, 204–217. [CrossRef] [PubMed]
83. Yin, J.; Guo, Y.; Li, Z. Platelet-derived growth factor-B signalling might promote epithelial-mesenchymal transition in gastric carcinoma cells through activation of the MAPK/ERK pathway. *Contemp. Oncol.* **2021**, *25*, 1–6. [CrossRef] [PubMed]
84. Zhang, H.; Sun, J.D.; Yan, L.J.; Zhao, X.P. PDGF-D/PDGFRbeta promotes tongue squamous carcinoma cell (TSCC) progression via activating p38/AKT/ERK/EMT signal pathway. *Biochem. Biophys. Res. Commun.* **2016**, *478*, 845–851. [CrossRef]
85. Dang, C.V. MYC on the path to cancer. *Cell* **2012**, *149*, 22–35. [CrossRef]
86. Chen, X.; Johns, D.C.; Geiman, D.E.; Marban, E.; Dang, D.T.; Hamlin, G.; Sun, R.; Yang, V.W. Kruppel-like factor 4 (gut-enriched Kruppel-like factor) inhibits cell proliferation by blocking G1/S progression of the cell cycle. *J. Biol. Chem.* **2001**, *276*, 30423–30428. [CrossRef] [PubMed]
87. Proença, C.; Freitas, M.; Ribeiro, D.; Rufino, A.T.; Fernandes, E.; Ferreira de Oliveira, J.M.P. The role of flavonoids in the regulation of epithelial-mesenchymal transition in cancer: A review on targeting signaling pathways and metastasis. *Med. Res. Rev.* **2023**, *43*, 1878–1945. [CrossRef]
88. Bushweller, J.H. Targeting transcription factors in cancer—From undruggable to reality. *Nat. Rev. Cancer* **2019**, *19*, 611–624. [CrossRef]
89. Liu, J.; Huang, Z.; Chen, H.N.; Qin, S.; Chen, Y.; Jiang, J.; Zhang, Z.; Luo, M.; Ye, Q.; Xie, N.; et al. ZNF37A promotes tumor metastasis through transcriptional control of THSD4/TGF-beta axis in colorectal cancer. *Oncogene* **2021**, *40*, 3394–3407. [CrossRef]

90. Kwon, W.; Choi, S.K.; Kim, D.; Kim, H.G.; Park, J.K.; Han, J.E.; Cho, G.J.; Yun, S.; Yu, W.; Han, S.H.; et al. ZNF507 affects TGF-beta signaling via TGFBR1 and MAP3K8 activation in the progression of prostate cancer to an aggressive state. *J. Exp. Clin. Cancer Res.* **2021**, *40*, 291. [CrossRef]
91. Li, K.; Xu, C.; Du, Y.; Junaid, M.; Kaushik, A.C.; Wei, D.Q. Comprehensive epigenetic analyses reveal master regulators driving lung metastasis of breast cancer. *J. Cell. Mol. Med.* **2019**, *23*, 5415–5431. [CrossRef] [PubMed]
92. Gao, P.; He, M.; Zhang, C.; Geng, C. Integrated analysis of gene expression signatures associated with colon cancer from three datasets. *Gene* **2018**, *654*, 95–102. [CrossRef] [PubMed]
93. Perumal, N.; Kanchan, R.K.; Doss, D.; Bastola, N.; Atri, P.; Chirravuri-Venkata, R.; Thapa, I.; Vengoji, R.; Maurya, S.K.; Klinkebiel, D.; et al. MiR-212-3p functions as a tumor suppressor gene in group 3 medulloblastoma via targeting nuclear factor I/B (NFIB). *Acta Neuropathol. Commun.* **2021**, *9*, 195. [CrossRef]
94. Zhong, Q.H.; Zha, S.W.; Lau, A.T.Y.; Xu, Y.M. Recent knowledge of NFATc4 in oncogenesis and cancer prognosis. *Cancer Cell Int.* **2022**, *22*, 212. [CrossRef]
95. Cheung, K.L.; Kim, C.; Zhou, M.M. The Functions of BET Proteins in Gene Transcription of Biology and Diseases. *Front. Mol. Biosci.* **2021**, *8*, 728777. [CrossRef]
96. Zhou, S.; Zhang, S.; Wang, L.; Huang, S.; Yuan, Y.; Yang, J.; Wang, H.; Li, X.; Wang, P.; Zhou, L.; et al. BET protein inhibitor JQ1 downregulates chromatin accessibility and suppresses metastasis of gastric cancer via inactivating RUNX2/NID1 signaling. *Oncogenesis* **2020**, *9*, 33. [CrossRef]
97. He, P.C.; Wei, J.; Dou, X.; Harada, B.T.; Zhang, Z.; Ge, R.; Liu, C.; Zhang, L.S.; Yu, X.; Wang, S.; et al. Exon architecture controls mRNA m^6A suppression and gene expression. *Science* **2023**, *379*, 677–682. [CrossRef]
98. Yang, Q.; Al-Hendy, A. The Functional Role and Regulatory Mechanism of FTO m^6A RNA Demethylase in Human Uterine Leiomyosarcoma. *Int. J. Mol. Sci.* **2023**, *24*, 7957. [CrossRef] [PubMed]

MDPI AG
Grosspeteranlage 5
4052 Basel
Switzerland
Tel.: +41 61 683 77 34

Cells Editorial Office
E-mail: cells@mdpi.com
www.mdpi.com/journal/cells

www.ingramcontent.com/pod-product-compliance
Lightning Source LLC
Chambersburg PA
CBHW041016270926
51850CB00024B/344
* 9 7 8 3 7 2 5 8 5 4 1 3 4 *